CIRCUIT ANALYSIS

THEORY AND PRACTICE

6.05
5.18

STRAIGHT LINE

$y = mx + b$ where m = slope = $\Delta y / \Delta x$. Δy is the amount y changes when x changes by Δx.

QUADRATIC EQUATION

Roots of $ax^2 + bx + c = 0$ are

$$x_{1,2} = \frac{-b \pm \sqrt{b^2 - 4ac}}{2a}$$

If $(b^2 - 4ac) \geq 0$, roots are real; if $(b^2 - 4ac) < 0$, roots are complex conjugates.

TRIGONOMETRIC IDENTITIES

$\sin(-\alpha) = -\sin \alpha$ $\cos(-\alpha) = \cos \alpha$

$\sin(\omega t \pm 90°) = \pm \cos \omega t$ $\sin(\omega t \pm 180°) = -\sin \omega t$

$\cos(\omega t \pm 90°) = \mp \sin \omega t$ $\cos(\omega t \pm 180°) = -\cos \omega t$

$$\sin^2 \omega t = \frac{1 - \cos 2\omega t}{2}$$

NATURAL LOGARITHMS

Given $e^x = y$. x may be found by noting that $\ln e^x = x$. For example, if $e^{-5t} = 0.7788$, then

$\ln e^{-5t} = \ln 0.7788$

$\quad -5t = -0.25$

$\qquad t = 0.05$ s

CALCULUS

$\dfrac{d}{dx}(ax) = a$ $\dfrac{d}{dx}(\sin ax) = a \cos ax$

$\dfrac{d}{dx}(e^{ax}) = a e^{ax}$ $\dfrac{d}{dx}(\cos ax) = -a \sin ax$

$\displaystyle\int x\,dx = \frac{x^2}{2}$ $\displaystyle\int \sin ax\,dx = -\frac{1}{a}\cos ax$

$\displaystyle\int \frac{dx}{x} = \ln x$ $\displaystyle\int \cos ax\,dx = \frac{1}{a}\sin ax$

Allan H. Robbins

Red River Community College, Manitoba

Wilhelm C. Miller

Red River Community College, Manitoba

 Delmar Publishers

I(T)P An International Thomson Publishing Company

Albany • Bonn • Boston • Cincinnati • Detroit • London • Madrid • Melbourne
Mexico City • New York • Pacific Grove • Paris • San Francisco • Singapore • Tokyo
Toronto • Washington

NOTICE TO THE READER

Cover photo courtesy of Mike Gallitelli
Cover Design: Lisa Bower

Delmar Staff

Publisher: Michael A. McDermott
Administrative Editor: Wendy Welch
Developmental Editor: Mary E. Clyne

Senior Project Editor: Christopher Chien
Production Manager: Larry Main
Art and Design Coordinator: Lisa Bower

COPYRIGHT © 1995
By Delmar Publishers
a division of International Thomson Publishing Inc.

The ITP logo is a trademark under license.

Printed in the United States of America

For more information, contact:

Delmar Publishers
3 Columbia Circle, Box 15015
Albany, New York 12212-5015

International Thomson Publishing Europe
Berkshire House 168-173
High Holborn
London, WC1V 7AA
England

Thomas Nelson Australia
102 Dodds Street
South Melbourne, 3205
Victoria, Australia

Nelson Canada
1120 Birchmont Road
Scarborough, Ontario
Canada, M1K 5G4

International Thomson Editores
Campos Eliseos 385, Piso 7
Col Polanco
11560 Mexico D F Mexico

International Thomson Publishing GmbH
Konigswinterer Strasse 418
53227 Bonn
Germany

International Thomson Publishing Asia
221 Henderson Road
#05-10 Henderson Building
Singapore 0315

International Thomson Publishing—Japan
Hirakawacho Kyowa Building, 3F
2-2-1 Hirakawacho
Chiyoda-ku, Tokyo 102
Japan

1 2 3 4 5 6 7 8 9 10 XXX 01 00 99 98 97 96 95

Library of Congress Cataloging-in-Publication Data:

Robbins, Allan.
  Circuit analysis: theory and practice / Allan Robbins, Wilhelm Miller.
     p. cm.
  Includes index.
  ISBN 0-8273-5414-2
    1. Electric circuit analysis. I. Miller, Wilhelm (Wilhelm C.)
  II. Title.
    TK454.R56 1994
    621.319′2—dc20

                      94-1022
                      CIP

Contents

Preface

This text is written primarily for students in electrical, electronic, and related engineering technology programs who are taking their first course in electric circuit theory. It covers fundamentals of dc and ac circuits, methods of analysis, capacitance, inductance, magnetism, simple transients, and other topics that are usually taught in an introductory circuit analysis course. While aimed primarily at students in two- and three-year technology programs, it may also serve as an introductory text in four-year university programs.

Features of the Book

- Clearly written, easy-to-understand style emphasizes principles and concepts.

- Hundreds of worked-out examples promote student understanding.

- Over 1200 diagrams and photographs illustrate the text. Full-color photographs and full-color diagrams with color-coded symbols and waveforms make visualizing important ideas easier.

- *In-Process Learning Checks* help identify learning gaps before students move on to new material.

- *Chapter Previews* provide a brief overview to prepare students for what is coming in the chapter.

- *Competency-based objectives* define the knowledge or skill that the student is expected to gain from each chapter.

- *Key terms* at the beginning of each chapter identify new terms to be introduced.

- A wide selection of end-of-chapter problems includes answers to odd-numbered problems.

- Computer-aided circuit analysis using PSpice illustrates how computer techniques may be used to analyze electric circuit problems.

- End-of-chapter problems are keyed to section numbers. Difficult problems are identified by having their number printed in red.

- Boxed articles and icons help locate important ideas.

Required Background

A working knowledge of basic algebra and trigonometry is required. In addition, students should be familiar with some linear algebra. Notably, they should be able to solve second-order linear equations. (However, a review of determinants and the solution of simultaneous equations is included in Appendix B.) Calculus is introduced gradually in later chapters to aid in the development of ideas. (This is in keeping with ABET guidelines, which require the use of some calculus in accredited engineering technology programs. It is expected that the student will be taking a concurrent calculus course.) Optional derivations using calculus are included for those colleges that stress calculus. They are identified by an $\int$ icon. These derivations may be omitted by those colleges that do not require calculus.

In terms of electrical background, no previous knowledge is assumed other than that gained through everyday exposure. That is, we expect that students will be generally familiar with the terms voltage, current, and power through their association with common electrical and electronic devices. In terms of physics, we expect that students will know the MKS (meter-kilogram-second) metric system and the atomic nature of matter.

Text Organization

The book is divided into five main parts: Foundation dc Concepts, Basic dc Analysis, Capacitance and Inductance, Foundation ac Concepts, and Impedance Networks. Each part begins with an overview that provides a context for the material to come, while individual chapters include previews that set the stage for the chapter itself.

Chapters 1 to 4 are introductory. They cover foundation concepts including voltage, current, resistance, Ohm's law, and power. Chapters 5 to 9 focus on dc analysis. Covered here are Kirchhoff's laws, series and parallel circuits, mesh and nodal analysis, wye and delta transformations, source transformations, superposition, Thévenin's theorem, the maximum power theorem, and so on. Chapters 10 to 14 cover basic concepts of capacitance, magnetism, and inductance as well as the analysis of simple transients in dc circuits. Chapters 15 to 17 cover foundation concepts of ac, ac voltage generation, the basic ideas of frequency, period, phase angle, and so on. Phasors and the impedance concept are introduced and used to solve simple problems. Lastly, power in ac circuits is investigated. Chapters 18 to 25 then apply these ideas to the analysis of ac circuits. Topics covered include ac equivalents of earlier dc techniques (e.g., mesh and nodal analysis, superposition, Thevénin's theorem, and so on), as well as new ideas such as resonance, filters and Bode plots, three-phase systems, transformers, and the analysis of nonsinusoidal waveforms.

Several appendices round out the book. Appendix A provides a minitutorial on PSpice and the solution of circuit problems by computer. Appendix B reviews determinants and the solution of simultaneous linear equations. Appendix C contains answers to odd-numbered end-of-chapter problems.

The Ancillary Support Package

A comprehensive set of ancillaries support this book.

Instructor's Solution Manual: Contains step-by-step solutions to all end-of-chapter problems.

Test Bank: Contains problems and multiple-choice questions in computerized form.

Student Laboratory Manual: Contains instructions for hands-on laboratory work. Includes a brief theory overview for each lab exercise.

Transparencies and Transparency Masters: Selected diagrams from the book have been reproduced as full-color overhead transparencies; others have been packaged in a form suitable for making overhead transparencies.

Suggestions for Use

The presentation and format are such that an instructor can either follow the sequence as laid out or design his or her own course by selecting topics and omitting others as needed. For example, the two chapters on basic transients (Chapters 11 and 14) can be postponed until later without loss of continuity. The sections on resonance (Chapter 21), filters and Bode plots (Chapter 22), and nonsinusoidal waveforms (Chapter 25) can also be covered in a different order or omitted without serious discontinuity. Topics omitted can then be covered in a later sequence of study as needed.

To The Student

Learning circuit theory should be challenging, interesting, and hopefully, fun. However, it is also hard work as knowledge and skills can only be acquired through practice. We offer a few guidelines.

1. As you go through the material, try to gain an appreciation of where circuit theory comes from—i.e., the basic experimental laws on which it is based. This will help you better understand the foundation ideas on which the theory is built.

2. Learn the terminology and definitions. Important new terms are introduced frequently. Learn what they mean and where they are used.

3. Study each section carefully and be sure you understand the basic ideas and how they are put together. Work your way through the examples with your calculator. Try the practice problems and end-of-chapter problems. (Answers are in Appendices C and D.)

4. When you are ready, test your understanding using the *In-Process Learning Checks* located in each chapter.

5. Finally, when you have mastered the material, move on to the next block. For those concepts with which you are having difficulty, consult your instructor or some other authoritative source.

Calculators for Circuit Analysis

You will need a good scientific calculator. A good calculator will permit you to more easily master the numerical aspect of problem solving and leave you more time to concentrate on circuit theory itself. (Many calculators are difficult/complex to use and thereby make the manipulation of numbers an impediment to learning. This is especially true for ac, where complex number work dominates.) There are some very inexpensive, nonprogrammable calculators on the market that handle complex-number arithmetic almost as easily as real-number arithmetic. Such calculators save an enormous amount of time. If your school permits their use in introductory courses, we strongly recommend that you acquire one.

Getting the Most Out of This Book

Circuit Analysis: Theory and Practice is a learning tool with many in-text learning features that you should find useful. It uses full-color photos and diagrams (many of which incorporate 3-D effects) to illustrate and clarify ideas. Voltages and currents are color-coded, and color is used to highlight definitions, terminology, and so on. Icons are used to help locate features such as In-Process Learning Checks and Practical Notes, while marginal notes are used to define key terms and expand on ideas. Some of these features are illustrated below.

But, when converting the source, we get

$$I = \frac{E}{R_S}$$

And so

$$I_L = \left(\frac{R_S}{R_S + R_L} \right) \left(\frac{E}{R_S} \right)$$

This result is equivalent to the current obtained in Equation 8–4. The voltage across the resistor is given as

$$V_L = I_L R_L$$

$$= \left(\frac{E}{R_S + R_L} \right) R_L$$

The voltage across the resistor is precisely the same as the result obtained in Equation 8–3. We therefore conclude that the load current and voltage drop are the same whether the source is a voltage source or an equivalent current source.

note

Although the sources are equivalent, currents and voltages within the sources may no longer be the same. The sources are only equivalent with respect to elements connected exernal to the terminals.

■ **EXAMPLE 8–4**   Convert the voltage source of Figure 8–9(a) into a current source and verify that the current, I_L, through the load is the same for each source.

(a)

(b)

Figure 8–9

Figure 11–15   The larger the time constant, the longer the capacitor takes to charge.

Figure 11–15 shows how voltage and current are affected by the time constant. The larger the resistance and capacitance, the larger the time constant and hence the longer it takes for the capacitor to charge.

practical notes

Since resistors and capacitors have tolerances, their actual values may be different from their marked values. This will make the actual time constant of the circuit different than the computed time constant. In many cases, this is not important. However, it may be. In critical applications, you should check the effect that R and C tolerances have on τ.

in process learning check I

1. If the capacitor of Figure 11–16 is uncharged, what is the current immediately after closing the switch?

Figure 11–16

2. Given $i_C = 50e^{-20t}$ mA.
   a. What is τ?
   b. Compute the current at $t = 0^+$ s, 25 ms, 50 ms, 75 ms, 100 ms, and 500 ms and sketch it.

3. Repeat Problem 2 for $v_C = 100(1 - e^{-50t})$ V.

4. For Figure 11–16, determine expressions for v_C and i_C.

5. Refer to Figure 11–10:
   a. What are $v_C(0^-)$ and $v_C(0^+)$?
   b. What are $i_C(0^-)$ and $i_C(0^+)$?
   c. What are the steady state voltage and current?

6. For the circuit of Figure 11–11, the current just after the switch is closed is 2 mA. The transient lasts 40 ms and the capacitor charges to 80 V. Determine E, R, and C.

Acknowledgements

Many people have contributed to the development of this book and we would now like to thank them. First, the reviewers.

- Clay Laster, San Antonio College, Texas,

- Martin Mayer, Red River Community College, Manitoba,

- Lyle McCurdy, Cal Poly Pomona, California,

- Jim Rhodes, Trident Technical College, North Carolina,

- Lee Rosenthal, Fairleigh Dickenson University, New Jersey,

- Richard Sturtevant, Springfield Technical College, Massachusetts.

The following firms and individuals supplied us with photographs, diagrams, or other useful information. To each of you, our sincere thanks.

Allen-Bradley
AT & T
AVX Corporation
B+K Precision
Bourns, Inc.
Butterworth & Co. Ltd.
Carte International Ltd.
Cognex Corporation
Condor DC Power Supplies Inc.
GE Reasearch and
   Development Center
Goguen Industries
Harris Corporation—Electronic
   Design Automation, Inc.
Harris Corporation—
   Farinon Division

Honeywell
Illinois Capacitor Inc.
JBL Professional
John Fluke Mfg. Co. Inc.
Manitoba Hydro
Martin Marietta
Mathsoft Inc.
MicroSim Corporation
Motorola
Siemens Solar Industries
Simpson Electric Company
Sprint
Tektronix
Transformer Manufacturers Inc.
Vansco Electronics

To the staff at Delmar for their tireless efforts in putting this book together, we express our deep appreciation. To our developmental editor,

Mary Clyne, for her encouragement and assistance, to Wendy Welch, our administrative editor, for her advice and valued help, and to Mark Huth, acquisition specialist, for initiating this project—a special thanks to all of you.

Lastly, we thank our wives and families for their support and patience during the preparation of this book.

Allan H. Robbins
Wilhelm C. Miller
November, 1993

P A R T I

Foundation dc Concepts

Circuit theory provides the tools and concepts needed to understand and analyze electrical and electronic circuits. The foundations of this theory were laid down well over a hundred years ago by a number of pioneer researchers. In 1780, Alessandro Volta of Italy developed an electric cell (battery) that provided the first source of what we now call dc voltage. Around the same time, the concept of current was evolved (even though nothing was known about the atomic structure of matter until much later). In 1826, Georg Simon Ohm of Germany brought the two ideas together and experimentally determined the relationship between voltage and current in a resistive circuit. This result, known as Ohm's law, set the stage for the development of modern-day circuit theory.

In Part I, we examine the foundation of this theory. We look at voltage, current, power, energy, and the relationships between them. The ideas developed here are used throughout the remainder of the book and in practice.

1

Introduction

CHAPTER PREVIEW

An electrical circuit is a system of interconnected components such as resistors, capacitors, inductors, voltage sources, and so on. The electrical behavior of these components is governed by a few basic experimental laws. These laws and the principles, concepts, mathematical relationships, and methods of analysis that have evolved from them are known as *circuit theory*.

Much of circuit theory deals with problem solving and numerical analysis. When you analyze a problem or design a circuit, for example, you are typically required to compute values for voltage, current, and power. In addition to a numerical value, your answer must include a unit. The system of units used for this purpose is the SI system (Système International). The SI system is a modernized metric system; it encompasses not only the familiar metric units for length, mass, and time, but also units for electrical and magnetic quantities as well.

Quite frequently, however, the SI units yield numbers that are either too large or too small for convenient use. To handle these, engineering notation and a set of standard prefixes have been developed. Their use in representation and computation is described and illustrated. The question of significant digits is also investigated.

Since circuit theory is somewhat abstract, diagrams are used to help present ideas. We look at several types—schematic, pictorial, and block diagrams—and show how to use them to represent circuits and systems.

We conclude the chapter with a brief look at computers and their use in circuit analysis and design. Several popular application packages and programming languages are described. Special emphasis is placed on PSpice, the software package used in this book.

KEY TERMS

Ampere
Block Diagram
Circuit
Conversion Factor
Current
Energy
Joule
Meter
Newton
Pictorial Diagram
Power of Ten Notation
Prefixes
Programming Language
Resistance
Schematic Diagram
Scientific Notation
SI Units
Significant Digits
SPICE
Volt
Watt

An electronic control panel from a vehicle control system. *(Courtesy Vansco Electronics. © Walter Kaiser/KKS/Wpg)*

OBJECTIVES

After studying this chapter, you will be able to

- describe the SI system of measurement,

- convert between various sets of units,

- use power of ten notation to simplify handling of large and small numbers,

- express electrical units using standard prefix notation such as μA, kV, mW, etc.,

- use a sensible number of significant digits in calculations,

- describe what block diagrams are and why they are used,

- convert a simple pictorial circuit to its schematic representation,

- describe generally how computers fit in the electrical circuit analysis picture.

Hints on Problem Solving

*D*uring the analysis of electric circuits, you will find yourself solving quite a few problems. An organized approach helps. Listed below are some useful guidelines:

1. *Make a sketch (e.g., a circuit diagram), mark on it what you know, then identify what it is that you are trying to determine. This should help you to organize your thoughts.*

2. *Think through the problem to identify the principles involved, then look for relationships that tie together the unknown and known quantities.*

3. *Substitute the known information into the selected equation(s) and solve for the unknown. (For complex problems, the solution may require a series of steps involving several concepts. If you cannot identify the complete set of steps before you start, start anyway. As each piece of the solution emerges, you are one step closer to the answer. You may make false starts. However, even experienced people do not get it right on the first try every time. Note also that there is seldom one "right" way to solve a problem. You may therefore come up with an entirely different correct solution than the authors do.)*

4. *Check the answer to see that it is sensible—that is, is it in the "right ballpark"? Does it have the correct sign? Do the units match?*

1.1 Introduction

Technology is rapidly changing the way we do things; we now have computers in our homes, electronic control systems in our cars, cellular phones that can be used just about anywhere, robots that produce products on production lines, and so on.

A first step to understanding these technologies is electric circuit theory. Circuit theory provides you with the knowledge of basic principles that you need to understand the behavior of electric and electronic devices, circuits, and systems. In this book, we develop and explore its basic ideas.

Before We Begin

Before we begin, let us look at a few examples of the technology at work. (As you go through these, you will see devices, components, and ideas that have not yet been discussed. You will learn about these later. For the moment, just concentrate on the general ideas.)

As a first example, consider Figure 1–1, which shows a VCR. Its design is based on electrical, electronic, and magnetic circuit principles. For example, resistors, capacitors, transistors, and integrated circuits are used to control the voltages and currents that operate its motors and amplify the audio and video signals that are the heart of the system. A magnetic circuit (the read/write system) performs the actual tape reads and writes. It creates, shapes, and controls the magnetic field that records audio and video signals on the tape. Another magnetic circuit, the power transformer, transforms the ac voltage from the 120-volt wall outlet voltage to the lower voltages required by the system.

Figure 1–2 shows another example. In this case, a designer, using a

Figure 1–1   A VCR is a familiar example of an electronic system.

Figure 1–2   A transformer designer using a 3-D electromagnetic analysis program to check the operation of a power transformer. Upper inset: Magnetic field pattern. *(Courtesy Carte International Inc.)*

personal computer, is analyzing the performance of a power transformer. The transformer must meet not only the voltage and current requirements of the application, but safety- and efficiency-related concerns as well. A software application package, programmed with basic electrical and magnetic circuit fundamentals, helps the user perform this task.

Figure 1–3 shows another application. It illustrates a manufacturing facility where electronic systems are assembled. Here, a "pick-and-place" robot is shown installing surface mount (SMT) components on a printed circuit (PC) card. On the bottom row of Figure 1–4, you can see some of

Figure 1–3   A "pick and place" robot installing surface mount components on a printed circuit board.

Surface mount
parts

Figure 1–4   Some typical electronic components. The small components at the bottom are surface mount parts.

Figure 1–5   The Cognex 5000, a machine vision card for a personal computer. *(Courtesy Cognex Corporation)*

TABLE 1–1 Common Quantities

1 meter = 100 centimeters = 39.37 inches
1 millimeter = 39.37 mils
1 inch = 2.54 centimeters
1 foot = 0.3048 meter
1 yard = 0.9144 meter
1 mile = 1.609 kilometers
1 kilogram = 1000 grams = 2.2 pounds
1 gallon (US) = 3.786 liters

these parts. Note how small they are. Electronic control provides the high precision needed to accurately position parts as small as these.

Figure 1–5 shows our final example, a machine vision card for a personal computer (PC). Machine vision systems are computer-based systems that use image processing techniques to give "sight" to machines. Typically, they use cameras or X-ray systems to collect visual information, which they then process to make decisions about what they "see." One important application of vision technology is in quality control; vision systems have been put to use to inspect the quality of solder joints on printed circuit boards during manufacture.

Before We Move On

Before we move on, we should note that, as diverse as these applications are, they all have one thing in common: all are rooted in the principles of circuit theory.

1.2 The SI System of Units

The solution of technical problems requires the use of units. At present, two major systems—the English (US Customary) and the metric—are in everyday use. For scientific and technical purposes, however, the English system has been largely superseded. In its place the SI system is used. Table 1–1 shows a few frequently encountered quantities with units expressed in both systems.

TABLE 1–2 Some SI Base Units

QUANTITY	SYMBOL	UNIT	ABBREVIATION
Length	ℓ	meter	m
Mass	m	kilogram	kg
Time	t	second	s
Electric current	I, i	ampere	A
Temperature	T	kelvin	K

The SI system combines the MKS metric units and the electrical units into one unified system: See Tables 1–2 and 1–3. (Do not worry about the electrical units yet. We define them later, starting in Chapter 2.) The units in Table 1–2 are defined units, while the units in Table 1–3 are derived units, obtained by combining units from Table 1–2. Note that some symbols and abbreviations use capital letters while others use lowercase letters.

A few non-SI units are still in use. For example, electric motors are commonly rated in horsepower, and wires are frequently specified in AWG sizes (American Wire Gage, Section 3.2). On occasion, you will need to convert non-SI units to SI units. Table 1–4 may be used for this purpose.

Definition of Units

When the metric system came into being in 1792, the meter was defined as one ten-millionth of the distance from the north pole to the equator and the second as $1/60 \times 1/60 \times 1/24$ of the mean solar day. Later, more accurate definitions based on physical laws of nature were adopted. The meter is now defined as the distance travelled by light in a vacuum in 1/299 792 458 of a second, while the second is defined in terms of the period of a cesium-based atomic clock. The definition of the kilogram is

TABLE 1–3 Some SI Derived Units*

QUANTITY	SYMBOL	UNIT	ABBREVIATION
Force	F	newton	N
Energy	W	joule	J
Power	P, p	watt	W
Voltage	V, v, E, e	volt	V
Charge	Q, q	coulomb	C
Resistance	R	ohm	Ω
Capacitance	C	farad	F
Inductance	L	henry	H
Frequency	f	hertz	Hz
Magnetic flux	Φ	weber	Wb
Magnetic flux density	B	tesla	T

*Electrical and magnetic quantities will be explained as you progress through the book. As in Table 1–2, the distinction between capitalized and lowercase letters is important.

TABLE 1-4 Conversions

	WHEN YOU KNOW	MULTIPLY BY	TO FIND
Length	inches (in)	0.0254	meters (m)
	feet (ft)	0.3048	meters (m)
	miles (mi)	1.609	kilometers (km)
Force	pounds (lb)	4.448	newtons (N)
Power	horsepower (hp)	746	watts (W)
Energy	kilowatthour (kWh)	3.6×10^6	joules* (J)
	foot-pound (ft-lb)	1.356	joules* (J)

Note: 1 joule = 1 newton-meter.

the mass of a specific platinum-iridium cylinder (the international proto-type), preserved at the International Bureau of Weights and Measures in France. This, however, may soon change; research is underway to develop an "electronic kilogram," based also on fundamental laws of nature.

Relative Size of the Units*

To gain a feel for the SI units and their relative size, refer to Tables 1–1 and 1–4. Note that 1 meter is equal to 39.37 inches; thus, 1 inch equals 1/39.37 = 0.0254 meter or 2.54 centimeters. A force of one pound is equal to 4.448 newtons; thus, 1 newton is equal to 1/4.448 = 0.225 pound of force, which is about the force required to lift a $^1/_4$-pound weight. One joule is the work done in moving a distance of one meter against a force of one newton. This is about equal to the work required to raise a quarter-pound weight one meter. Raising the weight one meter in one second requires about one watt of power.

The watt is also the SI unit for electrical power. A typical electric lamp, for example, consumes power at the rate of 60 watts, and a toaster at a rate of about 1000 watts, while a stereo system can deliver power to its speaker system at a rate of anywhere from a few watts to several hundred watts, depending on its design.

The link between electrical and mechanical units can be easily established. Consider an electrical generator. Mechanical power input produces electrical power output. If the generator were 100% efficient, then one watt of mechanical power input would yield one watt of electrical power output. This clearly ties the electrical and mechanical systems of units together.

However, just how big is a watt? While the above examples suggest that the watt is quite small in terms of the rate at which a human can work, it is actually quite large. For example, a person can do manual labor at a rate of about 60 watts when averaged over an 8-hour day—just enough to power a standard 60-watt electric lamp continuously over this time! A horse can do somewhat better. Based on experiment, Isaac Watt determined that a strong dray horse could average 746 watts. From this, he defined the horsepower (hp) as 1 horsepower = 746 watts. This is the figure that we still use today.

*Paraphrased from Edward C. Jordan and Keith Balmain, *Electromagnetic Waves and Radiating Systems,* Second Edition. (Englewood Cliffs, New Jersey: Prentice-Hall, Inc, 1968).

1.3 Converting Units

Often quantities expressed in one unit must be converted to another. For example, suppose you want to determine how many kilometers there are in ten miles. Given that 1 mile is equal to 1.609 kilometers, Table 1–1, you can write 1 mi = 1.609 km, using the abbreviations in Table 1–4. Now multiply both sides by 10. Thus, 10 mi = 16.09 km.

This procedure is quite adequate for simple conversions. However, for complex conversions, it may be difficult to keep track of units. The procedure outlined next helps. It involves writing units into the conversion sequence, cancelling where applicable, then gathering up the remaining units to ensure that the final result has the correct units.

To get at the idea, suppose you want to convert 12 centimeters to inches. From Table 1–1, 2.54 cm = 1 in. Since these are equivalent, you can write

$$\frac{2.54 \text{ cm}}{1 \text{ in}} = 1 \quad \text{or} \quad \frac{1 \text{ in}}{2.54 \text{ cm}} = 1 \tag{1–1}$$

Now multiply 12 cm by the second ratio and note that unwanted units cancel. Thus,

$$12 \text{ cm} \times \frac{1 \text{ in}}{2.54 \text{ cm}} = 4.72 \text{ in}$$

The quantities in equation 1–1 are called *conversion factors*. Conversion factors have a value of 1 and you can multiply by them without changing the value of an expression. When you have a chain of conversions, select factors so that all unwanted units cancel. This provides an automatic check on the final result as illustrated in part (b) of Example 1–1.

■ **EXAMPLE 1–1**   Given a speed of 60 miles per hour (mph),

   a. convert it to kilometers per hour,
   b. convert it to meters per second.

Solution

   a. Recall, 1 mi = 1.609 km. Thus,

$$1 = \frac{1.609 \text{ km}}{1 \text{ mi}}$$

Now multiply both sides by 60 mi/h and cancel units:

$$60 \text{ mi/h} = \frac{60 \text{ mi}}{\text{h}} \times \frac{1.609 \text{ km}}{1 \text{ mi}} = 96.54 \text{ km/h}$$

   b. Given that 1 mi = 1.609 km, 1 km = 1000 m, 1 h = 60 min, and 1 min = 60 s, define conversion factors

$$1 = \frac{1.609 \text{ km}}{1 \text{ mi}}, \quad 1 = \frac{1000 \text{ m}}{1 \text{ km}}, \quad 1 = \frac{1 \text{ h}}{60 \text{ min}}, \quad \text{and } 1 = \frac{1 \text{ min}}{60 \text{ s}}$$

Thus,

$$\frac{60 \text{ mi}}{\text{h}} = \frac{60 \text{ mi}}{\text{h}} \times \frac{1.609 \text{ km}}{1 \text{ mi}} \times \frac{1000 \text{ m}}{1 \text{ km}} \times \frac{1 \text{ h}}{60 \text{ min}} \times \frac{1 \text{ min}}{60 \text{ s}} = 26.82 \text{ m/s}$$

You can also solve this problem by treating the numerator and denominator separately. For example, you can convert miles to meters and hours to seconds, then divide (see Example 1–2). In the final analysis, both methods are equivalent.

■ **EXAMPLE 1–2**   Do Example 1–1(b) by expanding the top and bottom separately.

Solution

$$60 \text{ mi} = 60 \text{ mi} \times \frac{1.609 \text{ km}}{1 \text{ mi}} \times \frac{1000 \text{ m}}{1 \text{ km}} = 96\,540 \text{ m}$$

$$1 \text{ h} = 1 \text{ h} \times \frac{60 \text{ min}}{1 \text{ h}} \times \frac{60 \text{ s}}{1 \text{ min}} = 3600 \text{ s}$$

Thus, velocity = 96 540 m/3600 s = 26.82 m/s as above.

practice
problems
1

1. Area = πr^2. Given $r = 8$ inches, determine area in square meters (m^2).

2. A car travels 60 feet in 2 seconds. Determine
   a. its speed in meters per second,
   b. its speed in kilometers per hour.

For part (b), use the method of Example 1–1, then check using the method of Example 1–2. (Answers are in Appendix C.)

1.4 Power of Ten Notation

Electrical values vary tremendously in size. In electronic systems, for example, voltages may range from a few millionths of a volt to several thousand volts, while in power systems, voltages of up to several hundred thousand are common. To handle this large range, the *power of ten notation* (Table 1–5) is used.

To express a number in power of ten notation, move the decimal point to where you want it, then multiply the result by the power of ten needed to restore the number to its original value. Thus, 247 000 = 2.47 × 10^5. (The number 10 is called the *base*, and its power is called the *exponent*.) An easy way to determine the exponent is to count the number of places (right or left) that you moved the decimal point. Thus,

$$247\,000 = 2\,4\,7\,0\,0\,0 = 2.47 \times 10^5$$
$$5\,4\,3\,2\,1$$

TABLE 1–5 Common Power of Ten Multipliers

$1\ 000\ 000 = 10^6$	$0.000001 = 10^{-6}$
$100\ 000 = 10^5$	$0.00001 = 10^{-5}$
$10\ 000 = 10^4$	$0.0001 = 10^{-4}$
$1\ 000 = 10^3$	$0.001 = 10^{-3}$
$100 = 10^2$	$0.01 = 10^{-2}$
$10 = 10^1$	$0.1 = 10^{-1}$
$1 = 10^0$	$1 = 10^0$

Similarly, the number 0.003 69 may be expressed as 3.69×10^{-3} as illustrated below.

$$0.003\ 69 = 0.0\ 0\ 3\ 6\ 9 = 3.69 \times 10^{-3}$$

$$1\ 2\ 3$$

Multiplication and Division Using Powers of Ten

To multiply numbers in power of 10 notation, multiply their base numbers, then add their exponents. Thus,

$$(1.2 \times 10^3)(1.5 \times 10^4) = (1.2)(1.5) \times 10^{(3+4)} = 1.8 \times 10^7$$

For division, subtract the exponents in the denominator from those in the numerator. Thus,

$$\frac{4.5 \times 10^2}{3 \times 10^{-2}} = \frac{4.5}{3} \times 10^{2-(-2)} = 1.5 \times 10^4$$

■ **EXAMPLE 1–3**   Convert the following numbers to power of ten notation, then perform the operation indicated:

a. 276×0.009,

b. $98\ 200/20$.

Solution

a. $276 \times 0.009 = (2.76 \times 10^2)(9 \times 10^{-3}) = 24.8 \times 10^{-1} = 2.48$

b. $\dfrac{98\ 200}{20} = \dfrac{9.82 \times 10^4}{2 \times 10^1} = 4.91 \times 10^3$

Addition and Subtraction Using Powers of Ten

To add or subtract, first adjust all numbers to the same power of ten. It does not matter what exponent you choose, as long as they are all the same.

■ **EXAMPLE 1–4**   Add 3.25×10^2 and 5×10^3

a. using 10^2 representation,
b. using 10^3 representation.

Solution

a. $5 \times 10^3 = 50 \times 10^2$. Thus, $3.25 \times 10^2 + 50 \times 10^2 = 53.25 \times 10^2$

b. $3.25 \times 10^2 = 0.325 \times 10^3$. Thus, $0.325 \times 10^3 + 5 \times 10^3 = 5.325 \times 10^3$, which is the same as 53.25×10^2

Powers

Raising a number to a power is a form of multiplication (or division if the exponent is negative). For example,

$$(2 \times 10^3)^2 = (2 \times 10^3)(2 \times 10^3) = 4 \times 10^6$$

In general, $(N \times 10^n)^m = N^m \times 10^{nm}$. In this notation, $(2 \times 10^3)^2 = 2^2 \times 10^{3 \times 2} = 4 \times 10^6$ as before.

Integer fractional powers represent roots. Thus, $4^{1/2} = \sqrt{4}$ and $27^{1/3} = \sqrt[3]{27} = 3$.

■ **EXAMPLE 1–5**   Expand the following:

a. $(250)^3$ b. $(0.0056)^2$ c. $(141)^{-2}$ d. $(60)^{1/3}$

Solution

a. $(250)^3 = (2.5 \times 10^2)^3 = (2.5)^3 \times 10^{2 \times 3} = 15.625 \times 10^6$
b. $(0.0056)^2 = (5.6 \times 10^{-3})^2 = (5.6)^2 \times 10^{-6} = 31.36 \times 10^{-6}$
c. $(141)^{-2} = (1.41 \times 10^2)^{-2} = (1.41)^{-2} \times (10^2)^{-2} = 0.503 \times 10^{-4}$
d. $(60)^{1/3} = \sqrt[3]{60} = 3.91$

practice problems II

Determine the following:

a. $(6.9 \times 10^5)(0.392 \times 10^{-2})$
b. $(23.9 \times 10^{11})/(8.15 \times 10^5)$
c. $14.6 \times 10^2 + 11.2 \times 10^1$ (Express in 10^2 and 10^1 notation.)
d. $(29.6)^3$
e. $(0.385)^{-2}$

Use common sense when handling numbers. With modern calculators, for example, it may be easier to work with the numbers in their original form, then convert to a power of ten as the last step only if necessary. For example, it is more sensible to multiply 276×0.009 directly than to convert to power of ten notation first as we did in Example 1–3(a).

1.5 Prefixes

Scientific and Engineering Notation

TABLE 1–6 Common Prefixes

POWER OF 10	PREfiX	SYMBOL
10^{12}	tera	T
10^{9}	giga	G
10^{6}	mega	M
10^{3}	kilo	k
10^{-3}	milli	m
10^{-6}	micro	μ
10^{-9}	nano	n
10^{-12}	pico	p

If power of ten numbers are written with one digit to the left of the decimal place, they are said to be in *scientific notation*. Thus, 2.47×10^5 is in scientific notation, while 24.7×10^4 and 0.247×10^6 are not. However, we are more interested in *engineering notation*. In engineering notation, prefixes are used to represent certain powers of ten; see Table 1–6. Thus, a quantity such as 0.045 A (amperes) can be expressed as 45×10^{-3} A, but it is preferable to express it as 45 mA. Here, we have substituted the prefix *milli* for the multiplier 10^{-3}. It is usual to select a prefix that results in a base number between 0.1 and 999. Thus, 1.5×10^{-5} s would be expressed as 15 μs.

■ **EXAMPLE 1–6**   Express the following in prefix notation:

  a. 10×10^4 volts   **b.** 0.1×10^{-3} watts   **c.** 250×10^{-7} seconds

Solution

  a. 10×10^4 V $= 100 \times 10^3$ V $= 100$ kilovolts $= 100$ kV
  b. 0.1×10^{-3} W $= 0.1$ milliwatts $= 0.1$ mW
  c. 250×10^{-7} s $= 25 \times 10^{-6}$ s $= 25$ microseconds $= 25$ μs

■ **EXAMPLE 1–7**   Convert 0.1 MV to kilovolts (kV).

Solution

$$0.1 \text{ MV} = 0.1 \times 10^6 \text{ V} = (0.1 \times 10^3) \times 10^3 = 100 \text{ kV}$$

Remember that a prefix represents a power of ten and thus the rules for power of ten computation apply. For example, when adding or subtracting, adjust to a common base, as illustrated in Example 1–8.

■ **EXAMPLE 1–8**   Compute the sum of 1 ampere (amp) and 100 milliamperes.

  Solution   Adjust to a common base, either amps (A) or milliamps (mA). Thus,

$$1 \text{ A} + 100 \text{ mA} = 1 \text{ A} + 100 \times 10^{-3} \text{ A} = 1 \text{ A} + 0.1 \text{ A} = 1.1 \text{ A}$$

Alternatively, $1 \text{ A} + 100 \text{ mA} = 1000 \text{ mA} + 100 \text{ mA} = 1100 \text{ mA}$.

**practice
problems
III**

1. Convert 1800 kV to megavolts (MV).

2. In Chapter 4, we show that voltage is the product of current times resistance—that is, $V = I \times R$, where V is in volts, I is in amperes, and R is in ohms. Given $I = 25$ mA and $R = 4$ kΩ, convert these to power of ten notation, then show that $V = 100$ V.

**in process
learning
check I**

1. All conversion factors have a value of what?

2. Convert 14 yards to centimeters.

3. What units does the following reduce to?

$$\frac{km}{h} \times \frac{m}{km} \times \frac{h}{min} \times \frac{min}{s}$$

4. Express the following in engineering notation:
    a. 4270 ms   b. 0.001 53 V   c. 12.3×10^{-4} s

5. Express the result of each of the following computations as a number times 10 to the power indicated:
    a. 150×120 as a value times 10^4; as a value times 10^3.
    b. $300 \times 6/0.005$ as a value times 10^4; as a value times 10^5; as a value times 10^6.
    c. $430 + 15$ as a value times 10^2; as a value times 10^1.
    d. $(3 \times 10^{-2})^3$ as a value times 10^{-6}; as a value times 10^{-5}.

6. Express each of the following as indicated.
    a. 752 μA in mA.
    b. 0.98 mV in μV.
    c. 270 μs + 0.13 ms in μs and in ms.

(Answers are in Appendix C.)

1.6 Significant Digits and Numerical Accuracy

The number of digits in a number that carry actual information are termed *significant digits*. Thus, if we say a piece of wire is 3.57 meters long, we mean that its length is closer to 3.57 m than it is to 3.56 m or 3.58 m and we have three significant digits. (The number of significant digits includes the first estimated digit.) If we say that it is 3.570 m, we mean that it is closer to 3.570 m than to 3.569 m or 3.571 m and we have four significant digits. When determining significant digits, zeros used to locate the decimal point are not counted. Thus, 0.004 57 has three significant digits; this can be seen if you express it as 4.57×10^{-3}.

Most calculations that you will do in circuit theory will be done using a hand calculator. An error that has become quite common is to show more digits of "accuracy" in an answer than are warranted, simply because the numbers appear on the calculator display. The number of digits that you should show is related to the number of significant digits in the numbers used in the calculation.

To illustrate, suppose you have two numbers—$A = 3.76$ and $B = 3.7$ —to be multiplied. Their product is 13.912. If the numbers 3.76 and 3.7 are exact (i.e., have no uncertainty), this answer is correct. However, if the numbers have been obtained by measurement where values cannot be determined exactly, they will have some uncertainty and the product must reflect this uncertainty. For example, suppose A and B have an uncertainty of 1 in their first estimated digit—that is, $A = 3.76 \pm 0.01$ and $B = 3.7 \pm 0.1$. This means that A can be as small as 3.75 or as large as 3.77, while B can be as small as 3.6 or as large as 3.8. Thus, their product can be as small as $3.75 \times 3.6 = 13.50$ or as large as $3.77 \times 3.8 = 14.326$. The best that we can say about the product is that it is 14, i.e., that you know it only to the nearest whole number. You cannot even say that it is 14.0 since this implies that you know the answer to the nearest tenth, which, as you can see from the above, you do not.

We can now give a "rule of thumb" for determining significant digits. *The number of significant digits in a result due to multiplication or division is the same as the number of significant digits in the number with the least number of significant digits.* In the previous calculation, for example, 3.7 has two significant digits so that the answer can have only two significant digits as well. This agrees with our earlier observation that the answer is 14, not 14.0 (which has three).

When adding or subtracting, you must also use common sense. For example, suppose two currents are measured as 24.7 A (one place known after the decimal point) and 123 mA (i.e., 0.123 A). Their sum is 24.823 A. However, the right-hand digits 23 in the answer are not significant. They cannot be, since, if you don't know what the second digit after the decimal point is for the first current, it is senseless to claim that you know their sum to the third decimal place! The best that we can say about the sum is that it also has one significant digit after the decimal place, that is,

$$
\begin{array}{r}
24.7 \text{ A} \quad \text{(One place after decimal)} \\
+\ 0.123 \text{ A} \\
\hline
24.823 \text{ A}
\end{array}
\ \longrightarrow\ 24.8 \text{ A (One place after decimal)}
$$

Therefore, when adding numbers, add the given data, then round the result to the last column where all given numbers have significant digits. The process is similar for subtraction.

**practice
problems
IV**

1. Assume that only the digits shown in $8.75 \times 2.446 \times 9.15$ are significant. Determine their product and show it with the correct number of significant digits.

2. For the numbers of Problem 1, determine

$$
\frac{8.75 \times 2.446}{9.15}
$$

3. If the numbers in Problems 1 and 2 are exact, what are the answers to eight digits?

4. Three currents are measured as 2.36 A, 11.5 A, and 452 mA. Only the digits shown are significant. What is their sum shown to the correct number of significant digits?

1.7 Circuit Diagrams

Electric circuits are constructed using components such as batteries, switches, resistors, capacitors, transistors, interconnecting wires, etc. To represent these circuits on paper, diagrams are used. In this book, we use three types: *block diagrams*, *schematic diagrams*, and *pictorials*.

Block Diagrams

Block diagrams describe a circuit or system in simplified form. The overall problem is broken into blocks, each representing a portion of the system or circuit. Blocks are labelled to indicate what they do or what they contain, then interconnected to show their relationship to each other. General signal flow is usually from left to right and top to bottom. Figure 1–6, for example, represents an audio amplifier. Although you have not covered any of its circuits yet, you should be able to follow the general idea quite easily—sound is picked up by the microphone, converted to an electrical signal, amplified by a pair of amplifiers, then output to the speaker, where it is converted back to sound. A power supply energizes the system. The advantage of a block diagram is that it gives you the overall picture and helps you understand the general nature of a problem. However, it does not provide detail.

Pictorial Diagrams

Pictorial diagrams are one of the types of diagrams that provide detail. They help you visualize circuits and their operation by showing components as they actually appear. For example, the circuit of Figure 1–7 consists of a battery, a switch, and an electric lamp, all interconnected by wire. Operation is easy to visualize—when the switch is closed, the battery causes current in the circuit, which lights the lamp. The battery is referred to as the source and the lamp as the load.

Schematic Diagrams

While pictorial diagrams help you visualize circuits, they are cumbersome to draw. Schematic diagrams get around this by using simplified, standard symbols to represent components; see Table 1–7. (The meaning

Figure 1–6   An example block diagram. Pictured is a simplified representation of an audio amplification system.

Figure 1–7   A pictorial diagram. The battery is referred to as a *source* while the lamp is referred to as a *load*. (The + and − on the battery are discussed in Chapter 2.)

TABLE 1-7   Schematic Circuit Symbols

(a) Schematic using lamp symbol

(b) Schematic using resistance symbol

Figure 1–8   Schematic representation of Figure 1–7. The lamp has a circuit property called *resistance*; see Chapter 3.

of these symbols will be made clear as you progress through the book.) In Figure 1–8(a), for example, we have used some of these symbols to create a schematic for the circuit of Figure 1–7. Each component has been replaced by its corresponding circuit symbol.

When choosing symbols, choose those that are appropriate to the occasion. Consider the lamp of Figure 1–8(a). As we will show later, the lamp possesses a property called *resistance* that causes it to resist the passage of charge. When you wish to emphasize this property, use the resistance symbol rather than the lamp symbol, as in Figure 1–8(b).

When you draw schematic diagrams, draw them with horizontal and vertical lines joined at right angles as in Figure 1–8. This is standard practice. (At this point you should glance through some later chapters, e.g., Chapter 7, and study additional examples.)

1.8 Circuit Analysis Using Computers

Personal computers are used extensively for analysis and design. Software tools for such tasks fall into two broad categories: prepackaged application programs (*application packages*) and *programming languages*. The distinction between these two is quite fundamental. Application packages solve problems without requiring programming on the part of the user. Programming languages, by contrast, require the user to write code for each type of problem to be solved. There are some trade-offs. Application packages are much easier to use, but programming languages can be more flexible. Cost may also be a factor.

Prepackaged Simulation Software

Application packages used for circuit analysis fall mainly into the category of simulator packages and mathematics packages. (Simulation software solves circuits by simulating their behavior, rather than by solving a set of equations.) A popular package is SPICE. (SPICE is an acronym for *Simulation Program with Integrated Circuit Emphasis*.) It analyzes circuits from a physical description of the circuit rather than from equations. To describe the circuit to SPICE, you tell it: 1) what components the circuit consists of, 2) what their values are, and 3) how they are interconnected. From such a description, SPICE analyzes the circuit and determines voltages and currents as required.

One way to create a circuit description for SPICE is shown in Figure 1–9. First, you number nodes (points where two or more components connect) on the schematic, then prepare and enter into the computer a list summarizing how the various circuit elements are connected. For the circuit in Figure 1–9, for example, you tell SPICE that you have a 12-volt battery between nodes 1 and 0, a 10-ohm resistance between nodes 1 and 2, and a 15-ohm resistance between nodes 2 and 0. (A worked-out example is shown at the end of Chapter 4 if you care to look at it at this time.)

Many versions of SPICE also include a graphics entry (*schematic*

Figure 1–9   Schematic diagram annotated for data entry into SPICE.

Figure 1–10   Computer screen illustrating circuit analysis using simulation software. This circuit was created on the screen using Schematics, a schematic entry package. *(Courtesy MicroSim Corporation)*

capture) package that lets you create circuit schematics directly on your computer screen, Figure 1–10. (You will learn later what these symbols mean.)

The version of SPICE used in this book is the Design Center evaluation version of PSpice from MicroSim Corporation. (The evaluation version is a subset of their production version, but is capable of solving all examples and problems in this book. It is free to educational institutes.)

Prepackaged Math Software

Math packages are also useful. Again, no programming is required. A popular product is Mathcad from Mathsoft Inc. With Mathcad, you enter equations in standard mathematical notation. For example, to find the first root of a quadratic equation, you would use

$$x: = \frac{-b + \sqrt{b^2 - 4 \cdot a \cdot c}}{2 \cdot a}$$

Mathcad incorporates a built-in Electronic Handbook (software reference library) that contains hundreds of formulas plus some circuit diagrams. These can be selected by the user and positioned on the screen, eliminating a great deal of typing; see Figure 1–11. (You will learn later what these symbols mean.)

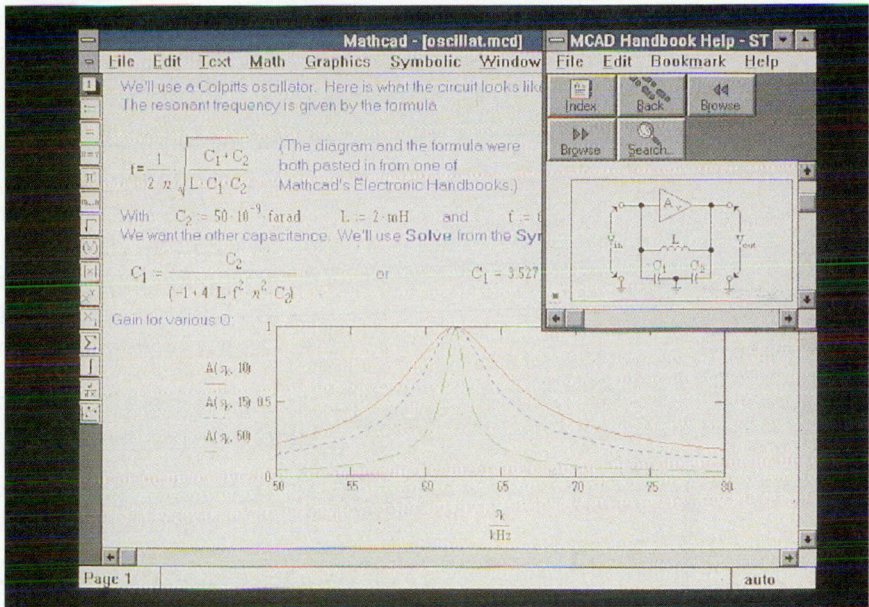

Figure 1–11   Computer screen showing circuit analysis using math software. *(Courtesy Mathsoft Inc.)*

Programming Languages

Many problems can also be solved using programming languages such as BASIC, C, or FORTRAN. To solve a problem using a programming language, you code its solution, step by step. While this can be tedious, you have the advantage of complete control. However, it generally takes considerably longer to create a program than to use an application package. In addition, you need programming skills. In contrast, application packages usually require only that you need to know how to enter data.

A Word of Caution

With the widespread availability of inexpensive software tools, you may wonder why you are asked to solve problems manually throughout this book. The reason is quite fundamental; as a student, your job is to learn principles and concepts. Getting correct answers using prepackaged software does not necessarily mean that you understand the theory—it may mean only that you know how to enter data. Software tools should always be used wisely. Before you use PSpice (or any other application package), be sure that you understand the basics of the subject that you are studying. This is why you should solve problems manually with your calculator first. Following this, try some of the application packages to explore ideas. Most chapters (starting with Chapter 4) include a selection of worked-out examples and PSpice problems to get you started.

Problems*

1.3   Converting Units

1. Make the following conversions:
   a.   100 square feet to square meters.
   b.   1 square inch to square meters.
   c.   A 1.2-m by 70-cm area to square meters (m^2).
   d.   47-pound force to newtons.
   e.   35 horsepower to watts.
   f.   1380 revolutions per hour to degrees per second.

2. If your car gets 27 miles per gallon (US), what is this in kilometers per liter?

3. Set up the appropriate conversion factors and compute the number of seconds in three years. (Assume no leap years.)

4. The equatorial radius of the earth is 3963 miles. What is the earth's circumference in kilometers at the equator?

5. The height of horses is sometimes measured in "hands," where 1 hand = 4 inches. How many meters tall is a 16-hand horse? How many centimeters?

6. Suppose $s = vt$ is given, where s is distance travelled, v is velocity, and t is time. If you travel at $v = 60$ mph for 500 seconds, you get upon unthinking substitution $s = vt = (60)(500) = 30,000$ miles. What is wrong with this calculation? What is the correct answer?

7. How long does it take for a pizza cutter traveling at 0.12 m/s to cut diagonally across a 15-in pizza?

8. Joe S. was asked to convert 2000 yd/h to meters per second. Here is Joe's work: velocity = $2000 \times 0.9144 \times 60/60 = 1828.8$ m/s. Determine conversion factors, write units into the conversion, and find the correct answer.

9. The mean distance from the earth to the moon is 238 857 miles. Radio signals travel at 299 792 458 m/s. How long does it take a radio signal to reach the moon?

10. Given that 1 hp = 550 ft-lb/s, 1 ft = 0.3048 m, 1 lb = 4.448 N, 1 J = 1 N-m, and 1 W = 1 J/s, show that 1 hp = 746 W.

1.4   Power of Ten Notation

11. Express each of the following in power of ten notation with one nonzero digit to the left of the decimal point:
   a.   8675
   b.   0.008 72
   c. 12.4×10^2
   d. 37.2×10^{-2}
   e. $0.003\ 48 \times 10^5$
   f. $0.000\ 215 \times 10^{-3}$
   g. 14.7×10^0

*Answers to odd numbers are in Appendix D.

12. For each of the following, convert the numbers to power of ten notation, then perform the indicated computations. Round your answer to four digits:

a. $(452)(6.73 \times 10^4)$

b. $(0.009\ 85)(4700)$

c. $(0.0892)/(0.000\ 067\ 3)$

d. $12.40 - 236 \times 10^{-2}$

e. $(1.27)^3 + 47.9/(0.8)^2$

f. $(-643 \times 10^{-3})^3$

g. $[(0.0025)^{1/2}][1.6 \times 10^4]$

h. $[(-0.027)^{1/3}]/[1.5 \times 10^{-4}]$

i. $\dfrac{(3.5 \times 10^4)^{-2} \times (0.0045)^2 \times (729)^{1/3}}{[(0.008\ 72) \times (47)^3] - 356}$

13. For the following,

a. convert numbers to power of ten notation, then perform the indicated computation,

b. perform the operation directly on your calculator without conversion. What is your conclusion?

i. 842×0.0014

ii. $\dfrac{0.0352}{0.007\ 91}$

14. Express each of the following in conventional notation:

a. 34.9×10^4

b. 15.1×10^0

c. 234.6×10^{-4}

d. 6.97×10^{-2}

e. $45\ 786.97 \times 10^{-1}$

f. 6.97×10^{-5}

15. One coulomb (Chapter 2) is the amount of charge represented by 6 240 000 000 000 000 000 electrons. Express this quantity in power of ten notation.

16. Compute the distance traveled in meters by light in a vacuum in 1.2×10^{-8} second.

17. How long does it take light to travel 3.47×10^5 km in a vacuum?

1.5   Prefixes

18. What is the appropriate prefix and its abbreviation for each of the following multipliers ?

a. 1000

b. 1 000 000

c. 10^9

d. 0.000 001

e. 10^{-3}

f. 10^{-12}

19. Express the following in terms of their abbreviations, e.g., microwatts as μW. Pay particular attention to capitalization (e.g., V, not v, for volts).

a. milliamperes

b. kilovolts

c. megawatts

d. microseconds

e. micrometers

f. milliseconds

g. nanoamps

20. Express the following in the most sensible engineering notation:

a. 0.0015 s   b. 0.000 027 s   c. 0.000 35 ms

21. Convert the following:

   a. 0.15 mV to microvolts
   b. 156 mV to volts
   c. 47 kW to watts
   d. 0.057 MW to kilowatts
   e. 3.5×10^4 volts to kilovolts
   f. 0.000 035 7 amps to microamps

22. Convert each of the following to a single prefix. For example, $15\ \mu\mu s = 15$ ps.

   a. 35 mmA
   b. 12×10^{-3} ns
   c. 0.049 kkV
   d. 9 kMW

23. A radio signal travels at 299 792.458 km/s and a telephone signal at 150 m/μs. If they originate at the same point, which arrives first at a destination 5000 km away? By how much?

1.6   Significant Digits and Numerical Accuracy

For each of the following, assume that the given digits are significant.

24. Determine the answer to three significant digits:

$$2.35 - 1.47 \times 10^{-6}$$

25. Given $V = IR$. If $I = 2.54$ and $R = 52.71$, determine V to the correct number of significant digits.

26. If $A = 4.05 \pm 0.01$ is divided by $B = 2.80 \pm 0.01$,

   a. What is the smallest that the result can be?
   b. What is the largest that the result can be?
   c. Based on this, give the result A/B to the correct number of significant digits.

27. The large black plastic component soldered onto the printed circuit board of Figure 1–12(a) is an electronic device known as an *integrated circuit*. As indicated in (b), the center-to-center spacing of its leads (commonly called pins) is 0.8 ± 0.1 mm. Pin diameters can vary from 0.25 to 0.45 mm. Considering these uncertainties,

   a. What is the minimum distance between pins due to manufacturing tolerances?
   b. What is the maximum distance ?

1.7   Circuit Diagrams

28. Consider the pictorial diagram of Figure 1–13. Using the appropriate symbols from Table 1–7, draw this in schematic form. Hint: In later chapters, there are many schematic circuits containing resistors, inductors, and capacitors. Use these as aids.

29. Draw the schematic diagram for a simple flashlight.

0.8 TYP ±0.1   0.25 / 0.45

(b)

Figure 1–12   (a)

Figure 1–13

1.8   Circuit Analysis Using Computers

30. Many electronic and computer magazines carry advertisements for computer software tools. Circuit and math tools are available under such names as PSpice, SpiceNet, Mathcad, MLAB, Matlab, Maple V, plus others. Investigate a few of these magazines in your school's library; by studying such advertisements, you can gain valuable insight into what modern software packages are able to do.

Voltage and Current

CHAPTER PREVIEW

A basic electric circuit consisting of a source, a switch, a load, and inter-connecting wire is shown in Figure 2–1. When the switch is closed, current in the circuit causes the light to come on. This circuit is representative of many common circuits found in practice, including those of flashlights and automobile headlight systems. We will use it to help develop an understanding of voltage and current.

Figure 2–1   A basic electric circuit.

KEY TERMS
Ampere
Atom
Battery
Cell
Circuit Breaker
Conductor
Coulomb
Coulomb's Law
Current
Electric Charge
Electron
Free Electrons
Fuse
Insulator
Ion
Neutron
Polarity
Potential Difference
Proton
Semiconductor
Shell
Switch
Valence
Volt

Elementary atomic theory shows that the current in Figure 2–1 is actually a flow of charges. The cause of their movement is the "voltage" of the source. While in Figure 2–1, this source is a battery, in practice it may be any one of a number of practical sources including generators, power supplies, solar cells, and so on.

In this chapter we look at the basic ideas of voltage and current. We begin with a discussion of atomic theory. This leads to the concept of free electrons and the idea of current as a movement of charge. The fundamental definitions of voltage and current are then developed. Following this, we look at a number of common voltage sources. The chapter concludes with a discussion of voltmeters and ammeters and the measurement of voltage and current in practice.

A familiar form of voltage source is the battery.

OBJECTIVES

After studying this chapter, you will be able to

- describe the makeup of an atom,

- explain the relationships between valence shells, free electrons, and conduction,

- describe the fundamental (coulomb) force within an atom, and the energy required to create free electrons,

- describe what ions are and how they are created,

- describe the characteristics of conductors, insulators, and semiconductors,

- describe the coulomb as a measure of charge,

- define voltage,

- describe how a battery "creates" voltage,

- explain current as a movement of charge and how voltage causes current in a conductor,

- describe important battery types and their characteristics,

- describe how to measure voltage and current.

The Equations of Circuit Theory

I n this chapter you meet the first of the equations and formulas that we use to describe the relationships of circuit theory. Remembering formulas is made easier if you clearly understand the principles and concepts on which they are based. As you may recall from high school physics, formulas can come about only through experiment, definition, or by mathematical manipulation.

Experimental Formulas

Circuit theory rests on a few basic experimental results. These are results that can be proven in no other way; they are valid solely because experiment has shown them to be true. The most fundamental of these are called "laws." Four examples are Ohm's law, Kirchhoff's current law, Kirchhoff's voltage law, and Faraday's law. (These laws will be met in various chapters throughout the book.) When you see a formula referred to as a law or an experimental result, remember that it is based on experiment and cannot be obtained in any other way.

Defined Formulas

Some formulas are created by definition, i.e., we make them up. For example, there are 60 seconds in a minute because we define the second as 1/60 of a minute. From this we get the formula $t_{sec} = 60 \times t_{min}$.

Derived Formulas

This type of formula or equation is created mathematically by combining or manipulating other formulas. In contrast to the other two types of formulas, the only way that a derived relationship can be proven is by mathematics.

An awareness of where circuit theory formulas come from is important to you. This awareness not only helps you understand and remember formulas, it helps you understand the very foundations of the theory—the basic experimental premises upon which it rests, the important definitions that have been made, and the methods by which these foundation ideas have been put together. This can help enormously in understanding and remembering concepts.

2.1 Atomic Theory

The basic structure of an atom is shown symbolically in Figure 2–2. It consists of a nucleus of protons and neutrons surrounded by a group of orbiting electrons. The electrons are negatively charged (−), while the protons are positively charged (+). Each atom (in its normal state) has an equal number of electrons and protons, and since their charges are equal and opposite, they cancel, leaving the atom electrically neutral, i.e., with zero net charge. The nucleus, however, has a net positive charge, since it consists of positively charged protons and uncharged neutrons.

The basic structure of Figure 2–2 applies to all elements, but each element has its own unique combination of electrons, protons, and neutrons. For example, the hydrogen atom, the simplest of all atoms, has one proton and one electron, while the copper atom has 29 electrons, 29 protons, and 35 neutrons. Silicon, which is important because of its use in transistors and other electronic devices, has 14 electrons, 14 protons, and 14 neutrons.

Electrons orbit the nucleus in discrete orbits called *shells*, designated by letters *K*, *L*, *M*, *N*, and so on (Figure 2–3). Only certain numbers of electrons can exist within any given shell. For example, there can be up to 2 electrons in the *K* shell, up to 8 in the *L* shell, up to 18 in the *M* shell, and up to 32 in the *N* shell. The number in any shell depends on the element. For instance, the copper atom, which has 29 electrons, has all three of its inner shells completely filled but its outer shell (shell *N*)

Electron
(negative charge)

Proton (positive charge)

Neutron (uncharged)

Figure 2–2   Bohr model of the atom. Electrons travel around the nucleus at incredible speeds, making billions of trips in a fraction of a second.

has only 1 electron, Figure 2–4. This electron (as noted below) is called its valence electron.

Valence Shells

The outermost shell of an atom is called its *valence shell* and electrons in this shell are called *valence electrons*. No element can have more than eight valence electrons; when a valence shell has eight electrons, it is filled. As we shall see, the number of valence electrons that an element has directly affects its electrical properties.

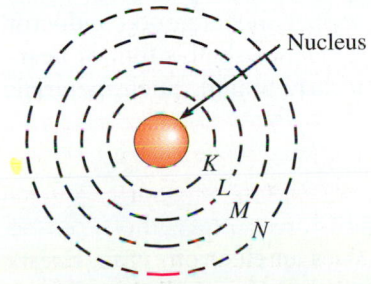

Nucleus

K
L
M
N

Figure 2–3   Simplified representation of the atom. Electrons travel in discrete orbits called "shells."

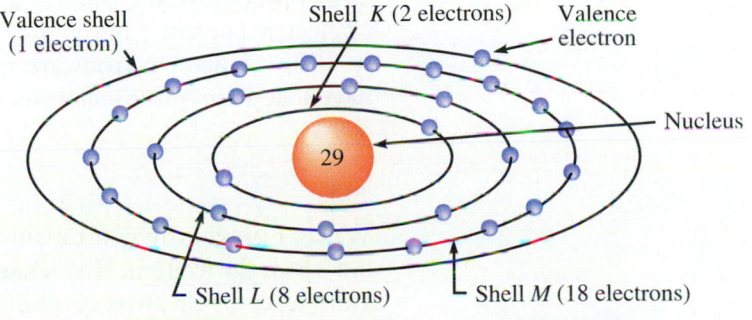

Valence shell
(1 electron)

Shell K (2 electrons)

Valence
electron

Nucleus

29

Shell L (8 electrons)

Shell M (18 electrons)

Figure 2–4   Copper atom. The valence electron is loosely bound.

(a) Like charges repel

(b) Unlike charges attract

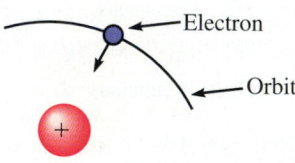

Electron

Orbit

(c) The force of attraction keeps electrons in orbit

Figure 2–5 Coulomb law forces.

Figure 2–6 Random motion of free electrons in a conductor.

Free Electrons

The electrons of Figure 2–2 travel around the nucleus at incredible speeds. What keeps them in orbit is the force of attraction that exists between unlike charges. This force was studied by Charles Coulomb (1736–1806). Coulomb determined experimentally that the force of attraction or repulsion between charged bodies is directly proportional to the product of their charges and inversely proportional to the square of the distance between them. Mathematically, Coulomb's law states

$$F = k\frac{Q_1 Q_2}{r^2} \quad \text{[newtons, N]} \tag{2–1}$$

where Q_1 and Q_2 are the charges on the two bodies, r is the center-to-center spacing between them in meters, and $k = 9 \times 10^9$. (Charges will be discussed in Section 2.2. For the moment, think of the charged bodies as being an electron and a proton respectively.) Unlike charges attract and like charges repel, as shown in Figure 2–5. Thus, electrons (−) are attracted to the positive charge (+) of the nucleus as indicated in Figure 2–5(c).

As Coulomb's law indicates, force decreases inversely as the square of distance; thus, if the distance between two charges is doubled, the force decreases to one quarter of its original value. Because of this relationship, electrons in outer orbits are less strongly attracted to the nucleus than those in inner orbits; that is, they are less *tightly bound* to the nucleus than those close by. Valence electrons are the least tightly bound and will, if they acquire sufficient energy, escape from their parent atoms.

The amount of energy required to escape depends on the number of electrons in the valence shell. If an atom has only a few valence electrons, only a small amount of additional energy is needed. For example, for a metal like copper, valence electrons can gain sufficient energy from heat alone (thermal energy), even at room temperature, to escape from their parent atoms and wander from atom to atom throughout the material as depicted in Figure 2–6. (Note that these electrons do not leave the substance, they simply wander from atom to atom. The material therefore remains electrically neutral.) Such electrons are called *free electrons*. In copper, there are of the order of 10^{23} free electrons per cubic centimeter at room temperature. As we shall see, it is the presence of this large number of free electrons that makes copper such a good conductor of electric current. On the other hand, if the valence shell is full (or nearly full), valence electrons are much more tightly bound. Such materials have few (if any) free electrons.

Ions

When a previously neutral atom gains or loses an electron, it acquires a net electrical charge. The charged atom is referred to as an *ion*. If the atom loses an electron, it is called a *positive ion*; if it gains an electron, it is called a *negative ion*.

Conductors, Insulators, and Semiconductors

The atomic structure of matter affects how easily charges move through a substance and hence how it is used electrically. Materials are classified as conductors, insulators, or semiconductors.

Conductors. Materials that have large numbers of free electrons that are able to move about easily are termed *conductors*. In general, metals are good conductors because they have only a few loosely bound valence electrons. In particular, silver, copper, gold, and aluminum are excellent conductors. Of these, copper is the most widely used. It is used in electrical and electronic equipment, in house wiring, and so on. Aluminum, although it is only about 60% as good a conductor as copper, is also used, mainly in applications where light weight is important, such as in overhead power transmission lines. Silver and gold are too expensive for general use. However, gold, because it oxidizes less than other materials, is used in specialized applications; for example, some critical electrical connectors use it because it makes a more reliable connection than other materials.

Insulators. Materials that do not conduct (e.g., glass, porcelain, plastic, rubber, and so on) are termed *insulators*. The covering on electric lamp cords, for example, is an insulator. It is used to prevent the wires from touching and to protect us from electric shock.

Insulators do not conduct because they have full or nearly full valence shells and thus their electrons are tightly bound. However, when high enough voltage is applied, electrons are literally torn from their parent atoms, causing the insulation to break down and conduct. In air, you see this as an arc or flashover. In solids, a small, carbonized burn path may result.

Semiconductors. Silicon and germanium (plus a few other materials) have half-filled valence shells and are thus neither good conductors nor good insulators. Known as *semiconductors*, they have unique electrical properties that make them important to the electronics industry. The most important material is silicon. It is used to make transistors, diodes, integrated circuits, and other electronic devices. Semiconductors have made possible personal computers, VCRs, portable CD players, calculators, and a host of other electronic products. You will study them in great detail in your electronics courses.

in process
l e a r n i n g
check I

1. Describe the basic structure of the atom in terms of its constituent particles: electrons, protons, and neutrons. Why is the nucleus positively charged? Why is the atom as a whole electrically neutral?

2. What are valence shells and valence electrons?

3. Describe Coulomb's law and use it to help explain why electrons far from the nucleus are loosely bound.

4. What are free electrons? Describe how they are created, using copper as an example. Explain what role thermal energy plays in the process.

(continued)

5. Briefly distinguish between a normal atom, a positive ion, and a negative ion.

6. Many atoms in Figure 2–6 have lost electrons and are thus positively charged, yet the substance as a whole is uncharged. Why?

7. In terms of free electrons, describe what makes a good conductor and a good insulator.

2.2 Electric Charge

Normally, substances as a whole have equal numbers of electrons and protons and are thus electrically neutral. Such substances are said to be *uncharged*. If charge is now removed from one body and transferred to another, both bodies become charged. When we refer to a body as being charged, we mean that it has an excess or deficiency of electrons. If it has an excess of electrons, it is negatively charged; if it has a deficiency, it is positively charged. Charge is represented by the letter Q.

A familiar example of charging occurs when you walk across a wool carpet. This action strips electrons from one material and transfers them to the other. Since both your shoes and the carpet are insulators, the displaced electrons cannot flow back, and hence, one material is left with an excess of electrons and the other a deficiency. Thus, your shoes (and you) become charged. If you now touch a metal object, you get a shock. A similar situation occurs when you tumble clothes in a dryer. Electrons are stripped from some garments and transferred to others. The force of attraction between unlike charges causes the garments to cling together, hence static cling occurs.

The Unit of Electrical Charge: The Coulomb

The unit of electrical charge is the coulomb (C). One coulomb is defined as the charge carried by 6.24×10^{18} electrons. Thus, if a neutral body acquires an excess of 6.24×10^{18} electrons, it will have a negative charge of 1 C (i.e., $Q = -1$ C), while if it loses 6.24×10^{18} electrons, it will have a positive charge of 1 C (i.e., $Q = +1$ C). The charge on one electron is $Q_e = 1/(6.24 \times 10^{18}) = 1.6 \times 10^{-19}$ C.

■ **EXAMPLE 2–1**   An initially neutral body has 1.7 μC of negative charge removed. Later, 18.7×10^{11} electrons are added. What is the body's final charge?

Solution   Initially the body is neutral, i.e., $Q_{\text{initial}} = 0$ C. When 1.7 μC of electrons is removed, the body is left with a positive charge of 1.7 μC. Now, 18.7×10^{11} electrons are added back. This is equivalent to

$$18.7 \times 10^{11} \text{ electrons} \times \frac{1 \text{ coulomb}}{6.24 \times 10^{18} \text{ electrons}} = 0.3 \ \mu\text{C}$$

of negative charge. The final charge on the body is therefore $Q_f = 1.7 \ \mu$C $- 0.3 \ \mu$C $= +1.4 \ \mu$C.

Now that charge has been defined, we can compute the force between charges. This force can be enormous. For example, if you could place 1 C of charge on two bodies separated by one third of a meter, the force between them would be 81×10^9 N, or about 9 million tons!

practice problems
1

1. Positive charges $Q_1 = 2\ \mu C$ and $Q_2 = 12\ \mu C$ are separated center to center by 10 mm. Compute the force between them. Is it attractive or repulsive?

2. Two equal charges are separated by 1 cm. If the force of repulsion between them is 9.7×10^{-2} N, what is their charge? What may the charges be, both positive, both negative, or one positive and one negative?

3. After 10.61×10^{13} electrons are added to a metal plate, it has a negative charge of 3 μC. What was its initial charge in coulombs?

2.3 VOLTAGE

When charges are detached from one body and transferred to another, a *potential difference* or *voltage* results between them. A familiar example is the voltage that develops when you walk across a carpet as noted in the previous section. Voltages in excess of ten thousand volts can be created in this way. This voltage is due entirely to the separation of positive and negative charges.

Figure 2–7 illustrates another example. During stormy weather, thunderclouds develop, with positive charges pushed to the top of the cloud and negative charges to the bottom. The force of repulsion between like charges then drives electrons away under the cloud, leaving the ground positively charged. Extremely high voltages can be created in this manner—voltages as high as several hundred million volts have been observed. These voltages are so great that the air breaks down and a lightning discharge occurs.

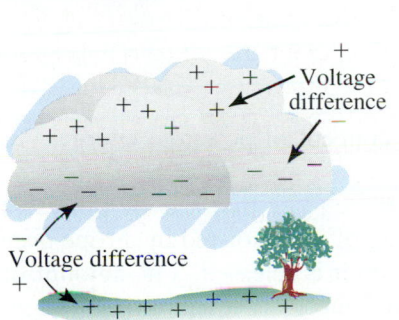

Figure 2–7   Voltages created by separation of charges in a thunder cloud. Charges of hundreds of millions of volts have been observed.

Practical Voltage Sources

As the preceding examples show, voltage is created by the separation of charges. We now look at practical voltage sources. A common example is the battery. In a battery, charges are separated by chemical action. An ordinary flashlight battery (dry cell) illustrates the concept in Figure 2–8. The inner electrode is a carbon rod and the outer electrode is a zinc case. The chemical reaction between the ammonium-chloride/manganese-dioxide paste and the zinc case creates an excess of electrons; hence, the zinc carries a negative charge. An alternate reaction leaves the carbon rod with a deficiency of electrons, causing it to be positively charged. These separated charges create a voltage (1.5 V in this case) between the two electrodes. The battery is useful as a source since its chemical action creates a continuous supply of energy that is able to do useful work, such as light a lamp or run a motor.

Metal cover and
positive terminal

Seal
Insulated Spacer
Carbon rod (+)

Ammonium
chloride
and manganese
dioxide mix

Zinc case (−)

Jacket

(a) Basic construction.
Voltage = 1.5 V

(b) C cell, commonly called a flashlight
battery.

Figure 2–8   Carbon-zinc cell. Voltage is created by separation of charge due to chemical action.

notes

The source of Figure 2–8 is more properly called a cell than a battery, since "cell" refers to a single cell while "battery" refers to a group of cells. However, through common usage, such cells are referred to as batteries. In what follows, we will also call them batteries.

Potential Energy

The concept of voltage is tied into the concept of potential energy. We therefore look briefly at energy.

In mechanics, potential energy is the energy that a body possesses because of its position. For example, a bag of sand hoisted by a rope over a pulley has the potential to do work when it is released. The amount of work that went into giving it this potential energy is equal to the product of force times the distance through which the bag was hoisted (i.e., work equals force times distance).

In a similar fashion, work is required to move positive and negative charges apart. This gives them potential energy. To understand why, consider again the cloud of Figure 2–7. Assume the cloud is initially uncharged. Now assume a charge of Q electrons is moved from the top of the cloud to the bottom. The positive charge left at the top of the cloud exerts a force on the electrons that tries to pull them back as they are being moved away. Since the electrons are being moved against this force, work (force times distance) is required. Since the separated charges experience a force to return to the top of the cloud, they have the potential to do work if released, i.e., they possess potential energy.

Definition of Voltage: The Volt

In electrical terms, a difference in potential energy is defined as *voltage*. In general, the amount of energy required to separate charges depends on the voltage developed and the amount of charge moved. By definition, *the voltage between two points is one volt if it requires one joule of ener-*

gy to move one coulomb of charge from one point to the other. In equation form,

$$V = \frac{W}{Q} \quad \text{[volts, V]} \tag{2-2}$$

where W is energy in joules, Q is charge in coulombs, and V is the resulting voltage in volts.

Note carefully that voltage is defined between points. For the case of the battery, voltage appears between terminals; for the cloud, voltage appears between points within the cloud or between points within the cloud and the ground. Thus, voltage does not exist at a point by itself; it is always determined with respect to some other point. Note also that, although we considered static electricity in developing the energy argument, the same conclusion results regardless of how you separate the charges; this may be by chemical means as in a battery, by mechanical means as in a generator, by photoelectric means as in a solar cell, and so on.

Alternate arrangements of Equation 2–2 are useful:

$$W = QV \quad \text{[joules, J]} \tag{2-3}$$

$$Q = \frac{W}{V} \quad \text{[coulombs, C]} \tag{2-4}$$

> ■ **EXAMPLE 2–2**   If it takes 35 J of energy to move a charge of 5 C from one point to another, what is the voltage between the two points?
>
> **Solution**
>
> $$V = \frac{W}{Q} = \frac{35 \text{ J}}{5 \text{ C}} = 7 \text{ J/C} = 7 \text{ V}$$

practice
problems
II

1. The voltage between two points is 19 V. How much energy is required to move 67×10^{18} electrons from one point to the other?

2. The voltage between two points is 140 mV. If 280 μJ of work are required to move a charge Q from one point to the other, what is Q?

Although Equation 2–2 is the formal definition of voltage, it does not provide a useful way to measure voltage (that is, Equation 2–2 is never actually used in practice to measure voltage). In practice, we use an instrument called a voltmeter (Section 2.6). Equation 2–2 is important, however, as it provides the theoretical foundation for the voltmeter as well as the foundation for many other important circuit relationships that you will soon encounter.

Symbol for DC Voltage Sources

Connecting a battery as in Figure 2–1 creates current in the circuit. Because one of its terminals is always positive and the other is always negative (i.e., the polarity of the battery never changes), the current is always in the same direction. Such a unidirectional current is called *dc* or *direct current*, and the battery is called a *dc source*. Symbols for dc sources are shown in Figure 2–9. The long bar denotes the positive terminal. On actual batteries, the positive terminal is usually marked POS (+) and the negative terminal NEG (–).

2.4 Current

Earlier, you learned that there are large numbers of free electrons in metals like copper. These electrons move randomly throughout the material (Figure 2–6) but their net movement in any given direction is zero.

Assume now that a battery is connected as in Figure 2–10. Since electrons are attracted by the positive pole of the battery and repelled by the negative pole, they move around the circuit, passing through the wire, the lamp, and the battery. This movement of charge is called an *electric current*. The more electrons per second that pass through the circuit, the greater is the current. Thus, current is the *rate of flow* (or *rate of movement*) of charge.

The Ampere

Since charge is measured in coulombs, its rate of flow is coulombs per second. In the SI system, one coulomb per second is defined as one *ampere* (commonly abbreviated amp or A). From this, we get that *one ampere is the current in a circuit when one coulomb of charge passes a given point in one*

$E \quad \substack{+ \\ -}$

(a) Symbol for a cell

$E \quad \substack{+ \\ -}$

(b) Symbol for a battery

1.5 V $\quad \substack{+ \\ -}$

(c) A 1.5 volt battery

Figure 2–9   Battery symbol. The long bar denotes the positive terminal and the short bar the negative terminal. Thus, it is not necessary to put + and – signs on the diagram.

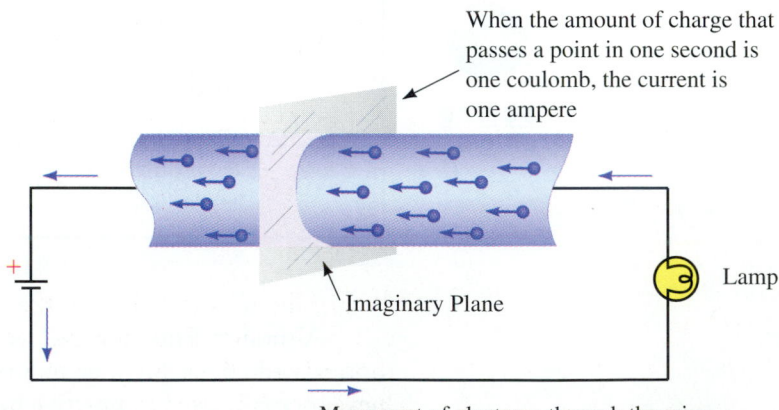

When the amount of charge that passes a point in one second is one coulomb, the current is one ampere

Imaginary Plane

Lamp

Movement of electrons through the wire

Figure 2–10   Electron flow in a conductor. Electrons (–) are attracted to the positive (+) pole of the battery. As electrons move around the circuit, they are replenished at the negative pole of the battery. This flow of charge is called an electric current.

second (Figure 2–10). Let the symbol for current be *I*. Expressed mathematically,

$$I = \frac{Q}{t} \quad \text{[amperes, A]} \qquad (2\text{--}5)$$

where *Q* is the charge (in coulombs) and *t* is the time interval (in seconds) over which it is measured. *In Equation 2–5, it is important to note that t does not represent a discrete point in time but rather it is the interval of time during which the transfer of charges is measured or computed.* Alternate forms of Equation 2–5 are

$$Q = It \quad \text{[coulombs, C]} \qquad (2\text{--}6)$$

and

$$t = \frac{Q}{I} \quad \text{[seconds, s]} \qquad (2\text{--}7)$$

■ **EXAMPLE 2–3**   If 49 coulombs of charge pass point *P* in Figure 2–10 in 7 seconds, what is the current?

Solution

$$I = \frac{Q}{t} = \frac{49 \text{ C}}{7 \text{ s}} = 7 \text{ C/s} = 7 \text{ A}$$

practice
problems
III

1. Between *t* = 1 ms and *t* = 14 ms, 8 μC of charge pass through a wire. What is the current?

2. After the switch of Figure 2–1 is closed, current *I* = 4 A. How much charge passes through the lamp between the time the switch is closed and the time that it is opened 3 minutes later?

Although Equation 2–5 is the theoretical definition of current, we never actually use it to measure current. In practice, we use an instrument called an ammeter, which is considered in Section 2.6. However, it is an extremely important equation that we will soon use to develop other relationships.

Current Direction

Originally it was believed that current was a movement of positive charge and that these charges moved around the circuit from the positive terminal of the battery to the negative as depicted in Figure 2–11(a). Based on this, all the laws, formulas, and symbols of circuit theory were developed. (We now refer to this direction as the *conventional current direction*.) After the discovery of the atomic nature of matter, it was learned that what actually moves in metalic conductors are electrons and that they move through the circuit as in Figure 2–11(b). This direction is

(a) Conventional current direction   (b) Electron flow direction

Figure 2–11   Conventional current versus electron flow. In this book we use conventional current.

called the *electron flow direction*. However, because the conventional current direction was so well established, most users stayed with it. We do likewise. Thus, **in this book, the conventional direction for current is used.**

Alternating Current (AC)

So far, we have considered only dc. Before we move on, we briefly mention *ac* or *alternating current*. Alternating current is current that changes direction cyclically, i.e., charges alternately flow in one direction, then in the other. The most common ac source is the commercial ac power system that supplies power to your home. We mention it here because you will encounter it briefly in Section 2.5. It is covered in detail in Chapter 15.

in process
l e a r n i n g
check II

1. Body A has a negative charge of 0.2 μC and body B has a charge of 0.37 μC (positive). If 87×10^{12} electrons are transferred from A to B, what are the charges in coulombs on A and on B after the transfer?

2. Briefly describe the mechanism of voltage creation using the carbon-zinc cell of Figure 2–8 to illustrate.

3. When the switch in Figure 2–1 is open, the current is zero, yet free electrons in the copper wire are moving about. Describe their motion. Why does their movement not constitute an electric current?

4. If 12.48×10^{20} electrons pass a certain point in a circuit in 2.5 s, what is the current in amperes?

5. For Figure 2–1, assume a 12-V battery. The switch is closed, then opened. If $I = 6$ A and the battery expends 230 040 J moving charge through the circuit, how long was the switch closed?

2.5 Practical dc Voltage Sources

Batteries

Batteries are the most common dc source. They are made in a variety of shapes, sizes, and ratings, from miniaturized button batteries capable of delivering only a few microamps to large automotive batteries capable of delivering hundreds of amps. Common sizes are the AAA, AA, C, and D as illustrated in the various photos of this chapter. All batteries use unlike conductive electrodes immersed in an electrolyte. Chemical interaction between the electrodes and the electrolyte creates the voltage of the battery.

Primary and Secondary Batteries

Batteries eventually become "discharged" and unable to produce usable voltage and current. Some types of batteries, however, can be "recharged." Such batteries are called *secondary* batteries. Other types, called *primary* batteries, cannot be recharged. A familiar example of a secondary battery is the automobile battery. It can be recharged by passing current through it opposite to its discharge direction. A familiar example of a primary cell is the flashlight battery.

Types of Batteries and Their Applications

The voltage of a battery, its service life, and other characteristics depend on the material from which it is made.

Alkaline.   This is one of the most widely used, general-purpose primary cells available. Alkaline batteries are used in flashlights, portable radios, TV remote controllers, cassette players, cameras, toys, and so on. They come in various sizes as depicted in Figure 2–12. Alkaline batteries provide 50% to 100% more total energy for the same size unit than carbon-zinc cells. Their cell voltage is 1.5 V.

Carbon-Zinc.   Also called a *dry cell*, the carbon-zinc battery was for many years the most widely used primary cell, but it is now giving way to other types. Its cell voltage is 1.5 volts. Popular sizes are the same as for the alkaline battery.

Lithium.   Lithium batteries (Figure 2–13) feature small size, low cost, and long life (e.g., shelf lives of 10 to 20 years). Applications include watches, pacemakers, cameras, and battery backup of computer memories. Several types of lithium cells are available, with voltages from of 2 V to 3.5 V and current ratings from the microampere to the ampere range.

Nickel Cadmium.   Commonly called "nicads," these are the most popular, general-purpose rechargeable batteries available. They have long service lives, operate over wide temperature ranges, and are manufactured in many styles and sizes, including C, D, AAA, and AA. Inexpensive chargers make it economically feasible to use nickel-cadmium batteries for home entertainment equipment.

Figure 2–12   Alkaline batteries. From left to right: 9-V rectangular battery, AAA cell, D cell, AA cell, and C cell.

Figure 2–13   Lithium batteries. Background: A 6-V battery from a camera; Right: A 3.6-V battery (the blue component) used for memory backup in a computer; Foregound: A selection of coin (button) batteries.

Lead Acid.   This is the familiar automotive battery. Its basic cell voltage is about 2 volts, but typically, six cells are connected internally to provide 12 volts at its terminals. Lead-acid batteries are capable of delivering large current (in excess of 100 A) for short periods as required, for example, to start an automobile.

Battery Capacity

Batteries run down under use. Their useful life, also called *capacity*, is specified in *ampere-hours* (Ah). The ampere-hour rating of a battery is equal to the product of its current drain times the length of time that you can expect to draw the specified current before the battery becomes unusable. For example, a battery rated at 200 Ah can theoretically supply 20 A for 10 h, or 5 A for 40 h, etc. The relationship between capacity, life, and current drain is

$$\text{life} = \frac{\text{capacity}}{\text{current drain}} \qquad (2\text{--}8)$$

The capacity of batteries is not a fixed value as suggested above but is affected by discharge rates, operating schedules, temperature, and other factors. At best, therefore, capacity is an estimate of expected life under certain conditions. Table 2–1 illustrates approximate service capacities for several sizes of carbon-zinc batteries at three values of current drain at 21°C. Under the conditions listed, the AA cell has a service life of (3 mA)(450 h) = 1350 mAh at a drain of 3 mA, but its service life decreases to (30 mA)(32 h) = 960 mAh at a drain of 30 mA. Figure 2–14 shows a typical variation of capacity of a nicad battery with changes in temperature.

TABLE 2–1 Capacity-Current Drain of Selected Carbon-Zinc Cells

CELL	STARTING DRAIN (mA)	SERVICE CAPACITY (h)
AA	3.0	450
	15.0	80
	30.0	32
C	5.0	520
	25.0	115
	50.0	53
D	10.0	525
	50.0	125
	100.0	57

Courtesy T. R. Crompton, *Battery Reference Book,* Butterworths & Co. (Publishers) Ltd, 1990.

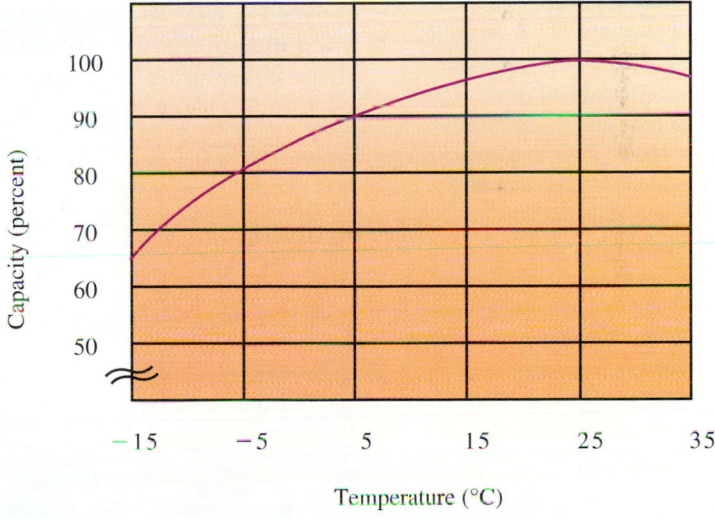

Figure 2–14   Typical variation of capacity versus temperature for a nicad battery.

■ **EXAMPLE 2–4**   Assume the battery of Figure 2–14 has a capacity of 240 Ah at 25°C. What is its capacity at –15°C?

Solution   From the graph, capacity at –15°C is down to 65%. Thus, capacity = 0.65 × 240 = 156 Ah.

(a) Total voltage is the sum of the cell voltages

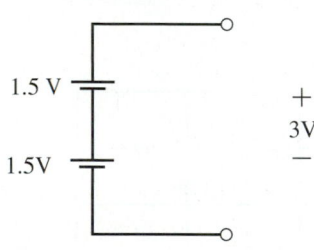

(b) Schematic representation

Figure 2–15   Cells connected in series.

Cells in Series and Parallel

Cells may be connected as in Figures 2–15 and 2–16 to increase their voltage and current capabilities.

Electronic Power Supplies

Electronic systems such as TV sets, VCRs, computers, and so on, require dc for their operation. Except for portable units which use batteries, they obtain their power from the commercial ac power lines by means of built-in power supplies (Figure 2–17). Such supplies convert the incoming ac to the dc voltages required by the equipment. Power supplies are also used in electronic laboratories. These are usually variable to provide the range of voltages needed for prototype development and circuit testing. Figure 2–18 shows a variable supply.

Solar Cells

Solar cells convert light energy to electrical energy using photovoltaic means. The basic cell consists of two layers of semiconductor material. When light strikes the cell, many electrons gain enough energy to cross from one layer to the other to create a dc voltage.

Solar energy has a number of practical applications. Figure 2–19, for example, shows an array of solar panels supplying power to a commercial ac network. (In this case, the dc is first converted to ac before it is fed into the power system.) In remote areas, solar panels are used to power communications systems and irrigation pumps. In space, they are used to power satellites. In everyday life, they are used to power hand-held calculators.

DC Generators

Direct current (dc) generators, which convert mechanical energy to electrical energy, are another source of dc. They create voltage by means of a

(a) Output voltage remains unchanged. However, available current is increased.

(b) Schematic representation

Figure 2–16   Cells connected in parallel.

coil of wire rotated through a magnetic field. Their principle of operation is similar to that of ac generators (discussed in Chapter 15).

Figure 2–17   Power supplies used to convert ac to dc. *(Photo courtesy of Condor DC Power Supplies Inc.)*

Figure 2–18   Variable laboratory power supply.

Figure 2–19   Davis, California Pacific Gas & Electric PVUSA (Photovoltaic for Utility Scale Applications). The plant is rated at 174 kilowatts ac. *(Courtesy Siemens Solar Industries, Camarillo, California)*

2.6 Measuring Voltage and Current

Voltage and current are measured in practice using instruments called *voltmeters* and *ammeters*. While voltmeters and ammeters are available as individual instruments, they are more commonly combined into a multi-purpose instrument called a *multimeter* or *VOM* (volt-ohm-milliamme-ter). Figure 2–20 shows both digital and analog multimeters. Analog instruments use a needle pointer to indicate measured values, while digital instruments use a numeric readout. Digital instruments are more popular than analog types because they are easier to use. We will use digital meters for most of our examples.

Setting the Multimeter for Voltage and Current Measurement

Multimeters typically have a set of terminals marked VΩ, A, and COM as can be seen in Figure 2–20, as well as a function selector switch or set of push buttons that permit you to select functions and ranges. Terminal VΩ is the terminal to use to measure voltage and resistance, while terminal A is used for current measurement. The terminal marked COM is the common terminal for all measurements. (Some multimeters combine the VΩ and A terminals into one terminal marked VΩA.) On many instruments the VΩ terminal is called the + terminal and the COM terminal is called the – terminal.

Voltage Select.   When set to voltage (V), the meter measures the voltage between its VΩ (or +) and COM (or –) terminals. In Figure 2–21(a), for example, with its leads placed across a 47.2-volt source, the instrument indicates 47.2 V.

(a) Analog multimeter.

(b) Hand-held digital multimeter (DMM).
(Reproduced with permission from the John Fluke Mfg. Co., Inc.)

Figure 2–20   Typical multimeters. On many multimeters, the VΩA terminal is also marked with a + sign and the COM terminal with a − sign.

(a) Set selector to V
to measure voltage

(b) Set selector to A
to measure current

Figure 2–21   Measuring voltage and current with a multimeter. By convention, the red lead is connected to the V- Ω -A (+) terminal and the black lead to the COM (−) terminal.

Figure 2–22   Place voltmeter leads across the circuit component whose voltage you wish to measure. If the voltmeter reads positive, the point where the red lead is connected is positive with respect to the point where the black lead is connected.

Current Select.   When set to current (A), the multimeter measures the current passing through it, i.e., the current entering its A (or +) terminal and leaving its COM (or −) terminal. In Figure 2–21(b), the meter measures and displays a current of 3.6 A.

How to Measure Voltage

Since voltage is the potential difference between two points, you measure voltage by placing the voltmeter leads **across** the component whose voltage you wish to determine as we did in Figure 2–21(a). Thus, to measure the voltage across the lamp of Figure 2–22, connect the leads as shown. If you have no idea how large the voltage is, set the meter to its highest range, then work your way down to avoid damage to the instrument.

Be sure to note the sign of the measured quantity. (Most digital instruments have an *autopolarity* feature that automatically determines the sign for you.) If the meter is connected as in Figure 2–21(a) with its + lead connected to the + terminal of the battery, the display will show 47.2 as indicated, while if the leads are reversed, the display will show −47.2. For an analog instrument, if the leads are connected correctly, the meter will read upscale; if they are connected backwards, the meter will try to read down scale. In this case, reverse the leads.

practical

notes

*B*y convention, voltmeters have one red lead and one black lead, with the red lead connected to the (+) or V Ω A terminal of the meter and the black connected to the (−) or COM terminal. Thus, if the voltmeter indicates a positive or upscale value, the point where the red lead is touching is positive with respect to the point where the black lead is touching; inversely, if the meter indicates negative or downscale, the point where the red lead is touching is negative with respect to the point where the black lead is connected.

How to Measure Current

As indicated by Figure 2–21(b), the current that you wish to measure must pass **through** the meter. Consider Figure 2–23(a). To measure this current, open the circuit as in (b) and insert the ammeter. Be sure to connect it so that current enters at the A (or +) terminal and exits at the COM (or −) terminal.

Reading Analog Multimeters

Consider the analog meter of Figure 2–24. Note that it has a selector switch for selecting dc volts, ac volts, dc current, and ohms plus a variety of scales to go with these functions and their ranges. To measure a quantity, set the selector switch to the desired function and range, then read the value from the appropriate scale.

(a) Current to be measured

(b) Ammeter inserted in the circuit to measure current

Figure 2–23   Open the circuit and insert the ammeter into the circuit so that the current that you wish to measure passes through the instrument, entering at the + or A terminal and exiting at the − or COM terminal.

Figure 2–24   Typical analog multimeter. The quantity being measured is indicated on the scale corresponding to the selector switch setting.

■ **EXAMPLE 2–5**   The meter of Figure 2–24 is set to the 100 volts dc range. This means that the instrument reads full scale when 100 volts is applied and proportionally less for other voltages. For the case shown, (expanded detail, Figure 2–25), the needle indicates 70 volts.

Figure 2–25   Meter indicates 70 volts on the 100-V scale.

practice
problems
IV

1. What is the current in the circuit of Figure 2–26?

Figure 2–26

2. In a new measurement, the selector switch is rotated one position counterclockwise. If the needle takes up the same position as in Figure 2–26, what is the new current?

Meter Symbols

In our examples so far, we have shown meters pictorially. Usually, however, they are shown schematically. The schematic symbol for a voltmeter is a circle with the letter V, while the symbol for an ammeter is a circle with the letter I or A. The circuits of Figures 2–22 and 2–23 have been redrawn (Figure 2–27) to indicate this.

(a) Voltmeter   (b) Use *I* or *A* in ammeter symbol

Figure 2–27   Schematic symbols for voltmeter and ammeter.

practical

notes

1. *B*eginners sometimes make statements such as ". . . the voltage through a resistor" or ". . . the current across a resistor." These statements are incorrect. Voltage does not pass through anything; voltage is a potential difference and appears across things. This is why we connect a voltmeter **across** components to measure their voltage. Similarly, current does not appear across anything; current is a flow of charge that passes **through** circuit elements. This is why we put the ammeter in the current path—to measure the current in it. Thus, the correct statements are ". . . voltage across the resistor . . ." and ". . . current through the resistor . . . ."*

2. *Be sure **not** to connect ammeters directly across a voltage source. Ammeters have nearly zero resistance and damage will probably result.*

2.7 Switches, Fuses, and Circuit Breakers

Switches

The most basic switch is a single-pole, single-throw (SPST) switch as shown in Figure 2–28. With the switch open, the current path is broken and the lamp is off; with it closed, the lamp is on. This type of switch is used, for example, for light switches in homes.

Figure 2–29 shows a single-pole, double-throw (SPDT) switch. Two of these switches may be used as in (b) for two-way control of a light. This type of arrangement is sometimes used for stairway lights; you can turn the light on or off from either the bottom or the top of the stairs.

Other types of switches are shown in Figure 2–30. A double-pole, single-throw (DPST) switch is shown in (a); it switches two lines simultaneously. Push button switches are also shown. Note that these can be either normally open (NO) or normally closed (NC). Push button switches are used in motor control circuits.

Fuses and Circuit Breakers

Fuses and circuit breakers are used to protect equipment or wiring against excessive current. For example, in your home, if you connect too many appliances to an outlet, the fuse or circuit breaker (depending on how your circuits are protected) "blows." This opens the circuit to protect against overloading and possible fire. Fuses and circuit breakers may

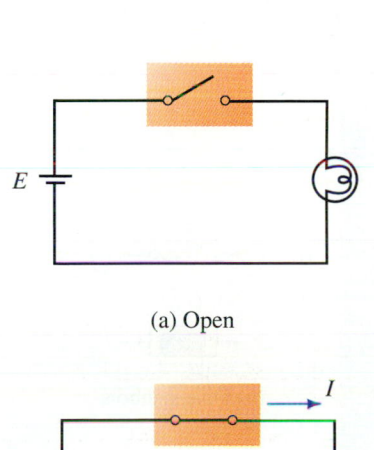

(a) Open

(b) Closed

Figure 2–28   Single-pole, single-throw (SPST) switch.

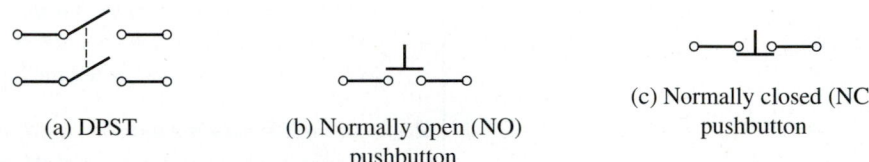

(a) SPDT switch   (b) Two-way switch control of a light

Figure 2–29   Single-pole, double-throw (SPDT) switch.

(a) DPST   (b) Normally open (NO)   (c) Normally closed (NC)
                      pushbutton   pushbutton

Figure 2–30   Other switch configurations.

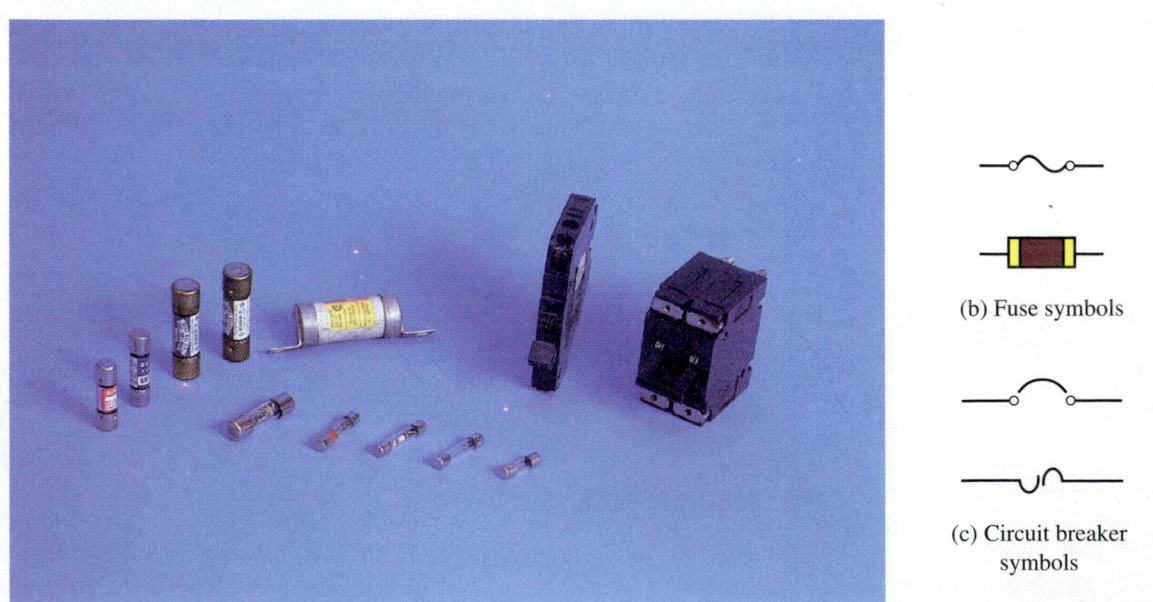

(a) A variety of fuses and circuit breakers.

(b) Fuse symbols

(c) Circuit breaker symbols

Figure 2–31   Fuses and circuit breakers.

also be installed in equipment to protect against internal faults. Figure 2–31 shows a variety of fuses and breakers.

Fuses use a metallic element that melts when current exceeds a pre-set value. Thus, if a fuse is rated at 3 A, it will "blow" if more than 3 amps passes through it. Fuses are made as fast-blow and slow-blow types. Fast-blow fuses are very fast; typically, they blow in a fraction of a second. Slow-blow fuses, on the other hand, operate more slowly so that they do not blow on small, momentary overloads.

Circuit breakers work on a different principle. When the current exceeds the rated value of a breaker, the magnetic field produced by the excessive current operates a mechanism that trips open a switch. After the fault or overload condition has been cleared, the breaker can be reset and used again. Since they are mechanical devices, their operation is slower than that of a fuse; thus, they do not "pop" on momentary overloads as, for example, when a motor is started.

Problems

2.1   Atomic Theory

1. In terms of the structure of an atom, briefly discuss the shell concept. Given that the maximum number of atoms that any shell can hold is $2n^2$, there n is the shell number, compute the maximum number of electrons that shells K, L, and M can hold. Compare these numbers to the values given in the text.

2. Two charges are separated by a certain distance. If the magnitude of both charges is doubled, how is the force affected?

3. Two charges are separated by a certain distance. If the distance between them is tripled, how is the force affected?

4. Two charges are separated by a certain distance. If the magnitude of one charge is doubled and the other tripled and the distance between them halved, how is the force affected?

5. Two materials have four and one electrons in their valence shells, respectively. Which is the better conductor?

6. a. Although gold is very expensive, it is sometimes used in electronics as a conductor. Why?
   b. Why is aluminum sometimes used when its conductivity is only about 60% as good as that of copper?

2.2   Electric Charge

7. What do we mean when we say that a body is "charged"?

8. Compute the force between the following charges and state whether it is attractive or repulsive. .
   a. A +1-μC charge and a +7-μC charge, separated 10 mm.
   b. $Q_1 = 8 \ \mu$C and $Q_2 = -4 \ \mu$C, separated 12 cm.
   c. Two electrons separated by 12×10^{-8} m.
   d. An electron and a proton separated by 5.3×10^{-11} m.
   e. An electron and a neutron separated by 5.7×10^{-11} m.

9. The force between a positive charge and a negative charge that are 2 cm apart is 180 N. If $Q_1 = 4 \ \mu$C, what is Q_2? Is the force attraction or repulsion?

10. If you could place a charge of 1 C on each of two bodies separated 25 cm center to center, what would be the force between them in newtons? In tons?

11. The force of repulsion between two charges separated by 50 cm is 0.02 N. If $Q_2 = 5Q_1$, determine the charges and their possible signs.

12. Standard copper house wire has a diameter of about 0.163 cm. Approximately how many free electrons are there per meter of length at room temperature?

13. Determine the charge possessed by 19×10^{13} electrons.

14. An electrically neutral metal plate acquires a negative charge of 47 μC. How many electrons were added to it?

15. A metal plate has 14.6×10^{11} electrons added. Later, $-0.93 \ \mu$C of charge is added. If the final charge on the plate is 0.13 μC, what was its initial charge?

16. The charge passing through a wire is given by $Q = 10t + 4$, where Q is in coulombs and t in seconds,
   a. How much charge has passed at $t = 5$ s?
   b. How much charge has passed at $t = 8$ s?
   c. What is the current in amps?

2.3 Voltage

17. Sliding off a chair and touching someone can result in a shock. Explain why.

18. How do we "create voltage"? Describe how voltage is created by a solar cell.

19. How much energy is gained by a charge of 0.5 μC as it moves through a potential difference of 8.5 kV?

20. If 1.2 kJ of energy are required to move 500 mC from one point to another, what is the voltage between the two points?

21. If 723 J are used to move 241 C of charge through a lamp, what is the voltage across the lamp?

22. The voltage between two points is 100 V.
  a. How much energy is required to move an electron between the two points?
  b. A neutron?

23. Given a voltage of 12 V for the battery in Figure 2–1, how much charge is moved through the lamp if it takes 57 J of energy to move it?

2.4   Current

24. For the circuit of Figure 2–1, if 27 C pass through the lamp in 9 seconds, what is the current in amperes?

25. If the current $I = 4$ A in Figure 2–1, how many coulombs pass through the lamp in 7 ms?

26. How long does it take for 100 μC to pass a point if the current is 25 mA?

27. If 93.6×10^{12} electrons pass through a lamp in 5 ms, what is the current ?

28. If 1353.6 J are required to move 47×10^{19} electrons through the lamp of Figure 2–1 in 1.3 min, what are V and I?

2.5   Practical Voltage Sources (DC)

29. What do we mean by dc? By ac?

30. How do you charge a secondary battery? Make a sketch. How do you charge a primary battery?

31. A battery rated 1400 mAh supplies 28 mA to a load. How long can it be expected to last?

32. The battery of Figure 2–14 is expected to last 17 h at a current drain of 1.5 A at 25°C. How long do you expect it to last at 5°C at a current drain of 0.8 A?

33. A lead-acid battery has 12 cells connected internally as in Figure 2–32. What is its terminal voltage?

2.6   Measuring Voltage and Current

34. The digital voltmeter of Figure 2–33 has autopolarity. For each case, determine its reading.

12 cells

Figure 2–32

25 V

(a)

25 V

(b)

10 V   4 V

(c)

Figure 2–33

4 V   10 V

(d)

35. The current in the circuit of Figure 2–34 is 9.17 mA. Which ammeter correctly indicates the current? (a) Meter 1, (b) Meter 2, (c) both.

Figure 2–34

36. What is the voltage indicated in Figure 2–35?

37. What is the current indicated in Figure 2–36?

Figure 2–35

Figure 2–36

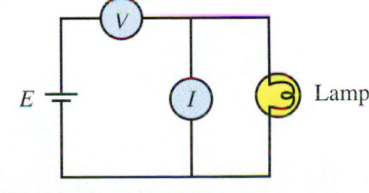

Figure 2–37   What is wrong here?

38. What is wrong with the statement that the voltage through the lamp of Figure 2–22 is 70.3 V?

39. What is wrong with the metering scheme shown in Figure 2–37? Fix it.

2.7   Control and Protective Devices

40. The two-way switch arrangement of Figure 2–29 has four possible combinations of settings, only one of which is shown. Draw the other three and note which combinations turn the light on.

41. For each of the combinations of switch positions shown in Figure 2–38, determine whether the lamp is on or off.

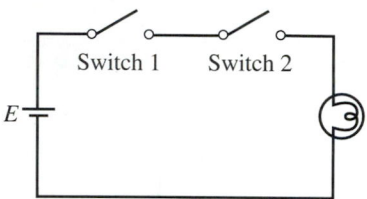

Switch 1	Switch 2	Lamp
Open	Open	
Open	Closed	
Closed	Open	
Closed	Closed	

Figure 2–38

TABLE 2–2

SWITCH 1	SWITCH 2	LAMP
Open	Open	Off
Open	Closed	On
Closed	Open	On
Closed	Closed	On

42. It is desired to control a light using two switches as indicated in Table 2–2. Draw the required circuit.

43. Fuses have a current rating so that you can select the proper size to protect a circuit against overcurrent. They also have a voltage rating. Why? Hint: Read the section on insulators, i.e., Section 2.1.

Resistance

CHAPTER PREVIEW

You have been introduced to the concepts of voltage and current in previous chapters and have found that current involves the movement of charge. In a conductor, the charge carriers are the free electrons which are moved due to the voltage of an externally applied source. As these electrons move through the material, they constantly collide with atoms and other electrons within the conductor. In a process similar to friction, the moving electrons give up some of their energy in the form of heat. These collisions represent an opposition to charge movement that is called *resistance*. The greater the opposition (i.e., the greater the resistance), the smaller will be the current for a given applied voltage.

Circuit components (called *resistors*) are specifically designed to possess resistance and are used in almost all electronic and electrical circuits. Although the resistor is the most simple component in any circuit, its effect is very important in determining the operation of a circuit.

Resistance is represented by the symbol R (Figure 3–1) and is measured in units of ohms (after Georg Simon Ohm). The symbol for ohms is the capital Greek letter omega (Ω).

Figure 3–1   Basic resistive circuit.

In this chapter, we examine resistance in its various forms. Beginning with metallic conductors, we study the factors which affect resistance in conductors. Following this, we look at commercial resistors, including both fixed and variable types. We then discuss important nonlinear resistance devices and conclude with an overview of superconductivity and its potential impact and use.

KEY TERMS

Color Codes
Conductance
Diode
Ohmmeter
Open Circuit
Photocell
Resistance
Resistivity
Short Circuit
Superconductance
Temperature Coefficient
Thermistor
Varistor
Wire gauge

Resistance is added to electric and electronic circuits to control voltage and current. *(Courtesy of Bourns, Inc.)*

OBJECTIVES

After studying this chapter, you will be able to

• calculate the resistance of a section of conductor, given its cross-sectional area and length,

• convert between areas measured in square mils, square meters, and circular mils,

• use tables of wire data to obtain the cross-sectional dimensions of various gauges of wire and predict the allowable current for a particular gauge of wire,

• use the temperature coefficient of a material to calculate the change in resistance as the temperature of the sample changes,

• use resistor color codes to determine the resistance and tolerance of a given fixed-composition resistor,

• demonstrate the procedure for using an ohmmeter to determine circuit continuity and to measure the resistance of both an isolated component and one which is located in a circuit,

• develop an understanding of various ohmic devices such as thermistors and photocells,

• develop an understanding of the resistance of nonlinear devices such as varistors and diodes,

• calculate the conductance of any resistive component.

Georg Simon Ohm and Resistance

One of the fundamental relationships of circuit theory is that between voltage, current, and resistance. This relationship and the properties of resistance were investigated by the German physicist Georg Simon Ohm (1787–1854) using a circuit similar to that of Figure 3–1. Working with Volta's recently developed battery and wires of different materials, lengths, and thicknesses, Ohm found that current depended on both voltage and resistance. For example, for a fixed resistance, he found that doubling the voltage doubled the current, tripling the voltage tripled the current, and so on. Also, for a fixed voltage, Ohm found that the opposition to current was directly proportional to the length of the wire and inversely proportional to its cross-sectional area. From this, he was able to define the resistance of a wire and show that current was inversely proportional to this resistance, e.g., when he doubled the resistance, he found that the current decreased to half of its former value.

These two results when combined together form what is known as Ohm's law. (You will study Ohm's law in great detail in Chapter 4.) Ohm's results are of such fundamental importance that they represent the real beginnings of what we now call electrical circuit analysis.

3.1 Resistance of Conductors

As mentioned in the chapter preview, conductors are materials which permit the flow of charge. However, conductors do not all behave the same way. Rather, we find that the resistance of a material is dependent upon several factors:

- Type of material
- Length of the conductor
- Cross-sectional area
- Temperature

If a certain length of wire is subjected to a current, the moving electrons will collide with other electrons within the material. Differences at the atomic level of various materials cause variation in how the collisions affect resistance. For example, silver has more free electrons than copper, and so the resistance of a silver wire will be less than the resistance of a copper wire having the identical dimensions. We may therefore conclude the following:

The resistance of a conductor is dependent upon the type of material.

If we were to double the length of the wire, we can expect that the number of collisions over the length of the wire would double, thereby causing the resistance to also double. This effect may be summarized as follows:

The resistance of a metallic conductor is directly proportional to the length of the conductor.

A somewhat less intuitive property of a conductor is the effect of cross-sectional area on the resistance. As the cross-sectional area is increased, the moving electrons are able to move more freely through the conductor, just as water moves more freely through a large-diameter pipe than a small-diameter pipe. If the cross-sectional area is doubled, the electrons would be involved in half as many collisions over the length of the wire. We may summarize this effect as follows:

The resistance of a metallic conductor is inversely proportional to the cross-sectional area of the conductor.

TABLE 3–1 Resistivity of Materials, ρ

MATERIAL	RESISTIVITY, ρ, AT 20°C (Ω-m)
Silver	1.645×10^{-8}
Copper	1.723×10^{-8}
Gold	2.443×10^{-8}
Aluminum	2.825×10^{-8}
Tungsten	5.485×10^{-8}
Iron	12.30×10^{-8}
Lead	22×10^{-8}
Mercury	95.8×10^{-8}
Nichrome	99.72×10^{-8}
Carbon	3500×10^{-8}
Germanium	$20–2300^*$
Silicon	$\cong 500^*$
Wood	$10^8–10^{14}$
Glass	$10^{10}–10^{14}$
Mica	$10^{11}–10^{15}$
Hard rubber	$10^{13}–10^{16}$
Amber	5×10^{14}
Sulphur	1×10^{15}
Teflon	1×10^{16}

*The resistivities of these materials are dependent upon the impurities within the materials.

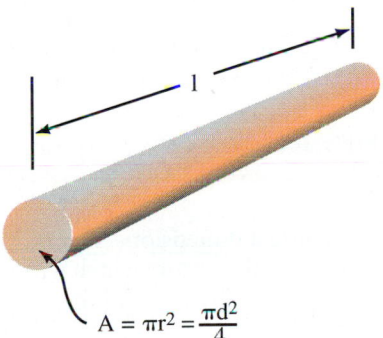

$$A = \pi r^2 = \frac{\pi d^2}{4}$$

Figure 3–2   Conductor with a circular cross-section.

The factors governing the resistance of a conductor at a given temperature may be summarized mathematically as follows:

$$R = \frac{\rho\ell}{A} \quad [\text{ohms, } \Omega] \tag{3–1}$$

where

ρ = resistivity, in ohm-meters (Ω-m)

ℓ = length, in meters (m)

A = cross-sectional area, in square meters (m^2).

In the above equation the lowercase Greek letter rho (ρ) is the constant of proportionality and is called the *resistivity* of the material. Resistivity is a physical property of a material and is measured in ohm-meters (Ω-m) in the SI system. Table 3–1 lists the resistivities of various materials at a temperature of 20°C. The effects on resistance due to changes in temperature will be examined in Section 3.4.

Since most conductors are circular, as shown in Figure 3–2, we may determine the cross-sectional area from either the radius or the diameter as follows:

$$A = \pi r^2 = \pi\left(\frac{d}{2}\right)^2 = \frac{\pi d^2}{4} \tag{3–2}$$

■ **EXAMPLE 3–1**   Most homes use solid copper wire having a diameter of 1.63 mm to provide electrical distribution to outlets and light sockets. Determine the resistance of 75 meters of a solid copper wire having the above diameter.

Solution   We will first calculate the cross-sectional area of the wire using equation 3–2.

$$A = \frac{\pi d^2}{4}$$

$$= \frac{\pi(1.63 \times 10^{-3}\ \text{m})^2}{4}$$

$$= 2.09 \times 10^{-6}\ \text{m}^2$$

Now, using Table 3–1, the resistance of the length of wire is found as

$$R = \frac{\rho\ell}{A}$$

$$= \frac{(1.723 \times 10^{-8}\ \Omega\text{-m})(75\ \text{m})}{2.09 \times 10^{-6}\ \text{m}^2}$$

$$= 0.619\ \Omega$$

practice problem 1

Find the resistance of a 100-m long tungsten wire which has a circular cross-section with a diameter of 0.1 mm ($T = 20°C$).

■ **EXAMPLE 3–2**   Bus bars are bare solid conductors (usually rectangular) used to carry large currents within buildings such as power generating stations, telephone exchanges, and large factories. Given a piece of aluminum bus bar as shown in Figure 3–3, determine the resistance between the ends of this bar at a temperature of 20°C.

Figure 3–3   Conductor with a rectangular cross-section.

Solution   The cross-sectional area is

$$A = (150 \text{ mm})(6 \text{ mm})$$
$$= (0.15 \text{ m})(0.006 \text{ m})$$
$$= 0.0009 \text{ m}^2$$
$$= 9.00 \times 10^{-4} \text{ m}^2$$

The resistance between the ends of the bus bar is determined as

$$R = \frac{\rho \ell}{A}$$
$$= \frac{(2.825 \times 10^{-8} \text{ } \Omega\text{-m})(270 \text{ m})}{9.00 \times 10^{-4} \text{ m}^2}$$
$$= 8.475 \times 10^{-3} \text{ } \Omega = 8.475 \text{ m}\Omega$$

in process
learning
check I

1. Given two lengths of wire having identical dimensions. If one wire is made of copper and the other is made of iron, which wire will have the greater resistance? How much greater will the resistance be?

2. Given two pieces of copper wire which have the same cross-sectional area, determine the relative resistance of a piece of wire which is twice as long as a control sample.

3. Given two pieces of copper wire which have the same length, determine the relative resistance of a piece of wire which has twice the diameter of a control sample.

3.2 Electrical Wire Tables

Although the SI system is the standard measurement for electrical and other physical quantities, the English system is still used extensively in the United States and to a lesser degree throughout the rest of the English-speaking world. One area which has been slow to convert to the SI system is the designation of cables and wires, where the American Wire Gauge (AWG) is the primary system used to denote wire diameters. In this system, each wire diameter is assigned a gauge number. The higher the AWG number, the smaller the diameter of the cable or wire, e.g., AWG 22 gauge wire is a smaller diameter than AWG 14 gauge. Since cross-sectional area is inversely proportional to the square of the diameter, a given length of 22-gauge wire will have more resistance than an equal length of 14-gauge wire. Because of the difference in resistance, we can intuitively deduce that large-diameter cables will be able to handle more current than smaller-diameter cables. Table 3–2 provides a listing of data for standard bare copper wire.

TABLE 3–2   Standard Solid Copper Wire at 20°C

SIZE (AWG)	DIAMETER		AREA		RESISTANCE (Ω/1000 ft)	CURRENT CAPACITY (A)
	(inches)	(mm)	(CM)	(mm²)		
56	0.0005	0.012	0.240	0.000122	43 200	
54	0.0006	0.016	0.384	0.000195	27 000	
52	0.0008	0.020	0.608	0.000308	17 000	
50	0.0010	0.025	0.980	0.000497	10 600	
48	0.0013	0.032	1.54	0.000779	6 750	
46	0.0016	0.040	2.46	0.00125	4 210	
45	0.0019	0.047	3.10	0.00157	3 350	
44	0.0020	0.051	4.00	0.00243	2 590	
43	0.0022	0.056	4.84	0.00245	2 140	
42	0.0025	0.064	6.25	0.00317	1 660	
41	0.0028	0.071	7.84	0.00397	1 320	
40	0.0031	0.079	9.61	0.00487	1 080	
39	0.0035	0.089	12.2	0.00621	847	
38	0.0040	0.102	16.0	0.00811	648	
37	0.0045	0.114	20.2	0.0103	521	
36	0.0050	0.127	25.0	0.0127	415	
35	0.0056	0.142	31.4	0.0159	331	
34	0.0063	0.160	39.7	0.0201	261	
33	0.0071	0.180	50.4	0.0255	206	
32	0.0080	0.203	64.0	0.0324	162	
31	0.0089	0.226	79.2	0.0401	131	
30	0.0100	0.254	100	0.0507	104	
29	0.0113	0.287	128	0.0647	81.2	
28	0.0126	0.320	159	0.0804	65.3	
27	0.0142	0.361	202	0.102	51.4	
26	0.0159	0.404	253	0.128	41.0	0.75*
25	0.0179	0.455	320	0.162	32.4	
24	0.0201	0.511	404	0.205	25.7	1.3*

*This current is suitable for single conductors and surface or loose wiring.
†This current may be accommodated in up to three wires in a sheathed cable. For four to six wires, the current in each wire must be reduced to 80% of the indicated value. For seven to nine wires, the current in each wire must be reduced to 70% of the indicated value.

(Continued)

TABLE 3–2   Standard Solid Copper Wire at 20°C   *(Continued)*

SIZE (AWG)	DIAMETER		AREA		RESISTANCE (Ω/1000 ft)	CURRENT CAPACITY (A)
	(inches)	(mm)	(CM)	(mm²)		
23	0.0226	0.574	511	0.259	20.3	
22	0.0253	0.643	640	0.324	16.2	2.0*
21	0.0285	0.724	812	0.412	12.8	
20	0.0320	0.813	1 020	0.519	10.1	3.0*
19	0.0359	0.912	1 290	0.653	8.05	
18	0.0403	1.02	1 620	0.823	6.39	5.0†
17	0.0453	1.15	2 050	1.04	5.05	
16	0.0508	1.29	2 580	1.31	4.02	10.0†
15	0.0571	1.45	3 260	1.65	3.18	
14	0.0641	1.63	4 110	2.08	2.52	15.0†
13	0.0720	1.83	5 180	2.63	2.00	
12	0.0808	2.05	6 530	3.31	1.59	20.0†
11	0.0907	2.30	8 230	4.17	1.26	
10	0.1019	2.588	10 380	5.261	0.998 8	30.0†
9	0.1144	2.906	13 090	6.632	0.792 5	
8	0.1285	3.264	16 510	8.367	0.628 1	
7	0.1443	3.665	20 820	10.55	0.498 1	
6	0.1620	4.115	26 240	13.30	0.395 2	
5	0.1819	4.620	33 090	16.77	0.313 4	
4	0.2043	5.189	41 740	21.15	0.248 5	
3	0.2294	5.827	52 620	26.67	0.197 1	
2	0.2576	6.543	66 360	33.62	0.156 3	
1	0.2893	7.348	83 690	42.41	0.123 9	
1/0	0.3249	8.252	105 600	53.49	0.098 25	
2/0	0.3648	9.266	133 100	67.43	0.077 93	
3/0	0.4096	10.40	167 800	85.01	0.061 82	
4/0	0.4600	11.68	211 600	107.2	0.049 01	

*This current is suitable for single conductors and surface or loose wiring.
†This current may be accommodated in up to three wires in a sheathed cable. For four to six wires, the current in each wire must be reduced to 80% of the indicated value. For seven to nine wires, the current in each wire must be reduced to 70% of the indicated value.

Even though Table 3–2 provides data for solid conductors up to AWG 4/0, most applications do not use solid conductor sizes beyond AWG 10. Solid conductors are difficult to bend and are easily damaged by mechanical flexing. For this reason, large-diameter cables are nearly always stranded rather than solid. Stranded wires and cables use anywhere from seven strands, as shown in Figure 3–4, to in excess of a hundred strands.

As one might expect, stranded wire uses the same AWG notation as solid wire. Consequently, AWG 10 stranded wire will have the same cross-sectional conductor area as AWG 10 solid wire. However, due to the additional space lost between the conductors, the stranded wire will have a larger overall diameter than the solid wire. Also, because the individual strands are coiled as a helix, the overall strand length will be slightly longer than the cable length.

Figure 3–4   Stranded wire (seven strands).

Wire tables similar to Table 3–2 are available for stranded copper cables and for cables constructed of other materials (notably aluminum).

■ **EXAMPLE 3–3**   Calculate the resistance of 200 feet of AWG 16 solid copper wire at 20°C.

Solution   From Table 3–2, we see that AWG 16 wire has a resistance of 4.02 Ω per 1000 feet. Since we are given a length of only 200 feet, the resistance will be determined as

$$R = \left(\frac{4.02 \ \Omega}{1000 \ \text{ft}} \right)(200 \ \text{ft}) = 0.804 \ \Omega$$

By examining Table 3–2, several important points may be observed:

- If the wire size increases by three gauge sizes, the cross-sectional area will approximately double. Since resistance is inversely proportional to cross-sectional area, a given length of larger-diameter cable will have a resistance which is approximately half as large as the resistance of a similar length of the smaller-diameter cable.

- If there is a difference of three gauge sizes between cables, then the larger-diameter cable will be able to handle approximately twice as much current as the smaller-diameter cable. The amount of current that a conductor can safely handle is directly proportional to the cross-sectional area.

- If the wire size increases by ten gauge sizes, the cross-sectional area will increase by a factor of about ten. Due to the inverse relationship between resistance, and cross-sectional area, the larger-diameter cable will have about one tenth the resistance of a similar length of the smaller-diameter cable.

- For a 10-gauge difference in cable sizes, the larger-diameter cable will have ten times the cross-sectional area of the smaller-diameter cable and so it will be able to handle approximately ten times more current.

■ **EXAMPLE 3–4**   If AWG 14 solid copper wire is able to handle 15 A of current, determine the expected current capacity of AWG 24 and AWG 8 copper wire at 20°C.

Solution   Since AWG 24 is ten sizes smaller than AWG 14, the smaller cable will be able to handle about one tenth the capacity of the larger-diameter cable.

AWG 24 will be able to handle approximately 1.5 A of current.

AWG 8 is six sizes larger than AWG 14. Since current capacity doubles for an increase of three sizes, AWG 11 would be able to handle 30 A and AWG 8 will be able to handle 60 A.

practice
problems
II

1. From Table 3–2 find the diameters in millimeters and the cross-sectional areas in square millimeters of AWG 19 and AWG 30 solid wire.

2. By using the cross-sectional areas for AWG 19 and AWG 30, determine the areas that AWG 16 and AWG 40 should have.

3. Compare the actual cross-sectional areas as listed in Table 3–2 to the areas found in Problems 1 and 2 above. (You will find a slight variation between your calculated values and the actual areas. This is because the actual diameters of the wires have been adjusted to provide optimum sizes for manufacturing.)

in process
learning
check II

1. AWG 12-gauge wire is able to safely handle 20 amps of current. How much current should an AWG 2-gauge cable be able to handle?

2. The electrical code actually permits up to 120 A for this cable. How does the actual value compare to your theoretical value? Why do you think there is a difference?

3.3 Resistance of Wires—Circular Mils

The American Wire Gauge system for specifying wire diameters was developed using a unit called the *circular mil* (CM), which is defined as the area contained within a circle having a diameter of 1 mil (1 mil = 0.001 inch). A *square mil* is defined as the area contained in a square having side dimensions of 1 mil. By referring to Figure 3–5, it is apparent that the area of a circular mil is smaller than the area of a square mil.

Because not all conductors have circular cross-sections, it is occasionally necessary to convert areas expressed in square mils into circular mils. We will now determine the relationship between the circular mil and the square mil.

Suppose that a wire has the circular cross section shown in Figure 3–5(a). By applying Equation 3–2, the area, in square mils, of the circular cross section is determined as follows:

$$A = \frac{\pi d^2}{4}$$

$$= \frac{\pi (1\ \text{mil})^2}{4}$$

$$= \frac{\pi}{4}\ \text{sq. mil}$$

From the above derivation the following relations must apply:

$$1\ \text{CM} = \frac{\pi}{4}\ \text{sq. mil} \qquad (3\text{–}3)$$

$$1\ \text{sq. mil} = \frac{4}{\pi}\ \text{CM} \qquad (3\text{–}4)$$

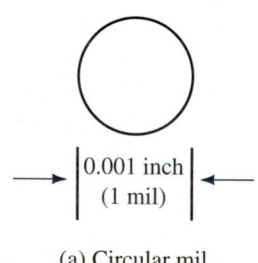

0.001 inch
(1 mil)

(a) Circular mil

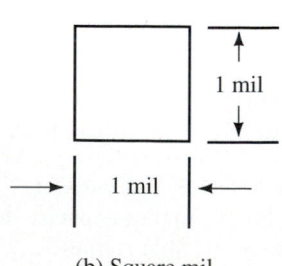

1 mil

1 mil

(b) Square mil

Figure 3–5

The greatest advantage of using the circular mil to express areas of wires is the simplicity with which calculations may be made. Unlike previous area calculations which involved the use of π, area calculations may be reduced to simply finding the square of the diameter. A wire which has a diameter d expressed in mils has an area in circular mils given as

$$A_{CM} = d_{mil}^2 \quad \text{[circular mils, CM]} \qquad (3-5)$$

■ **EXAMPLE 3–5**   Determine the cross-sectional area in circular mils of a wire having the following diameters:

a. 0.0159 inch (AWG 26 wire)

b. 0.500 inch

Solution

a. $d = 0.0159$ inch
   $= (0.0159 \text{ inch})(1000 \text{ mils/inch})$
   $= 15.9$ mils

Now, using Equation 3–5, we obtain

$$A_{CM} = (15.9)^2 = 253 \text{ CM.}$$

From Table 3–2, we see that the above result is precisely the area given for AWG 26 wire.

b. $d = 0.500$ inch
   $= (0.500 \text{ inch})(1000 \text{ mils/inch})$
   $= 500$ mils

$$A_{CM} = (500)^2 = 250\ 000 \text{ CM}$$

In Example 3–5(b) we see that the cross-sectional area of a cable may be a large number when it is expressed in circular mils. In order to simplify the units for area, the Roman numeral M is often used to represent 1000. If a wire has a cross-sectional area of 250 000 CM, it is more easily written as 250 MCM.

Clearly, this is a departure from the SI system, where M is used to represent one million. Since there is no simple way to overcome this conflict, the student working with cable areas expressed in MCM will need to remember that the M stands for one thousand and not for one million.

■ **EXAMPLE 3–6**

a. Determine the cross-sectional area in square mils and in circular mils of a copper bus bar having cross-sectional dimensions of 0.250 inch × 6.00 inch.

b. If this copper bus bar were to be replaced by AWG 2/0 cables, how many cables would be required?

Solution

a. $A_{\text{sq. mil}} = (250 \text{ mils})(6000 \text{ mils})$
$= 1\,500\,000 \text{ sq. mils}$

The area in circular mils is found by applying Equation 3–4, and this will be

$$A_{\text{CM}} = (250 \text{ mils})(6000 \text{ mils})$$

$$= (1\,500\,000 \text{ sq. mils})\left(\frac{4}{\pi} \text{ CM/sq. mil}\right)$$

$$= 1\,910\,000 \text{ CM}$$

$$= 1910 \text{ MCM}$$

b. From Table 3–2, we see that AWG 2/0 cable has a cross-sectional area of 133.1 MCM (133 100 CM) and so the bus bar is equivalent to the following number of cables:

$$n = \frac{1910 \text{ MCM}}{133.1 \text{ MCM}} = 14.35$$

This example illustrates that 15 cables would need to be installed to be equivalent to a single 6-inch by 0.25-inch bus bar. Due to the expense and awkwardness of using this many cables, we see the economy of using solid bus bar. The main disadvantage of using bus bar is that the conductor is not covered with an insulation, and so the bus bar does not offer the same protection as cable. However, since bus bar is generally used in locations where the only experienced technicians are permitted access, this disadvantage is a minor one.

As we have seen in Section 3.1, the resistance of a conductor was determined to be

$$R = \frac{\rho \ell}{A} \quad [\text{ohms}, \Omega] \tag{3–1}$$

Although the original equation used SI units, the equation will also apply if the units are expressed in any other convenient system. If cable length is generally expressed in feet and the area in circular mils, then the resistivity must be expressed in the appropriate units. Table 3–3 gives the resistivities of some conductors represented in circular mil-ohms per foot.

The following example illustrates how Table 3–3 may be used to determine the resistance of a given section of wire.

TABLE 3–3 Resistivity of Conductors, ρ

MATERIAL	RESISTIVITY, ρ, AT 20°C (CM-Ω/ft)
Silver	9.90
Copper	10.36
Gold	14.7
Aluminum	17.0
Tungsten	33.0
Iron	74.0
Lead	132.
Mercury	576.
Nichrome	600.

■ **EXAMPLE 3–7** Determine the resistance of an AWG 16 copper wire at 20°C if the wire has a diameter of 0.0508 inch and a length of 400 feet.

Solution
The diameter in mils is found as

$$d = 0.0508 \text{ inch} = 50.8 \text{ mils}$$

Therefore the cross-sectional area (in circular mils) of AWG 16 is

$$A_{CM} = 50.8^2 = 2580 \text{ CM}$$

Now, by applying Equation 3–1 and using the appropriate units, we obtain the following:

$$R = \frac{\rho \ell}{A_{CM}}$$

$$= \frac{\left(10.36 \, \frac{\text{CM-}\Omega}{\text{ft}}\right)(400 \text{ ft})}{2580 \text{ CM}}$$

$$= 1.61 \, \Omega$$

practice problems III

1. Determine the resistance of 1 mile (5280 feet) of AWG 19 copper wire at 20°C, if the cross-sectional area is 1290 CM.

2. Compare the above result with the value that would be obtained by using the resistance (in ohms per thousand feet) given in Table 3–2.

3. An aluminum conductor having a cross-sectional area of 1843 MCM is used to transmit power from a high-voltage dc (HVDC) generating station to a large urban center. If the city is 900 km from the generating station, determine the resistance of the conductor at a temperature of 20°C. (Use 1 ft = 0.3048 m.)

in process learning check III

A conductor has a cross-sectional area of 50 square mils. Determine the cross-sectional area in circular mils, square meters, and square millimeters.

3.4 Temperature Effects

Section 3.1 indicated that the resistance of a conductor will not be constant at all temperatures. As temperature increases, more electrons will escape their orbits, causing additional collisions within the conductor. For most conducting materials, the increase in the number of collisions translates into a relatively linear increase in resistance, as shown in Figure 3–6.

The rate at which the resistance of a material changes with a variation in temperature is called the *temperature coefficient* of the material and is assigned the Greek letter alpha (α). Some materials have only very slight changes in resistance, while other materials demonstrate dramatic changes in resistance with a change in temperature.

Any material for which resistance increases as temperature increases is said to have a *positive temperature coefficient*.

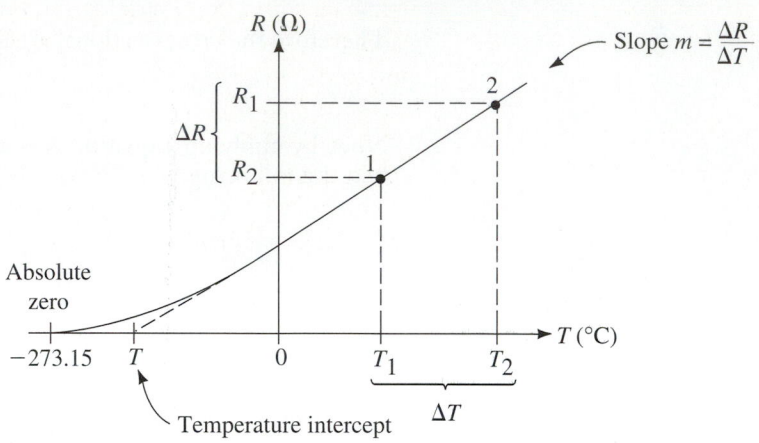

Figure 3–6   Temperature effects on the resistance of a conductor.

For semiconductor materials such as carbon, germanium, and silicon, increases in temperature allow electrons to escape their usually stable orbits and become free to move within the material. Although additional collisions do occur within the semiconductor, the effect of the collisions is minimal when compared with the contribution of the extra electrons to the overall flow of charge. As the temperature increases, the number of charge electrons increases, resulting in more current. Therefore, an increase in temperature results in a decrease in resistance. Consequently, these materials are referred to as having *negative temperature coefficients*.

Table 3–4 gives the temperature coefficients, α per degree Celsius, of various materials at 20°C and at 0°C.

If we consider that Figure 3–6 illustrates how the resistance of copper changes with temperature, we observe an almost linear increase in

TABLE 3–4   Temperature Intercepts and Coefficients for Common Materials

	T (°C)	α (°C)$^{-1}$ AT 20°C	α (°C)$^{-1}$ AT 0°C
Silver	−243	0.003 8	0.004 12
Copper	−234.5	0.003 93	0.004 27
Aluminum	−236	0.003 91	0.004 24
Tungsten	−202	0.004 50	0.004 95
Iron	−162	0.005 5	0.006 18
Lead	−224	0.004 26	0.004 66
Nichrome	−2270	0.000 44	0.000 44
Brass	−480	0.002 00	0.002 08
Platinum	−310	0.003 03	0.003 23
Carbon		−0.000 5	
Germanium		−0.048	
Silicon		−0.075	

resistance as the temperature increases. Further, we see that as the temperature is decreased to *absolute zero* ($T = -273.15°C$), the resistance approaches zero.

In Figure 3–6, the point at which the linear portion of the line is extrapolated to cross the abscissa (temperature axis) is referred to as the *temperature intercept* or the *inferred absolute temperature T* of the material.

By examining the straight-line portion of the graph, we may see that the slope of the line will be given as follows:

$$1/2m = \frac{\Delta R}{\Delta T}$$

$$= \frac{R_1 - 0}{T_1 - T} = \frac{R_1}{T_1 - T} \quad [\Omega/°C]$$

Now, we define the *temperature coefficient α* as

$$\alpha_1 = \frac{m}{R_1} \quad [(°C)^{-1}] \tag{3–6}$$

where

R_1 = the resistance, in ohms, at a temperature of T_1,

m = the slope of the linear portion of the curve ($\Delta R/\Delta T$), in ohms per degree Celsius.

As seen from the above definition, the temperature coefficient will not be constant everywhere on the straight line, but rather is dependent upon the resistance R_1 (and on the temperature T_1) of the conductor. You may again refer to Table 3–4, which provides the temperature coefficients of various materials at two temperatures, 20°C and 0°C.

By applying the equation of a straight line, we express the resistance anywhere on the linear portion of the curve as follows:

$$R_2 = R_1 + m\Delta T$$

$$= R_1 + m(T_2 - T_1)$$

$$= R_1[1 + \frac{m}{R_1}(T_2 - T_1)]$$

$$R_2 = R_1[1 + \alpha_1(T_2 - T_1)] \quad [\text{ohms}, \Omega] \tag{3–7}$$

where R_2 is the resistance, in ohms, at a temperature of T_2.

■ **EXAMPLE 3–8**   An aluminum wire has a resistance of 20 Ω at room temperature (20°C). Calculate the resistance of the same wire at temperatures of −40°C, 100°C, and 200°C.

Solution

At −40°C:
From Table 3–4, we see that aluminum has a temperature coefficient of $\alpha = 0.00391$. The resistance at −40°C is now determined to be

$$R_{-40°C} = (20\ \Omega)\{1 + [0.00391(°C)^{-1}][-40°C - 20°C]\}$$

$$= 15.3\ \Omega$$

At 100°C:

$$R_{100°C} = (20\ \Omega)\{1 + [0.00391(°C)^{-1}][100°C - 20°C]\}$$
$$= 26.3\ \Omega$$

At 200°C:

$$R_{200°C} = (20\ \Omega)\{1 + [0.00391(°C)^{-1}][200°C - 20°C]\}$$
$$= 34.1\ \Omega$$

The above phenomenon indicates that the resistance of conductors changes quite dramatically with changes in temperature. It is for this reason that manufacturers generally specify the range of temperatures over which a conductor may operate safely.

■ **EXAMPLE 3–9**   Tungsten wire is used as filaments in incandescent light bulbs. Current in the wire causes the wire to reach extremely high temperatures. Determine the temperature of the filament of a 100-W light bulb if the resistance at room temperature is measured to be 11.7 Ω and when the light is on, the resistance is determined to be 144 Ω.

Solution   In applying Equation 3–7, we set up the equation as follows:

$$144\ \Omega = (11.7\ \Omega)[1 + 0.0045(°C)^{-1}(T_1 - 20°C)]$$

Now, solving for the temperature T_1, we get

$$T_1 = \frac{\left(\dfrac{144\ \Omega}{11.7\ \Omega} - 1\right)}{0.0045(°C)^{-1}} + 20°C = 2530°C$$

practice
problem
IV

A HVDC (high voltage dc) transmission line must be able to operate over a wide temperature range. Calculate the resistance of 900 km of 1843 MCM aluminum conductor at temperatures of −40°C and +40°C.

in process
learning
check IV

Explain what is meant by the terms *positive temperature coefficient* and *negative temperature coefficient*. To which category does aluminum belong?

3.5 Types of Resistors

Virtually all electric and electronic circuits involve the control of voltage and/or current. The best way to provide such control is by inserting appropriate values of resistance into the circuit. Although various types and sizes of resistors are used in electrical and electronic applications, all resistors fall into two main categories: *fixed resistors* and *variable resistors*.

Fixed Resistors

As the name implies, fixed resistors are resistors having resistance values which are essentially constant. There are numerous types of fixed resistors, ranging in size from almost microscopic (as in integrated circuits) to high-power resistors which are capable of dissipating many watts of power. Figure 3–7 illustrates the basic structure of a *molded carbon composition* resistor.

As shown in the Figure 3–7, the molded carbon composition resistor consists of a carbon core mixed with an insulating filler. The ratio of carbon to filler determines the resistance value of the component: the higher the proportion of carbon, the lower the resistance. Metal leads are inserted into the carbon core, and then the entire resistor is encapsulated with an insulated coating. Carbon composition resistors are available in resistances from less than 1 Ω to 100 MΩ and typically have power ratings from ⅛ W to 2 W. Figure 3–8 shows various sizes of resistors, with the larger resistors being able to dissipate more power than the smaller resistors.

Although carbon-core resistors have the advantages of being inexpensive and easy to produce, they tend to have wide tolerances and are susceptible to large changes in resistance due to temperature variation. As shown in Figure 3–9, the resistance of a carbon composition resistor may change by as much as 5% when temperature is changed by 100°C.

Other types of fixed resistors include *carbon film, metal film, metal oxide, wire-wound,* and *integrated circuit packages.*

If fixed resistors are required in applications where precision is an important factor then film resistors are usually employed. These resistors consist of either carbon, metal, or metal-oxide film deposited onto a ceramic cylinder. The desired resistance is obtained by removing part of the resistive material, resulting in a helical pattern around the ceramic

Insulated coating

Carbon composition

Color coding

Leads imbedded into resistive material

Figure 3–7 Structure of a molded carbon composition resistor.

Figure 3–8   Actual size of carbon resistors (2 W, 1 W, ½ W, ¼ W, ⅛ W).

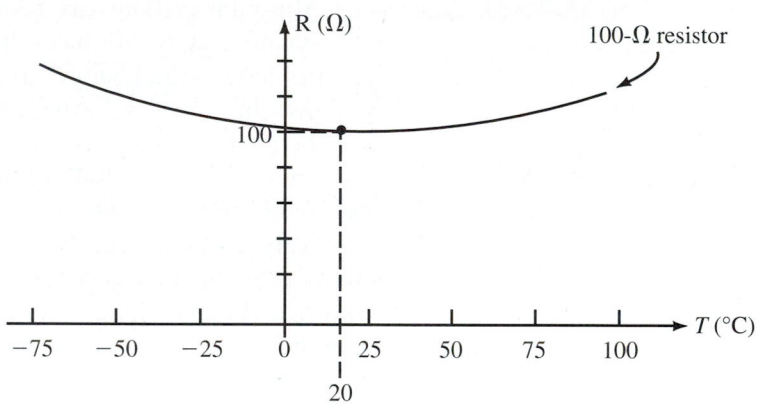

Figure 3–9   Variation in resistance of a carbon composition fixed resistor.

core. If variation of resistance due to temperature is not a major concern then low-cost carbon is used. However, if close tolerances are required over a wide temperature range then the resistors are made of films consisting of alloys such as nickel chromium, constantum, or manganin, which have very small temperature coefficients.

Occasionally a circuit requires a resistor to be able to dissipate large quantities of heat. In such cases, wire-wound resistors may be used. These resistors are constructed of a metal alloy wound around a hollow porcelain core which is then covered with a thin layer of porcelain to seal it in place. The porcelain is able to quickly dissipate heat generated due to current passing through the wire. Figure 3–10 shows a few of the various types of power resistors available.

In circuits where the dissipation of heat is not a major design consideration, fixed resistances may be constructed in miniature packages

Figure 3–10   Power resistors.

(called integrated circuits or ICs) capable of containing many individual resistors. The obvious advantage of such packages is their ability to conserve space on a circuit board. Figure 3–11 illustrates a typical resistor IC package.

Variable Resistors

Variable resistors provide indispensable functions which we use in one form or another almost daily. These components are used to adjust the volume of our radios, set the level of lighting in our homes, and adjust

(a) Internal resistor arrangement   (b) Integrated resistor network. *(Courtesy of Bourns, Inc.)*

Figure 3–11

the heat of our stoves and furnaces. Figure 3–12 shows the internal and the external view of typical variable resistors.

In Figure 3–13, we see that variable resistors have three terminals, two of which are fixed to the ends of the resistive material. The central terminal is connected to a wiper which moves over the resistive material when the shaft is rotated with either a knob or a screwdriver. The resis-

(a) External view of variable resistors.

(b) Internal view of a variable resistor.

Figure 3–12   Variable resistors. *(Courtesy of Bourns, Inc.)*

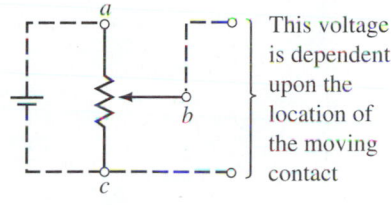

(b) Terminals of a variable resistor

This voltage
is dependent
upon the
location of
the moving
contact

(c) Variable resistor used
as a potentiometer

Figure 3–13   (a) Variable resistor. *(Courtesy of Bourns, Inc.)*

tance between the two outermost terminals will remain constant while the resistance between the central terminal and either terminal will change according to the position of the wiper.

If we examine the schematic of a variable resistor as shown in Figure 3–13(b), we see that the following relationship must apply:

$$R_{ac} = R_{ab} + R_{bc}$$

Variable resistors are used for two principal functions. *Potentiometers*, shown in Figure 3–13(c), are used to adjust the amount of potential (voltage) provided to a circuit. *Rheostats*, the connections and schematic of which are shown in Figure 3–14, are used to adjust the amount of current within a circuit. Applications of potentiometers and rheostats will be covered in later chapters.

The symbol used to represent a variable resistor used as a rheostat is shown in Figure 3–14(b).

$R_{ab} = 0$ when the
moving contact is at a

(No connections)

(a) Connections of a rheostat

(b) Symbol

Figure 3–14

3.6 Color Coding of Resistors

Large resistors such as the wire-wound resistors or the ceramic-encased power resistors have their resistor values and tolerances printed on their cases. Smaller resistors, whether constructed of a molded carbon composition or a metal film, may be too small to have their values printed on the component. Instead, these smaller resistors are usually covered by an epoxy or similar insulating coating over which several colored bands are printed radially as shown in Figure 3–15.

The colored bands provide a quickly recognizable code for determining the value of resistance, the tolerance (in percentage), and occasionally the expected reliability of the resistor. The colored bands are always read from left to right, left being defined as the side of the resistor with the band nearest to it.

The first two bands represent the first and second digits of the resistance value. The third band is called the multiplier band and represents the number of zeros following the first two digits; it is usually given as a power of ten. The fourth band indicates the tolerance of the resistor, and the fifth band (if present) is an indication of the expected reliability of the component. The reliability is a statistical indication of the expected number of components which will no longer have the indicated resistance value after 1000 hours of use. For example, if a particular resistor has a reliability of 1% it is expected that after 1000 hours of use, no more than one resistor in 100 is likely to be outside the specified range of resistance as indicated in the first four bands of the color codes. Table 3–5 shows the colors of the various bands and the corresponding values.

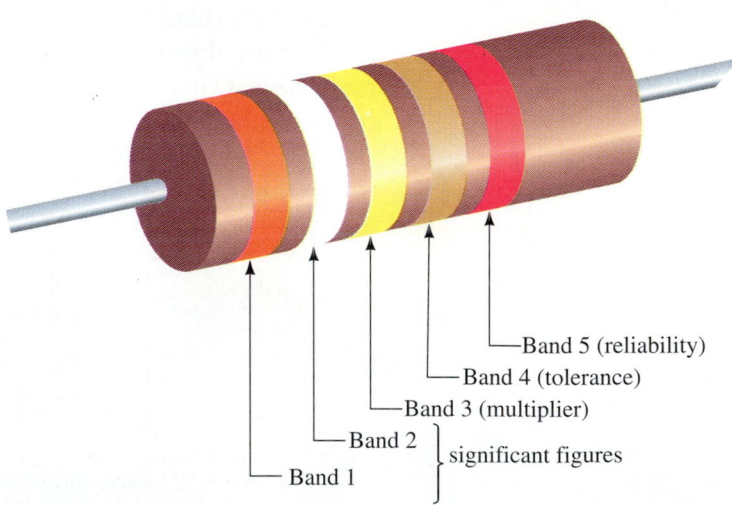

Band 5 (reliability)
Band 4 (tolerance)
Band 3 (multiplier)
Band 2
significant figures
Band 1

Figure 3–15   Resistor color codes.

TABLE 3–5 Resistor Color Codes

COLOR	BAND 1 SIG. FIG.	BAND 2 SIG. FIG.	BAND 3 MULTIPLIER	BAND 4 TOLERANCE	BAND 5 RELIABILITY
Black		0	$10^0 = 1$		
Brown	1	1	$10^1 = 10$		1%
Red	2	2	$10^2 = 100$		0.1%
Orange	3	3	$10^3 = 1\ 000$		0.01%
Yellow	4	4	$10^4 = 10\ 000$		0.001%
Green	5	5	$10^5 = 100\ 000$		
Blue	6	6	$10^6 = 1\ 000\ 000$		
Violet	7	7	$10^7 = 10\ 000\ 000$		
Gray	8	8			
White	9	9			
Gold			0.1	5%	
Silver			0.01	10%	
No color				20%	

■ **EXAMPLE 3–10**   Determine the resistance of a carbon film resistor having the color codes shown in Figure 3–16.

Red (0.1% reliability)
Gold (5% tolerance)
Orange ($\times 10^3$)
Gray (8)
Brown (1)

Figure 3–16

Solution   From Table 3–5, we see that the resistor will have a value determined as

$$R = 18 \times 10^3 \ \Omega \pm 5\%$$

$$= 18 \ \text{k}\Omega \pm 0.9 \ \text{k}\Omega \text{ with a reliability of } 0.1\%$$

This specification indicates that the resistance will fall between 171 kΩ and 189 kΩ. After 1000 hours, we would expect that no more than 1 resistor in 1000 would fall outside the specified range.

A resistor manufacturer produces carbon composition resistors of 100 MΩ, with a tolerance of ±5. What will be the color codes on the resistor? (Left to right)

3.7 Measuring Resistance—The Ohmmeter

The *ohmmeter* is an instrument which is generally part of a multimeter (usually including a voltmeter and an ammeter) and is used to measure the resistance of a component. Although it has limitations, the ohmmeter is used almost daily in service shops and laboratories to measure resistance of components and also to determine whether a circuit is faulty. In addition, the ohmmeter may also be used to determine the condition of semiconductor devices such as diodes and transistors. Figure 3–17 shows both an analog ohmmeter and the more modern digital ohmmeter.

(a) Analog ohmmeter.
Figure 3–17

(b) Digital ohmmeter. *(Reproduced with permission from the John Fluke Mfg. Co., Inc.)*

Figure 3–18   Ohmmeter used to measure an isolated component.

In order to measure the resistance of an isolated component or circuit, the ohmmeter is placed across the component under test, as shown in Figure 3–18. The resistance is then simply read from the meter display.

When using an ohmmeter to measure the resistance of a component which is located in an operating circuit, the following steps should be observed:

1. As shown in Figure 3–19(a), remove all power supplies from the circuit or component to be tested. If this step is not followed, the ohmmeter reading will, at best, be meaningless, and the ohmmeter may be severely damaged.

2. If you wish to measure the resistance of a particular component, it is necessary to isolate the component from the rest of the circuit. This is done by disconnecting at least one terminal of the component from the balance of the circuit as shown in Figure 3–19(b). If this step is not followed, in all likelihood the resistance reading indicated by the ohmmeter will not be the resistance of the desired resistor, but rather the resistance of the combination.

3. As shown in Figure 3–19(b), connect the two probes of the ohmmeter across the component to be measured. The black and red leads of the ohmmeter may be interchanged when

(a) Disconnect the circuit from the voltage/current source

(b) Isolate and measure the component

Figure 3–19   Using an ohmmeter to measure resistance in a circuit.

measuring resistance. When measuring resistance of other components, however, the measured resistance will be dependent upon the direction of the sensing current. Such devices are covered briefly in a later section of this chapter.

4. Ensure that the ohmmeter is on the correct range to provide the most accurate reading. For example, although a digital multimeter (DMM) can measure a reading for a 1.2-kΩ resistor on the 2-MΩ range, the same ohmmeter will provide additional significant digits (hence more precision) when it is switched to the 2-kΩ range. For analog meters, the best accuracy is obtained when the needle is approximately in the center of the scale.

5. When you are finished, turn the ohmmeter off. Because the ohmmeter uses an internal battery to provide a small sensing current, it is possible to drain the battery if the probes accidently connect together for an extended period.

In addition to measuring resistance, the ohmmeter may also be used to indicate the continuity of a circuit. Many modern digital ohmmeters have an audible tone which indicates that a circuit is unbroken from one point to another point. As demonstrated in Figure 3–20(a), the audible tone of a digital ohmmeter allows the user to determine continuity without having to look away from the circuit under test.

Ohmmeters are particularly useful instruments in determining whether a given circuit has been *short circuited* or *open circuited*.

A short circuit occurs when a low-resistance conductor such as a piece of wire or any other conductor is connected between two points in

(a) Short circuit

(b) Open circuit

Figure 3–20

a circuit. Due to the very low resistance of the short circuit, current will bypass the rest of the circuit and go through the short. An ohmmeter will indicate a very low (theoretically zero) resistance when used to measure across a short circuit.

An open circuit occurs when a conductor is broken between the points under test. An ohmmeter will indicate infinite resistance when used to measure the resistance of a circuit having an open circuit.

Figure 3–20 illustrates circuits having a short circuit and an open circuit.

practical

notes

W̶hen a digital ohmmeter measures an open circuit, the display on the meter will usually be the digit 1 at the left-hand side, with no following digits. This reading should not be confused with a reading of 1 Ω, 1 kΩ, or 1 MΩ, which would appear on the right-hand side of the display.

practice
problem
VI

An ohmmeter is used to measure across the terminals of a switch.

    a. What will the ohmmeter indicate when the switch is closed?

    b. What will the ohmmeter indicate when the switch is opened?

3.8 Thermistors

In Section 3.4 we saw how resistance changes with changes in temperature. While this effect is generally undesirable in resistors, there are many applications which use electronic components having characteristics which vary according to changes in temperature. Any device or component which causes an electrical change due to a physical change is referred to as a *transducer*.

A *thermistor* is a two-terminal transducer in which resistance changes significantly with changes in temperature (hence a thermistor is a "thermal resistor"). The resistance of thermistors may be changed either by external temperature changes or by changes in temperature caused by current through the component. By applying this principle, thermistors may be used in circuits to control current and to measure or control temperature. Typical applications include electronic thermometers and thermostatic control circuits for furnaces. Figure 3–21 shows a typical thermistor and its electrical symbol.

Thermistors are constructed of oxides of various materials such as cobalt, manganese, nickel, and strontium. As the temperature of the thermistor is increased, the outermost (valence) electrons in the atoms of the material become more active and break away from the atom. These extra electrons are now free to move within the circuit, thereby causing a reduction in the resistance of the component (*negative temperature coefficient*). Figure 3–22 shows how resistance of a thermistor varies with temperature effects.

(a) Thermistor   (b) Thermistor symbol

Figure 3–21

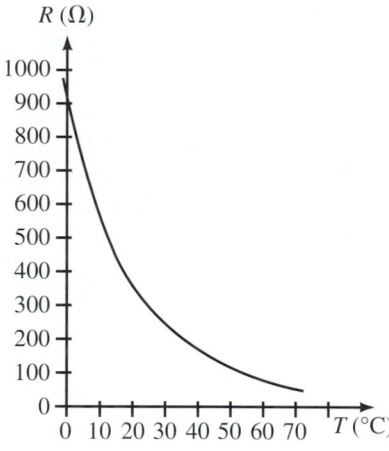

$R\ (\Omega)$

Figure 3–22   Thermistor resistance as a function of temperature.

practice
problem
VII

Referring to Figure 3–22, determine the approximate resistance of a thermistor at each of the following temperatures:

a. 10°C.

b. 30°C.

c. 50°C.

3.9 Photoconductive Cells

Photoconductive cells or *photocells* are two-terminal transducers which have a resistance determined by the amount of light falling on the cell. Most photocells are constructed of either cadmium sulfide (CdS) or cadmium selenide (CdSe) and are sensitive to light having wavelengths between 4000 Å (blue light) and 10 000 Å (infrared). The angstrom (Å) is a unit commonly used to measure the wavelength of light and has a dimension given as $1 \text{ Å} = 1 \times 10^{-10}$ m. Light, which is a form of energy, strikes the material of the photocell and causes the release of valence electrons, thereby reducing the resistance of the component. Figure 3–23 shows the structure, symbol, and resistance characteristics of a typical photocell.

Photocells may be used to measure light intensity and/or to control lighting. They are typically used as part of a security system.

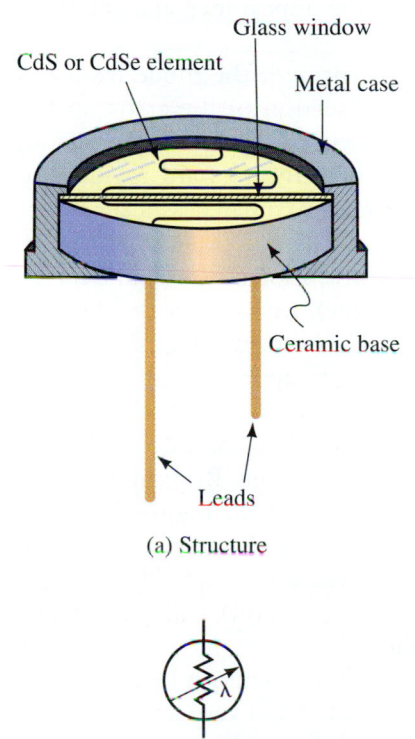

(a) Structure

(b) Symbol of a photocell

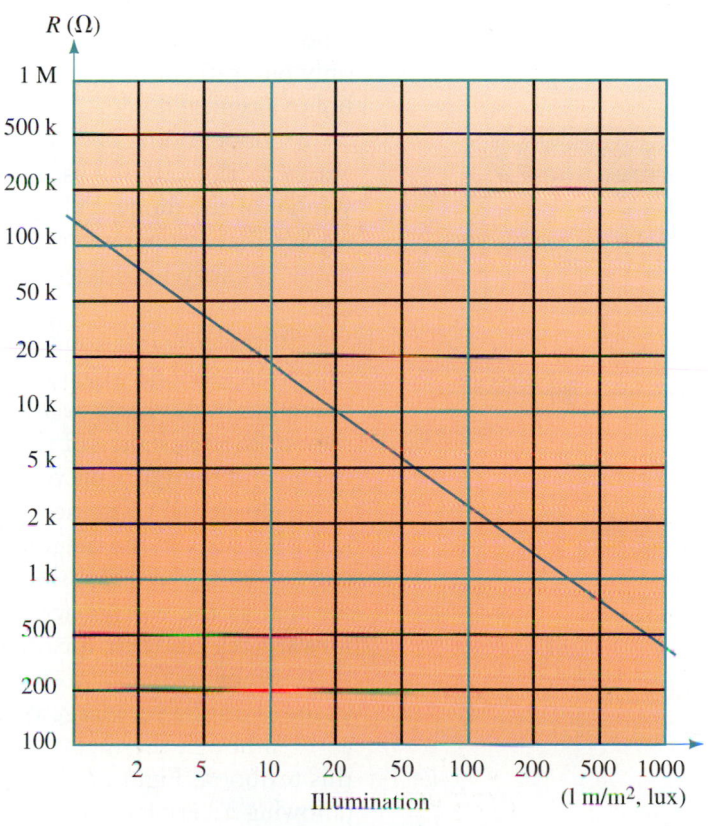

(c) Resistance versus illumination

Figure 3–23   Photocell.

3.10 Nonlinear Resistance

Up to this point, the components we have examined have had values of resistance which were essentially constant for a given temperature (or, in the case of a photocell, for a given amount of light). If we were to examine the current versus voltage relationship for these components, we would find that the relationship is linear, as shown in Figure 3–24.

If a device has a linear (straight-line) current-voltage relation then it is referred to as an *ohmic device*. (The linear current-voltage relationship will be will be covered in greater detail in the next chapter.) Often in electronics, we use components which do not have a linear current-voltage relationship and therefore these devices are referred to as *nonohmic devices*. On the other hand, some components, such as the thermistor, can be shown to have both an ohmic region and a nonohmic region. For large current through the thermistor, the component will get hotter. This increase in temperature will result in a decrease of resistance. Consequently, for large currents, the thermistor is a *nonohmic device*.

We will now briefly examine two common nonohmic devices.

Figure 3–24   Linear current-voltage relationship.

Diodes

The diode is a semiconductor device which permits charge to flow in only one direction. Figure 3–25 illustrates the appearance and the symbol of a typical diode.

Conventional current will easily pass through the diode from the anode to the cathode as indicated by the direction of the arrow in the symbol of the diode. When the diode conducts current in this direction, we say that it operates in its *forward region*. In the forward region, the diode has very little resistance and is often approximated as a short circuit.

If the circuit is connected such that conventional flow is in a direction from cathode to anode (against the arrow), we say that the diode is operating in its *reverse region*. When the diode is operating in the reverse region, the resistance of the diode is very high, approximating an open circuit and therefore preventing current.

The actual resistance of the diode is dependent not only on the polarity but also on the magnitude of the applied voltage. By examining the current-voltage relationship of the diode, as shown in Figure 3–26, we see that as the voltage across the device is increased, the current in the forward region (first quadrant) will also increase very quickly, whereas the current in the opposite direction (third quadrant) will remain unchanged until a very large voltage is applied.

Although the theory of operation of a diode will not be covered in this textbook, Figure 3–26 shows that the diode has a very low resistance (allowing a very large current) in the forward direction once the voltage across the diode is greater than V_F. In the reverse direction, the diode has a very large resistance (allowing almost no current) as long as the voltage is kept below V_R.

Figure 3–25   Diode.
(a) Typical structure.
(b) Symbol.

Figure 3–26   Current-voltage relation for a silicon diode.

practical

notes

Because an ohmmeter uses an internal voltage source to generate a small sensing current, the instrument may easily be used to determine the terminals (and hence the direction of conventional flow) of a diode. (See Figure 3–27.) If we measure the resis-

(a) Diode operating in its reverse region   (b) Diode operating in its forward region

Figure 3–27   Determining diode terminals with an ohmmeter.

tance of the diode in both directions, we will find that the resistance will be low when the positive terminal of the ohmmeter is connected to the anode of the diode. When the positive terminal is connected to the cathode, virtually no current will occur in the diode and so the indication on the ohmmeter will be a very high resistance (theoretically, R = ∞ Ω).

Varistors

Varistors, as shown in Figure 3–28, are semiconductor devices which have very high resistances when the voltage across the varistors is below the breakdown value. However, when the voltage across a varistor (either polarity) exceeds the rated value, the resistance of the device suddenly becomes very small, allowing charge to flow. Figure 3–29 shows the current-voltage relation for varistors.

(a) Varistors.
Figure 3–28

(b) Varistor symbols.

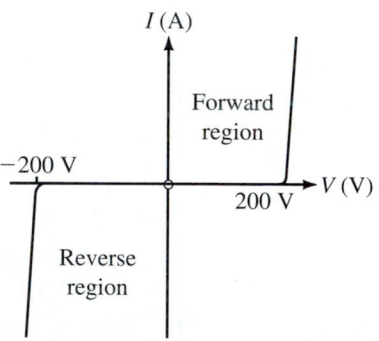

Figure 3–29   Current-voltage relation of a 200-V (peak) varistor.

Varistors are used in sensitive circuits, such as those in computers, to ensure that if the voltage suddenly exceeds a predetermined value, the varistor will effectively become a short circuit to the unwanted signal, thereby protecting the rest of the circuit from excessive voltage.

3.11 Conductance

Conductance, G, is defined as the measure of a material's ability to allow the flow of charge and is assigned the SI unit the siemens (S). A large conductance indicates that a material is able to conduct current well, whereas a low value of conductance indicates that a material does not readily permit the flow of charge. Mathematically, conductance is defined as the reciprocal of resistance. Thus

$$G = \frac{1}{R} \quad \text{[siemens, S]} \tag{3–8}$$

where R is resistance, in ohms (Ω).

■ **EXAMPLE 3-11**   Determine the conductance of following resistors:

a. 5 Ω.
b. 100 kΩ.
c. 50 mΩ.

Solution

a. $G = \dfrac{1}{5\ \Omega} = 0.2\ \text{S} = 200\ \text{mS}$

b. $G = \dfrac{1}{100\ \text{k}\Omega} = 0.01\ \text{mS} = 10\ \mu\text{S}$

c. $G = \dfrac{1}{50\ \text{m}\Omega} = 20\ \text{S}$

---

practice problems VIII

1. A given cable has a conductance given as 5.0 mS. Determine the value of the resistance, in ohms.

2. If the conductance is doubled, what happens to the resistance?

Although the SI unit of conductance (siemens) is almost universally accepted, older books and data sheets list conductance in the unit given as the mho (ohm spelled backwards) and having an upside-down omega, ℧, as the symbol. In such a case, the following relationship holds:

$$1\ ℧ = 1\ \text{S} \tag{3–9}$$

practice
problem
IX

A specification sheet for a radar transmitter indicates that one of the components has a conductance of 5 $\mu\mu\mho$.

    a. Express the conductance in the proper SI prefix and unit.
    b. Determine the resistance of the component, in ohms.

3.12 Superconductors

As you have seen, all power lines and distribution networks have internal resistance which results in energy loss due to heat as charge flows through the conductor. If there was some way of eliminating the resistance of the conductors, electricity could be transmitted farther and more economically. The idea that energy could be transmitted without losses along a "superconductor" transmission line was formerly a distant goal. However, recent discoveries in high-temperature superconductivity promise the almost magical ability to transmit and store energy with no loss in energy.

    In 1911, the Dutch physicist Heike Kamerlingh Onnes discovered the phenomenon of superconductivity. Studies of mercury, tin, and lead verified that the resistance of these materials decreases to no more than one ten-billionth of the room temperature resistance when subjected to temperatures of 4.6 K, 3.7 K, and 6 K respectively. Recall that the relationship between kelvins and degrees Celsius is given as follows:

$$T_K = T_{(°C)} + 273.15° \qquad (3\text{--}10)$$

The temperature at which a material becomes a superconductor is referred to as the *critical temperature*, T_C, of the material. Figure 3–30 shows how the resistance of a sample of mercury changes with temperature. Notice how the resistance suddenly drops to zero at a temperature of 4.6 K.

    Experiments with currents in supercooled loops of superconducting wire have determined that the induced currents will remain undiminished for many years within the conductor provided that the temperature is maintained below the critical temperature of the conductor.

    A peculiar, seemingly magical property of superconductors occurs when a permanent magnet is placed above the superconductor. The magnet will float above the surface of the conductor as if it is defying the law of gravity as shown in Figure 3–31.

    This principle, which is referred to as the *Meissner effect* (named after Walther Meissner), may be simply stated as follows:

    When a superconductor is cooled below its critical temperature, magnetic fields may surround but not enter the superconductor.

    The principle of superconductivity is explained in the behavior of electrons within the superconductor. Unlike conductors which have electrons moving randomly through the conductor and colliding with other electrons [Figure 3–32(a)], the electrons in superconductors form pairs which move through the material in a manner similar to a band marching in a parade. The orderly motion of electrons in a superconductor,

Figure 3–30 Critical temperature of mercury.

Figure 3–31   The Meissner effect: A magnetic cube hovers above a disk of ceramic superconductor. The disk is kept below its critical temperature in a bath of liquid nitrogen. *(Courtesy of AT&T Bell Laboratories/AT&T Archives)*

shown in Figure 3–32(b), results in an ideal conductor, since the electrons no longer collide.

The economy of having a high critical temperature has led to the search for high-temperature superconductors. In recent years, research at such institutes as the IBM Zurich Research Laboratory in Switzerland and the University of Houston in Texas have yielded superconducting materials which are able to operate at temperatures as high as 98 K (–175°C). While this temperature is still very low, it means that superconductivity can now be achieved by using the readily available liquid nitrogen rather than the much more expensive and more rare liquid helium.

Superconductivity has been found in seemingly unlikely materials such as ceramics consisting of barium, lanthanum, copper, and oxygen. Research is now centered on developing new materials which become

Figure 3–32   (a) In conductors, electrons are free to move in any direction through the conductor. Energy is lost due to collisions with atoms and other electrons, giving rise to the resistance of the conductor.
(b) In superconductors, electrons are bound in pairs and travel through the conductor in step, avoiding all collisions. Since there is no energy loss, the conductor has no resistance.

superconductors at ever higher temperatures and which are able to overcome the disadvantages of the early ceramic superconductors.

Although very expensive, low-temperature superconductivity is currently used in some giant particle accelerators and, to a limited degree, in electronic components (such as superfast Josephson junctions and SQUIDs, i.e., superconducting quantum interference devices, which are used to detect very small magnetic fields). Once research produces commercially viable, high-temperature superconductors, the possibilities of the applications will be virtually limitless. High-temperature superconductivity promises to yield improvements in transportation, energy storage and transmission, computers, and medical treatment and research. It is quite possible that high-temperature superconductivity will change electronics as much as the invention of the transistor.

Problems

3.1 Resistance of Conductors

1. Determine the resistance, at 20°C, of 100 m of solid aluminum wire having the following radii:
   a. 0.5 mm.
   b. 1.0 mm.
   c. 0.005 mm.
   d. 0.5 cm.

2. Determine the resistance, at 20°C, of 200 feet of iron conductors having the following cross sections:
   a. 0.25 inch by 0.25 inch square.
   b. 0.125 inch diameter round.
   c. 0.125 inch by 4.0 inch rectangle.

3. A 250-foot length of solid copper bus bar, shown in Figure 3–33, is used to connect a voltage source to a distribution panel. If the bar is to have a resistance of 0.02 Ω at 20°C, calculate the required height of the bus bar (in inches).

4. Nichrome wire is used to construct heating elements. Determine the length of 1.0-mm-diameter Nichrome wire needed to produce a heating element which has a resistance of 2.0 Ω at a temperature of 20°C.

5. A copper wire having a diameter of 0.80 mm is measured to have a resistance of 10.3 Ω at 20°C. How long is this wire in meters? How long is the wire in feet?

6. A piece of aluminum wire has a resistance, at 20°C, of 20 Ω. If this wire is melted down and used to produce a second wire having a length four times the original length, what will be the resistance of the new wire at 20°C? (Hint: The volume of the wire has not changed.)

Figure 3–33

7. Determine the resistivity (in ohm-meters) of a carbon graphite cylinder having a length of 6.00 cm, a diameter 0.50 mm, and a measured resistance of 3.0 Ω at 20°C. How does this value compare with the resistivity given for (carbon)?

8. A solid circular wire of length 200 m and diameter of 0.4 mm has a resistance measured to be 357 Ω at 20°C. Of what material is the wire constructed?

9. A 2500-m section of alloy wire has a resistance of 32 Ω. If the wire has a diameter of 1.5 mm, determine the resistivity of the material in ohm-meters. Is this alloy a better conductor than copper?

10. A section of iron wire having a diameter of 0.030 inch is measured to have a resistance of 2500 Ω (at a temperature of 20°C).
    a. Determine the cross-sectional area in square meters and in square millimeters. (Note: 1 inch = 2.54 cm = 25.4 mm)
    b. Calculate the length of the wire in meters.

3.2   Electrical Wire Tables

11. Use Table 3–2 to determine the resistance of 300 feet of AWG 22 and AWG 19 solid copper conductors. Compare the diameters and the resistances of the wires.

12. Use Table 3–2 to find the resistance of 250 m of AWG 8 and AWG 2 solid copper conductors. Compare the diameters and the resistances of the wires.

13. Determine the maximum current which could be handled by AWG 19 wire and by AWG 30 wire.

14. If AWG 8 is rated at a maximum of 40 A, how much current could AWG 2 handle safely?

15. A spool of AWG 36 copper transformer wire is measured to have a resistance of 550 Ω at a temperature of 20°C. How long is this wire in meters?

16. How much current should AWG 36 copper wire be able to handle?

3.3   Resistance of Wires—Circular Mils

17. Determine the area in circular mils of the following conductors ($T = 20°C$):
    a. Circular wire having a diameter of 0.016 inch.
    b. Circular wire having a diameter of 2.0 mm.
    c. Rectangular bus bar having dimensions 0.25 inch by 6.0 inch.

18. Express the cross-sectional areas of the conductors of Problem 17 in square mils and in square millimeters.

19. Calculate the resistance, at 20°C, of 400 feet of copper conductors having the cross-sectional areas given in Problem 17.

20. Determine the diameter in inches and in millimeters of circular cables having cross-sectional areas as given below: (Assume the cables to be solid conductors.)
    a. 250 CM.
    b. 1000 CM.
    c. 250 MCM.
    d. 750 MCM.

21. A 200-foot length of solid copper wire is measured to have a resistance of 0.500 Ω.
    a. Determine the cross-sectional area of the wire in both square mils and circular mils.
    b. Determine the diameter of the wire in mils and in inches.

22. Repeat Problem 21 if the wire had been made of Nichrome.

23. A spool of solid copper wire having a diameter of 0.040 inch is measured to have a resistance of 12.5 Ω (at a temperature of 20°C).
    a. Determine the cross-sectional area in both square mils and circular mils.
    b. Calculate the length of the wire in feet.

24. Iron wire having a diameter of 30 mils was occasionally used for telegraph transmission. A technician measures a section of telegraph line to have a resistance of 2500 Ω (at a temperature of 20°C).
    a. Determine the cross-sectional area in both square mils and circular mils.
    b. Calculate the length of the wire in feet and in meters. (Note: 1 ft = 0.3048 m.) Compare your answer to the answer obtained in Problem 10.

3.4   Temperature Effects

25. An aluminum conductor has a resistance of 50 Ω at room temperature. Find the resistance of the same conductor at −30°C, 0°C, and at 200°C.

26. AWG 14 solid copper house wire is designed to operate within a temperature range of −40°C to +90°C. Calculate the resistance of 200 circuit feet of wire at both temperatures. Note: A circuit foot is the length of cable needed for a current to travel to and from a load.

27. A given material has a resistance of 20 Ω at room temperature (20°C) and 25 Ω at a temperature of 85°C.

   a. Does the material have a positive or a negative temperature coefficient? Explain briefly.

   b. Determine the value of the temperature coefficient, α, at 20°C.

   c. Assuming the resistance versus temperature function to be linear, determine the expected resistance of the material at 0°C (the freezing point of water) and at 100°C (the boiling point of water).

28. A given material has a resistance of 100 Ω at room temperature (20°C) and 150 Ω at a temperature of –25°C.

   a. Does the material have a positive or a negative temperature coefficient? Explain briefly.

   b. Determine the value of the temperature coefficient, α, at 20°C.

   c. Assuming the resistance versus temperature function to be linear, determine the expected resistance of the material at 0°C (the freezing point of water) and at –40°C.

29. An electric heater is made of Nichrome wire. The wire has a resistance of 15.2 Ω at a temperature of 20°C. Determine the resistance of the Nichrome wire when the temperature of the wire is increased to 260°C.

30. A silicon diode is measured to have a resistance of 500 Ω at 20°C. Determine the resistance of the diode if the temperature of the component is increased with a soldering iron to 30°C. (Assume that the resistance versus temperature function is linear.)

3.5 Types of Resistors

31. A 10-kΩ variable resistor has its wiper (movable terminal b) initially at the bottom terminal, c. Determine the resistance R_{ab} between terminals a and b and the resistance R_{bc} between terminals b and c under the following conditions:

   a. The wiper is at c.

   b. The wiper is one fifth the way around the resistive surface.

   c. The wiper is four fifths the way around the resistive surface.

   d. The wiper is at a.

32. The resistance between wiper terminal b and bottom terminal c of a 200-kΩ variable resistor is measured to be 50 kΩ. Determine the resistance which would be measured between the top terminal, a and the wiper terminal, b.

3.6   Color Coding of Resistors

33. Given resistors having the following color codes (as read from left to right), determine the resistance, tolerance and reliability of each component. Express the uncertainty in both percentage and ohms.

a. Brown	Green	Yellow	Silver	
b. Red	Gray	Gold	Gold	Yellow
c. Yellow	Violet	Blue	Gold	
d. Orange	White	Black	Gold	Red

34. Determine the color codes required if you need the following resistors for a project:
  a. 33 kΩ ± 5% 0.1% reliability,
  b. 820 Ω ± 10%,
  c. 15 Ω ± 20%,
  d. 2.7 MΩ ± 5%.

3.7   Measuring Resistance—The Ohmmeter

35. Explain how an ohmmeter may be used to determine whether a light bulb is burned out.

36. If an ohmmeter were placed across the terminal of a switch, what resistance would you expect to measure when the contacts of the switch are closed? What resistance would you expect to measure when the contacts are opened?

37. Explain how you could use an ohmmeter to determine approximately how much wire is left on a spool of AWG 24 copper wire.

38. An analog ohmmeter is used to measure the resistance of a two-terminal component. The ohmmeter indicates a resistance of 1.5 kΩ. When the leads of the ohmmeter are reversed, the meter indicates that the resistance of the component is an open circuit. Is the component faulty? If not, what kind of component is being tested?

3.8   Thermistors

39. A thermistor has the characteristics shown in Figure 3–22.
  a. Determine the resistance of the device at room temperature, 20°C.
  b. Determine the resistance of the device at a temperature of 40°C.
  c. Does the thermistor have a positive or a negative temperature coefficient? Explain.

3.9   Photoconductive Cells

40. For the photocell having the characteristics as shown in Figure 3–23(c), determine the resistance

    a. in a dimly lit basement having an illuminance of 10 lux,

    b. in a home having an illuminance of 50 lux,

    c. in a classroom having an illuminance of 500 lux.

3.11   Conductance

41. Calculate the conductance of the following resistances:

    a. 0.25 Ω.

    b. 500 Ω.

    c. 250 kΩ.

    d. 12.5 MΩ.

42. Determine the resistance of components having the following conductances:

    a. 62.5 μS.

    b. 2500 mS.

    c. 5.75 mS.

    d. 25.0 S.

43. Determine the conductance of 1000 m of AWG 30 solid copper wire at a temperature of 20°C.

44. Determine the conductance of 200 feet of aluminum bus bar (at a temperature of 20°C) which has a cross-sectional dimension of 4.0 inches by 0.25 inch. If the temperature were to increase, what would happen to the conductance of the bus bar?

4

Ohm's Law, Power, and Energy

CHAPTER PREVIEW

In the previous two chapters, you studied voltage, current, and resistance separately. In this chapter, we consider them together. Beginning with Ohm's law, you will study the relationship between voltage and current in a resistive circuit, reference conventions, power, energy, efficiency, and the use of computers in solving simple Ohm's law problems.

(Photo by David Perry. Courtesy of Rockwell International)

OBJECTIVES

After studying this chapter, you will be able to

- compute voltage, current, and resistance in simple circuits using Ohm's law,

- use the voltage reference convention to determine polarity,

- describe how voltage, current, and power are related in a resistive circuit,

- compute power in dc circuits,

- use the power reference convention to describe the direction of power transfer,

- compute energy used by electrical loads,

- determine energy costs,

- determine the efficiency of machines and systems,

- use PSpice to solve Ohm's law problems.

Georg Simon Ohm

*I*n Chapter 3, we looked briefly at Ohm's experiments. We now take a look at Ohm the person.

Georg Simon Ohm was born in Erlangen, Bavaria, on March 16, 1787. His father was a master mechanic who determined that his son should obtain an education in science. Although Ohm became a teacher in a high school, he had aspirations to receive a university appointment. The only way that such an appointment could be realized would be if Ohm could produce important results through scientific research. Since the science of electricity was in its infancy, and because the electric cell had recently been invented by the Italian Conte Alessandro Volta, Ohm decided to study the behavior of current in resistive circuits. Because equipment was expensive and hard to come by, Ohm made much of his own, thanks, in large part, to his father's training. Using this equipment, Ohm determined experimentally that the amount of current transmitted along a wire was directly proportional to its cross-sectional area and inversely proportional to its length. From these results, Ohm was able to define resistance and show that there was a simple relationship between voltage, resistance, and current. This result, now known as Ohm's law, is probably the most fundamental relationship in circuit theory. However, when published in 1827, Ohm's results were met with ridicule. As a result, not only did Ohm miss out on a university appointment, he was forced to resign from his high-school teaching position. While Ohm was living in poverty and shame, his work became known and appreciated outside Germany. In 1842, Ohm was appointed a member of the Royal Society. Finally, in 1849, he was appointed as a professor at the University of Munich, where he was at last recognized for his important contributions.

4.1 Ohm's Law

Consider the circuit of Figure 4–1. Using a circuit similar in concept to this, Ohm determined experimentally that *current in a resistive circuit is directly proportional to its applied voltage and inversely proportional to its resistance*. In equation form, Ohm's law states

$$I = \frac{E}{R} \quad \text{[amps, A]} \tag{4–1}$$

where

E is the voltage in volts,

R is the resistance in ohms,

I is the current in amperes.

From this you can see that the larger the applied voltage, the larger the current, and the larger the resistance, the smaller the current.

The proportional relationship between voltage and current described by Equation 4–1 may be demonstrated by direct substitution as indicated in Figure 4–2. For a fixed resistance, doubling the voltage as shown in (b) doubles the current, while tripling the voltage as shown in (c) triples the current, and so on.

The inverse relationship between resistance and current is demonstrated in Figure 4–3. For a fixed voltage, doubling the resistance as shown in (b) halves the current, while tripling the resistance as shown in (c) reduces the current to one third of its original value, and so on.

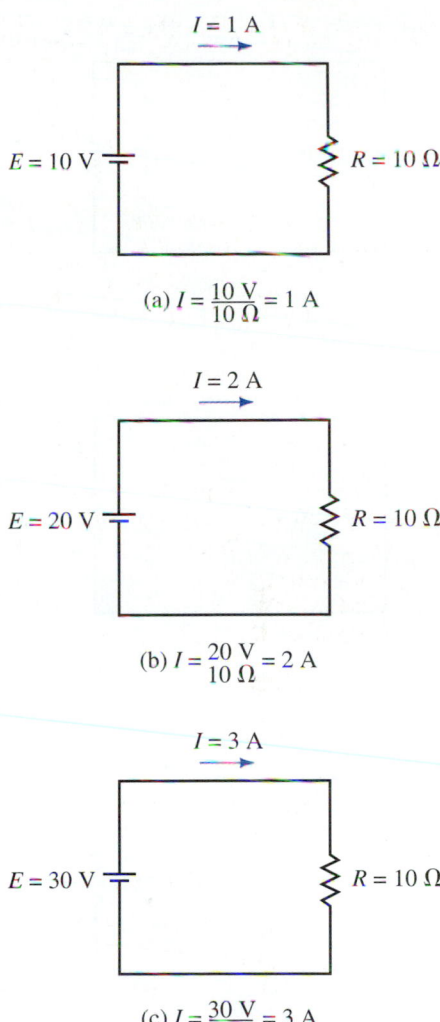

(a) $I = \dfrac{10\ V}{10\ \Omega} = 1\ A$

(b) $I = \dfrac{20\ V}{10\ \Omega} = 2\ A$

(c) $I = \dfrac{30\ V}{10\ \Omega} = 3\ A$

Figure 4–2   For a fixed resistance, current is directly proportional to voltage; thus, doubling the voltage as in (b) doubles the current, while tripling the voltage as in (c) triples the current, and so on.

R

Resistor

E

+

Jolt

Battery

(a) Test circuit

I

E

R

$I = \dfrac{E}{R}$

(b) Schematic. Meters not shown

Figure 4–1   Circuit for illustrating Ohm's law.

(a) $I = \dfrac{36\ V}{12\ \Omega} = 3\ A$

(b) $I = \dfrac{36\ V}{24\ \Omega} = 1.5\ A$

(c) $I = \dfrac{36\ V}{36\ \Omega} = 1\ A$

Figure 4–3 For a fixed voltage, current is inversely proportional to resistance; thus, doubling the resistance as in (b) halves the current, while tripling the resistance as in (c) results in one third the current, and so on.

Ohm's law may also be expressed as

$$E = IR \quad \text{[volts, V]} \qquad (4\text{--}2)$$

and

$$R = \frac{E}{I} \quad \text{[ohms, } \Omega] \qquad (4\text{--}3)$$

These are obtained from Equation 4–1 by simple algebra. When using Ohm's law, be sure to express all quantities in base units of volts, ohms, and amps.

■ **EXAMPLE 4–1**   A 27-Ω resistor is connected to a 12-V battery. What is the current?

Solution   Substituting the resistance and voltage values into Ohm's law yields

$$I = \frac{E}{R} = \frac{12\ V}{27\ \Omega} = 0.444\ A$$

■ **EXAMPLE 4–2**   The lamp of Figure 4–4 draws 25 mA when connected to a 6-V battery. What is its resistance?

Figure 4–4

Solution   Using Equation 4–3,

$$R = \frac{E}{I} = \frac{6\ V}{25 \times 10^{-3}\ A} = 240\ \Omega$$

■ **EXAMPLE 4–3**   A resistor with the color code brown, red, yellow is connected to a 30-V source. What is I?

Solution   Using the color code of Chapter 3, $R = 120$ kΩ. Thus,

$$I = \frac{E}{R} = \frac{30\ V}{120 \times 10^{3}\ \Omega} = 0.25 \times 10^{-3} = 0.25\ \text{mA}$$

■ **EXAMPLE 4–4**   If 125 μA is the current in a resistor with color bands red, red, yellow, what is the voltage across the resistor?

Solution   Using the color code, $R = 220$ kΩ. From Ohm's law, $E = IR = (125 \times 10^{-6}\ A)(220 \times 10^{3}\ \Omega) = 27.5$ V.

**practice
problems
1**

1. a. For the circuit of Figure 4–2(a), show that halving the voltage halves the current.
   b. For the circuit of Figure 4–3(a), show that halving the resistance doubles the current.
   Are these results consistent with the verbal statement of Ohm's law?

2. For each of the following, draw the circuit with values marked, then solve for the unknown.
   a. A 10 000-milliohm resistor is connected to a 24-V battery. What is the resistor current?
   b. How many volts are required to establish a current of 20 μA in a 100-kΩ resistor?
   c. If 125 V is applied to a resistor and 5 mA results, what is the resistance?

Ohm's Law in Graphical Form

The relationship between current and voltage described by Equation 4–1 may be shown graphically as in Figure 4–5. The graph, which is a straight line, shows clearly that the relationship is linear. Graphs such as this are important in electronics; we look at them in more detail in Section 4.7.

Open Circuits

Current can only exist where there is a conductive path (e.g., a length of wire). For the circuit of Figure 4–6, I equals zero since there is no conductor between points a and b. We refer to this as an *open circuit*. Since $I = 0$, substitution of this into Equation 4–3 yields

$$R = \frac{E}{I} = \frac{E}{0} \Rightarrow \infty \text{ ohms}$$

Thus, an open circuit has infinite resistance.

Voltage Symbols

Two different symbols are used to represent voltage. For sources, use uppercase E; for loads (and other components), use uppercase V. This is illustrated in Figure 4–7.

Using the symbol V, Ohm's law may be written in its several forms as

$$I = \frac{V}{R} \quad \text{[amps]} \tag{4–4}$$

$$V = IR \quad \text{[volts]} \tag{4–5}$$

$$R = \frac{V}{I} \quad \text{[ohms]} \tag{4–6}$$

These relationships hold for every resistor in a circuit, no matter how complex the circuit. Since $V = IR$, these voltages are sometimes referred to as *IR drops*.

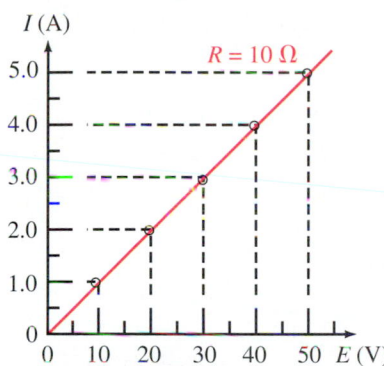

Figure 4–5   Graph showing the relationship between current and voltage for a linear resistance. Plot is for a 10-Ω resistor.

Figure 4–6   An open circuit has infinite resistance.

Figure 4–7   Symbols used to represent voltages. E is used for source voltages while V is used for voltages across circuit components such as resistors.

■ **EXAMPLE 4–5**   The current in each resistor of Figure 4–8 is $I = 0.5$ A. Compute V_1 and V_2.

$I = 0.5$ A

$R_1 = 20\ \Omega$ V_1

$R_2 = 100\ \Omega$ V_2

Figure 4–8   Ohm's law applies to each resistor.

Solution $V_1 = IR_1 = (0.5\ \text{A})(20\ \Omega) = 10$ V. Note, I is also the current in R_2. Thus, $V_2 = IR_2 = (0.5\ \text{A})(100\ \Omega) = 50$ V.

4.2   Voltage and Current Conventions

Voltage Polarities

So far, we have paid little attention to the polarity of voltages across resistors. However, polarity is of extreme importance; fortunately, there is a simple relationship between current direction and voltage polarity. To get at the idea, consider Figure 4–9. In (a), the polarity of V is obvious

Figure 4–9   Convention for voltage polarity. Place the plus sign for V at the tail of the current direction arrow.

$E = 12$ V

I V

$+ \quad V \quad -$
$I \longrightarrow$

$- \quad V \quad +$
$\longleftarrow I$

(a) Defining convention   (b) Examples

since the resistor is connected directly to the source. This makes the top end of the resistor positive with respect to the bottom end, and $V = E = 12$ V as indicated by the meters.

Now consider current. The direction of I is from top to bottom as we indicate by the current direction arrow. Examining voltage polarity, we see that the plus sign for V is at the tail of this arrow. This observation turns out to be true in general and gives us a convention for marking voltage polarity on circuit diagrams. *For voltage across a resistor, place the plus sign at the tail of the current arrow.* Two additional examples are shown in Figure 4–9(b).

practice problem II

For each resistor of Figure 4–10, compute V and show its polarity.

(a) $R = 10$ kΩ

$I = 0.1$ A

(b) $R = 3$ MΩ

$I = 0.15$ mA

(c) $R = 400$ Ω

$I = 0.3$ A

(d) $R = 0.4$ Ω

$I = 2.5$ A

Figure 4–10

(a)

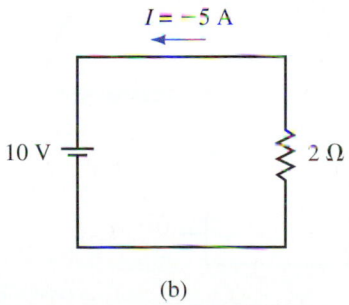

(b)

Figure 4–11   Two representations of the same current.

Current Direction

We also need to look at current direction in more detail. Normally, we show current out of the plus (+) terminal of a source as in Figure 4–11(a). (Here, $I = E/R = 10$ V/2 Ω = 5 A in the direction shown.) But what if you do not know the direction of a current in advance, as frequently is the case when you are analyzing complex circuits? What direction do you choose?

To get at the answer, consider Figure 4–11(b). Here, current is shown going into the plus terminal of the source. To compensate for the reversed direction, we have changed the sign of I. The interpretation placed on this is that a negative current going into an element is the same as a positive current coming out. Therefore, (a) and (b) are the same, i.e., they are two representations of the same current. This gives us the following convention. *If the actual current is in the direction of its reference arrow, it will have a positive value, while if it is opposite to its reference arrow, it will have a negative value.* In practice, this means that you can arbitrarily assign current directions, since, if you obtain a negative value when you solve for a current, you know that it is actually opposite in direction to what you have assumed.

(a) General

(b) Point x is 6 volts positive
with respect to point y

Figure 4–12   The +, – symbology
means that you are looking at the
voltage at the point marked + with
respect to the point marked –.

The Voltage Polarity Convention Revisited

The +, – sign convention of Figure 4–9 has a deeper meaning than so far considered. *Voltage exists between points, and when we place a + at one point and a – at another point, we define this to mean that we are looking at the voltage at the point marked + with respect to the point marked –.* Thus, in Figure 4–12(b), $V = 6$ volts means that point x is 6 V positive with respect to point y. Since the red lead of the meter is placed at point x and the black at point y, the meter will indicate +6 V.

Now consider Figure 4–13(b). [Part (a) has been repeated from Figure 4–12(b) for reference.] Here, we have placed the plus sign at point y, meaning that you are looking at the voltage at y with respect to x. Since point y is 6 V negative with respect to point x, V will have a value of minus 6 volts, i.e., $V = -6$ volts. Note also that the meter indicates –6 volts since we have reversed its leads and the red lead is now at point y and the black at point x. *It is important to realize here that the voltage across R has not changed. What has changed is how we are looking at it and how we have connected the meter to measure it.* Thus, since the actual voltage is the same in both cases, (a) and (b) are equivalent representations.

(a) Voltage at x with respect to y is 6 volts.   (b) Voltage at y with respect to x is –6 volts.

Figure 4–13   Two representations of the same voltage.

■ **EXAMPLE 4–6**   Consider Figure 4–14. In (a), $I = 3$ A, while in (b), $I = -3$ A. Using the voltage polarity convention, determine the voltages across the two resistors and show that they are equal.

Figure 4–14

Solution   In each case, place the plus sign at the tail of the current direction arrow. Then, for (a), you are looking at the polarity of x with respect to y and you get $V = IR = (3$ A$)(12$ Ω$) = 6$ V as expected. Now consider (b). The polarity markings mean that you are looking at the polarity of y with respect to x and you get $V = IR = (-3$ A$)(2$ Ω$) = -6$ V. This means that point y is 6 V negative with respect to x, or equivalently, x is 6 V positive with respect to y. Thus, the two voltages are equal.

In later chapters where you encounter complex circuits with unknown voltage polarities and current directions, proper application of the above principles will always yield correct results.

in process
learning
check I

1. For each of the circuits of Figure 4–15, compute the current and show its direction. Express the answer in the most appropriate unit, e.g., amps, milliamps, microamps, etc.

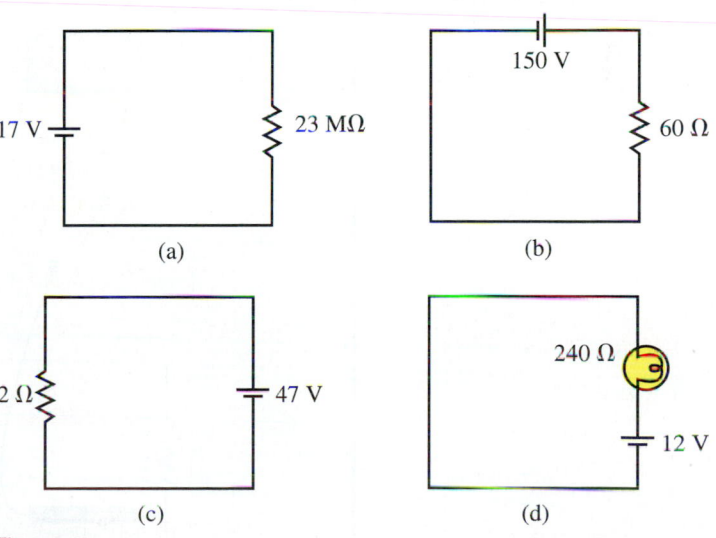

Figure 4–15

2. A resistor has color bands brown, black, and red and a current of 25 mA. Determine the voltage across it.

3. For a resistive circuit, what is I if $E = 500$ V and R is open circuited? Will the current change if the voltage is doubled?

4. A certain resistive circuit has voltage E. If $I = 2.5$ A, what will be the current if:
   a. R is doubled?
   b. R is quadrupled?
   c. R is reduced to 20% of its original value?
   d. R is doubled and E is quadrupled?

5. The voltmeters of Figure 4–16 have autopolarity. Determine the reading of each meter, magnitude, and sign.

(a) $R = 10 \ \Omega$

(c) $R = 15 \ \Omega$

(b) $R = 36 \ \Omega$

(d) $R = 40 \ \Omega$

Figure 4–16

4.3 Power

Power is familiar to all of us, at least in a general sort of way. We know, for example, that electric heaters and light bulbs are rated in watts (W) and that motors are rated in horsepower (or watts), both being units of power as discussed in Chapter 1. We also know that the higher the watt rating of a device, the more energy we can get out of it per unit time. Figure 4–17 illustrates the idea. In (a), the greater the power rating of the light, the more light energy that it can produce per second. In (b), the greater the power rating of the heater, the more heat energy it can produce per second. In (c), the larger the power rating of the motor, the more mechanical work that it can do per second.

As you can see, power is related to energy, which is the capacity to do work. Formally, power is defined as the rate of doing work or, equivalently, as the rate of transfer of energy. The symbol for power is P. By definition,

$$P = \frac{W}{t} \quad \text{[watts, W]} \tag{4–7}$$

where W is the work (or energy) in joules and t is the corresponding time interval of t seconds.

As noted in Chapter 1, the SI unit of power is the watt. From Equation 4–7, we see that P also has units of joules per second. Thus, *one watt equals one joule per second.*

Power in Electrical Systems

Since our interest is in electrical power, we need expressions for P in terms of electrical quantities. Recall from Chapter 2 that voltage is de-

notes

1. *Time t in Equation 4–7 is a time interval, not an instantaneous point in time.*

2. *The symbol for energy is W and the abbreviation for watts is W. Multiple use of symbols is common in technology. In such cases, you have to look at the context in which a symbol is used to determine its meaning.*

(a) A 100 W lamp produces more light energy per second than a 40 W lamp

(b) Hair dryer

(c) A 10 hp motor can do more work in a given time than a $\frac{1}{2}$ hp motor

Figure 4–17   Energy conversion. Power P is a measure of the rate of conversion.

fined as work per unit charge and current as the rate of transfer of charge, i.e.,

$$V = \frac{W}{Q} \qquad (4\text{–}8)$$

and

$$I = \frac{Q}{t} \qquad (4\text{–}9)$$

From Equation 4–8, $W = QV$. Substituting this into Equation 4–7 yields $P = W/t = (QV)/t = V(Q/t)$. Replacing Q/t with I, we get

$$P = VI \quad [\text{watts, W}] \qquad (4\text{–}10)$$

and, for a source,

$$P = EI \quad [\text{watts, W}] \qquad (4\text{–}11)$$

Additional relationships are obtained by substituting $V = IR$ and $I = V/R$ into Equation 4–10:

$$P = I^2 R \quad [\text{watts, W}] \qquad (4\text{–}12)$$

and

$$P = \frac{V^2}{R} \quad [\text{watts, W}] \qquad (4\text{–}13)$$

■ EXAMPLE 4–7   Compute the power supplied to the electric heater of Figure 4–18 using all three formulas.

Resistance of heater = 12 Ω

Figure 4–18   Power to the load (i.e., the heater) can be computed from any of the power formulas.

Solution $I = V/R = 120\ V/12\ \Omega = 10\ A$. Thus, the power may be calculated as follows:

a. $P = VI = (120\ V)(10\ A) = 1200\ W$

b. $P = I^2R = (10\ A)^2(12\ \Omega) = 1200\ W$

c. $P = V^2/R = (120\ V)^2/12\ \Omega = 1200\ W$

Note that all give the same answer, as they must.

■ EXAMPLE 4–8 Compute the power to each resistor in Figure 4–19 using Equation 4–13.

I

$R_1 = 20\ \Omega$ $V_1 = 10\ V$

$R_2 = 100\ \Omega$ $V_2 = 50\ V$

Figure 4–19

Solution You must use the appropriate voltage in the power equation. For resistor R_1, use V_1; for resistor R_2, use V_2.

a. $P_1 = V_1^2/R_1 = (10\ V)^2/20\ \Omega = 5\ W$

b. $P_2 = V_2^2/R_2 = (50\ V)^2/100\ \Omega = 25\ W$

■ EXAMPLE 4–9 If the motor of Figure 4–17(c) draws 6 A from a 120-V source,

a. Compute its power input in watts.

b. Assuming the motor is 100% efficient (i.e., that all electrical power supplied to it is output as mechanical power), compute its power output in horsepower.

Solution

a. $P_{in} = VI = (120\ V)(6\ A) = 720\ W$

b. $P_{out} = P_{in} = 720\ W$. Converting to horsepower, $P_{out} = (720\ W)/(746\ W/hp) = 0.965\ hp$.

practice problem III

a. Show that $I = \sqrt{\dfrac{P}{R}}$

b. Show that $V = \sqrt{PR}$

c. A 100-Ω resistor dissipates 169 W. What is its current?

d. A 3-Ω resistor dissipates 243 W. What is the voltage across it?

e. For Figure 4–19, $I = 0.5$ A. Use Equations 4–10 and 4–12 to compute power to each resistor. Compare your answers to the answers of Example 4–8.

Power Rating of Resistors

Resistors must be able to safely dissipate their heat without damage. For this reason, resistors are rated in watts. (For example, composition resistors of the type used in electronics are made with standard ratings of ⅛, ¼, ½, 1, and 2 W as you saw in Figure 3–8.) To provide a safety margin, it is customary to select a resistor that is capable of dissipating two or more times its computed power. By overrating a resistor, it will run a little cooler.

practical

notes

A properly chosen resistor is able to dissipate its heat safely without becoming excessively hot. However, if through bad design or subsequent component failure its current becomes excessive, it will overheat and damage may result, as shown in Figure 4–20. One of the symptoms of overheating is that the resistor becomes noticeably hotter than other resistors in the circuit. (Be careful, however, as you might get burned if you try to check by touch.) Component failure may also be detected by smell. Burned components have a characteristic odor that you will soon come to recognize. If you detect any of these symptoms, turn the equipment off and look for the source of the problem. Note, however, an overheated component is often the symptom of a problem, rather than its cause.

Figure 4–20   Damage to resistor due to overheating.

4.4 Power Direction Convention

For circuits with one source and one load, energy flows from the source to the load and the direction of power transfer is obvious. For circuits with multiple sources and loads, however, the direction of energy flow in some parts of the network may not be at all apparent. We therefore need to establish a clearly understood power transfer direction convention.

(a) Power transfer

(b) Source end

(c) Load end

(d) Generalization

Figure 4–21   Reference convention for power.

A resistive load (Figure 4–21) may be used to illustrate the idea. Since the direction of energy flow can only be into a resistor, never out of it (since resistors do not produce power), we define the positive direction of power transfer as from the source to the load as in (a) and indicate this by means of an arrow: $P\rightarrow$. We then adopt the convention that, *for the relative voltage polarities and current and power directions shown in Figure 4–21(a), when power transfer is in the direction of the arrow, it is positive, whereas when it is in the direction opposite to the arrow, it is negative.*

To help interpret the convention, consider Figure 4–21(b), which highlights the source end. From this we see that *power out of a source is positive when both the current and power arrows point out from the source, both I and P have positive values, and the source voltage has the polarity indicated.*

Now consider Figure 4–21(c), which highlights the load end. Note the relative polarity of the load voltage and the direction of the current and power arrows. From this, we see that power to a load is positive when both the current and power direction arrows point into the load, both have positive values, and the load voltage has the polarity indicated.

In Figure 4–21(d), we have generalized the concept. The box may contain either a source or a load. If P has a positive value, power transfer is into the box; if P has a negative value, its direction is out.

■ **EXAMPLE 4–10**   Use the above convention to describe power transfer for the electric vehicle of Figure 4–22.

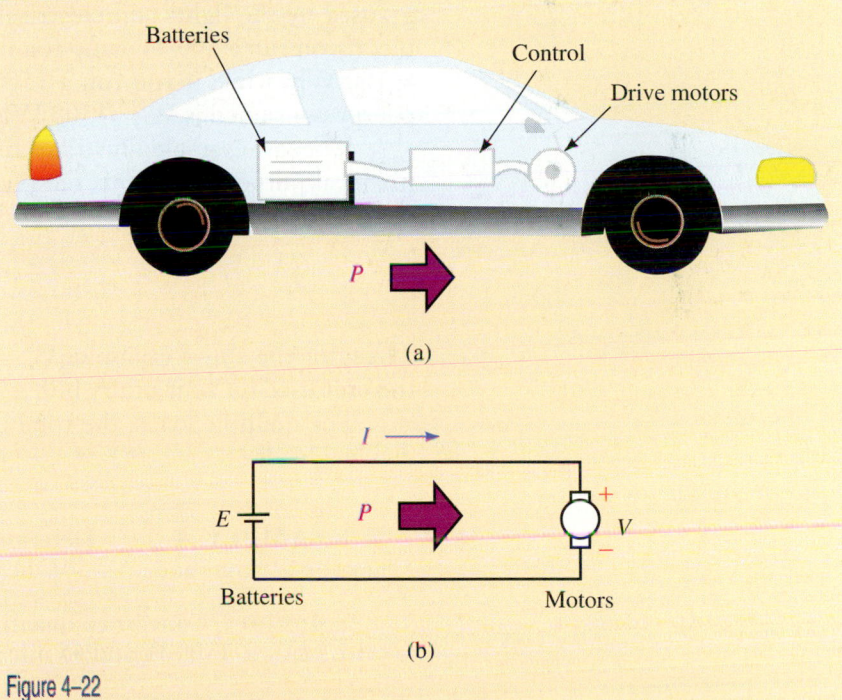

Figure 4–22

Solution   During normal operation, the batteries supply power to the motors, and current and power are both positive according to the conventions shown in Fig. 4–22(b). However, when the vehicle is going downhill, its motors are driven by the weight of the car and they act as generators, charging the batteries. Now, since I is in the direction opposite to its reference, it is negative. Thus, $P = VI$ is negative. The interpretation is therefore that power transfer is in the direction opposite to the power reference arrow. For example, if $V = 48$ volts and $I = -10$ A, then $P = VI = (48 \text{ V})(-10 \text{ A}) = -480$ W. This is consistent with what is happening, since minus 480 W *into* the motors is the same as plus 480 W *out*.

4.5 Energy

Earlier (Equation 4–7), we defined power as the rate of doing work. When you transpose this equation, you get the formula for energy:

$$W = Pt \tag{4–14}$$

If t is measured in seconds, W has units of watt-seconds (i.e., joules, J), while if t is measured in hours, W has units of watthours (Wh). Note that in Equation 4–14, P must be constant over the time interval under consideration. If it is not, apply Equation 4–14 to each interval over which P is constant as described later in this section. (For the more general case, you need calculus, which is beyond the scope of this book.)

The most familiar example of energy usage is the energy that we use in our homes and pay for on our utility bills. This energy is the energy used by the lights and electrical appliances in our homes. For example, if you run a 100-W lamp for 1 hour, the energy consumed is $W = Pt = 100$ Wh, while if you run a 1500-W electric heater for 12 hours, the energy consumed is $W = (1500 \text{ W})(12 \text{ h}) = 18\ 000$ Wh.

The last example illustrates that the watthour is too small a unit for practical purposes. For this reason, we use *kilowatthours* (kWh). By definition,

$$\text{energy}_{(kWh)} = \frac{\text{energy}_{(Wh)}}{1000} \tag{4–15}$$

Thus, for the above example, $W = 18$ kWh. The kilowatthour (kWh) is the unit used on your utility bill.

For multiple loads, the total energy is the sum of the energy of individual loads.

■ **EXAMPLE 4–11**   Determine the total energy used by a 100-W lamp for 12 hours and a 1.5-kW heater for 45 minutes.

Solution   Convert all quantities to the same set of units, e.g., convert 1.5 kW to 1500 W and 45 minutes to 0.75 h. Then,

$$W = (100 \text{ W})(12 \text{ h}) + (1500 \text{ W})(0.75 \text{ h}) = 2325 \text{ Wh} = 2.325 \text{ kWh}$$

Alternatively, convert all power to kilowatts first. Thus,

$$W = (0.1 \text{ kW})(12 \text{ h}) + (1.5 \text{ kW})(0.75 \text{ h}) = 2.325 \text{ kWh}$$

■ **EXAMPLE 4–12**   Suppose you use the following electrical appliances: a 1.5-kW heater for 7½ hours; a 3.6-kW broiler for 17 minutes; three 100-W lamps for 4 hours; a 900-W toaster for 6 minutes. At $0.09 per kilowatthour, how much will this cost you?

Solution   Convert time in minutes to hours. Thus,

$$W = (1500)(7½) + (3600)\left(\frac{17}{60}\right) + (3)(100)(4) + (900)\left(\frac{6}{60}\right)$$

$$= 13\ 560 \text{ Wh} = 13.56 \text{ kWh}$$

$$\text{cost} = (13.56 \text{ kWh})(\$0.09/\text{kWh}) = \$1.22$$

Watthour Meter

In practice, energy is measured by watthour meters, most of which are electromechanical devices that incorporate a small electric motor whose speed is proportional to power to the load. This motor drives a set of dials through a gear train (Figure 4–23). Since the angle through with the dials rotate depends on the speed of rotation (i.e., power consumed)

Figure 4–23   Watthour meter. Most meters still in use use a gear train to drive the dials. Newer meters are electronic.

and the length of time that this power flows, the dial position indicates energy used. To illustrate, assume a load of 1 kW for 1 hour. The meter's motor rotates at the corresponding speed, turning the dials through a corresponding angle. Now, assume a 2-kW load for 1 hour. Since the motor rotates at twice the speed, the dials rotate through twice the previous angle, yielding a reading twice as large. Newer meters use electronics rather than motors and gear trains.

Law of Conservation of Energy

Before leaving this section, we consider the law of conservation of energy. It states that energy can neither be created nor destroyed, but is instead converted from one form to another. You saw examples of this above—for example, the conversion of electrical energy into heat energy by a resistor, and the conversion of electrical energy into mechanical energy by a motor. In fact, several types of energy may be produced simultaneously. For example, electrical energy is converted to mechanical energy by a motor, but some heat is also produced. This results in a lowering of efficiency, a topic we consider next.

4.6 Efficiency

Efficiency is important to all of us. Poor efficiency results in wasted energy and higher costs. For example, an inefficient motor costs more to run than an efficient one for the same output. An inefficient piece of electronic gear generates more heat than an efficient one, and this heat must be removed, resulting in increased costs for fans, heat sinks, and the like.

Efficiency can be expressed in terms of either energy or power. Power is generally easier to measure, so we usually use power. The efficiency of a device or system (Figure 4–24) is defined as the ratio of useful power output P_{out} to total power input P_{in}, and it is usually expressed in percent and denoted by the Greek letter η (eta). Thus,

$$\eta = \frac{P_{out}}{P_{in}} \times 100\% \tag{4–16}$$

In terms of energy,

$$\eta = \frac{W_{out}}{W_{in}} \times 100\% \tag{4–17}$$

Since $P_{in} = P_{out} + P_{losses}$, efficiency can also be expressed as

$$\eta = \frac{P_{out}}{P_{out} + P_{losses}} \times 100\% = \frac{1}{1 + \frac{P_{losses}}{P_{out}}} \times 100\% \tag{4–18}$$

The efficiency of equipment and machines varies greatly. Large power transformers, for example, have efficiencies of 98% or better, while electronic amplifiers have efficiencies below 50%. Note that efficiency will always be less than 100%.

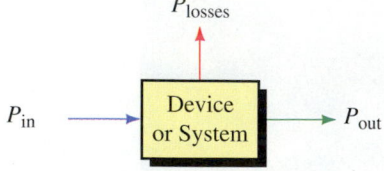

Figure 4–24   Input power equals output power plus losses.

■ **EXAMPLE 4–13**   A 120-V electric motor draws 12 A and develops an output power of 1.6 hp.

a. What is its efficiency?

b. How much power is wasted?

Solution

a. $P_{in} = EI = (120 \text{ V})(12 \text{ A}) = 1440 \text{ W}$, and $P_{out} = 1.6 \text{ hp} \times 746 \text{ W/hp} = 1194 \text{ W}$. Thus,

$$\eta = \frac{P_{out}}{P_{in}} = \frac{1194 \text{ W}}{1440 \text{ W}} \times 100 = 82.9\%$$

b. $P_{losses} = P_{in} - P_{out} = 1440 - 1194 = 246 \text{ W}$

■ **EXAMPLE 4–14**   A transformer delivering 40 kW has losses of 800 W. Compute its efficiency.

Solution

$$P_{in} = P_{out} + P_{losses} = 40{,}800 \text{ W}$$

$$\eta = \frac{P_{out}}{P_{in}} \times 100\% = \frac{40{,}000}{40{,}800} \times 100\% = 98\%$$

For systems with subsystems or components in cascade (Figure 4–25), overall efficiency is the product of the efficiencies of each individual part:

$$\eta_T = \eta_1 \times \eta_2 \times \eta_3 \times \cdots \times \eta_n \qquad (4\text{–}19)$$

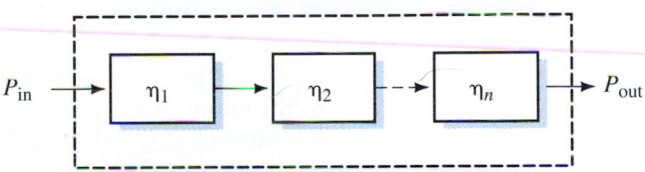

(a) Cascaded system

(b) Equivalent of (a)

Figure 4–25   For systems in cascade, the resultant efficiency is the product of the efficiencies of the individual stages.

■ **EXAMPLE 4–15**

a. For a certain system, $\eta_1 = 95\%$, $\eta_2 = 85\%$, and $\eta_3 = 75\%$. What is η_T?

b. If $\eta_T = 65\%$, $\eta_2 = 80\%$, and $\eta_3 = 90\%$, what is η_1?

Solution

a. Convert all efficiencies to a decimal value, then multiply. Thus, $\eta_T = \eta_1 \eta_2 \eta_3 = (0.95)(0.85)(0.75) = 0.61$ or 61%.

b. $\eta_1 = \eta_T / (\eta_2 \eta_3) = (0.65)/(0.80 \times 0.90) = 0.903$ or 90.3%

■ **EXAMPLE 4–16**   A motor drives a pump through a gearbox (Figure 4–26). Power input to the motor is 1200 W. How many horsepower are delivered to the pump?

P_{in} P_{out}
To pump

Gear box
($\eta_2 = 70\%$)

Motor
($\eta_1 = 90\%$)

(a) Physical system

P_{in} → η_1 η_2 → P_{out}

(b) Block diagram

Figure 4–26   Motor driving pump through a gear box.

Solution   The efficiency of the motor-gearbox combination is $\eta_T = (0.90)(0.70) = 0.63$. The output of the gearbox (and hence the input to the pump) is $P_{out} = \eta_T \times P_{in} = (0.63)(1200 \text{ W}) = 756 \text{ W}$. Converting to horsepower, $P_{out} = (756 \text{ W})/(746 \text{ W/hp}) = 1.01 \text{ hp}$.

■ **EXAMPLE 4–17**   The motor of Figure 4–26 is operated from 9:00 a.m. to 12:00 noon and from 1:00 p.m. to 5:00 p.m. each day, for 5 days a week, outputting 7 hp to a load. At \$0.085/kWh, it costs \$22.19 per week for electricity. What is the efficiency of the motor/gearbox combination?

Solution

$$W_{in} = \frac{\$22.19/\text{wk}}{\$0.085/\text{kWh}} = 261.1 \text{ kWh/wk}$$

The motor operates 35 h/wk. Thus,

$$P_{in} = \frac{W_{in}}{t} = \frac{261.1 \text{ kWh/wk}}{35 \text{ h/wk}} = 7460 \text{ W}$$

$$\eta_T = \frac{P_{out}}{P_{in}} = \frac{(7 \text{ hp} \times 746 \text{ W/hp})}{7460 \text{ W}} = 0.7$$

Thus, the motor/gearbox efficiency is 70%.

4.7 Nonlinear and Dynamic Resistances

All resistors considered so far have constant values that do not change with voltage or current. Such resistors are termed *linear* since their current-voltage *(I-V)* plot is a straight line. However, the resistance of some materials changes with voltage or current. These materials are termed *nonlinear* because their *I-V* plot is curved (Figure 4–27).

Since the resistance of all materials changes with temperature, all resistors are to some extent nonlinear, since they all produce heat and this heat changes their resistance. For most resistors, however, this effect is small over their normal operating range, and such resistors are considered to be linear. (The commercial resistors shown in Figure 3–8 and most others that you will encounter in this book are linear.)

Since an *I-V* plot is a graph of Ohm's law, resistance can be computed from the ratio *V/I*. First, consider the linear plot of Figure 4–27. Because the slope is constant, the resistance is constant and you can compute *R* at any point. For example, at $V = 10$ V, $I = 20$ mA and $R = 10$ V/20 mA = 500 Ω. Similarly, at $V = 20$ V, $I = 40$ mA and $R = 20$ V/40 mA = 500 Ω, which is the same as before. This is true at all points on this linear curve. The resulting resistance is referred to as dc resistance, R_{dc}. Thus, $R_{dc} = 500 \Omega$.

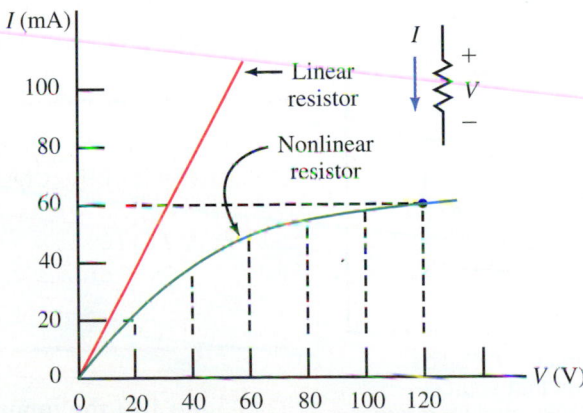

Figure 4–27   Linear and nonlinear resistance characteristics.

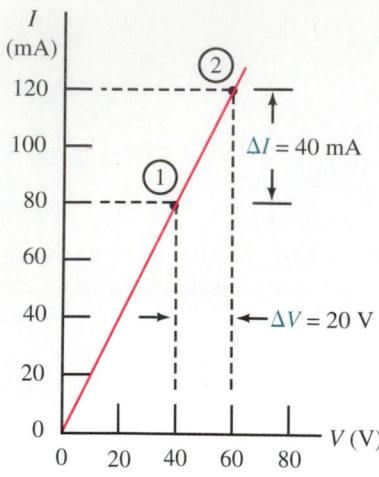

Figure 4–28 $R = \Delta V/\Delta I = 20$ V/40 mA $= 500\ \Omega$.

An alternate way to compute resistance is illustrated in Figure 4–28. At point 1, $V_1 = I_1 R$. At point 2, $V_2 = I_2 R$. Subtracting voltages and solving for R yields

$$R = \frac{V_2 - V_1}{I_2 - I_1} = \frac{\Delta V}{\Delta I} \quad \text{[ohms, } \Omega] \qquad (4\text{--}20)$$

where $\Delta V/\Delta I$ is the slope of the line. (Here, Δ is the Greek letter delta. It is used to represent a change or increment in value.) To illustrate, if you select ΔV to be 20 V, you find that the corresponding ΔI from Figure 4–28 is 40 mA. Thus, $R = \Delta V/\Delta I = 20$ V/40 mA $= 500\ \Omega$ as before. Resistance calculated as in Figure 4–28 is called *ac* or *dynamic resistance*. For linear resistors, $R_{ac} = R_{dc}$.

Now consider the nonlinear resistance plot of Figure 4–27. At $V = 20$ V, $I = 20$ mA. Therefore, $R_{dc} = 20$ V/20 mA $= 1.0$ kΩ; at $V = 120$ V, $I = 60$ mA, and $R_{dc} = 120$ V/60 mA $= 2.0$ kΩ. This resistance therefore increases with applied voltage. However, for small variations about a fixed point on the curve, the ac resistance will be constant. This is an important concept in electronics. However, since dynamic resistance is beyond the scope of this book, we must leave it for later courses to explore.

4.8 Computer-Aided Circuit Analysis

We end our introduction to Ohm's law by solving several simple problems using PSpice. As noted in Chapter 1, PSpice is a circuit analysis program that works from a physical description of a circuit. (In this book, we use the Design Center Evaluation version from MicroSim Corporation.) We enter circuit descriptions as data files.

Preparing a Circuit Description

Consider the circuit of Figure 4–29. The total resistance is 10 Ω + 15 Ω = 25 Ω and the current is $I = 12$ V/25 Ω = 0.48 A. (This type of circuit is known as a series circuit and is covered in detail in Chapter 5.) To describe the circuit to PSpice, number all nodes (i.e., points where two or more components connect), starting with 0 as the reference, identify all components with an appropriate symbol, then prepare a circuit description listing the components, what nodes they connect to, and what their numerical values are. The input file is shown below.

Figure 4–29   Circuit for PSpice. PSpice requires that source voltages be designated by *V* rather than by *E*.

```
Input File:                               Line
* Figure 4-29: Ohm's Law Problem     1
Vs   1   0   12V                      2
R1   1   2   10ohms                   3
R2   2   0   15ohms                   4
.END                                  5
```

Line 1 of the input file is a comment line. It is used by PSpice as a title on printouts and screen displays. Line 2 tells PSpice that a 12-V source, designated Vs, is connected between nodes 1 and 0. (Voltage source designations must always start with the letter V.) Lines 3 and 4 tell PSpice that resistors R1 and R2, with values of 10 Ω and 15 Ω, are connected between nodes 1 and 2 and 2 and 0 respectively. (Resistor designations must always start with the letter R.) Line 5 designates the end of the input file.

Details on how to enter and run this program are contained in Appendix A. You should now key it in and try it out. To view results, select FILE from the Control Shell menu, then Browse Output. The calculated source current is shown below (Output File).

```
Output File:
* Figure 4-29: Ohm's Law Problem
**** SMALL SIGNAL BIAS SOLUTION TEMPERATURE = 27 DEG C
*****************************************************
NODE VOLTAGE   NODE VOLTAGE   NODE VOLTAGE   NODE VOLTAGE
( 1) 12.000    ( 2)  7.200
VOLTAGE SOURCE CURRENTS
NAME            CURRENT
Vs              -4.800E-01
```

The answer appears under the heading VOLTAGE SOURCE CURRENTS (highlighted in blue). According to this, source current is -4.8×10^{-1} A = -0.48 A (rather than $I = 0.48$ A as we computed manually). The reason for the negative sign is that PSpice computes current into the node named first in a statement. Since the first-named node of V_s is node 1, PSpice computes current into Vs, rather than current out of it. Since -0.48 A into node 1 is the same as $+0.48$ A out, the answer is correct. (A way around this problem is shown in Appendix A, Example A-2), where this problem is solved in a slightly different fashion.)

Plotting Ohm's Law

PSpice can also be used to plot computed results. By varying V_S and plotting I, for example, you can get a graph of Ohm's law. To do this, we need two new PSpice commands. (Both are described in detail in Appendix A.)

The .DC Statement.   The first command is the .DC statement. It causes PSpice to vary the source voltage from a minimum value to a maximum value in a specified step size. For example, the statement .DC Vs 0 25V 5V causes V_s to step from 0 V to 25 V in 5-V increments. The current is calculated at each step.

The .PROBE Statement.   The second command is the .PROBE statement. (.PROBE is PSpice's graphic postprocessor.) The .PROBE command displays a set of empty axes at the end of analysis. By following the directions in Appendix A, you can call up and display computed results as indicated in Figure 4-30.

Figure 4–30   Graph of Ohm's law.

Now consider the input file below. Following the procedure detailed in Appendix A, key in the file and run PSpice. At the end of the run, a set of empty axes should appear on the screen. Using the procedure for displaying graphic output (Appendix A), add trace I(R1) to the display. You should then get the result shown in Figure 4–30.

```
Input File:
* Figure 4-30: Graph of Ohm's Law
Vs   1   0
R1   1   0   12.5ohms
.DC Vs   0   25V   5V
.PROBE
.END
```

Problems

4.1   Ohm's Law

1. Determine the current I for each of the following. Express your answer in the most appropriate unit: amps, milliamps, microamps, etc.

a. $E = 40$ V, $R = 20\ \Omega$.

b. $E = 35$ mV, $R = 5\ m\Omega$.

c. $E = 200$ V, $R = 40\ k\Omega$.

d. $E = 10$ V, $R = 2.5\ M\Omega$.

e. $E = 7.5$ V, $R = 2.5 \times 10^{3}\ \Omega$.

f. $E = 12$ kV, $R = 2\ M\Omega$.

2. Determine R for each of the following. Express your answer in the most appropriate unit: ohms, kilohms, megohms, etc.
   a. $E = 50$ V, $I = 2.5$ A.   c. $E = 2$ kV, $I = 0.1$ kA.
   b. $E = 37.5$ V, $I = 1$ mA.   d. $E = 4$ kV, $I = 8 \times 10^{-4}$ A.

3. Compute V for each of the following:
   a. 1 mA, 40 kΩ.   c. 10 mA, 4×10^4
   b. 10 μA, 30 kΩ. Ω.   d. 12 A, 3×10^{-2} Ω.

4. How much voltage can be applied to a 560-Ω resistor if its current must not exceed 50 mA?

5. What is the resistance of an electric heater that draws 12.5 A from a 120-V source?

6. If $I = 36$ mA when $E = 12$ V, what is I if the 12-V source is
   a. replaced by an 18-V source?
   b. replaced by a 4-V source?

7. A 20-V source is applied to a resistor with color bands brown, black, red, and silver
   a. Compute the nominal current in the circuit.
   b. Compute the minimum and maximum currents based on the tolerance of the resistor.

8. For Figure 4–31, if $E = 312$ V, what will the ammeter indicate?

9. Consider the I-V graph in Figure 4–32.
   a. What is R?
   b. If V is increased to 100 V, what is I?

Ammeter

1 A

Fuse

Green →
Blue →
Black →

E

Figure 4–31

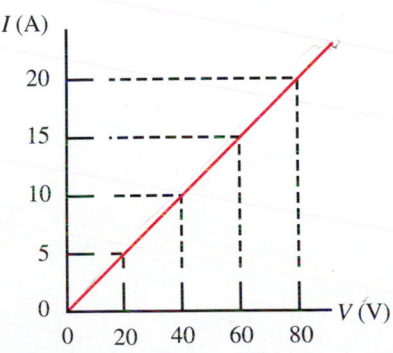

I (A)

20

15

10

5

0

0   20   40   60   80

V (V)

Figure 4–32

$+$ V $-$

$I \longrightarrow$

(a) $I = 3$ A

$+$ V $-$

$I \longrightarrow$

(b) $I = -4$ A

$-$ V $+$

$\longleftarrow$ I

(c) $I = 6$ A

$-$ V $+$

$\longleftarrow$ I

(d) $I = -7$ A

Figure 4–33   All resistors are 15 Ω.

10. For a resistive circuit, E is quadrupled and R is halved. If the new current is 24 A, what was the original current?

11. For a resistive circuit, $E = 100$ V. If R is doubled and E is changed so that the new current is double the original current, what is the new value of E?

12. You need to measure the resistance of a certain component, but have only a 12-V battery and an ammeter. Describe how you would determine its resistance. Include a sketch of your circuit.

13. If 25 m of 0.1-mm-diameter Nichrome wire is connected to a 12-V battery, what is the current?

14. If the current is 0.5 A when a length of No. 40 copper wire is connected to 48 V, what is the length of the wire?

4.2   Voltage and Current Conventions

15. All resistors in Figure 4–33 are 15 Ω. For each case, determine the magnitude and polarity of voltage V.

16. The ammeters of Figure 4–34 have autopolarity. Determine their readings, magnitude, and polarity. Hint: Recall that an ammeter gives a positive reading when current enters its A terminal.

(a)

(b)

Figure 4–34

4.3   Power

17. A resistor dissipates 723 joules of energy in 3 minutes and 47 seconds. Calculate the rate at which energy is being transferred to this resistor in joules per second. What is its power dissipation in watts?

18. How long does it take for a resistor dissipating 100 W to dissipate 1470 J?

19. A resistor draws 3 A from a 12-V battery. How much power does the battery deliver to the resistor?

20. A 120-V electric coffee maker is rated 960 W. Determine its resistance and rated current.

21. A 1.2-kW electric heater has a resistance of 6 Ω. How much current does it draw?

22. A resistor draws 1.5 mA when dissipating 2.25 W. What is its resistance?

23. How many volts must be applied to a 3-Ω resistor to result in a power dissipation of 752 W?

24. A resistor with color bands brown, black, and orange dissipates 0.25 W. Compute its voltage and current.

25. A portable radio transmitter has an input power of 0.455 kW. How much current does it draw from a 12-V battery?

26. For a resistive circuit, $E = 12$ V.
   a. If the load dissipates 8 W, what is the current in the circuit?
   b. If the load dissipates 36 W, what is the load resistance?

27. A motor delivers 3.56 hp to a load. How many watts is this?

28. The load on a 120-V circuit consists of six 100-W lamps, a 1.2-kW electric heater, and an electric motor drawing 1500 W. If the circuit is fused at 30 A, what happens when a 900-W toaster is plugged in? Justify your answer.

29. A 0.27-kΩ resistor is rated 2 W. Compute the maximum voltage that can be applied and the maximum current that it can carry without exceeding its ability to safely dissipate the heat generated.

30. Determine which, if any, of the following resistors may have been damaged by overheating.
   a. 560 Ω, ½ W, with 75 V across it.
   b. 3 Ω, 20 W, with 4 A through it.
   c. ¼ W, with 0.25 mA through it and 40 V across it.

4.4   Power Direction Convention

31. Each block of Figure 4–35 may be a source or a load. For each, determine power and its direction.

(a)

(b)

(c)

(d)

Figure 4–35

4.5   Energy

32. If a heater dissipates power at the rate of 800 W, how long does it take to dissipate 24 kJ?

33. A 40-W night safety light burns for 9 hours.
   a. Determine the energy used in joules.
   b. Determine the energy used in watthours.
   c. At $0.08/kWh, how much does it cost to run this light for 9 hours?

34. An indicator light on a control panel operates continuously, drawing 20 mA from a 120-V supply. At $0.09 per kilowatthour, how much does it cost per year to operate the light?

35. An electric device with a cycle time of 1 hour operates at full power (400 W) for 15 minutes, at half power for 30 minutes, then cuts off for the remainder of the hour. The cycle repeats continuously. At $0.10/kWh, determine the yearly cost of operating this device.

36. While the device of Problem 35 is operating, two other loads as follows are also operating:
   a. a 4-kW heater, continuously,
   b. a 3.6-kW heater, on for 12 hours per day.
   Calculate the yearly cost of running all loads.

37. Determine the total cost of using the following at $0.11 per kWh:
   a. a 900-W toaster for 5 minutes,
   b. a 120-V, 8-A heater for 1.7 hours,
   c. an 1100-W dishwasher for 36 minutes,
   d. a 120-V, 288-Ω soldering iron for 24 minutes.

38. At $0.08 per kilowatthour, it costs $1.20 to run a heater for 50 hours from a 120-V source. How much current does the heater draw?

39. If there are 23 slices in a loaf of bread and you have a two-slice, 1100-W toaster that takes 1 minute and 45 seconds to toast a pair of slices, at $0.13/kWh, how much does it cost to toast a loaf?

4.6   Efficiency

40. For a certain device, $\eta = 94\%$. If losses are 18 W, what are P_{in} and P_{out}?

41. The power input to a device is 1100 W. If the power lost due to various inefficiencies is 190 W, what is the efficiency of the device?

42. A 240-V, 4.5-A water heater produces heat energy at the rate of 3.6 MJ per hour. Compute
   a. the efficiency of the water heater,
   b. the annual cost of operation at $0.09/kWh if the heater is on for 6 h/day.

43. A 120-V electric motor with an efficiency of 89% draws 15 A from the source. What is its output horsepower?

44. A 120-V electric motor develops an output of 3.8 hp. If its efficiency is 87%, how much current does it draw?

45. Show that the efficiency of n devices or systems in cascade is the product of their individual efficiencies, i.e., that $\eta_T = \eta_1 \times \eta_2 \times \cdots \times \eta_n$.

46. The power/control system of an electric car consists of a 48-V onboard battery pack, electronic control/drive unit, and motor (Figure 4–36). If 20 A are drawn from the batteries, how many horsepower are delivered to the drive wheels?

Efficiency: 95%   Efficiency: 80%

Figure 4–36

47. A 120-V motor drives a pump through a gearbox (Figure 4–26). If the power input to the pump is 1100 W, the gearbox has an efficiency of 75%, and the power input to the motor is 1600 W, determine the horsepower output of the motor.

48. If the motor of Problem 47 is protected by a 15-A circuit breaker, will it open? Compute the current to find out.

49. If the overall efficiency of a radio transmitting station is 55% and it transmits at 35 kW for 24 h/day, compute the cost of energy used per day at $0.09/kWh.

50. In a factory, two machines, each delivering 27 kW are used on an average for 8.7 h/day, 320 days/year. If the efficiency of the newer machine is 87% and that of the older machine is 72%, compute the difference in cost per year of operating them at $0.10 per kilowatthour.

4.7   Nonlinear and Dynamic Resistances

51. A voltage-dependent resistor has the *I-V* characteristic of Figure 4–37.
   a. At $V = 25$ V, what is I? What is R_{dc}?
   b. At $V = 60$ V, what is I? What is R_{dc}?
   c. Why are the two values different?

Figure 4–37

52. For the resistor of Figure 4–37:
    a. Determine $R_{dynamic}$ for V between 0 and 40 V.
    b. Determine $R_{dynamic}$ for V greater than 40 V.
    c. If V changes from 20 V to 30 V, how much does I change?
    d. If V changes from 50 V to 70 V, how much does I change?

4.8   Computer-Aided Circuit Analysis

53. The 12-V source of Figure 4–29 is replaced by a –75-V source. You can handle this in one of two ways:
    a. Specify the source voltage as –75 V in the PSpice program, connected to nodes 1 and 0.
    b. Specify the source voltage as 75 V, but reverse the nodes to 0 and 1.

    Prepare PSpice programs for each of these two approaches, then run the programs and verify.

54. For the circuit of Figure 4–29, let $R_1 = 12$ kΩ and $R_2 = 18$ kΩ. Change the program to express resistance in kilohms rather than in ohms. Run the program and verify it.

55. Consult Appendix A and study the .PRINT command. Then, for the Ohm's law graphing example of Section 4.8, replace the .PROBE statement with a .PRINT statement so that a table of values is computed. Run the program, then use Browse Output to view the results. Plot them.

56. For the Ohm's law graphing example of Section 4.8, change R to 5 kΩ, then modify the program to graph current versus voltage from –20 V to 20 V in 2-V steps.

PART

II

Basic dc Analysis

Part I provided the foundation upon which proficient circuit analysis is built. The terms, units, and definitions used in the previous chapters have developed the necessary vocabulary which is used throughout electrical and electronics technology.

In Part II, we build on this foundation, applying the concepts developed in Part I to initially analyze series dc circuits. Several very important rules, laws, and theorems will be developed in the following chapters, providing additional tools needed to extend circuit theory to the analysis of parallel and series-parallel circuits. Circuit theorems will be used to simplify the most complex circuit into an equivalent circuit which is represented as a single source and a single resistor. Although most circuits are much more complex than the circuits used in this text, all circuits, even the most complex, follow the same laws.

Several different techniques are often available to analyze a given circuit. In order to give a broad overview, this textbook uses the various methods for illustrative purposes. You are encouraged to use different techniques where possible, so that you develop the skills to become proficient in circuit analysis.

Series Circuits

CHAPTER PREVIEW

In the previous chapter we examined the interrelation of current, voltage, resistance, and power in a single resistor circuit. In this chapter we will expand on these basic concepts to examine the behavior of circuits having several resistors in series.

We will use Ohm's law to derive the voltage divider rule and to verify Kirchhoff's voltage law. A good understanding of these important principles provides an important base upon which further circuit analysis techniques are built. Kirchhoff's voltage law and Kirchhoff's current law, which will be covered in the next chapter, are fundamental in understanding **all** electrical and electronic circuits.

After developing the basic framework of series circuit analysis, we will apply the ideas to analyze and design simple voltmeters and ohmmeters. While meters are usually covered in a separate instruments or measurements course, we examine these circuits merely as an application of the concepts of circuit analysis.

Similarly, we will observe how circuit principles are used to explain the loading effect of an ammeter placed in series with a circuit.

KEY TERMS

Electric Circuit
Ground
Kirchhoff's Voltage Law
Loading Effect (Ammeter)
Ohmmeter Design
Point Sources
Series Connection
Total Equivalent Resistance
Voltage Divider Rule
Voltage Subscripts
Voltmeter Design

(Courtesy of Martin Marietta Corp.)

OBJECTIVES

After studying this chapter, you will be able to

• determine the total resistance in a series circuit and calculate circuit current,

• use Ohm's law and the voltage divider rule to solve for the voltage across all resistors in the circuit,

• express Kirchhoff's voltage law and use it to analyze a given circuit,

• solve for the power dissipated by any resistor in a series circuit and show that the total power dissipated is exactly equal to the power delivered by the voltage source,

• solve for the voltage between any two points in a series or parallel circuit,

• design a simple voltmeter or ohmmeter given a particular meter movement,

• calculate the loading effect of an ammeter in a circuit.

Gustav Robert Kirchhoff

Kirchhoff was a German physicist born on March 12, 1824, in Königsberg, Prussia. His first research was on the conduction of electricity, which led to his presentation of the laws of closed electric circuits in 1845. Kirchhoff's current law and Kirchhoff's voltage law apply to all electrical circuits and therefore are fundamentally important in understanding circuit operation. Kirchhoff was also the first to verify that an electrical impulse travelled at the speed of light.

Although these discoveries have immortalized Kirchhoff's name in electrical science, he is better known for his work with R. W. Bunsen in which he made major contributions in the study of spectroscopy and advanced the research into blackbody radiation.

Kirchhoff died in Berlin on October 17, 1887.

5.1 Series Circuits

An *electric circuit* is the combination of any number of sources and loads connected in any manner which allows charge to flow. The electric circuit may be simple, such as a circuit consisting of a battery and a light bulb. Or the circuit may be very complex, such as the circuits contained within a television set, microwave oven, or computer. However, no matter how complicated, each circuit follows fairly simple rules in a predictable manner. Once these rules are understood, any circuit may be analyzed to determine the operation under various conditions.

All electric circuits obtain their energy either from a direct current (dc) source or from an alternating current (ac) source. In the next few chapters, we examine the operation of circuits when they are supplied by dc sources. Although ac circuits have fundamental differences when compared with dc circuits, the laws, theorems, and rules that you learn in dc circuits apply directly to ac circuits as well.

In the previous chapter, you were introduced to a simple dc circuit consisting of a single voltage source (such as a chemical battery) and a single load resistance. The schematic representation of such a simple circuit was covered in Chapter 4 and is shown again in Figure 5–1.

While the circuit of Figure 5–1 is useful in deriving some important concepts, very few practical circuits are this simple. However, we will find that even the most complicated dc circuits can generally be simplified to the circuit shown.

We begin by examining the most simple connection, the *series connection*. In Figure 5–2, we have two resistors, R_1 and R_2, connected at a single point in what is said to be a series connection.

Two elements are said to be in series if they are connected at a single point and if there are no other current-carrying connections at this point.

A *series circuit* is constructed by combining various elements in series, as shown in Figure 5–3. Current will leave the positive terminal of the voltage source, move through the resistors, and return to the negative terminal of the source.

In the circuit of Figure 5–3, we see that the voltage source, E, is in series with R_1, R_1 is in series with R_2, and R_2 is in series with E. By examining this circuit, another important characteristic of a series circuit becomes evident. In an analogy similar to water flowing in a pipe, current

Conventional flow

Figure 5–1

Single point of connection

Figure 5–2   Resistors in series.

Figure 5–3   Series circuit.

entering an element must be the same as the current leaving the element. Now, since current does not leave at any of the connections, we conclude that the following must be true.

The identical current occurs everywhere in a series circuit.

While the above statement seems self-evident, we will find that this will help to explain many of the other characteristics of a series circuit.

What two conditions determine whether two elements are connected in series?

in process
learning
check I

5.2 Kirchhoff's Voltage Law

Next to Ohm's law, one of the most important laws of electricity is Kirchhoff's voltage law (KVL) which states the following:

The summation of voltage rises and voltage drops around a closed loop is equal to zero. Symbolically, this may be stated as follows:

$$\sum V = 0 \quad \text{for a closed loop} \tag{5–1}$$

In the above symbolic representation, the uppercase Greek letter sigma (Σ) stands for summation and V stands for voltage rises and drops. A closed loop is any path that is chosen such as to leave a certain point, travel around the circuit, and return to the original point from a different direction.

An alternate way of stating Kirchhoff's voltage law is as follows:

The summation of voltage rises is equal to the summation of voltage drops around a closed loop.

$$\sum E_{\text{rises}} = \sum V_{\text{drops}} \quad \text{for a closed loop} \tag{5–2}$$

If we consider the circuit of Figure 5–4, we may begin at point *a* in the lower left-hand corner. By arbitrarily following the direction of the current, *I*, we move through the voltage source, which represents a rise in potential from point *a* to point *b*. Next, in moving from point *b* to point *c*, we pass through resistor R_1, which presents a potential drop of V_1. Continuing through resistors R_2 and R_3, we have additional drops of V_2 and V_3 respectively. By applying Kirchhoff's voltage law around the closed loop, we arrive at the following mathematical statement for the given circuit:

$$E - V_1 - V_2 - V_3 = 0$$

Although we chose to follow the direction of current in writing Kirchhoff's voltage law equation, it would be just as correct to move around the circuit in the opposite direction. In this case the equation would appear as follows:

$$V_3 + V_2 + V_1 - E = 0$$

By simple manipulation, it is quite easy to show that the two equations are identical.

Figure 5–4   Kirchhoff's voltage law.

■ **EXAMPLE 5–1**   Verify Kirchhoff's voltage law for the circuit of Figure 5–5.

Figure 5–5

Solution   If we follow the direction of the current, we write the loop equation as

$$15\,V - 2\,V - 3\,V - 6\,V - 3\,V - 1\,V = 0\,V$$

practice problem I

Verify Kirchhoff's voltage law for the circuit of Figure 5–6.

Figure 5–6

in process learning check II

Define Kirchhoff's voltage law.

5.3 Resistors in Series

Almost all complicated circuits can be simplified. We will now examine how to simplify a circuit consisting of a voltage source in series with several resistors. Consider the circuit shown in Figure 5–7.

Figure 5–7

Figure 5–8

Since the circuit is closed, the voltage source will cause a current I in the circuit. This current in turn produces a voltage drop across each resistor, where

$$V_x = IR_x$$

Applying Kirchhoff's voltage law to the closed loop gives

$$E = V_1 + V_2 + \cdots + V_n$$
$$= IR_1 + IR_2 + \cdots + IR_n$$
$$= I(R_1 + R_2 + \cdots + R_n)$$

If we were to replace all the resistors with an equivalent total resistance, R_T, then the circuit would appear as shown in Figure 5–8.

However, applying Ohm's law to the circuit of Figure 5–8 gives

$$E = IR_T \qquad (5\text{--}3)$$

Since the circuit of Figure 5–8 is equivalent to the circuit of Figure 5–7, we conclude that this can only occur if the total resistance of the n series resistors is given as

$$R_T = R_1 + R_2 + \cdots + R_n \quad [\text{ohms, } \Omega] \qquad (5\text{--}4)$$

If each of the n resistors has the same value, then the total resistance is determined as

$$R_T = nR \quad [\text{ohms, } \Omega] \qquad (5\text{--}5)$$

■ **EXAMPLE 5–2**   Determine the total resistance for each of the networks shown in Figure 5–9.

(a)   (b)

Figure 5–9

Solution

a. $R_T = 5\ \Omega + 10\ \Omega + 20\ \Omega + 15\ \Omega = 50\ \Omega$

b. $R_T = 4(10\ \text{k}\Omega) = 40\ \text{k}\Omega$

Any voltage source connected to the terminals of a network of series resistors will provide the same current as if a single resistance, having a value of R_T, were connected between the open terminals. From Ohm's law we get

$$I = \frac{E}{R_T} \quad [\text{amps, A}] \tag{5-6}$$

The power dissipated by each resistor is determined as

$$P_1 = V_1 I = \frac{V_1^{\,2}}{R_1} = I^2 R_1 \quad [\text{watts, W}]$$

$$P_2 = V_2 I = \frac{V_2^{\,2}}{R_2} = I^2 R_2 \quad [\text{watts, W}] \tag{5-7}$$

$$\vdots$$

$$P_n = V_n I = \frac{V_n^{\,2}}{R_n} = I^2 R_n \quad [\text{watts, W}]$$

In Chapter 4, we showed that the power delivered by a voltage source to a circuit is given as

$$P_T = EI \quad [\text{watts, W}] \tag{5-8}$$

Since energy must be conserved, the power delivered by the voltage source is equal to the total power dissipated by all the resistors. Hence

$$P_T = P_1 + P_2 + \cdots + P_n \quad [\text{watts, W}] \tag{5-9}$$

■ **EXAMPLE 5–3**

$R_1 = 2\ \Omega$

$E = 24\ \text{V}$

$R_2 = 6\ \Omega$

$R_3 = 4\ \Omega$

Figure 5–10

For the series circuit shown in Figure 5–10, find the following quantities:

a. Total resistance, R_T.

b. Circuit current, I.

c. Voltage across each resistor.

d. Power dissipated by each resistor.

e. Power delivered to the circuit by the voltage source.

f. Verify that the power dissipated by the resistors is equal to the power delivered to the circuit by the voltage source.

Solution

a. $R_T = 2\ \Omega + 6\ \Omega + 4\ \Omega = 12\ \Omega$

b. $I = (24\ \text{V})/(12\ \Omega) = 2.0\ \text{A}$

c. $V_1 = (2\ \text{A})(2\ \Omega) = 4.0\ \text{V}$

$V_2 = (2\ \text{A})(6\ \Omega) = 12.0\ \text{V}$

$V_3 = (2\ \text{A})(4\ \Omega) = 8.0\ \text{V}$

d. $P_1 = (2\ \text{A})^2(2\ \Omega) = (4\ \text{V})^2/2\ \Omega = 8.0\ \text{W}$

$P_2 = (2\ \text{A})^2(6\ \Omega) = 24.0\ \text{W}$

$P_3 = (2\ \text{A})^2(4\ \Omega) = 16.0\ \text{W}$

e. $P_T = (24\ \text{V})(2\ \text{A}) = 48.0\ \text{W}$

f. $P_T = 8\ \text{W} + 24\ \text{W} + 16\ \text{W} = 48.0\ \text{W}$

practice problem II

$R_1 = 20\ \Omega$

$E = 120\ \text{V}$ $R_2 = 40\ \Omega$

$R_3 = 30\ \Omega$

Figure 5–11

For the series circuit shown in Figure 5–11, find the following quantities:

a. Total resistance, R_T.
b. The direction and magnitude of the current, I.
c. Polarity and magnitude of the voltage across each resistor.
d. Power dissipated by each resistor.
e. Power delivered to the circuit by the voltage source.
f. Show that the power dissipated is equal to the power delivered.

in process **learning** *check III*

Three resistors, R_1, R_2, and R_3, are in series. Determine the value of each resistor if $R_T = 42\ \text{k}\Omega$, $R_2 = 3R_1$, and $R_3 = 2R_2$.

5.4 Voltage Sources in Series

If a circuit has more than one voltage source in series, then the voltage sources may effectively be replaced by a single source having a value that is the sum or difference of the individual sources. Since the sources may have different polarities, it is necessary to consider polarities in determining the resulting magnitude and polarity of the equivalent voltage source.

If the polarities of all the voltage sources are such that the sources appear as voltage rises in given direction, then the resultant source is determined by simple addition, as shown in Figure 5–12.

If the polarities of the voltage sources do not result in voltage rises in the same direction, then we must compare the rises in one direction to the rises in the other direction. The magnitude of the resultant source will be the sum of the rises in one direction minus the sum of the rises in the opposite direction. The polarity of the equivalent voltage source will be the same as the polarity of whichever direction has the greater rise. Consider the voltage sources shown in Figure 5–13.

If the rises in one direction were equal to the rises in the opposite direction, then the resultant voltage source would be equal to zero.

Figure 5–12

Figure 5–13

A typical lead-acid automobile battery consists of six cells connected in series. If the voltage between the battery terminals is measured to be 13.06 V, what is the average voltage of each cell within the battery?

5.5 Interchanging Series Components

The order of series components may be changed without affecting the operation of the circuit.

The two circuits in Figure 5–14 are equivalent.

(a)   (b)

Figure 5–14

Very often, once the circuits have been redrawn it becomes easier to visualize the circuit operation. Therefore, we will regularly use the technique of interchanging components to simplify circuits before we analyze them.

■ **EXAMPLE 5–4**   Simplify the circuit of Figure 5–15 into a single source in series with the four resistors. Determine the direction and magnitude of the current in the resulting circuit.

Figure 5–15

Solution   We may redraw the circuit by the two steps shown in Figure 5–16. It is necessary to ensure that the voltage sources are correctly moved since it is quite easy to assign the wrong polarity. Perhaps the easiest way to is to imagine that we slide the voltage source around the circuit to the new location.

(a)   (b)

Figure 5–16

The current in the resulting circuit will be in a counterclockwise direction around the circuit and will have a magnitude determined as

$$I = \frac{E_T}{R_T} = \frac{6\,\text{V} + 1\,\text{V} - 2\,\text{V}}{2\,\Omega + 4\,\Omega + 3\,\Omega + 1\,\Omega} = \frac{5\,\text{V}}{10\,\Omega} = 0.5\,\text{A}$$

Because the circuits are in fact equivalent, the current direction determined in Figure 5–16 also represents the direction for the current in the circuit of Figure 5–15.

5.6 The Voltage Divider Rule

The voltage applied to a series circuit will be dropped across all the resistors in proportion to the magnitude of the individual resistors. The total voltage dropped across the resistors must equal the applied voltage source(s) by KVL.

Consider the circuit of Figure 5–17.

After examining the circuit in Figure 5–17, we see that the total resistance $R_T = 10\,\text{k}\Omega$ results in a circuit current of $I = 1\,\text{mA}$. From Ohm's law, R_1 has a voltage drop of $V_1 = 2.0\,\text{V}$, while R_2, which is four times as large as R_1, has four times as much voltage drop, $V_2 = 8.0\,\text{V}$.

We also see that the summation of the voltage drops across the resistors is exactly equal to the voltage rise of the source, namely,

$$E = 10\,\text{V} = 2\,\text{V} + 8\,\text{V}$$

Figure 5–17

The voltage divider rule allows us to determine the voltage across any series resistance in a single step, without first calculating the current. We have seen that for any number of resistors in series the current in the circuit is determined by Ohm's law as

$$I = \frac{E}{R_T} \quad [\text{Amps, A}] \tag{5–10}$$

where the two resistors in Figure 5–17 result in a total resistance of

$$R_T = R_1 + R_2$$

By again applying Ohm's law, the voltage drop across any resistor in the series circuit is calculated as

$$V_x = IR_x$$

Now, by substituting Equation 5–4 into the above equation we write the *voltage divider rule* for two resistors as a simple equation:

$$V_x = \frac{R_x}{R_T} E = \frac{R_x}{R_1 + R_2} E$$

In general, for any number of resistors the voltage drop across any resistor may be found as

$$V_x = \frac{R_x}{R_T} E \qquad\qquad (5\text{–}11)$$

■ **EXAMPLE 5–5** Use the voltage divider rule to determine the voltage across each of the resistors in the circuit shown in Figure 5–18. Show that the summation of voltage drops is equal to the applied voltage rise in the circuit.

Figure 5–18

Solution

$$R_T = 6\ \Omega + 12\ \Omega + 7\ \Omega = 25\ \Omega$$

$$V_1 = \left(\frac{6\ \Omega}{25\ \Omega}\right)(18\ \text{V}) = 4.32\ \text{V}$$

$$V_2 = \left(\frac{12\ \Omega}{25\ \Omega}\right)(18\ \text{V}) = 8.64\ \text{V}$$

$$V_3 = \left(\frac{7\ \Omega}{25\ \Omega}\right)(18\ \text{V}) = 5.04\ \text{V}$$

The total voltage drop is the summation

$$V_T = 4.32\ \text{V} + 8.64\ \text{V} + 5.04\ \text{V} = 18.0\ \text{V} = E$$

■ **EXAMPLE 5–6** Using the voltage divider rule, determine the voltage across each of the resistors of the circuit shown in Figure 5–19.

Figure 5–19

Solution

$$R_T = 2 \ \Omega + 1 \ 000 \ 000 \ \Omega = 1 \ 000 \ 002 \ \Omega$$

$$V_1 = \left(\frac{2 \ \Omega}{1 \ 000 \ 002 \ \Omega} \right)(20 \ \text{V}) \approx 40 \ \mu\text{V}$$

$$V_2 = \left(\frac{1.0 \ \text{M}\Omega}{1.000 \ 002 \ \text{M}\Omega} \right)(20 \ \text{V}) = 19.999 \ 86 \ \text{V}$$

$$\approx 20.0 \text{V}$$

The previous example illustrates two important points which occur regularly in electronic circuits. If a single series resistance is very large in comparison with the other series resistances, then the voltage across that resistor will be essentially the total applied voltage. On the other hand, if a single resistance is very small in comparison with the other series resistances, then the voltage drop across the small resistor will be essentially zero. As a general rule, if a series resistor is more than 100 times larger than another series resistor, then the effect of the smaller resistor(s) may be effectively neglected.

practice
problem
III

Figure 5–20

For the circuits shown in Figure 5–20, determine the approximate voltage drop across each resistor without using a calculator. Compare your approximations to the actual values obtained with a calculator.

in process
learning
check V

The voltage drops across three resistors are measured to be 10.0 V, 15.0 V, and 25.0 V. If the largest resistor is 47.0 kΩ, determine the sizes of the other two resistors.

5.7 Circuit Ground

Perhaps one of the most misunderstood concepts in electronics is that of *ground*. This misunderstanding leads to many problems when circuits are designed and analyzed. The standard symbol for *circuit ground* is shown in Figure 5–21(a), while the symbol for *chassis ground* is shown in Figure 5–21(b).

In its most simple definition, ground is simply an "arbitrary electrical point of reference" or "common point" in a circuit. Using the ground symbol in this manner usually allows the circuit to be sketched more simply. When the ground symbol is used arbitrarily to designate a point of reference, it would be just as correct to redraw the circuit schematic showing all the ground points connected together or indeed to redraw the circuit using an entirely different point of reference. The circuits shown in Figure 5–22 are exactly equivalent circuits even though the circuits of Figure 5–22(a) and 5–22(c) use different points of reference.

While the ground symbol is used to designate a common point of reference within a circuit, it usually has a greater meaning to the technologist or engineer. Very often, the metal chassis of an appliance is connected to the circuit ground. Such a connection is referred to as a *chassis ground* and is usually designated as shown in Figure 5–21(b).

In order to help prevent electrocution, the chassis ground is usually further connected to the *earth ground* through a connection provided at the electrical outlet box. In the event of a failure within the circuit, the chassis would redirect current to ground (tripping a breaker or fuse), rather than presenting a hazard to an unsuspecting operator.

As the name implies, the earth ground is a connection which is bonded to the earth, either through water pipes or by a connection to ground rods. Everyone is familiar with the typical 120-Vac electrical outlet shown in Figure 5–23. The rounded terminal of the outlet is always the ground terminal and is used not only in ac circuits by may also be used to provide a common point for dc circuits. When a circuit is bonded

(a) Circuit ground or reference

(b) Chassis ground

Figure 5–21 Ground symbols.

(a)   (b)   (c)

Figure 5–22

Neutral   Line

Figure 5–23   Ground connection in a typical 120-Vac outlet.

Ground

to the earth through the ground terminal, then the ground symbol no longer represents an arbitrary connection, but rather represents a very specific type of connection.

in process
learning
check VI

If you measure the resistance between the ground terminal of the 120-Vac plug and the metal chassis of a microwave oven to be zero ohms, what does this tell you about the chassis of the oven?

5.8 Voltage Subscripts

Double Subscripts

As you have already seen, voltages are always expressed as the potential difference between two points. In a 9-V battery, there is a 9-volt rise in potential from the negative terminal to the positive terminal. Current through a resistor results in a voltage drop across the resistor such that the terminal from which charge leaves is at a lower potential than the terminal into which charge enters. We now examine how voltages within any circuit may be easily described as the voltage between two points. If we wish to express the voltage between two points (say points a and b in a circuit) then we express such a voltage in a subscripted form (e.g., V_{ab}), where the first term in the subscript is the point of interest and the second term is the point of reference.

Consider the series circuit of Figure 5–24.

If we label the points within the circuit a, b, c, and d, then we can see that the point b is at a higher potential than point a by an amount equal to the supply voltage. We may write this mathematically

Figure 5–24

as $V_{ba} = +50$ V. Although the plus sign is redundant, we show it here to indicated that point b is at a higher potential than point a. If we examine the voltage at point a with respect to point b, we see that a is at a lower potential than b. This may be written mathematically as $V_{ab} = -50$ V.

From the above illustration, we may make the following general statement:

$$V_{ab} = -V_{ba}$$

for any two points a and b within a circuit.

Current through the circuit results in voltage drops across the resistors as shown in Figure 5–24. If we determine the voltage drops on all resistors and show the correct polarities, then we see that the following must also apply:

$$V_{bc} = +10 \text{ V} \qquad V_{cb} = -10 \text{ V}$$
$$V_{cd} = +25 \text{ V} \qquad V_{dc} = -25 \text{ V}$$
$$V_{da} = +15 \text{ V} \qquad V_{ad} = -15 \text{ V}$$

If we wish to determine the voltage between any other two points within the circuit, it is a simple matter of adding all the voltages between the two points, taking into account the polarities of the voltages. The voltage between points b and d would be determined as follows:

$$V_{bd} = V_{bc} + V_{cd} = 10 \text{ V} + 25 \text{ V} = +35 \text{ V}$$

Similarly, the voltage between points b and a could be determined by using the voltage drops of the resistors:

$$V_{ba} = V_{bc} + V_{cd} + V_{da} = 10 \text{ V} + 25 \text{ V} + 15 \text{ V} = +50 \text{ V}$$

Notice that the above result is precisely the same as when we determined V_{ba} using only the voltage source. This result indicates that the voltage between two points is not dependent upon the path taken.

■ **EXAMPLE 5–7**   For the circuit of Figure 5–25, find the voltages V_{ac}, V_{ad}, V_{cf} and V_{eb}.

Figure 5–25

Solution   First, we determine that the equivalent supply voltage for the circuit is

$$E_{eq} = 3\text{ V} + 4\text{ V} - 2\text{ V} = 5.0\text{ V}$$

with a polarity such that current will move in a counterclockwise direction within the circuit.

Next, we determine the voltages on all resistors by using the voltage divider rule and assigning polarities based on the direction of the current.

$$V_1 = \frac{R_1}{R_T}E_{eq}$$

$$= \left(\frac{10\ \Omega}{10\ \Omega + 30\ \Omega + 40\ \Omega}\right)(5.0\text{ V}) = 0.625\text{ V}$$

$$V_2 = \frac{R_2}{R_T}E_{eq}$$

$$= \left(\frac{30\ \Omega}{10\ \Omega + 30\ \Omega + 40\ \Omega}\right)(5.0\text{ V}) = 1.875\text{ V}$$

$$V_3 = \frac{R_3}{R_T}E_{eq}$$

$$= \left(\frac{40\ \Omega}{10\ \Omega + 30\ \Omega + 40\ \Omega}\right)(5.0\text{ V}) = 2.50\text{ V}$$

The voltages appearing across the resistors are as shown in Figure 5–26.

Figure 5–26

Finally, we solve for the voltages between the indicated points:

$$V_{ac} = -2.0\text{ V} - 0.625\text{ V} = -2.625\text{ V}$$

$$V_{ad} = -2.0\text{ V} - 0.625\text{ V} + 3.0\text{ V} = +0.375\text{ V}$$

$$V_{cf} = +3.0\text{ V} - 1.875\text{ V} + 4.0\text{ V} = +5.125\text{ V}$$

$$V_{eb} = +1.875\text{ V} - 3.0\text{ V} + 0.625\text{ V} = -0.500\text{ V}$$

Or, selecting the opposite path, we get

$$V_{eb} = +4.0\text{ V} - 2.5\text{ V} - 2.0\text{ V} = -0.500\text{ V}$$

practical

notes

*S*ince most students initially find it difficult to determine the correct polarity for the voltage between two points, we present a simplified method to correctly determine the polarity and voltage between any two points within a circuit.

1. Determine the circuit current. Calculate the voltage drop across all components.

2. Polarize all resistors based upon the direction of the current. The terminal at which the current enters is assigned to be positive, while the terminal at which the current leaves is assigned to be negative.

3. In order to determine the voltage at point a with respect to point b, start at point a. Refer to Figure 5–27. Now, imagine that you walk around the circuit to the reference point b.

$$V_{ab} = \;+\; 5\ V \;-\; 3\ V \;+\; 6\ V = +\,8\ V$$

Figure 5–27

4. As you "walk" around the circuit, add the voltage drops and rises as you get to them. The assigned polarity of the voltage at any component (whether it is a source or a resistor) is the first polarity that you arrive at for the component.

5. The resulting voltage, V_{ab}, is the numerical sum of all the voltages between a and b.

For Figure 5–27, the voltage V_{ab} will be determined as

$$V_{ab} = +5\ V - 3\ V + 6\ V = 8\ V$$

practice problem IV

Find the voltage V_{ab} in the circuit of Figure 5–28.

Figure 5–28

Single Subscripts

In a circuit which has a reference point (or ground point), most voltages will be expressed with respect to the reference point. In such a case it is no longer necessary to express a voltage using a dual subscript. Rather if we wish to express the voltage at point *a* with respect to ground, we simply refer to this as V_a. Similarly, the voltage at point *b* would be referred to as V_b. Therefore, any voltage which has only a single subscript is always referenced to the ground point of the circuit.

■ **EXAMPLE 5–8**   For the circuit of Figure 5–29, determine the voltages V_a, V_b, V_c, and V_d.

Figure 5–29

Solution   Applying the voltage divider rule, we determine the voltage across each resistor as follows:

$$V_1 = \frac{2 \text{ k}\Omega}{2 \text{ k}\Omega + 3 \text{ k}\Omega + 5 \text{ k}\Omega} (20 \text{ V}) = 4 \text{ V}$$

$$V_2 = \frac{3 \text{ k}\Omega}{2 \text{ k}\Omega + 3 \text{ k}\Omega + 5 \text{ k}\Omega} (20 \text{ V}) = 6 \text{ V}$$

$$V_3 = \frac{5 \text{ k}\Omega}{2 \text{ k}\Omega + 3 \text{ k}\Omega + 5 \text{ k}\Omega} (20 \text{ V}) = 10 \text{ V}$$

Now we solve for the voltage at each of the points as follows:

$$V_a = 4 \text{ V} + 6 \text{ V} + 10 \text{ V} = +20 \text{ V} = E$$

$$V_b = 6 \text{ V} + 10 \text{ V} = +16 \text{ V}$$

$$V_c = +10 \text{ V}$$

$$V_d = 0 \text{ V}$$

If the voltage at various points in a circuit is known with respect to ground, then the voltage between the points may be easily determined as follows:

$$V_{ab} = V_a - V_b \quad \text{[volts, V]} \qquad (5\text{–}12)$$

■ **EXAMPLE 5–9** For the circuit of Figure 5–30, determine the voltages V_{ab} and V_{cb} given that $V_a = +5$ V, $V_b = +3$ V, and $V_c = -8$ V.

$V_a = +5$ V
$V_b = +3$ V
$V_c = -8$ V

Figure 5–30

Solution

$$V_{ab} = +5 \text{ V} - (+3 \text{ V}) = +2 \text{ V}$$

$$V_{cb} = -8 \text{ V} - (+3 \text{ V}) = -11 \text{ V}$$

Point Sources

Figure 5–31

The idea of voltages with respect to ground may be easily extended to include voltage sources. When a voltage source is given with respect to ground, it may be simplified in the circuit as a *point source* as shown in Figure 5–31.

Point sources are often used to simplify the representation of circuits. We need to remember that in all such cases the corresponding points always represent voltages with respect to ground (even if ground is not shown).

■ **EXAMPLE 5–10** Determine the current and direction in the circuit of Figure 5–32.

$E_1 = +5$ V $E_2 = -8$ V

$R_1 = 52$ kΩ

Figure 5–32

Solution The circuit may be redrawn showing the reference point and converting the voltage point sources into the more common schematic representation. The resulting circuit is shown in Figure 5–33.

Figure 5–33

Now, we easily calculate the current in the circuit as

$$I = \frac{E_T}{R_1} = \frac{5\ V + 8\ V}{52\ k\Omega} = 0.25\ mA$$

in process
l e a r n i n g
check VII

Voltage measurements are taken at three locations in a circuit. They are $V_a = +5.0\ V$, $V_b = -2.5\ V$, and $V_c = -5\ V$. Determine the voltages V_{ab}, V_{ca}, and V_{bc}.

5.9 Internal Resistance of Voltage Sources

So far we have worked only with ideal voltage sources, which maintain constant voltages regardless of the loads connected across the terminals. Consider a typical lead-acid automobile battery, which has a voltage of approximately 12 V. Similarly, four C-cell batteries, when connected in series, have a combined voltage of 12 V. Why, then, can we not use four C-cell batteries to operate the car? The answer, in part, is because the lead-acid battery has a much lower internal resistance than the low-energy C-cells. In practice, all voltage sources contain some internal resistance which will reduce the efficiency of the voltage source. We may symbolize any voltage source schematically as an ideal voltage source in series with an internal resistance. Figure 5–34 shows both an ideal voltage source and a practical or actual voltage source.

The voltage which appears between the positive and negative terminals is called the *terminal voltage*. In an ideal voltage source, the terminal voltage will remain constant regardless of the load connected. An ideal voltage source will be able to provide as much current as the circuit demands. In a practical voltage source, the terminal voltage is dependent upon the value of the load connected across the voltage source. As expected, the practical voltage source sometimes is not able to provide as much current as the load demands. Rather the current in the circuit is limited by the combination of the internal resistance and the load resistance.

Under a no-load condition ($R_L = \infty\ \Omega$), there is no current in the circuit and so the terminal voltage will be equal to the voltage appearing across the ideal voltage source. If the output terminals are shorted together ($R_L = 0\ \Omega$), the current in the circuit will be a maximum and the terminal voltage will be equal to approximately zero. In such a situation, the voltage dropped across the internal resistance will be equal to the voltage of the ideal source.

The following example helps to illustrate the above principles.

(a) Ideal voltage source

(b) Actual voltage source

Figure 5–34

■ **EXAMPLE 5–11**   Two batteries having an open-terminal voltage of 12 V are used to provide current to a load having a resistance of 0.10 Ω. If one battery has an internal resistance of 0.02 Ω and the second battery has an internal resistance of 100 Ω, calculate the current through the load and the resulting terminal voltage for each of the batteries.

Solution   The circuit for each of the batteries is shown in Figure 5–35.

(a) Low internal resistance

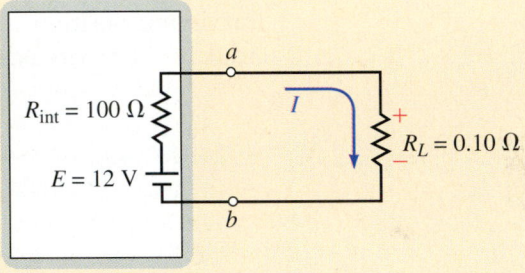

(b) High internal resistance

Figure 5–35

$R_{int} = 0.02\ \Omega$:

$$I = \frac{12\ V}{0.02\ \Omega + 0.10\ \Omega} = 100\ A$$

$$V_{ab} = (100\ A)(0.10\ \Omega) = 10.0\ V$$

$R_{int} = 100\ \Omega$:

$$I = \frac{12\ V}{100\ \Omega + 0.10\ \Omega} = 0.120\ A$$

$$V_{ab} = (0.120\ A)(0.10\ \Omega) = 0.012\ V$$

Although very simple, this example helps to illustrate why a 12-V automobile battery (which is actually 14.4 V) is able to start a car while nine 1.5 V-flashlight batteries connected in series will have virtually no measurable effect when connected to the same circuit.

5.10 Voltmeter Design

We now use the concept of a series circuit to analyze a functional circuit, namely a voltmeter. Although most technologists will seldom need to design a voltmeter, the principles presented here will help you understand the limitations of the instrument and hopefully eliminate some of its mystery.

The typical voltmeter consists of a meter movement in series with a current-limiting resistance. The meter may be either a permanent-magnet moving coil (PMMC) as shown in Figure 5–36, or a digital panel meter (DPM).

Although the digital meter is now more common, we will examine the circuits using the PMMC movement since this type of display is very simple and easily understood with concepts you have learned. The PMMC consists of an electromagnet mounted on a spring. When an external voltage is applied to the terminals of the voltmeter, a small current will occur in the voltmeter. As charge flows through the coils of the electromagnet, a magnetic field is developed. Since this movable coil is located inside a permanent magnet, the magnets will interact causing the coil to deflect proportional to the current within the movement. (In later chapters you will learn more about how magnetic fields may be created by electric currents.)

The amount of current which causes the movement to deflect to its maximum position is referred to as the *full-scale deflection current* and is usually abbreviated as I_{fsd}. The I_{fsd} of an analog meter can be determined

Figure 5–36   Permanent-magnet moving-coil (PMMC) movement. *(Courtesy of Simpson Electric Co.)*

from the *sensitivity* of the meter, S, which is generally printed on the meter face and is given in volts per ohm. The sensitivity is defined to be

$$S = \frac{1}{I_{fsd}}$$

If the current in a circuit is less than I_{fsd}, then the movement will deflect an amount proportional to the current as follows:

$$\text{deflection} = \frac{I}{I_{fsd}} \times 100\% \qquad (5\text{--}13)$$

If excessive current is applied to the meter movement, the fine needle may be bent or the movement itself may be destroyed.

Due to the extreme length of very fine wire in the electromagnet, the PMMC usually has a resistance on the order of several thousand ohms. This resistance, called the *meter resistance*, is abbreviated as R_m. The schematic representation of a typical PMMC movement is shown in Figure 5–37.

By combining the PMMC movement with a single series resistor as shown in Figure 5–38, it is possible to build a simple circuit capable of measuring external voltages.

In the schematic of Figure 5–38, the resistor R_s is used to limit the amount of current so that the meter movement never receives more than I_{fsd}. The value of this resistor is dependent not only on the type of meter movement used, but also on the voltage range of the voltmeter. Clearly we would like to have maximum meter deflection when the meter detects the maximum voltage for the particular range. Once we have decided the voltage range that the meter will be measuring, it is a simple task to scale the meter with numbers and graticules to help in interpreting a measurement. Such a scale is shown in Figure 5–39 for a voltmeter having a 10-V range. In the meter of Figure 5–39, the meter will indicate the maximum voltage (10 V) when the current through the meter movement is equal to I_{fsd}.

Figure 5–37   Schematic representation of a PMMC movement.

Figure 5–38   Simple voltmeter.

Full-scale deflection

Each graticule represents a variation of 0.2 V

Figure 5–39 Typical scaling of a voltmeter.

Using Ohm's law, we determine that the total circuit resistance is determined as

$$R_T = \frac{V_{range}}{I_{fsd}} = V_{range}S$$

Since the meter movement has a resistance of R_m, we may determine the required series resistance as

$$R_s = R_T - R_m$$

or

$$R_s = \frac{V_{range}}{I_{fsd}} - R_m \qquad (5-14)$$

By adding a selector switch, it is possible to design a multirange voltmeter, shown in Figure 5–40, which is able to select various ranges.

For a multirange voltmeter, the scaling is generally adjusted to allow a simple interpretation of the reading. The following example illustrates how a multirange voltmeter would be designed and how the scale might be indicated on the meter face.

Figure 5–40 Multirange voltmeter.

■ **EXAMPLE 5–12** Design a voltmeter which has 20-V, 50-V, and 100-V ranges and uses a meter movement having $I_{fsd} = 1$ mA and $R_m = 2$ kΩ.

Solution 20-V range: Using Equation 5–12, we determine the required series resistance as

$$R_1 = \frac{20\ V}{1\ mA} - 2\ k\Omega = 18\ k\Omega$$

Similarly, the other two ranges would be designed as follows:

50-V range:

$$R_2 = \frac{50 \text{ V}}{1 \text{ mA}} - 2 \text{ k}\Omega = 48 \text{ k}\Omega$$

100-V range:

$$R_3 = \frac{100 \text{ V}}{1 \text{ mA}} - 2 \text{ k}\Omega = 98 \text{ k}\Omega$$

The resulting circuit and corresponding scaling is shown in Figure 5–41. Now, by selecting between the various ranges and using the corresponding scale, we are able to use the meter to measure voltages up to 100 V.

(a) Voltmeter design

(b) Scaling of the meter face

Figure 5–41

■ **EXAMPLE 5–13** If the voltmeter of Example 5–12 is used to measure the voltage across a 40-V voltage source, determine the range(s) which could be used. For the range(s), calculate the current through the meter movement and find the percent of deflection.

Solution The voltmeter could not be used on its 20-V range, since to do so would cause the current to exceed I_{fsd}. However, both the 50-V range and the 100-V range could be used.

50-V range: If this range is used, then the current in the circuit is as shown in Figure 5–42.

Figure 5–42

By applying Ohm's law, we determine the current in the circuit to be

$$I = \frac{40 \text{ V}}{48 \text{ k}\Omega + 2 \text{ k}\Omega} = 0.80 \text{ mA}$$

The deflection of the movement is

$$\text{deflection} = \frac{0.80 \text{ mA}}{1.0 \text{ mA}} \times 100\% = 80\%$$

100-V range: If this range is used, then current in the circuit is as shown in Figure 5–43.

Figure 5–43

By applying Ohm's law, we determine the current in the circuit to be

$$I = \frac{40 \text{ V}}{98 \text{ k}\Omega + 2 \text{ k}\Omega} = 0.40 \text{ mA}$$

The deflection when measuring the voltage on this range is

$$\text{deflection} = \frac{0.40 \text{ mA}}{1.0 \text{ mA}} \times 100\% = 40\%$$

Although we are occasionally able to measure voltages on more than one range, it is generally more acceptable to measure on the range which gives the greatest nonmaximum deflection. In this example, the better range would therefore be the 50-V range since it provides the greater deflection.

practice problem V

Design a voltmeter which has a 250-V range and a 1000-V range. Use a meter movement having a sensitivity $S = 5$ kΩ/V ($I_{fsd} = 200$ μA) and $R_m = 5$ kΩ in your design.

in process learning check VIII

A voltmeter has a sensitivity of 20 kΩ/V. Determine the resistance that you would expect to measure between the terminals when the voltmeter is on its 100-V range.

5.11 Ohmmeter Design

In the previous section, we saw how a voltmeter is essentially constructed from a meter movement in series with a resistance. The movement deflects an amount which is proportional to the amount of current passing through it. By employing a similar principle, it is possible to use the same meter movement to construct an ohmmeter.

Unlike the voltmeter, which uses an external voltage source to provide the necessary current to cause a deflection within the PMMC movement, an ohmmeter must have an internal voltage source (usually a battery) to provide the required sensing current. The schematic of a simple ohmmeter is shown in Figure 5–44.

In the circuit of Figure 5–44, we can see that no current will be present until an unknown resistance, R_x, is connected across the open terminals of the ohmmeter. The ohmmeter is designed so that maximum current will pass through the meter movement when the resistance connected across the terminals is equal to zero (i.e., a short circuit ($R_x = 0$). The scaling of the meter face plate is determined according to the movement deflection for various values of unknown resistance.

Now, since we want maximum deflection when the terminals are shorted, the value of R_s is calculated in a manner similar to how the voltmeter was designed, namely

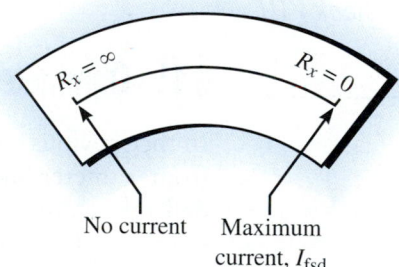

Figure 5–44   Simple ohmmeter.

$$R_s = \frac{E}{I_{fsd}} - R_m \tag{5–15}$$

It is now apparent that when the resistance being measured is a minimum ($R = 0$), current will be maximum. Conversely, when resistance is maximum ($R = \infty$), current will be minimum, namely zero. The scaling for an ohmmeter is shown in Figure 5–45.

Figure 5–45   Scaling of an ohmmeter.

Since current is inversely proportional to the resistance of the circuit, we find that the scale will not be linear. The following example demonstrates this principle.

■ **EXAMPLE 5–14**   Design an ohmmeter using a 9-V battery and a PMMC meter movement having $I_{fsd} = 1$ mA and $R_m = 2$ kΩ. Determine the value of R_x when the movement shows 25%, 50%, and 75% deflection.

Solution   The value of the series resistance is

$$R_s = \frac{9\ V}{1\ mA} - 2\ k\Omega = 7\ k\Omega$$

The resulting circuit is as shown in Figure 5–46(a).

(a) Ohmmeter circuit

(b) Ohmmeter scaling

Figure 5–46

By analyzing the series circuit, we can see that the current is the required $I_{fsd} = 1$ mA when $R_x = 0\Omega$.

At 25% deflection, the current in the circuit is

$$I = (0.25)(1 \text{ mA}) = 0.25 \text{ mA}$$

From Ohm's law, the total resistance in the circuit must be

$$R_T = \frac{9 \text{ V}}{0.25 \text{ mA}} = 36 \text{ k}\Omega$$

For the given circuit, only the load resistance, R_x, may change. Its value is determined as

$$R_x = R_T - R_s - R_m$$
$$= 36 \text{ k}\Omega - 7 \text{ k}\Omega - 2 \text{ k}\Omega$$
$$= 27 \text{ k}\Omega$$

Similarly, at 50% deflection, the current in the circuit is $I = 0.50$ mA and the total resistance is $R_T = 18$ kΩ. So, the unknown resistance must be $R_x = 9$ kΩ.

Finally, at 75% deflection, the current in the circuit will be $I = 0.75$ mA, resulting in a total circuit resistance of 12 kΩ. Therefore, for 75% deflection, the unknown resistance is $R_x = 3$ kΩ.

The resulting ohmmeter scale appears as shown in Figure 5–46(b).

5.12 Ammeter Loading Effects

As you have already learned, ammeters are instruments which measure the current in a circuit. In order to use an ammeter, the circuit must be disconnected and the ammeter placed in series with the branch for which the current is to be determined. Since an ammeter uses the current in the circuit to provide a reading, it will affect the circuit under measurement. This effect is referred to as *meter loading*. All instruments, regardless of type, will load the circuit to some degree. The amount of loading is dependent upon both the instrument and by the circuit being measured. For any meter, we define the loading effect as follows:

$$\text{loading effect} = \frac{\text{theoretical value} - \text{measured value}}{\text{theoretical value}} \times 100\%$$

■ **EXAMPLE 5–15**   For the series circuits of Figure 5–47, determine the current in each circuit. If an ammeter having an internal resistance of 250 Ω is used to measure the current in the circuits, determine the current through the ammeter and calculate the loading effect for each circuit.

$R_1 = 2$ kΩ

E_1 ⊥ 10 V

I_1

(a) Circuit #1

$R_2 = 100$ Ω

E_2 ⊥ 50 mV

I_2

(b) Circuit #2

Figure 5–47

Solution   Circuit No. 1: The current in the circuit is

$$I_1 = \frac{10 \text{ V}}{20 \text{ k}\Omega} = 0.50 \text{ mA}$$

Now, by placing the ammeter into the circuit as shown in Figure 5–48(a) the resistance of the ammeter will slightly affect the operation of the circuit.

(a) Circuit #1   (b) Circuit #2

Figure 5–48

The resulting current in the circuit will be reduced to

$$I_1 = \frac{10 \text{ V}}{20 \text{ k}\Omega + 0.25 \text{ k}\Omega} = 0.494 \text{ mA}$$

We see that by placing the ammeter into the circuit, the resistance of the meter slightly affects the operation of the circuit. The loading effect is

$$\text{loading effect} = \frac{0.500 \text{ mA} - 0.494 \text{ mA}}{0.500 \text{ mA}} \times 100\%$$

$$= 1.23\%$$

Circuit No. 2: The current in the circuit is

$$I_2 = \frac{50 \text{ mV}}{100 \text{ }\Omega} = 0.50 \text{ mA}$$

Now, by placing the ammeter into the circuit as shown in Figure 5–48(b), the resistance of the ammeter will greatly affect the operation of the circuit.

The resulting current in the circuit will be reduced to

$$I_2 = \frac{50 \text{ mV}}{100 \text{ }\Omega + 250 \text{ k}\Omega} = 0.143 \text{ mA}$$

We see that by placing the ammeter into the circuit, the resistance of the meter will load the circuit. The loading effect will be

$$\text{loading effect} = \frac{0.500 \text{ mA} - 0.143 \text{ mA}}{0.500 \text{ mA}} \times 100\%$$

$$= 71.4\%$$

The results of this example indicate that an ammeter, which usually has fairly low resistance, will not significantly load a circuit having a resistance of several thousand ohms. However, if the same meter is used to measure current in a circuit having low values of resistance, then the loading effect will be substantial.

5.13 Circuit Analysis Using Computers

PSpice software provides a simple means of determining the voltage, current, and total power in a circuit. The example given in this section is simple and does not demonstrate the full power of PSpice. However, it does provide the student with the necessary background for later analysis of more complex circuits.

■ **EXAMPLE 5–16**   Use PSpice to solve for the circuit current and the voltage across each of the resistors in Figure 5–49. This circuit was previously analyzed in Example 5–1.

Figure 5–49

Solution   First, number all nodes, with the reference node designated as node 0. Next, prepare the input file using the PSpice syntax as outlined in Appendix A. The input file for the circuit of Figure 5–49 is listed as follows.

```
Input File:
Example 5-16: Series Circuit, Figure 5-49
Vs    1    0     24V
R1    1    2     2ohmsR2   2    3    6ohms
R3    3    0     4ohms
.DC   Vs   24V   24V   1V
.PRINT DC V(1,2) V(2,3) V(3,0) I(Vs)
.OPTIONS NOPAGE
.END
```

In order to have PSpice print the values for voltage and current, it is first necessary to execute the .DC control statement. The indicated statement causes the supply voltage, V_s to start at 24 V and stop at 24 V with the increment between steps to be 1 V. Although we are analyzing this circuit for a given voltage, the .DC statement is particularly useful in determining how a circuit would operate under various supply voltages.

After running PSpice, the output file may be viewed by selecting the Browse menu item in the control shell or by printing the output file. For instance, if the file is saved as EX5-15.CIR, then the output file generated by PSpice will have the .OUT extension. The output file, EX5-16.OUT, appears as shown in Figure 5–50.

```
OUTPUT

***       DC TRANSFER CURVES      TEMPERATURE = 27.000 DEG C

Vs             I(Vs)       V(1,2)        V(2,3)         V(3,0)
2.400W+01   -2.000E+00   4.000E+00   1.200E+01   8.000E+00

            JOB CONCLUDED
            TOTAL JOB TIME        1.81
```

Figure 5–50

As expected, the results of this analysis are consistent with the calculations of Example 5–1. The assumed direction for the current through the voltage source is from node 1 to node 0. The negative sign indicates that the current is actually in the opposite direction.

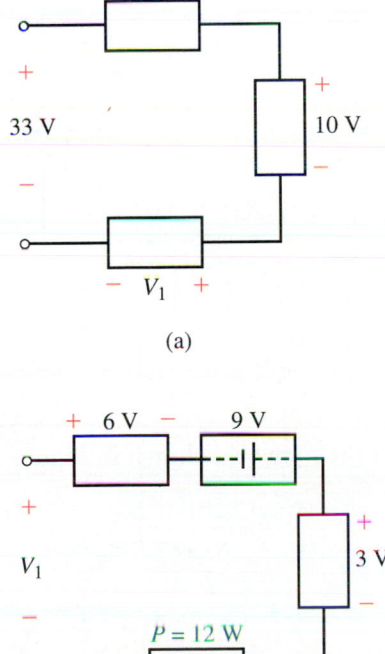

(a)

(b)

Figure 5–51

Problems

5.2 Kirchhoff's Voltage Law

1. Determine the unknown voltages in the networks of Figure 5–51.

2. Determine the unknown voltages in the networks of Figure 5–52.

(a)   (b)

Figure 5–52

3. Solve for the unknown voltages in the circuit of Figure 5–53.

4. Solve for the unknown voltages in the circuit of Figure 5–54.

Figure 5–53   Figure 5–54

5.3 Resistors in Series

5. Determine the total resistance of the networks shown in Figure 5–55.

(a)   (b)   (c)

Figure 5–55

6. Determine the unknown resistance in each of the networks in Figure 5–56.

(a)

(b) Each resistor band code is brown, red, orange

(c)

Figure 5–56

7. For the circuits shown in Figure 5–57, determine the total resistance, R_T, and the current, I.

(a)

(b)

Figure 5–57

8. The circuits of Figure 5–58 have the total resistance, R_T, as shown. For each of the circuits find the following:

   a. The magnitude of current in the circuit.
   b. The total power delivered by the voltage source.

(a)

(b)

Figure 5–58

c. The direction of current through each resistor in the circuit.
d. The value of the unknown resistance, R.
e. The voltage drop across each resistor.
f. The power dissipated by each resistor. Verify that the summation of powers dissipated by the resistors is equal to the power delivered by the voltage source.

9. For the circuit of Figure 5–59, find the following quantities:
   a. The circuit current.
   b. The total resistance of the circuit.
   c. The value of the unknown resistance, R.
   d. The voltage drop across all resistors in the circuit.
   e. The power dissipated by all resistors.

10. The circuit of Figure 5–60 has a current of 2.5 mA. Find the following quantities:
   a. The total resistance of the circuit.
   b. The value of the unknown resistance, R_2.
   c. The voltage drop across each resistor in the circuit.
   d. The power dissipated by each resistor in the circuit.

Figure 5–59   Figure 5–60

11. For the circuit of Figure 5–61, find the following quantities:
   a. The current, I.
   b. The voltage drop across each resistor.
   c. The voltage across the open terminals a and b.

12. Refer to the circuit of Figure 5–62:
   a. Use Kirchhoff's voltage law to find the voltage drops across R_2 and R_3.
   b. Determine the magnitude of the current, I.
   c. Solve for the unknown resistance, R_1.

Figure 5–61

Figure 5–62

Figure 5–63

13. Repeat Problem 12 for the circuit of Figure 5–63.

14. Refer to the circuit of Figure 5–64:

  a. Find R_T.
  b. Solve for the current, I.
  c. Determine the voltage drop across each resistor.
  d. Verify Kirchhoff's voltage law around the closed loop.
  e. Find the power dissipated by each resistor.
  f. Determine the minimum power rating of each resistor, if resistors are available with the following power ratings: ⅛ W, ¼ W, ½ W, 1 W, and 2 W.
  g. Show that the power delivered by the voltage source is equal to the summation of the powers dissipated by the resistors.

15. Repeat Problem 14 for the circuit of Figure 5–65.

Figure 5–64   Figure 5–65

16. Refer to the circuit of Figure 5–66.

  a. Calculate the voltage across each resistor.
  b. Determine the values of the resistors R_1 and R_2.
  c. Solve for the power dissipated by each of the resistors.

Figure 5–66

5.5   Interchanging Series Components

17. Redraw the circuits of Figure 5–67, showing a single voltage source for each circuit. Solve for the current in each circuit.

(a)   (b)

Figure 5–67

18. Use the information given to determine the polarity and magnitude of the unknown voltage source in each of the circuits of Figure 5–68.

(a)

Figure 5–68

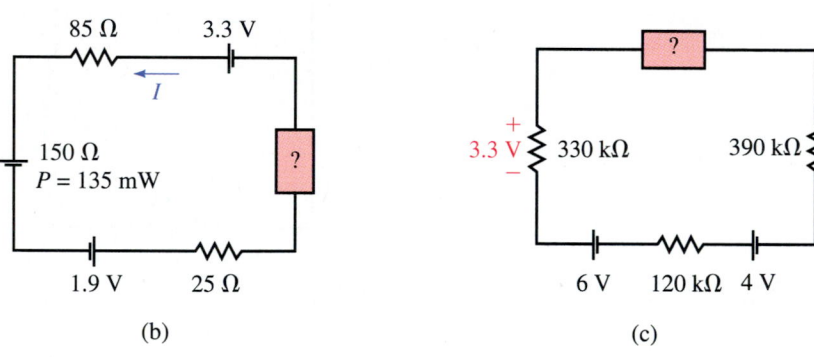

(b)   (c)

5.6   Voltage Divider Rule

19. Use the voltage divider rule to determine the voltage across each resistor in the circuits of Figure 5–69. Use your results to verify Kirchhoff's voltage law for each circuit.

(a)   (b)

Figure 5–69

20. Repeat Problem 19 for the circuits of Figure 5–70.

(a)   (b)

Figure 5–70

21. Refer to the circuits of Figure 5–71:
a. Find the values of the unknown resistors.
b. Calculate the voltage across each resistor.
c. Determine the power dissipated by each resistor.

(a)   (b)

Figure 5–71

22. Refer to the circuits of Figure 5–72:
a. Find the values of the unknown resistors using the voltage divider rule.
b. Calculate the voltage across R_1 and R_3.
c. Determine the power dissipated by each resistor.

(a)   (b)

Figure 5–72

$R = 25 \, \Omega$/light bulb

Figure 5–73

23. A string of 24 series light bulbs is connnected to a 120-V supply as shown in Figure 5–73.

   a.   Solve for the current in the circuit.

   b.   Use the voltage divider rule to find the voltage across each light bulb.

   c.   Calculate the power dissipated by each bulb.

   d.   If a single light bulb were to become an open circuit, the entire string would stop working. To prevent this from occurring, each light bulb has a small metal strip which shorts the light bulb when the filament fails. If two bulbs in the string were to burn out, repeat Steps a) through c).

   e.   Based on your calculations of Step d), what do you think would happen to the life expectancy of the remaining light bulbs if the two faulty bulbs were not replaced?

24. Repeat Problem 23 for a string consisting of 36 light bulbs.

5.8   Voltage Subscripts

25. Solve for the voltages V_{ab} and V_{bc} in the circuits of Figure 5–71.

26. Repeat Problem 25 for the circuits of Figure 5–72.

27. For the circuits of Figure 5–74, determine the voltage across each resistor and calculate the voltage V_a.

(a)

(b)

Figure 5–74

(a)

(b)

Figure 5–75

28. Given the circuits of Figure 5–75:

   a.   Determine the voltage across each resistor.

   b.   Find the magnitude and direction of the current in the 180-kΩ resistor.

   c.   Solve for the voltage V_a.

5.9   Internal Resistance of Voltage Sources

29. A battery is measured to have an open-terminal voltage of 14.2 V. When this voltage is connected to a 100-Ω load, the voltage measured between the terminals of the battery drops to 6.8 V.

   a. Determine the internal resistance of the battery.
   b. If the 100-Ω load were replaced with a 200-Ω load, what voltage would be measured across the terminals of the battery?

Figure 5–76

(a)   (b)

30. The voltage source shown in Figure 5–76 is measured to have an open-circuit voltage of 24 V. When a 10-Ω load is connected across the terminals, the voltage measured with a voltmeter drops to 22.8 V.

   a. Determine the internal resistance of the voltage source.
   b. If the source had only half the resistance determined in a), what voltage would be measured across the terminals with the 10-Ω resistor connected?

5.10   Voltmeter Design

31. Given a meter movement having $I_{fsd} = 2$ mA and $R_m = 1$ kΩ, design a voltmeter having the following ranges:

a. 20-V range.
b. 100-V range.
c. 500-V range.
d. Is it possible to have a 1-V range using the given meter movement? Explain.

32. Repeat Problem 31 using a meter movement having $I_{fsd} = 50$ μA and $R_m = 5$ kΩ.

5.11   Ohmmeter Design

33. Given a meter movement having $I_{fsd} = 2$ mA and $R_m = 1$ kΩ, design an ohmmeter which uses a 9-V battery to provide the sensing current. Determine the values of the unknown resistances which would result in meter deflections of 25%, 50%, 75%, and 100%.

34. Given a meter movement having $I_{fsd} = 50$ μA and $R_m = 5$ kΩ, design an ohmmeter which uses a 6-V battery to provide the sensing current. Determine the percentage meter deflection for a 10-kΩ resistor.

5.12   Ammeter Loading Effects

35. For the series circuits of Figure 5–77, determine the current in each circuit. If an ammeter having an internal resistance of 50 Ω is used to measure the current in the circuits, determine the current through the ammeter and calculate the loading effect for each circuit.

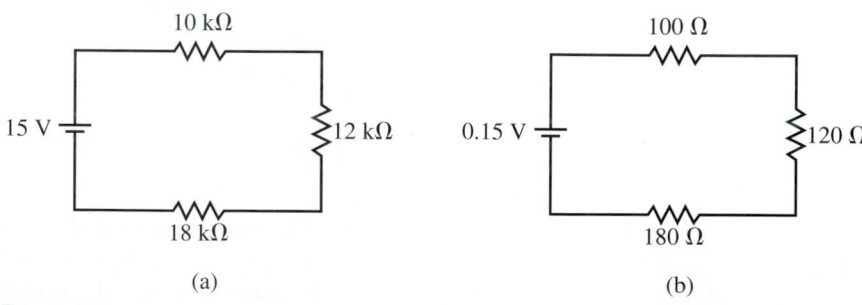

(a)   (b)

Figure 5–77

36. Repeat Problem 35 if the ammeter has a resistance of 10 Ω.

5.13   Circuit Analysis Using Computers

37. Refer to the circuits of Figure 5–69. Use PSpice to find the following:

a. The current in each circuit.
b. The voltage across each resistor in the circuit.

Figure 5–78

38. Given the circuit of Figure 5–78, use PSpice to determine the following:

a. The current through the voltage source, I(Vs).
b. The voltage across each resistor.
c. The voltage between terminals 1 and 3.
d. The voltage, with respect to ground, at terminal 2.

6

Parallel Circuits

CHAPTER PREVIEW

Two fundamental circuits form the basis of all electrical circuits. They are the *series circuit* and the *parallel circuit*. The previous chapter examined the principles and rules which applies to series circuits. In this chapter we study the *parallel* (or *shunt*) circuit and examine the rules governing the operation of these circuits.

Figure 6–1 illustrates a simple example of several light bulbs connected in parallel with one another and a battery supplying voltage to the bulbs.

This illustration shows one of the important differences between the series circuit and the parallel circuit. The parallel circuit will continue to operate even though one of the light bulbs may have defective (open) filament. Only the defective light bulb will no longer glow. If a circuit were made up of several light bulbs in series, however, the defective light bulb would prevent any current in the circuit, and so all the light bulbs would be off.

Figure 6–1   Simple parallel circuit.

(Courtesy of Siemens Solar Industries, Camarillo, CA 93011)

OBJECTIVES

After studying this chapter you will be able to

- recognize which elements and branches in a given circuit are connected in parallel and which are connected in series,

- calculate the total resistance and conductance of a network of parallel resistances,

- determine the current in any resistor in a parallel circuit,

- solve for the voltage across any parallel combinations of resistors,

- apply Kirchhoff's current law to solve for unknown currents in a circuit,

- explain why voltage sources of different magnitudes must never be connected in parallel,

- use the current divider rule to solve for the current through any resistor of a parallel combination,

- design a simple ammeter using a permanent-magnet, moving-coil, meter movement,

- identify and calculate the loading effects of a voltmeter connected into a circuit.

Galvani and the Discovery of Nerve Excitation

*L*uigi Galvani was born in Bologna, Italy, on September 9, 1737.

Galvani's main expertise was in anatomy, in which subject he was appointed lecturer at the university in Bologna.

Galvani discovered that when the nerves of frogs were connected to sources of electricity, the muscles twitched. Although he was unable to determine where the electrical pulses originated within the animal, Galvani's work was significant and helped to open further discoveries in nerve impulses.

Galvani's name has been adopted for the instrument called the galvanometer, *which is used for detecting very small currents.*

Luigi Galvani died in Bologna on December 4, 1798. Although he made many contributions to science, Galvani died poor and in controversy.

6.1 Parallel Circuits

The illustration of Figure 6–1 shows that one terminal of each light bulb is connected to the positive terminal of the battery and that the other terminal of the light bulb is connected to the negative terminal of the battery. These points of connection are often referred to as *nodes.*

Elements or branches are said to be in a parallel connection when they have exactly two nodes in common.

Figure 6–2 shows several different ways of sketching parallel elements. The elements between the nodes may be any two-terminal devices such as voltage sources, resistors, light bulbs, and the like.

In the illustrations of Figure 6–2, notice that every element has two terminals and that each of the terminals is connected to one of the two nodes.

Very often, circuits contain a combination of series and parallel components. Although we will study these circuits in greater depth in

Figure 6–2   Parallel elements.

Series

Parallel

(a)

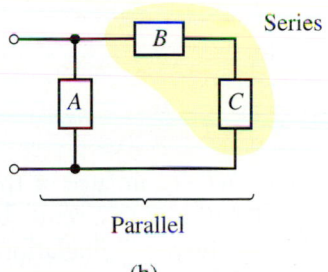

Series

Parallel

(b)

Figure 6–3   Series-parallel combinations.

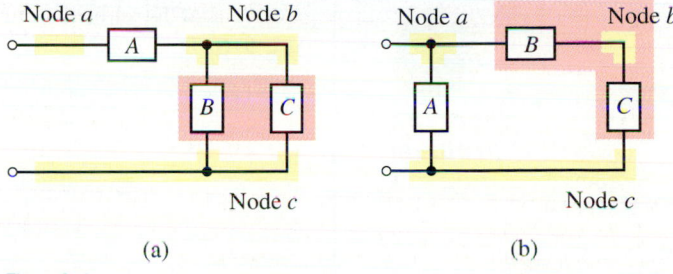

(a)

(b)

Figure 6–4

later chapters, it is important at this point to be able to recognize the various connections in a given network. Consider the networks shown in Figure 6–3.

When analyzing a particular circuit, it is usually easiest to first designate the nodes (we will use lowercase letters) and then to identify the types of connections. Figure 6–4 shows the nodes for the networks of Figure 6–3.

In the circuit of Figure 6–4(a), we see that element B is in parallel with element C since they each have nodes b and c in common. This parallel combination is now seen to be in series with element A.

In the circuit of Figure 6–4(b), element B is in series with element C since these elements have a single common node: node b. The branch consisting of the series combination of elements B and C is then determined to be in parallel with element A.

6.2 Kirchhoff's Current Law

Recall that Kirchhoff's voltage law was extremely useful in understanding the operation of the series circuit. In a similar manner, Kirchhoff's current law is the underlying principle which is used to explain the operation of a parallel circuit. Kirchhoff's current law states the following:

The summation of currents entering a node is equal to the summation of currents leaving the node.

An analogy which helps to understand the principle of Kirchhoff's current law is the flow of water. When water flows in a closed pipe, the amount of water entering a particular point in the pipe is exactly equal to the amount of water leaving, since there is no loss. In mathematical form, Kirchhoff's current law is stated as follows:

$$\sum I_{\text{entering node}} = \sum I_{\text{leaving node}} \qquad (6–1)$$

Figure 6–5 is an illustration of Kirchhoff's current law. Here we see that the node has two currents entering, $I_1 = 5$ A and $I_5 = 3$ A, and three currents leaving, $I_2 = 2$ A, $I_3 = 4$ A, and $I_4 = 2$ A. Now we can see that Equation 6–1 applies in the illustration, namely.

$$\sum I_{\text{in}} = \sum I_{\text{out}}$$
$$5\,\text{A} + 3\,\text{A} = 2\,\text{A} + 4\,\text{A} + 8\,\text{A}$$
$$8\,\text{A} = 8\,\text{A} \quad \text{(checks!)}$$

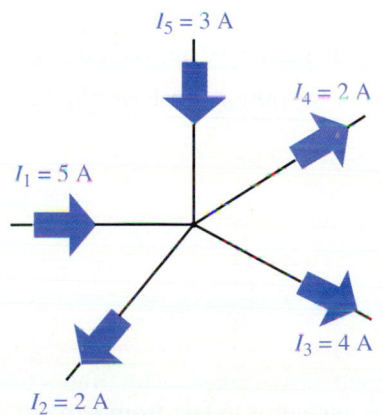

$I_5 = 3$ A

$I_4 = 2$ A

$I_1 = 5$ A

$I_3 = 4$ A

$I_2 = 2$ A

Figure 6–5   Kirchhoff's current law.

practice
problem
1

Verify that Kirchhoff's current law applies at the node shown in Figure 6–6.

Figure 6–6

Quite often, when we analyze a given circuit, we are unsure of the direction of current through a particular element within the circuit. In such cases, we assume a reference direction and base further calculations on this assumption. If our assumption is incorrect, calculations will show that the current has a negative sign. The negative sign simply indicates that the current is in fact opposite to the direction selected as the reference. The following example illustrates this very important concept.

■ **EXAMPLE 6–1**   Determine the magnitude and correct direction of the currents I_3 and I_5 for the network of Figure 6–7.

Figure 6–7

Solution   Although points a and b are in fact the same node, we treat the points as two separate nodes with 0 Ω resistance between them.

Since Kirchhoff's current law must be valid at point a, we have the following expression for this node:

$$I_1 = I_2 + I_3$$

and so

$$I_3 = I_1 - I_2$$
$$= 2\text{ A} - 3\text{ A} = -1\text{ A}$$

Notice that the reference direction of current I_3 was taken to be from a to b, while the negative sign indicates that the current is in fact from b to a.

Similarly, using Kirchhoff's current law at point b gives

$$I_3 = I_4 + I_5$$

which gives current I_5 as

$$I_5 = I_3 - I_4$$
$$= -1\,A - 6\,A = -7\,A$$

The negative sign indicates that the current I_5 is actually towards node b rather than away from the node. The actual directions and magnitudes of the currents are illustrated in Figure 6–8.

Figure 6–8

■ **EXAMPLE 6–2**   Find the magnitudes of the unknown currents for the circuit of Figure 6–9.

Figure 6–9

Solution   If we consider point a, we see that there are two unknown currents, I_1 and I_3. Since there is no way to solve for these values, we examine the currents at point b, where we again have two unknown currents, I_3 and I_4. Finally we observe that at point c there is only one unknown, I_4. Using Kirchhoff's current law we solve for the unknown current as follows:

$$I_4 + 3\,A + 2\,A = 10\,A$$

Therefore,

$$I_4 = 10\,A - 3\,A - 2\,A = 5\,A$$

Now we can see that at point b the current entering is

$$I_3 = 5\,A + 3\,A + 2\,A = 10\,A$$

And finally, by applying Kirchhoff's current law at point a, we determine that the current I_1 is

$$I_1 = 10\,\text{A} - 3\,\text{A} = 7\,\text{A}$$

■ **EXAMPLE 6–3**   Determine the unknown currents in the network of Figure 6–10.

Figure 6-10

Solution   We first assume reference directions for the unknown currents in the network.

Since we may use the analogy of water moving through conduits, we can easily assign directions for the currents I_3, I_5, and I_7. However, the direction for the current I_4 is not as easily determined, so we arbitrarily assume that its direction is to the right. Figure 6–10(b) shows the various nodes and the assumed current directions.

By examining the network, we see that there is only a single source of current, $I_1 = 24$ A. Using the analogy of water pipes, we conclude that the current leaving the network is $I_7 = I_1 = 24$ A.

Now, applying Kirchhoff's current law to node a, we calculate the current I_3 as follows:

$$I_1 = I_2 + I_3$$

Therefore,

$$I_3 = I_1 - I_2 = 24\,\text{A} - 11\,\text{A} = 13\,\text{A}$$

Similarily, at node c, we have

$$I_3 + I_4 = I_6$$

Therefore,

$$I_4 = I_6 - I_3 = 6\,\text{A} - 13\,\text{A} = -7\,\text{A}$$

Although the current I_4 is opposite to the assumed reference direction, we do not change its direction for further calculations. We use the original direction together with the negative sign; otherwise the calculations would be needlessy complicated.

Applying Kirchhoff's current law at node b, we get

$$I_2 = I_4 + I_5$$

which gives

$$I_5 = I_2 - I_4 = 11 \text{ A} - (-7 \text{ A}) = 18 \text{ A}$$

Finally, applying Kirchhoff's current law at node d gives

$$I_5 + I_6 = I_7$$

resulting in

$$I_7 = I_5 + I_6 = 18 \text{ A} + 6 \text{ A} = 24 \text{ A}$$

practice
problem
II

Determine the unknown currents in the network of Figure 6–11.

Figure 6–11

6.3 Resistors in Parallel

A simple parallel circuit is constructed by combining a voltage source with several resistors as shown in Figure 6–12.

The voltage source will result in current from the positive terminal of the source toward node a. At this point the current will split between

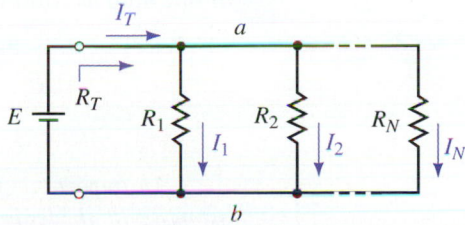

Figure 6–12

the various resistors and then recombine at node b before continuing to the negative terminal of the voltage source.

This circuit illustrates a very important concept of parallel circuits. If we were to apply Kirchhoff's voltage law around each closed loop in the parallel circuit of Figure 6–12, we would find that the voltage across all parallel resistors is exactly equal, namely $V_{R_1} = V_{R_2} = V_{R_3} = E$. Therefore, by applying Kirchhoff's voltage law, we make the following statement:

The voltage across all parallel elements in a circuit will be the same.

The above principle allows us to determine the equivalent resistance, R_T, of any number of resistors connected in parallel. The equivalent resistance, R_T, is the effective resistance "seen" by the source and determines the total current, I_T, provided to the circuit. Applying Kirchhoff's current law to the circuit of Figure 6–11, we have the following expression:

$$I_T = I_1 + I_2 + \cdots + I_N$$

However, since Kirchhoff's voltage law also applies to the parallel circuit, the voltage across each resistor must be equal to the supply voltage, E. The total current in the circuit, which is determined by the supply voltage and the equivalent resistance, may now be written as

$$\frac{E}{R_T} = \frac{E}{R_1} + \frac{E}{R_2} + \cdots + \frac{E}{R_N}$$

Simplifying the above expression gives us the general expression for total resistance of a parallel circuit as

$$\frac{1}{R_T} = \frac{1}{R_1} + \frac{1}{R_2} + \cdots + \frac{1}{R_N} \quad \text{(siemens, S)} \tag{6–2}$$

Since conductance was defined as the reciprocal of resistance, we may write the above equation in terms of conductance, namely,

$$G_T = G_1 + G_2 + \cdots + G_N \quad \text{(S)} \tag{6–3}$$

Whereas series resistors had a total resistance determined by the summation of the particular resistances, we see that any number of parallel resistors have a total conductance determined by the summation of the individual conductances.

The equivalent resistance of N parallel resistors may be determined in one step as follows:

$$R_T = \frac{1}{\dfrac{1}{R_1} + \dfrac{1}{R_2} + \cdots + \dfrac{1}{R_N}} \quad (\Omega) \tag{6–4}$$

An important effect of combining parallel resistors is that the resultant resistance will always be smaller than the smallest resistor in the combination.

■ **EXAMPLE 6–4**   Solve for the total conductance and total equivalent resistance of the circuit shown in Figure 6–13.

Figure 6–13

Solution   The total conductance is

$$G_T = G_1 + G_2 = \frac{1}{4\ \Omega} + \frac{1}{1\ \Omega} = 1.25\ \text{S}$$

The total equivalent resistance of the circuit is

$$R_T = \frac{1}{G_T} = \frac{1}{1.25\ \text{mS}} = 0.8\ \Omega$$

Notice that the equivalent resistance of the parallel resistors is indeed less than the value of each resistor.

■ **EXAMPLE 6–5**   Determine the conductance and resistance of the network of Figure 6–14.

Figure 6–14

Solution   The total conductance is

$$G_T = G_1 + G_2 + G_3$$

$$= \frac{1}{18\ \Omega} + \frac{1}{9\ \Omega} + \frac{1}{6\ \Omega}$$

$$= 0.0\overline{5}\ \text{S} + 0.1\overline{1}\ \text{S} + 0.1\overline{6}\ \text{S}$$

$$= 0.3\overline{3}\ \text{S}$$

where the overbar indicates that the number under it is repeated infinitely to the right.
    The total resistance is

$$R_T = \frac{1}{0.3\overline{3}\ \text{S}} = 3.0\ \Omega$$

Figure 6–15

For the parallel network of resistors shown in Figure 6–15, find the total conductance, G_T and the total resistance, R_T.

N Equal Resistors in Parallel

If we have N equal resistors in parallel, each resistor, R, has the same conductance, G. By applying Equation 6–3, the total conductance is found:

$$G_T = NG$$

The total resistance is now easily determined as

$$R_T = \frac{1}{G_T} = \frac{1}{NG} = \frac{R}{N} \qquad (6\text{–}5)$$

■ **EXAMPLE 6–6**   For the networks of Figure 6–16, calculate the total resistance.

(a)

(b)

Figure 6–16

Solution

a. $R_T = \dfrac{18\ \text{k}\Omega}{3} = 6\ \text{k}\Omega$

b. $R_T = \dfrac{200\ \Omega}{4} = 50\ \Omega$

Two Resistors in Parallel

Very often circuits have only two resistors in parallel. In such a case, the total resistance of the combination may be determined without the necessity of determining the conductance.

For two resistors, Equation 6–4 is written

$$R_T = \cfrac{1}{\dfrac{1}{R_1} + \dfrac{1}{R_2}}$$

By cross multiplying the terms in the denominator, the expression becomes

$$R_T = \cfrac{1}{\dfrac{R_1 + R_2}{R_1 R_2}}$$

And, so, for two resistors in parallel we have the following expression:

$$R_T = \frac{R_1 R_2}{R_1 + R_2} \tag{6-6}$$

For two resistors connected in parallel, the equivalent resistance is found by the product of the two values divided by the sum.

■ **EXAMPLE 6–7**   Determine the total resistance of the resistor combinations of Figure 6–17.

(a)

(b)

(c)

Figure 6–17

Solution

a. $R_T = \dfrac{(3\ M\Omega)(1\ M\Omega)}{3\ M\Omega + 1\ M\Omega} = 0.75\ M\Omega = 750\ k\Omega$

b. $R_T = \dfrac{(36\ \Omega)(24\ \Omega)}{36\ \Omega + 24\ \Omega} = 14.4\ \Omega$

c. $R_T = \dfrac{(98\ k\Omega)(2\ k\Omega)}{98\ k\Omega + 2\ k\Omega} = 1.96\ k\Omega$

Although Equation 6–6 is intended primarily to solve for two resistors in parallel, the approach may also be used to solve for any number of resistors by examining only two resistors at a time.

■ **EXAMPLE 6–8**   Calculate the total resistance of the resistor combination of Figure 6–18.

Figure 6–18

Solution   By grouping the resistors into combinations of two, the circuit may be simplified as shown in Figure 6–19.

Figure 6–19

The equivalent resistance of each of the indicated combinations is determined as follows:

$$R_A = \frac{(180\ \Omega)(90\ \Omega)}{180\ \Omega + 90\ \Omega} = 60\ \Omega$$

$$R_B = \frac{(60\ \Omega)(60\ \Omega)}{60\ \Omega + 60\ \Omega} = 30\ \Omega$$

The circuit can be further simplified as a combination of two resistors shown in Figure 6–20.

Figure 6–20

The resultant equivalent resistance is

$$R_T = \frac{(60\ \Omega)(30\ \Omega)}{60\ \Omega + 30\ \Omega} = 20\ \Omega$$

Three Resistors in Parallel

Using an approach similar to the derivation of Equation 6–6, we may arrive at an equation which solves for three resistors in parallel. Indeed, it is possible to write a general equation to solve for four resistors, five resistors, etc. Although such an equation is certainly useful, students are discouraged from memorizing such lengthy expressions. You will generally find that it is much more efficient to remember the principles upon which the equation is constructed. Consequently, the derivation of Equation 6–7 is left up to the student.

$$R_T = \frac{R_1 R_2 R_3}{R_1 R_2 + R_1 R_2 + R_2 R_3} \qquad (6\text{–}7)$$

practice
problem
IV

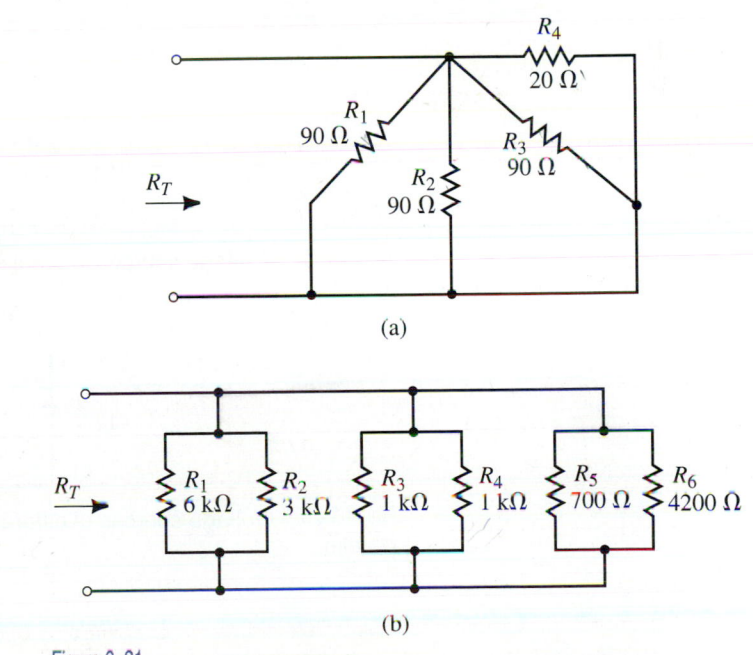

Figure 6–21

Find the total equivalent resistance for each network in Figure 6–21.

If the circuit of Figure 6–21(a) is connected to a 24-V voltage source, determine the following quantities:

a. The total current provided by the voltage source.
b. The current through each resistor of the network.
c. Verify Kirchhoff's current law at one of the voltage source terminals.

6.4 Voltage Sources in Parallel

Voltage sources of different potentials should never be connected in parallel, since to do so would contradict Kirchhoff's voltage law. However, when two equal potential sources are connected in parallel, each source will deliver half the required circuit current. For this reason automobile batteries are sometimes connected in parallel to assist in starting a car with a "weak" battery. Figure 6–22 illustrates this principle.

Figure 6–23 shows that if voltage sources of two different potentials are placed in parallel, Kirchhoff's voltage law would be violated around the closed loop. In practice, if voltage sources of different potentials are placed in parallel, a series circuit results in which a very large current may be present in the sources. The current will occur even though there may not be a load connected across the sources. The following example illustrates the large currents that can occur when two parallel batteries of different potential are connected.

Figure 6–22   Voltage sources in parallel.

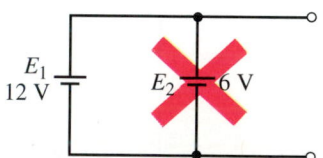

Figure 6–23   Voltage sources of different potentials must **never** be placed in parallel.

■ **EXAMPLE 6–9**   A 12-V battery and a 6-V battery (each having an internal resistance of 0.05 Ω) are inadvertently placed in parallel as shown in Figure 6–24. Determine the current through the batteries.

Figure 6–24

Solution   From Ohm's law,

$$I = \frac{E_T}{R_T} = \frac{12\text{ V} - 6\text{ V}}{0.05\ \Omega + 0.05\ \Omega} = 60\text{ A}$$

This example illustrates why batteries of different potential must never be connected in parallel. Tremendous currents will occur within the sources resulting in the possibility of a fire or explosion.

6.5 Current Divider Rule

When we examined series circuits we determined that the current in the series circuit was the same everywhere in the circuit, whereas the voltages across the series elements were typically different. The voltage divider rule (VDR) was used to determine the voltage across all resistors within a series network.

In parallel networks, the voltage across all parallel elements is the same. However, the currents through the various elements are typically different. The current divider rule (CDR) is used to determine how current entering a node is split between the various parallel resistors connected to the node.

Consider the network of parallel resistors shown in Figure 6–25.

If this network of resistors is supplied by a voltage source, the total current in the circuit is

$$I_T = \frac{E}{R_T} \qquad\qquad (6\text{–}8)$$

$$I_x = \frac{R_T}{R_x} I_T$$

Figure 6–25   Current divider rule.

Since each of the N parallel resistors has the same voltage, E, across its terminals, the current through any resistor in the network is given as

$$I_x = \frac{E}{R_x} \qquad (6\text{–}9)$$

By rewriting Equation 6–8 as $E = I_T R_T$ and then substituting this into Equation 6–9, we obtain the current divider rule as follows:

$$I_x = \frac{R_T}{R_x} I_T \qquad (6\text{–}10)$$

The current divider rule allows us to calculate the current in any resistor of a parallel network if we know the total current entering the network. Notice, though, that while the current divider rule is similar to the voltage divider rule, the portion of total current through an unknown parallel resistance is determined by dividing the total parallel resistance by the unknown resistance.

If the network consists of only two parallel resistors, then the current through each resistor may be found in a slightly different way. Recall that for two resistors in parallel, the total parallel resistance is given as

$$R_T = \frac{R_1 R_2}{R_1 + R_2}$$

Now, by substituting this expression for total resistance into Equation 6–10, we obtain

$$I_1 = \frac{I_T R_T}{R_1}$$

$$= \frac{I_T \left(\dfrac{R_1 R_2}{R_1 + R_2} \right)}{R_1}$$

which simplifies to

$$I_1 = \frac{R_2}{R_1 + R_2} I_T \qquad (6\text{–}11)$$

Similarly,

$$I_2 = \frac{R_1}{R_1 + R_2} I_T \qquad (6\text{–}12)$$

Notice the similarity between the above equations and the expression for the voltage divider rule. However, unlike the voltage across R_1 given by the voltage divider rule, the current through R_1 is determined by R_2 over the sum, rather than R_1 over the sum.

Several other important characteristics of parallel networks become evident.

If current enters a parallel network consisting of any number of equal resistors, then the current entering the network will split equally between all of the resistors.

If current enters a parallel network consisting of several values of resistance, then the smallest value of resistor in the network will have the largest amount of current. Inversely, the largest value of resistance will have the small-

est amount of current. By examining Equations 6–11 and 6–12, we see that if R_1 is the larger resistor, R_2 will have the greater amount of current.

The latter characteristic may be simplified by saying that *most of the current will follow the path of least resistance.*

■ **EXAMPLE 6–10**   For the network of Figure 6–26, determine the currents I_1, I_2, and I_3.

Figure 6–26

Solution   First, we calculate the total resistance of the network. Although we may use Equation 6–7 to find the total resistance of the three resistors, we use the more general expression given by Equation 6–4 to solve for R_T:

$$R_T = \frac{1}{\dfrac{1}{1\,\Omega} + \dfrac{1}{2\,\Omega} + \dfrac{1}{4\,\Omega}} = 0.5714\,\Omega$$

Now the currents are evaluated as follows:

$$I_1 = \frac{R_T}{R_1}I_T = \left(\frac{0.5714\,\Omega}{1\,\Omega}\right)(14\text{ A}) = 8\text{ A}$$

$$I_2 = \frac{R_T}{R_2}I_T = \left(\frac{0.5714\,\Omega}{2\,\Omega}\right)(14\text{ A}) = 4\text{ A}$$

$$I_3 = \frac{R_T}{R_3}I_T = \left(\frac{0.5714\,\Omega}{4\,\Omega}\right)(14\text{ A}) = 2\text{ A}$$

■ **EXAMPLE 6–11**   For the network of Figure 6–27, determine the currents I_1, I_2, and I_3.

Figure 6–27

Solution   Since all the resistors have the same value, the incoming current will split equally between the resistances. Therefore,

$$I_1 = I_2 = I_3 = \frac{12 \text{ mA}}{3} = 4 \text{ mA}$$

■ **EXAMPLE 6–12**   Determine the currents I_1 and I_2 in the network of Figure 6–28.

Figure 6–28

Solution   Because we have only two resistors in the given network, we use Equations 6–11 and 6–12:

$$I_1 = \frac{R_2}{R_1 + R_2} I_T = \left(\frac{200 \text{ }\Omega}{300 \text{ }\Omega + 200 \text{ }\Omega} \right)(20 \text{ mA}) = 8 \text{ mA}$$

$$I_2 = \frac{R_1}{R_1 + R_2} I_T = \left(\frac{300 \text{ }\Omega}{300 \text{ }\Omega + 200 \text{ }\Omega} \right)(20 \text{ mA}) = 12 \text{ mA}$$

■ **EXAMPLE 6–13**   Design the resistance R_1 so that current will divide as shown in the network of Figure 6–29.

Figure 6–29

Solution   There are several methods which may be used to solve this problem. We will examine only two of the possibilities.

Method I: Since we have two resistors in parallel, we may use Equation 6–12 to solve for the unknown resistor:

$$I_2 = \frac{R_1}{R_1 + R_2} I_T$$

$$5\ \text{A} = \left(\frac{R_1}{R_1 + 30\ \Omega}\right)(25\ \text{A})$$

Using algebra, we get

$$(5\ \text{A})R_1 + (5\ \text{A})(30\ \Omega) = (25\ \text{A})R_1$$

$$(20\ \text{A})R_1 = 150\ \text{V}$$

$$R_1 = \frac{150\ \text{V}}{20\ \text{A}} = 7.5\ \Omega$$

Method II: By applying Kirchhoff's current law, we see that the current in R_1 must be

$$I_1 = 25\ \text{A} - 5\ \text{A} = 20\ \text{A}$$

Now, since elements in parallel must have the same voltage across their terminals, the voltage across R_1 must be exactly the same as the voltage across R_2. By Ohm's law, the voltage across R_2 is

$$V_2 = (5\ \text{A})(30\ \Omega) = 150\ \text{V}$$

And so

$$R_1 = \frac{150\ \text{V}}{20\ \text{A}} = 7.5\ \Omega$$

As expected, the results are identical. This example illustrates that there is usually more than one method to solving a given problem. Although the methods are equally correct, we see that the second method in this example is less involved.

practice
problem
V

Use the current divider rule to calculate the unknown currents for the networks of Figure 6–30.

(a)

(b)

Figure 6–30   *(continues)*

Figure 6–30   (*continued*)

in process
learning
check II

Four resistors are connected in parallel. The values of the resistors are
1 Ω, 3 Ω, 4 Ω, and 5 Ω.

a. Using only a pencil and a piece of paper (no calculator),
   determine the current through each resistor if the current
   through the 5-Ω resistor is 6 A.

b. Again, without a calculator, solve for the total current applied to
   the parallel combination.

c. Use a calculator to determine the total parallel resistance of the
   four resistors. Use the current divider rule and the total current
   obtained in part b) to calculate the current through each
   resistor.

6.6 Analysis of Parallel Circuits

We will now examine how to use the principles developed in this chapter
when analyzing parallel circuits. In the examples to follow, we find that
the laws of conservation of energy apply equally well to parallel circuits
as to series circuits. Although we choose to analyze circuits a certain way,
remember that there is usually more than one way to arrive at the correct answer. As you become more proficient at circuit analysis you will
generally use the most efficient method. For now, however, use the
method with which you feel most comfortable.

■ EXAMPLE 6–14

Figure 6–31

For the circuit of Figure 6–31, determine the following quantities:

a. R_T.

b. I_T.

c. Power delivered by the voltage source.

d. I_1 and I_2 using the current divider rule.

e. Powers dissipated by the resistors.

Solution

a. $R_T = \dfrac{R_1 R_2}{R_1 + R_2} = \dfrac{(2 \text{ k}\Omega)(8 \text{ k}\Omega)}{2 \text{ k}\Omega + 8 \text{ k}\Omega} = 1.6 \text{ k}\Omega$

b. $I_T = \dfrac{E}{R_T} = \dfrac{36 \text{ V}}{1.6 \text{ k}\Omega} = 22.5 \text{ mA}$

c. $P_T = E I_T = (36 \text{ V})(22.5 \text{ mA}) = 810 \text{ mW}$

d. $I_2 = \dfrac{R_1}{R_1 + R_2} I_T = \left(\dfrac{2 \text{ k}\Omega}{2 \text{ k}\Omega + 8 \text{ k}\Omega} \right)(22.5 \text{ mA}) = 4.5 \text{ mA}$

$I_1 = \dfrac{R_2}{R_1 + R_2} I_T = \left(\dfrac{8 \text{ k}\Omega}{2 \text{ k}\Omega + 8 \text{ k}\Omega} \right)(22.5 \text{ mA}) = 18.0 \text{ mA}$

e. Since we know the voltage across each of the parallel resistors must be 36 V, we use this voltage to determine the power dissipated by each resistor. It would be equally correct to use the current through each resistor to calculate the power. However, it is generally best to use given information rather than calculated values to perform further calculations since it is then less likely that an error is carried through.

$$P_1 = \dfrac{E^2}{R_1} = \dfrac{(36 \text{ V})^2}{2 \text{ k}\Omega} = 648 \text{ mW}$$

$$P_2 = \dfrac{E^2}{R_2} = \dfrac{(36 \text{ V})^2}{8 \text{ k}\Omega} = 162 \text{ mW}$$

Notice that the power delivered by the voltage source is exactly equal to the total power dissipated by the resistors, namely $P_T = P_1 + P_2$.

■ **EXAMPLE 6–15**   Refer to the circuit of Figure 6–32:

Figure 6–32

a. Solve for the total power delivered by the voltage source.
b. Find the currents I_1, I_2, and I_3.
c. Determine the values of the unknown resistors R_2 and R_3.
d. Calculate the power dissipated by each resistor.
e. Verify that the power dissipated is equal to the power delivered by the voltage source.

Solution

a. $P_T = EI_T = (120 \text{ V})(2.2 \text{ A}) = 264 \text{ W}$

b. Since the three resistors of the circuit are in parallel, we know that the voltage across all resistors must be equal to $E = 120$ V.

$$I_1 = \frac{V_1}{R_1} = \frac{120 \text{ V}}{300 \text{ }\Omega} = 0.4 \text{ A}$$

$$I_3 = \frac{P_3}{V_3} = \frac{144 \text{ W}}{120 \text{ V}} = 1.2 \text{ A}$$

Because KCL must be maintained at each node, we may determine the current I_2 as

$$I_2 = I_T - I_1 - I_3$$
$$= 2.2 \text{ A} - 0.4 \text{ A} - 1.2 \text{ A} = 0.6 \text{ A}$$

c. $R_2 = \dfrac{V_2}{I_2} = \dfrac{120 \text{ V}}{0.6 \text{ A}} = 200 \text{ }\Omega$

Although we could use the calculated current I_3 to determine the resistance, it is best to use the given data in calculations rather than calculated values.

$$R_3 = \frac{V_3{}^2}{P_3} = \frac{(120 \text{ V})^2}{144 \text{ W}} = 100 \text{ }\Omega$$

d. $P_1 = \dfrac{V_1{}^2}{R_1} = \dfrac{(120 \text{ V})^2}{300 \text{ }\Omega} = 48 \text{ W}$

$P_2 = I_2 E_2 = (0.6 \text{ A})(120 \text{ V}) = 72 \text{ W}$

e. $\quad P_{\text{in}} = P_{\text{out}}$

$264 \text{ W} = P_1 + P_2 + P_3$

$264 \text{ W} = 48 \text{ W} + 72 \text{ W} + 144 \text{ W}$

$264 \text{ W} = 264 \text{ W}$ (checks!)

6.7 Ammeter Design

The ammeter (which is used to measure current in a circuit) is a very good practical application of a parallel circuit. Recall that in order to use the ammeter correctly, the circuit under test must be opened and the ammeter inserted between the open terminals. Although you will rarely be required to design an ammeter, it is important to understand the in-

Figure 6–33   Simple ammeter.

ternal operation of the ammeter. Once you understand the operation, you will appreciate the limitations of the instrument.

The schematic of a simple ammeter using a PMMC (permanent-magnet, moving-coil) movement is shown in Figure 6–33.

Since the PMMC movement is able to handle only very small currents, a precisely engineered resistance is placed in parallel with the resistance of the meter movement, R_m. The shunt resistance, R_{shunt}, ensures that the current entering the sensitive meter movement is kept below the I_{fsd} (full-scale deflection current) of the meter movement.

The ammeter is designed to provide full-scale deflection of the meter movement when the current entering the instrument is at the desired range current. The shunt resistance ensures that the excess current bypasses the meter movement. The following example shows how the value of shunt resistance is determined to result in a given ammeter.

The voltage across the parallel combination of R_m and R_{shunt} when the meter movement is at its maximum deflection is determined from Ohm's law as

$$V_m = I_{fsd}R_m$$

From Kirchhoff's current law, the current through the shunt resistor (at full-scale deflection) is

$$I_{shunt} = I_{range} - I_{fsd}$$

And so, from Ohm's law, the shunt resistor has a value determined as

$$R_{shunt} = \frac{I_{fsd}R_m}{I_{range} - I_{fsd}} = \frac{1}{\left(\dfrac{I_{range}}{I_{fsd}} - 1\right)}R_m \qquad (6–13)$$

Since the above equation is quite complicated, it is generally very difficult to memorize the expression. It is much easier to remember the principles from which the expression is conceived.

■ **EXAMPLE 6–16**   Determine the value of shunt resistance required to build a 100-mA ammeter using a meter movement having $I_{\text{fsd}} = 1$ mA and $R_m = 2$ kΩ.

Solution   When the ammeter is detecting the maximum current, the voltage across the meter movement (and the shunt resistance) is

$$V_m = I_{\text{fsd}}R_m = (1 \text{ mA})(2 \text{ k}\Omega) = 2.0 \text{ V}$$

The current in the shunt resistance is

$$I_{\text{shunt}} = I_{\text{range}} - I_{\text{fsd}} = 100 \text{ mA} - 1 \text{ mA} = 99 \text{ mA}$$

And so the shunt resistance must have a value of

$$R_{\text{shunt}} = \frac{2.0 \text{ V}}{99 \text{ mA}} = 20.2 \ \Omega$$

The resulting circuit appears in Figure 6–34.

Figure 6–34

Figure 6–35 shows the schematic of a simple multi-range ammeter. In order to have a multirange ammeter, it is necessary to have a different shunt resistor for each range.

To prevent damaging the sensitive meter movement, the ammeter of Figure 6–35 uses a "make-before-break" switch. As the name implies, this type of switch makes contact with the new position before breaking contact with the previous position. This feature prevents a large current from damaging the sensitive meter movement while a switch is in transition from one range to the next.

The selector switch must be a "make–before–break"
switch to prevent applying excessive current to
the meter movement when changing ranges.

Figure 6–35   Multirange ammeter.

practice
problem
VI

Design the shunt resistance needed to construct an ammeter which is able to measure up to 10 mA. Use a meter movement having I_{fsd} = 50 μA and R_m = 5 kΩ. Show a sketch of your design.

6.8 Voltmeter Loading Effects

In the previous chapter, we observed that a voltmeter is essentially a meter movement in series with a current-limiting resistance. When a voltmeter is placed across two terminals to provide a voltage reading, the circuit is affected in the same manner as if a resistance were placed across the two terminals. The effect is shown in Figure 6–36.

If the resistance of the voltmeter is very large in comparison with the resistance across which the voltage is to be measured, the meter will indicate essentially the same voltage which was present before the meter was connected. On the other hand, if the meter has an internal resistance which is near in value to the resistance across which the measurement is taken, then the meter will adversely load the circuit, resulting in an erroneous reading. Generally, if the meter resistance is more than ten times larger than the resistance across which the voltage is taken, then the loading effect is considered negligible and may be ignored.

In the circuit of Figure 6–37, there is no current in the circuit since the terminals *a* and *b* are open circuited. The voltage appearing between

Figure 6–36

$R = 100\ \Omega$

$E = 10\ V$

Figure 6–37

$R = 100\ \Omega$ a

$E = 10\ V$

b

Figure 6–38

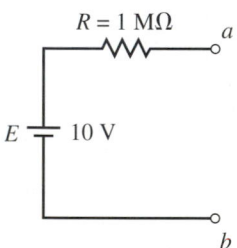

$R = 1\ M\Omega$ a

E $10\ V$

b

Figure 6–39

$R = 1\ M\Omega$ a

I

$E = 10\ V$

b

Figure 6–40

the open terminals must be $V_{ab} = 10$ V. Now, if we place a voltmeter having an internal resistance of 200 kΩ between the terminals, the circuit is closed, resulting in a small current. The complete circuit appears as shown in Figure 6–38.

The reading indicated on the face of the meter is the voltage which occurs across the internal resistance of the meter. Applying Kirchhoff's voltage law to the circuit, this voltage is

$$V_{ab} = \frac{200\ k\Omega}{200\ k\Omega + 100\ \Omega}(10\ V) = 9.995\ V$$

Clearly, the reading on the face of the meter is essentially equal to the expected value of 10 V. Recall from the previous chapter that we defined the loading effect of a meter as follows:

$$\text{loading effect} = \frac{\text{actual value} - \text{reading}}{\text{actual value}} \times 100\%$$

For the circuit of Figure 6–38, the voltmeter has a loading effect of

$$\text{loading effect} = \frac{10\ V - 9.995\ V}{10\ V} \times 100\% = 0.05\%$$

This loading error is virtually undetectable for the circuit given. The same would not be true if we had a circuit as shown in Figure 6–39 and used the same voltmeter to provide a reading.

Again, if the circuit were left open circuited, we would expect that $V_{ab} = 10$ V.

By connecting the 200-kΩ voltmeter between the terminals, as shown in Figure 6–40, we see that the voltage detected between terminals a and b will no longer be the desired voltage; rather,

$$V_{ab} = \frac{200\ k\Omega}{200\ k\Omega + 1\ M\Omega}(10\ V) = 1.667\ V$$

The loading effect of the meter in this circuit is

$$\text{loading effect} = \frac{10\text{ V} - 1.667\text{ V}}{10\text{ V}} \times 100\% = 83.33\%$$

The previous illustration is an example of a problem which can occur when taking measurements in electronic circuits. When an inexperienced technician or technologist obtains an unforeseen result, he or she assumes that something is wrong with either the circuit or the instrument. In fact, both the circuit and the instrument are behaving in a perfectly predictable manner. He or she merely forgot to take into account the meter's loading effect. All instruments have limitations and we must always be aware of these limitations.

■ **EXAMPLE 6–17**   A digital voltmeter having an internal resistance of 5 MΩ is used to measure the voltage across terminals *a* and *b* in the circuit of Figure 6–40.

    a. Determine the reading on the meter.
    b. Calculate the loading effect of the meter.

Solution

    a. The voltage applied to the meter terminals is

$$V_{ab} = \left(\frac{5\text{ M}\Omega}{1\text{ M}\Omega + 5\text{ M}\Omega} \right)(10\text{ V}) = 8.33\text{ V}$$

    b. The loading effect is

$$\text{loading error} = \frac{10\text{ V} - 8.33\text{ V}}{10\text{ V}} \times 100\% = 16.7\%$$

in process
l e a r n i n g
check III

All instruments have a loading effect on the circuit in which a measurement is taken. If you were given two voltmeters, one with an internal resistance of 200 kΩ and another with an internal resistance of 1 MΩ, which meter would load a circuit more? Explain.

6.9 Computer Analysis

Although PSpice is most useful for complex circuits, we can use the software as an illustrative tool to verify concepts learned in this chapter. The practice of solving simple circuits gives the student confidence and helps one learn techniques which provide the basis to undertake more difficult tasks.

■ **EXAMPLE 6–18**   Use PSpice to solve for the circuit current and the current through each resistor in the circuit of Figure 6–41. This circuit was previously analyzed in Example 6–15.

Figure 6–41

Solution   The voltage source is designated using the PSpice format, and the nodes of the given circuit are labelled as shown in Figure 6–42.

Figure 6–42

The PSpice input file is listed below:

```
Input File:
Example 6-18 - Parallel Circuit, Figure 6-41
Vs    1    0    36V
R1    1    0    2kohms
R2    1    0    8kohms
.DC   Vs   36V  36V   1V
.PRINT DC I(Vs)  I(R1)   I(R2)
.OPTIONS NOPAGE
.END
```

After running the PSpice program, the output file may be viewed by selecting the Browse menu item or by printing the output file with the .OUT extension. The output file, titled EX6-18.OUT, appears in Figure 6–43.

```
OUTPUT

***      DC TRANSFER CURVES    TEMPERATURE = 27.000 DEG C

Vs            I(Vs)         I(R1)        I(R2)
3.600E+01    -2.250E-02    1.800E-02    4.500E-03


      JOB CONCLUDED
      TOTAL JOB TIME       2.58
```

Figure 6–43

The results of the analysis are consistent with those obtained in Example 6–15. Again, notice that the current through the voltage source is given a negative sign, indicating that the current is not from node 1 to node 0, but rather from node 0 to node 1.

practice
problem
VII

Use PSpice to determine the current in each resistor of the circuit of Figure 6–21(a) if a 24-V voltage source is connected across the terminals of the resistor network.

Problems

6.1   Parallel Circuits

1. Indicate which of the elements in Figure 6–44 are connected in parallel and which elements are connected in series.

Figure 6–44

2. For the networks of Figure 6–45, indicate which resistors are connected in series and which resistors are connected in parallel.

Figure 6–45

3. Without changing the component positions, show at least one way of connecting all the elements of Figure 6–46 in parallel.

Figure 6–46

4. Repeat Problem 3 for the elements shown in Figure 6–47.

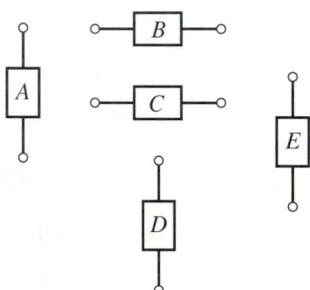

Figure 6–47

6.2   Kirchhoff's Current Law

5. Use Kirchhoff's current law to determine the magnitudes and directions of the indicated currents in each of the networks shown in Figure 6–48.

(a)

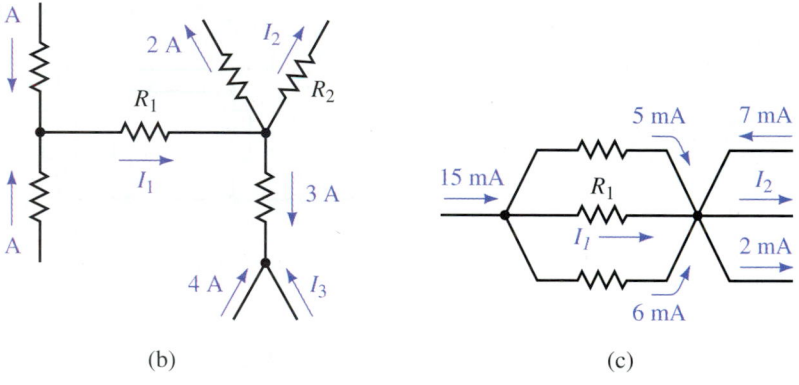

(b)   (c)

Figure 6–48

6. For the circuit of Figure 6–49, determine the magnitude and direction of each of the indicated currents.

Figure 6–49

7. Consider the network of Figure 6–50:
   a. Find the direction and magnitudes of the currents I_1, I_2, and I_3.
   b. Determine the value of the resistance R_3.

Figure 6–50

8. Find each of the unknown currents in the networks of Figure 6–51.

(a)

(b)

Figure 6–51

9. Refer to the network of Figure 6–52:

Figure 6–52

   a. Use Kirchhoff's current law to solve for the unknown currents, I_1, I_2, I_3, and I_4.
   b. Calculate the voltage, V, across the network.
   c. Determine the values of the unknown resistors, R_1, R_3, and R_4.

10. Refer to the network of Figure 6–53:

Figure 6–53

a. Use Kirchhoff's current law to solve for the unknown currents.
b. Calculate the voltage, *V*, across the network.
c. Determine the required value of the voltage source, *E*. (Hint: Use Kirchhoff's voltage law.)

6.3   Resistors in Parallel

11. Calculate the total conductance and total resistance of each of the networks shown in Figure 6–54.

Figure 6–54

12. For the networks of Figure 6–55, determine the value of the unknown resistance(s) to result in the given total conductance.

Figure 6–55

13. For the networks of Figure 6–56, determine the value of the unknown resistance(s) to result in the total resistances given.

(a)   (b)

Figure 6-56

14. Determine the value of each unknown resistor in the network of Figure 6–57, so that the total resistance is 100 kΩ.

Figure 6-57

15. Refer to the network of Figure 6–58:

Figure 6-58

a. Calculate the values of R_1, R_2, and R_3 so that the total resistance of the network is 200 Ω.
b. If R_3 has a current of 2 A, determine the current through each of the other resistors.
c. How much current must be applied to the entire network?

16. Refer to the network of Figure 6–59:

Figure 6-59

a. Calculate the values of R_1, R_2, R_3, and R_4 so that the total resistance of the network is 100 kΩ.
b. If R_4 has a current of 2 mA, determine the current through each of the other resistors.
c. How much current must be applied to the entire network?

17. Refer to the network of Figure 6–60:

Figure 6–60

    a. Find the voltages across R_1 and R_2.
    b. Determine the current I_2.

18. Refer to the network of Figure 6–61:

Figure 6–61

    a. Find the voltages across R_1, R_2 and R_3.
    b. Calculate the current I_2
    c. Calculate the current I_3.

19. Determine the total resistance of each network of Figure 6–62.

(a)   (b)   (c)

Figure 6–62

20. Determine the total resistance of each network of Figure 6–63.

(a)

(b)   (c)

Figure 6–63

21. Determine the values of the resistors in the circuit of Figure 6–64, given the indicated conditions.

22. Given the indicated conditions, calculate all currents and determine all resistor values for the circuit of Figure 6–65.

$I_2 = 3I_1$
$I_3 = 1.5I_2$
$R_T = 16 \text{ k}\Omega$

Figure 6–64

$R_1 = 36 \text{ }\Omega$
$I_3 = 500 \text{ mA}$
$R_2 = 4R_1$

Figure 6–65

23. Without using a pencil, paper, or a calculator, determine the resistance of each network of Figure 6–66.

(a)   (b)   (c)

Figure 6–66

Figure 6–67

24. Without using a pencil, paper, or calculator, determine the approximate resistance of the network of Figure 6–67.

6.4   Voltage Sources in Parallel

25. Two 20-V batteries are connected in parallel to provide current to a 100-Ω load as shown in Figure 6–68. Determine the current in the load and the current in each battery.

26. Two lead-acid automobile batteries are connected in parallel, as shown in Figure 6–69, to provide additional starting current. One of the batteries is fully charged at 14.2 V and the other battery has discharged to 9 V. If the internal resistance of each battery is 0.01 Ω, determine the current in the batteries. If each battery is intended to provide a maximum current of 150 A, should this method be used to start a car?

Figure 6–68

Figure 6–69

6.5   Current Divider Rule

27. Use the current divider rule to find the currents I_1 and I_2 in the networks of Figure 6–70.

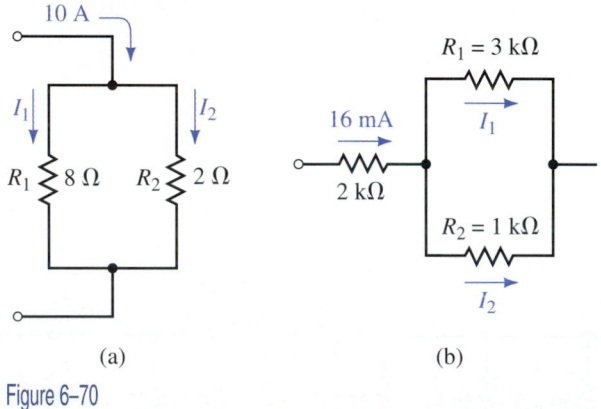

(a)

(b)

Figure 6–70

28. Repeat Problem 27 for the networks of Figure 6–71.

29. Use the current divider rule to determine all unknown currents for the networks of Figure 6–72.

Figure 6–71

Figure 6–72

30. Repeat Problem 29 for the networks of Figure 6–73.

Figure 6–73

31. Use the current divider rule to determine the unknown resistance in the network of Figure 6–74.

32. Use the current divider rule to determine the unknown resistance in the network of Figure 6–75.

Figure 6–74   Figure 6–75

33. Refer to the circuit of Figure 6–76:

Figure 6–76

   a.   Determine the equivalent resistance, R_T, of the circuit.
   b.   Solve for the current I.
   c.   Use the current divider rule to determine the current in each resistor.
   d.   Verify Kirchhoff's current law at node a.

34. Repeat Problem 33 for the circuit of Figure 6–77.

Figure 6–77

240 V, R_1 60 Ω, R_2 100 Ω, R_3 75 Ω

Node a

Figure 6–78

6.6 Analysis of Parallel Circuits

35. Refer to the circuit of Figure 6–78:

a. Find the total resistance, R_T, and solve for the current, I, through the voltage source.

b. Find all of the unknown currents in the circuit.

c. Verify Kirchhoff's current law at node a.

d. Determine the power dissipated by each resistor. Verify that the total power dissipated by the resistors is equal to the power delivered by the voltage source.

36. Repeat Problem 35 for the circuit of Figure 6–79.

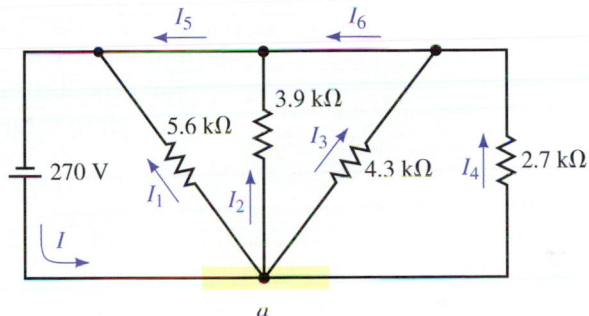

Figure 6–79

37. Refer to the circuit of Figure 6–80:

Figure 6–80

a. Calculate the current through each resistor in the circuit.

b. Determine the total current supplied by the voltage source.

c. Find the power dissipated by each resistor.

38. Refer to the circuit of Figure 6–81.

a. Solve for the indicated currents.

b. Find the power dissipated by each resistor.

c. Verify that the power delivered by the voltage source is equal to the total power dissipated by the resistors.

39. Given the circuit of Figure 6–82:

a. Determine the values of all resistors.

b. Calculate the currents through R_1, R_2, and R_4.

c. Find the currents I_1 and I_2.

d. Find the power dissipated by resistors R_2, R_3, and R_4.

Figure 6–81

Figure 6–82

40. A circuit consists of four resistors connected in parallel and connected to a 20-V source as shown in Figure 6–83. Determine the minimum power rating of each resistor if resistors are available with the following power ratings: ⅛ W, ¼ W, ½ W, 1 W, and 2 W.

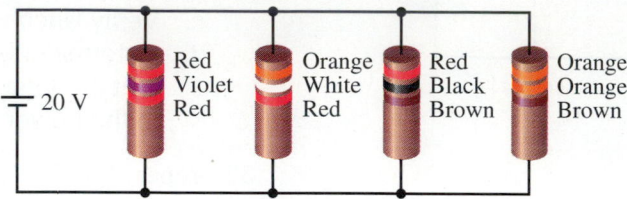

Figure 6–83

41. For the circuit of Figure 6–84, determine each of the indicated currents. If the circuit has a 15-A fuse as shown, is the current enough to cause the fuse to open?

Figure 6–84

42. a. For the circuit of Figure 6–84, calculate the value of R_3 which will result in a circuit current of exactly $I_T = 15$ A.
  b. If the value of R_3 is increased above the value found in part a), what will happen to the circuit current, I_T?

6.7 Ammeter Design

43. A common ammeter uses a meter movement having $I_{fsd} = 50$ μA and $R_m = 5$ kΩ. If the meter has a 10-mA range, determine the value of the shunt resistor needed for this range.

44. Using a meter movement having $I_{fsd} = 50$ μA and $R_m = 2$ kΩ, design an ammeter having a 10-mA range, a 100-mA range, and a 250-mA range. Sketch your design.

45. An ammeter has a 0.5-Ω shunt and a meter movement with $I_{fsd} = 1$ mA and $R_m = 5$ kΩ. Determine the maximum current which can be measured with this ammeter. If the meter deflects to 62% of its full-scale deflection, how much current is being applied to the ammeter?

46. Using a meter movement having $I_{fsd} = 1$ mA and $R_m = 5$ kΩ, design an ammeter having a 5-mA range, a 20-mA range, and a 100-mA range. Sketch your design.

6.8 Voltmeter Loading Effects

47. A voltmeter having a 1-MΩ internal resistance is used to measure the indicated voltage in the circuit shown in Figure 6–85.

    a. Determine the voltage reading which will be indicated by the meter.

    b. Calculate the voltmeter's loading effect when used to measure the indicated voltage.

48. Repeat Problem 47 if the 500-kΩ resistor of Figure 6–85 is replaced with a 2-MΩ resistor.

Figure 6–85   Figure 6–86

49. An inexpensive analog voltmeter is used to measure the voltage across terminals a and b of the circuit shown in Figure 6–86. If the voltmeter indicates that the voltage $V_{ab} = 1.2$ V, what is the actual voltage of the source if the resistance of the meter is 50 kΩ?

50. What would be the reading if a digital meter having an internal resistance of 10 MΩ is used instead of the analog meter of Problem 49?

6.9 Computer Analysis

51. Write an input file for PSpice to solve for the current through each resistor in the circuit of Figure 6–78.

52. Write an input file for PSpice to solve for the current through each resistor in the circuit of Figure 6–79.

Series-Parallel Circuits

CHAPTER PREVIEW

Most circuits encountered in electronics are not simple series circuits nor simple parallel circuits, but rather a combination of the two. Although series-parallel circuits appear to be more complicated than either of the previous types of circuits analyzed to this point, we find that the same principles apply.

This chapter examines how the Kirchhoff's voltage current laws are applied to the analysis of series-parallel circuits. We will also observe that voltage and current divider rules apply to the more complex circuits. In the analysis of series-parallel circuits, we often simplify the given circuit to enable us to more clearly see how the rules and laws of circuit analysis apply. Students are encouraged to redraw circuits whenever the solution of a problem is not immediately apparent. This technique is used by even the most experienced engineers, technologists, and technicians.

In this chapter, we begin by examining simple resistor circuits. The principles of analysis are then applied to more practical circuits such as those containing zener diodes and transistors. The same principles are then applied to determine the loading effects of voltmeters and ammeters in more complex circuits.

After analyzing a complex circuit, we want to know whether the solutions are in fact correct. As you have already seen, electrical circuits usually may be studied in more than one way to arrive at a solution. Once currents and voltages for a circuit have been found, it is very easy to determine whether the resultant solution verifies the law of conservation of energy, Kirchhoff's current law, and Kirchhoff's voltage law. If there is any discrepancy (other than rounding error), there is an error in the calculation!

KEY TERMS

Branch Currents
Parallel Branches
Potentiometer Circuits
Series-Parallel Connection
Transistor Bias
Zener Diode

Primus 2000 system on the Cessna Citation III aircraft. *(Courtesy of Honeywell Inc.)*

OBJECTIVES

After studying this chapter, you will be able to

- find the total resistance of a network consisting of resistors connected in various series-parallel configurations,

- solve for the current through any branch or component of a series-parallel circuit,

- determine the difference in potential between any two points in a series-parallel circuit,

- calculate the voltage drop across a resistor connected to a potentiometer,

- describe how the size of load resistor connected to a potentiometer affects the output voltage,

- calculate the loading effects of using a voltmeter or ammeter to measure the voltage or current in any circuit,

- use PSpice to solve for voltages and currents in series-parallel circuits.

Benjamin Franklin

*B*enjamin Franklin was born in Boston, Massachusetts, in 1706. Although Franklin is best known as a great statesman and diplomat, he also furthered the cause of science with his experiments in electricity. This particularly includes his work with the Leyden jar, which was used to store electric charge. In his famous experiment of 1752, he used a kite to demonstrate that lightning is an electrical event. It was Franklin who postulated that positive and negative electricity are in fact a single "fluid."

Although Franklin's major accomplishments came as a result of his work in achieving independence of the Thirteen Colonies, he was nonetheless a notable scientist.

Benjamin Franklin died in his Philadelphia home on February 12, 1790, at the age of eighty-four.

7.1 The Series-Parallel Network

Figure 7–1

In electric circuits, we define a *branch* as any circuit which can be simplified as having two terminals. The components between the two terminals may be any combination of resistors, voltage sources, or other elements. Many complex circuits may be separated into a combination of both series and/or parallel elements, while other circuits consist of even more elaborate combinations which are neither series nor parallel.

In order to analyze a complicated circuit, it is important to be able to recognize which elements are in series and which elements or branches are in parallel. Consider the network of resistors shown in Figure 7–1.

We immediately recognize that the resistors R_2, R_3, and R_4 are in parallel. This parallel combination is in series with the resistors R_1 and R_5. The total resistance may now be written as follows:

$$R_T = R_1 + (R_2 \| R_3 \| R_4) + R_5$$

■ **EXAMPLE 7–1**   For the network of Figure 7–2, determine which resistors and branches are in series and which are in parallel. Write an expression for the total equivalent resistance, R_T.

Figure 7–2

Solution   First, we recognize that the resistors R_3 and R_4 are in parallel: $(R_3 \| R_4)$.

Next, we see that this combination is in series with the resistor R_2: $[R_2 + (R_3 \| R_4)]$.

Finally, the entire combination is in parallel with the resistor R_1. The total resistance of the circuit may now be written as follows:

$$R_T = R_1 \parallel [R_2 + (R_3 \parallel R_4)]$$

practice
problem
1

For the network of Figure 7–3, determine which resistors and branches are in series and which are in parallel. Write an expression for the total resistance, R_T.

Figure 7–3

7.2 Analysis of Series-Parallel Circuits

Series-parallel networks are often difficult to analyze because they initially appear confusing. However, the analysis of even the most complex circuit is simplified by following some fairly basic steps. By practicing (not memorizing) the techniques outlined in this section, you will find that most circuits can be reduced to groupings of series and parallel combinations. In analyzing such circuits, it is imperative to remember that the rules for analyzing series and parallel elements still apply.

The same current occurs through all series elements.
The same voltage occurs across all parallel elements.

In addition, remember that Kirchhoff's voltage law and Kirchhoff's current law apply for all circuits regardless of whether the circuits are series, parallel, or series-parallel. The following steps will help to simplify the analysis of series-parallel circuits:

1. Whenever necessary, redraw complicated circuits showing the source connection at the left-hand side. All nodes should be labelled to ensure that the new circuit is equivalent to the original circuit. You will find that as you become more experienced at analyzing circuits, this step will no longer be as important and may therefore be omitted.

2. Examine the circuit to determine the strategy which will work best in analyzing the circuit for the required quantities. You will usually find it best to begin the analysis of the circuit at the components most distant to the source.

3. Simplify recognizable combinations of components wherever possible, redrawing the resulting circuit as often as necessary. Keep the same labels for corresponding nodes.

4. Determine the equivalent circuit resistance, R_T.

5. Solve for the total circuit current. Indicate the directions of all currents and label the correct polarities of the voltage drops on all components.

6. Calculate how currents and voltages split between the elements of the circuit.

7. Since there are usually several possible ways at arriving at solutions, verify the answers by using a different approach. The extra time taken in this step will usually ensure that the correct answer has been found.

■ **EXAMPLE 7–2**   Consider the circuit of Figure 7–4.

Figure 7–4

a. Find R_T.
b. Calculate I_1, I_2, and I_3.
c. Determine the voltages V_1 and V_2.

Solution   By examining the circuit of Figure 7–4, we see that resistors R_2 and R_3 are in parallel. This parallel combination is in series with the resistor R_1.

The combination of resistors may be represented by a simple series network shown in Figure 7–5. Notice that the nodes have been labelled using the same notation.

Figure 7–5

a. The total resistance of the circuit may be determined from the combination

$$R_T = R_1 + R_2 \| R_3$$

$$R_T = 12 \text{ k}\Omega + \frac{(10 \text{ k}\Omega)(40 \text{ k}\Omega)}{10 \text{ k}\Omega + 40 \text{ k}\Omega}$$

$$= 12 \text{ k}\Omega + 8 \text{ k}\Omega = 20 \text{ k}\Omega$$

b. From Ohm's law, the total current is

$$I_T = I_1 = \frac{48 \text{ V}}{20 \text{ k}\Omega} = 2.4 \text{ mA}$$

The current I_1 will enter node b and then split between the two resistors R_2 and R_3. This current divider may be simplified as shown in the partial circuit of Figure 7–6.

Figure 7–6

Applying the current divider rule gives

$$I_2 = \frac{(40 \text{ k}\Omega)(2.4 \text{ mA})}{10 \text{ k}\Omega + 40 \text{ k}\Omega} = 1.92 \text{ mA}$$

$$I_3 = \frac{(10 \text{ k}\Omega)(2.4 \text{ mA})}{10 \text{ k}\Omega + 40 \text{ k}\Omega} = 0.48 \text{ A}$$

c. Using the above currents, we determine voltages using Ohm's law:

$$V_1 = (2.4 \text{ mA})(12 \text{ k}\Omega) = 28.8 \text{ V}$$

$$V_3 = (0.48 \text{ mA})(40 \text{ k}\Omega) = 19.2 \text{ V}$$

In order to check the answers, we may simply apply Kirchhoff's voltage law around any closed loop which includes the voltage source:

$$\sum V = E - V_1 - V_3$$

$$= 48 \text{ V} - 28.8 \text{ V} - 19.2 \text{ V}$$

$$= 0 \text{ V (checks!)}$$

The solution may be verified by ensuring that the power delivered by the voltage source is equal to the summation of powers dissipated by the resistors.

in process
l e a r n i n g
check I

Use the results of Example 7–2 to verify that the law of conservation of energy applies to the circuit of Figure 7–4 by showing that the voltage source delivers the same power as the total power dissipated by all resistors.

■ **EXAMPLE 7–3** Find the voltage V_{ab} for the circuit of Figure 7–7.

Figure 7–7

Solution We begin by redrawing the circuit in a more simple representation as shown in Figure 7–8.

Figure 7–8

From Figure 7–8, we see that the original circuit consists of two parallel branches, where each branch is a series combination of two resistors.

If we take a moment to examine the circuit, we see that the voltage V_{ab} may be determined from the combination of voltages across R_1 and R_2. Alternatively, the voltage may be found from the combination of voltages across R_3 and R_4.

As usual, several methods of analysis are possible. Because the two branches are in parallel, the voltage across each branch must be 40 V. Using the voltage divider rules allows us to quickly calculate the voltage across each resistor. Although equally correct, other methods of calculating the voltages would be more lengthy.

$$V_2 = \frac{R_2}{R_2 + R_3} E$$

$$= \left(\frac{50\ \Omega}{50\ \Omega + 200\ \Omega} \right)(40\ V) = 8.0\ V$$

$$V_1 = \frac{R_1}{R_1 + R_4} E$$

$$= \left(\frac{100\ \Omega}{100\ \Omega + 300\ \Omega} \right)(40\ V) = 10.0\ V$$

As shown in Figure 7–9, we apply Kirchhoff's voltage law to determine the voltage between terminals a and b.

Figure 7–9

$$V_{ab} = -10.0\ V + 8.0\ V = -2.0\ V$$

■ **EXAMPLE 7–4**   Consider the circuit of Figure 7–10:

Figure 7–10

a. Find the total resistance R_T "seen" by the source E.

b. Calculate I_T, I_1, and I_2.

c. Determine the voltages V_2 and V_4.

Solution   We begin the analysis by redrawing the circuit. Since we generally like to see the source on the left-hand side, one possible way of redrawing the resultant circuit is shown in Figure 7–11. Notice that the polarities of voltages across all resistors have been shown.

$$R'_T = (4 \text{ k}\Omega + 6 \text{ k}\Omega) \parallel (15 \text{ k}\Omega)$$
$$= 6 \text{ k}\Omega$$

Figure 7–11

a. From the redrawn circuit, we see that the total resistance of the circuit is

$$R_T = R_3 + [(R_1 + R_2) \| R_4]$$

$$= 3 \text{ k}\Omega + \frac{(4 \text{ k}\Omega + 6 \text{ k}\Omega)(15 \text{ k}\Omega)}{(4 \text{ k}\Omega + 6 \text{ k}\Omega) + 15 \text{ k}\Omega}$$

$$= 3 \text{ k}\Omega + 6 \text{ k}\Omega = 9 \text{ k}\Omega$$

b. The current supplied by the voltage source is

$$I_T = \frac{E}{R_T} = \frac{45 \text{ V}}{9 \text{ k}\Omega} = 5 \text{ mA}$$

We see that the supply current divides between the parallel branches as shown in Figure 7–12.

Figure 7–12

Applying the current divider rule, we calculate the branch currents as

$$I_1 = I_T \frac{R'_T}{(R_1 + R_2)} = \frac{(5 \text{ mA})(6 \text{ k}\Omega)}{4 \text{ k}\Omega + 6 \text{ k}\Omega} = 3 \text{ mA}$$

$$I_2 = I_T \frac{R'_T}{R_4} = \frac{(5 \text{ mA})(6 \text{ k}\Omega)}{15 \text{ k}\Omega} = 2 \text{ mA}$$

Notice: When determining the branch currents, the resistance R'_T is used in the calculations rather the the total circuit resistance. This is because the current $I_T = 5$ mA splits between the two branches of R'_T and the split is not affected by the value of R_3.

   c. The voltages V_2 and V_4 are now easily calculated by using Ohm's law:

$$V_2 = I_1 R_2 = (3 \text{ mA})(4 \text{ k}\Omega) = 12 \text{ V}$$

$$V_4 = I_2 R_4 = (2 \text{ mA})(15 \text{ k}\Omega) = 30 \text{ V}$$

■ **EXAMPLE 7–5** For the circuit of Figure 7–13, find the indicated currents and voltages.

Figure 7–13

Solution Because the above circuit contains voltage point sources, it is necessary to redraw the circuit to help visualize the operation.

The point sources are voltages with respect to ground, and so we begin by drawing a circuit with the reference point as shown in Figure 7–14.

Figure 7–14

Now, we can see that the circuit may be further simplified by combining the voltage sources ($E = E_1 + E_2$) and by showing the resistors in a more suitable location. The simplified circuit is shown in Figure 7–15.

Figure 7–15

The total resistance "seen" by the equivalent voltage source is

$$R_T = R_1 + [R_4 \| (R_2 + R_3)]$$

$$= 10\ \Omega + \frac{(30\ \Omega)(10\ \Omega + 50\ \Omega)}{30\ \Omega + (10\ \Omega + 50\ \Omega)} = 30\ \Omega$$

And so the total current provided into the circuit is

$$I_1 = \frac{E}{R_T} = \frac{18\ \text{V}}{30\ \Omega} = 600\ \text{mA}$$

At node b this current divides between the two branches as follows:

$$I_3 = \frac{(R_2 + R_3)I_1}{R_4 + R_2 + R_3} = \frac{(60\ \Omega)(600\ \text{mA})}{30\ \Omega + 10\ \Omega + 50\ \Omega} = 400\ \text{mA}$$

$$I_2 = \frac{R_4 I_1}{R_4 + R_2 + R_3} = \frac{(30\ \Omega)(600\ \text{mA})}{30\ \Omega + 10\ \Omega + 50\ \Omega} = 200\ \text{mA}$$

The voltage V_{ab} has the same magnitude as the voltage across the resistor R_2, but with a negative polarity (since b is at a higher potential than a):

$$V_{ab} = -I_2 R_2 = -(200\ \text{mA})(10\ \Omega) = -2.0\ \text{V}$$

practice
problem
___*II*___

Consider the circuit of Figure 7–16:

Figure 7–16

a. Find the total circuit resistance, R_T.
b. Determine the current I_T through the voltage sources.
c. Solve for the currents I_1 and I_2.
d. Calculate the voltage V_{ab}.

7.3 Applications of Series-Parallel Circuits

We now examine how the methods developed in the first two sections of this chapter are applied when analyzing practical circuits. You may find that some of the circuits introduce you to unfamiliar devices. For now, you do not need to know precisely how these devices operate, simply that the voltages and currents in the circuits follow the same rules and laws that you have used up to now.

■ **EXAMPLE 7–6**   The circuit of Figure 7–17 is referred to as a *bridge circuit* and is used extensively in electronic and scientific instruments.

Figure 7–17

Calculate the current I and the voltage V_{ab} when

a. $R_x = 0 \ \Omega$ (short circuit),
b. $R_x = 15 \ k\Omega$,
c. $R_x = \infty$ (open circuit).

Solution

a. $R_x = 0 \ \Omega$:

The circuit is redrawn as shown in Figure 7–18.

Figure 7–18

The voltage source "sees" a total resistance of

$$R_T = (R_1 + R_3) \| R_2 = 250 \ \Omega \| 5 \ k\Omega = 238 \ \Omega$$

resulting in a source current of

$$I = \frac{10 \ V}{238 \ \Omega} = 42.0 \ mA$$

The voltage V_{ab} may be determined by voltage division between R_1 and R_2.

By examining the circuit, we see that the voltage across R_1 will be constant regardless of the value of the variable resistor R_x. Hence

$$V_1 = \left(\frac{(50 \ \Omega}{50 \ \Omega + 200 \ \Omega)} \right)(10 \ V) = 2 \ V$$

Now, since the variable resistor is a short circuit, the entire source voltage will appear across the resistor R_2, giving

$$V_2 = 10.0 \ V$$

And so

$$V_{ab} = -V_1 + V_2 = -2.0 \ V + 10.0 \ V = +8.0 \ V$$

b. $R_x = 15 \ k\Omega$:

The circuit is redrawn in Figure 7–19.

Figure 7–19

The voltage source "sees" a circuit resistance of

$$R_T = (R_1 + R_3)\|(R_2 + R_x)$$
$$= 250\ \Omega\|20\ \text{k}\Omega = 247\ \Omega$$

which results in a source current of

$$I = \frac{10\ \text{V}}{247\Omega} = 40.5\ \text{mA}$$

The voltages across R_1 and R_2 are

$$V_1 = 2\ \text{V}\quad\text{(as before)}$$

$$V_2 = \frac{R_2}{R_2 + R_x}E$$

$$= \left(\frac{5\ \text{k}\Omega}{5\ \text{k}\Omega + 15\ \text{k}\Omega}\right)(10\ \text{V}) = 2.5\ \text{V}$$

Now the voltage between terminals a and b is found as

$$V_{ab} = -V_1 + V_2$$
$$= -2.0\ \text{V} + 2.5\ \text{V} = +0.5\ \text{V}$$

c. $R_x = \infty$:

The circuit is redrawn in Figure 7–20.

Figure 7–20

Because the second branch is an open circuit due to the resistor R_x, the total resistance "seen" by the source is

$$R_T = R_1 + R_3 = 250\ \Omega$$

resulting in a source current of

$$I = \frac{10\ \text{V}}{250\ \Omega} = 40.0\ \text{mA}$$

The voltages across R_1 and R_2 are

$$V_1 = 2\ \text{V} \quad \text{(as before)}$$

$$V_2 = 0\ \text{V} \quad \text{(since the branch is open)}$$

And so the resulting voltage between terminals a and b is

$$V_{ab} = -V_1 + V_2$$

$$= -2.0\ \text{V} + 0\ \text{V} = -2.0\ \text{V}$$

The previous example illustrates how voltages and currents within a circuit are affected by changes elsewhere in the circuit. In the example, we saw that the voltage V_{ab} varied from −2 V to +8 V, while the total circuit current varied from a minimum value of 40 mA to a maximum value of 42 mA. These changes occurred even though the resistor R_x varied from 0 Ω to ∞.

A *transistor* is a three-terminal device which may be used to amplify small signals. In order for the transistor to operate as an amplifier, however, certain dc conditions must be met. These conditions set the "bias point" of the transistor. The bias current of a transistor circuit is determined by a dc voltage source and several resistors. Although the operation of the transistor is outside the scope of this textbook, we can analyze the bias circuit of a transistor using elementary circuit theory.

■ **EXAMPLE 7–7**　Use the given conditions to determine I_C, I_E, V_{CE}, and V_B for the transistor circuit of Figure 7–21.

Figure 7–21

Solution   In order to simplify the work, the circuit of Figure 7–21 is separated into two circuits: one circuit containing the known voltage V_{BE} and the other containing the unknown voltage V_{CE}.

Since we always start will the given information, we redraw the circuit containing the known voltage V_{BE} as illustrated in Figure 7–22.

Figure 7–22

Although the circuit of Figure 7–22 initially appears to be a series circuit, we see that this cannot be the case, since we are given that $I_E \cong I_C = 100I_B$. We know that the current everywhere in a series circuit must be the same. However, Kirchhoff's voltage law may still be applied around the closed loop, resulting in the following:

$$V_{BB} = R_B I_B + V_{BE} + R_E I_E$$

The above expression contains two unknowns. From the given information we have the current $I_E = 100I_B$. This allows the expression to be rewritten as

$$V_{BB} = R_B I_B + V_{BE} + R_E (100I_B)$$

Solving for the unknown current I_B, we have

$$5.0 \text{ V} = (200 \text{ k}\Omega)I_B + 0.7 \text{ V} + (1 \text{ k}\Omega)(100I_B)$$

$$(300 \text{ k}\Omega)I_B = 5.0 \text{ V} - 0.7 \text{ V} = 4.3 \text{ V}$$

$$I_B = \frac{4.3 \text{ V}}{300 \text{ k}\Omega} = 14.3 \text{ }\mu\text{A}$$

The current $I_E \cong I_C = 100I_B = 1.43$ mA.

As mentioned previously, the circuit can be redrawn as two separate circuits. The circuit containing the unknown voltage V_{CE} is illustrated in Figure 7–23. Notice that the resistor R_E appears in both Figure 7–22 and Figure 7–23.

Figure 7–23

Applying Kirchhoff's voltage law around the closed loop of Figure 7–23, we have the following:

$$V_{R_C} + V_{CE} + V_{R_E} = V_{CC}$$

The voltage V_{CE} is found as

$$
\begin{aligned}
V_{CE} &= V_{CC} - V_{R_C} - V_{R_E} \\
&= V_{CC} - R_C I_C - R_E I_E \\
&= 20.0\ \text{V} - (4\ \text{k}\Omega)(1.43\ \text{mA}) - (1\ \text{k}\Omega)(1.43\ \text{mA}) \\
&= 20.0\ \text{V} - 5.73\ \text{V} - 1.43\ \text{V} = 12.8\ \text{V}
\end{aligned}
$$

Finally, applying Kirchhoff's voltage law from B to ground, we have

$$
\begin{aligned}
V_B &= V_{BE} + V_{R_E} \\
&= 0.7\ \text{V} + 1.43\ \text{V} \\
&= 2.13\ \text{V}
\end{aligned}
$$

practice problem III

Use the given information to find V_G, I_D, and V_{DS} for the circuit of Figure 7–24.

Figure 7–24

A *zener diode* is a two-terminal device similar to a varistor (refer to Chapter 3). When the voltage across the zener diode attempts to go above the rated voltage for the device, the zener diode provides a low-resistance path for the extra current. Due to this action, a relatively constant voltage, V_Z, is maintained across the zener diode. This characteristic is referred to as *voltage regulation* and has many applications in electronic and electrical circuits. Once again, although the theory of operation of the zener diode is outside the scope of this textbook, we are able to apply simple circuit theory to examine how the circuit operates.

■ **EXAMPLE 7–8**   For the circuit of Figure 7–25, calculate the indicated quantities.

Figure 7–25

Solution   If we take a moment to examine the circuit, we see that the zener diode is placed in parallel with the resistor R_2. This parallel combination is in series with the resistor R_1 and the voltage source, E.

In order for the zener diode to operate, the voltage across the diode would have to be above the zener voltage without the diode present. If we remove the zener diode the circuit would appear as shown in Figure 7–26.

Figure 7–26

From Figure 7–26, we may determine the voltage V_2 which would be present without the zener diode in the circuit. Since the circuit is a simple series circuit, the voltage divider rule may be used to determine V_2:

$$V_2 = \frac{R_2}{R_1 + R_2} E = \left(\frac{10 \text{ k}\Omega}{5 \text{ k}\Omega + 10 \text{ k}\Omega} \right)(15 \text{ V}) = 10 \text{ V}$$

When the zener diode is placed across the resistor R_2 the device will operate to limit the voltage to $V_Z = 5$ V.

Because the zener diode is operating, the voltage across both the diode and the resistor R_2 must be the same, namely 5 V. Because the parallel combination of D_Z and R_2 is in series with the resistor R_1, the voltage across R_1 is easily determined from Kirchhoff's voltage law as

$$V_1 = E - V_Z = 15\text{ V} - 5\text{V} = 10\text{ V}$$

Now, from Ohm's law, the currents I_1 and I_2 are easily found to be

$$I_2 = \frac{V_2}{R_2} = \frac{5\text{ V}}{10\text{ k}\Omega} = 0.5\text{ mA}$$

$$I_1 = \frac{V_1}{R_1} = \frac{10\text{ V}}{5\text{ k}\Omega} = 2.0\text{ mA}$$

And applying Kirchoff's current law at node a, we get the zener diode current as

$$I_Z = I_1 - I_2 = 2.0\text{ mA} - 0.5\text{ mA} = 1.5\text{ mA}$$

Finally, the power dissipated by the zener diode must be

$$P_Z = V_Z I_Z = (5\text{ V})(1.5\text{ mA}) = 7.5\text{ mW}$$

in process
learning
check II

Use the results of Example 7–8 to show that the power delivered to the circuit, by the voltage source of Figure 7–25, is equal to the total power dissipated by the resistors and the zener diode.

practice
problems
IV

1. Determine I_1, I_Z, and I_2 for the circuit of Figure 7–25 if the resistor R_1 is increased to 10 kΩ.

2. Repeat Problem 1 if R_1 is increased to 30 kΩ.

7.4 Potentiometers

As mentioned in Chapter 3, variable resistors may be used as potentiometers as shown in Figure 7–27 to control voltages within circuits.

Figure 7–27

An example of a variable resistor used as a potentiometer is the volume control on a receiver or amplifier. When the movable terminal is at the uppermost position, the voltage appearing between terminals b and c is simply calculated by using the voltage divider rule as

$$V_{bc} = \left(\frac{50 \text{ k}\Omega}{50 \text{ k}\Omega + 50 \text{ k}\Omega} \right)(120 \text{ V}) = 60 \text{ V}$$

Alternatively, when the movable terminal is at the lowermost position, the voltage between terminals b and c is $V_{bc} = 0$ V, since the two terminals are effectively shorted and the voltage across a short circuit is always zero.

The circuit of Figure 7–27 represents a potentiometer having an output voltage which is adjustable between 0 and 60 V. This output is referred to as the *nonloaded output*, since there is no load resistance connected between the terminals b and c. If a load resistance were connected between these terminals, the output voltage, called the *loaded output*, will no longer be the same. The following example is an illustration of circuit loading.

■ **EXAMPLE 7–9**   For the circuit of Figure 7–28, determine the range of the voltage V_{bc} as the potentiometer varies between its minimum and maximum values.

Figure 7–28

Solution   The minimum voltage between terminals b and c will occur when the movable contact is at the lowermost contact of the variable resistor. In this position, the voltage $V_{bc} = 0$ V, since the terminals b and c are shorted.

The maximum voltage V_{bc} occurs when the movable contact is at the uppermost contact of the variable resistor. In this position, the circuit may be represented as shown in Figure 7–29.

Figure 7–29

In Figure 7–29, we see that the resistance R_2 is in parallel with the load resistor R_L. The voltage between terminals b and c is easily determined from the voltage divider rule, as follows:

$$V_{bc} = \frac{R_2 \| R_L}{(R_2 \| R_L) + R_1} E$$

$$= \left(\frac{25 \text{ k}\Omega}{25 \text{ k}\Omega + 50 \text{ k}\Omega} \right)(120 \text{ V}) = 40 \text{ V}$$

We conclude that the voltage at the output of the potentiometer is adjustable from 0 V to 40 V for a load resistance of $R_L = 50$ kΩ.

practice
problem
V

Refer to the circuit of Figure 7–28.

a. Determine the output voltage range of the potentiometer if the load resistor is $R_L = 5$ kΩ.

b. Repeat a) if the load resistor is $R_L = 500$ kΩ.

c. What conclusion may be made about the output voltage of a potentiometer when the load resistance is large in comparison with the potentiometer resistance?

in process
learning
check III

A 20-kΩ potentiometer is connected across a voltage source with a 2-kΩ load resistor connected between the wiper (center terminal) and the negative terminal of the voltage source.

a. What percentage of the source voltage will appear across the load when the wiper is one fourth of the way from the bottom?

b. Determine the percentage of the source voltage which appears across the load when the wiper is one half and three fourths of the way from the bottom.

c. Repeat the calculations of a) and b) for a load resistor of 200 kΩ.

d. From the above results, what conclusion can you make about the effect of placing a load across a potentiometer?

7.5 Loading Effects of Instruments

In Chapters 5 and 6 we examined how ammeters and voltmeters affect the operation of simple series circuits. The degree to which the circuits are affected is called the *loading effect* of the instrument. Recall that, in order for an instrument to provide an accurate indication of how a circuit operates, the loading effect should ideally be zero. In practice, it is impossible for any instrument to have zero loading effect, since all instruments absorb some energy from the circuit under test, thereby affecting circuit operation.

In this section, we will determine how instrument loading affects more complex circuits.

■ **EXAMPLE 7–10**   Calculate the loading effects if a digital multimeter, having an internal resistance of 10 MΩ, is used to measure V_1 and V_2 in the circuit of Figure 7–30.

Figure 7–30

Solution   In order to determine the loading effect for a particular reading, we need to calculate both the non-loaded voltage and the loaded voltage.

For the circuit given in Figure 7–30, the nonloaded voltages across each resistor is

$$V_1 = \left(\frac{5\ \text{M}\Omega}{5\ \text{M}\Omega + 10\ \text{M}\Omega}\right)(27\ \text{V}) = 9.0\ \text{V}$$

$$V_2 = \left(\frac{10\ \text{M}\Omega}{5\ \text{M}\Omega + 10\ \text{M}\Omega}\right)(27\ \text{V}) = 18.0\ \text{V}$$

When the voltmeter is used to measure V_1, the result is equivalent to connecting a 10-MΩ resistor across resistor R_1, as shown in Figure 7–31.

Figure 7–31

The voltage appearing across the parallel combination of R_1 and resistance of the voltmeter is calculated as

$$V_1 = \left(\frac{5 \text{ M}\Omega \| 10 \text{ M}\Omega}{(5 \text{ M}\Omega \| 10 \text{ M}\Omega) + 10 \text{ M}\Omega} \right)(27 \text{ V})$$

$$= \left(\frac{3.33 \text{ M}\Omega}{13.3 \text{ M}\Omega} \right)(27 \text{ V})$$

$$= 6.75 \text{ V}$$

Notice that the measured voltage is less than the 9 V that we had expected to measure.

With the voltmeter connected across resistor R_2, the circuit appears as shown in Figure 7–32.

Figure 7–32

The voltage appearing across the parallel combination of R_2 and resistance of the voltmeter is calculated as

$$V_2 = \left(\frac{10 \text{ M}\Omega \| 10 \text{ M}\Omega}{5 \text{ M}\Omega + (10 \text{ M}\Omega \| 10 \text{ M}\Omega)} \right)(27 \text{ V})$$

$$= \left(\frac{5.0 \text{ M}\Omega}{10.0 \text{ M}\Omega} \right)(27 \text{ V})$$

$$= 13.5 \text{ V}$$

Again, we notice that the measured voltage is less than the 18 V that we had expected.

Now the loading effects are calculated as follows.

When measuring V_1:

$$\text{loading effect} = \frac{9.0\text{ V} - 6.75\text{ V}}{9.0\text{ V}} \times 100\%$$

$$= 25\%$$

When measuring V_2:

$$\text{loading effect} = \frac{18.0\text{ V} - 13.5\text{ V}}{18.0\text{ V}} \times 100\%$$

$$= 25\%$$

practice problem VI

Calculate the loading effects if an analog voltmeter, having an internal resistance of 200 kΩ, is used to measure V_1 and V_2 in the circuit of Figure 7–30.

■ **EXAMPLE 7–11**   For the circuit of Figure 7–33, calculate the loading effect if a 5.00-Ω ammeter is used to measure the currents I_T, I_1, and I_2.

Figure 7–33

Solution   We begin by determining the nonloaded currents in the circuit. Using Ohm's law, we solve for the currents I_1 and I_2:

$$I_1 = \frac{100\text{ mV}}{25\ \Omega} = 4.0\text{ mA}$$

$$I_2 = \frac{100\text{ mV}}{5\ \Omega} = 20.0\text{ mA}$$

Now, by Kirchhoff's current law,

$$I_T = 4.0\text{ mA} + 20.0\text{ mA} = 24.0\text{ mA}$$

If we were to insert the ammeter into the branch with resistor R_1, the circuit would appear as shown in Figure 7–34.

Figure 7-34

The current through the ammeter would be

$$I_1 = \frac{100 \text{ mV}}{25 \text{ } \Omega + 5 \text{ } \Omega} = 3.33 \text{ mA}$$

If we were to insert the ammeter into the branch with resistor R_2, the circuit would appear as shown in Figure 7–35.

Figure 7-35

The current through the ammeter would be

$$I_2 = \frac{100 \text{ mV}}{5\,\Omega + 5\,\Omega} = 10.0 \text{ mA}$$

If the ammeter were inserted into the circuit to measure the current I_T, the equivalent circuit would appear as shown in Figure 7–36.

Figure 7–36

The total resistance of the circuit would be

$$R_T = 5\,\Omega + 25\,\Omega \| 5\,\Omega = 9.17\,\Omega$$

This would result in a current I_T determined by Ohm's law as

$$I_T = \frac{100 \text{ mV}}{9.1\overline{6}\,\Omega}$$

$$= 10.9 \text{ mA}$$

The loading effects for the various current measurements are as follows:

When measuring I_1,

$$\text{loading effect} = \frac{4.0 \text{ mA} - 3.33 \text{ mA}}{4.0 \text{ mA}} \times 100\%$$

$$= 16.7\%$$

When measuring I_2,

$$\text{loading effect} = \frac{20 \text{ mA} - 10 \text{ mA}}{20 \text{ mA}} \times 100\%$$

$$= 50\%$$

When measuring I_T,

$$\text{loading effect} = \frac{24 \text{ mA} - 10.9 \text{ mA}}{24 \text{ mA}} \times 100\%$$

$$= 54.5\%$$

Notice that the loading effect for an ammeter is most pronounced when it is used to measure current in a branch having a resistance on the same order of magnitude as the meter.

You will also notice that if this ammeter were used in a circuit to verify the correctness of Kirchhoff's current law, the loading effect of the meter would produce an apparent contradiction. From KCL,

$$I_T = I_1 + I_2$$

By substituting the measured values of current into the above equation, we have

$$10.91 \text{ mA} = 3.33 \text{ mA} + 10.0 \text{ mA}$$

$$10.91 \text{ mA} \neq 13.33 \text{ mA (contradiction)}$$

This example illustrates that the loading effect of a meter may severely affect the current in a circuit, giving results which seem to contradict the laws of circuit theory. Therefore, wherever an instrument is used to measure a particular quantity, we must always take into account the limitations of the instrument and question the validity of a resulting reading.

practice
problem
VII

Calculate the readings and the loading error if an ammeter having an internal resistance of 1 Ω is used to measure the currents in the circuit of Figure 7–33.

7.6 Circuit Analysis Using Computers

The analysis of series-parallel circuits using PSpice is very similar to the methods used to analyze series and parallel circuits in previous chapters. The following example illustrates how PSpice may be used to determine the required quantities in a given circuit. The example analyzes the same circuit as that used in Example 7–4.

■ EXAMPLE 7–12

Figure 7–37

Given the circuit of Figure 7–37, use PSpice to find the following quantities:

    a. Voltages V_2 and V_4.
    b. Currents I_T, I_1, and I_2.
    c. Total resistance, R_T.

Solution   The nodes for the given circuit may be labelled as shown in Figure 7–38. You will notice that the voltage source has been renamed from E to V_s to correspond to the notation required by the PSpice program.

Figure 7–38

The voltages V_2 and V_4 in the circuit of Figure 7–37 correspond respectively to the nodal voltages V(3,2) and V(3,0) in PSpice notation. The input file for the above circuit is given below:

```
Input File:
Example 7-12: Series/parallel Circuit, Figure 7-38
Vs     1     0      45V
R1     2     0      6kohms
R2     2     3      4kohms
R3     3     1      3kohms
R4     3     0      15kohms
.DC   Vs    45V    45V   1V
.PRINT DC V(3,2) V(3,0) I(R1) I(R4) I(Vs)
.OPTIONS NOPAGE
.END
```

The input file provides for the analysis using an incremented dc voltage source (Vs). The required voltages and currents are calculated and then printed in the output file. The results of the analysis may be viewed from within the in the Control Shell by using the Browse menu. The pertinent data from the output file for the given circuit is shown next.

```
Output File:
Vs          V(3,2)      V(3,0)      I(R1)       I(R4)
I(Vs)
4.500E+01 1.200E+01 3.000E+01 3.000E-03 2.000E-03 -
5.000E-03
```

From the PSpice output file, we see that the circuit has the following currents and voltages:

a. $V_2 = 12$ V and $V_4 = 30$ V.

b. $I_T = 5.0$ mA, $I_1 = 3.0$ mA, and $I_2 = 2.0$ mA.

These results are consistent with the results obtained in Example 7–4.

c. In order to determine the total resistance external to the 45-V supply, we use the results obtained from PSpice as follows:

$$R_T = \frac{E}{I_T} = \frac{45 \text{ V}}{5 \text{ mA}} = 9 \text{ k}\Omega$$

As expected, this result is consistent with the result obtained in Example 7–4. If we had wanted to determine the power dissipated by a particular resistor (or delivered by the voltage source), we simply would have used any of the power equations and performed the necessary calculation.

Very often we would like to know how a circuit will behave when a particular component value is allowed to change. PSpice is particularily useful in solving such circuits, since it eliminates the time-consuming process of solving the same circuit under several different parameters. In order to allow for a variation in a given component, we will need to use the .MODEL and .STEP statements. The syntax for these statements is outlined below:

`.MODEL    mname    mtype    [(par = value)]`
mname is the assigned name of the component to be modelled.
mtype is the type of model used. Some of the models available are RES(resistors), IND(inductors), and CAP(capacitors).
[(par = value)] is an optionally assigned value. The default values for resistors, inductors, and capacitors are (R = 1), (L = 1), and (C = 1).

`.STEP    [type]    name    start    end    incr`
[type] is an optional specification which allows the user to specify the type of sweep. The sweep types which are available are LIN for a linear variation, OCT for a variation in octaves, and DEC for a variation in decades. The default type is LIN.
name is the name of the element to be varied. When used to step through resistors the name must correspond to *mname* used in the .MODEL statement (e.g., RES)

start is the value at which the variation is to start. The value must be a finite, nonzero value.

end is the value at which the variation is to stop. The end value may be smaller than the start value but must also be finite and nonzero.

incr is the step size for linear variation and the number of values in either each octave or decade.

■ **EXAMPLE 7–13**   Write the PSpice input file for the circuit of Figure 7–39. The output shall display the source current and the voltage V_{ab} as the resistor as R_x is varied in 5-kΩ increments between 5 kΩ and 50 kΩ. Compare the results to those obtained when the circuit was analyzed in Example 7–6.

Figure 7–39

Solution   The nodes and the voltage source are labelled as shown in Figure 7–40. The voltage V_{ab} is the same as the voltage between terminals 2 and 3, V(2,3). The input file for the circuit is given as follows.

Figure 7–40

```
Input file:
Example 7-13: Bridge circuit, Figure 7-39
Vs    1    0    10V
R1    1    2    50ohms
```

```
R2      1     3      5kohms
R3      2     0      200ohms
Rx      3     0      Rvar        1
.MODEL       Rvar  RES(R=1)
.STEP        RES   Rvar(R)  5kohms     50kohms     5kohms
.DC    Vs    10V   10V      1V
.PRINT       DC    I(Vs)    V(2,3)
.OPTIONS NOPAGE
.END
```

After PSpice is run, the output file may be viewed using the Browse menu item in the control shell or by printing the output file. The actual output file contains a large amount of information relating to the analysis of the circuit. However, since we are only interested in a single voltage and current, the following edited version of the output file is shown.

```
Output file:
RVAR R                    Vs              I(Vs)        V(2,3)
  5.0000E+03            1.000E+01      -4.100E-02    3.000E+00
 10.0000E+03            1.000E+01      -4.067E-02    1.333E+00
 15.0000E+03            1.000E+01      -4.050E-02    5.000E-01
 20.0000E+03            1.000E+01      -4.040E-02    0.000E+00
 25.0000E+03            1.000E+01      -4.033E-02   -3.333E-01
 30.0000E+03            1.000E+01      -4.029E-02   -5.714E-01
 35.0000E+03            1.000E+01      -4.025E-02   -7.500E-01
 40.0000E+03            1.000E+01      -4.022E-02   -8.889E-01
 45.0000E+03            1.000E+01      -4.020E-02   -1.000E+00
 50.0000E+03            1.000E+01      -4.018E-02   -1.091E+00
```

Although we are not able to determine the results when $R_x = 0$ nor when $R_x = \infty$, we see that when $R_x = 15$ kΩ, the current through the source is $I = 40.5$ mA and the voltage $V_{ab} = 0.50$ V. These results are consistent with those obtained in Example 7–6.

practice problem VIII

Write the PSpice input file for the circuit of Figure 7–41. The output shall display the source current, I, the load current, I_L, and the voltage V_{ab} as the resistor R_L is varied in 100-Ω increments between 100 Ω and 1000 Ω.

Figure 7–41

Problems

7.1   The Series-Parallel Network

1. For the network of Figure 7–42, determine which resistors and branches are in series and which are in parallel. Write an expression for the total resistance, R_T.

[handwritten:] $R_T \{ R_4 \| (R_2 + R_3) \} + R_1 + R_5$

[handwritten:] $(R_1 \| R_2) + R_3 \| (R_4 \cdots$

(a)   (b)

Figure 7–42

2. For each of the networks of Figure 7–43, write an expression for the total resistance, R_T.

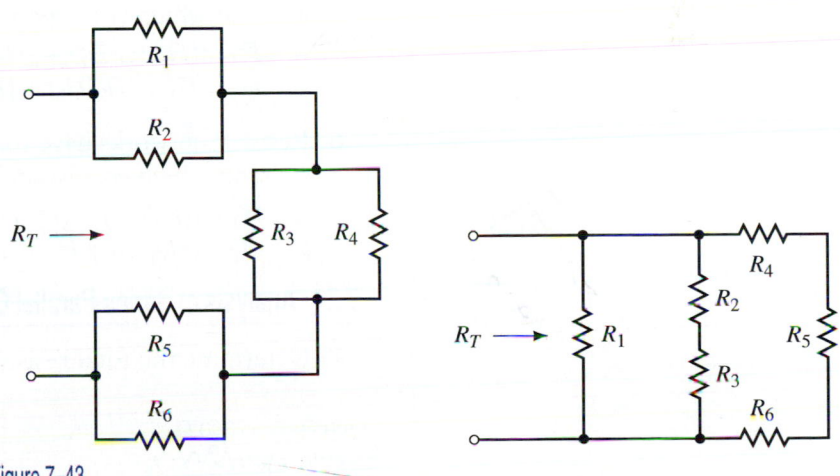

Figure 7–43

3. Write an expression for both R_{T_1} and R_{T_2} for the networks of Figure 7–44.

Figure 7–44

4. Write an expression for both R_{T_1} and R_{T_2} for the networks of Figure 7–45.

Figure 7–45

5. Resistor networks have total resistances as given below. Sketch a circuit which corresponds to each expression.
   a. $R_T = (R_1\|R_2\|R_3) + (R_4\|R_5)$.
   b. $R_T = R_1 + (R_2\|R_3) + [R_4\|(R_5 + R_6)]$.

6. Resistor networks have total resistances as given below. Sketch a circuit which corresponds to each expression.
   a. $R_T = [(R_1\|R_2) + (R_3\|R_4)]\|R_5$.
   b. $R_T = (R_1\|R_2) + R_3 + [(R_4 + R_5)\|R_6]$.

7.2   Analysis of Series-Parallel Circuits

7. Determine the total resistance of each network in Figure 7–46.

Figure 7–46

$R_1 \left(R_2 \| R_3 \| R_5 \right)$

$R_5 \| R_3 \| R_2$

r_1

r_2

8. Determine the total resistance of each network in Figure 7–47.

All resistors are 1 kΩ.

All resistors are 1 kΩ.

$R_T \longrightarrow$

$R_T \longrightarrow$

(a)   (b)

Figure 7–47

9. Calculate the resistances R_{ab} and R_{cd} in the circuit of Figure 7–48.

Figure 7–48

10. Calculate the resistances R_{ab}, R_{bc}, and R_{ac} in the circuit of Figure 7–49.

Figure 7–49

Figure 7-50

11. Refer to the circuit of Figure 7–50:

Find the following quantities:
 a. R_T.
 b. I_T, I_1, I_2, I_3, I_4.
 c. V_{ab}, V_{bc}.

12. Refer to the circuit of Figure 7–51:

Figure 7–51

Find the following quantities:
 a. R_T (equivalent resistance "seen" by the voltage source).
 b. I_T, I_1, I_2, I_3, I_4.
 c. V_{ab}, V_{bc}, V_{cd}.

13. Refer to the circuit of Figure 7–52:

Figure 7–52

 a. Find the currents I_1, I_2, I_3, I_4, I_5, and I_6.
 b. Solve for the voltages V_{ab} and V_{cd}.
 c. Verify that the power delivered to the circuit is equal to the summation of powers dissipated by the resistors.

14. Refer to the circuit of Figure 7–53:
 a. Find the currents I_1, I_2, I_3, I_4, and I_5.
 b. Solve for the voltages V_{ab} and V_{bc}.
 c. Verify that the power delivered to the circuit is equal to the summation of powers dissipated by the resistors.

Figure 7–53

15. Refer to the circuits of Figure 7–54:

(a)   (b)

Figure 7–54

a. Find the indicated currents.
b. Solve for the voltage V_{ab}.
c. Verify that the power delivered to the circuit is equal to the summation of powers dissipated by the resistors.

16. Refer to the circuit of Figure 7–55:

Figure 7–55

a. Solve for the currents I_1, I_2, and I_3 when $R_x = 0\ \Omega$ and when $R_x = 5\ \text{k}\Omega$.
b. Calculate the voltage V_{ab} when $R_x = 0\ \Omega$ and when $R_x = 5\ \text{k}\Omega$.

Figure 7–56

Figure 7–57

Figure 7–58

Figure 7–59

7.3   Applications of Series-Parallel Circuits

17. Solve for all currents and voltage drops in the circuit of Figure 7–56. Verify that the power delivered by the voltage source is equal to the power dissipated by the resistors and the zener diode.

18. Refer to the circuit of Figure 7–57:
   a. Determine the power dissipated by the 6.2-V zener diode. If the zener diode is rated for a maximum power of 1/4 W, is it likely to be destroyed?
   b. Repeat Part a) if the resistance R_1 is doubled.

19. Given the circuit of Figure 7–58, determine the range of R (maximum and minimum values) which will ensure that the output voltage $V_L = 5.6$ V while the maximum power rating of the zener diode is not exceeded.

20. Given the circuit of Figure 7–59, determine the range of R (maximum and minimum values) which will ensure that the output voltage $V_L = 5.6$ V while the maximum power rating of the zener diode is not exceeded.

21. Given the circuit of Figure 7–60, determine V_B, I_C, and V_{CE}.

22. Repeat Problem 21 if R_B is increased to 10 kΩ. (All other quantities remain unchanged.)

23. Consider the circuit of Figure 7–61 and the indicated values:
   a. Determine I_D.
   b. Calculate the required value for R_S.
   c. Solve for V_{DS}.

Figure 7–60

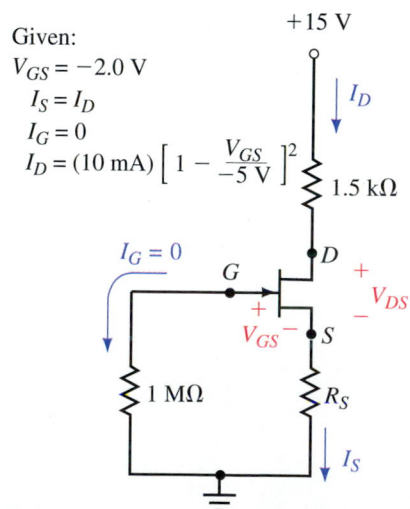

Figure 7–61

24. Consider the circuit of Figure 7–62 and the indicated values:

   a. Determine I_D and V_G.

   b. Design the required values for R_S and R_D.

Given:
$V_{GS} = -2.5$ V
$I_S = I_D$
$I_G = 0$
$I_D = (10 \text{ mA}) \left[1 - \dfrac{V_{GS}}{-5 \text{ V}} \right]^2$
$V_{DS} = +6.0$ V

Figure 7–62

7.4 Potentiometers

25. Refer to the circuit of Figure 7–63:

   a. Determine the range of voltages which will appear across R_L as the potentiometer is varied between its minimum and maximum values.

   b. If R_2 is adjusted to be 2.5 kΩ, what will be the voltage V_L? If the load resistor is now removed, what voltage would appear between terminals a and b?

26. Repeat Problem 25 using the load resistor $R_L = 30$ kΩ.

27. If the potentiometer of Figure 7–64 is adjusted so that $R_2 = 200$ Ω, determine the voltages V_{ab} and V_{bc}.

Figure 7–63

Figure 7–64

28. Calculate the values of R_1 and R_2 required in the potentiometer of Figure 7–64 if the voltage V_L across the 50-Ω load resistor is to be 6.0 V.

29. Refer to the circuit of Figure 7–65:

    a. Determine the range of output voltage (minimum to maximum) which can be expected as the potentiometer is adjusted from minimum to maximum.

    b. Calculate R_2 when $V_o = 20$ V.

Figure 7–65

30. In the circuit of Figure 7–65, what value of R_2 results in an output voltage of 40 V?

7.5   Loading Effects of Instruments

31. A voltmeter having a sensitivity $S = 20$ kΩ/V is used on the 10-V range (200-kΩ total internal resistance) to measure voltage across the 750-kΩ resistor of Figure 7–66. The voltage indicated by the meter is 5.00 V.

    a. Determine the value of the supply voltage, *E*.

    b. What voltage will be present across the 750-kΩ resistor when the voltmeter is removed from the circuit?

    c. Calculate the loading effect of the meter when used as shown.

    d. If the same voltmeter is used to measure the voltage across the 200-kΩ resistor, what voltage would be indicated?

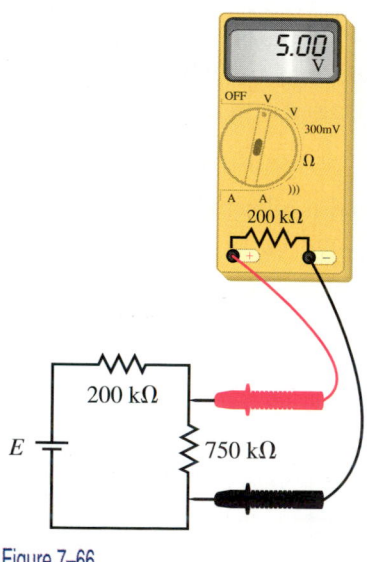

Figure 7–66

32. The voltmeter of Figure 7–67 has a sensitivity $S = 2$ kΩ/V.

    a. If the meter is used on its 50-V range ($R = 100$ kΩ) to measure the voltage across R_2, what will be the meter reading and the loading error?

    b. If the meter is changed to its 20-V range ($R = 40$ kΩ) determine the reading on this range and the loading error. Will the meter be damaged on this range? Will the loading error be less or more than the error in Part a)?

R_1

300 kΩ

60 V R_2 1.2 MΩ

Figure 7–67

33. An ammeter is used to measure curren[t]
Figure 7–68.

   a.   Explain how to correctly connect
        measure the current I_1.
   b.   Determine the values indicated w
        to measure each of the indicated
   c.   Calculate the loading effect of th
        measuring each of the currents.

34. Suppose the ammeter in Figure 7–68 has an internal resistance
of 0.5 Ω:

   a.   Determine the values indicated when the ammeter is
        used to measure the indicated currents in the circuit.
   b.   Calculate the loading effect of the meter when
        measuring each of the currents.

5.6 Ω

I_1 I_2 I_3

0.2 V   6.8 Ω   3.9 Ω

Figure 7–68

7.6   Circuit Analysis Using Computers

35. Using PSpice, write the input file to solve for V_{ab}, I_T, I_1, I_2, I_3,
and I_4 of the circuit of Figure 7–51.

36. Use PSpice to write the input file for the circuits of Figure
7–54. Print out the currents I_1, I_2, and I_3 and the voltage V_{ab} for
each of the circuits.

37. Use PSpice to write the input file for the circuit of Figure 7–55.
Print out the currents I_1, I_2, and I_3 and the voltage V_{ab} as the
resistor R_x is varied from 500 Ω to 5 kΩ in 500-Ω increments.

Methods of Analysis

CHAPTER PREVIEW

The networks you have worked with so far have generally had a single voltage source and could be easily analyzed using techniques such as Kirchhoff's voltage law and Kirchhoff's current law. In this chapter, you will examine circuits which have more than one voltage source or which cannot be easily analyzed using techniques studied in previous chapters.

The methods used in the determining the operation of complex networks will include *branch-current analysis, mesh (or loop) analysis,* and *nodal analysis*. Although any of the above methods may be used, you will find that certain circuits are more easily analyzed using one particular approach. The advantages of each method will be discussed in the appropriate section.

In using the techniques outlined above, it is assumed that the networks are *linear bilateral networks*. The term *linear* indicates that the components used in the circuit have voltage-current characteristics which follow a straight line. Refer to Figure 8–1.

The term *bilateral* indicates that the components in the network will have characteristics which are independent of the direction of the current through the element or the voltage across the element. A resistor is an example of a linear bilateral component since the voltage across a resistor is directly proportional to the current through it and the operation of the resistor is the same regardless of the direction of the current.

In this chapter you will be introduced to the conversion of a network from a delta (Δ) configuration to an equivalent wye (Y) configuration. Conversely, we will examine the transformation from a Y configuration to an equivalent Δ configuration. You will use these conversions to examine the operation of an unbalanced bridge network.

KEY TERMS
Branch-Current Analysis
Bridge Networks
Constant-Current Sources
Delta-Wye Conversions
Linear Bilateral Networks
Mesh Analysis
Nodal Analysis
Wye-Delta Conversions

(a) Linear V-I characteristics

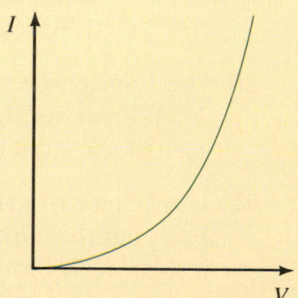

(b) Non-linear V-I characteristics

Figure 8–1

Fiber optic cables. *(Courtesy of Sprint; Shawnee, KS 66205)*

OBJECTIVES

After studying this chapter you will be able to

• convert a voltage source into an equivalent current source,

• convert a current source into an equivalent voltage source,

• analyze circuits having two or more current sources in parallel,

• write and solve *branch* equations for a network,

• write and solve *mesh* equations for a network,

• write and solve *nodal* equations for a network,

• convert a resistive delta to an equivalent wye circuit or a wye to its equivalent delta circuit and solve the resulting simplified circuit,

• determine the voltage across or current through any portion of a bridge network,

• use PSpice to analyze multiloop circuits.

Sir Charles Wheatstone

Charles Wheatstone was born in Gloucester, England, on February 6, 1802. Wheatstone's original interest was in the study of acoustics and musical instruments. However, he gained fame and a knighthood as a result of inventing the telegraph and improving the electric generator.

Although he did not invent the bridge circuit, Wheatstone used one for measuring resistance very precisely. He found that when the currents in the Wheatstone bridge are exactly balanced, the unknown resistance can be compared to a known standard.

Sir Charles died in Paris, France, on October 19, 1875.

8.1 Constant-Current Sources

All the circuits presented so far have used voltage sources as the means of providing power. However, the analysis of certain circuits is easier if you work with current rather than with voltage. Unlike a voltage source, a constant-current source maintains the same current in its branch of the circuit regardless of how components are connected external to the source. The symbol for a constant-current source is shown in Figure 8–2.

The direction of the current source arrow indicates the direction of current in the branch. In previous chapters you learned that the magnitude and the direction of current through a voltage source varies according to the size of the circuit resistances and how other voltage sources are connected in the circuit. For current sources, the *voltage* across the current source depends on how the other components are connected.

Figure 8–2 Ideal constant-current source.

■ **EXAMPLE 8–1** Refer to the circuit of Figure 8–3:

Figure 8–3

a. Calculate the voltage V_S across the current source if the resistor is 100 Ω.

b. Calculate the voltage if the resistor is 2 kΩ.

Solution   The current source maintains a constant current of 2 A through the circuit. Therefore,

a. $V_S = V_R = (2\text{ A})(100\ \Omega) = 200$ V.

b. $V_S = V_R = (2\text{ A})(2\text{ k}\Omega) = 4000$ V.

If the current source is the only source in the circuit, then the polarity of voltage across the source will be as shown in Figure 8–3. This, however, may not be the case if there is more than one source.

■ **EXAMPLE 8–2**   Determine the voltages V_1, V_2, and V_S and the current I_S for the circuit of Figure 8–4.

Figure 8–4

Solution   Since the given circuit is a series circuit, the current everywhere in the circuit must be the same, namely

$$I_S = 2 \text{ mA}$$

Using Ohm's law,

$$V_1 = (2\text{ mA})(1\text{ k}\Omega) = 2 \text{ V}$$
$$V_2 = (2\text{ mA})(2\text{ k}\Omega) = 4 \text{ V}$$

Applying Kirchhoff's voltage law around the closed loop,

$$\sum V = V_S - V_1 - V_2 + E = 0$$
$$V_S = V_1 + V_2 - E$$
$$= 2 \text{ V} + 4 \text{ V} - 10 \text{ V} = -4 \text{ V}$$

From the above result, you see that the actual polarity of V_S is opposite to that assumed.

■ **EXAMPLE 8–3**   Calculate the currents I_1 and I_2 and the voltage V_S for the circuit of Figure 8–5.

Figure 8–5

Solution   Because the –5-V supply is effectively across the load resistor,

$$I_1 = \frac{5\text{ V}}{10\text{ }\Omega} = 0.5\text{ A}\quad\text{(in the direction assumed)}$$

Applying Kirchhoff's current law at point a,

$$I_2 = 0.5\text{ A} + 2.0\text{ A} = 2.5\text{ A}$$

From Kirchhoff's voltage law,

$$\Sigma V = -10\text{ V} + V_S + 5\text{ V} = 0\text{ V}$$
$$V_S = 10\text{ V} - 5\text{ V} = +5\text{ V}$$

From examining the previous examples, the following conclusions may be made regarding current sources:

The constant-current source determines the current in its branch of the circuit.

The magnitude and polarity of voltage appearing across a constant-current source is dependent upon the network in which the source is connected.

8.2 Source Conversions

In the previous section you were introduced to the ideal constant-current source. This is a source which has no internal resistance included as part of the circuit. As you recall, voltage sources always have some series resistance, although in some cases this resistance is so small in comparison with other circuit resistance that it may effectively be ignored when determining the operation of the circuit. Similarly, a constant-current source will always have some shunt (or parallel) resistance. If this resistance is very large in comparison with the other circuit resistance, the internal resistance of the source may once again be ignored. *An ideal current source has an infinite shunt resistance.*

Figure 8–6 shows equivalent voltage and current sources.

If the internal resistance of a source is considered, the source, whether it is a voltage source or a current source, is easily converted to

$$E = IR_S$$

Figure 8-6

the other type. The current source of Figure 8–6 is equivalent to the voltage source if

$$I = \frac{E}{R_S} \tag{8–1}$$

and the resistance in both sources is R_S.

Similarly, a current source may be converted to an equivalent voltage source by letting

$$E = IR_S \tag{8–2}$$

These results may be easily verified by connecting an external resistance, R_L, across each source. The sources can be equivalent only if the voltage across R_L is the same for both sources. Similarly, the sources are equivalent only if the current through R_L is the same when connected to either source.

Consider the circuit shown in Figure 8–7.

The voltage across the load resistor is given as

$$V_L = \frac{R_L}{R_L + R_S} E \tag{8–3}$$

The current through the resistor R_L is given as

$$I_L = \frac{E}{R_L + R_S} \tag{8–4}$$

Next, consider an equivalent current source connected to the same load as shown in Figure 8–8. The current through the resistor R_L is given by

$$I_L = \frac{R_S}{R_S + R_L} I$$

Figure 8-7

Figure 8-8

But, when converting the source, we get

$$I = \frac{E}{R_S}$$

And so

$$I_L = \left(\frac{R_S}{R_S + R_L}\right)\left(\frac{E}{R_S}\right)$$

This result is equivalent to the current obtained in Equation 8–4. The voltage across the resistor is given as

$$V_L = I_L R_L$$

$$= \left(\frac{E}{R_S + R_L}\right)R_L$$

The voltage across the resistor is precisely the same as the result obtained in Equation 8–3. We therefore conclude that the load current and voltage drop are the same whether the source is a voltage source or an equivalent current source.

■ **EXAMPLE 8–4**   Convert the voltage source of Figure 8–9(a) into a current source and verify that the current, I_L, through the load is the same for each source.

(a)

(b)

Figure 8–9

Solution   The equivalent current source will have a current magnitude given as

$$I = \frac{48 \text{ V}}{10 \text{ }\Omega} = 4.8 \text{ A}$$

The resulting circuit is shown in Figure 8–9(b).

For the circuit of Figure 8–9(a), the current through the load is found as

$$I_L = \frac{48 \text{ V}}{10 \text{ }\Omega + 40 \text{ }\Omega} = 0.96 \text{ A}$$

For the equivalent circuit of Figure 8–9(b), the current through the load is

$$I_L = \frac{(4.8 \text{ A})(10 \text{ }\Omega)}{10 \text{ }\Omega + 40 \text{ }\Omega} = 0.96 \text{ A}$$

Clearly the results are the same.

■ **EXAMPLE 8–5**   Convert the current source of Figure 8–10(a) into a voltage source and verify that the voltage, V_L, across the load is the same for each source.

(a)

$$E = (30 \text{ mA})(30 \text{ k}\Omega) = 900 \text{ V}$$

(b)

Figure 8–10

Solution   The equivalent voltage source will have a magnitude given as

$$E = (30 \text{ mA})(30 \text{ k}\Omega) = 900 \text{ V}$$

The resulting circuit is shown in Figure 8–10(b).

For the circuit of Figure 8–10(a), the voltage across the load is determined as

$$I_L = \frac{(30 \text{ k}\Omega)(30 \text{ mA})}{30 \text{ k}\Omega + 10 \text{ k}\Omega} = 22.5 \text{ mA}$$

$$V_L = I_L R_L = (22.5 \text{ mA})(10 \text{ k}\Omega) = 225 \text{ V}$$

For the equivalent circuit of Figure 8–10(b), the voltage across the load is

$$V_L = \frac{10 \text{ k}\Omega}{10 \text{ k}\Omega + 30 \text{ k}\Omega}(900 \text{ V}) = 225 \text{ V}$$

Once again, we see that the circuits are equivalent.

practice problems 1

1. Convert the voltage sources of Figure 8–11 into equivalent current sources.

Figure 8–11

2. Convert the current sources of Figure 8–12 into equivalent voltage sources.

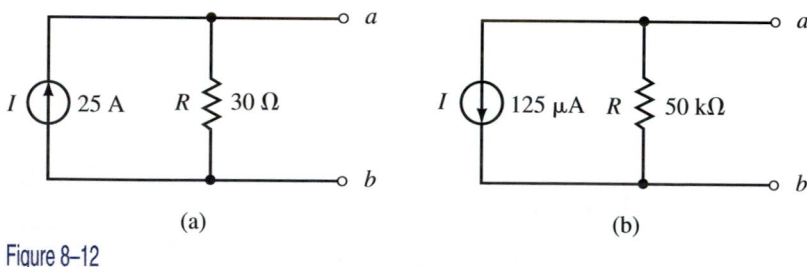

Figure 8–12

8.3   Current Sources in Parallel and Series

When several current sources are placed in parallel, then the circuit may be simplified by replacing the current sources with a single current source. The magnitude and direction of this resultant source is determined by adding the currents in one direction and then subtracting the currents in the opposite direction.

■ **EXAMPLE 8–6**   Simplify the circuit of Figure 8–13 and determine the voltage V_{ab}.

Figure 8–13

Solution   Since all of the current sources are in parallel, they can be replaced by a single current source. The equivalent current source will have a direction which is the same as both I_2 and I_3, since the magnitude of current in the downward direction is greater than the current in the upward direction. The equivalent current source has a magnitude of

$$I = 2\,A + 6\,A - 3\,A = 5\,A$$

as shown in Figure 8–14(a).

The circuit is further simplified by combining the resistors into a single value:

$$R_T = 6\,\Omega \| 3\,\Omega \| 6\,\Omega = 1.5\,\Omega$$

The equivalent circuit is shown in Figure 8–14(b).

Figure 8–14

The voltage V_{ab} is found as

$$V_{ab} = -(5\,A)(1.5\,\Omega) = -7.5\,V$$

■ **EXAMPLE 8–7**   Reduce the circuit of Figure 8–15 into a single current source and solve for the current through the resistor R_L.

Figure 8–15

Solution   The voltage source in this circuit is converted to an equivalent current source as shown. The resulting circuit may then be simpified to a single current source where

$$I_S = 200 \text{ mA} + 50 \text{ mA} = 250 \text{ mA}$$

and

$$R_S = 400 \text{ }\Omega\|100 \text{ }\Omega = 80 \text{ }\Omega$$

The simplified circuit is shown in Figure 8–16.

Figure 8–16

The current through R_L is now easily calculated as

$$I_L = \left(\frac{80 \text{ }\Omega}{80 \text{ }\Omega + 20 \text{ }\Omega} \right)(250 \text{ mA}) = 200 \text{ mA}$$

Kirchhoff's current law
is violated at this node
since $I_{in} \neq I_{out}$.

$I_1 = 3 \text{ A}$ $I_2 = 5 \text{ A}$

Figure 8–17

Current sources should never be placed in series. If a node is chosen between the current sources, it becomes immediately apparent that the current entering the node is not the same as the current leaving the node. Clearly, this cannot not occur since there would then be a violation of Kirchhoff's current law (see Figure 8–17).

in process
l e a r n i n g
check *I*

1. Briefly explain the procedure for converting a voltage source into an equivalent current source.

2. What is the most important rule determining how current sources are connected into a circuit?

8.4   Branch-Current Analysis

In previous chapters we used Kirchhoff's circuit law and Kirchhoff's voltage law to solve equations for circuits having a single voltage source. In this section, you will use these powerful tools to analyze circuits having more than one source.

Branch-current analysis allows us to directly calculate the current in each branch of a circuit. Since the method involves the analysis of several simultaneous linear equations, you may find that a review of determinants is in order. Appendix B has been included to provide a review of the mechanics of solving simultaneous linear equations.

When applying branch-current analysis, you will find the technique listed below useful.

1. Arbitrarily assign current directions to each branch in the network. If a particular branch has a current source, then this step is not necessary since you already know the magnitude and direction of the current in this branch.

2. Using the assigned currents, label the polarities of the voltage drops across all resistors in the circuit.

3. Apply Kirchhoff's voltage law around each of the closed loops. Write just enough equations to include all branches in the loop equations. If a branch has only a current source and no series resistance, it is not necessary to include it in the KVL equations.

4. Apply Kirchhoff's current law at enough nodes to ensure that all branch currents have been included. In the event that a branch has only a current source, it will need to be included in this step.

5. Solve the resulting simultaneous linear equations.

■ **EXAMPLE 8–8**   Find the current in each branch in the circuit of Figure 8–18.

Figure 8–18

Solution

Step 1: Assign currents as shown in Figure 8–18.

Step 2: Indicate the polarities of the voltage drops on all resistors in the circuit, using the assumed current directions.

Step 3: Write the Kirchhoff voltage law equations.

Loop *abcda*: $6\text{ V} - (2\ \Omega)I_1 + (2\ \Omega)I_2 - 4\text{ V} = 0\text{ V}$

Notice that the circuit still has one branch which has not been included in the KVL equations, namely the branch *cefd*. This branch would be included if a loop equation for *cefdc* or for *abcefda* were written. There is no reason for choosing one loop over another, since the overall result will remain unchanged even though the intermediate steps will not give the same results.

Loop *cefdc*: $4\text{ V} - (2\ \Omega)I_2 - (4\ \Omega)I_3 + 2\text{ V} = 0\text{ V}$

Now that all branches have been included in the loop equations, there is no need to write any more. Although more loops exist, writing more loop equations would needlessly complicate the calculations.

Step 4: Write the Kirchhoff current law equation(s).

By applying KCL at node *c*, all branch currents in the network are included.

Node *c*: $I_3 = I_1 + I_2$

To simplify the solution of the simultaneous linear equations we write them as follows:

$$2I_1 - 2I_2 + 0I_3 = 2$$

$$0I_1 - 2I_2 - 4I_3 = -6$$

$$1I_1 + 1I_2 - 1I_3 = 0$$

The principles of linear algebra (Appendix B) allow us to solve for the determinant of the denominator as follows:

$$D = \begin{vmatrix} 2 & -2 & 0 \\ 0 & -2 & -4 \\ 1 & 1 & -1 \end{vmatrix}$$

$$= 2\begin{vmatrix} -2 & -4 \\ 1 & -1 \end{vmatrix} - 0\begin{vmatrix} -2 & 0 \\ 1 & -1 \end{vmatrix} + 1\begin{vmatrix} -2 & 0 \\ -2 & -4 \end{vmatrix}$$

$$= 2(2 + 4) - 0 + 1(8) = 20$$

Now, solving for the currents, we have the following:

$$I_1 = \frac{\begin{vmatrix} 2 & -2 & 0 \\ -6 & -2 & -4 \\ 0 & 1 & -1 \end{vmatrix}}{D}$$

$$= \frac{2\begin{vmatrix} -2 & -4 \\ 1 & -1 \end{vmatrix} - (-6)\begin{vmatrix} -2 & 0 \\ 1 & -1 \end{vmatrix} + 0\begin{vmatrix} -2 & 0 \\ -2 & -4 \end{vmatrix}}{20}$$

$$= \frac{2(2+4) + 6(2) + 0}{20} = \frac{24}{20} = 1.2 \text{ A}$$

$$I_2 = \frac{\begin{vmatrix} 2 & 2 & 0 \\ 0 & -6 & -4 \\ 1 & 0 & -1 \end{vmatrix}}{D}$$

$$= \frac{2\begin{vmatrix} -6 & -4 \\ 0 & -1 \end{vmatrix} - 0\begin{vmatrix} 2 & 0 \\ 0 & -1 \end{vmatrix} + 1\begin{vmatrix} 2 & 0 \\ -6 & -4 \end{vmatrix}}{20}$$

$$= \frac{2(6) + 0 + 1(-8)}{20} = \frac{4}{20} = 0.2 \text{ A}$$

$$I_3 = \frac{\begin{vmatrix} 2 & -2 & 2 \\ 0 & -2 & -6 \\ 1 & 1 & 0 \end{vmatrix}}{D}$$

$$= \frac{2\begin{vmatrix} -2 & -6 \\ 1 & 0 \end{vmatrix} - 0\begin{vmatrix} -2 & 2 \\ 1 & 0 \end{vmatrix} + 1\begin{vmatrix} -2 & 2 \\ -2 & -6 \end{vmatrix}}{20}$$

$$= \frac{2(6) - 0 + 1(12 + 4)}{20} = \frac{28}{20} = 1.4 \text{ A}$$

■ **EXAMPLE 8–9**   Find the currents in each branch of the circuit shown in Figure 8–19. Solve for the voltage V_{ab}.

Figure 8–19

Solution   Notice that although the above circuit has four currents, there are only three **unknown** currents: I_2, I_3, and I_4. The current I_1 is given by the value of the constant-current source. In order to solve this network we will need three linear equations. As before, the equations are determined by Kirchhoff's voltage and current laws.

Step 1: The currents are indicated in the given circuit.

Step 2: The polarities of the voltages across all resistors are shown.

Step 3: Kirchhoff's voltage law is applied at the indicated loops:

Loop *badb:* $-(2\ \Omega)(I_2) + (3\ \Omega)(I_3) - 8\ \text{V} = 0\ \text{V}$

Loop *bacb:* $-(2\ \Omega)(I_2) + (1\ \Omega)(I_4) - 6\ \text{V} = 0\ \text{V}$

Step 4: Kirchhoff's current law is applied as follows:

Node *a:* $I_2 + I_3 + I_4 = 5\ \text{A}$

Rewriting the linear equations,

$$-2I_2 + 3I_3 + 0I_4 = 8$$

$$-2I_2 + 0I_3 + 1I_4 = 6$$

$$1I_2 + 1I_3 + 1I_4 = 5$$

The determinant of the denominator is evaluated as

$$D = \begin{vmatrix} -2 & 3 & 0 \\ -2 & 0 & 1 \\ 1 & 1 & 1 \end{vmatrix} = 11$$

Now solving for the currents, we have

$$I_2 = \frac{\begin{vmatrix} 8 & 3 & 0 \\ 6 & 0 & 1 \\ 5 & 1 & 1 \end{vmatrix}}{D} = \frac{11}{11} = -1.0\ \text{A}$$

$$I_3 = \frac{\begin{vmatrix} -2 & 8 & 0 \\ -2 & 6 & 1 \\ 1 & 5 & 1 \end{vmatrix}}{D} = \frac{22}{11} = 2.0\ \text{A}$$

$$I_4 = \frac{\begin{vmatrix} -2 & 3 & 8 \\ -2 & 0 & 6 \\ 1 & 1 & 5 \end{vmatrix}}{D} = \frac{44}{11} = 4.0\ \text{A}$$

The current I_2 is negative, which simply means that the actual direction of the current is opposite to the chosen direction.

Although the network may be further analyzed using the assumed current directions, it is easier to understand the circuit operation by showing the correct current directions as in Figure 8–20.

$$I_2 = 1.0 \text{ A}$$
$$I_3 = 2.0 \text{ A}$$
$$I_4 = 4.0 \text{ A}$$

Figure 8–20

Using the correct direction for I_2,

$$V_{ab} = +(2 \text{ } \Omega)(1 \text{ A}) = +2 \text{ V}$$

practice
problem
II

Use branch-current analysis to solve for the indicated currents in the circuit of Figure 8–21.

Figure 8–21

8.5 Mesh (Loop) Analysis

In the previous section you used Kirchhoff's laws to solve for the current in each branch of a given network. While the methods used were relatively simple, branch-current analysis is awkward to use because it generally involves solving several simultaneous linear equations. It is not difficult to see that the number of equations may be prohibitively large even for a relatively simple circuit.

A better approach and one which is used extensively in analyzing linear bilateral networks is called mesh (or loop) analysis. While the technique is similar to branch-current analysis, the number of simultaneous linear equations tends to be less. The principal difference between mesh analysis and branch-current analysis is that we simply need to apply Kirchhoff's voltage law around closed loops without the need for applying Kirchhoff's current law.

The steps used in solving a circuit using mesh analysis are as follows:

1. Arbitrarily assign a clockwise current to each interior closed loop in the network. Although the assigned current may be in any direction, a clockwise direction is used to make later work simpler.

2. Using the assigned loop currents, indicate the voltage polarities across all resistors in the circuit. For a resistor which is common to two loops, the polarities of the voltage drop due to each loop current should be indicated on the appropriate side of the component.

3. Applying Kirchhoff's voltage law, write the loop equations for each loop in the network. Do not forget that resistors which are common to two loops will have two voltage drops, one due to each loop.

4. Solve the resultant simultaneous linear equations.

5. Branch currents are determined by algebraically combining the loop currents which are common to the branch.

■ **EXAMPLE 8–10** Find the current in each branch for the circuit of Figure 8–22.

Figure 8–22

Solution

Step 1: Loop currents are assigned as shown in Figure 8–22. These currents are designated I_1 and I_2.

Step 2: Voltage polarities are assigned according to the loop currents. Notice that the resistor R_2 has two different voltage polarities due to the different loop currents.

Step 3: The loop equations are written by applying Kirchhoff's voltage law in each of the loops. The equations are as follows:

Loop 1: $6 \text{ V} - (2 \text{ }\Omega)I_1 - (2 \text{ }\Omega)I_1 + (2 \text{ }\Omega)I_2 - 4 \text{ V} = 0$

Loop 2: $4 \text{ V} - (2 \text{ }\Omega)I_2 + (2 \text{ }\Omega)I_1 - (4 \text{ }\Omega)I_2 + 2 \text{ V} = 0$

Note that the voltage across R_2 due to the currents I_1 and I_2 is indicated as two separated terms, where one term represents a voltage drop in the direction of I_1 and the other term represents a voltage rise in the same direction. The magnitude and polarity of the voltage across R_2 is determined by the actual size and directions of the loop currents. The above loop equations may be simplified as follows:

Loop 1: $(4\ \Omega)I_1 - (2\ \Omega)I_2 = 2\ V$

Loop 2: $-(2\ \Omega)I_1 + (6\ \Omega)I_2 = 6\ V$

Using determinants, the loop equations are easily solved as

$$I_1 = \frac{\begin{vmatrix} 2 & -2 \\ 6 & 6 \end{vmatrix}}{\begin{vmatrix} 4 & -2 \\ -2 & 6 \end{vmatrix}} = \frac{12 + 12}{24 - 4} = \frac{24}{20} = 1.2\ A$$

and

$$I_1 = \frac{\begin{vmatrix} 4 & 2 \\ -2 & 6 \end{vmatrix}}{\begin{vmatrix} 4 & -2 \\ -2 & 6 \end{vmatrix}} = \frac{24 + 4}{24 - 4} = \frac{28}{20} = 1.4\ A$$

From the above results, we see that the currents through resistors R_1 and R_3 are I_1 and I_2 respectively.

The branch current for R_2 is found by combining the loop currents through this resistor:

$$I_{R_2} = 1.4\ A - 1.2\ A = 0.2\ A \quad \text{(upward)}$$

Notice that the results obtained by using mesh analysis are exactly the same as those obtained by branch-current analysis. Unlike branch-current analysis, this approach required the solution of only two simultaneous linear equations, whereas branch-current analysis required three equations. Mesh analysis also requires that only Kirchhoff's voltage law be applied and clearly illustrates why mesh analysis is preferred to branch-current analysis.

If the circuit being analyzed contains current sources, the procedure is a bit more complicated. The circuit may be simplified by converting the current source(s) to voltage sources and then solving the resulting network using the procedure shown in the previous example. Alternatively, you may not wish to alter the circuit, in which case the current source will provide one of the loop currents.

■ **EXAMPLE 8–11**   Determine the current through the 8-V battery for the circuit shown in Figure 8–23.

Figure 8–23

Solution   Convert the current source into an equivalent voltage source. The equivalent circuit may now be analyzed by using the loop currents shown in Figure 8–24.

Figure 8–24

Loop 1: $-10\text{ V} - (2\ \Omega)I_1 - (3\ \Omega)I_1 + (3\ \Omega)I_2 - 8\text{ V} = 0$

Loop 2: $8\text{ V} - (3\ \Omega)I_2 + (3\ \Omega)I_1 - (1\ \Omega)I_2 - 6\text{ V} = 0$

Rewriting the linear equations, you get the following:

Loop 1: $(5\ \Omega)I_1 - (3\ \Omega)I_2 = -18\text{ V}$

Loop 2: $-(3\ \Omega)I_1 + (4\ \Omega)I_2 = 2\text{ V}$

Solving the equations using determinants, we have the following:

$$I_1 = \frac{\begin{vmatrix} -18 & -3 \\ 2 & 4 \end{vmatrix}}{\begin{vmatrix} 5 & -3 \\ -3 & 4 \end{vmatrix}} = -\frac{66}{11} = -6.0\text{ A}$$

$$I_1 = \frac{\begin{vmatrix} 5 & -18 \\ -3 & 2 \end{vmatrix}}{\begin{vmatrix} 5 & -3 \\ -3 & 4 \end{vmatrix}} = -\frac{44}{11} = -4.0\text{ A}$$

If the assumed direction of current in the 8-V battery is taken to be I_2, then

$$I = I_2 - I_1 = -4.0\text{ A} - (-6.0\text{ A}) = 2.0\text{ A}$$

The direction of the resultant current is the same as I_2 (upward).

The circuit of Figure 8–23 may also be analyzed without converting the current source to a voltage source. Although the approach is generally not used, the following example illustrates the above-mentioned technique.

■ **EXAMPLE 8–12**   Determine the current through R_1 for the circuit shown in Figure 8–25.

Figure 8–25

Solution   By inspection, we see that the loop current $I_1 = -5$ A. The mesh equations for the other two loops are as follows:

Loop 2: $-(2\ \Omega)I_2 + (2\ \Omega)I_1 - (3\ \Omega)I_2 + (3\ \Omega)I_3 - 8\ V = 0$

Loop 3: $8\ V - (3\ \Omega)I_3 + (3\ \Omega)I_2 - (1\ \Omega)I_3 - 6\ V = 0$

Although it is possible to analyze the circuit by solving three linear equations, it is easier to substitute the known value $I_1 = -5V$ into the mesh equation for loop 2, which may now be written as

Loop 2: $-(2\ \Omega)I_2 - 10\ V - (3\ \Omega)I_2 + (3\ \Omega)I_3 - 8\ V = 0$

The loop equations may now be simplified as

Loop 2: $(5\ \Omega)I_2 - (3\ \Omega)I_3 = -18\ V$

Loop 3: $-(3\ \Omega)I_2 + (4\ \Omega)I_3 = 2\ V$

The simultaneous linear equations are solved as follows:

$$I_1 = \frac{\begin{vmatrix} -18 & -3 \\ 2 & 4 \end{vmatrix}}{\begin{vmatrix} 5 & -3 \\ -3 & 4 \end{vmatrix}} = -\frac{66}{11} = -6.0\ A$$

$$I_1 = \frac{\begin{vmatrix} 5 & -18 \\ -3 & 2 \end{vmatrix}}{\begin{vmatrix} 5 & -3 \\ -3 & 4 \end{vmatrix}} = -\frac{44}{11} = -4.0\ A$$

The calculated values of the assumed reference currents allow us to determine the actual current through the various resistors as follows:

$$I_{R_1} = I_1 - I_2 = -5\ A - (-6\ A) = 1\ A \quad \text{downward}$$
$$I_{R_2} = I_3 - I_2 = -4\ A - (-6\ A) = 2\ A \quad \text{upward}$$
$$I_{R_3} = -I_3 = 4\ A \quad \text{left}$$

These results are consistent with those obtained in Example 8–9.

Format Approach for Mesh Analysis

A very simple technique may be used to write the mesh equations for any linear bilateral network. When this format approach is used, the simultaneous linear equations for a network having n independent loops will appear as follows:

$$R_{11}I_1 - R_{12}I_2 - R_{13} - \ldots - R_{1n} = E_1$$
$$-R_{21}I_1 + R_{22}I_2 - R_{23} - \ldots - R_{2n} = E_2$$
$$\vdots$$
$$-R_{n1}I_1 - R_{n2}I_2 - R_{n3} - \ldots + R_{nn} = E_n$$

The terms R_{11}, R_{22}, R_{33}, . . . , R_{nn} represent the total resistance in each loop and are found by simply adding all the resistances in a particular loop. The remaining resistance terms are called the *mutual resistance* terms. These resistances represent resistance which is shared between two loops. For example, the mutual resistance R_{12} is the resistance in loop 1 which is located in the branch between loop 1 and loop 2. If there is no resistance between two loops, this term will be zero.

Notice that the terms containing R_{11}, R_{22}, R_{33}, . . . , R_{nn} are positive and that all of the mutual resistance terms are negative. This characteristic occurs because all currents are assumed to be clockwise.

If the linear equations are correctly written, you will find that the coefficients along the principal diagonal (R_{11}, R_{22}, R_{33}, . . . , R_{nn}) will be positive. All other coefficients will be negative. Also, if the equations are correctly written, the terms will be symmetrical about the principal diagonal, i.e., $R_{12} = R_{21}$.

The terms E_1, E_2, E_3, . . . , E_n are the summation of the voltage rises in the direction of the loop currents. If a voltage source appears in the branch shared by two loops, it will be included in the calculation of the voltage rise for each loop.

The method used in applying the format approach of mesh analysis is as follows:

1. Convert current sources into equivalent voltage sources.

2. Assign clockwise currents to each independent closed loop in the network.

3. Write the simultaneous linear equations in the format outlined.

4. Solve the resulting simultaneous linear equations.

■ **EXAMPLE 8–13**   Solve for the currents through R_2 and R_3 in the circuit of Figure 8–26.

Figure 8–26

Solution

Step 1: Although we see that the circuit has a current source, it may not be immediately evident how the source can be converted into an equivalent voltage source. Redrawing the circuit into a more recognizable form, as shown in Figure 8–27, we see that the 2-mA current source is in parallel with a 6-kΩ resistor. The source conversion is also illustrated in Figure 8–27.

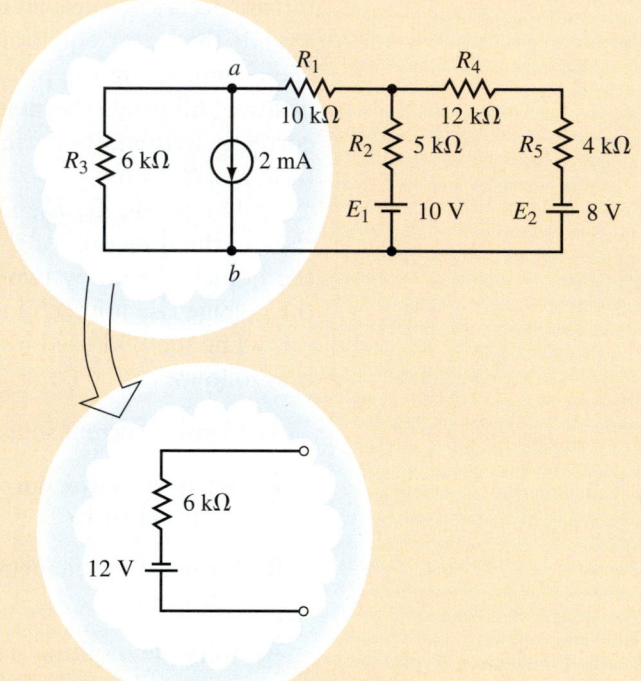

Figure 8–27

Step 2: Redrawing the circuit is further simplified by labelling some of the nodes, in this case *a* and *b*. After performing a source conversion, we have the two-loop circuit shown in Figure 8–28. The current directions for I_1 and I_2 are also illustrated.

Figure 8–28

Step 3: The loop equations are

Loop 1: $(6\text{ k}\Omega + 10\text{ k}\Omega + 5\text{ k}\Omega)I_1 - (5\text{ k}\Omega)I_2 = -12\text{ V} - 10\text{ V}$

Loop 2: $-(5\text{ k}\Omega)I_1 + (5\text{ k}\Omega + 12\text{ k}\Omega + 4\text{ k}\Omega)I_2 = 10\text{ V} + 8\text{ V}$

In loop 1, both voltages are negative since they appear as voltage drops when following the direction of the loop current.

These equations are rewritten as

$$(21\text{ k}\Omega)I_1 - (5\text{ k}\Omega)I_2 = -22\text{ V}$$

$$-(5\text{ k}\Omega)I_1 + (21\text{ k}\Omega)I_2 = 18\text{ V}$$

Step 4: Solving these equations gives

$$I_1 = \dfrac{\begin{vmatrix} -22 & -5 \\ 18 & 21 \end{vmatrix}}{\begin{vmatrix} 21 & -5 \\ -5 & 21 \end{vmatrix}} = -\dfrac{-462 + 90}{441 - 25} = \dfrac{-372}{416} = -0.894\text{ mA}$$

$$I_2 = \dfrac{\begin{vmatrix} 21 & -22 \\ -5 & 18 \end{vmatrix}}{\begin{vmatrix} 21 & -5 \\ -5 & 21 \end{vmatrix}} = \dfrac{378 + 110}{441 - 25} = \dfrac{268}{416} = 0.644\text{ mA}$$

The current through the resistor R_2 is easily determined to be

$$I_2 - I_1 = 0.644\text{ mA} - (-0.894\text{ mA}) = 1.54\text{ mA}$$

The current through R_3 is not found as easily. A common mistake is to say that the current in R_3 is the same as the current through the 6-kΩ resistor of the circuit in Figure 8–28. **This is not the case.** Since this resistor was part of the source conversion it is no longer in the same location as in the original circuit.

Although there are several ways of finding the required current, the easiest method involves the application of Kirchhoff's current law at node a. If we reexamine the circuit of Figure 8–26 and make the assumption that the current through R_3 enters the node, we see that the following expression must apply at node a:

$$\sum I_{\text{entering}} = \sum I_{\text{leaving}}$$

$$I_{R_3} = 2\text{ mA} + I_1$$

$$= 2\text{ mA} + (-0.894\text{ mA})$$

$$= 1.106\text{ mA}$$

practice problem III

Use mesh analysis to find the loop currents in the circuit of Figure 8–29.

Figure 8–29

8.6 Nodal Analysis

In the previous section we applied Kirchhoff's voltage law to arrive at loop currents in a network. In this section we will apply Kirchhoff's current law to determine the potential difference (voltage) at any node with respect to some arbitrary reference point in a network. Once the potentials of all nodes are known, it is a simple matter to determine other quantities such as current and power within the network.

The steps used in solving a circuit using nodal analysis are as follows:

1. Arbitrarily assign a reference node within the circuit and indicate this node as *ground*. The reference node is usually located at the bottom of the circuit, although it may be located anywhere.

2. Convert each voltage source in the network to its equivalent current source. This step, although not absolutely necessary, makes further calculations easier to understand.

3. Arbitrarily assign voltages $(V_1, V_2, \ldots, V_n)$ to the remaining nodes in the circuit. (Remember that you have already assigned a reference node, so these voltages will all be with respect to the chosen reference.)

4. Arbitrarily assign a current direction to each branch in which there is no current source. Using the assigned current directions, indicate the corresponding polarities of the voltage drops on all resistors.

5. With the exception of the reference node (ground), apply Kirchhoff's current law at each of the nodes. If a circuit has a total of $n + 1$ nodes (including the reference node), there will be n simultaneous linear equations.

6. Rewrite each of the arbitrarily assigned currents in terms of the potential difference across a known resistance.

7. Solve the resulting simultaneous linear equations for the voltages $(V_1, V_2, \ldots, V_n)$.

■ **EXAMPLE 8–14**   Given the circuit of Figure 8–30, use nodal analysis to solve for the voltage V_{ab}.

Figure 8–30

Solution

Step 1: Select a convenient reference node.

Step 2: Convert the voltage sources into equivalent current sources. The equivalent circuit is shown in Figure 8–31.

Figure 8–31

Steps 3 and 4: Arbitrarily assign node voltages and branch currents. Indicate the voltage polarities across all resistors according to the assumed current directions.

Step 5: We now apply Kirchhoff's current law at the nodes labelled as V_1 and V_2:

Node V_1:
$$\sum I_{\text{entering}} = \sum I_{\text{leaving}}$$
$$200 \text{ mA} + 50 \text{ mA} = I_1 + I_2$$

Node V_2:
$$\sum I_{\text{entering}} = \sum I_{\text{leaving}}$$
$$200 \text{ mA} + I_2 = 50 \text{ mA} + I_3$$

Step 6: The currents are rewritten in terms of the voltages across the resistors as follows:

$$I_1 = \frac{V_1}{20 \ \Omega}$$

$$I_2 = \frac{V_1 - V_2}{40 \ \Omega}$$

$$I_3 = \frac{V_2}{30 \ \Omega}$$

The nodal equations become

$$200 \text{ mA} + 50 \text{ mA} = \frac{V_1}{20 \ \Omega} + \frac{V_1 - V_2}{40 \ \Omega}$$

$$200 \text{ mA} + \frac{V_1 - V_2}{40 \ \Omega} = 50 \text{ mA} + \frac{V_2}{30 \ \Omega}$$

Substituting the voltage expressions into the original nodal equations, we have the following simultaneous linear equations:

$$\left(\frac{1}{20 \ \Omega} + \frac{1}{40 \ \Omega}\right)V_1 - \left(\frac{1}{40 \ \Omega}\right)V_2 = 250 \text{ mA}$$

$$-\left(\frac{1}{40 \ \Omega}\right)V_1 + \left(\frac{1}{30 \ \Omega} + \frac{1}{40 \ \Omega}\right)V_2 = 150 \text{ mA}$$

These may be further simplified as

$$(0.075 \text{ S})V_1 - (0.025 \text{ S})V_2 = 250 \text{ mA}$$

$$-(0.025 \text{ S})V_1 + (0.0583)V_2 = 150 \text{ mA}$$

Step 7: Use determinants to solve for the nodal voltages as

$$V_1 = \frac{\begin{vmatrix} 0.250 & -0.025 \\ 0.150 & 0.058\bar{3} \end{vmatrix}}{\begin{vmatrix} 0.075 & -0.025 \\ -0.025 & 0.058\bar{3} \end{vmatrix}}$$

$$= \frac{(0.250)(0.058\bar{3}) - (0.150)(-0.025)}{(0.075(0.058\bar{3}) - (-0.025)(-0.025)}$$

$$= \frac{0.018\bar{3}}{0.00375} = 4.889 \text{ V}$$

and

$$V_2 = \cfrac{\begin{vmatrix} 0.075 & 0.250 \\ -0.025 & 0.150 \end{vmatrix}}{\begin{vmatrix} 0.075 & -0.025 \\ -0.025 & 0.058\bar{3} \end{vmatrix}}$$

$$= \frac{(0.075)(0.150) - (-0.025)(0.250)}{0.00375}$$

$$= \frac{0.0175}{0.00375} = 4.667 \text{ V}$$

If we go back to the original circuit of Figure 8–30, we see that the voltage V_2 is the same as the voltage V_a, namely

$$V_a = 4.667 \text{ V} = 6.0 \text{ V} + V_{ab}$$

Therefore, the voltage V_{ab} is simply found as

$$V_{ab} = 4.667 \text{ V} - 6.0 \text{ V} = -1.333 \text{ V}$$

■ **EXAMPLE 8–15**   Determine the nodal voltages for the circuit shown in Figure 8–32.

Figure 8–32

Solution   By following the steps outlined, the circuit may be re-drawn as shown in Figure 8–33.

Figure 8–33

Applying Kirchhoff's current law to the nodes corresponding to V_1 and V_2, the following nodal equations are obtained:

$$\sum I_{\text{leaving}} = \sum I_{\text{entering}}$$

Node V_1: $\qquad\qquad I_1 + I_2 = 2 \text{ A}$

Node V_2: $\qquad\qquad I_3 + I_4 = I_2 + 3 \text{ A}$

The currents may once again be written in terms of the voltages across the resistors:

$$I_1 = \frac{V_1}{5 \,\Omega}$$

$$I_2 = \frac{V_1 - V_2}{3 \,\Omega}$$

$$I_3 = \frac{V_2}{4 \,\Omega}$$

$$I_4 = \frac{V_2}{6 \,\Omega}$$

The nodal equations become

Node V_1: $\qquad\qquad \dfrac{V_1}{5 \,\Omega} + \dfrac{(V_1 - V_2)}{3 \,\Omega} = 2 \text{ A}$

Node V_2: $\qquad\qquad \dfrac{V_2}{4 \,\Omega} + \dfrac{V_2}{6 \,\Omega} = \dfrac{(V_1 - V_2)}{3 \,\Omega} + 3 \text{ A}$

These equations may now be simplified as

Node V_1: $\qquad\qquad \left(\dfrac{1}{5 \,\Omega} + \dfrac{1}{3 \,\Omega}\right) V_1 - \left(\dfrac{1}{3 \,\Omega}\right) V_2 = 2 \text{ A}$

Node V_2: $\qquad\qquad -\left(\dfrac{1}{3 \,\Omega}\right) V_1 + \left(\dfrac{1}{4 \,\Omega} + \dfrac{1}{6 \,\Omega} + \dfrac{1}{3 \,\Omega}\right) V_2 = 3 \text{ A}$

The solutions for V_1 and V_2 are found using determinants:

$$V_1 = \frac{\begin{vmatrix} 2 & -0.333 \\ 3 & 0.750 \end{vmatrix}}{\begin{vmatrix} 0.533 & -0.333 \\ -0.333 & 0.750 \end{vmatrix}} = \frac{2.500}{0.289} = 8.654 \text{ V}$$

$$V_2 = \frac{\begin{vmatrix} 0.533 & 2 \\ -0.333 & 3 \end{vmatrix}}{\begin{vmatrix} 0.533 & -0.333 \\ -0.333 & 0.750 \end{vmatrix}} = \frac{2.267}{0.289} = 7.846 \text{ V}$$

In the previous two examples, you may have noticed that the simultaneous linear equations have a format similar to that developed for mesh analysis. When we wrote the nodal equation for node V_1 the coefficient for the variable V_1 was positive, and it had a magnitude given by the summation of the conductance attached to this node. The coefficient for the variable V_2 was negative and had a magnitude given by the mutual conductance between nodes V_1 and V_2.

Format Approach

A simple format approach may be used to write the nodal equations for any network having $n + 1$ nodes. Where one of these nodes is denoted as the reference node, there will be n simultaneous linear equations which will appear as follows:

$$G_{11}V_1 - G_{12}V_2 - G_{13}V_3 - \ldots - R_{1n}V_n = I_1$$
$$-G_{21}V_1 + G_{22}V_2 - G_{23}V_3 - \ldots - R_{2n}V_n = I_2$$
$$\vdots$$
$$-G_{n1}V_1 - G_{n2}V_2 - G_{n3}V_3 - \ldots + R_{nn}V_n = I_n$$

The coefficients (constants) G_{11}, G_{22}, G_{33}, . . . , G_{nn} represent the summation of the conductances attached to the particular node. The remaining coefficients are called the *mutual conductance* terms. For example, the mutual conductance G_{23} is the conductance attached to node V_2, which is common to node V_3. If there is no conductance that is common to two nodes, then this term would be zero. Notice that the terms G_{11}, G_{22}, G_{33}, . . . , G_{nn} are positive and that the mutual conductance terms are negative. Further, if the equations are written correctly, then the terms will be symmetrical about the principal diagonal, i.e., $G_{23} = G_{32}$.

The terms V_1, V_2, . . . , V_n are the unknown node voltages. Each voltage represents the potential difference between the node in question and the reference node.

The terms I_1, I_2, . . . , I_n are the summation of current sources entering the node. If a current source has a current such that it is leaving the node, then the current is simply assigned as negative. If a particular current source is shared between two nodes, then this current must be included in both nodal equations.

The method used in applying the format approach of nodal analysis is as follows:

1. Convert voltage sources into equivalent current sources.

2. Label the reference node as $\perp$. Label the remaining nodes as V_1, V_2, . . . , V_n.

3. Write the linear equation for each node using the format outlined.

4. Solve the resulting simultaneous linear equations for V_1, V_2, . . . , V_n.

The next examples illustrate how the format approach is used to solve circuit problems.

■ **EXAMPLE 8–16** Determine the nodal voltages for the circuit shown in Figure 8–34.

Figure 8–34

Solution   The circuit has a total of three nodes: the reference node (at a potential of zero volts) and two other nodes, V_1 and V_2.

By applying the format approach for writing the nodal equations, we get two equations:

Node V_1: $\left(\dfrac{1}{3\,\Omega} + \dfrac{1}{5\,\Omega}\right)V_1 - \left(\dfrac{1}{5\,\Omega}\right)V_2 = -6\,\text{A} + 1\,\text{A}$

Node V_2: $-\left(\dfrac{1}{5\,\Omega}\right)V_1 + \left(\dfrac{1}{5\,\Omega} + \dfrac{1}{4\,\Omega}\right)V_2 = -1\,\text{A} - 2\,\text{A}$

On the right-hand sides of the above, those currents that are leaving the nodes and not entering are given a negative sign.

These equations may be rewritten as

Node V_1: $(0.533\ \text{S})V_1 - (0.200\ \text{S})V_2 = -5\ \text{A}$

Node V_2: $-(0.200\ \text{S})V_1 + (0.450\ \text{S})V_2 = -3\ \text{A}$

Using determinants to solve these equations, we have

$$V_1 = \frac{\begin{vmatrix} -5 & -0.200 \\ -3 & 0.450 \end{vmatrix}}{\begin{vmatrix} 0.533 & -0.200 \\ -0.200 & 0.450 \end{vmatrix}} = \frac{-2.85}{0.200} = -14.25\ \text{V}$$

$$V_2 = \frac{\begin{vmatrix} 0.533 & -5 \\ -0.200 & -3 \end{vmatrix}}{\begin{vmatrix} 0.533 & -0.200 \\ -0.200 & 0.450 \end{vmatrix}} = \frac{-2.60}{0.200} = -13.0\ \text{V}$$

■ **EXAMPLE 8–17**   Use nodal analysis to find the nodal voltages for the circuit of Figure 8–35. Use the answers to solve for the current through R_1.

Figure 8–35

Solution   In order to apply nodal analysis, we must first convert the voltage source into its equivalent current source. The resulting circuit is shown in Figure 8–36.

Figure 8–36

Labelling the nodes and writing the nodal equations, we obtain the following:

Node V_1: $\left(\dfrac{1}{5\ k\Omega} + \dfrac{1}{3\ k\Omega} + \dfrac{1}{4\ k\Omega}\right)V_1 - \left(\dfrac{1}{4\ k\Omega}\right)V_2 = 2\ \text{mA} - 3\ \text{mA}$

Node V_2: $-\left(\dfrac{1}{4\ k\Omega}\right)V_1 + \left(\dfrac{1}{4\ k\Omega} + \dfrac{1}{2\ k\Omega}\right)V_2 = 2\ \text{mA}$

Because it is inconvenient to use kilohms and milliamps throughout our calculations, we may eliminate these units in our calculations. You have already seen that any voltage obtained by using these quantities will result in the units being "volts." Therefore the nodal equations may be simplified as

Node V_1: $(0.7833)V_1 - (0.2500)V_2 = -1$

Node V_2: $-(0.2500)V_1 + (0.750)V_2 = 2$

The solutions are as follows:

$$V_1 = \frac{\begin{vmatrix} -1 & -0.250 \\ 2 & 0.750 \end{vmatrix}}{\begin{vmatrix} 0.7833 & -0.250 \\ -0.250 & 0.750 \end{vmatrix}} = \frac{-0.250}{0.525} = -0.476 \text{ V}$$

$$V_2 = \frac{\begin{vmatrix} 0.7833 & -1 \\ -0.250 & 2 \end{vmatrix}}{\begin{vmatrix} 0.7833 & -0.250 \\ -0.250 & 0.750 \end{vmatrix}} = \frac{1.3167}{0.525} = 2.508 \text{ V}$$

Using the values derived for the nodal voltages, it is now possible to solve for any other quantities in the circuit. To determine the current through resistor $R_1 = 5$ kΩ, we first reassemble the circuit as it appeared originally. Since the node voltage V_1 is the same in both circuits, we use it in determining the desired current. The resistor may be isolated as shown in Figure 8–37.

Figure 8–37

The current is easily found as

$$I = 10 \text{ V} - \frac{(-0.476 \text{ V})}{5 \text{ k}\Omega} = 2.095 \text{ mA}   \text{(upward)}$$

practice problem
IV

Use nodal analysis to determine the node voltages for the circuit of Figure 8–38.

Figure 8-38

8.7 Delta-Wye (Pi-Tee) Conversion

Delta-Wye Conversion

Figure 8–39

You have previously examined resistor networks involving series, parallel, and series-parallel combinations. We will next examine networks which cannot be placed into any of the above categories. While these circuits may be analyzed using techniques developed earlier in this chapter, there is an easier approach. For example, consider the circuit shown in Figure 8–39.

This circuit could be analyzed using mesh analysis. However, you see that the analysis would involve solving four simultaneous linear equations, since there are four separate loops in the circuit. If we were to use nodal analysis, the solution would require determining three node voltages, since there are three nodes in addition to a reference node. Unless a computer is used, both techniques are very time-consuming and prone to error.

As you have already seen, it is occasionally easier to examine a circuit after it has been converted to some equivalent form. We will now develop a technique for converting a circuit from a delta (or pi) into an equivalent wye (or tee) circuit. Consider the circuits shown in Figure 8–40. We start by making the assumption that the networks shown in Figure 8–40(a) are equivalent to those shown in Figure 8–40(b). Then, using this assumption, we will determine the mathematical relationships between the various resistors in the equivalent circuits.

The circuit of Figure 8–40(a) can be equivalent to the circuit of Figure 8–40(b) only if the resistance "seen" between any two terminals is

(a) Wye ("Y") or Tee ("T") network   (b) Delta ("Δ") or Pi ("Π") network

Figure 8–40

exactly the same. If we were to connect a source between terminals a and b of the "Y," the resistance between the terminals would be

$$R_{ab} = R_1 + R_2 \qquad (8\text{--}5)$$

But the resistance between terminals a and b of the "Δ" is

$$R_{ab} = R_C \| (R_A + R_B) \qquad (8\text{--}6)$$

Combining Equations 8–5 and 8–6, we get

$$R_1 + R_2 = \frac{R_C(R_A + R_B)}{R_A + R_B + R_C}$$

$$R_1 + R_2 = \frac{R_A R_C + R_B R_C}{R_A + R_B + R_C} \qquad (8\text{--}7)$$

Using a similar approach between terminals b and c, we get

$$R_2 + R_3 = \frac{R_A R_B + R_A R_C}{R_A + R_B + R_C} \qquad (8\text{--}8)$$

and between terminals c and a we get

$$R_1 + R_3 = \frac{R_A R_B + R_B R_C}{R_A + R_B + R_C} \qquad (8\text{--}9)$$

If Equation 8–8 is subtracted from Equation 8–7, then

$$R_1 + R_2 - (R_2 + R_3) = \frac{R_A R_C + R_B R_C}{R_A + R_B + R_C} - \frac{R_A R_B + R_A R_C}{R_A + R_B + R_C}$$

$$R_1 - R_3 = \frac{R_B R_C - R_A R_B}{R_A + R_B + R_C} \qquad (8\text{--}10)$$

Adding Equations 8–9 and 8–10, we get

$$R_1 + R_3 + R_1 - R_3 = \frac{R_A R_B + R_B R_C}{R_A + R_B + R_C} + \frac{R_B R_C - R_A R_B}{R_A + R_B + R_C}$$

$$2R_1 = \frac{2R_B R_C}{R_A + R_B + R_C}$$

$$R_1 = \frac{R_B R_C}{R_A + R_B + R_C} \qquad (8\text{--}11)$$

Using a similar approach, we obtain

$$R_2 = \frac{R_A R_C}{R_A + R_B + R_C} \qquad (8\text{--}12)$$

$$R_3 = \frac{R_A R_B}{R_A + R_B + R_C} \qquad (8\text{--}13)$$

Notice that any resistor connected to a point of the "Y" is obtained by finding the product of the resistors connected to the same point in the "Δ" and then dividing by the sum of all the "Δ" resistances.

If all the resistors in a Δ circuit have the same value, R_Δ, then the resulting resistors in the equivalent Y network will also be equal and have a value given as

$$R_Y = \frac{R_\Delta}{3} \qquad (8\text{--}14)$$

■ **EXAMPLE 8–18**   Find the equivalent Y circuit for the Δ circuit shown in Figure 8–41.

Figure 8–41

Solution   From the circuit of Figure 8–41, we see that we have the following resistor values:

$$R_A = 90 \ \Omega$$

$$R_B = 60 \ \Omega$$

$$R_C = 30 \ \Omega$$

Applying Equations 8–11 through 8–13 we have the following equivalent "Y" resistor values:

$$R_1 = \frac{(30 \ \Omega)(60 \ \Omega)}{30 \ \Omega + 60 \ \Omega + 90 \ \Omega}$$

$$= \frac{1800 \ \Omega}{180} = 10 \ \Omega$$

$$R_2 = \frac{(30 \ \Omega)(90 \ \Omega)}{30 \ \Omega + 60 \ \Omega + 90 \ \Omega}$$

$$= \frac{2700 \ \Omega}{180} = 15 \ \Omega$$

$$R_3 = \frac{(60 \ \Omega)(90 \ \Omega)}{30 \ \Omega + 60 \ \Omega + 90 \ \Omega}$$

$$= \frac{5400 \ \Omega}{180} = 30 \ \Omega$$

The resulting circuit is shown in Figure 8–42.

Figure 8–42

Wye-Delta Conversion

By using Equations 8–11 to 8–13, it is possible to derive another set of equations which allow the conversion from a "Y" into an equivalent "Δ." Examining Equations 8–11 through 8–13, we see that the following must be true:

$$R_A + R_B + R_C = \frac{R_A R_B}{R_3} = \frac{R_A R_C}{R_2} = \frac{R_B R_C}{R_1}$$

From the above expression we may write the following two equations:

$$R_B = \frac{R_A R_1}{R_2} \tag{8–15}$$

$$R_C\ 3 = \frac{R_A R_1}{R_3} \tag{8–16}$$

Now, substituting Equations 8–15 and 8–16 into Equation 8–11, we have the following:

$$R_1 = \frac{\left(\dfrac{R_A R_1}{R_2}\right)\left(\dfrac{R_A R_1}{R_3}\right)}{R_A + \left(\dfrac{R_A R_1}{R_2}\right) + \left(\dfrac{R_A R_1}{R_3}\right)}$$

By factoring R_A out of each term in the denominator, we are able to arrive at

$$R_1 = \frac{\left(\dfrac{R_A R_1}{R_2}\right)\left(\dfrac{R_A R_1}{R_3}\right)}{R_A\left[1 + \left(\dfrac{R_1}{R_2}\right) + \left(\dfrac{R_1}{R_3}\right)\right]}$$

$$R_1 = \frac{\left(\dfrac{R_A R_1 R_1}{R_2 R_3}\right)}{\left[1 + \left(\dfrac{R_1}{R_2}\right) + \left(\dfrac{R_1}{R_3}\right)\right]}$$

$$= \frac{\left(\dfrac{R_A R_1 R_1}{R_2 R_3}\right)}{\left(\dfrac{R_1 R_2 + R_1 R_3 + R_2 R_3}{R_2 R_3}\right)}$$

$$= \frac{R_A R_1 R_1}{R_1 R_2 + R_1 R_3 + R_2 R_3}$$

Rewriting the above expression gives

$$R_A = \frac{R_1 R_2 + R_1 R_3 + R_2 R_3}{R_1} \tag{8–17}$$

Similarly,

$$R_B = \frac{R_1 R_2 + R_1 R_3 + R_2 R_3}{R_2} \tag{8–18}$$

and

$$R_C = \frac{R_1 R_2 + R_1 R_3 + R_2 R_3}{R_3} \qquad (8\text{--}19)$$

In general, we see that the resistor in any side of a "Δ" is found by taking the sum of all **two-product** combinations of "Y" resistor values and then dividing by the resistance in the "Y" which is located directly opposite to the resistor being calculated.

If the resistors in a Y network are all equal, then the resultant resistors in the equivalent Δ circuit will also be equal and given as

$$R_\Delta = 3R_Y \qquad (8\text{--}20)$$

■ **EXAMPLE 8–19**   Find the Δ network equivalent of the Y network shown in Figure 8–43.

Figure 8–43

Solution   The equivalent Δ network is shown in Figure 8–44.

Figure 8–44

The values of the resistors are determined as follows:

$$R_A = \frac{(4.8 \text{ k}\Omega)(2.4 \text{ k}\Omega) + (4.8 \text{ k}\Omega)(3.6 \text{ k}\Omega) + (2.4 \text{ k}\Omega)(3.6 \text{ k}\Omega)}{4.8 \text{ k}\Omega}$$

$$= 7.8 \text{ k}\Omega$$

$$R_B = \frac{(4.8 \text{ k}\Omega)(2.4 \text{ k}\Omega) + (4.8 \text{ k}\Omega)(3.6 \text{ k}\Omega) + (2.4 \text{ k}\Omega)(3.6 \text{ k}\Omega)}{3.6 \text{ k}\Omega}$$

$$= 10.4 \text{ k}\Omega$$

$$R_C = \frac{(4.8 \text{ k}\Omega)(2.4 \text{ k}\Omega) + (4.8 \text{ k}\Omega)(3.6 \text{ k}\Omega) + (2.4 \text{ k}\Omega)(3.6 \text{ k}\Omega)}{2.4 \text{ k}\Omega}$$

$$= 15.6 \text{ k}\Omega$$

■ **EXAMPLE 8–20**   Given the circuit of Figure 8–45, find the total resistance, R_T, and the total current, I.

Figure 8–45

Solution   As is often the case, the given circuit may be solved in one of two ways. We may convert the "Δ" into its equivalent "Y," and solve the circuit by placing the resultant branches in parallel, or we may convert the "Y" into its equivalent "Δ." We choose to use the latter conversion since the resistors in the "Y" have the same value. The equivalent "Δ" will have all resistors given as

$$R_\Delta = 3(10\ \Omega) = 30\ \Omega$$

The resulting circuit is shown in Figure 8–46(a).

(a)

(b)

Figure 8–46

We see that the sides of the resulting "Δ" are in parallel, which allows us to simplify the circuit even further as shown in Figure 8–46(b). The total restance of the circuit is now easily determined as

$$R_T = 15 \ \Omega \| (20 \ \Omega + 22.5 \ \Omega)$$

$$= 11.09 \ k\Omega$$

This results in a circuit current of

$$I = \frac{30 \ V}{11.09 \ k\Omega} = 2.706 \ mA$$

practice problem V

Convert the Δ network of Figure 8–44 into an equivalent Y network. Verify that the result you obtain is the same as that found in Figure 8–43.

8.8 Bridge Networks

In this section you will be introduced to the *bridge network*. Bridge networks are used in electronic measuring equipment to precisely measure resistance in dc circuits and similar quantities in ac circuits. The bridge circuit was originally used by Sir Charles Wheatstone in the mid-nineteenth century to measure resistance by balancing small currents. The Wheatstone bridge circuit is still used to measure resistance very precisely. The digital bridge, shown in Figure 8–47, is an example of one such instrument.

Figure 8–47   Digital bridge used for precisely measuring resistance, inductance, and capacitance.

You will use the techniques developed earlier in the chapter to analyze the operation of these networks. Bridge circuits may be shown in various configurations as seen in Figure 8–48.

(a)

(b)

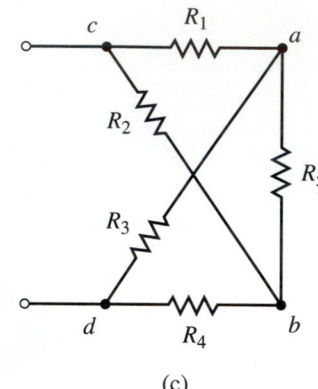

(c)

Figure 8–48

Although a bridge circuit may appear in one of three forms, you can see that they are equivalent. There are, however, two different states of bridges: the *balanced bridge* and the *unbalanced bridge*.

A *balanced bridge* is one in which the current through the resistance R_5 is equal to zero. In practical circuits, R_5 is generally a variable resistor in series with a sensitive galvanometer. When the current through R_5 is zero, then it follows that

$$V_{ab} = (R_5)(0 \text{ A}) = 0 \text{ V}$$

$$I_{R_1} = I_{R_3} = \frac{V_{cd}}{R_1 + R_3}$$

$$I_{R_2} = I_{R_4} = \frac{V_{cd}}{R_2 + R_4}$$

But the voltage V_{ab} is found as

$$V_{ab} = V_{ad} - V_{bd} = 0$$

Therefore, $V_{ad} = V_{bd}$ and

$$R_3 I_{R_3} = R_4 I_{R_4}$$

$$R_3 \left(\frac{V_{cd}}{R_1 + R_3} \right) = R_4 \left(\frac{V_{cd}}{R_2 + R_4} \right)$$

which simplifies to

$$\frac{R_3}{R_1 + R_3} = \frac{R_4}{R_2 + R_4}$$

Now, if we invert both sides of the equation and simplify, we get the following:

$$\frac{R_1 + R_3}{R_3} = \frac{R_2 + R_4}{R_4}$$

$$\frac{R_1}{R_3} + 1 = \frac{R_2}{R_4} + 1$$

Finally, by subtracting 1 from each side, we obtain the following ratio for a balanced bridge:

$$\frac{R_1}{R_3} = \frac{R_2}{R_4} \tag{8-21}$$

From Equation 8–21, we notice that a bridge network is balanced whenever the ratios of the resistors in the two arms are the same.

An *unbalanced bridge* is one in which the current through R_5 is not zero and so the above ratio does not apply to an unbalanced bridge network. Figure 8–49 illustrates each condition of a bridge network.

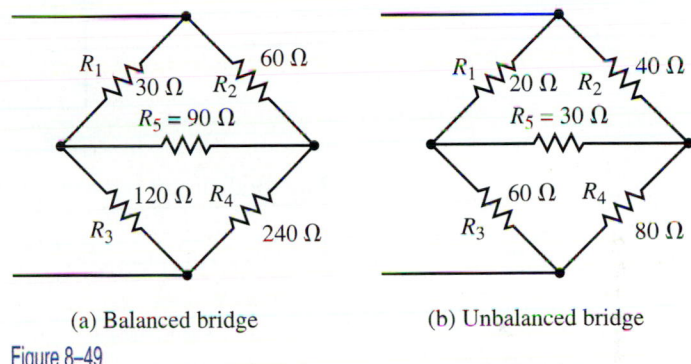

(a) Balanced bridge   (b) Unbalanced bridge

Figure 8-49

If a balanced bridge appears as part of a complete circuit, its analysis is very simple since the resistor R_5 may be removed and replaced with either a short (since $V_{R_5} = 0$) or an open (since $I_{R_5} = 0$).

However, if a circuit contains an unbalanced bridge, the analysis is more complicated. In such cases, it is possible to determine currents and voltages by using mesh analysis, nodal analysis, or by using Δ to Y conversion. The following examples illustrate how bridges may be analyzed.

■ **EXAMPLE 8–21**   Solve for the currents through R_1 and R_4 in the circuit of Figure 8–50.

Figure 8-50

Solution   We see that the bridge of the above circuit is balanced (since $R_1/R_3 = R_2/R_4$). Because the circuit is balanced, we may remove R_5 and replace it with either a short circuit (since the voltage across a short circuit is zero) or an open circuit (since the current through an open circuit is zero). The remaining circuit is then solved by one of the methods developed in previous chapters. Both methods will be illustrated to show that the results are exactly the same.

Method 1: If R_5 is replaced by an open, the result is the circuit shown in Figure 8–51.

$R_S = 10\ \Omega$

I_{R_1}

R_1 $3\ \Omega$

R_2 $6\ \Omega$

60 V

R_3 $12\ \Omega$

I_{R_4} R_4 $24\ \Omega$

Figure 8–51

The total circuit resistance is found as

$$R_T = 10\ \Omega + (3\ \Omega + 12\ \Omega)\|(6\ \Omega + 24\ \Omega)$$

$$= 10\ \Omega + 15\ \Omega\|30\ \Omega$$

$$= 20\ \Omega$$

The circuit current is

$$I_T = \frac{60\ \text{V}}{20\ \Omega} = 3.0\ \text{A}$$

The current in each branch is then found by using the current divider rule:

$$I_{R_1} = \left(\frac{30\ \Omega}{30\ \Omega + 15\ \Omega}\right)(3.0\ \text{A}) = 2.0\ \text{A}$$

$$I_{R_4} = \frac{10\ \Omega}{24\ \Omega + 6\ \Omega}(3.0\ \text{A}) = 1.0\ \text{A}$$

Method 2: If R_5 is replaced with a short circuit, the result is the circuit shown in Figure 8–52.

$R_S = 10\ \Omega$

I_{R_1}

R_1 $3\ \Omega$

R_2 $6\ \Omega$

60 V

R_3 $12\ \Omega$

R_4 $24\ \Omega$

I_{R_4}

Figure 8–52

The total circuit resistance is found as

$$R_T = 10\ \Omega + (3\ \Omega\|6\ \Omega) + (12\ \Omega\|24\ \Omega)$$

$$= 10\ \Omega + 2\ \Omega + 8\ \Omega$$

$$= 20\ \Omega$$

The above result is precisely the same as that found using Method 1. Therefore the circuit current will remain as $I_T = 3.0$ A.

The currents through R_1 and R_4 may be found by the current divider rule as

$$I_{R_1} = \left(\frac{6\,\Omega}{6\,\Omega + 3\,\Omega} \right)(3.0\text{ A}) = 2.0\text{ A}$$

and

$$I_{R_4} = \left(\frac{12\,\Omega}{12\,\Omega + 24\,\Omega} \right)(3.0\text{ A}) = 1.0\text{ A}$$

Clearly, theses results are precisely those obtained in Method 1, illustrating that the methods are equivalent. Remember, though, R_5 can be replaced with a short circuit or an open circuit only when the bridge is balanced.

■ **EXAMPLE 8–22**   Use mesh analysis to find the currents through R_1 and R_5 in the circuit of Figure 8–53.

$R_S = 6\ \Omega$

R_1 $6\ \Omega$

R_2 $12\ \Omega$

R_5 $18\ \Omega$

30 V

R_3 $3\ \Omega$

R_4 $3\ \Omega$

I_1 I_2 I_3

Figure 8–53

Solution   After assigning loop currents as shown, we write the loop equations as

Loop 1: $(15\,\Omega)I_1 - (6\,\Omega)I_2 - (3\,\Omega)I_3 = 30$ V

Loop 2: $-(6\,\Omega)I_1 + (36\,\Omega)I_2 - (18\,\Omega)I_3 = 0$

Loop 3: $-(3\,\Omega)I_1 - (18\,\Omega)I_2 + (24\,\Omega)I_3 = 0$

The determinant for the denominator will is

$$D = \begin{vmatrix} 15 & -6 & -6 \\ -6 & 36 & -18 \\ 3 & -18 & 24 \end{vmatrix} = 6264$$

Notice that, as expected, the elements in the principal diagonal are positive and that the determinant is symmetrical around the principal diagonal.

The loop currents are now evaluated as

$$I_1 = \frac{\begin{vmatrix} 30 & -6 & -3 \\ 0 & 36 & -18 \\ 0 & -18 & 24 \end{vmatrix}}{D} = \frac{16\,200}{6264} = 2.586\text{ A}$$

$$I_2 = \frac{\begin{vmatrix} 15 & 30 & -3 \\ -6 & 0 & -18 \\ -3 & 0 & 24 \end{vmatrix}}{D} = \frac{5940}{6264} = 0.948 \text{ A}$$

$$I_3 = \frac{\begin{vmatrix} 15 & -6 & 30 \\ -6 & 36 & 0 \\ -3 & -18 & 0 \end{vmatrix}}{D} = \frac{6480}{6264} = 1.034 \text{ A}$$

The current through R_1 is found as

$$I_{R_1} = I_1 - I_2 = 2.586 \text{ A} - 0.948 \text{ A} = 1.638 \text{ A}$$

The current through R_5 is found as

$$I_{R_5} = I_3 - I_2 = 1.034 \text{ A} - 0.948 \text{ A}$$
$$= 0.086 \text{ A} \quad \text{to the right}$$

The previous example illustrates that if the bridge is not balanced, there will always be some current through resistor R_5. The unbalanced circuit may also be easily analyzed using nodal analysis, as in the following example.

■ **EXAMPLE 8–23**   Determine the node voltages and the voltage V_{R_5} for the circuit of Figure 8–54.

Figure 8–54

Solution   By converting the voltage source into an equivalent current source, we obtain the circuit shown in Figure 8–55.

Figure 8–55

The nodal equations for the circuit are as follows:

Node 1: $\left(\dfrac{1}{6\,\Omega}+\dfrac{1}{6\,\Omega}+\dfrac{1}{12\,\Omega}\right)V_1-\left(\dfrac{1}{6\,\Omega}\right)V_2-\left(\dfrac{1}{12\,\Omega}\right)V_3=5\ \text{A}$

Node 2: $-\left(\dfrac{1}{6\,\Omega}\right)V_1+\left(\dfrac{1}{6\,\Omega}+\dfrac{1}{3\,\Omega}+\dfrac{1}{18\,\Omega}\right)V_2-\left(\dfrac{1}{18\,\Omega}\right)V_3=0\ \text{A}$

Node 3: $-\left(\dfrac{1}{12\,\Omega}\right)V_1-\left(\dfrac{1}{18\,\Omega}\right)V_2+\left(\dfrac{1}{3\,\Omega}+\dfrac{1}{12\,\Omega}+\dfrac{1}{18\,\Omega}\right)V_3=0\ \text{A}$

The linear equations are

Node 1: $0.4167V_1-0.1667V_2-0.0833V_3=5\ \text{A}$

Node 2: $-0.1667V_1+0.5556V_2-0.0556V_3=0$

Node 3: $-0.0833V_1-0.0556V_2+0.4722V_3=0$

The determinant of the denominator is

$$D=\begin{vmatrix} 0.4167 & -0.1667 & -0.0833 \\ -0.1667 & 0.5556 & -0.0556 \\ -0.0833 & -0.0556 & 0.4722 \end{vmatrix}=0.08951\ \text{A}$$

Again notice that the elements on the principal diagonal are positive and that the determinant is symmetrical about the principal diagonal.

The node voltages are calculated to be

$$V_1=\frac{\begin{vmatrix} 5 & -0.1667 & -0.0833 \\ 0 & 0.5556 & -0.0556 \\ 0 & -0.0556 & 0.4722 \end{vmatrix}}{D}=\frac{1.2963}{0.08951}=14.48\ \text{A}$$

$$V_2=\begin{vmatrix} 0.4167 & 5 & -0.0833 \\ -0.1667 & 0 & -0.0556 \\ -0.0833 & 0 & 0.4722 \end{vmatrix}=\frac{0.41667}{0.08951}=4.655\ \text{V}$$

$$V_3=\begin{vmatrix} 0.4167 & -0.1667 & 5 \\ -0.1667 & 0.0556 & 0 \\ -0.0833 & -0.0556 & 0 \end{vmatrix}=\frac{0.2778}{0.08951}=3.103\ \text{V}$$

Using the above results, we find the voltage across R_5:

$$V_{R_5} = V_2 - V_3 = 4.655 \text{ V} - 3.103 \text{ V} = 1.552 \text{ V}$$

and the current through R_5 is

$$I_{R_5} = \frac{1.552 \text{ V}}{18 \text{ }\Omega} = 0.086 \text{ A} \quad \text{to the right}$$

As expected, the results are the same whether we use mesh analysis or nodal analysis. It is therefore a matter of personal preference as to which approach should be used.

A final method for analyzing bridge networks involves the use of Δ to Y conversion. The following example illustrates the method used.

■ **EXAMPLE 8–24**   Find the current through R_5 for the circuit shown in Figure 8–56.

Figure 8–56

Solution   By inspection we see that this circuit is not balanced, since

$$\frac{R_1}{R_3} \neq \frac{R_2}{R_4}$$

Therefore, the current through R_5 cannot be zero. Notice, also, that the circuit contains two possible Δ configurations. If we choose to convert the top Δ to its equivalent Y, we get the circuit shown in Figure 8–57.

$$\frac{(6)(18)}{6 + 12 + 18} = 3\ \Omega$$

$$\frac{(6)(12)}{6 + 12 + 18} = 2\ \Omega$$

$$\frac{(12)(18)}{6 + 12 + 18} = 6\ \Omega$$

Figure 8–57

By combining resistors, it is possible to reduce the complicated circuit to the simple series circuit shown in Figure 8–58.

$$2 + (3 + 3)\ \|\ (6 + 3) = 3.6\ \Omega$$

Figure 8–58

The circuit of Figure 8–58 is easily analyzed to give a total circuit current of

$$I = \frac{30\ \text{V}}{6\ \Omega + 2\ \Omega + 3.6\ \Omega} = 2.586\ \text{A}$$

Using the calculated current, it is possible to work back to the original circuit. The currents in the resistors R_3 and R_4 are found by using the current divider rule for the corresponding resistor branches, as shown in Figure 8–57.

$$I_{R_3} = \frac{(6\ \Omega + 3\ \Omega)}{(6\ \Omega + 3\ \Omega) + (3\ \Omega + 3\ \Omega)}(2.586\ \text{A}) = 1.552\ \text{A}$$

$$I_{R_4} = \frac{(3\ \Omega + 3\ \Omega)}{(6\ \Omega + 3\ \Omega) + (3\ \Omega + 3\ \Omega)}(2.586\ \text{A}) = 1.034\ \text{A}$$

These results are exactly the same as those found in Examples 8–21 and 8–22. Using these currents, it is now possible to determine the voltage V_{bc} as

$$V_{bc} = -(3\ \Omega)I_{R_4} + (3\ \Omega)I_{R_3}$$

$$= (-3\ \Omega)(1.034\ \text{A}) + (3\ \Omega)(3.103\ \text{A})$$

$$= 1.554\ \text{V}$$

The current through R_5 is determined to be

$$I_{R_5} = \frac{1.554\ \text{V}}{18\ \Omega} = 0.086\ \text{A} \quad \text{to the right}$$

**in process
learning
check II**

1. For a balanced bridge, what will be the value of voltage between the midpoints of the arms of the bridge?

2. If a resistor or sensitive galvanometer is placed between the arms of a balanced bridge, what will be the current through the resistor?

3. In order to simplify the analysis of a balanced bridge, how may the resistance R_5, between the arms of the bridge, be replaced?

**practice
problems
VI**

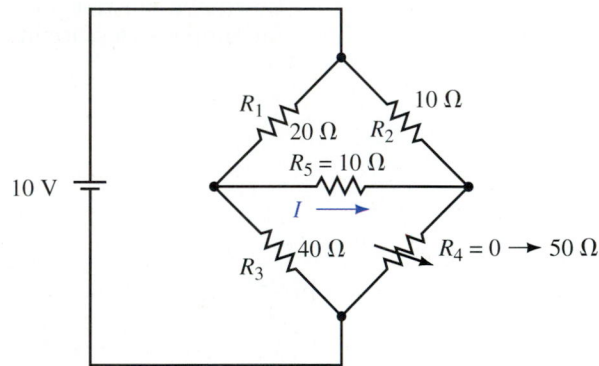

Figure 8–59

1. For the circuit shown in Figure 8–59, what value of R_4 will ensure that the bridge is balanced?

2. Determine the current I through R_5 in Figure 8–59 when $R_4 = 0\ \Omega$ and when $R_4 = 50\ \Omega$.

8.9 Circuit Analysis Using Computers

PSpice is able to analyze a circuit without the need to convert between voltage and current sources or having to write lengthy linear equations. It is possible to have the program output the value of voltage across or current through any element in a given circuit. The following example was previously analyzed using several other methods throughout this chapter.

■ **EXAMPLE 8–25**   Given the circuit of Figure 8–60, use PSpice to find the voltage V_{ab} and the current through each resistor.

Figure 8–60

Solution   The nodes and voltage sources are relabelled using the required PSpice syntax. The resultant circuit is shown in Figure 8–61.

Figure 8–61

The current source is identified in a manner similar to that used to describe a voltage source, namely *Ixxx*, where the current source has a direction from the first node to the second node. The input and output files for this circuit are given below. Once again, the output file has been edited for clarity.

```
Input file:
Example 8-25: Mesh/Nodal Analysis Circuit, Figure 8-61
I1       1    0    5A
R1       1    0    2ohms
R2       1    2    3ohms
Vs1      2    0    8V
R3       1    3    1ohm
Vs2      3    0    6V
.DC    Vs1   8V   8V     1V
.PRINT       DC    V(1)   I(R1)    I(R2)    I(R3)
.OPTIONS NOPAGE
```

```
Output file:
Vs1          V(1)          V(2)          I(R1)          I(R2)          I(R3)
8.000E+00   2.000E+00   8.000E+00   1.000E+00   -2.000E+00   -4.000E+00
```

From the PSpice output file we have the following results:

$$V_{ab} = V(1) = 2.0 \text{ V}$$

$$I_{R_1} = 1.0 \text{ A} \quad \text{(downward)}$$

$$I_{R_2} = 2.0 \text{ A} \quad \text{(upward)}$$

$$I_{R_3} = 4.0 \text{ A} \quad \text{(to the left)}$$

practice problem VII

Write the PSpice input file for the circuit of Figure 8–54, so that the output file will provide the currents through R_1, R_2, and R_5. Compare your results to those obtained in Example 8–23.

Problems

8.1   Constant-Current Sources

1. Find the voltage V_S for the circuit shown in Figure 8–62.

Figure 8–62   Figure 8–63

2. Find the voltage V_S for the circuit shown in Figure 8–63.

3. Refer to the circuit of Figure 8–64:

Figure 8–64

a. Find the current I_3.
b. Determine the voltages V_S and V_1.

4. Consider the circuit of Figure 8–65:
   a. Calculate the voltages V_2 and V_S.
   b. Find the currents I and I_3.

Figure 8–65

5. For the circuit of Figure 8–66, find the currents I_1 and I_2.

Figure 8–66   Figure 8–67

6. Refer to the circuit of Figure 8–67:
   a. Find the voltages V_S and V_2.
   b. Determine the current I_4.

7. Verify that the power supplied by the sources is equal to the summation of the powers dissipated by the resistors in the circuit of Figure 8–66.

8. Verify that the power supplied by the source in the circuit of Figure 8–67 is equal to the summation of the powers dissipated by the resistors.

8.2   Source Conversions

9. Convert each of the voltage sources of Figure 8–63 into its equivalent current source.

(a)   (b)

Figure 8–68

10. Convert each of the current sources of Figure 8–69 into its equivalent voltage source.

(a)   (b)

Figure 8–69

11. Refer to the circuit of Figure 8–70:
  a. Solve for the current through the load resistor using the current divider rule.
  b. Convert the current source into its equivalent voltage source and again determine the current through the load.

Figure 8–70   Figure 8–71

12. Find V_{ab} and I_2 for the network of Figure 8–71.

13. Refer to the circuit of Figure 8–72:

Figure 8–72

  a. Convert the current source and the 330-Ω resistor into an equivalent voltage source.
  b. Solve for the current I through R_L.
  c. Determine the voltage V_{ab}.

14. Refer to the circuit of Figure 8–73:

Figure 8–73

a. Convert the voltage source and the 36-Ω resistor into an equivalent current source.
b. Solve for the current I through R_L.
c. Determine the voltage V_{ab}.

8.3 Current Sources in Parallel and Series

15. Find the voltage V_2 and the current I_1 for the circuit of Figure 8–74.

Figure 8–74

16. Convert the voltage sources of Figure 8–75 into current sources and solve for the current I_1 and the voltage V_{ab}.

Figure 8–75

17. For the network of Figure 8–76 convert the current source and the 2.4-kΩ resistors into a voltage source and find the voltage V_{ab} and the current I_3.

Figure 8–76

18. For the circuit of Figure 8–76, convert the voltage source and the series resistors into an equivalent current source.
  a. Determine the current I_2.
  b. Solve for the voltage V_{ab}.

8.4   Branch-Current Analysis

19. Write the branch-current equations for the networks shown in Figure 8–77 and solve for the branch currents using determinants.

(a)   (b)

Figure 8–77

20. Refer to the circuit of Figure 8–78:
  a. Solve for the current I_1 using branch-current analysis.
  b. Determine the voltage V_{ab}.

Figure 8–78

21. Write the branch-current equations for the circuit shown in Figure 8–79 and solve for the current I_2.

Figure 8–79

22. Refer to the circuit shown in Figure 8–80:

Figure 8–80

a. Write the branch-current equations.
b. Solve for the currents I_1 and I_2.
c. Determine the voltage V_{ab}.

23. Refer to the circuit shown in Figure 8–81:

Figure 8–81

a. Write the branch-current equations.
b. Solve for the current I_2.
c. Determine the voltage V_{ab}.

24. Refer to the circuit shown in Figure 8–82:

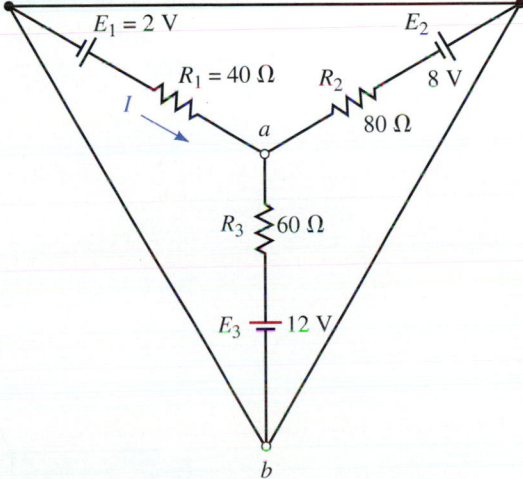

Figure 8–82

a. Write the branch-current equations.
b. Solve for the current I.
c. Determine the voltage V_{ab}.

8.5 Mesh (Loop) Analysis

25. Write the mesh equations for the networks shown in Figure 8–77 and solve for the loop currents.

26. Use mesh analysis for the network of Figure 8–78 to solve for the current I_1.

27. Use mesh analysis to solve for the current I_2 in the network of Figure 8–79.

28. Use mesh analysis to solve for the loop currents in the circuit of Figure 8–81. Use your results to determine I_2 and V_{ab}.

29. Use mesh analysis to solve for the loop currents in the circuit of Figure 8–82. Use your results to determine I and V_{ab}.

30. Using mesh analysis, determine the current through the 6-Ω resistor in each of the networks in Figure 8–83.

(a)   (b)

Figure 8–83

31. Write the mesh equations for the network in Figure 8–84. Solve for the loop currents using determinants.

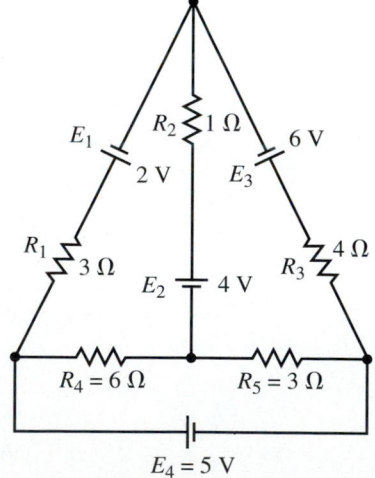

Figure 8–84

32. Repeat Problem 31 for the network in Figure 8–85.

Figure 8–85

8.6 Nodal Analysis

33. Write the nodal equations for each of the networks of Figure 8–86 and solve for the nodal voltages.

(a)   (b)

Figure 8–86

34. Write the nodal equations for each of the networks of Figure 8–87 and determine the voltage V_{ab}.

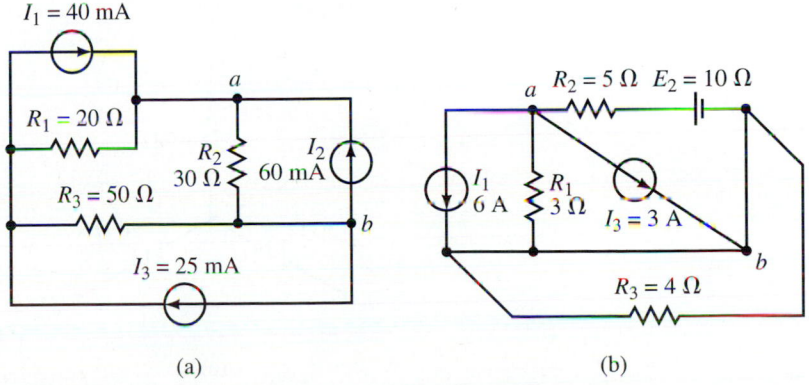

(a)   (b)

Figure 8–87

35. Write the nodal equations for each of the networks of Figure 8–83 and solve for $V_{6\,\Omega}$.

36. Write the nodal equations for the circuit of Figure 8–84 and solve for $V_{6\,\Omega}$.

8.7   Delta-Wye (Pi-Tee) Conversion

37. Convert each of the Δ networks of Figure 8–88 into its equivalent Y configuration.

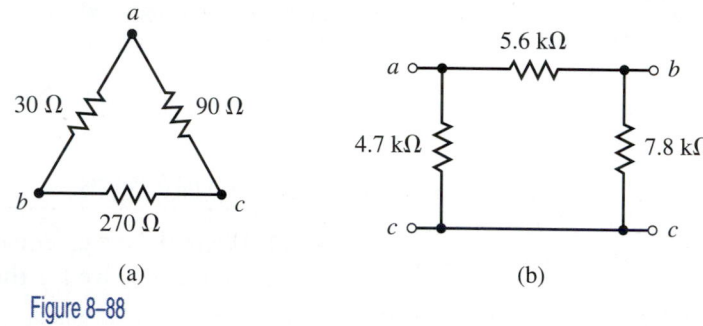

(a)   (b)

Figure 8–88

38. Convert each of the Δ networks of Figure 8–89 into its equivalent Y configuration.

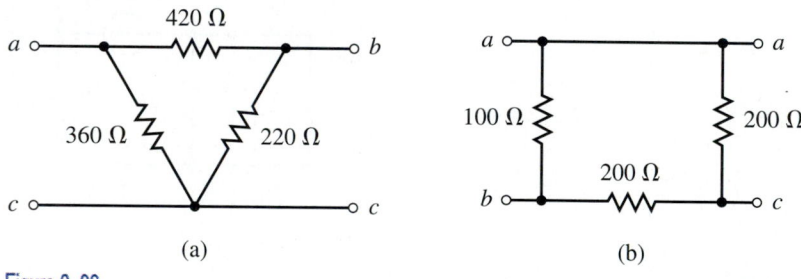

(a)   (b)

Figure 8–89

39. Convert each of the Y networks of Figure 8–90 into its equivalent Δ configuration.

(a)   (b)

Figure 8–90

40. Convert each of the Y networks of Figure 8–91 into its equivalent Δ configuration.

(a)

(b)

Figure 8–91

41. Using Δ-Y or Y-Δ conversion, find the current I for each of the circuits in Figure 8–92.

(a) All resistors are 4.5 kΩ

(b)

Figure 8–92

42. Using Δ-Y or Y-Δ conversion, find the current I and the voltage V_{ab} for each of the circuits in Figure 8–93.

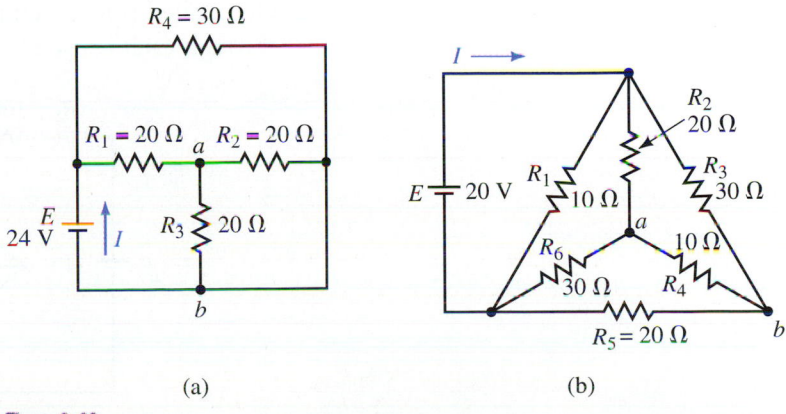

(a)

(b)

Figure 8–93

8.8   Bridge Networks

43. Refer to the bridge network of Figure 8–94:

Figure 8–94

    a. Is the bridge balanced? Explain.
    b. Write the mesh equations.
    c. Calculate the current through R_5.
    d. Determine the voltage across R_5.

44. Consider the bridge network of Figure 8–95:

Figure 8–95

    a. Is the bridge balanced? Explain.
    b. Write the mesh equations.
    c. Determine the current through R_5.
    d. Calculate the voltage across R_5.

45. Given the bridge circuit of Figure 8–96, find the current through each resistor.

Figure 8–96

46. Refer to the bridge circuit of Figure 8–97:

Figure 8–97

a. Determine the value of resistance R_x such that the bridge is balanced.

b. Calculate the current through R_5 when $R_x = 0\ \Omega$ and when $R_x = 10\ k\Omega$.

8.9   Circuit Analysis Using Computers

47. Use PSpice to solve for the currents through all resistors of the circuit shown in Figure 8–84.

48. Use PSpice to solve for the voltage across the 5-kΩ resistor in the circuit of Figure 8–86(b).

Network Theorems

CHAPTER PREVIEW

In this chapter you will learn how to use some basic theorems which will allow the analysis of even the most complex resistive networks. The theorems which are most useful in analyzing networks are the *superposition, Thévenin, Norton,* and *maximum power transfer theorems.*

You will also be introduced to other theorems which, while useful for providing a well-rounded appreciation of circuit analysis, have limited use in the analysis of circuits. These theorems, which apply to specific types of circuits, are the *substitution, reciprocity,* and *Millman* theorems. Your instructor may choose to omit the latter theorems without any loss in continuity.

IRIDIUM will connect users to their local cellular telephone system or to satellites overhead. *(Courtesy of Motorola, Inc.)*

OBJECTIVES

After studying this chapter you will be able to

• apply the superposition theorem to determine the current through or voltage across any resistance in a given network,

• state Thévenin's theorem and determine the Thévenin equivalent circuit of any resistive network,

• state Norton's theorem and determine the Norton equivalent circuit of any resistive network,

• determine the required load resistance of any circuit to ensure that the load receives maximum power from the circuit,

• apply Millman's theorem to determine the current through or voltage across any resistor sup-

plied by any number of sources in parallel,

• state the reciprocity theorem and demonstrate that it applies for a given single-source circuit,

• state the substitution theorem and apply the theorem in simplifying the operation of a given circuit.

9.1 Superposition Theorem

The superposition theorem is a method which allows us to determine the current through or the voltage across any resistor or branch in a network. The advantage of using this approach instead of mesh analysis or nodal analysis is that it is not necessary to use determinants or matrix algebra to analyze a given circuit. The theorem states the following:

The total current through or voltage across a resistor or branch may be determined by summing the effects due to each independent source.

In order to apply the superposition theorem it is necessary to remove all sources other than the one being examined. In order to "zero" a voltage source, we *replace it with a short circuit*, since the voltage across a short circuit is zero volts. A current source is zeroed by *replacing it with an open circuit*, since the current through an open circuit is zero amps.

If we wish to determine the power dissipated by any resistor, we must first find either the voltage across the resistor or the current through the resistor:

$$P = I^2 R = \frac{V^2}{R}$$

notes

T_{he} *superposition theorem does not apply to power, since power is not a linear quantity, but rather is found as the square of either current or voltage.*

■ **EXAMPLE 9–1**   Consider the circuit of Figure 9–1:

Figure 9–1

a. Determine the current in the load resistor, R_L.

b. Verify that the superposition theorem does not apply to power.

Solution

a. We first determine the current through R_L due to the voltage source by removing the current source and replacing in with an open circuit (zero amps) as shown in Figure 9–2.

Current source
replaced with an
open circuit

Figure 9–2

The resulting current through R_L is determined from Ohm's law as

$$I_L(1) = \frac{20\text{ V}}{16\text{ }\Omega + 24\text{ }\Omega} = 0.5\text{ A}$$

Next, we determine the current through R_L due to the current source by removing the voltage source and replacing it with a short circuit (zero volts) as shown in Figure 9–3.

Voltage source
replaced with a
short circuit

Figure 9–3

The resulting current through R_L is found with the current divider rule as

$$I_L(2) = -\left(\frac{24\text{ }\Omega}{24\text{ }\Omega + 16\text{ }\Omega} \right)(2\text{ A}) = -1.2\text{ A}$$

The resultant current through R_L is found by applying the superposition theorem:

$$I_L = 0.5 \text{ A} - 1.2 \text{ A} = -0.7 \text{ A}$$

The negative sign indicates that the current through R_L is opposite to the assumed reference direction. Consequently, the current through R_L will, in fact, be upward with a magnitude of 0.7 A.

    b. If we assume that the superposition theorem applies for power, we would have the power due the first source given as

$$P_1 = I_L{}^2(1)R_L = (0.5 \text{ A})^2(16 \text{ }\Omega) = 4.0 \text{ W}$$

and the power due the second source as

$$P_2 = I_L{}^2(2)R_L = (1.2 \text{ A})^2(16 \text{ }\Omega) = 23.04 \text{ W}$$

The total power, if superposition applies, would be

$$P_T = P_1 + P_2 = 4.0 \text{ W} + 23.04 \text{ W} = 27.04 \text{ W}$$

Clearly, this result is wrong, since the actual power dissipated by the load resistor is correctly given as

$$P_L = I_L{}^2 R_L = (0.7 \text{ A})^2(16 \text{ }\Omega) = 7.84 \text{ W}$$

The superposition theorem may also be used to determine the voltage across any component or branch within the circuit.

■ **EXAMPLE 9–2**   Determine the voltage drop across the resistor R_2 of the circuit shown in Figure 9–4.

Figure 9–4

Solution   Since this circuit has three separate sources, it is necessary to determine the voltage across R_2 due to each individual source.

First, we consider the voltage across R_2 due to the 16-V source as shown in Figure 9–5.

Figure 9–5

The voltage across R_2 will be the same as for the voltage across the parallel combination of $R_2 \| R_3 = 0.8\ \text{k}\Omega$. Therefore,

$$V_{R_2}(1) = -\left(\frac{0.8\ \text{k}\Omega}{0.8\ \text{k}\Omega + 2.4\ \text{k}\Omega}\right)(16\ \text{V}) = -4.0\ \text{V}$$

The negative sign in the above calculation simply indicates that the voltage across the resistor due to the first source is opposite to the assumed reference polarity.

Next, we consider the current source. The resulting circuit will be as shown in Figure 9–6.

Figure 9-6

From this circuit, you can observe that the total resistance "seen" by the current source is

$$R_T = R_1 \| R_2 \| R_3 = 0.6\ \text{k}\Omega$$

The resulting voltage across R_2 is

$$V_{R_2}(2) = (0.6\ \text{k}\Omega)(5\ \text{mA}) = 3.0\ \text{V}$$

Finally, the voltage due to the 32-V source is found by analyzing the circuit of Figure 9–7.

The voltage across R_2 is

$$V_{R_2}(3) = \left(\frac{0.96 \text{ k}\Omega}{0.96 \text{ k}\Omega + 1.6 \text{ k}\Omega} \right)(32 \text{ V}) = 12.0 \text{ V}$$

By superposition, the resulting voltage is

$$V_{R_2} = -4.0 \text{ V} + 3.0 \text{ V} + 12.0 \text{ V} = 11.0 \text{ V}$$

Figure 9–7

practice problem I

Use the superposition theorem to determine the voltage across R_1 and R_3 in the circuit of Figure 9–4.

in process learning check I

Use the final results of Example 9–2 and Practice Problem I to determine the power dissipated by the resistors in the circuit of Figure 9–4. Verify that the superposition theorem does not apply to power.

9.2 Thévenin's Theorem

In this section, we will apply one of the most important theorems of electric circuits. Thévenin's theorem allows even the most complicated circuit to be reduced to a single voltage source and a single resistance. The importance of such a theorem becomes evident when we try to analyze a circuit as shown in Figure 9–8.

Figure 9–8

If we wanted to find the current through the variable load resistor when $R_L = 0$, $R_L = 2$ kΩ, and $R_L = 5$ kΩ using existing methods, we would need to analyze the entire circuit three separate times. However, if we could reduce the entire circuit external to the load resistor to a single voltage source in series with a resistor, the solution becomes very easy.

Thévenin's theorem is a circuit analysis technique which reduces any linear bilateral network to an equivalent circuit having only one voltage source and one series resistor. The resulting two-terminal circuit is equivalent to the original circuit when connected to any external branch or component. In summary, Thévenin's theorem is simplified as follows:

Any linear bilateral network may be reduced to a simplified two-terminal circuit consisting of a single voltage source in series with a single resistor as shown in Figure 9–9.

Figure 9–9 Thévenin equivalent circuit.

The following steps provide a technique which converts any circuit into its Thévenin equivalent:

1. *Remove the load from the circuit.*

2. *Label the resulting two terminals. We will label them as a and b, although any notation may be used.*

3. *Set all sources in the circuit to zero.*

   Voltage sources are set to zero by replacing them with short circuits (zero volts).

   Current sources are set to zero by replacing them with open circuits (zero amps).

4. *Determine the Thévenin equivalent resistance, R_{Th}, by calculating the resistance "seen" between the terminals a and b. It may be necessary to redraw the circuit to simplify this step.*

5. *Replace the sources removed in Step 3, and determine the open-circuit voltage between the terminals. If the circuit has more than one source, it may be necessary to use the superposition theorem. In that case, it will be necessary to determine the open-circuit voltage due to each source separately and then determine the combined effect. The resulting open-circuit voltage will be the value of the Thévenin voltage, E_{Th}.*

6. *Draw the Thévenin equivalent circuit using the resistance determined in Step 4 and the voltage calculated in Step 5. As part of the resulting circuit, include that portion of the network removed in Step 1.*

■ **EXAMPLE 9–3**   Determine the Thévenin equivalent circuit external to the resistor R_L for the circuit of Figure 9–10. Use the Thévenin equivalent circuit to calculate the current through R_L.

Figure 9–10

Solution

Steps 1 and 2: Removing the load resistor from the circuit and labelling the remaining terminals, we obtain the circuit shown in Figure 9–11.

Figure 9–11

Step 3: Setting the sources to zero, we have the circuit shown in Figure 9–12.

Voltage source replaced with a short circuit

Current source replaced with an open circuit

Figure 9–12

Step 4: The Thévenin resistance between the terminals is $R_{Th} = 24\ \Omega$.

Step 5: From Figure 9–11, the open-circuit voltage between terminals a and b is found as

$$V_{ab} = 20\ V - (24\ \Omega)(2\ A) = -28\ V$$

Step 6: The resulting Thévenin equivalent circuit is shown in Figure 9–13.

Figure 9–13

Using this Thévenin equivalent circuit, we easily find the current through R_L as

$$I_L = \left(\frac{28\ V}{24\ \Omega + 16\ \Omega} \right) = 0.700\ A \quad \text{(upward)}$$

This result is the same as that obtained by using the superposition theorem in Example 9–1.

■ **EXAMPLE 9–4**   Find the Thévenin equivalent circuit of the indicated area in Figure 9–14. Using the equivalent circuit, determine the current through the load resistor when $R_L = 0$, $R_L = 2\ k\Omega$, and $R_L = 5\ k\Omega$.

Figure 9–14

Solution

Steps 1, 2, and 3: After removing the load, labelling the terminals, and setting the sources to zero, we have the circuit shown in Figure 9–15.

Voltage source replaced with a short circuit

Current source replaced with an open circuit

Figure 9–15

Step 4: The Thévenin resistance of the circuit is

$$R_{Th} = 6 \text{ k}\Omega \| 2 \text{ k}\Omega = 1.5 \text{ k}\Omega$$

Step 5: Although several methods are possible, we will use the superposition theorem to find the open-circuit voltage V_{ab}. Figure 9–16 shows the circuit for determining the contribution due to the 15-V source.

Figure 9–16

$$V_{ab}(1) = \left(\frac{2 \text{ k}\Omega}{2 \text{ k}\Omega + 6 \text{ k}\Omega} \right)(15 \text{ V}) = +3.75 \text{ V}$$

Figure 9–17 shows the circuit for determining the contribution due to the 5-mA source.

Figure 9–17

$$V_{ab}(2) = \left(\frac{(2\ k\Omega)(6\ k\Omega)}{2\ k\Omega + 6\ k\Omega} \right)(5\ mA) = +7.5\ V$$

The Thévenin equivalent voltage is

$$E_{Th} = V_{ab}(1) + V_{ab}(2) = +3.75\ V + 7.5\ V = 11.25\ V$$

Step 6: The resulting Thévenin equivalent circuit is shown in Figure 9–18.

Figure 9–18

From this circuit, it is now an easy matter to determine the current for any value of load resistor:

$R_L = 0\ \Omega:$ $I_L = \dfrac{11.25\ V}{1.5\ k\Omega} = 7.5\ mA$

$R_L = 2\ k\Omega:$ $I_L = \dfrac{11.25\ V}{1.5\ k\Omega + 2\ k\Omega} = 3.214\ mA$

$R_L = 5\ k\Omega:$ $I_L = \dfrac{11.25\ V}{1.5\ k\Omega + 5\ k\Omega} = 1.731\ mA$

■ **EXAMPLE 9–5**   Find the Thévenin equivalent circuit external to R_5 in the circuit in Figure 9–19. Use the equivalent circuit to determine the current through the resistor.

Figure 9–19

Solution   Notice that the circuit is an unbalanced bridge circuit. If the techniques of the previous chapter had to be used, we would need to solve either three mesh equations or three nodal equations.

Steps 1 and 2: Removing the resistor R_5 from the circuit and labelling the two terminals a and b, we obtain the circuit shown in Figure 9–20.

Figure 9–20

By examining the circuit shown in Figure 9–20, we see that it is no simple task to determine the equivalent circuit between terminals a and b. The process is simplified by redrawing the circuit as illustrated in Figure 9–21.

Figure 9–21

Notice that the circuit of Figure 9–21 has nodes a and b conveniently shown at the top and bottom of the circuit. Additional nodes (node c and node d) are added to simplify the task of correctly placing resistors between the nodes.

After simplifying a circuit, it is always a good idea to ensure that the resulting circuit is indeed an equivalent circuit. You may verify the equivalence of the two circuits by confirming that each component is connected between the same nodes for each circuit.

Now that we have a circuit which is easier to analyze, we find the Thévenin equivalent of the resultant.

Step 3: Setting the voltage source to zero by replacing it with a short, we obtain the circuit shown in Figure 9–22.

Figure 9–22

Step 4: The resulting Thévenin resistance is

$$R_{Th} = 10\ \Omega \| 20\ \Omega + 20\ \Omega \| 50\ \Omega$$

$$= 6.67\ \Omega + 14.29\ \Omega = 20.95\ \Omega$$

Step 5: The open-circuit voltage between the terminals a and b is found by first evaluating the loop currents I_1 and I_2 in the circuit of Figure 9–23.

Figure 9–23

Because the voltage source, E, provides a constant voltage across the resistor combinations R_1-R_3 and R_2-R_4, we simply use the voltage divider rule to determine the voltage across the various components:

$$V_{ab} = -V_{R_1} + V_{R_2}$$

$$= -\frac{(10\ \Omega)(10\ \text{V})}{30\ \Omega} + \frac{(20\ \Omega)(10\ \text{V})}{70\ \Omega}$$

$$= -0.476\ \text{V}$$

Note: The previous technique could not be used if the source had some series resistance, since then the voltage provided to resistor combinations R_1-R_3 and R_2-R_4 would no longer be the entire supply voltage but rather would be dependent upon the value of the series resistance of the source.

Step 6: The resulting Thévenin circuit is shown in Figure 9–24.

Figure 9–24

From the circuit of Figure 9–24, it is now possible to calculate the current through the resistor R_5 as

$$I = \frac{0.476 \text{ V}}{20.95 \ \Omega + 30 \ \Omega} = 9.34 \text{ mA}$$

This example illustrates the importance of labelling the terminals which remain after a component or branch is removed. If we had not labelled the terminals and drawn an equivalent circuit, the current through R_5 would not have been found as easily.

practice
problem
II

Find the Thévenin equivalent circuit external to resistor R_1 in the circuit of Figure 9–1.

practice
problem
III

Use Thévenin's theorem to determine the current through load resistor R_L for the circuit of Figure 9–25.

Figure 9–25

in process
learning
check II

In the circuit of Figure 9–25, what would the value of R_1 need to be in order that the Thévenin resistance is equal to $R_L = 80 \ \Omega$?

9.3 Norton's Theorem

Norton's theorem is a circuit analysis technique which is similar to Thévenin's theorem. By using this theorem the circuit is reduced to a single current source and one parallel resistor. As with the Thévenin equivalent circuit, the resulting two-terminal circuit is equivalent to the original circuit when connected to any external branch or component. In summary, Norton's theorem may be simplified as follows:

Any linear bilateral network may be reduced to a simplified two-terminal circuit consisting of a single current source and a single shunt resistor as shown in Figure 9–26.

Figure 9–26   Norton equivalent circuit.

The following steps provide a technique which allows the conversion of any circuit into its Norton equivalent:

1. *Remove the load from the circuit.*

2. *Label the resulting two terminals. We will label them as a and b, although any notation may be used.*

3. *Set all sources to zero. As before, voltage sources are set to zero by replacing them with short circuits and current sources are set to zero by replacing them with open circuits.*

4. *Determine the Norton equivalent resistance, R_N, by calculating the resistance seen between the terminals a and b. It may be necessary to redraw the circuit to simplify this step.*

5. *Replace the sources removed in Step 3, and determine the current which would occur in a short if the short were connected between terminals a and b. If the original circuit has more than one source, it may be necessary to use the superposition theorem. In this case, it will be necessary to determine the short-circuit current due to each source separately and then determine the combined effect. The resulting short-circuit current will be the value of the Norton current I_N.*

6. *Sketch the Norton equivalent circuit using the resistance determined in Step 4 and the current calculated in Step 5. As part of the resulting circuit, include that portion of the network removed in Step 1.*

The Norton equivalent circuit may also be determined directly from the Thévenin equivalent circuit by using the source conversion technique developed in Chapter 8. As a result, the Thévenin and Norton circuits shown in Figure 9–27 are equivalent.

Thévenin equivalent circuit   Norton equivalent circuit

Figure 9–27

From Figure 9–27 we see that the relationship between the circuits is as follows:

$$E_{Th} = I_N R_N \tag{9–1}$$

$$I_N = \frac{E_{Th}}{R_{Th}} \tag{9–2}$$

■ **EXAMPLE 9–6**   Determine the Norton equivalent circuit external to the resistor R_L for the circuit of Figure 9–28. Use the Norton equivalent circuit to calculate the current through R_L. Compare the results to those obtained using Thévenin's theorem in Example 9–3.

Figure 9–28

Solution

Steps 1 and 2: Remove load resistor R_L from the circuit and label the remaining terminals as a and b. The resulting circuit is shown in Figure 9–29.

Figure 9–29

Step 3: Zero the voltage and current sources as shown in the circuit of Figure 9–30.

Figure 9–30

Step 4: The resulting Norton resistance between the terminals is

$$R_N = R_{ab} = 24 \ \Omega$$

Step 5: The short-circuit current is determined by first calculating the current through the short due to each source. The circuit for each calculation is illustrated in Figure 9–31.

Voltage Source, E: The current in the short between terminals a and b [Figure 9–31(a)] is found from Ohm's law as

$$I_{ab}(1) = \frac{20 \text{ V}}{24 \ \Omega} = 0.833 \text{ A}$$

(a) Voltage source

Figure 9–31   (*Continues*)

Current Source, I: By examining the circuit for the current source [Figure 9–31(b)] we see that the short between terminals a and b effectively removes R_1 from the circuit. Therefore, the current through the short will be

$$I_{ab}(2) = -2.0 \text{ A}$$

Notice that R_1 is shorted by the short circuit between a and b

(b) Current source

Figure 9–31   (*Continued*)

Notice that the current I_{ab} is indicated as being a negative quantity. As we have seen before, this result merely indicates that the actual current is opposite to the assumed reference direction.

Now, applying the superposition theorem, we find the Norton current as

$$I_N = I_{ab}(1) + I_{ab}(2) = 0.833 \text{ A} - 2.0 \text{ A} = -1.167 \text{ A}$$

As before, the negative sign indicates that the short-circuit current is actually from terminal b toward terminal a.

Step 6: The resultant Norton equivalent circuit is shown in Figure 9–32.

Figure 9–32

Now we can easily find the current through load resistor R_L by using the current divider rule:

$$I_L = \left(\frac{24 \ \Omega}{24 \ \Omega + 16 \ \Omega} \right)(1.167 \text{ A}) = 0.700 \text{ A}   \text{ (upward)}$$

By referring to Example 9–3, we see that the same result was obtained by finding the Thévenin equivalent circuit. An alternate method of finding the Norton equivalent circuit is by converting the Thévenin circuit found in Example 9–3 into an equivalent Norton circuit as shown in Figure 9–33.

Figure 9–33

■ **EXAMPLE 9–7**   Find the Norton equivalent of the circuit external to resistor R_L in the circuit in Figure 9–34. Use the equivalent circuit to determine the load current I_L when $R_L = 0$, 2 kΩ, and 5 kΩ.

Figure 9–34

Solution

Steps 1, 2, and 3: After removing the load resistor, labelling the remaining two terminals a and b, and setting the sources to zero, we have the circuit of Figure 9–35.

Voltage source replaced with a short circuit

Current source replaced with an open circuit

Figure 9–35

Step 4: The Norton resistance of the circuit is found as

$$R_N = 6 \text{ k}\Omega \| 2 \text{ k}\Omega = 1.5 \text{ k}\Omega$$

Step 5: The value of the Norton constant-current source is found by determining the current effects due to each independent source acting on a short circuit between terminals a and b.

Voltage Source, E: Referring to Figure 9–36(a), a short circuit between terminals a and b short-circuits resistor R_2. The short-circuit current due to the voltage source is

$$I_{ab}(1) = \frac{15 \text{ V}}{6 \text{ k}\Omega} = 2.50 \text{ mA}$$

(a)

(b)

Figure 9–36

Current Source, I: Referring to Figure 9–36(b), the short circuit between terminals a and b eliminates both resistors R_1 and R_2. The short-circuit current due to the current source is therefore

$$I_{ab}(2) = 5.00 \text{ mA}$$

The resultant Norton current is found from superposition as

$$I_N = I_{ab}(1) + I_{ab}(2) = 2.50 \text{ mA} + 5.00 \text{ mA} = 7.50 \text{ mA}$$

Step 6: The Norton equivalent circuit is shown in Figure 9–37.

Figure 9–37

Let $R_L = 0$: The current I_L must equal the source current, and so

$$I_L = 7.50 \text{ mA}$$

Let $R_L = 2\ k\Omega$: The current I_L is found from the current divider rule as

$$I_L = \left(\frac{1.5 \text{ k}\Omega}{1.5 \text{ k}\Omega + 2 \text{ k}\Omega} \right)(7.50 \text{ mA}) = 3.214 \text{ mA}$$

Let $R_L = 5\ k\Omega$: Using the current divider rule again, the current I_L is found as

$$I_L = \left(\frac{1.5 \text{ k}\Omega}{1.5 \text{ k}\Omega + 5 \text{ k}\Omega} \right)(7.50 \text{ mA}) = 1.731 \text{ mA}$$

Comparing the above results to those obtained in Example 9–4, we see that they are precisely the same.

■ EXAMPLE 9–8   Consider the circuit of Figure 9–38:

Figure 9–38

a. Find the Norton equivalent circuit external to terminals *a* and *b*.

b. Determine the current through R_L.

Solution

a. Steps 1 and 2: After removing the load (which consists of a current source in parallel with a resistor), we have the circuit of Figure 9–39.

Figure 9–39

Step 3: After zeroing the sources, we have the network shown in Figure 9–40.

Figure 9–40

Step 4: The Norton equivalent resistance is found as

$$R_N = 120\ \Omega \| 280\ \Omega = 84\ \Omega$$

Step 5: In order to determine the Norton current we must again determine the short-circuit current due to each source separately and then combine the results using the superposition theorem.

Voltage Source, E: Referring to Figure 9–41(a), notice that the resistor R_2 is shorted by the short circuit between terminals a and b and so the current in the short circuit is

$$I_{ab}(1) = \frac{24\ \text{V}}{120\ \Omega} = 0.2\ \text{A} = 200\ \text{mA}$$

This resistor is shorted by the short circuit between terminals a and b

Figure 9–41   (a)

Current Source, I: Referring to Figure 9–41(b), the short circuit between terminals a and b will now eliminate both resistors. The current through the short will simply be the source current. However, since the current will not be from a to b but rather in the opposite direction, we write

$$I_{ab}(2) = -560\ \text{mA}$$

Now the Norton current is found as the summation of the short-circuit currents due to each source:

$$I_N = I_{ab}(1) + I_{ab}(2) = 200 \text{ mA} + (-560 \text{ mA}) = -360 \text{ mA}$$

Both resistors are shorted by the short circuit between terminals a and b

Figure 9–41   (b)

The negative sign in the above calculation for current indicates that if a short circuit were placed between terminals a and b, current would actually be in the direction from b to a. The Norton equivalent circuit is shown in Figure 9–42.

Figure 9–42

b. The current through the load resistor is found by applying the current divider rule:

$$I_L = \left(\frac{84 \text{ }\Omega}{84 \text{ }\Omega + 168 \text{ }\Omega} \right)(360 \text{ mA} - 180 \text{ mA}) = 60 \text{ mA} \quad \text{(upward)}$$

practice
problem
IV

Find the Norton equivalent of the circuit shown in Figure 9–43. Use the source conversion technique to determine the Thévenin equivalent of the circuit.

Figure 9–43

practice
problem
V

Find the Norton equivalent external to R_L in the circuit of Figure 9–44. Solve for the current I_L when $R_L = 0$, 10 kΩ, 50 kΩ, and 100 kΩ.

Figure 9–44

in process
learning
check III

1. Show the relationship between the Thévenin equivalent circuit and the Norton equivalent circuit. Sketch each circuit.

2. If a Thévenin equivalent circuit has $E_{Th} = 100$ mV and $R_{Th} = 500$ Ω, draw the corresponding Norton equivalent circuit.

3. If a Norton equivalent circuit has $I_N = 10$ μA and $R_N = 20$ kΩ, draw the corresponding Thévenin equivalent circuit.

9.4   Maximum Power Transfer Theorem

In amplifiers and in most communication circuits such as radio receivers and transmitters, it is often desired that the load receives the maximum amount of power from a source.

The maximum power transfer theorem states the following:

A load resistance will receive maximum power from a circuit when the resistance of the load is exactly the same as the Thévenin (Norton) resistance looking back at the circuit.

(a)

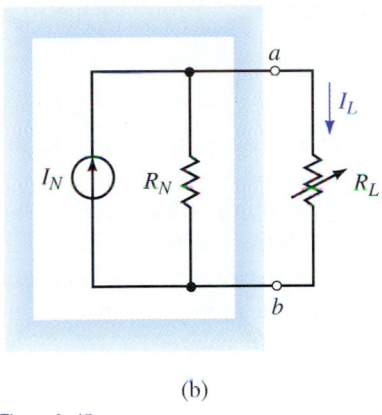

(b)

Figure 9–45

From Figure 9–45 we see that once the network has been simplified using either Thévenin's or Norton's theorem, maximum power will occur when

$$R_L = R_{Th} = R_N \qquad (9\text{–}3)$$

By examining the equivalent circuits of Figure 9–45, we see that the following equations determine the power delivered to the load:

$$P_L = \frac{\left(\dfrac{R_L}{R_L + R_{Th}} \times E_{Th}\right)^2}{R_L}$$

which gives

$$P_L = R_L \times \frac{E_{Th}^2}{(R_L + R_{Th})^2} \qquad (9\text{–}4)$$

Similarly,

$$P_L = \left(I_N \times \frac{R_N}{R_L + R_N}\right)^2 \times R_L \qquad (9\text{–}5)$$

Under maximum power conditions ($R_L = R_{Th} = R_N$), the above equations may be used to determine the maximum power delivered to the load and may therefore be written as

$$P_{max} = \frac{E_{Th}^2}{4R_{Th}} \qquad (9\text{–}6)$$

$$P_{max} = \frac{I_N^2 R_N}{4} \qquad (9\text{–}7)$$

TABLE 9–1

R_L (Ω)	V_L (V)	I_L (A)	P_L (W)
0	0	2.000	0
1	1.667	1.667	2.778
2	2.857	1.429	4.082
3	3.750	1.250	4.688
4	4.444	1.111	4.938
5	5.000	1.000	5.000
6	5.455	0.909	4.959
7	5.833	0.833	4.861
8	0.615	0.769	4.734
9	6.429	0.714	4.592
10	6.667	0.667	4.444

■ **EXAMPLE 9–9**   For the circuit of Figure 9–46, sketch graphs of V_L, I_L, and P_L as functions of R_L.

Figure 9–46

Solution   We may first set up a table of data for various values of resistance, R_L. See Table 9–1. Voltage and current values are determined by using voltage divider rule and current divider rule respectively. The power P_L for each value of resistance is determined by finding the product $P_L = V_L I_L$, or by using Equation 9–4.

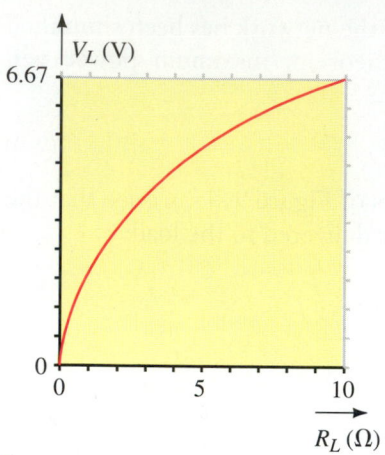

Figure 9–47   Voltage versus R_L.

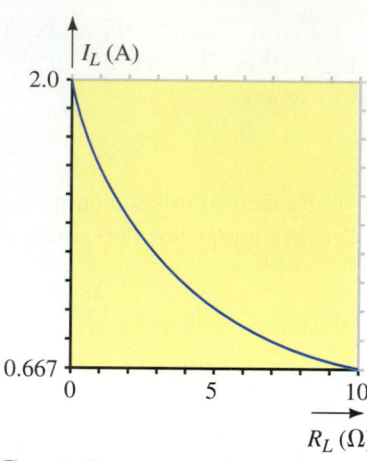

Figure 9–48   Current versus R_L.

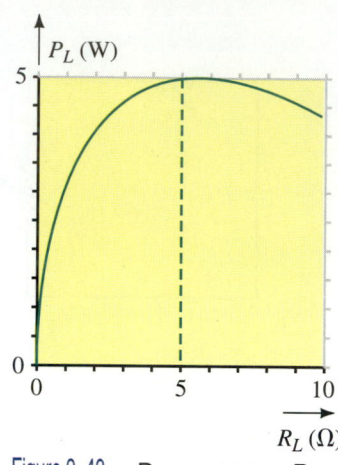

Figure 9–49   Power versus R_L.

If the data from Table 9–1 are plotted on a linear graphs, the graphs will appear as shown in Figures 9–47, 9–48, and 9–49.

Notice in the previous graphs that although voltage across the load increases as R_L increases, the power delivered to the load will be a maximum when $R_L = R_{Th} = 5 \ \Omega$. The reason for this apparent contradiction is because, as R_L increases, the reduction in current more than offsets the corresponding increase in voltage.

■ **EXAMPLE 9–10**   Consider the circuit of Figure 9–50:

Figure 9–50

a. Determine the value of load resistance required to ensure that maximum power is transferred to the load.

b. Find V_L, I_L, and P_L when maximum power is delivered to the load.

Solution

a. In order to determine the conditions for maximum power transfer, it is first necessary to determine the equivalent circuit external to the load. We may determine either the Thévenin equivalent circuit or the Norton equivalent circuit. This circuit was analyzed in Example 9–4 using Thévenin's theorem, and we determined the equivalent circuit to be as shown in Figure 9–51.

Figure 9–51

Maximum power will occur when $R_L = 1.5$ kΩ.

b. Letting $R_L = 1.5$ kΩ, we see that half of the Thévenin voltage will appear across the load resistor and half will appear across the Thévenin resistance. So, at maximum power,

$$V_L = \frac{E_{Th}}{2} = \frac{11.25 \text{ V}}{2} = 5.625 \text{ V}$$

$$I_L = \frac{5.625 \text{ V}}{1.5 \text{ k}\Omega} = 3.750 \text{ mA}$$

The power delivered to the load is found as

$$P_L = \frac{V_L^2}{R_L} = \frac{(5.625 \text{ V})^2}{1.5 \text{ k}\Omega} = 21.1 \text{ mW}$$

Or, alternatively using current, we calculate the power as

$$P_L = I_L^2 R_L = (3.75 \text{ mA})^2 (1.5 \text{ k}\Omega) = 21.1 \text{ mW}$$

In solving this problem, we could just as easily have used the Norton equivalent circuit to determine required values.

Recall that efficiency was defined as the ratio of output power to input power:

$$\eta = \frac{P_{out}}{P_{in}}$$

or as a percentage:

$$\eta = \frac{P_{out}}{P_{in}} \times 100\%$$

By using the maximum power transfer theorem, we see that under the condition of maximum power the efficiency of the circuit is

$$\eta = \frac{P_{out}}{P_{in}} \times 100\%$$

$$= \frac{\dfrac{E_{Th}^2}{4R_{Th}}}{\dfrac{E_{Th}^2}{2R_{Th}}} \times 100\% = 0.500 \times 100\% = 50\% \qquad (9\text{--}8)$$

For communication circuits and for many amplifier circuits, 50% represents the maximum possible efficiency. At this efficiency level, the voltage presented to the following stage would only be half of the maximum terminal voltage.

In power transmission such as the 115-Vac, 60-Hz power in your home, the condition of maximum power is not a requirement. This is because when the source delivers maximum power, the voltage across the load will be reduced to half of the maximum available terminal voltage. Clearly, if we are working with power supplies, we would like to ensure that efficiency is brought as close to 100% as possible. In such cases, load resistance R_L is kept much larger than the internal resistance of the voltage source (typically $R_L \geq 10R_{int}$), ensuring that the voltage appearing across the load will be very nearly equal to the maximum terminal voltage of the voltage source.

■ **EXAMPLE 9–11**   Refer to the circuit of Figure 9–52, which represents a typical dc power supply.

Figure 9–52

a. Determine the value of R_L needed for maximum power transfer.

b. Determine terminal voltage V_L and the efficiency when the value of the load resistor is $R_L = 50\ \Omega$.

c. Determine terminal voltage V_L and the efficiency when the value of the load resistor is $R_L = 100\ \Omega$.

Solution

a. For maximum power transfer, the load resistor will be given as $R_L = 0.05\ \Omega$. At this value of load resistance, the efficiency will be only 50%.

b. For $R_L = 50\ \Omega$, the voltage appearing across the output terminals of the voltage source is

$$V_L = \left(\frac{50\ \Omega}{50\ \Omega + 0.05\ \Omega} \right)(9.0\ \text{V}) = 8.991\ \text{V}$$

The efficiency is

$$\eta = \frac{P_{out}}{P_{in}} \times 100\%$$

$$= \frac{\dfrac{(8.991 \text{ V})^2}{50 \text{ }\Omega}}{\dfrac{(9.0 \text{ V})^2}{50.05 \text{ }\Omega}} \times 100\%$$

$$= \frac{1.6168 \text{ W}}{1.6184 \text{ W}} \times 100\% = 99.90\%$$

c. For $R_L = 100$ Ω, the voltage appearing across the output terminals of the voltage source is

$$V_L = \left(\frac{100 \text{ }\Omega}{100 \text{ }\Omega + 0.05 \text{ }\Omega}\right)(9.0 \text{ V}) = 8.995 \text{ 50 V}$$

The efficiency is

$$\eta = \frac{P_{out}}{P_{in}} \times 100\%$$

$$= \frac{\dfrac{(8.9955 \text{ V})^2}{100 \text{ }\Omega}}{\dfrac{(9.0 \text{ V})^2}{100.05 \text{ }\Omega}} \times 100\%$$

$$= \frac{1.6168 \text{ W}}{1.6184 \text{ W}} \times 100\% = 99.95\%$$

From this example, we see that if efficiency is important, as it is in power transmission, then the load resistance should be much larger than the resistance of the source (typically $R_L \geq 10R_{int}$). If on the other hand, it is more important to ensure maximum power transfer, then the load resistance should equal the source resistance ($R_L = R_{int}$).

practice problem VI

Refer to the circuit of Figure 9–44. For what value of R_L will the load receive maximum power? Determine the power when $R_L = R_N$, when $R_L = 25$ kΩ, and when $R_L = 50$ kΩ.

In process learning check IV

A Thévenin equivalent circuit consists of $E_{Th} = 10$ V and $R_{Th} = 2$ kΩ. Determine the efficiency of the circuit when

a. $R_L = R_{Th}$,
b. $R_L = 0.5R_{Th}$,
c. $R_L = 2R_{Th}$.

Figure 9–53

9.5 Substitution Theorem

The substitution theorem states the following:

Any branch within a circuit may be replaced by an equivalent branch, provided the replacement branch has the same current through it and voltage across it as the original branch.

This theorem is best illustrated by examining the operation of a circuit. Consider the circuit of Figure 9–53.

The voltage V_{ab} and the current I in the circuit of Figure 9–53 are given as

$$V_{ab} = \left(\frac{6 \text{ k}\Omega}{4 \text{ k}\Omega + 6 \text{ k}\Omega} \right)(10 \text{ V}) = +6.0 \text{ V}$$

and

$$I = \frac{10 \text{ V}}{4 \text{ k}\Omega + 6 \text{ k}\Omega} = 1 \text{ mA}$$

The resistor R_2 may be replaced with any combination of components, provided that the resulting components maintain the above conditions. We see that the branches of Figure 9–54 are each equivalent to the original branch between terminals a and b of the circuit in Figure 9–53.

Although each of the branches in Figure 9–54 is different, the current entering or leaving each branch will be same as that in the original branch. Similarly, the voltage across each branch will be the same. If any of these branches is substituted into the original circuit, the balance of the circuit will operate in the same way as the original. This theorem allows us to replace any branch within a given circuit with an equivalent branch, thereby simplifying the analysis of the remaining circuit.

(a)   (b)   (c)   (d)   (e)

Figure 9–54

■ **EXAMPLE 9–12**   If the indicated portion in the circuit of Figure 9–55 is to be replaced with a current source and a 240-Ω shunt resistor, determine the magnitude and direction of the required current source.

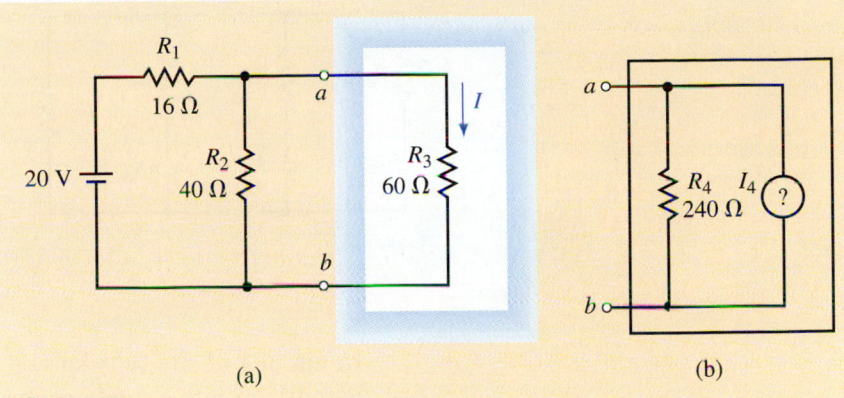

(a)   (b)

Figure 9–55

Solution   The voltage across the branch in the original circuit is

$$V_{ab} = \left(\frac{40\ \Omega \| 60\ \Omega}{16\ \Omega + (40\ \Omega \| 60\ \Omega)} \right)(20\ \text{V}) = \left(\frac{24\ \Omega}{16\ \Omega + 24\ \Omega} \right)(20\ \text{V}) = 12.0\ \text{V}$$

which results in a current of

$$I = \frac{12.0\ \text{V}}{60\ \Omega} = 0.200\ \text{A} = 200\ \text{mA}$$

In order to maintain the same terminal voltage, $V_{ab} = 12.0$ V, the current through resistor $R_4 = 240\ \Omega$ must be

$$I_{R4} = \frac{12.0\ \text{V}}{240\ \Omega} = 0.050\ \text{A} = 50\ \text{mA}$$

Finally, we know that the current entering terminal a is $I = 200$ mA. In order for Kirchhoff's current law to be satisfied at this node, the current source must have a magnitude of 150 mA and the direction must be downward, as shown in Figure 9–56.

Figure 9–56

9.6 Millman's Theorem

Millman's theorem is used to simplify circuits having several parallel voltage sources as illustrated in Figure 9–57. Although any of the other theorems developed in this chapter will work in this case, Millman's theorem provides a much simpler and more direct equivalent.

Figure 9–57

Figure 9–58

In circuits of the type shown in Figure 9–57, the voltage sources may be replaced with a single equivalent source as shown in Figure 9–58.

To find the values of the equivalent voltage source E_{eq} and series resistance R_{eq}, we need to convert each of the voltage sources of Figure 9–57 into its equivalent current source using the technique developed in Chapter 8. The value of each current source would be determined by using Ohm's law (i.e., $I_1 = E_1/R_1$, $I_2 = E_2/R_2$, etc.). After the source conversions are completed, the circuit would appear as shown in Figure 9–59.

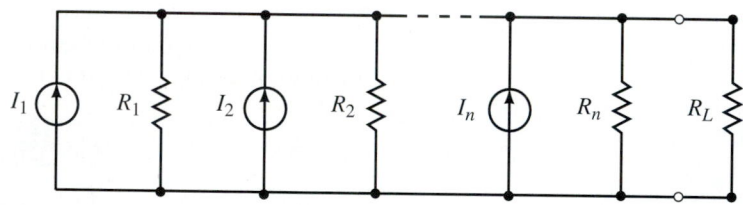

Figure 9–59

From the circuit of Figure 9–59 we see that all of the current sources have the same direction. Clearly, this will not always be the case since the direction of each current source will be determined by the initial polarity of the corresponding voltage source. It is now possible to replace the n current sources with a single current source having a magnitude given as

$$I_{eq} = \sum_{x=0}^{n} I_x = I_1 + I_2 + I_3 + \cdots + I_n \tag{9–9}$$

which may be written as

$$I_{eq} = \frac{E_1}{R_1} + \frac{E_2}{R_2} + \frac{E_3}{R_3} + \cdots + \frac{E_n}{R_n} \tag{9–10}$$

If the direction of any current source is opposite to the direction shown, then the corresponding magnitude would be subtracted, rather than added. From Figure 9–59, we can see that by removing the current sources, the circuit corresponding to the sources would have an equivalent resistance given as

$$R_{eq} = R_1 \| R_2 \| R_3 \| \cdots \| R_n \tag{9–11}$$

which may be determined as

$$R_{eq} = \frac{1}{G_{eq}} = \frac{1}{\dfrac{1}{R_1} + \dfrac{1}{R_2} + \dfrac{1}{R_3} + \cdots \dfrac{1}{R_n}} \qquad (9\text{–}12)$$

The general expression for the equivalent voltage is

$$E_{eq} = I_{eq}R_{eq} = \frac{\dfrac{E_1}{R_1} + \dfrac{E_2}{R_2} + \dfrac{E_3}{R_3} + \cdots + \dfrac{E_n}{R_n}}{\dfrac{1}{R_1} + \dfrac{1}{R_2} + \dfrac{1}{R_3} + \cdots + \dfrac{1}{R_n}} \qquad (9\text{–}13)$$

■ **EXAMPLE 9–13**   Use Millman's Theorem to simplify the circuit of Figure 9–60 so that it has only a single source. Use the simplified circuit to find the current in the load resistor, R_L.

Figure 9–60

Solution   From Equation 9–13, we express the equivalent voltage source as

$$E_{eq} = \frac{\dfrac{-96\ \text{V}}{240\ \Omega} + \dfrac{40\ \text{V}}{200\ \Omega} + \dfrac{-80\ \text{V}}{800\ \Omega}}{\dfrac{1}{240\ \Omega} + \dfrac{1}{200\ \Omega} + \dfrac{1}{800\ \Omega}}$$

$$= \frac{-0.300}{10.41\overline{6}\ \text{mS}} = -28.8\ \text{V}$$

The equivalent resistance is

$$R_{eq} = \frac{1}{\dfrac{1}{240\ \Omega} + \dfrac{1}{200\ \Omega} + \dfrac{1}{800\ \Omega}} = \frac{1}{10.41\overline{6}\ \text{mS}} = 96\ \Omega$$

The equivalent circuit using Millman's theorem is shown in Figure 9–61. Notice that the equivalent voltage source has a polarity which is opposite to the originally assumed polarity. This is because the voltage sources E_1 and E_3 have magnitudes which overcome the polarity and magnitude of the source E_2.

Figure 9–61

From the equivalent circuit of Figure 9–61, it is a simple matter to determine the current through the load resistor:

$$I_L = \frac{28.8 \text{ V}}{96 \ \Omega + 192 \ \Omega} = 0.100 \text{ A} = 100 \text{ mA} \quad \text{(upward)}$$

9.7 Reciprocity Theorem

The reciprocity theorem is a theorem which can only be used with single-source circuits. This theorem, however, may be applied to either voltage sources or current sources. The theorem states the following:

Voltage Sources

A voltage source causing a current I in any branch of a circuit may be removed from the original location and placed into that branch. The voltage source in the new location will produce a current in the original source location which is exactly equal to the originally calculated current, I.

When applying the reciprocity theorem for a voltage source, the following steps must be followed:

1. The voltage source is replaced by a short circuit in the original location.

2. The polarity of the source in the new location must ensure that the current direction in that branch remains unchanged.

Current Sources

A current source causing a voltage V at any node of a circuit may be removed from the original location and connected to that node. The current source in the new location will produce a voltage in the original source location which is exactly equal to the originally calculated voltage, V.

When applying the reciprocity theorem for a current source, the following conditions must be met:

1. The current source is replaced by an open circuit in the original location.

2. The direction of the source in the new location must ensure that the polarity of the voltage at the node to which the current source is now connected remains unchanged.

The following examples illustrate how the reciprocity theorem is used within a circuit.

■ EXAMPLE 9–14   Consider the circuit of Figure 9–62:

Figure 9–62

a. Calculate the current I.

b. Remove voltage source E and place it into the branch with R_3. Show that the current through the branch which formerly had E is now the same as the current I.

Solution

a. $V_{12\,\Omega} = \left(\dfrac{8\,\Omega \| 12\,\Omega}{4\,\Omega + (8\,\Omega \| 12\,\Omega)} \right)(22\ \text{V}) = \left(\dfrac{4.8}{8.8} \right)(22\ \text{V}) = 12.0\ \text{V}$

$I = \dfrac{V_{12\,\Omega}}{12\,\Omega} = \dfrac{12.0\ \text{V}}{12\,\Omega} = 1.00\ \text{A}$

b. Now removing the voltage source from its original location and moving it into the branch containing the current I, we obtain the circuit shown in Figure 9–63.

Polarity of the source is such that the current direction remains unchanged.

When E is removed it is replaced by a short circuit.

Figure 9–63

For the circuit of Figure 9–63, we determine the current I as follows:

$$V_{4\,\Omega} = \left(\frac{4\,\Omega \| 8\,\Omega}{12\,\Omega + (4\,\Omega \| 8\,\Omega)} \right)(22\text{ V}) = \left(\frac{2.\overline{6}}{14.6} \right)(22\text{ V}) = 4.00\text{ V}$$

$$I = \frac{V_{4\,\Omega}}{4\,\Omega} = \frac{4.00\text{ V}}{4\,\Omega} = 1.00\text{ A}$$

From this example, we see that the reciprocity theorem does indeed apply.

■ **EXAMPLE 9–15**   Consider the circuit shown in Figure 9–64:

Figure 9–64

a. Determine the voltage V across resistor R_3.

b. Remove the current source I and place it between node b and the reference node. Show that the voltage across the former location of the current source (node a) is now the same as the voltage V.

Solution

a. The node voltages for the circuit of Figure 9–64 are determined as follows:

$$R_T = 6\text{ k}\Omega \| (9\text{ k}\Omega + 3\text{ k}\Omega) = 4\text{ k}\Omega$$

$$V_a = (2\text{ mA})(4\text{ k}\Omega) = 8.0\text{ V}$$

$$V_b = \left(\frac{3\text{ k}\Omega}{3\text{ k}\Omega + 9\text{ k}\Omega} \right)(8.0\text{ V}) = 2.0\text{ V}$$

b. After relocating the current source from the original location, and connecting it between node b and ground, we obtain the circuit shown in Figure 9–65.

Figure 9–65

The resulting node voltages are now found as follows:

$$R_T = 3 \text{ k}\Omega \| (6 \text{ k}\Omega + 9 \text{ k}\Omega) = 2.5 \text{ k}\Omega$$

$$V_b = (2 \text{ mA})(2.5 \text{ k}\Omega) = 5.0 \text{ V}$$

$$V_a = (\frac{6 \text{ k}\Omega}{6 \text{ k}\Omega + 9 \text{ k}\Omega})(5.0 \text{ V}) = 2.0 \text{ V}$$

From the above results, we conclude that the reciprocity theorem again applies for the given circuit.

9.8 Circuit Analysis Using Computers

PSpice is easily used to find the Thévenin equivalent of any dc circuit. However, in order to find the equivalent, we must be able to use the .TF (transfer function) statement. This statement is used by PSpice to generate three quantities for the circuit under analysis. First, it provides the ratio of the output variable to the input variable. This ratio, called the *transfer function* of the circuit, can be one of four possible combinations of currents or voltage. The second quantity generated is the input resistance at the input source. The final quantity is the resistance at the output terminals of the circuit. This last quantity represents the Thévenin resistance.

The format for the transfer function statement is as follows:

.TF   Output   Input

■ **EXAMPLE 9–16** Write the PSpice input file to determine the Thévenin equivalent external to the load resistor of the circuit in Figure 9–66.

Figure 9–66

Solution As before, we must remove the load resistor from the circuit before we can determine the Thévenin equivalent circuit. If we did not follow this step, the load resistor would be included in the calculation of the Thévenin equivalent. The resulting circuit which corresponds to the required PSpice format is shown in Figure 9–67.

Figure 9–67

The PSpice input file is given as follows:

```
Input File:
Example 9-16: Thévenin equivalent circuit, Figure 9-
67
Vs    1    0    dc       15V
R1    1    2    6kohms
R2    2    0    2kohms
Is    0    2    dc       5mA
.TF   V(2,0)   Vs
.OPTIONS NOPAGE
.END
```

The pertinent information from the output file is given as follows:

```
Output File:
NODE    VOLTAGE        NODE    VOLTAGE        NODE    VOLTAGE
(1)    15.0000        (2)    11.2500


      VOLTAGE SOURCE CURRENTS
      NAME            CURRENT


      Vs              -6.250E-04


      TOTAL POWER DISSIPATION   9.38E-03   WATTS


        V(2,0)/Vs =   2.500E-01


        INPUT RESISTANCE AT Vs =   8.000E+03


        OUTPUT RESISTANCE AT V(2,0) =   1.500E+03
```

From the above information, we see that the Thévenin voltage is the voltage which appears at node 2, $V(2,0)$, which PSpice has determined as 11.25 V. In the output file, the Thévenin resistance is called the output resistance and was found to be 1.500 kΩ.

From the PSpice results we have the following values:

$$E_{Th} = 11.25 \text{ V}$$

and

$$R_{Th} = 1.50 \text{ k}\Omega$$

The equivalent circuit is shown in Figure 9–68.

Figure 9–68

These results are consistent with those obtained in Example 9–4.

practice
problem
VII

Write the input file to determine the Thévenin equivalent of the circuit shown in Figure 9–25. Convert the Thévenin circuit to its Norton equivalent. Compare the PSpice results to those obtained in Practice Problem III.

PSpice has an additional postprocessor called PROBE, which is able to provide a graphical display of numerous variables. We will use the PROBE postprocessor to examine how output power, voltage, and current change as a function of load resistance.

■ **EXAMPLE 9–17**   Consider the circuit of Figure 9–69:

Figure 9–69

Write a PSpice input file using the PROBE postprocessor to observe how the output power changes as a function of the load resistance.

Solution The circuit in the PSpice format is shown in Figure 9–70.

Figure 9–70

PSpice will not easily permit us to have a resistance of 0 Ω between two nodes and so we select the load resistance to vary between 1 Ω and 10 Ω. The size of the increment determines how fast the computer is able to provide a graph. We select an increment of 0.1 Ω as a reasonable step. The input file is given below. Notice that we have not specified which variables are to be plotted. PSpice calculates all voltages and currents for each value of R_L and generates a data file with all the information. The PROBE postprocessor uses the data to plot any variables.

```
Input File:
Example 9-17: Maximum power transfer theorem, Figure
9-70
VTH    1    0     dc     10V
RTH    1    2     5ohms
RL     2    0     Rvar   1
.MODEL      Rvar RES(R=1)
.DC    RES Rvar(R)     1     10     0.1
.PROBE
.OPTIONS NOPAGE
.END
```

If the program is run from the PSpice shell, then the PROBE program will run automatically. Once the blank graph with the corresponding menu appears on the screen, we may control the display by indicating which variables we wish to observe.

For instance, if we wish to observe the output power as a function of R, we use the *Add_trace* menu item. Output power is given as V(2)*V(2)/R. Up to a total of three traces may be shown on the screen at a time. The other traces may be added by using the *Y_axis*, *Add_axis*, and *Add_trace* menu items. The voltage and current may be displayed by typing V(2) and I(RL) at the appropriate command lines.

The display will appear as shown in Figure 9–71. Notice that the display indicates that the output power will be a maximum when $R_L = R_{Th} = 5$ Ω.

Figure 9–71 PSpice displays of voltage, current, and power.

Problems

9.1 Superposition Theorem

1. Given the circuit of Figure 9–72, use superposition to calculate the current through each of the resistors.

Figure 9–72

2. Use superposition to determine the voltage drop across each of the resistors of the circuit in Figure 9–73.

Figure 9–73

3. Use superposition to solve for the voltage V_a and the current I in the circuit of Figure 9–74.

Figure 9–74   Figure 9–75

4. Using superposition, find the current through the 480-Ω resistor in the circuit of Figure 9–75:

5. Given the circuit of Figure 9–76, what must be the value of the unknown voltage source to ensure that the current through the load is $I_L = 5$ mA as shown. Verify the results using superposition.

Figure 9–76

6. If the load resistor in the circuit of Figure 9–77 is to dissipate 120 W, determine the value of the unknown voltage source. Verify the results using superposition.

Figure 9–77

9.2   Thévenin's Theorem

7. Find the Thévenin equivalent external to R_L in circuits of Figure 9–78. Use the equivalent circuit to find V_{ab} for each circuit.

(a)   (b)

Figure 9–78

8. Repeat Problem 7 for the circuits of Figure 9–79.

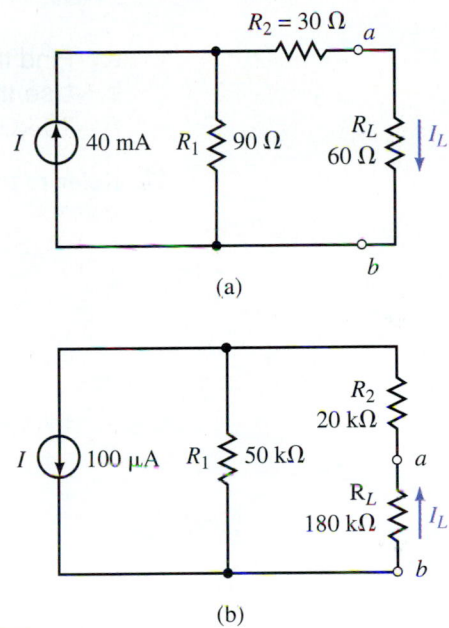

(a)

(b)

Figure 9–79

9. Refer to the circuit of Figure 9–80:

Figure 9–80

a. Find the Thévenin equivalent circuit external to R_L.
b. Use the equivalent circuit to determine V_{ab} when $R_L = 20\ \Omega$ and when $R_L = 50\ \Omega$.

10. Refer to the circuit of Figure 9–81:

Figure 9–81

a. Find the Thévenin equivalent circuit external to R_L.
b. Use the equivalent circuit to determine V_{ab} when $R_L = $ 10 kΩ and when $R_L = 20$ kΩ.

11. Refer to the circuits of Figure 9–82:

(a)

(b)

Figure 9–82

a. Find the Thévenin equivalent circuit external to the indicated terminals.
b. Use the Thévenin equivalent circuit to determine the current through each of the indicated branches.

12. Refer to the circuit of Figure 9–83:

All resistors are 3.3 kΩ.

Figure 9-83

    a. Find the Thévenin equivalent circuit external to R_L.
    b. Use the Thévenin equivalent circuit to find V_L.

13. Refer to the circuits of Figure 9–84:

(a)

Figure 9-84

    a. Find the Thévenin equivalent circuit external to R_L.
    b. Use the Thévenin equivalent circuit to find the current I when $R_L = 0$, 10 kΩ, and 50 kΩ.

14. Refer to the circuit of Figure 9–85:

    a. Find the Thévenin equivalent circuit external to R_L.

    b. Use the Thévenin equivalent circuit to find the power dissipated by R_L.

Figure 9–85   Figure 9–86

15. Find the Thévenin equivalent circuit of the network external to the indicated branch as shown in Figure 9–86.

16. Repeat Problem 11 for each of the circuits of Figure 9–87.

(a)   (b)

Figure 9–87

9.3   Norton's Theorem

17. Find the Norton equivalent circuit external to R_L in circuits of Figure 9–78. Use the equivalent circuit to find I_L for each circuit.

18. Repeat Problem 17 for the circuits of Figure 9–79.

19. Refer to the circuit of Figure 9–80:

    a. Find the Norton equivalent circuit external to R_L.

    b. Use the equivalent circuit to determine I_L when $R_L = 20\ \Omega$ and when $R_L = 50\ \Omega$.

20. Refer to the circuit of Figure 9–81:

    a. Find the Norton equivalent circuit external to R_L.

    b. Use the equivalent circuit to determine I_L when $R_L = 10\ k\Omega$ and when $R_L = 20\ k\Omega$.

21. a. Find the Norton equivalent circuit external to the indicated terminals of Figure 9–82.

    b. Convert the Thévenin equivalent circuit of Problem 11 to its Norton equivalent.

22. a. Find the Norton equivalent circuit external to R_L in the circuit of Figure 9–83.

    b. Convert the Thévenin equivalent circuit of Problem 12 to its Norton equivalent.

23. Repeat Problem 21 for each of the circuits of Figure 9–84.

24. Repeat Problem 21 for the circuit of Figure 9–85.

25. Repeat Problem 21 for the circuit of Figure 9–86.

26. Repeat Problem 21 for each of the circuits of Figure 9–87.

9.4 Maximum Power Transfer Theorem

27. a. For the circuit of Figure 9–84(a), determine the value of R_L so that maximum power is delivered to the load.

    b. Calculate the value of the maximum power which can be delivered to the load.

    c. Sketch the curve of power versus resistance as R_L is adjusted from $0\ \Omega$ to $50\ k\Omega$ in increments of $5\ k\Omega$.

28. Repeat Problem 27 for the circuit of Figure 9–84(b).

29. a. For each of the circuits of Figure 9–88, find the value of R so that $R_L = R_{Th}$.

    b. Calculate the maximum power dissipated by R_L.

(a)

(b)

Figure 9–88

30. a. For the circuit of Figure 9–89, determine the values of R_1 and R_2 so that the 25-kΩ load receives maximum power.
  b. Calculate the maximum power delivered to R_L.

Figure 9–89

9.5 Substitution Theorem

31. If the indicated portion of the circuit in Figure 9–90 is to be replaced with a voltage source and a 50-Ω series resistor, determine the magnitude and polarity of the resulting voltage source.

32. If the indicated portion of the circuit in Figure 9–90 is to be replaced with a current source and a 200-Ω shunt resistor, determine the magnitude and direction of the resulting current source.

Figure 9–90

9.6 Millman's Theorem

33. Use Millman's theorem to find the current through and the power dissipated by R_L in each of the circuits of Figure 9–91.

Figure 9–91

34. Repeat Problem 33 for the circuit of Figure 9–92.

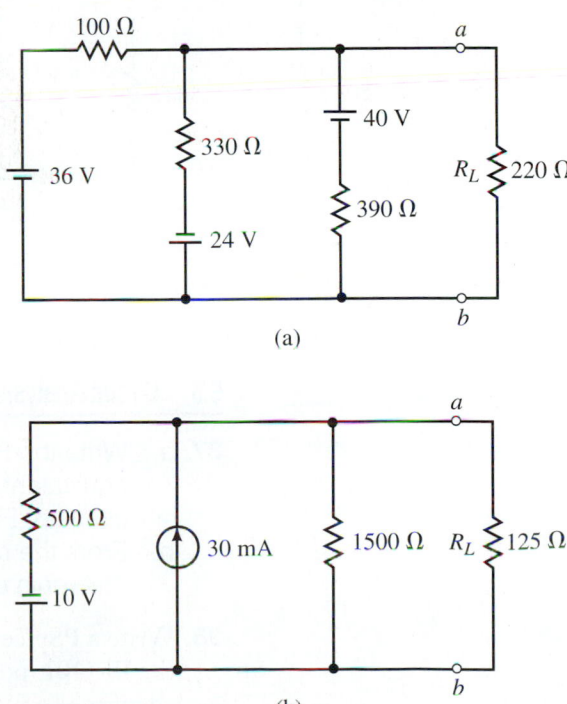

Figure 9–92

9.7   Reciprocity Theorem

35. a.   Determine the current I in the circuit of Figure 9–93(a).
   b.   Determine the current I in the circuit of Figure 9–93(b).
   c.   Show that reciprocity applies for the circuits given.

(a)   (b)

Figure 9–93

36. a.   Determine the voltage V in the circuit of Figure 9–94(a).
   b.   Determine the voltage V in the circuit of Figure 9–94(b).
   c.   Show that reciprocity applies for the circuits given.

(a)   (b)

Figure 9–94

9.8   Circuit Analysis Using Computers

37. a.   Write the PSpice input file to determine the Thévenin equivalent circuit of the network external to R_L in the circuit of Figure 9–80.
   b.   From the results obtained in PSpice, sketch the Thévenin equivalent circuit.

38. Write a PSpice input file for the circuit of Figure 9–84(a). Use the PROBE postprocessor to observe how the output power changes as a function of load resistance R_L. Allow the resistor to vary from 200 Ω to 50 kΩ in increments of 200 Ω.

PART III

Capacitance and Inductance

Resistance, inductance, and capacitance are the three basic circuit properties that we use to control voltages and currents in electric circuits. However, each behaves in a fundamentally different way. Resistance, for example, opposes current, while inductance opposes any change in current, and capacitance opposes any change in voltage. In addition, resistance dissipates energy, while inductance and capacitance both store energy—inductance in its magnetic field and capacitance in its electric field.

Circuit elements that are built to possess capacitance are called capacitors, while elements built to possess inductance are called inductors. In Part III of this book, we explore these elements, their properties, and their behavior in electric circuits.

Note: Because many institutions cover *ac* principles before transient analysis, the material in Part III has been organized so that Chapters 11 and 14 can be delayed until after the *ac* topics without loss of continuity.

Capacitors and Capacitance

CHAPTER PREVIEW

A *capacitor* is a circuit component designed to store electrical charge. If you connect a dc voltage source to a capacitor, for example, the capacitor will "charge" to the voltage of the source. If you then disconnect the source, the capacitor's voltage will remain at the value to which it was charged (assuming no leakage). Because of this tendency to hold voltage, *a capacitor opposes changes in voltage*. It is this characteristic that gives capacitors their unique properties.

Capacitors are widely used in electrical and electronic applications. They are used in radio and TV systems, for example, to tune in signals, in cameras to store the charge that fires the photoflash, on pump and refrigeration motors to increase starting torque, in electric power systems to increase operating efficiency, and so on.

Capacitance is the electrical property of capacitors: it is a measure of how much charge a capacitor can hold. In this chapter, we look at capacitance and its basic properties. In Chapter 11, we look at capacitors in dc and pulse circuits; in later chapters, we look at capacitors in ac applications.

KEY TERMS

Capacitance
Capacitor
Dielectric
Dielectric Constant
Dielectric Absorption
Electric Flux
Electric Field Intensity
Electric Flux Density
Electric Field
Electrolytic
Farad
Leakage
Permittivity
Voltage Gradient
Voltage Breakdown
Working Voltage

A montage of capacitors. *(Courtesy Illinois Capacitor Inc.)*

OBJECTIVES

After studying this chapter, you will be able to

- describe the basic construction of capacitors,

- explain how capacitors store charge,

- define capacitance,

- describe what factors affect capacitance and in what way,

- describe the electric field of a capacitor,

- compute the breakdown voltages of various materials,

- describe various types of commercial capacitors,

- compute the capacitance of capacitors in series and in parallel combinations,

- compute capacitor voltage and current for simple time-varying waveforms,

- determine stored energy,

- describe capacitor faults and the basic troubleshooting of capacitors.

371

Michael Faraday and The Field Concept

The unit of capacitance, the farad, *is named after Michael Faraday (1791–1867). Born in England to a working class family, Faraday received limited education. Nonetheless, he was responsible for many of the fundamental discoveries of electricity and magnetism. Lacking mathematical skills, he used his intuitive ability to develop conceptual pictures of basic phenomenon rather than mathematical models. It was his development of the* field concept *that made it possible to map out the fields that exist around electrical charges.*

To get at this idea, recall from Chapter 2 that unlike charges attract and like charges repel, i.e. a force exists between charges. We call the region where this force acts an electric field. *To visualize this field, we use Faraday's* field concept *and draw* lines of force *(or* flux lines*) that show at every point in space the magnitude and direction of the force. Now, rather than supposing that one charge exerts a force on another, we instead visualize that the original charges create a field in space and that other charges introduced into this field experience a force due to the field. This concept is helpful in studying certain aspects of capacitors as you will see in this chapter.*

The development of the field concept had a significant impact on science. We now picture several important phenomena in terms of fields, including electric fields, gravitation, and magnetism. When Faraday published his theory in 1844, however, it was not taken seriously, much like Ohm's work two decades earlier. It is also interesting to note that the development of the field concept grew out of Faraday's research into magnetism, not electric charge.

10.1 Capacitance

A capacitor consists of two conductors separated by an insulator. One of its basic forms is the parallel-plate capacitor shown in Figure 10–1. It consists of two metal plates separated by a nonconducting material called a *dielectric*. The dielectric may be air, oil, mica, plastic, ceramic, or other suitable insulating material.

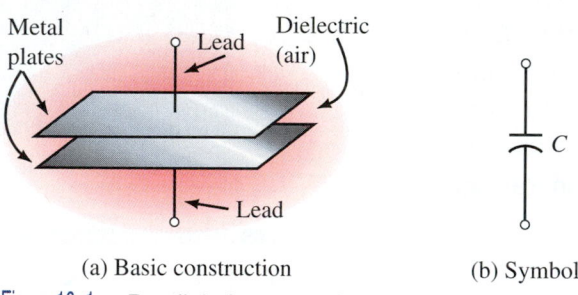

(a) Basic construction   (b) Symbol

Figure 10–1   Parallel-plate capacitor.

Figure 10–2 When a source is connected, electrons are removed from plate *A* and an equal number deposited on plate *B*. This leaves the top plate positively charged and the bottom plate negatively charged.

Since the plates of the capacitor are metal, they contain huge numbers of free electrons. In their normal state, however, they are uncharged, that is, there is no excess or deficiency of electrons on either plate. If a dc source is now connected (Figure 10–2), electrons are pulled from the top plate by the positive potential of the battery and the same number deposited on the bottom plate. This leaves the top plate with a deficiency of electrons (i.e., positive charge) and the bottom plate with an excess (i.e., negative charge). In this state, the capacitor is said to be

Figure 10–3   Capacitor after charging. When the source is disconnected, electrons are trapped on the bottom plate. Thus, charge is stored.

charged. Note, however, that the capacitor's net charge is still zero. All that we have done is move charge from one plate to the other.

If we now disconnect the source (Figure 10–3), the excess electrons remain trapped on the bottom plate as they have no way to return to the top plate. The capacitor therefore remains charged even though no source is present. Because of this, we say that *a capacitor can store charge.* Capacitors with little leakage (Section 10.5) can hold their charge for a considerable time.

Large capacitors charged to high voltages contain a great deal of energy and can give you a bad shock. Always *discharge* capacitors after power has been removed if you intend to handle them. You can do this by shorting a wire across their leads. Electrons then return to the top plate, restoring the charge balance and reducing the capacitor voltage to zero. (However, you also need to be concerned about residual voltage due to dielectric absorption. This is discussed in Section 10.5.)

Definition of Capacitance

The amount of charge (Q) that a capacitor can store depends on the applied voltage. Experiments show that for a given capacitor, Q is proportional to voltage. Let the constant of proportionality be C. Then,

$$Q = CV \qquad (10\text{–}1)$$

Rearranging terms yields

$$C = \frac{Q}{V} \quad \text{(farads, F)} \qquad (10\text{–}2)$$

The term C is defined as the capacitance of the capacitor. As indicated, its unit is the *farad.* By definition, *the capacitance of a capacitor is one farad if it stores one coulomb of charge when the voltage across its terminals is one volt.* The farad, however, is a very large unit. Most practical capacitors range in size from picofarads (pF or 10^{-12} F) to microfarads (μF or 10^{-6} F). The larger the value of C, the more charge that the capacitor can hold for a given voltage.

■ **EXAMPLE 10–1**

a. How much charge is stored on a 10-μF capacitor when it is connected to a 24-volt source?

b. The charge on a 20-nF capacitor is 1.7 μC. What is its voltage?

Solution

a. From Equation 10–1, $Q = CV$. Thus, $Q = (10 \times 10^{-6} \text{ F})(24 \text{ V}) = 240\ \mu$C.

b. Rearranging Equation 10–1, $V = Q/C = (1.7 \times 10^{-6} \text{ C})/(20 \times 10^{-9} \text{ F}) = 85$ V.

10.2 Factors Affecting Capacitance

Effect of Area

As shown by Equation 10–2, capacitance is directly proportional to charge. This means that the more charge you can put on a capacitor's plates for a given voltage, the greater will be its capacitance. Consider Figure 10–4. The capacitor of (b) has four times the area of (a). Since it has the same number of free electrons per unit area, it has four times the total charge and hence four times the capacitance. This turns out to be true in general, that is, *capacitance is directly proportional to plate area.*

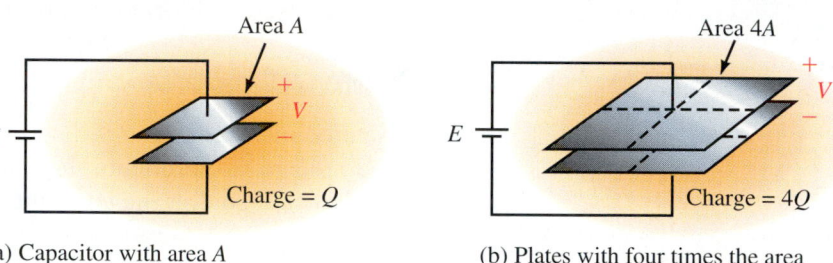

(a) Capacitor with area *A*
    and charge *Q*

(b) Plates with four times the area
    have four times the charge and
    therefore, four times the capacitance

Figure 10–4   Capacitance is proportional to area.

Effect of Spacing

Now consider Figure 10–5. Since the top plate has a deficiency of electrons and the bottom plate an excess, a force of attraction exists across the gap. For a fixed spacing as in (a), the charges are in equilibrium. Now move the plates closer together as in (b). As spacing decreases, the force

(a) Charges in
    equilibrium

(b) As spacing decreases, more
    electrons move from A to B,
    increasing the charge on the
    plates and hence, their
    capacitance

Figure 10–5   Decreasing plate spacing increases capacitance.

of attraction increases, pulling more electrons from within plate B to its surface, where they congregate. This creates a deficiency of electrons in the lower levels of B. To replenish these, the source moves additional electrons from A around the circuit, leaving A with an even greater deficiency and B with an even greater excess. The charge on the plates therefore increases and hence, according to Equation 10–2, so does the capacitance. We therefore conclude that decreasing spacing increases capacitance, and vice versa. In fact, as we will show later, *capacitance is inversely proportional to plate spacing.*

Effect of Dielectric

Capacitance also depends on the dielectric. Consider Figure 10–6(a), which shows an air-dielectric capacitor. If you substitute different materials for air, the capacitance increases. Table 10–1 shows the factor by which capacitance increases for a number of different materials. For example, if Teflon® is used instead of air, capacitance is increased by a factor of 2.1. This factor is called the *relative dielectric constant* or *relative permittivity* of the material. (Permittivity is a measure of how easy it is to establish electric flux in a material.) Note that high-permittivity ceramic increases capacitance by as much as 7500, as indicated in Figure 10–6(b).

TABLE 10–1 Relative Dielectric Constants (Also Called Relative Permittivities

MATERIAL	ϵ_r (NOMINAL VALUES)
Vacuum	1
Air	1.0006
Ceramic	30–7500
Mica	5.5
Mylar	3
Oil	4
Paper (dry)	2.2
Polystyrene	2.6
Teflon	2.1

(a) $C = 200$ pF with air dielectric

(b) $C = 1.5$ µF with high permittivity ceramic dielectric

Figure 10–6   The factor by which a material increases capacitance is termed its relative dielectric constant. The ceramic used here has a value of 7500.

Capacitance of a Parallel-Plate Capacitor

From the above observations, we see that capacitance is directly proportional to plate area, inversely proportional to plate separation, and dependent on the dielectric. In equation form,

$$C = \epsilon \frac{A}{d} \quad \text{(F)} \tag{10–3}$$

where area A is in square meters and spacing d is in meters.

Dielectric Constant

The constant ϵ in Equation 10–3 is the *absolute dielectric constant* of the insulating material. Its units are farads per meter (F/m). For air or vacuum, ϵ has a value of $\epsilon_o = 8.85 \times 10^{-12}$ F/m. For other materials, ϵ is expressed as the product of the relative dielectric constant, ϵ_r (shown in Table 10–1), times ϵ_o. That is,

$$\epsilon = \epsilon_r \epsilon_o \qquad (10\text{--}4)$$

Consider, again, Equation 10–3: $C = \epsilon A/d = \epsilon_r \epsilon_o A/d$. Note that $\epsilon_o A/d$ is the capacitance of a vacuum- (or air-) dielectric capacitor. Denote it by C_o. Then, for any other dielectric,

$$C = \epsilon_r C_o \qquad (10\text{--}5)$$

■ **EXAMPLE 10–2**   Compute the capacitance for a parallel-plate capacitor that has plates 10 cm by 20 cm, has a plate separation of 5 mm, and

   a. an air dielectric,

   b. a ceramic dielectric with permittivity of 7500.

Solution   Convert all dimensions to meters. Thus, $A = (0.1 \text{ m})(0.2 \text{ m}) = 0.02 \text{ m}^2$, and $d = 5 \times 10^{-3}$ m.

   a. For air, $C = \epsilon_o A/d = (8.85 \times 10^{-12})(2 \times 10^{-2})/(5 \times 10^{-3}) = 35.4 \times 10^{-12}$ F = 35.4 pF.

   b. For ceramic with $\epsilon_r = 7500$ $C = 7500(35.4 \text{ pF}) = 0.266 \ \mu\text{F}$.

■ **EXAMPLE 10–3**   A parallel-plate capacitor with air dielectric has a value of $C = 12$ pF. What is the capacitance of a capacitor that has the following:

   a. The same separation and dielectric but five times the plate area?

   b. The same dielectric but four times the area and one fifth the plate spacing?

   c. A dry paper dielectric, six times the plate area, and twice the plate spacing?

Solution

   a. Since the plate area has increased by a factor of five and everything else remains the same, C increases by a factor of five. Thus, $C = 5(12 \text{ pF}) = 60$ pF.

   b. With four times the plate area, C increases by a factor of four. With one fifth the plate spacing, C increases by a factor of five. Thus, $C = (4)(5)(12 \text{ pF}) = 240$ pF.

   c. Dry paper increases C by a factor of 2.2. The increase in plate area increases C by a factor of six. Doubling the plate spacing reduces C by one half. Thus, $C = (2.2)(6)(\frac{1}{2})(12 \text{ pF}) = 79.2$ pF.

Earlier we stated that the farad is a very large unit. To gain some insight about this, consider how big a 1-farad parallel-plate capacitor must be. For a spacing of 1 mm in a vacuum, you would need over 10^8 m^2 of area, i.e., a set of plates about 6 miles by 7 miles! Modern technology, however, can package a great deal of capacitance in a much smaller volume. For example, an experimental capacitor has been built with a capacitance of nearly 1000 farads in a package about the size of a flashlight battery! Most practical capacitors, however, are far from either of these two extremes.

in process
l e a r n i n g
check I

1. For a parallel-plate capacitor, if you triple the plate area and halve the plate spacing, how does capacitance change?

2. For the capacitor of Figure 10–6, if you use mica instead of ceramic, what will be the capacitance?

3. What is the dielectric for the capacitor of Figure 10–7(b)?

(a) C = 24pF with air dielectric   (b) C = 66 pF

Figure 10–7

10.3   Electric Fields

Electric Flux

Electric fields are force fields that exist in the region surrounding charged bodies. Some familiarity with electric fields is necessary to understand dielectrics and their effect on capacitance. We now look briefly at the key ideas.

Consider Figure 10–8(a). As noted in Chapter 2, unlike charges attract and like charges repel, i.e. a force exists between them. The region

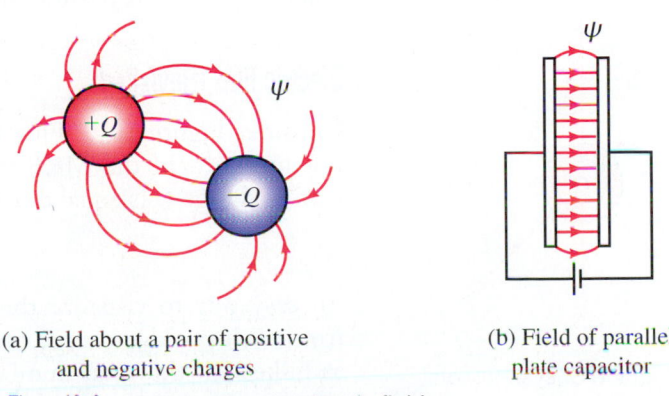

(a) Field about a pair of positive   (b) Field of parallel
and negative charges   plate capacitor

Figure 10–8   Some example electric fields.

where this force exists is called an *electric field*. To visualize this field, we use Faraday's field concept. The direction of the field is defined as the direction of force on a positive charge. It is therefore directed outward from the positive charge and inward toward the negative charge as shown. Field lines never cross, and the density of the lines indicates the strength of the field; the closer the lines are, the stronger the field. Figure 10–8(b) shows the field of a parallel-plate capacitor. In this case, the field is uniform across the gap, with some fringing near its edges. Electric flux lines are represented by the Greek letter ψ (psi).

Electric Field Intensity

The strength of an electric field, also called its *electric field intensity*, is the force that the field exerts on a small, positive test charge, Q_t. Let the field strength be denoted by $\mathscr{E}$. Then, by definition,

$$\mathscr{E} = F/Q_t \quad \text{(newtons/coulomb, N/C)} \tag{10–6}$$

To illustrate, let us determine the field about a point charge, Q. When the test charge is placed near Q, it experiences a force of $F = kQQ_t/r^2$ (Coulomb's law, Chapter 2). The constant in Coulomb's law is actually equal to $1/4\pi\epsilon$. Thus, $F = QQ_t/4\pi\epsilon r^2$, and from Equation 10–6,

$$\mathscr{E} = \frac{F}{Q_t} = \frac{Q}{4\pi\epsilon r^2} \quad \text{(N/C)} \tag{10–7}$$

Electric Flux Density

Because of the presence of ϵ in Equation 10–7, the electric field intensity depends on the medium in which the charge is located. Let us define a new quantity, D, that is independent of the medium. Let

$$D = \epsilon \mathscr{E} \tag{10–8}$$

D is known as electric *flux density*. Although not apparent here, D represents the density of flux lines in space, that is,

$$D = \frac{\text{total flux}}{\text{area}} = \frac{\psi}{A} \tag{10–9}$$

where ψ is the flux passing through area A.

Electric Flux (Revisited)

Consider Figure 10–9. Flux ψ is due to the charge Q. Although we will not prove it, in the MKS system the number of flux lines emanating from a charge Q is equal to the charge itself, that is,

$$\psi = Q \quad \text{(C)} \tag{10–10}$$

An easy way to visualize this is to think of one flux line as emanating from each positive charge on the body as shown in Figure 10–9. Then, as indicated, the total number of lines is equal to the total number of charges.

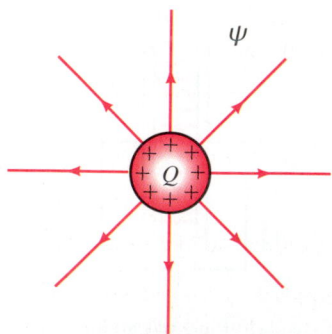

ψ

Figure 10–9   In the SI system, total flux ψ equals charge Q.

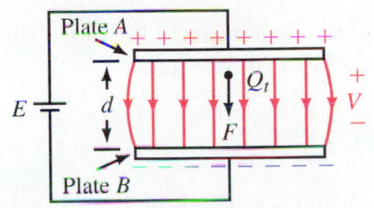

Figure 10–10   Work moving test charge Q_t is force times distance.

Field of a Parallel-Plate Capacitor

Now consider a parallel-plate capacitor (Figure 10–10). The field here is created by the charge distributed over its plates. Since plate A has a deficiency of electrons, it looks like a sheet of positive charge, while plate B looks like a sheet of negative charge. A positive test charge Q_t between these sheets is therefore repelled by the positive sheet and attracted by the negative sheet.

Now move the test charge from plate B to plate A. The work W required to move the charge against the force F is force times distance. Thus,

$$W = Fd \quad \text{(J)} \tag{10–11}$$

In Chapter 2, we defined voltage as work divided by charge, i.e., $V = W/Q$. Since the charge here is the test charge, Q_t, the voltage between plates A and B is

$$V = W/Q_t = (Fd)/Q_t \quad \text{(V)} \tag{10–12}$$

Now divide both sides by d. This yields $V/d = F/Q_t$. But $F/Q_t = \mathscr{E}$, from Equation 10–6. Thus,

$$\mathscr{E} = V/d \quad \text{(V/m)} \tag{10–13}$$

Equation 10–13 shows that the electric field strength between capacitor plates is equal to the voltage across the plates divided by the distance between them.

■ **EXAMPLE 10–4**   Suppose that the electric field intensity between the plates of a capacitor is 50 000 V/m when 80 V is applied:

a. What is the plate spacing if the dielectric is air? If the dielectric is ceramic?

b. What is $\mathscr{E}$ if the plate spacing is halved?

Solution

a. $\mathscr{E} = V/d$, independent of dielectric. Thus,

$$d = \frac{V}{\mathscr{E}} = \frac{80 \text{ V}}{50 \times 10^3 \text{ V/m}} = 1.6 \times 10^{-3} \text{ m}$$

b. Since $\mathscr{E} = V/d$, $\mathscr{E}$ will double to 100 000 V/m.

---

practice problems 1

1. What happens to the electric field intensity of a capacitor if you do the following:
   a. Double the applied voltage?
   b. Triple the applied voltage and double the plate spacing?

2. If the electric field intensity of a capacitor with polystyrene dielectric and plate size 2 cm by 4 cm is 100 kV/m when 50 V is applied, what is its capacitance?

Capacitance (Revisited)

With the above background, we can examine capacitance a bit more rigorously. Recall, $C = Q/V$. Using the above relationships yields

$$C = \frac{Q}{V} = \frac{\psi}{V} = \frac{AD}{\mathcal{E}d} = \frac{D}{\mathcal{E}}\left(\frac{A}{d}\right) = \epsilon\frac{A}{d}$$

This is the same equation (Equation 10–3) that we developed intuitively in Section 10.2.

10.4 Dielectrics

As you saw in Figure 10–6, a dielectric increases capacitance. We now examine why. Consider Figure 10–11. For a charged capacitor, electron orbits (which are normally circular) become elongated as electrons are attracted toward the positive (+) plate and repelled from the negative (−) plate. This makes the end of the atom nearest the positive plate appear negative while its other end appears positive. Such atoms are *polarized*. Throughout the bulk of the dielectric, the negative end of a polarized atom is adjacent to the positive end of another atom, and the effects cancel. However, at the surfaces of the dielectric, there are no atoms to cancel, and the net effect is as if a layer of negative charge exists on the surface of the dielectric at the positive plate and a layer of positive charge at the negative plate. This makes the plates appear closer, thus increasing capacitance. Materials for which the effect is largest result in the greatest increase in capacitance.

Figure 10–11   Effect of the capacitor's electric field on an atom of its dielectric.

TABLE 10–2 Dielectric Strength*

MATERIAL	kV/mm
Air	3
Ceramic (high ϵ_r)	3
Mica	40
Mylar	16
Oil	15
Polystyrene	24
Rubber	18
Teflon®	60

*Values depend on the composition of the material. These are the values we use in this book.

Voltage Breakdown

If the voltage of Figure 10–11 is increased beyond a critical value, the force on the electrons is so great that they are literally torn from orbit. This is called *dielectric breakdown* and the electric field intensity (Equation 10–13) at breakdown is called the *dielectric strength* of the material. For air, breakdown occurs when the voltage gradient exceeds 3 kV/mm. The breakdown strengths for other materials are shown in Table 10–2. Since the quality of a dielectric depends on many factors, dielectric strength varies from sample to sample. Solid dielectrics are usually damaged by breakdown.

Breakdown is not limited to capacitors; it can occur with any type of electrical apparatus whose insulation is stressed beyond safe limits. (For example, air breaks down and flashovers occur on high-voltage transmission lines when they arc struck by lightning.) The shape of conductors also affects breakdown voltage. Breakdown occurs at lower voltages at sharp points than at blunt points. This effect is made use of in lightning arresters.

■ **EXAMPLE 10–5**   A capacitor with plate dimensions of 2.5 cm by 2.5 cm and a ceramic dielectric with $\epsilon_r = 7500$ experiences breakdown at 2400 V. What is C?

Solution   From Table 10–2 dielectric strength = 3 kV/mm. Thus, $d = 2400$ V/3000 V/mm $= 0.8$ mm $= 8 \times 10^{-4}$ m. So

$$C = \epsilon_r \epsilon_o A/d$$

$$= (7500)(8.85 \times 10^{-12})(0.025 \text{ m})^2/(8 \times 10^{-4} \text{ m})$$

$$= 51.9 \text{ nF}$$

---

*practice
problem
II*

An air-dielectric capacitor breaks down at 500 V. If the plate spacing is doubled and the capacitor is filled with oil, at what voltage will breakdown occur?

Voltage Rating

Because of dielectric breakdown, capacitors are rated for maximum operating voltage (called *working voltage*) by their manufacturer (indicated on the capacitor as WVDC or *working voltage dc*). If you operate a capacitor beyond its working voltage, you may damage it.

10.5 Nonideal Effects

So far, we have assumed ideal capacitors. However, real capacitors have nonideal characteristics.

Leakage Current

When a charged capacitor is disconnected from its source, it will eventually discharge. This is because no insulator is perfect and a small amount of charge "leaks" through the dielectric. Similarly, a small leakage current will pass through its dielectric when a capacitor is connected to a source.

The effect of leakage is modeled by a resistor in Figure 10–12. Since leakage is very small, R is very large, typically hundreds of megohms. The larger R is, the longer a capacitor can hold its charge. For most applications, leakage can be neglected.

Figure 10–12   Leakage current.

Equivalent Series Resistance (ESR)

As a capacitor ages, resistance may develop in its leads as its internal connections begin to fail. This resistance is in series with the capacitor and can cause problems in high-frequency circuits.

Dielectric Absorption

When a capacitor is discharged by temporarily shorting its leads, it should have zero volts when the short is removed. However, atoms sometimes remain partially polarized, and when the short is removed, they cause a residual voltage to appear across the capacitor. This effect is known as *dielectric absorption*. In electronic circuits, the voltage due to dielectric absorption can upset circuit voltage levels; in TV tubes and electrical power apparatus, it can result in large and potentially dangerous voltages.

Temperature Coefficient

Because dielectrics are affected by temperature, capacitance may change with temperature. If capacitance increases with increasing temperature, the capacitor is said to have a *positive temperature coefficient*; if it decreases, the capacitor has a *negative temperature coefficient*; if it remains essentially constant, the capacitor has a *zero temperature coefficient*.

The temperature coefficient is specified as a change in capacitance in parts per million (ppm) per degree Celsius. Consider a 1-μF capacitor. Since 1 μF = 1 million pF, 1 ppm is 1 pF. Thus, a 1-μF capacitor with a temperature coefficient of 200 ppm/°C could change as much as 200 pF per degree Celsius.

10.6   Types of Capacitors

No single capacitor type suits all applications, and thus capacitors are made in a wide variety of types and sizes. Among these are fixed and variable types with differing dielectrics and recommended areas of application.

Fixed Capacitors

Fixed capacitors are often identified by their dielectric. Common dielectric materials include ceramic, plastic, and mica, plus, for electrolytic capacitors, aluminum and tantalum oxide. Design variations include tubular and interleaved plates. The interleave design (Figure 10–13) uses multiple plates to increase effective plate area. A layer of insulation separates plates, and alternate plates are connected together. The tubular design (Figure 10–14) uses sheets of metal foil separated by an insulator such as plastic film. Fixed capacitors are encapsulated in plastic, epoxy resin, or other insulating material and identified with value, tolerance, and other appropriate data either via body markings or color coding. Electrical characteristics and physical size depend on the dielectric used.

Figure 10–13   Stacked capacitor construction. The stack is compressed, leads attached, and the unit coated with epoxy resin or other insulating material.

Figure 10–14   Tubular capacitor with axial leads.

Ceramic Capacitors.   First, consider ceramic. The permittivity of ceramic varies widely (as indicated in Table 10–1). At one end are ceramics with extremely high permittivity. These permit packaging a great deal of capacitance in a small space, but yield capacitors whose characteristics vary widely with temperature and operating voltage. However, they are popular in limited temperature applications where small size and cost are important. At the other end are ceramics with highly stable characteristics. They yield capacitors whose values change little with temperature, voltage, or aging. However, since their dielectric constants are relatively low (typically 30 to 80), these capacitors are physically larger than those made using high-permittivity ceramic. Many surface mount capacitors (considered later in this section) use ceramic dielectrics.

Plastic Film Capacitors.   Plastic-film capacitors are of two basic types: film/foil or metalized film. *Film/foil* capacitors use metal foil separated by plastic film as in Figure 10–14, while *metallized-film* capacitors have their foil material vacuum deposited directly onto plastic film. Film/foil capacitors are generally larger than metallized-foil units, but have better capacitance stability and higher insulation resistance. Typical film materials are polyester, Mylar, polypropylene, and polycarbonate. Figure 10–15 shows a selection of plastic-film capacitors.

Figure 10–15   Radial lead film capacitors.   *(Courtesy Illinois Capacitor Inc.)*

Metalized-film capacitors are *self-healing*. Thus, if voltage stress at an imperfection exceeds breakdown, an arc occurs which evaporates the metallized area around the fault, isolating the defect. (Film/foil capacitors are not self-healing.)

Mica Capacitors.   Mica capacitors are low in cost with low leakage and good stability. Available values range from a few picofarads to about 0.1 μF.

Electrolytic Capacitors.   Electrolytic capacitors provide large capacitance (i.e., up to several hundred thousand microfarads) at a relatively low cost. (Their capacitance is large because they have a very thin layer of oxide as their dielectric.) However, their leakage is relatively high and breakdown voltage relatively low. Electrolytics have either aluminum or tantalum as their plate material. Tantalum devices are smaller than aluminum devices, have less leakage, and are more stable.

The basic aluminum electrolytic capacitor construction is similar to that of Figure 10–14, with strips of aluminum foil separated by gauze saturated with an electrolyte. During manufacture, chemical action creates a thin oxide layer that acts as the dielectric. This layer must be maintained during use. For this reason, electrolytic capacitors are polarized (marked with a + and − sign), and the plus (+) terminal must always be kept positive with respect to the minus (−) terminal. Electrolytic capacitors have a *shelf life*, that is, if they are not used for an extended period (typically two years), they may fail when powered up again. Figure 10–16 shows a selection of aluminum electrolytic devices.

Tantalum capacitors come in two basic types: wet slug and solid dielectric. Figure 10–17 shows a cutaway view of a solid tantalum unit. The slug, made from powdered tantalum, is highly porous and provides a large internal surface area that is coated with an oxide to form the dieiectric. Tantalum capacitors are polarized and must be inserted into a circuit properly.

Surface Mount Capacitors.   Many electronic products now use *surface mount devices* (SMDs). (SMDs do not have connection leads but are soldered directly onto printed circuit boards.) Figure 10–18 shows a surface

Figure 10–16   Radial lead aluminum electrolytic capacitor. *(Courtesy Illinois Capacitor Inc.)*

Figure 10–17   Cutaway view of a solid tantalum capacitor. *(Courtesy AVX Corporation)*

mount, ceramic chip capacitor. Such devices are extremely small and provide high packaging density.

Variable Capacitors

The most common variable capacitor is that used in radio tuning circuits (Figure 10–19). It has a set of stationary plates and a set of movable plates which are ganged together and mounted on a shaft. As the shaft is

(a) Typical size is 2 mm × 1 mm
(See Figure 1-4)

(b) Cutaway View
(Courtesy AVX Corporation)

Figure 10–18   Surface mount, ceramic chip capacitor.

Figure 10–19   Variable capacitor of type used in radios.

rotated, the movable plates mesh with the stationary plates, changing the effective surface area (and hence the capacitance).

Another adjustable type is the *trimmer* or *padder* capacitor, which is used for fine adjustments, usually over a very small range. In contrast to the variable capacitor (which is frequently varied by the user), a trimmer is usually set to its required value, then never touched again (unless further trimming is needed.)

practice
problem
III

A 2.5-μF capacitor has a tolerance of +80% and −20%. Determine what its maximum and minimum values could be.

10.7 Capacitors in Parallel and Series

Capacitors may be connected in various combinations. Equivalent values are easily computed.

Capacitors in Parallel

For capacitors in parallel, the effective plate area is the sum of the individual plate areas; thus, the total capacitance is the sum of the individual capacitances. This is easily shown. Consider Figure 10–20. The charge on each capacitor is given by Equation 10–1. Thus, $Q_1 = C_1V$ and $Q_2 = C_2V$. Since $Q_T = Q_1 + Q_2$, $Q_T = C_1V + C_2V = (C_1 + C_2)V$. But $Q_T = C_TV$. Thus, $C_T = C_1 + C_2$. If there are more than two capacitors, then

$$C_T = C_1 + C_2 + \cdots + C_N \qquad (10\text{–}14)$$

(a) $Q_T = Q_1 + Q_2$

(b)

(c) $C_T = C_1 + C_2$

Figure 10–20   Capacitors in parallel. Voltage is the same across each.

That is, *the total capacitance of capacitors in parallel is the sum of their individual capacitances.* (Note that Equation 10–14 is similar to the formula for resistors in series.)

■ **EXAMPLE 10–6**   A 10-μF, a 15-μF, and a 100-μF capacitor are connected in parallel across a 50-V source. Determine the following:

   a. Total capacitance.
   b. Total charge stored.
   c. Charge on each capacitor.

Solution

   a. $C_T = C_1 + C_2 + C_3 = 10\ \mu F + 15\ \mu F + 100\ \mu F = 125\ \mu F$

   b. $Q_T = C_T V = (125\ \mu F)(50\ V) = 6.25\ mC$

   c. $Q_1 = C_1 V = (10\ \mu F)(50\ V) = 0.5\ mC$
      $Q_2 = C_2 V = (15\ \mu F)(50\ V) = 0.75\ mC$
      $Q_3 = C_3 V = (100\ \mu F)(50\ V) = 5.0\ mC$

Note $Q_1 + Q_2 + Q_3 = (0.5 + 0.75 + 5.0)\ mC = 6.25\ mC$, which checks with (b).

---

practice
problems
IV

1. Three capacitors are paralleled across an 80-V source, with $Q_T = 0.12$ C. If $C_1 = 200\ \mu F$ and $C_2 = 300\ \mu F$, what is C_3?

2. Three capacitors are paralleled. If the value of the second capacitor is twice that of the first and the value of the third is one quarter that of the second and the total capacitance is 70 μF, what are the values of each capacitor?

Capacitors in Series

For capacitors in series (Figure 10–21), the same charge appears on each. Thus, $Q = C_1 V_1$, $Q = C_2 V_2$, etc. Solving for voltages yields $V_1 = Q/C_1$, $V_2 = Q/C_2$, and so on. Applying KVL, we get $V = V_1 + V_2 + \cdots + V_N$. Therefore,

$$V = \frac{Q}{C_1} + \frac{Q}{C_2} + \cdots + \frac{Q}{C_N} = Q\left(\frac{1}{C_1} + \frac{1}{C_2} + \cdots + \frac{1}{C_N}\right)$$

But $V = Q/C_T$. Equating this with the right side above and cancelling Q yields

$$\frac{1}{C_T} = \frac{1}{C_1} + \frac{1}{C_2} + \cdots + \frac{1}{C_N} \tag{10–15}$$

(This is similar to the formula for resistors in parallel.) For two capacitors in series, the equation reduces to

$$C_T = \frac{C_1 C_2}{C_1 + C_2} \tag{10–16}$$

For N equal capacitors in series, Equation 10–15 yields $C_T = C/N$.

(a)

(b) Equivalent

Figure 10–21   Capacitors in series. Charge is the same on each.

■ **EXAMPLE 10–7**   Refer to Figure 10–22(a):

a. Determine C_T.

b. If 50 V is applied across the capacitors, determine Q.

c. Determine the voltage on each capacitor.

30 μF   60 μF   20 μF   10 μF

C_1 C_2 C_3 C_T

(a)   (b)

Figure 10–22

Solution

a. $$\frac{1}{C_T} = \frac{1}{C_1} + \frac{1}{C_2} + \frac{1}{C_3} = \frac{1}{30\ \mu F} + \frac{1}{60\ \mu F} + \frac{1}{20\ \mu F}$$

$$= 0.0333 \times 10^6 + 0.0167 \times 10^6 + 0.05 \times 10^6 = 0.1 \times 10^6$$

Therefore,

$$C_T = \frac{1}{0.1 \times 10^6} = 10\ \mu F$$

b. $Q = C_T V = (10 \times 10^{-6}\ \text{F})(50\ \text{V}) = 0.5\ \text{mC}$

c. $V_1 = Q/C_1 = (0.5 \times 10^{-3}\ \text{C})/(30 \times 10^{-6}\ \text{F}) = 16.7\ \text{V}$

$V_2 = Q/C_2 = (0.5 \times 10^{-3}\ \text{C})/(60 \times 10^{-6}\ \text{F}) = 8.3\ \text{V}$

$V_3 = Q/C_3 = (0.5 \times 10^{-3}\ \text{C})/(20 \times 10^{-6}\ \text{F}) = 25.0\ \text{V}$

Check: $V_1 + V_2 + V_3 = 16.7 + 8.3 + 25 = 50\ \text{V}$.

■ **EXAMPLE 10–8** For the circuit of Figure 10–23(a), determine C_T.

(a)

(b)

(c)

Figure 10–23 Systematic reduction.

Solution The problem is easily solved through step-by-step reduction. C_2 and C_3 in parallel yield $45\ \mu\text{F} + 15\ \mu\text{F} = 60\ \mu\text{F}$. C_4 and C_5 in parallel total $20\ \mu\text{F}$. The reduced circuit is shown in (b). The two $60\text{-}\mu\text{F}$ capacitances in series reduce to $30\ \mu\text{F}$. The series combination of $30\ \mu\text{F}$ and $20\ \mu\text{F}$ can be found from Equation 10–16. Thus,

$$C_T = \frac{30\ \mu\text{F} \times 20\ \mu\text{F}}{30\ \mu\text{F} + 20\ \mu\text{F}} = 12\ \mu\text{F}$$

Alternately, you can reduce (b) directly using Equation 10–15. Try it.

Voltage Divider Rule for Series Capacitors

For capacitors in series (Figure 10–24) a simple voltage divider rule can be developed. Recall, for individual capacitors, $Q_1 = C_1 V_1$, $Q_2 = C_2 V_2$, etc., and for the complete string, $Q_T = C_T V_T$. As noted earlier, $Q_1 = Q_2 = \cdots = Q_T$. Thus, $C_1 V_1 = C_T V_T$. Solving for V_1 yields

$$V_1 = \left(\frac{C_T}{C_1}\right)V_T$$

This relationship holds for all capacitors. Thus,

$$V_x = \left(\frac{C_T}{C_x}\right)V_T \qquad (10–17)$$

From this, you can see that the voltage across a capacitor is inversely proportional to its capacitance, that is, the smaller the capacitance, the larger the voltage, and vice versa. Other useful variations are

$$V_1 = \left(\frac{C_2}{C_1}\right)V_2, \qquad V_1 = \left(\frac{C_3}{C_1}\right)V_3, \qquad V_2 = \left(\frac{C_3}{C_2}\right)V_3, \qquad \text{etc.}$$

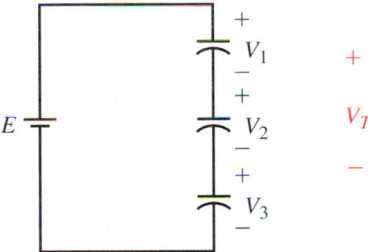

Figure 10–24   Capacitive voltage divider.

notes

For capacitors in parallel, total capacitance is always larger than the largest capacitance; for capacitors in series, total capacitance is always smaller than the smallest capacitance.

practice problems
V

1. Compute the voltages of Example 10–7 using Equation 10–17.

2. Determine the voltage across each capacitor of Figure 10–23 if the voltage across C_5 is 30 V.

10.8 Capacitor Current and Voltage

As noted earlier (Figure 10–2), during charging, electrons are moved from one plate of a capacitor to the other plate. Several points should be noted. 1) This movement of electrons constitutes a current. 2) This current lasts only long enough for the capacitor to charge. When the capacitor is fully charged, current is zero. 3) Current in the circuit during charging is due solely to the movement of electrons from one plate to the other around the external circuit; no current passes through the dielectric between the plates. 4) As charge is deposited on the plates, the capacitor voltage builds. However, this voltage does not jump to full value immediately since it takes time to move electrons from one plate to the other. (Billions of electrons must be moved.) 5) Since voltage builds up as charging progresses, the difference in voltage between the source and the capacitor decreases and hence the rate of movement of electrons (i.e., the current) decreases as the capacitor approaches full charge.

Figure 10–25 shows what the voltage and current look like during the charging process. As indicated, the current starts out with an initial surge, then decays to zero while the capacitor voltage gradually climbs from zero to full voltage. The charging time typically ranges from nanoseconds to milliseconds, depending on the resistance and capacitance of the circuit. (We study these relationships in detail in Chapter 11.) A similar surge (but in the opposite direction) occurs during discharge.

(a)

(b) Current surge during charging. Current is zero when fully charged.

(c) Capacitor voltage. $v_C = E$ when fully charged.

Figure 10–25   The capacitor does not charge instantaneously, as a finite amount of time is required to move electrons around the circuit.

In Figure 10–25, note that current exists only while the capacitor voltage is changing. This observation turns out to be true in general, that is, *current in a capacitor exists only while capacitor voltage is changing.* The reason is not hard to understand. As you saw before, a capacitor's dielectric is an insulator and consequently no current can pass through it (assuming zero leakage). The only charges that can move, therefore, are the free electrons that exist on the capacitor's plates. When capacitor voltage is constant, these charges are in equilibrium, no net movement of charge occurs, and the current is thus zero. However, if the source voltage is increased, additional electrons are pulled from the positive plate; inversely, if the source voltage is decreased, excess electrons on the negative plate are returned to the positive plate. Thus, in both cases, capacitor current results when capacitor voltage is changed. As we now show, this current is proportional to the rate of change of voltage. Before we do this, however, we need to look at symbols.

Symbols for Time-Varying Voltages and Currents

Quantities that vary with time are called *instantaneous* quantities. *Standard industry practice requires that we use lowercase letters for time-varying quantities, rather than capital letters as for dc.* Thus, we use v_C and i_C to represent changing capacitor voltage and current rather than V_C and I_C. (Often we drop the subscripts and just use v and i.) Since these quantities are functions of time, they may also be shown as $v_C(t)$ and $i_C(t)$.

Capacitor v-i Relationship

The relationship between charge and voltage for a capacitor is given by Equation 10–1. For the time-varying case, it is

$$q = Cv_C \tag{10–18}$$

But current is the rate of movement of charge. In calculus notation, this is $i_C = dq/dt$. Differentiating Equation 10–18 yields

$$i_C = \frac{dq}{dt} = \frac{d}{dt}(Cv_c) \tag{10–19}$$

Since C is constant, we get

$$i_C = C\frac{dv_C}{dt} \quad \text{(A)} \tag{10–20}$$

Equation 10–20 shows that *current through a capacitor is equal to C times the rate of change of voltage across it.* This means that the faster the voltage changes, the larger the current, and vice versa. It also means that if the voltage is constant, the current is zero (as we noted earlier).

Reference conventions for voltage and current are shown in Figure 10–26. As usual, the plus sign goes at the tail of the arrow. If the voltage is increasing, dv_C/dt is positive and the current is in the direction of the reference arrow; if the voltage is decreasing, dv_C/dt is negative and the current is opposite to the arrow.

Equation 10–20 is the fundamental voltage-current relationship for capacitance but requires the use of calculus. For now, we will look only at situations where voltage changes linearly with time as in Figure 10–27(a). For this case, dv_C/dt can be found as the *slope* of the line, that is, $dv_C/dt = \text{rise/run} = \Delta v/\Delta t$. We therefore do not need calculus.

Figure 10–26   The + sign for v_C goes at the tail of the current arrow.

■ **EXAMPLE 10–9**   A signal generator applies voltage to a 5-μF capacitor with a waveform as in Figure 10–27(a). The voltage rises linearly from 0 to 10 V in 1 ms, falls linearly to –10 V at $t = 3$ ms, remains constant until $t = 4$ ms, rises to 10 V at $t = 5$ ms, and remains constant thereafter.

    a. Determine the slope of v_C.

    b. Determine the current and sketch its graph.

(a)

(b)

Figure 10–27

Solution

a. We need the slope of v_C during each time interval where slope = rise/run = $\Delta v / \Delta t$.

0 ms to 1 ms: $\Delta v = 10$ V; $\Delta t = 1$ ms; Therefore, slope = 10 V/1 ms = 10 000 V/s.

1 ms to 3 ms: Slope = –20 V/2 ms = –10 000 V/s.

3 ms to 4 ms: Slope = 0 V/s.

4 ms to 5 ms: Slope = 20 V/1 ms = 20 000 V/s.

b. $i_C = C dv_C / dt = C$ times slope. Thus,

0 ms to 1 ms: $i = (5 \times 10^{-6} \text{ F})(10\ 000 \text{ V/s}) = 50$ mA.

1 ms to 3 ms: $i = -(5 \times 10^{-6} \text{ F})(10\ 000 \text{ V/s}) = -50$ mA.

3 ms to 4 ms: $i = (5 \times 10^{-6} \text{ F})(0 \text{ V/s}) = 0$ A.

4 ms to 5 ms: $i = (5 \times 10^{-6} \text{ F})(20\ 000 \text{ V/s}) = 100$ mA.

The current is plotted in Figure 10–27(b).

10.9   Energy Stored by a Capacitor

An ideal capacitor does not dissipate power. When power is transferred to a capacitor, all of it is stored as energy in the capacitor's electric field. When the capacitor is discharged, this stored energy is returned to the circuit.

To determine the stored energy, consider Figure 10–28. Power is given by $p = vi$ watts. Using calculus (see ∫), it can be shown that the stored energy is given by

$$W = \frac{1}{2} CV^2 \quad \text{(J)} \tag{10–21}$$

where V is the voltage across the capacitor. This means that the energy at any time depends on the value of the capacitor's voltage at that time.

Figure 10–28   Storing energy in a capacitor.

Deriving Equation 10–21

Power to the capacitor (Figure 10–28) is given by $p = vi$, where $i = Cdv/dt$. Therefore, $p = Cdv/dt$. However, $p = dW/dt$. Integrating both sides yields

$$W = \int_0^t p\,dt = C\int_0^t v\frac{dv}{dt}\,dt = C\int_0^V v\,dv = \frac{1}{2}CV^2$$

10.10   Capacitor Failures and Troubleshooting

Capacitors fail because of misapplication, excessive voltage, current, temperature, or simply because they age. They can short internally, leads may become open, dielectrics may become excessively leaky, and they may fail catastrophically due to incorrect use. (If an electrolytic capacitor is connected with its polarity reversed, for example, it may explode.) Capacitors should be used well within their rating limits. Excessive voltage can lead to dielectric puncture creating pinholes that short the plates together. High temperatures may cause an increase in leakage and/or a permanent shift in capacitance. High temperatures may be caused by inadequate heat removal, excessive current, lossy dielectrics, or an operating frequency beyond the capacitor's rated limit.

Basic Testing with an Ohmmeter

Some basic (out-of-circuit) tests can be made with an analog ohmmeter. This ohmmeter can detect opens and shorts and, to a certain extent, leaky dielectrics. First, ensure that the capacitor is discharged, then set the ohmmeter to its highest range and connect it to the capacitor. (For electrolytic devices, ensure that the plus (+) side of the ohmmeter is connected to the plus (+) side of the capacitor.)

Initially, the ohmmeter reading should be low, then for a good capacitor gradually increase to infinity as the capacitor charges through the ohmmeter circuit. (Or at least a very high value, since most good capacitors, except electrolytics, have a resistance of hundreds of megohms.) For small capacitors, however, the time to charge may be too short to yield useful results.

Faulty capacitors respond differently. If a capacitor is shorted, the meter resistance reading will stay low. If it is leaky, the reading will be lower than normal. If it is open circuited, the meter will indicate infinity immediately, without dipping to zero when first connected.

Capacitor Testers

Ohmmeter testing of capacitors has its limitations and other tools may be needed. Figure 10–29 shows several. The DMM in (a) can measure capacitance and display it directly on its readout. The LCR (inductance, capacitance, resistance) analyzer in (b) can determine capacitance as well as detect opens and shorts. More sophisticated testers are available that determine capacitance value, leakage at rated voltage, dielectric absorption, and so on.

(a) Measuring *C* with a DMM. (Not all DMMs can measure capacitance.)

(b) Capacitor/inductor analyzer. *(Courtesy B + K Precision)*

Figure 10–29   Capacitor testing.

Problems

10.1   Capacitance

1. The voltage across a 20-μF capacitor is 80 V. What is the charge on the capacitor?

2. The charge on a 50-μF capacitor is 10×10^{-3} C. What is the voltage?

3. When 10 μC of charge is placed on a capacitor, its voltage is 25 V. What is the capacitance?

10.2   Factors Affecting Capacitance

4. A parallel-plate capacitor has a capacitance of 120 pF. If the distance between its plates is tripled, what is its new capacitance?

5. A parallel-plate capacitor with a mica dielectric has dimensions of 1 cm × 1.5 cm and separation of 0.1 mm. What is its capacitance?

6. For the capacitor of Problem 5, if the mica is removed, what is its new capacitance?

7. The capacitance of an oil-filled capacitor is 200 pF. If the separation between its plates is 0.1 mm, what is the area of its plates?

8. A 0.01-μF capacitor has ceramic with a dielectric constant of 7500. If the ceramic is removed, the plate separation doubled, and the spacing between plates filled with oil, what is the new value for C?

9. The plate area of a capacitor is 4.5 in². and the plate separation is 5 mils. If the relative permittivity of the dielectric is 80, what is C?

10.3   Electric Fields

10. a.   What is the electric field strength $\mathscr{E}$ at a distance of 1 cm from a 100-mC charge in transformer oil?
    b.   What is $\mathscr{E}$ at twice the distance?

11. Suppose that 150 V is applied across a 100-pF parallel-plate capacitor whose plates are separated by 1 mm. What is the electric field intensity $\mathscr{E}$ between the plates?

10.4   Dielectrics

12. An air-dielectric capacitor has plate spacing of 1.5 mm. How much voltage can be applied before breakdown occurs?

13. Repeat Problem 12 if the dielectric is mica and the spacing is 2 mils.

14. A mica-dielectric capacitor breaks down when E volts is applied. The mica is removed and the spacing between plates doubled. If breakdown now occurs at 500 V, what is E?

15. Determine at what voltage the dielectric of a 200 nF Mylar capacitor with a plate area of 0.625 m² will break down.

16. Figure 10–30 shows several gaps, including a parallel-plate capacitor, a set of small spherical points, and a pair of sharp points. The spacing is the same for each. As the voltage is increased, which gap breaks down for each case?

Plates   Points

(a)

Spheres   Plates

(b)

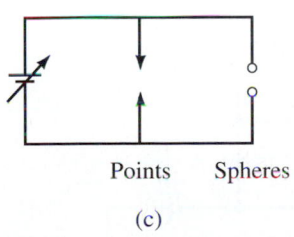

Points   Spheres

(c)

Figure 10–30   Voltage is increased until one of the gaps breaks down. (The source has high internal resistance to limit current following breakdown.)

17. If you continue to increase the source voltage of Figures 10–30(a), (b), and (c) after a gap breaks down, will the second gap also break down? Justify your answer.

10.5 Nonideal Effects

18. A 25-μF capacitor has a negative temperature coefficient of 175 ppm/°C. By how much and in what direction might it vary if the temperature rises by 50°C? What would be its new value?

19. If a 4.7-μF capacitor changes to 4.8 μF when the temperature rises 40°C, what is its temperature coefficient?

10.7 Capacitors in Parallel and Series

20. Determine C_T for each circuit of Figure 10–31.

(a)

(b)

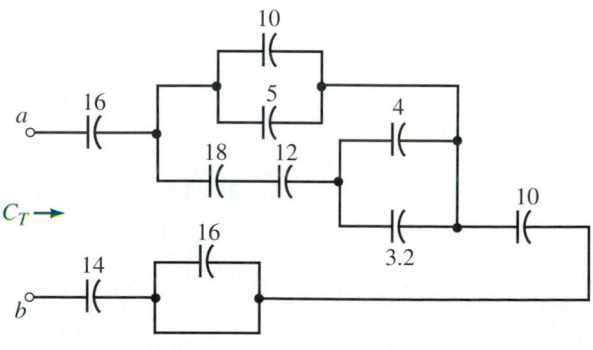

(c) All values in μF

Figure 10–31

21. For Figure 10–32, determine C_x.

Figure 10–32

22. For Figure 10–33, determine C_3 and C_4.

Figure 10–33

23. For Figure 10–34, determine C_T.

Figure 10–34

24. A 10-μF and 4.7-μF capacitor are connnected in parallel. After a third capacitor is added to the circuit, $C_T = 2.695\ \mu$F. What is the value of the third capacitor? How is it connected?

25. Consider capacitors of 1 μF, 1.5 μF, and 10 μF. If $C_T = 10.6\ \mu$F, how are the capacitors connected?

26. For the capacitors of Problem 25, if $C_T = 2.304\ \mu$F, how are the capacitors connected?

27. For Figures 10–31(a) and (b), find the voltage on each capacitor if 100 V is applied to terminals a-b.

28. Use the voltage divider rule to find the voltage across each capacitor of Figure 10–35.

(a)

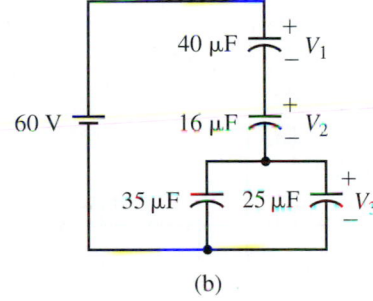

(b)

Figure 10–35

29. For Figure 10–36, $V_x = 50$ V. Determine C_x.

Figure 10–36

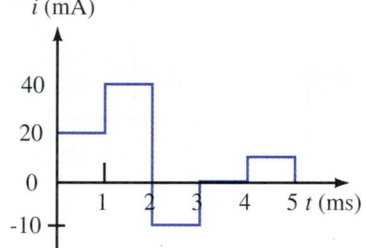

Figure 10–37

30. For Figure 10–37, determine C_x.

10.8   Capacitor Current and Voltage

31. The voltage across the capacitor of Figure 10–38(a) is shown in (b). Sketch current i_C scaled with numerical values.

(a)

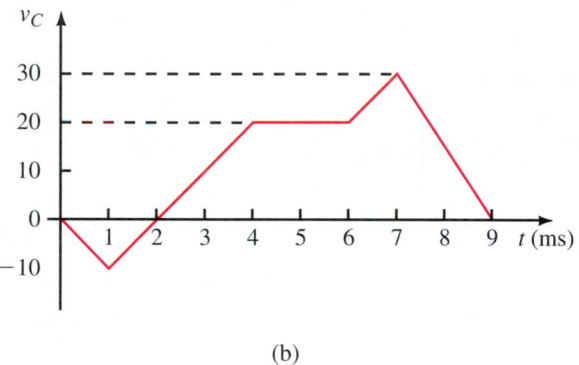

(b)

Figure 10–38

32. The current in a 1-μF capacitor is shown in Figure 10–39. Sketch voltage v_C scaled with numerical values. Voltage at $t = 0$ s is 0 V.

33. If the voltage across a 4.7-μF capacitor is $v_C = 100e^{-0.05t}$ V, what is i_C? (This is a calculus problem.)

10.9   Energy Stored by a Capacitor

34. For the circuit of Figure 10–35(a), determine the energy stored in each capacitor.

35. For Figure 10–38, determine the capacitor's energy at each of the following times: $t = 0$, 1 ms, 4 ms, 5 ms, 7 ms, and 9 ms.

10.10   Capacitor Failures and Troubleshooting

36. For each case shown in Figure 10–40, what is the likely fault?

Figure 10–39

(a)   (b)   (c)

Figure 10–40

Capacitive Transients, Pulse, and Waveshaping

CHAPTER PREVIEW

As you saw in Chapter 10, capacitors do not charge or discharge instantaneously. Instead, as illustrated in Figure 10–25, voltages and currents take time to reach their new values. As we will show in this chapter, these voltages and currents are exponential in form, and the length of time that it takes to charge or discharge depends on the resistance and capacitance of the circuit. Since the voltages and currents that exist during charging and discharging are transitory in nature, they are called *transients*. Transients do not last very long, typically only a fraction of a second. However, they are important to us for a number of reasons, some of which you will learn in this chapter.

   Transients occur in both capacitive and inductive circuits. In capacitive circuits, they occur because capacitor voltage cannot change instantaneously; in inductive circuits, they occur because inductor current cannot change instantaneously. In this chapter, we look at capacitive transients; in Chapter 14, we look at inductive transients. As you will see, many of the basic principles are the same.

   Note: Optional derivations using calculus are marked by a ∫ icon. They may be omitted without loss of continuity by colleges that do not stress calculus.

OUTLINE

Performing a transient analysis on an electronic circuit using SPICE.
(Photographer/Artist—L. G. Meares. Courtesy of Intusoft)

Introduction
- Capacitor Charging
- Steady State Conditions
- Capacitor Discharging
- The Meaning of Time in Transient Analysis
- Exponential Functions

Capacitor Charging Equations
- The Time Constant
- Duration of a Transient

Capacitor with an Initial Voltage

Capacitor Discharging Equations

More Complex Circuits

An *RC* Timing Application

Pulse Response of *RC* Circuits
- Pulse Basics
- The Effect of Pulse Width
- Capacitive Loading

Transient Analysis Using PSpice
- Creating a Step Voltage
- Creating a Switched Input

OBJECTIVES

After studying this chapter, you will be able to

- explain why transients occur in *RC* circuits,
- explain why an uncharged capacitor looks like a short circuit when first energized,
- describe why a capacitor looks like an open circuit to steady state dc,
- describe charging and discharging of simple *RC* circuits with dc excitation,
- determine voltages and currents in simple *RC* circuits during charging and discharging,
- plot voltage and current transients,
- understand the part that time constants play in determining the duration of transients,
- compute time constants,
- describe the use of charging and discharging waveforms in simple timing applications,
- calculate the pulse response of simple *RC* circuits,
- solve simple *RC* transient problems using PSpice.

403

Desirable and Undesirable Transients

*T*ransients occur in capacitive and inductive circuits whenever circuit conditions are changed, for example, by the sudden application of a voltage, the switching in or out of a circuit element, or the malfunctioning of a circuit component. Some transients are desirable and useful; others occur under abnormal conditions and are potentially destructive in nature.

An example of the latter is the transient that results when lightning strikes a power line. Following a strike, the line voltage, which may have been only a few thousand volts before the strike, suddenly rises to many hundreds of thousands of volts or higher, then rapidly decays, while the current, which may have been only a few hundred amps, suddenly rises to many times its normal value. Although these transients do not last very long, they can cause serious damage. While this is a rather severe example of a transient, it nonetheless illustrates that during transient conditions, some of a circuit or system's most difficult problems may arise.

Many transient effects, on the other hand, are useful. For example, a number of important electronic devices and circuits depend on transient effects; these include timers, oscillators, and waveshaping circuits. As you will see in this chapter and in later electronics courses, the charge/discharge characteristic of RC circuits is fundamental to their operation.

Figure 11–1 A transient results when an *RC* circuit is switched.

11.1 Introduction

A basic *RC* circuit is shown in Figure 11–1. Most of the key ideas concerning charging and discharging and dc transients in *RC* circuits can be developed from it.

Capacitor Charging

First, assume the capacitor is uncharged and that the switch is open. Now move the switch to the charge position. At the instant the switch is closed the current jumps to E/R amps, then decays to zero, while the voltage, which is zero at the instant the switch is closed, gradually climbs to E volts. This is shown in Figure 11–2. The shapes of these curves are easily explained.

First, consider voltage. In order to change capacitor voltage, electrons must be moved from one plate to the other. Even for a relatively

(a)

(b)

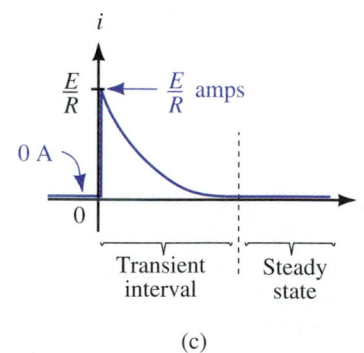

(c)

Figure 11–2 Voltage and current during charging. Time $t = 0$ s is defined as the instant the switch is moved to the charge position.

(a) Circuit as it looks just after the switch is moved to the charge position; v_C is still zero.

Looks like a short

(b) Since $v_C = 0$,
$i = E/R$.

Figure 11–3   An uncharged capacitor initially looks like a short circuit.

small capacitor, billions of electrons must be moved. This takes time. Consequently, *capacitor voltage cannot change instantaneously, i.e., it cannot jump abruptly from one value to another*. Instead, it climbs gradually and smoothly as illustrated in Figure 11–2(b).

Now consider current. As indicated in Figure 11–2(c), the current is discontinuous, that is, it jumps from 0 to E/R amps. To understand why, note that since capacitor voltage cannot change instantaneously, its value just after switching must be the same as just before. Initially the capacitor is uncharged. Thus, as indicated in Figure 11–3(a), the capacitor voltage just after switching will still be zero. Since the capacitor has current through it (because electrons start to move immediately) and zero voltage across it, it looks momentarily like a short circuit, (b). (This is an important observation and one that is true in general, that is, *an uncharged capacitor looks like a short circuit at the instant of switching*.) The circuit current therefore jumps to $i = E/R$ amps. It then gradually decays to zero.

Steady State Conditions

When the capacitor voltage and current reach their final values and stop changing, the circuit is said to be in *steady state*. Figure 11–4(a) shows the circuit after it has reached steady state. Since the capacitor has voltage across it but no current through it, it looks like an open circuit as indicated in (b). This is also an important observation and one that is true in general, that is, *a capacitor looks like an open circuit to steady state dc*.

(a) Steady state:
$v_C = E$ and $i_C = 0$.

(b) Since the capacitor has voltage across it but zero current, it looks like an open circuit.

Figure 11–4   A capacitor looks like an open circuit to dc in the steady state.

Capacitor Discharging

Now consider the discharge case shown in Figure 11–5. Since the capacitor has E volts across it before the switch is moved to discharge, it will still have E volts across it just after. This is indicated in (a). The capacitor therefore looks momentarily like a voltage source as shown in (b), and the current thus jumps to $-E/R$ amps. (The current is negative since it is opposite in direction to the reference arrow.) The voltage and current then decay to zero as indicated in Figure 11–6.

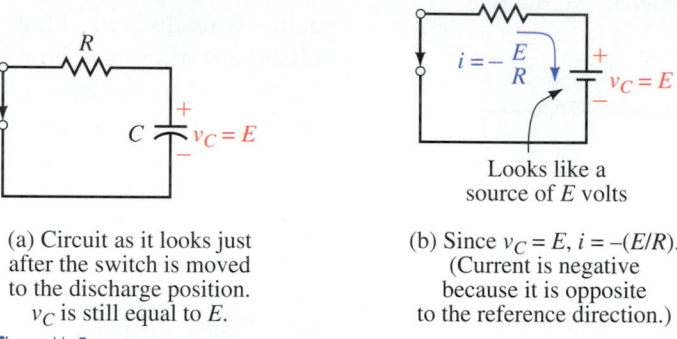

(a) Circuit as it looks just after the switch is moved to the discharge position. v_C is still equal to E.

(b) Since $v_C = E$, $i = -(E/R)$. (Current is negative because it is opposite to the reference direction.)

Figure 11–5   A charged capacitor looks initially like a voltage source.

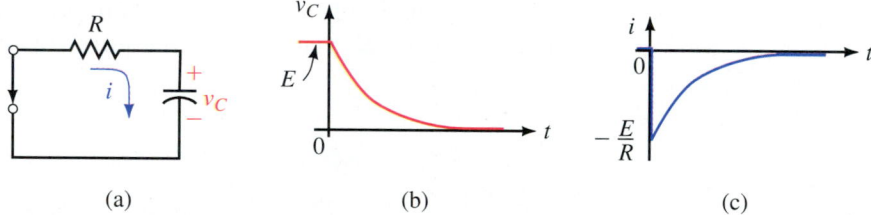

(a)   (b)   (c)

Figure 11–6   Voltage and current during discharge. Time $t = 0$ s is defined as the instant the switch is moved to the discharge position.

■ **EXAMPLE 11–1**   For Figure 11–1, $E = 40$ V, $R = 10\ \Omega$, and the capacitor is initially uncharged. The switch is moved to the charge position and the capacitor allowed to charge. Then the switch is moved to the discharge position and the capacitor allowed to discharge. Sketch the voltages and currents and determine the values at switching and steady state.

Solution   The current and voltage curves are shown in Figure 11–7. Initially, $i = 0$ A since the switch is open. Immediately after it is moved to the charge position, the current jumps to $E/R = 40$ V/10 Ω = 4 A; then it decays to zero. At the same time, v_C starts at 0 V and climbs to 40 V. When the switch is moved to the discharge position, the capacitor looks momentarily like a 40-V source and the current jumps to -40 V/10 Ω = -4 A; then it decays to zero. At the same time, v_C also decays to zero.

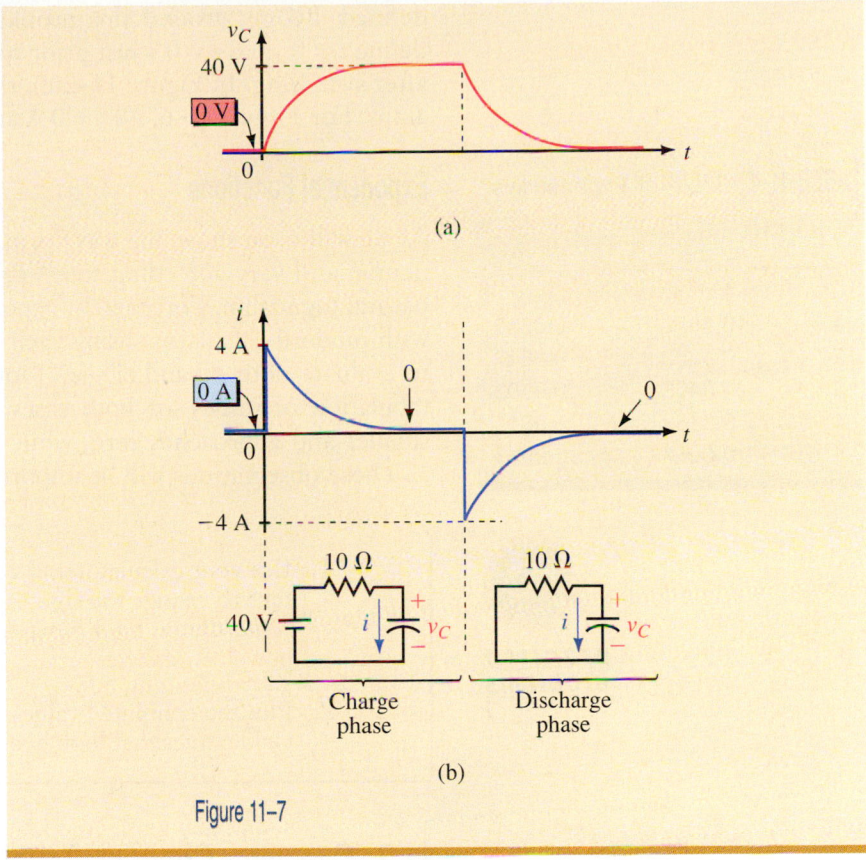

Figure 11–7

S*ome key points to remember are the following:*

1. Voltage across a capacitor cannot change instantaneously, but current can.

2. To determine currents and voltages in a circuit at the instant of switching, replace uncharged capacitors with short circuits and charged capacitors with dc sources equal to their respective voltages at the instant of switching.

3. To determine currents and voltages in a dc circuit after it has reached steady state, replace capacitors with open circuits.

The Meaning of Time in Transient Analysis

The time t used in transient analysis is measured from the instant of switching. Thus, $t = 0$ in Figure 11–2 is defined as the instant the switch is moved to charge, while in Figure 11–6, it is defined as the instant the switch is moved to discharge. Voltages and currents are then represented in terms of this time as $v_C(t)$ and $i(t)$. For example, the voltage across a capacitor at $t = 0$ s is denoted as $v_C(0)$, while the voltage at $t = 10$ ms is denoted as $v_C(10 \text{ ms})$, and so on.

A problem arises when a quantity is discontinuous as is the current of Figure 11–2(c). Since its value is changing at $t = 0$ s, $i(0)$ cannot be defined. To get around this problem, we define two values for 0. We define $t = 0^-$ s as $t = 0$ s just prior to switching and $t = 0^+$ s as $t = 0$ s just after switching. In Figure 11–2(c), therefore, $i(0^-) = 0$ A while $i(0^+) = E/R$ amps. For Figure 11–6, $i(0^-) = 0$ A and $i(0^+) = -E/R$ amps.

Exponential Functions

As we will soon show, the waveforms of Figures 11–2 and 11–6 are exponential and vary according to e^{-x} or $(1 - e^{-x})$, where e is the base of the natural logarithm. Fortunately, exponential functions are easy to evaluate with modern calculators using their e^x function. You will need to be able to evaluate both e^{-x} and $(1 - e^{-x})$ for any value of x. Table 11–1 shows a tabulation of values for both cases. Note that as x gets larger, e^{-x} gets smaller and approaches zero, while $(1 - e^{-x})$ gets larger and approaches 1. These observations will be important to you later.

TABLE 11–1 Table of Exponentials

x	e^{-x}	$1 - e^{-x}$
0	1	0
1	0.3679	0.6321
2	0.1353	0.8647
3	0.0498	0.9502
4	0.0183	0.9817
5	0.0067	0.9933

practice problems 1

1. Use your calculator and verify the entries in Table 11–1. Be sure to change the sign of x before using the e^x function. Note that $e^{-0} = 1$ since any quantity raised to the zeroth power is one.

2. Plot the computed values on graph paper and verify that they yield curves that look like those shown in Figure 11–2.

11.2 Capacitor Charging Equations

We will now develop equations for voltages and current during charging. Consider Figure 11–8. KVL yields

$$v_R + v_C = E \tag{11–1}$$

But $v_R = Ri_C$ and $i_C = Cdv_C/dt$ (Equation 10–20). Thus, $v_R = RCdv_C/dt$. Substituting this into Equation 11–1 yields

$$RC\frac{dv_c}{dt} + v_C = E \tag{11–2}$$

Figure 11–8 Circuit for the charging case. $v_R + v_C = E$.

Equation 11–2 can be solved for v_C using basic calculus (see). The result is

$$v_C = E(1 - e^{-t/RC}) \tag{11–3}$$

where R is in ohms, C is in farads, t is in seconds and $e^{-t/RC}$ is the exponential function discussed earlier. The product RC has units of seconds. (This is left as an exercise for the student to show.)

SOLVING EQUATION 11–2

First, rearrange Equation 11–2:

$$\frac{dv_C}{dt} = \frac{1}{RC}(E - v_C)$$

Thus,

$$\frac{dv_C}{E - v_C} = \frac{dt}{RC}$$

Now multiply both sides by −1 and integrate.

$$\int_0^{v_C} \frac{dv_C}{v_C - E} = -\frac{1}{RC} \int_0^t dt$$

$$\ln(v_C - E) \Big]_0^{v_C} = -\frac{t}{RC} \Big]_0^t$$

Next, substitute integration limits,

$$\ln(v_C - E) - \ln(-E) = -\frac{t}{RC}$$

$$\ln\left(\frac{v_C - E}{-E}\right) = -\frac{t}{RC}$$

Finally, take the inverse log of both sides. Thus,

$$\frac{v_C - E}{-E} = e^{-t/RC}$$

When you rearrange this, you get Equation 11–3. That is,

$$v_C = E(1 - e^{-t/RC})$$

Now consider the resistor voltage. From Equation 11–1, $v_R = E - v_C$. Substituting v_C from Equation 11–3 yields $v_R = E - E(1 - e^{-t/RC}) = E - E + Ee^{-t/RC}$. After cancellation, you get

$$v_R = Ee^{-t/RC} \tag{11–4}$$

Now divide both sides by R. Since $i_C = i_R = v_R/R$, this yields

$$i_C = \frac{E}{R} e^{-t/RC} \tag{11–5}$$

The waveforms are shown in Figure 11–9. Values at any time may be determined by substitution.

Figure 11–9   Curves for the circuit of Figure 11–8.

■ **EXAMPLE 11–2**   Suppose $E = 100$ V, $R = 10$ kΩ, and $C = 10$ μF:

a. Determine the expression for v_C.
b. Determine the expression for i_C.
c. Compute the capacitor voltage at $t = 150$ ms.
d. Compute the capacitor current at $t = 150$ ms.
e. Locate the computed points on the curves.

Solution

a. $RC = (10 \times 10^3\ \Omega)(10 \times 10^{-6}\ F) = 0.1\ s$. From Equation 11–3,
$v_C = E(1 - e^{-t/RC}) = 100(1 - e^{-t/0.1}) = 100(1 - e^{-10\,t})$ V.

b. From Equation 11–5, $i_C = (E/R)e^{-t/RC} = (100\ V/10\ k\Omega)e^{-10\,t} = 10e^{-10\,t}$ mA.

c. At $t = 0.15$ s, $v_C = 100(1 - e^{-10\,t}) = 100(1 - e^{-10(0.15)}) = 100(1 - e^{-1.5}) = 100(1 - 0.223) = 77.7$ V.

d. $i_C = 10e^{-10\,t}$ mA $= 10e^{-10(0.15)}$ mA $= 10e^{-1.5}$ mA $= 2.23$ mA.

e. The corresponding points are shown in Figure 11–10.

Figure 11–10   Plotting the computed points on the v_C and i_C curves.

In the above example, we expressed voltage as $v_C = 100(1 - e^{-t/0.1})$ and as $100(1 - e^{-10t})$ V. Similarly, current can be expressed as $i_C = 10e^{-t/0.1}$ or as $10e^{-10t}$ mA. Although some authors prefer one notation over the other, both are correct and we will use them interchangeably.

practice problems II

1. Determine additional voltage and current points for Figure 11–10 by computing values of v_C and i_C at values of time from $t = 0$ s to $t = 500$ ms at 100-ms intervals. Plot the results.

2. The switch of Figure 11–11 is closed at $t = 0$ s. If $E = 80$ V, $R = 4$ kΩ, and $C = 5\ \mu F$, determine expressions for v_C and i_C. Plot the results from $t = 0$ s to $t = 100$ ms at 20-ms intervals. Note that charging takes less time here than for Problem 1.

Figure 11–11

■ **EXAMPLE 11–3**   For the circuit of Figure 11–11, $E = 60$ V, $R = 2$ kΩ, and $C = 25$ μF. The switch is closed at $t = 0$ s, opened 40 ms later and left open. Determine equations for capacitor voltage and current and plot.

Solution $RC = (2\text{ k}\Omega)(25\text{ }\mu\text{F}) = 50$ ms. From $t = 0$ s to 40 ms, the following equations hold:

$$v_C = E(1 - e^{-t/RC}) = 60(1 - e^{-t/50\text{ ms}})\text{ V}$$

$$i_C = (E/R)e^{-t/RC} = 30e^{-t/50\text{ ms}}\text{ mA}$$

Voltage starts at 0 V and rises exponentially. At $t = 40$ ms, the switch is opened, interrupting charging. At this instant, $v_C = 60(1 - e^{-(40/50)}) = 60(1 - e^{-0.8}) = 33.0$ V. Since the switch is left open, the voltage remains constant at 33 V thereafter as indicated in Figure 11–12. (The dotted curve shows how the voltage would have kept rising if the switch had remained closed.)

Now consider current. The current starts at 30 mA and decays to $i_C = 30e^{-(40/50)}$ mA = 13.5 mA at $t = 40$ ms. At this point, the switch is opened, and the current drops instantly to zero. (The dotted line shows how the current would have decayed if the switch had not been opened.)

(a)   (b)

Figure 11–12   Incomplete charging. The switch of Figure 11–11 was opened at $t = 40$ ms, causing charging to cease.

The Time Constant

The rate at which a capacitor charges depends on the product of R and C. This product is known as the *time constant* of the circuit and is given the symbol τ (the Greek letter tau). As noted earlier, RC has units of seconds. Thus,

$$\tau = RC \quad \text{(seconds, s)} \tag{11–6}$$

Using τ, Equations 11–3 to 11–5 can be written as

$$v_C = E(1 - e^{-t/\tau}) \tag{11–7}$$

$$i_C = \frac{E}{R}e^{-t/\tau} \tag{11–8}$$

and

$$v_R = Ee^{-t/\tau} \tag{11–9}$$

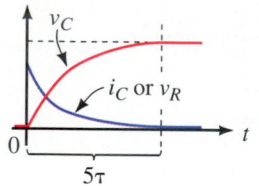

Figure 11–13   Transients last five time constants.

Duration of a Transient

The length of time that a transient lasts depends on the exponential function $e^{-t/\tau}$. As t increases, $e^{-t/\tau}$ decreases, and when it reaches zero, the transient is gone. Theoretically, this takes infinite time. In practice, however, over 99% of the transition takes place during the first five time constants (i.e., transients are within 1% of their final value at $t = 5\tau$). This can be verified by direct substitution. At $t = 5\tau$, $v_C = E(1 - e^{-t/\tau}) = E(1 - e^{-5}) = E(1 - 0.0067) = 0.993E$, meaning that the transient has achieved 99.3% of its final value. Similarly, the current falls to within 1% of its final value in five time constants. Thus, *for all practical purposes, transients can be considered to last for only five time constants* (Figure 11–13).

■ **EXAMPLE 11–4**   For the circuit of Figure 11–11, how long will it take for the capacitor to charge if $R = 2\ k\Omega$ and $C = 10\ \mu F$?

Solution $\tau = RC = (2\ k\Omega)(10\ \mu F) = 20$ ms. Therefore, the capacitor charges in $5\tau = 100$ ms.

■ **EXAMPLE 11–5**   The transient in a circuit with $C = 40\ \mu F$ lasts 0.5 s. What is R?

Solution $5\tau = 0.5$ s. Thus, $\tau = 0.1$ s and $R = \tau/C = 0.1\ s/(40 \times 10^{-6}\ F) = 2.5\ k\Omega$.

Figure 11–14 shows percent capacitor voltage and current plotted versus multiples of time constant. (Points are computed from $v_C = 100(1 - e^{-t/\tau})$ and $i_C = 100e^{-t/\tau}$. For example, at $t = \tau$, $v_C = 100(1 - e^{-t/\tau}) = 100(1 - e^{-\tau/\tau}) = 100(1 - e^{-1}) = 63.2$ V, i.e., 63.2%, and $i_C = 100e^{-\tau/\tau} = 100e^{-1} = 36.8$ A, which is 36.8%, and so on.) These curves are referred to as universal time constant curves. They are useful as a guide for checking your work.

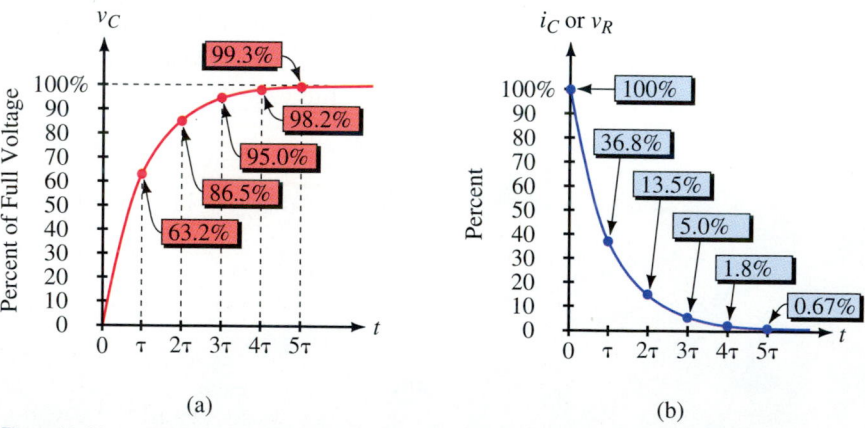

(a)   (b)

Figure 11–14   Universal voltage and current curves for the charging case.

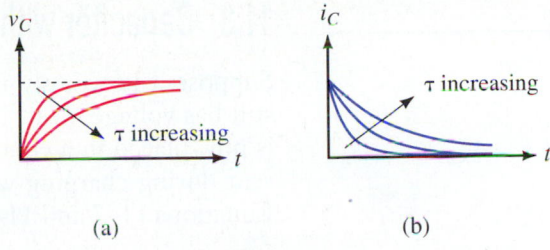

Figure 11–15   The larger the time constant, the longer the capacitor takes to charge.

Figure 11–15 shows how voltage and current are affected by the time constant. The larger the resistance and capacitance, the larger the time constant and hence the longer it takes for the capacitor to charge.

practical

notes

*S*ince *resistors and capacitors have tolerances, their actual values may be different from their marked values. This will make the actual time constant of the circuit different than the computed time constant. In many cases, this is not important. However, it may be. In critical applications, you should check the effect that R and C tolerances have on* τ.

in process
learning
check I

1. If the capacitor of Figure 11–16 is uncharged, what is the current immediately after closing the switch?

Figure 11–16

2. Given $i_C = 50e^{-20t}$ mA.
    a. What is τ?
    b. Compute the current at $t = 0^+$ s, 25 ms, 50 ms, 75 ms, 100 ms, and 500 ms and sketch it.

3. Repeat Problem 2 for $v_C = 100(1 - e^{-50t})$ V.

4. For Figure 11–16, determine expressions for v_C and i_C.

5. Refer to Figure 11–10:
    a. What are $v_C(0^-)$ and $v_C(0^+)$?
    b. What are $i_C(0^-)$ and $i_C(0^+)$?
    c. What are the steady state voltage and current?

6. For the circuit of Figure 11–11, the current just after the switch is closed is 2 mA. The transient lasts 40 ms and the capacitor charges to 80 V. Determine E, R, and C.

11.3 Capacitor with an Initial Voltage

Suppose a previously used capacitor has not been discharged and thus still has voltage on it. Let this voltage be denoted as V_0. If the capacitor is now placed in a circuit like that in Figure 11–16, the voltage and current during charging will be affected by the initial voltage. In this case, Equations 11–7 and 11–8 become

$$v_C = E + (V_0 - E)e^{-t/\tau} \tag{11–10}$$

$$i_C = \frac{E - V_0}{R}e^{-t/\tau} \tag{11–11}$$

■ **EXAMPLE 11–6** Suppose the capacitor of Figure 11–16 has 25 volts on it with polarity shown at the time the switch is closed.

a. Determine the expression for v_C.
b. Determine the expression for i_C.
c. Compute v_C and i_C at $t = 0.1$ s.
d. Sketch v_C and i_C.

Solution $\tau = RC = (200\ \Omega)(1000\ \mu F) = 0.2$ s

a. From Equation 11–10,

$$v_C = E + (V_0 - E)e^{-t/\tau} = 40 + (25 - 40)e^{-t/0.2} = 40 - 15e^{-5t}\ \text{V}$$

b. From Equation 11–11,

$$i_C = \frac{E - V_0}{R}e^{-t/\tau} = \frac{40 - 25}{200}e^{-5t} = 75e^{-5t}\ \text{mA}$$

c. At $t = 0.1$ s,

$$v_C = 40 - 15e^{-5t} = 40 - 15e^{-0.5} = 30.9\ \text{V}$$

$$i_C = 75e^{-5t}\ \text{mA} = 75e^{-0.5}\ \text{mA} = 45.5\ \text{mA}$$

d. The waveforms are shown in Figure 11–17 with the above points plotted.

(a) $V_0 = 25$V.

(b)

Figure 11–17   Capacitor with an initial voltage.

practice
problem
III

Repeat Example 11–6 for the circuit of Figure 11–16 if $V_0 = -150$ V.

Figure 11–18   Discharge case. Initial capacitor voltage is V_0. Note the reference direction for i_C. To maintain consistency with the voltage-current reference convention discussed in Section 10.8, i_C must be chosen in this direction so that the + sign for v_C is at the tail of the current arrow. This means that when you solve for i_C, it will have a negative value. You can see this in Figure 11–19(b).

11.4 Capacitor Discharging Equations

To determine the discharge equations, move the switch to the discharge position (Figure 11–18), KVL yields $v_R + v_C = 0$. Substituting $v_R = RC\,dv_C/dt$ from Section 11.2 yields

$$RC\frac{dv_C}{dt} + v_C = 0 \qquad (11\text{–}12)$$

This can be solved for v_C using basic calculus. The result is

$$v_C = V_0 e^{-t/RC} \qquad (11\text{–}13)$$

where V_0 is the voltage on the capacitor at the instant the switch is moved to discharge. Now consider the resistor voltage. Since $v_R + v_C = 0$, $v_R = -v_C$ and

$$v_R = -V_0 e^{-t/RC} \qquad (11\text{–}14)$$

Now divide both sides by R. Since $i_C = i_R = v_R/R$,

$$i_C = -\frac{V_0}{R} e^{-t/RC} \qquad (11\text{–}15)$$

Note that this is negative, since, during discharge, the current is opposite in direction to the reference arrow of Figure 11–18. Voltage v_C and current i_C are shown in Figure 11–19. As for the charging case, *discharge transients last five time constants.*

In Equations 11–13 to 11–15, V_0 represents the voltage on the capacitor at the instant the switch is moved to the discharge position. If the switch has been in the charge position long enough for the capacitor to fully charge, $V_0 = E$ and Equations 11–13 and 11–15 become $v_C = Ee^{-t/RC}$ and $i_C = -(E/R)e^{-t/RC}$ respectively. If the charging is stopped early, as in Figure 11–12, V_0 will be the voltage reached by the capacitor when charging is interrupted.

(a)

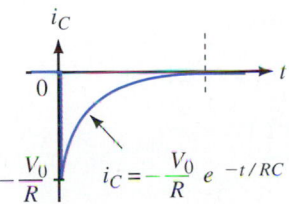

(b) Discharge current is opposite to the reference for i_C.

Figure 11–19   Waveforms for the discharge case.

■ **EXAMPLE 11–7**   For the circuit of Figure 11–18, assume the capacitor is charged to 100 V before the switch is moved to the discharge position. Suppose $R = 5\ \text{k}\Omega$ and $C = 25\ \mu F$:

a. Determine the expression for v_C.

b. Determine the expression for i_C.

c. Compute the voltage and current at 0.1 s.

Solution $RC = (5 \text{ k}\Omega)(25 \ \mu\text{F}) = 0.125$ s and $V_0 = 100$ V. Therefore,

a. $v_C = V_0 e^{-t/RC} = 100 e^{-t/0.125} = 100 e^{-8t}$ V.

b. $i_C = -(V_0/R)e^{-t/RC} = -20 e^{-8t}$ mA.

c. At $t = 0.1$ s,

$$v_C = 100 e^{-8t} = 100 e^{-0.8} = 44.9 \text{ V}$$

$$i_C = -20 e^{-8t} \text{ mA} = -20 e^{-0.8} \text{ mA} = -8.99 \text{ mA}$$

11.5 More Complex Circuits

The charge and discharge equations described previously apply only to circuits of the form shown in Figures 11–2 and 11–6 respectively. Fortunately, many circuits can be reduced to these forms using standard circuit reduction techniques such as series and parallel combinations, source conversions, Thévenin's theorem, and so on. Once a circuit has been reduced to its series equivalent, you can use any of the equations that we have developed so far.

■ **EXAMPLE 11–8**   For the circuit of Figure 11–20(a), determine expressions for v_C and i_C. Capacitors are uncharged.

(a)   (b)

Figure 11–20

Solution   Reduce circuit (a) to circuit (b).

$$R_{eq} = R_1 \| R_2 = 2.0 \text{ k}\Omega; \qquad C_{eq} = C_1 + C_2 = 10 \ \mu\text{F}.$$
$$R_{eq}C_{eq} = (2 \text{ k}\Omega)(10 \times 10^{-6} \text{ F}) = 0.020 \text{ s}$$

Thus,

$$v_C = E(1 - e^{-t/R_{eq}C_{eq}}) = 100(1 - e^{-t/0.02}) = 100(1 - e^{-50t}) \text{ V}$$

$$i_C = \frac{E}{R_{eq}} e^{-t/R_{eq}C_{eq}} = \frac{100}{2000} e^{-t/0.02} = 50 \ e^{-50t} \text{ mA}$$

■ **EXAMPLE 11–9** The capacitor of Figure 11–21 is initially uncharged. Close the switch at $t = 0$ s.

a. Determine the expression for v_C.

b. Determine the expression for i_C.

Figure 11–21

Solution Reduce the circuit to its series equivalent using Thévenin's theorem:

$$R'_2 = R_2 \| R_3 = 160 \ \Omega$$

From Figure 11–22(a),

$$R_{Th} = R_1 \| R'_2 + R_4 = 240 \| 160 + 104 = 96 + 104 = 200 \ \Omega$$

From Figure 11–22(b),

$$V'_2 = \left(\frac{R'_2}{R_1 + R'_2} \right) E = \left(\frac{160}{240 + 160} \right) \times 100 \ V = 40 \ V$$

(a) Finding R_{Th} (b) Finding E_{Th}

Figure 11–22   Thévenizing the circuit of Figure 11–21 following switch closure.

From KVL, $E_{Th} = V'_2 = 40$ V. The resultant equivalent circuit is shown in Figure 11–23.

$$\tau = R_{Th}C = (200 \ \Omega)(50 \ \mu F) = 10 \ ms$$

a. $v_C = E_{Th}(1 - e^{-t/\tau}) = 40(1 - e^{-100t})$ V

b. $i_C = \dfrac{E_{Th}}{R_{Th}} e^{-t/\tau} = \dfrac{40}{200} e^{-t/0.01} = 200e^{-100t}$ mA

Figure 11–23   Thévenizing the circuit of Figure 11–21 (continued).

1. For Figure 11–21, if $R_1 = 400\ \Omega$, $R_2 = 1200\ \Omega$, $R_3 = 300\ \Omega$, $R_4 = 50\ \Omega$, $C = 20\ \mu F$, and $E = 200$ V, determine v_C and i_C.

2. Using the values shown in Figure 11–21, determine v_C and i_C if the capacitor has an initial voltage of 60 V.

3. Using the values of Problem 1, determine v_C and i_C if the capacitor has an initial voltage of −50 V.

Notes About Time References

1. *So far, we have dealt with charging and discharging problems separately. For these, we define t = 0 s as the instant the switch is moved to the charge position for charging problems and to the discharge position for discharging problems.*

2. *When you have both charge and discharge cases in the same example, you need to establish clearly what you mean by "time." We use the following procedure:*
   a. *Define t = 0 s as the instant the switch is moved to the first position, then determine corresponding expressions for v_C and i_C. These expressions and the corresponding time scale are valid until the switch is moved to its new position.*
   b. *When the switch is moved to its new position, shift the time reference and make t = 0 s the time at which the switch is moved to its new position, then determine corresponding expressions for v_C and i_C. These new expressions are only valid from the new t = 0 s reference point. The old expressions remain valid on the old time scale.*
   c. *We now have two time scales for the same graph. However, we generally only show the first scale explicitly; the second scale is implied rather than shown.*
   d. *Use τ_c to represent the time constant for charging and τ_d to represent the time constant for discharging. Since the equivalent resistance and capacitance for discharging may be different than that for charging, the time constants may be different for the two cases.*

■ **EXAMPLE 11–10**   The capacitor of Figure 11–24(a) is uncharged. The switch is moved to position 1 for 10 ms, then to position 2, where it remains.

   a. Determine v_C during charge.
   b. Determine i_C during charge.
   c. Determine v_C during discharge.

d. Determine i_C during discharge.

e. Sketch the charge and discharge waveforms.

(a) Circuit

(b) Charging circuit

(c) Discharging circuit
$V_0 = 100$ V at $t = 0$ s

Figure 11-24

Solution   Figure 11–24(b) shows the equivalent charging circuit. Here,

$$\tau_c = (R_1 + R_2)C = (1 \text{ k}\Omega)(2 \text{ }\mu\text{F}) = 2.0 \text{ ms.}$$

a. $v_C = E(1 - e^{-t/\tau_c}) = 100(1 - e^{-500t})$

b. $i_C = \dfrac{E}{R_{T_c}} e^{-t/\tau_c} = \dfrac{100}{1000} e^{-500t} = 100e^{-500t} \text{ mA}$

Since $5\tau_c = 10$ ms, charging is complete by the time the switch is moved to discharge.

c. Figure 11–24(c) shows the equivalent discharge circuit. Note $V_0 = 100$ V.

$$\tau_d = (500 \text{ }\Omega)(2 \text{ }\mu\text{F}) = 1.0 \text{ ms}$$
$$v_C = V_0 e^{-t/\tau_d} = 100e^{-1000t} \text{ V}$$

where $t = 0$ s has been redefined for discharge as noted above.

d. $i_C = -\dfrac{E}{R_2 + R_3} e^{-t/\tau_d} = -\dfrac{100}{500} e^{-1000t} = -200e^{-1000t} \text{ mA}$

e. See Figure 11–25. Note that discharge is more rapid than charge since $\tau_d < \tau_c$.

Figure 11–25   Waveforms for the circuit of Figure 11–24. Note that τ_d is shorter than τ_c.

■ **EXAMPLE 11–11**   The capacitor of Figure 11–26 is uncharged. The switch is moved to position 1 for 5 ms, then to position 2 and left there.

a. Determine v_C while the switch is in position 1.
b. Determine i_C while the switch is in position 1.
c. Compute v_C and i_C at $t = 5$ ms.
d. Determine v_C while the switch is in position 2.
e. Determine i_C while the switch is in position 2.
f. Sketch the voltage and current waveforms.

Figure 11–26

Solution

$$\tau_c = \tau_d = RC = (1 \text{ k}\Omega)(4 \text{ }\mu\text{F}) = 4 \text{ ms}$$

a. $v_C = E_1(1 - e^{-t/\tau_c}) = 10(1 - e^{-250t})$ V

b. $i_C = \dfrac{E_1}{R}e^{-t/\tau_c} = \dfrac{10}{1000}e^{-250t} = 10e^{-250t}$ mA

c. At $t = 5$ ms,

$$v_C = 10(1 - e^{-250 \times 0.005}) = 7.14 \text{ V}$$

$$i_C = 10e^{-250 \times 0.005} \text{ mA} = 2.87 \text{ mA}$$

d. In position 2, $E_2 = 30$ V, and $V_0 = 7.14$ V. Use Equation 11–10:

$$v_C = E_2 + (V_0 - E_2)e^{-t/\tau_d} = 30 + (7.14 - 30)e^{-250t}$$

$$= 30 - 22.86e^{-250t} \text{ V}$$

where $t = 0$ s has been redefined for position 2.

e. $\quad i_C = \dfrac{E_2 - V_0}{R} e^{-t/\tau_d} = \dfrac{30 - 7.14}{1000} e^{-250t} = 22.86 e^{-250t}$ mA

f. See Figure 11–27.

Figure 11–27   Waveforms for the circuit of Figure 11–26.

■ **EXAMPLE 11–12**   In Figure 11–28(a), the capacitor is initially uncharged. The switch is moved to the charge position, then to the discharge position, yielding the current shown in (b). The capacitor discharges in 1.75 ms. Determine the following:

a. E.   b. R_1.   c. C.

Figure 11–28

Solution

a. Since the capacitor charges fully, it has a value of E volts when switched to discharge. The discharge current spike is

$$-\frac{E}{10\,\Omega + 25\,\Omega} = -3 \text{ A}$$

Thus, $E = 105$ V.

b. The charging current spike has a value of

$$\frac{E}{10\,\Omega + R_1} = 7 \text{ A}$$

Since $E = 105$ V, this yields $R_1 = 5\ \Omega$.

c. $5\tau_d = 1.75$ ms. Therefore $\tau_d = 350\ \mu s$. *But* $\tau_d = (R_2 + R_3)C$. Thus, $C = 350\ \mu s/35\ \Omega = 10\ \mu F$.

■ **EXAMPLE 11–13**   The circuit of Figure 11–29(a) has reached steady state. Determine the capacitor voltages.

(a)

(b)

Figure 11–29

Solution   Since the circuit is in steady state, a transient analysis is not required. Replace all capacitors with open circuits as in (b). Then,

$$I_1 = \frac{200\text{ V}}{40\ \Omega + 60\ \Omega} = 2\text{ A}, \qquad I_2 = \frac{90\text{ V}}{40\ \Omega + 8\ \Omega + 12\ \Omega} = 1.5\text{ A}$$

KVL: $V_1 - 120 - 18 = 0$. Therefore, $V_1 = 138$ V. Further,

$$V_2 = (8\ \Omega)(1.5\text{ A}) = 12\text{ V}$$

practice
problems
V

1. For the circuit of Figure 11–30(a), the switch is moved to position 1 at $t = 0$ s and to position 2 100 ms later. Determine v_C and i_C for position 2.

2. Repeat for Figure 11–30(b).

(a) *C* = 500 μF

(b) *C* = 20 μF

Figure 11–30

3. The circuit of Figure 11–31 has reached steady state. Determine source currents I_1 and I_2.

Figure 11–31

11.6 An *RC* Timing Application

RC circuits are used to create delays for alarm, motor control, and timing applications. Figure 11–32 shows an alarm application. The alarm unit contains a threshold detector, and when the input to this detector exceeds a preset value, the alarm is turned on.

(a) Delay circuit

(b)

Figure 11–32   Creating a delay with an *RC* circuit.

■ **EXAMPLE 11–14**   The circuit of Figure 11–32 is part of a building security system. When an armed door is opened, you have a specified number of seconds to disarm the system before the alarm goes off. If $E = 20$ V, $C = 40$ μF, the alarm is activated when v_C reaches 16 V, and you want a delay of 25 s, what value of R is needed?

Solution   Now, $v_C = E(1 - e^{-t/RC})$. After a bit of manipulation, you get

$$e^{-t/RC} = \frac{E - v_C}{E}$$

Taking the natural log of both sides yields

$$-\frac{t}{RC} = \ln\left(\frac{E - v_C}{E}\right)$$

At $t = 25$ s, $v_C = 16$ V. Thus,

$$-\frac{t}{RC} = \ln\left(\frac{20 - 16}{20}\right) = \ln 0.2 = -1.6094$$

Substituting $t = 25$ s and $C = 40$ μF yields

$$R = \frac{t}{1.6094C} = \frac{25 \text{ s}}{1.6094 \times 40 \times 10^{-6}} = 388 \text{ k}\Omega$$

practice
problems
VI

1. Suppose you want to increase the disarm time of Example 11–14 to 35 s. Compute the new value of R.

2. If, in Example 11–14, the threshold is 15 V and $R = 1$ MΩ, what is the disarm time?

in process
learning
check II

1. Refer to Figure 11–16:
   a. Determine the expression for v_C when $V_0 = 80$ V. Sketch v_C.
   b. Repeat (a) if $V_0 = 40$ V. Why is there no transient?
   c. Repeat (a) if $V_0 = -60$ V.

2. For Part (c) of Question 1, determine how long it takes for v_C to reach zero volts using the technique of Example 11–14.

3. For the circuit of Figure 11–18, suppose $R = 10$ kΩ and $C = 10$ μF:
   a. Determine the expressions for v_C and i_C when $V_0 = 100$ V. Sketch v_C and i_C.
   b. Repeat (a) if $V_0 = -100$ V.

4. Repeat Example 11–11 if voltage source 2 is reversed, i.e., $E_2 = -30$ V.

5. The switch of Figure 11–33(a) is closed at $t = 0$ s. The Norton equivalent of the circuit in the box is shown in (b). Determine expressions for v_C and i_C. The capacitor is initially uncharged.

(a)   (b) Norton equivalent

Figure 11–33

11.7 Pulse Response of *RC* Circuits

In previous sections, we looked at the response of *RC* circuits to switched dc inputs. In this section, we consider the effect that *RC* circuits have on pulse waveforms. Since many electronic devices and systems utilize pulse or rectangular waveforms, including computers, communications systems, and motor control circuits, these are important considerations.

Pulse Basics

A *pulse* is a voltage or current that changes from one level to the other and back again as in Figure 11–34(a) and (b). A *pulse train* is a repetitive stream of pulses, as in (c). If a waveform's high time equals its low time, as in (d), it is called a *square wave*.

The length of each cycle of a pulse train is termed its *period, T,* and the number of pulses per second is defined as its *pulse repetition rate*

(a) Positive pulse

(b) Negative pulse

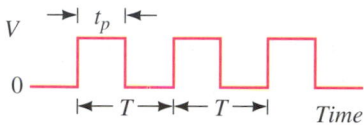

(c) Pulse train. T is referred to as the period of the pulse train

(d) Square wave

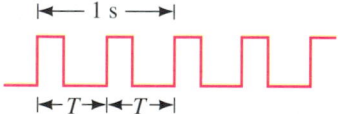

(e) PRR = 2 pulses/s

Figure 11–34   Ideal pulses and pulse waveforms.

(PRR) or *pulse repetition frequency* (PRF). For example, in (e), there are two complete cycles in one second; therefore, the PRR = 2 pulses/s. With two cycles every second, the time for one cycle is $T = 1/2$ s. Note that this is 1/PRR. This is true in general. That is,

$$T = \frac{1}{PRR} \quad s \tag{11–16}$$

The width, t_p, of a pulse relative to its period, [Figure 11–34(c)] is its *duty cycle*. Thus,

$$\text{duty cycle} = \frac{t_p}{T} \times 100\% \tag{11–17}$$

A square wave [Figure 11–34(d)] therefore has a 50% duty cycle, while a waveform with $t_p = 1.5\ \mu s$ and a period of 10 μs has a duty cycle of 15%.

In practice, waveforms are not ideal, that is, they do not change from low to high or high to low instantaneously. Instead, they have finite *rise* and *fall times*. Rise and fall times are denoted as t_r and t_f and are measured between the 10% and 90% points as indicated in Figure 11–35(a). *Pulse width* is measured at the 50% point. The difference between a real waveform and an ideal waveform is often slight. For example, rise and fall times of real pulses may be only a few nanoseconds and when viewed on an oscilloscope, as in Figure 11–35(b), appear to be ideal. In what follows, we will assume ideal waveforms.

(a) Pulse definitions

(b) Pulse waveform viewed on an oscilloscope.

Figure 11–35   Practical waveforms.

The Effect of Pulse Width

The width of a pulse relative to a circuit's time constant determines how it is affected by an *RC* circuit. Consider Figure 11–36. In (a), the circuit has been drawn to focus on the voltage across *C*; in (b), it has been drawn to focus on the voltage across *R*. (Otherwise, the circuits are identical.)

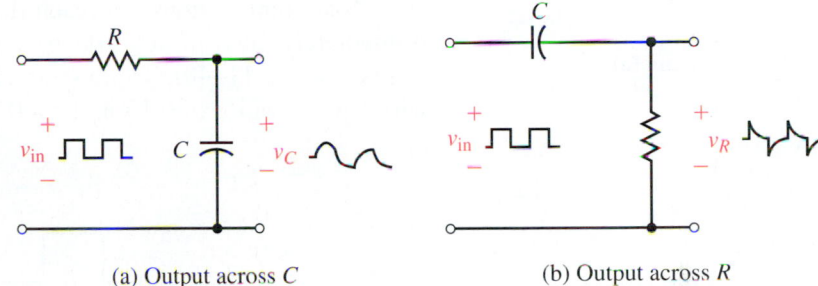

(a) Output across *C*   (b) Output across *R*

Figure 11–36   *RC* circuits with pulse input.

(a)

(b)

(c)

Figure 11–37   Pulse width much greater than 5τ. Note the shaded areas; they indicate where the capacitor is charging and discharging.

Pulse Width $t_p \gg 5\tau$.   First, consider the ouput of circuit (a). When the pulse width and time between pulses are very long compared with the circuit time constant, the capacitor charges and discharges fully. Note, however, that charging and discharging occur at the transitions of the pulse. This increases the rise and fall times of the output [Figure 11–37(b)]. In high-speed circuits, this may be a problem. (We look at this in more detail at the end of this section under the heading Capacitive Loading.

■ **EXAMPLE 11–15**   A square wave is applied to the input of Figure 11–36(a). If $R = 1$ kΩ and $C = 100$ pF, estimate the rise and fall time of the output signal.

Solution   Here, $\tau = RC = (1 \times 10^3)(100 \times 10^{-12}) = 100$ ns. Now consider the universal time constant curve in Figure 11–14(a). Note that v_C reaches the 10% point at about 0.1τ, which is $(0.1)(100$ ns$) = 10$ ns. The 90% point is reached at about 2.3τ, which is $(2.3)(100$ ns$) = 230$ ns. The rise time is therefore approximately 220 ns. The fall time will be the same.

Figure 11–38   Pulse width equal to 5τ.

Now consider the circuit in (b). The output here is a series of short, sharp spikes that occur at input transitions. Under the conditions here (i.e., pulse width much greater than the circuit time constant), v_R is an approximation to the derivative of v_{in} and the circuit is called a *differentiator circuit*. Such circuits have important practical uses.

Pulse Width $t_p = 5\tau$.   These waveforms are shown in Figure 11–38. Since the pulse width is 5τ, the capacitor fully charges and discharges during each pulse.

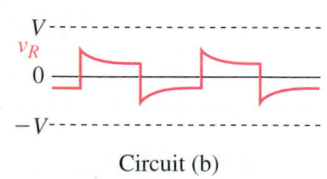

Circuit (a)

Circuit (b)

Figure 11–39   Pulse width much less than 5τ.

Pulse Width $t_p \ll 5\tau$.   For circuit (a), when the pulse width is much shorter than 5τ, the capacitor does not have time to charge significantly before it is switched to discharge, and vice versa. Since charging and discharging occur on the linear portion of the charge and discharge curves, v_C is nearly triangular in shape (Figure 11–39). As shown below, it has an average value of $V/2$. Under the conditions here, v_C is the approximate integral of v_{in} and the circuit is called an *integrator circuit*.

Note that v_C does not reach the steady state shown in Figure 11–39 immediately. Instead, it works its way up over a period of five time constants (Figure 11–40). To illustrate, assume an input square wave of 5 V with a pulse width of 0.1 s and $\tau = 0.1$ s.

(a) Input waveform

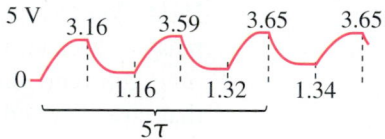

(b) Output voltage v_C

Figure 11–40   Circuit takes five time constants to stabilize.

Pulse 1: $v_C = E(1 - e^{-t/\tau})$. At the end of the first pulse ($t = 0.1$ s), v_C climbs to $v_C = 5(1 - e^{-0.1/0.1}) = 5(1 - e^{-1}) = 3.16$ V. From the end of pulse 1 to the beginning of pulse 2 (i.e., over an interval of 0.1 s), v_C decays from 3.16 V to $3.16e^{-0.1/0.1} = 3.16e^{-1} = 1.16$ V.

Pulse 2: v_C starts at 1.16 V and 0.1 s later has a value of $v_C = E + (V_0 - E)e^{-t/\tau} = 5 + (1.16 - 5)e^{-0.1/0.1} = 5 - 3.84e^{-1} = 3.59$ V. It then decays to $3.59e^{-1} = 1.32$ V over the next 0.1 s.

Continuing in this manner, the remaining values for Figure 11–40(b) are determined. After 5τ, v_C cycles between 1.34 and 3.65 V, with an average of $(1.34 + 3.65)/2 = 2.5$ V, or half the input pulse amplitude.

practice problem VII

Verify the remaining points of Figure 11–40(b).

Capacitive Loading

(a) Unloaded driver

Load

(b) Distorted signal

Figure 11–41   Distortion caused by capacitive loading.

Capacitance occurs whenever conductors are separated by insulating material. This means that capacitance exists between wires in cables, between traces on printed circuit boards, and so on. In general, this capacitance is undesirable but it cannot be avoided. It is called *stray capacitance*. Fortunately, stray capacitance is often so small that it can be neglected. However, in high-speed circuits, it may cause problems.

To illustrate, consider Figure 11–41. The electronic driver of (a) produces square pulses. However, when it is connected as in (b), stray capacitance loads it and increases the signal's rise and fall times (since capacitance takes time to charge and discharge). If the rise and fall times become excessively long, the signal reaching the load may be so degraded that the system malfunctions. Capacitive loading can create serious problems in computers (and other high-speed systems), where signal rise and fall times of a few nanoseconds or less may be required.

11.8   Transient Analysis Using PSpice

Transient analysis requires a few new PSpice commands. In this section, we introduce the .TRAN statement, the PULSE declaration, and the .PROBE statement. (All are described in Appendix A.) The PULSE command allows us to apply pulses to circuits, the .TRAN statement allows us to compute the transient response of a network, and the .PROBE statement permits us to display computed waveforms on the computer screen.

Creating a Step Voltage

Step voltages may be create in several ways. In this chapter, we use the .TRAN statement with the UIC (Use Initial Condition) keyword. In Chapter 14, we use a voltage-controlled switch. You can also use a wide pulse.

An Example.   To illustrate, consider the circuit of Figure 11–42. The capacitor is initially uncharged and the switch is open. At $t = 0$ s, the switch is moved to the charge position.

The effect of closing the switch is to suddenly apply 20 volts across R and C. The easiest way to do this is to use the .TRAN statement with keyword UIC. As noted in Appendix A, this causes PSpice to apply the dc source as a step voltage. The *Input File* is shown below.

Figure 11–42   PSpice example.

```
Input File:
* Figure 11-42: Step Response, Charge Position
* Circuit Description
Vs   1   0   20V
R    1   2   5kohms
C    2   0   40uF
```

```
* Transient Analysis Specification
.TRAN 200ms  2s   UIC
.PROBE
.END
```

Analysis of the Program.

1. The combination of the Vs statement and the .TRAN statement (with UIC keyword) creates a 20-volt step at $t = 0$ s.

2. The statement .TRAN 200ms 2s causes PSpice to compute values over the time interval 0 to 2 s. The 200-ms specification is the plotting increment.

3. The .PROBE statement permits you to display results in graphic form on your terminal.

Key in and run this program. Use the procedure entitled *Displaying Multiple Traces* in Section A.1 to display v_C and i_C [denoted respectively as v(2) and i(C)] to get the display of Figure 11–43. Note that since $5\tau = 1$ s, the transient should last 1 s (which it does). The initial current should be $E/R = (20\ V)/(5\ k\Omega) = 4$ mA and the final voltage 20 V (which they are). Thus, the solution checks.

Figure 11–43   Waveforms for the circuit of Figure 11–42. Charge case.

If you do not have a graphics monitor, you can graph results manually. Replace the .PROBE statement with the .PRINT TRAN V(2) I(C) statement. To access results, use pull-down menu *Files* and select

Browse Output. To get a hard copy, print the output file as discussed in Appendix A.

Creating a Switched Input

Let us now move the switch to charge for a time, then to discharge. This is equivalent to placing 20 volts across the *RC* combination for the charging time, then 0 volts thereafter. You can do this using a PULSE command. The pulse declaration has the form

   PULSE(V1 V2 td tr tf pw per)

where V1 is the initial value of the pulse, V2 is its amplitude, td is its starting delay, tr and tf are rise and fall times, pw is the pulse width, and per is its period. The following example illustrates its use. It also illustrates how to place an initial voltage on a capacitor.

■ **EXAMPLE 11–16**   The capacitor of Figure 11–42 has an initial voltage $V_0 = -10$ V. The switch is moved to the charge position for 1 s, then to the discharge position, where it remains. Determine v_C and i_C.

Solution   The *Input File* is shown below. Vs has an initial value of 0 V, an amplitude of 20 V, a delay of 0 s, 1-μs rise and fall times, a 1-s width, and a 5-s period. Thus, 20 V is applied to the circuit for 1 s (the charge position), then 0 V thereafter (the discharge position). (Theoretically, we would like zero rise and fall times, but PSpice requires nonzero values. One microsecond was chosen as a workable compromise.) The specification IC = –10V gives the capacitor an initial value of –10 V. UIC (Use Initial Condition) must be used with .TRAN when you have initial conditions. The waveforms are shown in Figure 11–44.

```
Input File:
* Figure 11-42: Charge/Discharge Response
Vs  1   0   PULSE (0 20 0 1us 1us 1s 5s)
R   1   2   5kohms
C   2   0   40uF   IC = -10V
.TRAN 200ms   2s    UIC
.PROBE
.END
```

Note that voltage starts at –10 V and climbs to 20 V while current starts at $(E - V_0)/R = 30$ V/5 kΩ = 6 mA and decays to zero. When the switch is turned to the discharge position, the current jumps from 0 A to –20 V/5 kΩ = –4 mA and then decays to zero while the voltage decays to zero. Thus the solution checks.

Figure 11–44   Waveforms for the circuit of Figure 11–42. Charge/discharge case.

Problems

11.1   Introduction

1. The capacitor of Figure 11–45 is uncharged.
   a. What are the capacitor voltage and current just after the switch is closed?
   b. What are the capacitor voltage and current after the capacitor is fully charged?

2. Repeat Problem 1 if the 20-V source is replaced by a −60-V source.

3. a. What does an uncharged capacitor look like at the instant of switching?
   b. What does a charged capacitor look like at the instant of switching?
   c. What does a capacitor look like to steady state dc?
   d. What do we mean by $i(0^-)$? By $i(0^+)$?

11.2   Capacitor Charging Equations

4. The switch of Figure 11–45 is closed at $t = 0$ s.
   a. Determine the equation for charging voltage v_C.
   b. Determine the equation for charging current i_C.

Figure 11–45

c. By direct substitution, compute v_C and i_C at $t = 0^+$ s, 40 μs, 80 μs, 120 μs, 160 μs, and 200 μs.

d. Plot v_C and i_C on graph paper using the results of (c). Hint: See Example 11–2.

5. Repeat Problem 4 if $R = 500\ \Omega$, $C = 25\ \mu$F, and $E = 45$ V, except compute and plot values at $t = 0^+$ s, 20 ms, 40 ms, 60 ms, 80 ms, and 100 ms.

6. The capacitor of Figure 11–2 is uncharged at the instant the switched is closed. If $E = 80$ V, $C = 10\ \mu$F, and $i_C(0^+) = 20$ mA, determine the equations for v_C and i_C.

7. Show that $\tau = RC$ has units of seconds.

8. Determine the time constant for the circuit of Figure 11–45. How long (in seconds) will it take for the capacitor to charge?

9. A capacitor takes 200 ms to charge. If $R = 5$ kΩ, what is C?

10. If $i_C = 25e^{-40t}$ A, how long will the transient last?

11. For Figure 11–2, the current jumps to 3 mA when the switch is closed. The capacitor takes 1 s to charge. If $E = 75$ V, determine R and C.

12. For Figure 11–2, if $v_C = 100(1 - e^{-50t})$ V and $i_C = 25e^{-50t}$ mA, what are E, R, and C?

13. For Figure 11–2, determine E, R, and C if the capacitor takes 5 ms to charge, the current at 1 time constant after the switch is closed is 3.678 mA, and the capacitor charges to 45 volts.

11.3 Capacitor with an Initial Voltage

14. The capacitor of Figure 11–45 has an initial voltage.
    a. If $V_0 = 10$ V, what is the current just after the switch is closed?
    b. Repeat (a) if $V_0 = 20$ V and $V_0 = -10$ V.

15. For the capacitor of Figure 11–45, $V_0 = 30$ V.
    a. Determine the expression for charging voltage v_C.
    b. Determine the expression for current i_C.
    c. Sketch v_C and i_C.

16. Repeat Problem 15 if $V_0 = -5$ V.

11.4 Capacitor Discharging Equations

17. For the circuit of Figure 11–46, assume the capacitor is charged to 50 V before the switch is closed.
    a. Determine the equation for discharge voltage v_C.
    b. Determine the equation for discharge current i_C.
    c. Determine the time constant of the circuit.

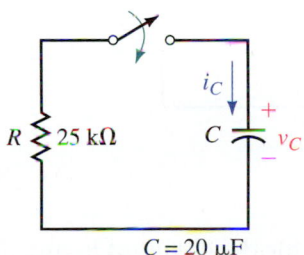

R ⩾ 25 kΩ C ⏚ v_C

$C = 20\ \mu$F

Figure 11–46

    d. Compute v_C and i_C at $t = 0^+$ s, $t = \tau$, 2τ, 3τ, 4τ, and 5τ.
    e. Plot the results of (d) with the time axis scaled in seconds and time constants.

18. For Figure 11–47, let $E = 200$ V, $R_2 = 1$ kΩ, and $C = 0.5$ μF. After the capacitor has fully charged in position 1, the switch is moved to position 2.
    a. What is the capacitor voltage immediately after the switch is moved to position 2? What is its current?
    b. What is the discharge time constant?
    c. Determine discharge equations for v_C and i_C.

Figure 11–47

19. For Figure 11–47, C is fully charged before the switch is moved to discharge. Current just after it is moved is $i_C = -4$ mA and C takes 20 ms to discharge. If $E = 80$ V, what are R_2 and C?

11.5   More Complex Circuits

20. For Figure 11–48, the switch is closed at $t = 0$. Given $V_0 = 0$V.
    a. Determine the equations for v_C and i_C.
    b. Compute the capacitor voltage at $t = 0^+$, 2, 4, 6, 8, 10, and 12 ms.
    c. Repeat (b) for the capacitor current.
    d. Why does 225 V/30 Ω also yield $i(0^+)$?

Figure 11–48

21. Repeat Problem 20, parts (a) to (c) for the circuit of Figure 11–49.

Figure 11–49

22. Consider again Figure 11–47. Suppose $E = 80$ V, $R_2 = 25$ kΩ, and $C = 0.5$ μF:
   a. What is the charge time constant?
   b. What is the discharge time constant?
   c. With the capacitor initially discharged, move the switch to position 1 and determine equations for v_C and i_C during charge.
   d. Move the switch to the discharge position. How long does it take for the capacitor to discharge?
   e. Sketch v_C and i_C from the time the switch is placed in charge to the time that the capacitor is fully discharged. Assume the capacitor is in the charge position for 80 ms.

23. For the circuit of Figure 11–47, the capacitor is initially uncharged. The switch is first moved to charge, then to discharge, yielding the current shown in Figure 11–50. The capacitor fully charges in 12.5 s. Determine E, R_2, and C.

24. Refer to the circuit of Figure 11–51:
   a. What is the charge time constant?
   b. What is the discharge time constant?
   c. With the capacitor initially discharged, move the switch to position 1 and determine equations for v_C and i_C during charge.

Figure 11–50

Figure 11–51

d. After the capacitor has charged for two time constants, move the switch to position 2 and determine equations for v_C and i_C during discharge.

e. Sketch v_C and i_C.

25. Determine the capacitor voltages and the source current for the circuit of Figure 11–52 after it has reached steady state.

Figure 11–52

26. A black box containing dc sources and resistors has open-circuit voltage of 45 volts as in Figure 11–53(a). When the output is shorted as in (b), the short-circuit current is 1.5 mA. A switch and an uncharged 500-μF capacitor are connected as in (c). Determine the capacitor voltage and current 25 s after the switch is closed.

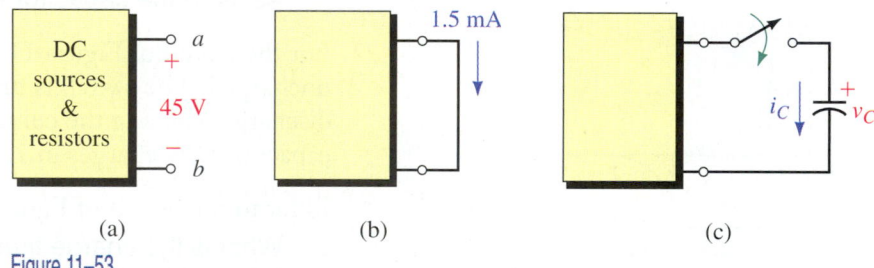

(a)   (b)   (c)

Figure 11–53

11.6 An *RC* Timing Application

27. For the alarm circuit of Figure 11–32, if the input from the sensor is 5 V, $R = 750$ kΩ, and the alarm is activated at 15 s when $v_C = 3.8$ V, what is C?

28. For the alarm circuit of Figure 11–32, the input from the sensor is 5 V, $C = 47$ μF, and the alarm is activated when $v_C = 4.2$ V. Choose the nearest standard resistor value to achieve a delay of at least 37 s.

29. For the alarm circuit of Figure 11–32, the input from the sensor is 5 V, $R = 750$ kΩ, $C = 15$ μF, and the alarm is activated when $v_C = 3.8$ V.

a. Determine the delay time.
b. If the tolerance on R is ±1%, the tolerance on C is ±5%, the tolerance on the threshold detector is ±0.1 V, and the input from the sensor is 5 V ± 0.5 V, determine the minimum and maximum delays which might occur.

11.7   Pulse Response of *RC* Circuits

30. Consider the waveform of Figure 11–54.
   a. What is the period?
   b. What is the duty cycle?
   c. What is the PRR?

Figure 11–54

Figure 11–55

31. Repeat Problem 30 for the waveform of Figure 11–55.

32. Determine the rise time, fall time, and pulse width for the pulse in Figure 11–56.

Figure 11–56

Figure 11–57

33. A single pulse is input to the circuit of Figure 11–57. Assuming that the capacitor is initially unchanged, sketch the output for each set of values below:
   a. $R = 2\ k\Omega$, $C = 1\ \mu F$.
   b. $R = 2\ k\Omega$, $C = 0.1\ \mu F$.

34. A pulse train is input to the circuit of Figure 11–58. Assuming that the capacitor is initially unchanged, sketch the output for each set of values below:
   a. $R = 2\ k\Omega$, $C = 0.1\ \mu F$.
   b. $R = 20\ k\Omega$, $C = 1.0\ \mu F$.

35. A step is applied to the circuit of Figure 11–58. If $R = 150\ \Omega$ and $C = 20\ pF$, estimate the rise time of the output voltage.

Figure 11–58

11.8   Transient Analysis Using PSpice

36. Redo Example 11–16 with the switch in the charge position for 0.5 s and everything else the same. With your calculator, compute v_C and i_C at 0.5 s and compare them to PSpice's plot. Repeat for i_C just after moving the switch to the discharge position.

37. We solved the circuit of Figure 11–21 by reducing it to its Thévenin equivalent. Using PSpice, solve the circuit in its original form (Figure 11–21) and compute and plot capacitor voltage and current. Now, using the solution of Example 11–9, compute v_C and i_C at several values of time and compare them with values from the PSpice output.

38. Use PSpice to solve for voltages and currents in the circuit of Figure 11–52. From this, determine the final (steady state) voltages and currents and compare them to the answers of Problem 25.

39. Apply a 100-V step to the circuit of Figure 11–58 with $R = 150\ \Omega$ and $C = 20$ pF, and view v_C on the screen. Use the cursor to measure the rise time. Compare this to the answer to Problem 35.

Magnetism and Magnetic Circuits

CHAPTER PREVIEW

Many common devices rely on magnetism. Familiar examples include computer disk drives, tape recorders, VCRs, transformers, motors, generators, and so on. To understand their operation, you need a knowledge of magnetism and magnetic circuit principles. In this chapter, we look at fundamentals of magnetism, relationships between electrical and magnetic quantities, magnetic circuit concepts, and methods of analysis. In Chapter 13, we look at electromagnetic induction and inductance, and in Chapter 24, we apply magnetic principles to the study of transformers.

rg_new.grd step 1 time 0.0000E+00 16-Sep-93 09:23:26

Vector Potentials

Magnetic flux for a dc motor, armature only excited. *(Courtesy GE Research and Development Center)*

OBJECTIVES

After studying this chapter, you will be able to

- represent magnetic fields using Faraday's flux concept,

- describe magnetic fields quantitatively in terms of flux and flux density,

- explain what magnetic circuits are and why they are used,

- determine magnetic field intensity or magnetic flux density from a B-H curve,

- solve series magnetic circuits,

- solve series-parallel magnetic circuits,

- compute the attractive force of an electromagnet,

- explain the domain theory of magnetism,

- describe the demagnetization process.

Magnetism and Electromagnetism

*W*hile the basic facts about magnetism have been known since ancient times, it was not until the early 1800s that the connection between electricity and magnetism was made and the foundations of modern electromagnetic theory laid down.

In 1819, Hans Christian Oersted, a Danish scientist, demonstrated that electricity and magnetism were related when he showed that a compass needle was deflected by a current-carrying conductor. The following year, Andre Ampere (1775–1836) showed that current-carrying conductors attract or repel each other just like magnets. However, it was Michael Faraday (recall Chapter 10) who developed our present concept of the magnetic field as a collection of flux lines in space that conceptually represent both the intensity and the direction of the field. It was this concept that led to an understanding of magnetism and the development of important practical devices such as the transformer and the electric generator.

In 1873, James Clerk Maxwell, a Scottish scientist, tied the then known theoretical and experimental concepts together and developed a unified theory of electromagnetism that predicted the existence of radio waves. Some 30 years later, Heinrich Hertz, a German physcist, showed experimentally that such waves existed, thus verifying Maxwell's theories and paving the way for modern radio and television.

12.1 The Nature of a Magnetic Field

Magnetism refers to the force that acts between magnets and magnetic materials. We know, for example, that magnets attract pieces of iron, deflect compass needles, attract or repel other magnets, and so on. This force acts at a distance and without the need for direct physical contact. The region where the force is felt is called the "field of the magnet" or simply, its *magnetic field*. Thus, **a magnetic field is a force field.**

Magnetic Flux

Faraday's flux concept helps us visualize this field. Using Faraday's representation, magnetic fields are shown as lines in space. These lines, called *flux lines* or *lines of force*, show the direction and intensity of the field at all points. This is illustrated in Figure 12–1 for the field of a bar magnet. As indicated, the field is strongest at the *poles* of the magnet (where flux lines are most dense), its direction is from north (N) to south (S) external to the magnet, and flux lines never cross.

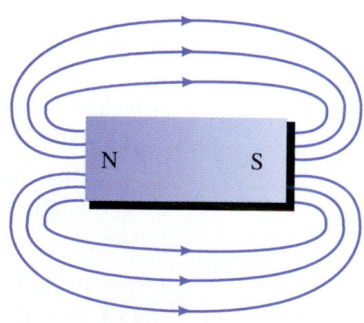

Figure 12–1   Field of a bar magnet.

Figure 12–2 shows what happens when two magnets are brought close together. In (a), unlike poles attract, and flux lines reach from one magnet to the other. In (b), like poles repel, and the flux lines are pushed back as indicated by the flattening of the field between the two magnets.

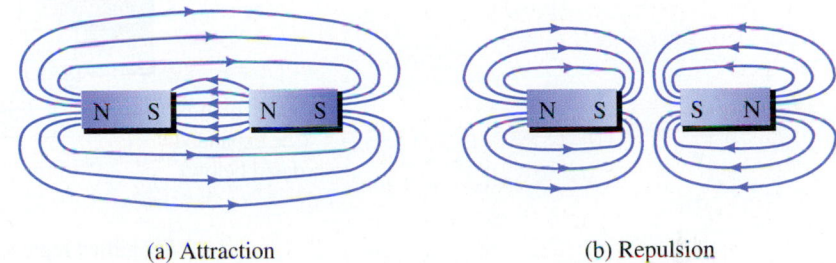

(a) Attraction   (b) Repulsion

Figure 12–2   Field patterns due to attraction and repulsion.

Ferromagnetic Materials

Magnetic materials (materials that are attracted by magnets such as iron, nickel, cobalt, and their alloys) are called *ferromagnetic* materials. Ferromagnetic materials provide an easy path for magnetic flux. Thus, if a ferromagnetic object is placed in a magnetic field as in Figure 12–3, a high density of flux passes through it. Note, however, that nonmagnetic materials (plastic, wood, glass, and so on) have no effect on the field.

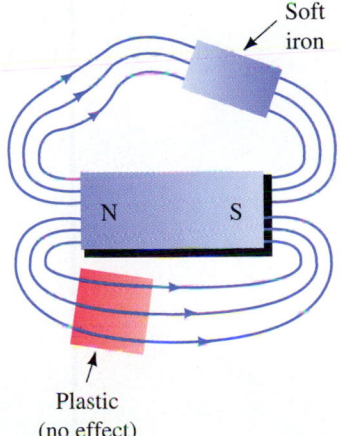

Soft
iron

N   S

Plastic
(no effect)

Figure 12–3   Magnetic field follows the path through the iron. It is not affected by the plastic.

Figure 12–4 shows an application of these principles. Part (a) shows a simplified representation of a loudspeaker, and part (b) shows expanded detail of its magnetic field. (Since the speaker is symmetrical, only half its structure is shown in (b).) The field is created by a permanent magnet, and the iron pole pieces guide the field and concentrate it in the gap where the speaker coil is placed. (For a description of how the speaker works, see Section 12.4.) Note how the flux crowds together at sharp interior corners, spreads apart at exterior corners, and is essentially uniform elsewhere. This is characteristic of magnetic fields in iron.

Voice coils
(see Figure 12–11)

Magnet

Magnetic flux
(see expanded
detail below)

Speaker
cone

Magnet

(a) Simplified representation of a loudspeaker. The
magnetic structure and voice coil are called a
"speaker motor."

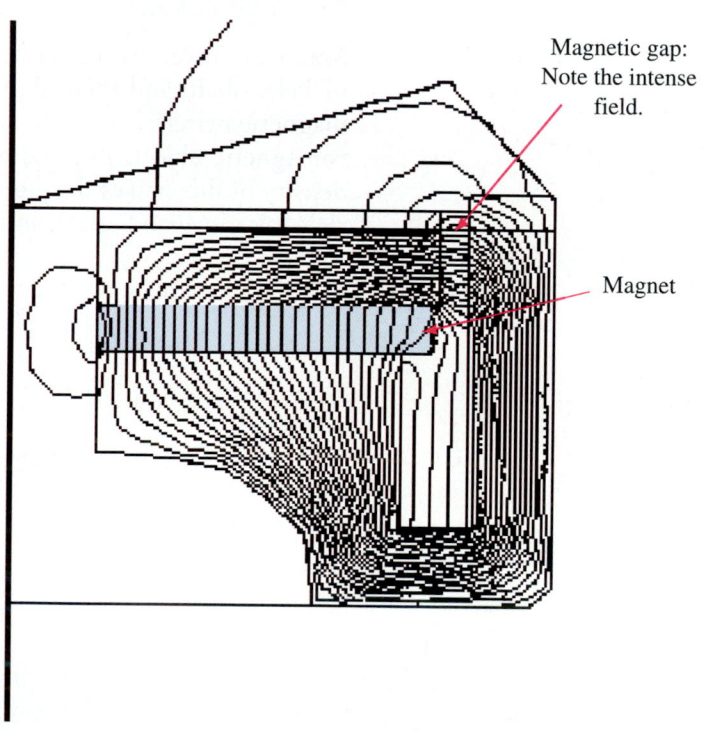

Magnetic gap:
Note the intense
field.

Magnet

(b) Magnetic field pattern (expanded detail).
(Courtesy JBL Professional)

Figure 12–4   Magnetic field pattern for a loudspeaker.

12.2 Electromagnetism

Most applications of magnetism involve magnetic effects due to electric
currents. We look first at some basic principles. Consider Figure 12–5.
The current, I, creates a magnetic field that is concentric about the conductor, uniform along its length, and whose strength is directly proportional to I. Note the direction of the field. It may be remembered with the

aid of the *right-hand rule*. As indicated in (b), place your right hand around the conductor with your thumb pointing in the direction of current. Your fingers then point in the direction of the field. If you reverse the direction of the current, the direction of the field reverses. If the conductor is wound into a coil, the fields of its individual turns combine, producing a resultant field as in Figure 12–6. The direction of the coil flux can also be remembered by means of a simple rule: curl the fingers of your right hand around the coil in the direction of the current and your thumb will point in the direction of the field. If the direction of the current is reversed, the field also reverses. Provided no ferromagnetic material is present, the strength of the coil's field is also proportional to its current.

(a) Magnetic field produced by current. Field is proportional to *I*.

(b) Right-hand rule.

Figure 12–5   Field about a current-carrying conductor.

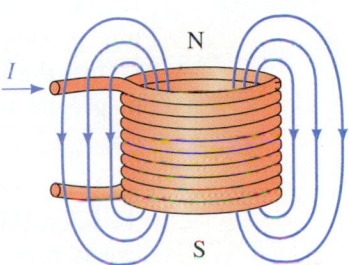

Figure 12–6   Field produced by a coil. If the current direction is reversed, the field direction will reverse.

If the coil is wound on a ferromagnetic core as in Figure 12–7 (transformers are built this way), almost all flux is confined to the core, although a small amount (called stray or leakage flux) returns through the surrounding air. However, now that ferromagnetic material is present, the core flux is no longer proportional to current. The reason for this is discussed in Section 12.14.

Core flux (simplified representation) Flux is denoted by the symbol Φ.

Leakage flux

Iron core

Figure 12–7   For ferromagnetic materials, almost all of the flux is confined to the core.

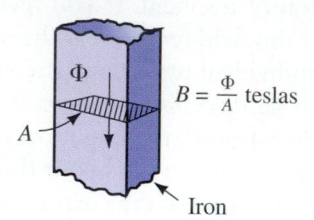

Figure 12–8   Concept of flux density. $1\ T = 1\ Wb/m^2$.

12.3 Flux and Flux Density

Magnetic fields are described in terms of *flux*, Φ (Greek letter phi), and *flux density*, B. Flux, Φ, refers to the total number of lines of force in a field, while flux density refers to their number per unit area. In the SI system, flux is measured in units of *webers* (Wb, in honor of pioneer researcher Wilhelm Eduard Weber, 1804–1891), while flux density is measured in units of *teslas* (T, in honor of Nikola Tesla, 1856–1943). By definition, *1 tesla = 1 Wb/m²*. As illustrated in Figure 12–8, flux density is found by dividing total flux passing perpendicularly through an area by the size of the area. That is,

$$B = \Phi/A \quad \text{(teslas, T)} \tag{12–1}$$

Thus, if 600 μWb of flux pass perpendicularly through an area $A = 20 \times 10^{-4}\ m^2$, the flux density is $B = (600 \times 10^{-6}\ \text{Wb})/(20 \times 10^{-4}\ m^2) = 0.3$ T. The greater the flux density, the stronger the field.

■ **EXAMPLE 12–1**   For the magnetic core of Figure 12–9, the flux density at cross section 1 is $B_1 = 0.4$ T. Determine B_2.

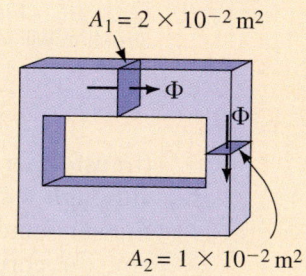

$A_1 = 2 \times 10^{-2}\ m^2$

$A_2 = 1 \times 10^{-2}\ m^2$

Figure 12–9

Solution $\Phi = B_1 \times A_1 = (0.4\ \text{T})(2 \times 10^{-2}\ m^2) = 0.8 \times 10^{-2}$ Wb. Since all flux is confined to the core, the flux at cross section 2 is the same as at cross section 1. Therefore,

$$B_2 = \Phi/A_2 = (0.8 \times 10^{-2}\ \text{Wb})/(1 \times 10^{-2}\ m^2) = 0.8\ \text{T}$$

practice problems 1

1. Refer to the core of Figure 12–8:
    a. If A is 2 cm × 2.5 cm and $B = 0.4$ T, compute Φ in webers.
    b. If A is 0.5 inch by 0.8 inch and $B = 0.35$ T, compute Φ in webers.

2. In Figure 12–9, if $\Phi = 100 \times 10^{-4}$ Wb, compute B_1 and B_2.

To gain a feeling for the size of magnetic units, note that the strength of the earth's field is approximately 50 μT near the earth's surface, the field of a large generator or motor is on the order of 1 or 2 T, and the largest fields yet produced (using superconducting magnets) are on the order of 25 T.

Other systems of units (now largely superseded) are the CGS system and the English systems. In the CGS system, flux is measured in maxwells and flux density in gauss. In the English system, flux is measured in lines and flux density in lines per square inch. Conversion factors are given in Table 12–1. We use only the SI system in this book.

TABLE 12–1 Magnetic Units Conversion Table

SYSTEM	FLUX (Φ)	FLUX DENSITY (B)
SI	webers (Wb)	teslas (T) $1\ T = 1\ Wb/m^2$
English	lines $1\ Wb = 10^8$ lines	lines/in^2 $1\ T = 6.452 \times 10^4$ lines/in^2
CGS	maxwells $1\ Wb = 10^8$ maxwells	gauss 1 gauss $= 1$ maxwell/cm^2 $1\ T = 10^4$ gauss

in process
l e a r n i n g
check I

1. A magnetic field is a _____ field.

2. With Faraday's flux concept, the density of lines represents the _____ of the field and their direction represents the _____ of the field.

3. Three ferromagnetic materials are _____, _____, and _____.

4. The direction of a magnetic field is from _____ to _____ outside a magnet.

5. For Figures 12–5 and 12–6, if the direction of current is reversed, sketch what the fields look like.

6. If the core shown in Figure 12–7 is plastic, sketch what the field will look like.

7. Flux density B is defined as the ratio Φ/A, where A is the area (parallel, perpendicular) to Φ.

8. For Figure 12–9, if A_1 is 2 cm × 2.5 cm, B_1 is 0.5 T, and $B_2 = 0.25$ T, what is A_2?

12.4 Magnetic Circuits

Most practical applications of magnetism use magnetic structures to guide and shape magnetic fields by providing a well-defined path for flux. Such structures are called *magnetic circuits*. Magnetic circuits are found in motors, generators, computer disk drives, tape recorders, and so on. The speaker of Figure 12–4 illustrates the concept. It uses a powerful magnet to create flux and an iron circuit to guide the flux to the air gap to provide the intense field required by the voice coil. Note how effectively it does its job; almost the entire flux produced by the magnet is confined to the iron path with little leakage into the air.

A second example is shown in Figures 12–10 and 12–11. Tape recorders, VCRs, and computer disk drives all store information magnetically on iron oxide coated surfaces for later retrieval and use. The basic tape recorder scheme is shown symbolically in Figure 12–10. Sound picked up by a microphone is converted to an electrical signal, amplified, and the output applied to the record head. The record head is a small magnetic circuit. Current from the amplifier passes through its coil, creating a magnetic field that magnetizes the moving tape. The magnetized patterns on the tape correspond to the original sound input.

(a) Recording System

Figure 12–10   The recording head of a tape recorder is a magnetic circuit.

During playback the magnetized tape is passed by a playback head, as shown in Figure 12–11(a). Voltages induced in the playback coil are amplified and applied to a speaker. The speaker (b) utilizes a flexible cone to reproduce sound. A coil of fine wire attached at the apex of this cone is placed in the field of the speaker air gap. Current from the amplifier passes through this coil, creating a varying field that interacts with the fixed field of the speaker magnet, causing the cone to vibrate. Since these vibrations correspond to the magnetized patterns on the tape, the original sound is reproduced. Computer disk drives use a similar record/playback scheme; in this case, binary logic patterns are stored and retrieved rather than music and voice.

12.5 Air Gaps, Fringing, and Laminated Cores

For magnetic circuits with air gaps, *fringing* occurs, causing a decrease in flux density in the gap as in Figure 12–12(a). For short gaps, fringing can usually be neglected. Alternatively, correction can be made by increasing each cross-sectional dimension of the gap by the size of the gap to approximate the decrease in flux density.

(a) Playback System

(b) Speaker

Figure 12–11   Both the playback system and the speaker use magnetic circuits.

(a) Fringing at gap

(b) Laminated section.
Effective magnetic area is
less than the physical area

Figure 12–12   Fringing and
laminations.

■ **EXAMPLE 12–2**   A core with cross-sectional dimensions of 2.5 cm by 3 cm has a 0.1-mm gap. If flux density $B = 0.86$ T in the iron, what is the approximate flux density in the gap?

Solution

$$\Phi = BA = (0.86 \text{ T})(2.5 \times 10^{-2} \text{ m})(3 \times 10^{-2} \text{ m}) = 0.645 \text{ mWb}$$
$$A_g \simeq (2.51 \times 10^{-2} \text{ m})(3.01 \times 10^{-2} \text{ m}) = 7.555 \times 10^{-4} \text{ m}^2$$

Thus,

$$B_g \simeq 0.645 \text{ mWb}/7.555 \times 10^{-4} \text{ m}^2 = 0.854 \text{ T}$$

Now consider laminations. Many practical magnetic circuits (such as transformers) use thin sheets of stacked iron or steel as in Figure 12–12(b). Since the core is not a solid block, its effective cross-sectional area (i.e., the actual area of iron) is less than its physical area. A *stacking factor*, defined as the ratio of the actual area of ferrous material to the physical area of the core, permits you to determine the core's effective area.

practice problem II

A laminated section of core has cross-sectional dimensions of 0.03 m by 0.05 m and a stacking factor of 0.9.

a. What is the effective area of the core?

b. Given $\Phi = 1.4 \times 10^{-3}$ Wb, what is the flux density, B?

12.6 Series Elements and Parallel Elements

Magnetic circuits may have sections of different materials. For example, the circuit of Figure 12–13 has sections of cast iron, sheet steel, and an air gap. For this circuit, flux Φ is the same in all sections. Such a circuit is called a *series magnetic circuit*. Although the flux is the same in all sections, the flux density in each section depends on its effective cross-sectional area as you saw earlier.

A circuit may also have elements in parallel (Figure 12–14). At each junction, the sum of fluxes entering is equal to the sum leaving. This is the counterpart of Kirchhoff's current law. Thus, for Figure 12–14, if $\Phi_1 = 25\ \mu$Wb and $\Phi_2 = 15\ \mu$Wb, then $\Phi_3 = 10\ \mu$Wb.

Figure 12–13   Series magnetic circuit. Flux Φ is the same throughout.

Figure 12–14   The sum of the flux entering a junction equals the sum leaving. Thus, $\Phi_1 = \Phi_2 + \Phi_3$.

in process **learning** *check II*

1. Why is the flux density in each section of Figure 12–13 different?

2. For Figure 12–13, $\Phi = 1.32$ mWb, the cross section of the core is 3 cm by 4 cm, the laminated section has a stacking factor of 0.8, and the gap is 1 mm. Determine the flux density in each section, taking fringing into account.

3. If core of Figure 12–14 is symmetrical about its center leg, $B_1 = 0.4$ T, and the cross-sectional area of the center leg is 25 cm^2, what are Φ_2 and Φ_3?

12.7 Magnetic Circuits With dc Excitation

We now look at the analysis of magnetic circuits with dc excitation. There are two basic problems to consider: 1) given the flux, to determine the current required to produce it and 2) given the current, to compute the flux produced. To help visualize how to solve such problems, we first establish an analogy between magnetic circuits and electric circuits.

MMF: The Source of Magnetic Flux

Current through a coil creates magnetic flux. The greater the current or the greater the number of turns, the greater will be the flux. This flux-producing ability of a coil is called its *magnetomotive force* (mmf). Magnetomotive force is given the symbol $\mathscr{F}$ and is defined as

$$\mathscr{F} = NI \quad \text{(ampere-turns, At)} \tag{12–2}$$

Thus, a coil with 100 turns and 2.5 amps will have an mmf of 250 ampere-turns, while a coil with 500 turns and 4 amps will have an mmf of 2000 ampere-turns.

Reluctance, $\mathscr{R}$: Opposition to Magnetic Flux

Flux in a magnetic circuit also depends on the opposition that the circuit presents to it. Termed *reluctance*, this opposition depends on the dimensions of the core and the material of which it is made. Like the resistance of a wire, reluctance is directly proportional to length and inversely proportional to cross-sectional area. In equation form,

$$\mathscr{R} = \frac{\ell}{\mu A} \quad \text{(At/Wb)} \tag{12–3}$$

where μ is a property of the core material called its *permeability* (discussed in Section 12.8). Permeability is a measure of how easy it is to establish flux in a material. Ferromagnetic materials have high permeability and hence low $\mathscr{R}$, while nonmagnetic materials have low permeability and high $\mathscr{R}$.

Ohm's Law for Magnetic Circuits

The relationship between flux, mmf, and reluctance is

$$\Phi = \mathscr{F}/\mathscr{R} \quad \text{(Wb)} \tag{12–4}$$

This relationship is similar to Ohm's law and is depicted symbolically in Figure 12–15.

Figure 12–15   Electric circuit analogy of a magnetic circuit. $\Phi = \mathscr{F}/\mathscr{R}$.

■ **EXAMPLE 12–3**   For Figure 12–16, if the reluctance of the magnetic circuit is $\mathfrak{R} = 12 \times 10^4$ At/Wb, what is the flux in the circuit?

N = 300 turns

Figure 12–16

Solution

$$\mathfrak{F} = NI = (300)(0.5\ \text{A}) = 150\ \text{At}$$

$$\Phi = \mathfrak{F}/\mathfrak{R} = (150\ \text{At})/(12 \times 10^4\ \text{At/Wb}) = 12.5 \times 10^{-4}\ \text{Wb}$$

In Example 12–3, we assumed that the reluctance of the core was constant. This is only approximately true under certain conditions. In general, it is not true, since $\mathfrak{R}$ is a function of flux density. Thus Equation 12–4 is not really very useful, since for ferromagnetic material, $\mathfrak{R}$ depends on flux, the very quantity that you are trying to find. The main use of Equations 12–3 and 12–4 is to provide an analogy between electric and magnetic circuit analysis.

12.8   Magnetic Field Intensity and Magnetization Curves

We now look at a more practical approach to analyzing magnetic circuits. First, we require a quantity called *magnetic field intensity*, H (also known as *magnetizing force*). It is a measure of the mmf per unit length of a circuit.

To get at the idea, suppose you apply the same mmf (say 600 At) to two circuits with different path lengths (Figure 12–17). In (a), you have 600 ampere-turns of mmf to "drive" flux through 0.6 m of core; in (b), you have the same mmf but it is spread across only 0.15 m of path length. Thus the mmf per unit length in the second case is more intense. Based on this idea, one can define magnetic field intensity as the ratio of applied mmf to the length of path that it acts over. Thus,

(a) A long path

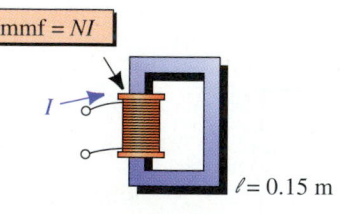

(b) A short path

Figure 12–17   By definition, $H =$ mmf/length $= NI/\ell$.

$$H = \mathfrak{F}/\ell = NI/\ell \quad (\text{At/m}) \tag{12–5}$$

For the circuit of Figure 12–17(a), $H = 600/0.6 = 1000$ At/m, while for the circuit of (b), $H = 600/0.15 = 4000$ At/m. Thus, in (a) you have 1000 ampere-turns of "driving force" per meter of length to create flux, whereas in (b) you have four times as much. (However, you won't get four times as much flux, since the opposition to flux varies with the density of the flux.)

Rearranging Equation 12–5 yields an important result:

$$NI = H\ell \quad (\text{At}) \tag{12–6}$$

$$NI = H\ell$$

Figure 12–18   Circuit analogy, $H\ell$ model.

In an analogy with electric circuits (Figure 12–18), the NI product is an *mmf source* while the $H\ell$ product is an *mmf drop*.

The Relationship between *B* and *H*

From Equation 12–5, you can see that magnetizing force, H, is a measure of the flux-producing ability of the coil (since it depends on NI). You also know that B is a measure of the resulting flux (since $B = \Phi/A$). Thus, B and H are related. The relationship is

$$B = \mu H \tag{12–7}$$

where μ is the permeability of the core (recall Equation 12–3).

It was stated earlier that permeability is a measure of how easy it is to establish flux in a material. To see why, note from Equation 12–7 that the larger the value of μ, the larger the flux density for a given H. However, H is proportional to current; therefore, the larger the value of μ, the larger the flux density for a given magnetizing current. From this, it follows that the larger the permeability, the more flux you get for a given magnetizing current.

In the SI system, μ has units of webers per ampere-turn-meter. The permeability of free space is $\mu_0 = 4\pi \times 10^{-7}$. For all practical purposes, the permeability of air and other nonmagnetic materials is the same as for a vacuum. Thus, in air gaps,

$$B_g = \mu_0 H_g = 4\pi \times 10^{-7} \times H_g \tag{12–8}$$

Rearranging Equation 12–8 yields

$$H_g = \frac{B_g}{4\pi \times 10^{-7}} = 7.96 \times 10^5 B_g \quad \text{(At/m)} \tag{12–9}$$

practice problem III

For Figure 12–16, the core cross section is 0.05 m × 0.08 m. If a gap is cut in the core and H in the gap is 3.6×10^5 At/m, what is the flux in the core? Neglect fringing.

B-H Curves

For ferromagnetic materials, μ is not constant but varies with flux density and there is no easy way to compute it. In reality, however, it isn't μ that you are interested in: What you really want to know is, given B, what is H, and vice versa. A set of curves, called *B-H* or *magnetization curves*, provides this information. (These curves are obtained experimentally and are available in handbooks. A separate curve is required for each material.) Figure 12–19 shows typical curves for cast iron, cast steel, and sheet steel.

Figure 12–19   *B-H* curves for selected materials.

■ **EXAMPLE 12–4**   If $B = 1.4$ T for sheet steel, what is H?

Solution   Enter Figure 12–19 on the axis at $B = 1.4$ T, continue across until you encounter the curve for sheet steel, then read the corresponding value for H as indicated in Figure 12–20: $H = 1000$ At/m.

Figure 12–20   For sheet steel, $H = 1000$ At/m when $B = 1.4$ T.

practice problem IV

The cross section of a sheet steel core is 0.1 m × 0.1 m and its stacking factor is 0.93. If $H = 1500$ At/m, compute the magnetic flux, Φ.

12.9 Ampere's Circuital Law

One of the key relationships in magnetic circuit theory is *Ampere's circuital law*. Ampere's law was determined experimentally and is a generalization of the relationship $\mathscr{F} = NI = H\ell$ that we developed earlier. Ampere showed that the algebraic sum of mmfs around a closed loop in a magnetic circuit is zero, regardless of the number of sections or coils. That is,

$$\sum_{\circlearrowleft} \mathscr{F} = 0 \qquad (12\text{--}10)$$

This can be rewrittten as

$$\sum_{\circlearrowleft} NI = \sum_{\circlearrowleft} H\ell \quad \text{At} \qquad (12\text{--}11)$$

which states that the sum of applied mmfs around a closed loop equals the sum of the mmf drops. The summation is algebraic and terms are additive or subtractive, depending on the direction of flux and how the coils are wound. To illustrate, consider again Figure 12–13. Here,

$$NI - H_{\text{iron}}\ell_{\text{iron}} - H_{\text{steel}}\ell_{\text{steel}} - H_g\ell_g = 0$$

Thus,

$$\underbrace{NI}_{\substack{\text{Impressed}\\\text{mmf}}} = \underbrace{H_{\text{iron}}\ell_{\text{iron}} + H_{\text{steel}}\ell_{\text{steel}} + H_g\ell_g}_{\text{sum of mmf drops}}$$

which states that the applied mmf NI is equal to the sum of the $H\ell$ drops around the loop. The path to use for the $H\ell$ terms is the mean (average) path.

You now have two magnetic circuit models (Figure 12–21). While the reluctance model (a) is not very useful for solving problems, it helps relate magnetic circuit problems to familiar electrical circuit concepts. The Ampere's law model, on the other hand, permits us to solve practical problems. We look at how to do this in the next section.

(a) Reluctance model   (b) Ampere's circuital law model

Figure 12–21   Two models for the circuit of Figure 12–13.

1. If the mmf of a 200-turn coil is 700 At, the current in the coil is _____ amps.

2. For Figure 12–17, if $H = 3500$ At/m and $N = 1000$ turns, then for (a), I is _____ A, while for (b), I is _____ A.

3. For cast iron, if $B = 0.5$ T, then $H =$ _____ At/m.

4. A circuit consists of one coil, a section of iron, a section of steel, and two air gaps (of different sizes). Draw the Ampere's law model.

5. Which is the correct answer for the circuit of Figure 12–22?

    a. Ampere's law around loop 1 yields ($NI = H_1\ell_1 + H_2\ell_2$, $NI = H_1\ell_1 - H_2\ell_2$).

    b. Ampere's law around loop 2 yields ($0 = H_2\ell_2 + H_3\ell_3$, $0 = H_2\ell_2 - H_3\ell_3$).

Path $\ell_2 = \ell_{cda}$
$H = H_2$

Path $\ell_1 = \ell_{abc}$ Path $\ell_3 = \ell_{cda}$
$H = H_1$ $H = H_3$

Figure 12–22

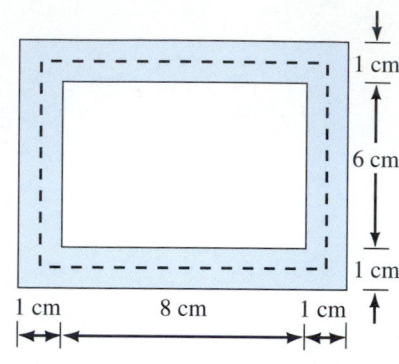

1 cm

6 cm

1 cm

1 cm   8 cm   1 cm

Figure 12–23

6. For the circuit of Figure 12–23, the length ℓ to use in Ampere's law is (0.36 m, 0.32 m, 0.28 m). Why?

12.10 Series Magnetic Circuits: Given Φ, Find *NI*

You now have the tools needed to solve basic magnetic circuit problems. We will begin with series circuits where Φ is known and we want to find the excitation to produce it. Problems of this type can be solved using four basic steps:

1. Compute B for each section using $B = \Phi/A$.

2. Determine H for each magnetic section from the *B-H* curves. Use $H_g = 7.96 \times 10^5 B_g$ for air gaps.

3. Compute *NI* using Ampere's circuital law.

4. Use the computed *NI* to determine coil current or turns as required. (Circuits with more than one coil are handled as in Example 12–6.)

Be sure to use the mean path through the circuit when applying Ampere's law.

practical

*M*agnetic circuit analysis is not as precise as electric circuit analysis because 1) the assumption of uniform flux density breaks down at sharp corners as you saw in Figure 12–4, 2) it is difficult to scale accurate values from B-H curves, and 3) the B-H curve is a mean curve and has considerable uncertainty as discussed later (Section 12–14).

Although the answers are approximate, they are adequate for most purposes.

■ **EXAMPLE 12–5**   If the core of Figure 12–24 is cast iron and Φ $= 0.1 \times 10^{-3}$ Wb, what is the coil current?

Mean length *abcda* = 0.25 m
N = 500 turns
$A = 0.2 \times 10^{-3}$ m²

Figure 12–24

Solution   Following the four steps outlined above:

1. The flux density is

$$B = \frac{\Phi}{A} = \frac{0.1 \times 10^{-3}}{0.2 \times 10^{-3}} = 0.5 \text{ T}$$

2. From the *B-H* curve (cast iron), Figure 12–19, $H = 1575$ At/m.

3. Apply Ampere's law. There is only one coil and one core section. Length = 0.25 m. Thus,

$$NI = H\ell = 1575 \times 0.25 = 393.8 \text{ At}$$

4. Divide by N:

$$I = 393.8/500 = 0.788 \text{ amps}$$

■ **EXAMPLE 12–6**   A second coil is added as shown in Figure 12–25. If Φ $= 0.1 \times 10^{-3}$ Wb as before, but $I_1 = 1.5$ amps, what is I_2?

$N_1 = 500$ $N_2 = 200$
Figure 12–25

Solution From the previous example, you know that a current of 0.788 amps in coil 1 produces $\Phi = 0.1 \times 10^{-3}$ Wb. But you already have 1.5 amps in coil 1. Thus, coil 2 must be wound in opposition so that its mmf is subtractive. Applying Ampere's law yields $N_1I_1 - N_2I_2 = H\ell$. Hence,

$$(500)(1.5\ A) - 200I_2 = 393.8\ \text{At}$$

and so $I_2 = 1.78$ amps.

Note. Since magnetic circuits are nonlinear, you cannot use superposition, that is, you cannot consider each coil of Figure 12–25 by itself, then sum the results. You must consider them simultaneously as we did in Example 12–6.

More Examples

If a magnetic circuit contains an air gap, add another element to the conceptual models (recall Figure 12–21). Since air represents a poor path, its reluctance will be high compared with that of iron. Recalling our analogy to electric circuits, this suggests that the mmf drop across the gap will be large compared with that of the iron. You can see this in the following example.

■ **EXAMPLE 12–7** The core of Figure 12–24 has a 0.008-m gap cut as shown in Figure 12–26. Determine how much the current must increase to maintain the original core flux. Neglect fringing.

Figure 12–26

Solution

Iron
$\ell_{\text{iron}} = 0.25 - 0.008 = 0.242$ m. Since Φ does not change, B and H will be the same as before.

Air Gap
B_g is the same as B_{iron}. Thus, $B_g = 0.5$ T and $H_g = 7.96 \times 10^5 B_g = 3.98 \times 10^5$ At/m.

Ampere's Law
$NI = H_{\text{iron}}\ell_{\text{iron}} + H_g\ell_g = (1575)(0.242) + (3.98 \times 10^5)(0.008) = 381 + 3184 = 3565$ At. Thus, $I = 3565/500 = 7.13$ amps. Note that the current had to increase from 0.788 amp to 7.13 amps in order to maintain the same flux, almost a ninefold increase.

■ **EXAMPLE 12–8**   The laminated sheet steel section of Figure 12–27 has a stacking factor of 0.9. Compute the current required to establish a flux of $\Phi = 1.4 \times 10^{-4}$ Wb.

Cross section = 0.5" × 0.8" (all members)
$\Phi = 1.4 \times 10^{-4}$ Wb

Figure 12–27

Solution   Convert all dimensions to metric.
Cast Iron

$$\ell_{\text{iron}} = \ell_{ab} + \ell_{cdef} = 2.5 + 2 + 2.5 - 0.2 = 6.8 \text{ in} = 0.173 \text{ m}$$

$$A_{\text{iron}} = (0.5 \text{ in})(0.8 \text{ in}) = 0.4 \text{ in}^2 = 0.258 \times 10^{-3} \text{ m}^2$$

$$B_{\text{iron}} = \Phi/A_{\text{iron}} = (1.4 \times 10^{-4})/(0.258 \times 10^{-3}) = 0.54 \text{ T}$$

$$H_{\text{iron}} = 1850 \text{ At/m} \quad \text{(from Figure 12–19)}$$

Sheet Steel

$$\ell_{\text{steel}} = \ell_{fg} + \ell_{gb} + \ell_{ba} = 0.25 + 2 + 0.25 = 2.5 \text{ in} = 6.35 \times 10^{-2} \text{ m}$$

$$A_{\text{steel}} = (0.9)(0.258 \times 10^{-3}) = 0.232 \times 10^{-3} \text{ m}^2$$

$$B_{\text{steel}} = \Phi/A_{\text{steel}} = (1.4 \times 10^{-4})/(0.232 \times 10^{-3}) = 0.60 \text{ T}$$

$$H_{\text{steel}} = 125 \text{ At/m} \quad \text{(from Figure 12–19)}$$

Air Gap

$$\ell_g = 0.2 \text{ in} = 5.08 \times 10^{-3} \text{ m}$$

$$B_g = B_{\text{iron}} = 0.54 \text{ T}$$

$$H_g = (7.96 \times 10^5)(0.54) = 4.3 \times 10^5 \text{ At/m}$$

Ampere's Law

$$NI = H_{\text{iron}}\ell_{\text{iron}} + H_{\text{steel}}\ell_{\text{steel}} + H_g\ell_g$$

$$= (1850)(0.173) + (125)(6.35 \times 10^{-2}) + (4.3 \times 10^5)(5.08 \times 10^{-3})$$

$$= 320 + 7.9 + 2184 = 2512 \text{ At}$$

$$I = 2512/N = 2512/150 = 16.7 \text{ amps}$$

■ **EXAMPLE 12–9**   Figure 12–28 shows a portion of a solenoid. Flux $\Phi = 4 \times 10^{-4}$ Wb when $I = 2.5$ amps. Find the number of turns on the coil.

Figure 12–28   Solenoid. All parts are cast steel.

Solution

Yoke

$$A_{\text{yoke}} = 2.5 \text{ cm} \times 2.5 \text{ cm} = 6.25 \text{ cm}^2 = 6.25 \times 10^{-4} \text{ m}^2$$

$$B_{\text{yoke}} = \frac{\Phi}{A_{\text{yoke}}} = \frac{4 \times 10^{-4}}{6.25 \times 10^{-4}} = 0.64 \text{ T}$$

$$H_{\text{yoke}} = 450 \text{ At/m} \quad \text{(from Figure 12–19)}$$

Plunger

$$A_{\text{plunger}} = 2.0 \text{ cm} \times 2.5 \text{ cm} = 5.0 \text{ cm}^2 = 5.0 \times 10^{-4} \text{ m}^2$$

$$B_{\text{plunger}} = \frac{\Phi}{A_{\text{plunger}}} = \frac{4 \times 10^{-4}}{5.0 \times 10^{-4}} = 0.8 \text{ T}$$

$$H_{\text{plunger}} = 500 \text{ At/m} \quad \text{(from Figure 12–19)}$$

Air Gap

There are two identical gaps. For each,

$$B_g = B_{\text{yoke}} = 0.64 \text{ T}$$

Thus,

$$H_g = (7.96 \times 10^5)(0.64) = 5.09 \times 10^5 \text{ At/m}$$

The results are summarized in Table 12–2.

Ampere's Law

$$NI = H_{\text{yoke}}\ell_{\text{yoke}} + H_{\text{plunger}}\ell_{\text{plunger}} + 2H_g\ell_g = 90 + 50 + 2(2036) = 4212 \text{ At}$$

$$N = 4212/2.5 = 1685 \text{ turns}$$

TABLE 12–2

MATERIAL	SECTION	LENGTH (m)	A (m²)	B (T)	H (At/m)	$H\ell$ (At)
Cast steel	yoke	0.2	6.25×10^{-4}	0.64	400	80
Cast steel	plunger	0.1	5×10^{-4}	0.8	500	50
Air	gap	0.4×10^{-2}	6.25×10^{-4}	0.64	5.09×10^5	2036

12.11 Series-Parallel Magnetic Circuits

Series-parallel magnetic circuits are handled using the sum of fluxes principle (Figure 12–14) and Ampere's law.

■ **EXAMPLE 12–10** The core of Figure 12–29 is cast steel. Determine the current to establish an air-gap flux $\Phi_g = 6 \times 10^{-3}$ Wb. Neglect fringing.

Cast steel
$A = 2 \times 10^{-2} \, m^2$
$\ell_g = \ell_{bc} = 0.25 \times 10^{-3}$ m

$\Phi_g = \Phi_3$

$\ell_{ab} = \ell_{cd} = 0.25$ m $\ell_{da} = 0.2$ m
$\ell_{dea} = 0.35$ m

Figure 12–29

Solution Consider each section in turn.
Air Gap

$$B_g = \Phi_g/A_g = (6 \times 10^{-3})/(2 \times 10^{-2}) = 0.3 \text{ T}$$
$$H_g = (7.96 \times 10^5)(0.3) = 2.388 \times 10^5 \text{ At/m}$$

Sections ab and cd

$$B_{ab} = B_{cd} = B_g = 0.3 \text{ T}$$
$$H_{ab} = H_{cd} = 250 \text{ At/m} \quad \text{(from Figure 12–19)}$$

Ampere's Law (Loop 2)
$\Sigma_{\circlearrowleft} NI = \Sigma_{\circlearrowleft} H\ell$. Since you are going opposite to flux in leg *da*, the corresponding term (i.e., $H_{da}\ell_{da}$) will be subtractive. Also, $NI = 0$ for loop 2. Thus,

$$0 = \Sigma_{\circlearrowleft \text{ loop2}} H\ell$$
$$0 = H_{ab}\ell_{ab} + H_g\ell_g + H_{cd}\ell_{cd} - H_{da}\ell_{da}$$
$$= (250)(0.25) + (2.388 \times 10^5)(0.25 \times 10^{-3}) + (250)(0.25) - 0.2H_{da}$$
$$= 62.5 + 59.7 + 62.5 - 0.2H_{da} = 184.7 - 0.2H_{da}$$

Thus, $0.2H_{da} = 184.7$ and $H_{da} = 925$ At/m. From Figure 12–19, $B_{da} = 1.12$ T.

$$\Phi_2 = B_{da}A = 1.12 \times 0.02 = 2.24 \times 10^{-2} \text{ Wb}$$
$$\Phi_1 = \Phi_2 + \Phi_3 = 2.84 \times 10^{-2} \text{ Wb.}$$
$$B_{dea} = \Phi_1/A = (2.84 \times 10^{-2})/0.02 = 1.42 \text{ T}$$
$$H_{dea} = 2125 \text{ At/m} \quad \text{(from Figure 12–19)}$$

Ampere's Law (Loop 1)

$$NI = H_{dea}\ell_{dea} + H_{ad}\ell_{ad} = (2125)(0.35) + 184.7 = 929 \text{ At}$$

$$I = 929/200 = 4.65 \text{ A}$$

practice problem V

The cast-iron core of Figure 12–30 is symmetrical. Determine current *I*. Hint: To find *NI*, you can write Ampere's law around either loop. (Circuit analogies are shown in Figure 12–31.)

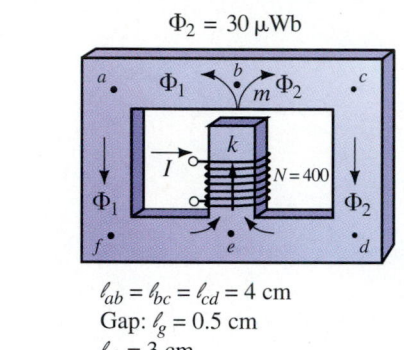

$\ell_{ab} = \ell_{bc} = \ell_{cd} = 4$ cm
Gap: $\ell_g = 0.5$ cm
$\ell_{ek} = 3$ cm

Core dimensions: 1 cm × 1 cm

Figure 12–30

(a)   (b)

Figure 12–31   Circuit analogies for Figure 12–30.

12.12 Series Magnetic Circuits: Given *NI*, Find Φ

In previous problems, you were given the flux and asked to find the current. We now look at the converse problem: given *NI*, to find the resultant flux. For the special case of a core of one material and constant cross section (Example 12–11) this is straightforward. For all other cases, trial and error must be used.

■ **EXAMPLE 12–11**   For the circuit of Figure 12–32, $NI = 250$ At. Determine Φ.

Figure 12-32

Solution $H\ell = NI$. Thus, $H = NI/\ell = 250/0.2 = 1250$ At/m. From the *B-H* curve of Figure 12–19, $B = 1.22$ T. Therefore, $Φ = BA = 1.22 \times 0.01 = 1.22 \times 10^{-2}$ Wb.

For circuits with two or more sections, the process is not so simple. Before you can find *H* in any section, for example, you need to know the flux density. However, in order to determine flux density, you need to know *H*. Thus, neither Φ nor *H* can be found without knowing the other first.

To get around this problem, use trial and error. First, take a guess at the value for flux, compute *NI* as in the examples of Section 12.11, then compare the computed *NI* against the given *NI*. If they agree, the problem is solved. If they don't, adjust your guess and try again. Repeat the procedure until you are within 5% of the given *NI*.

The problem is how to come up with a good first guess. For circuits of the type of Figure 12–33, note that $NI = H_{steel}\ell_{steel} + H_g\ell_g$. As a first guess, assume that the reluctance of the steel is so high that the full mmf drop appears across the gap. Thus, $NI \simeq H_g\ell_g$, and

$$H_g \simeq NI/\ell_g \qquad (12\text{--}12)$$

You can now apply Ampere's law to see how close to the given *NI* your trial guess is.

$\mathcal{F} = H_{steel} + H_g\,\ell_g$

$\qquad \simeq H_g\,\ell_g$ if $H_g\,\ell_g \gg H_{steel}\,\ell_{steel}$

$\ell_g = 0.002$ m
$\ell_{steel} = 0.2$ m

Figure 12–33

practical

note

$\mathcal{S}$*ince we know that some of the mmf drop appears across the steel, we will start at* *than 100% for the gap. Common sense and a bit of experience helps. If the gap is large* *pared with the length of the steel, a very high percentage appears across the gap. The* *tive size of the mmf drops also depends on the core material. For cast iron, the percent-* *drop across the iron is larger than the percentage across a similar piece of sheet steel or* *steel. This is illustrated in the next two examples.*

■ **EXAMPLE 12–12**   The core of Figure 12–33 is cast steel, $NI =$ 1100 At, and the cross-sectional area everywhere is 0.0025 m². Determine the flux in the core.

Solution

Initial Guess
Assume that 90% of the mmf appears across the gap. The applied mmf is 1100 At. Ninety percent of this is 990 At. Thus, $H_g \simeq 0.9NI/\ell = 990/0.002$ $= 4.95 \times 10^5$ At/m and $B_g = \mu_0 H_g = (4\pi \times 10^{-7})(4.95 \times 10^5) = 0.62$ T.

Trial 1
Since the area of the steel is the same as that of the gap, the flux density is the same, neglecting fringing. Thus, $B_{steel} = B_g = 0.62$ T. From the B-H curve, $H_{steel} = 400$ At/m. Now apply Ampere's law:

$$NI = H_{steel}\ell_{steel} + H_g\ell_g = (400)(0.2) + (4.95 \times 10^5)(0.002)$$

$$= 80 + 990 = 1070 \text{ At}$$

This answer is 2.7% lower than the given NI of 1100 At and is therefore acceptable. Thus, $\Phi = BA = 0.62 \times 0.0025 = 1.55 \times 10^{-3}$ Wb.

The initial guess in Example 12–12 yielded an acceptable answer on the first trial. (You are seldom this lucky.)

■ **EXAMPLE 12–13**   If the core of Figure 12–33 is cast iron instead of steel, compute Φ.

Solution   Because cast iron has a larger H for a given flux density (Figure 12–19), it will have a larger $H\ell$ drop and less will appear across the gap. Assume 75% across the gap.

Initial Guess
$H_g \simeq 0.75\ NI/\ell = (0.75)(1100)/0.002 = 4.125 \times 10^5$ At/m.
$B_g = \mu_0 H_g = (4\mu \times 10^{-7})(4.125 \times 10^5) = 0.52$ T.

Trial 1
$B_{iron} = B_g$. Thus, $B_{iron} = 0.52$ T. From the B-H curve, $H_{iron} = 1675$ At/m.

Ampere's Law

$$NI = H_{iron}\ell_{iron} + H_g\ell_g = (1675)(0.2) + (4.125 \times 10^5)(0.002)$$

$$= 335 + 825 = 1160 \text{ At} \quad \text{(high by 5.45\%)}$$

Trial 2
Reduce the guess by 5.45% to $B_{iron} = 0.49$ T. Thus, $H_{iron} = 1500$ At/m (from the *B-H* curve) and $H_g = 7.96 \times 10^5 \, B_g = 3.90 \times 10^5$ At/m.

Ampere's Law

$$NI = H_{iron}\ell_{iron} + H_g\ell_g = (1500)(0.2) + (3.90 \times 10^5)(0.002)$$

$$= 300 + 780 = 1080 \text{ At}$$

The error is now 1.82%, which is excellent. Thus, $\Phi = BA = (0.49)(2.5 \times 10^{-3}) = 1.23 \times 10^{-3}$ Wb. If the error had been larger than 5%, a new trial would have been needed.

12.13 Force Due to an Electromagnet

Electromagnets are used in relays, door bells, lifting magnets, and so on. For an electromagnetic relay as in Figure 12–34, it can be shown that the force created by the magnetic field is

$$F = \frac{B_g^2 A_g}{2\mu_0} \tag{12–13}$$

where B_g is flux density in the gap in teslas, A_g is gap area in square meters, and F is force in newtons.

■ **EXAMPLE 12–14**   Figure 12–34 shows a typical relay. The force due to the current-carrying coil pulls the pivoted arm against spring tension to close the contacts and energize the load. If the pole face is 1/4 inch square and $\Phi = 0.5 \times 10^{-4}$ Wb, what is the pull on the armature in pounds?

(a) A typical relay

Figure 12–34

(b) Typical application of a relay

Figure 12–34

Solution   Convert to metric units.

$$A_g = (0.25 \text{ in})(0.25 \text{ in}) = 0.0625 \text{ in}^2 = 0.403 \times 10^{-4} \text{ m}^2$$
$$B_g = \Phi/A_g = (0.5 \times 10^{-4})/(0.403 \times 10^{-4}) = 1.24 \text{ T}$$

Thus,

$$F = \frac{B_g^2 A}{2\mu_0} = \frac{(1.24)^2(0.403 \times 10^{-4})}{2(4\pi \times 10^{-7})} = 24.66 \text{ N} = 5.54 \text{ lb}$$

12.14 Properties of Magnetic Materials

Magnetic properties are related to atomic structure. Each atom of a substance for example produces a tiny, atomic level, magnetic field because its moving electrons constitute an atomic level current and currents create magnetic fields. For nonmagnetic materials, these fields are randomly oriented and cancel. However, for ferromagnetic materials, the fields in small regions, called *domains* (Figure 12–35), do not cancel. (Domains are of microscopic size, but are large enough to hold from 10^{17} to 10^{21} atoms.) If the domain fields in a ferromagnetic material line up, the material is magnetized; if they are randomly oriented, the material is not magnetized.

Magnetizing a Specimen

A nonmagnetized specimen can be magnetized by making its domain fields line up. Figure 12–36 shows how this can be done. As current through the coil is increased, the field strength increases and more and more domains align themselves in the direction of the field. If the field is made strong enough, almost all domain fields line up and the material is said to be in *saturation* (the almost flat portion of the *B-H* curve). In satu-

Ferromagnetic material

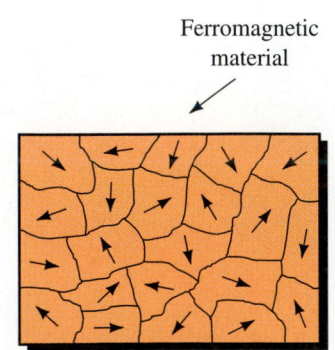

Figure 12–35   Random orientation of microscopic fields in a nonmagnetized ferromagnetic material. The small regions are called domains.

(a) Set-up

(b) Progressive change in the domains as the external field is increased. H is proportional to current I.

Figure 12–36   The magnetization process.

ration, the flux density increases slowly as magnetization intensity increases. This means that once the material is in saturation, you cannot magnetize it much further no matter how hard you try. Path 0-a traced from the nonmagnetized state to the saturated state is termed the *dc* or *normal magnetization curve*. (This is the *B-H* curve that you used earlier when you solved magnetic circuit problems.)

Hysteresis

If you now reduce the current to zero, you will find that the material still retains some magnetism, called *residual magnetism* (Figure 12–37, point b). If now, you reverse the current, the flux reverses and the bottom part of the curve can be traced. By reversing the current again at d, the curve can be traced back to point a. The result is called a *hysteresis loop*. A major source of uncertainty in magnetic circuit behavior should now be apparent: As you can see, flux density depends not just on current, it also depends on which arm of the curve the sample is magnetized on, i.e., it depends on the circuit's past history. For this reason, *B-H* curves are the average of the two arms of the hysteresis loop, i.e., the dc curve of Figure 12–36.

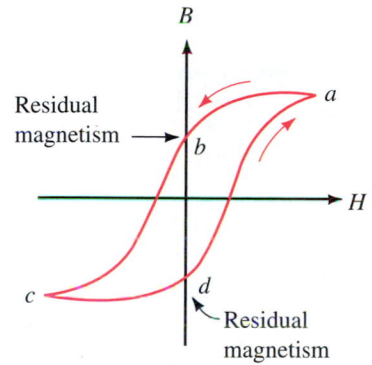

Figure 12–37   Hysteresis loop.

The Demagnetization Process

As indicated above, simply turning the current off does not demagnetize ferromagnetic material. To demagnetize it, you must successively decrease its hysteresis loop to zero as in Figure 12–38. You can place the specimen inside a coil that is driven by a variable ac source and gradually decrease the coil current to zero, or you can use a fixed ac supply and gradually withdraw the specimen from the field. Such procedures are used by service personnel to "degauss" TV picture tubes.

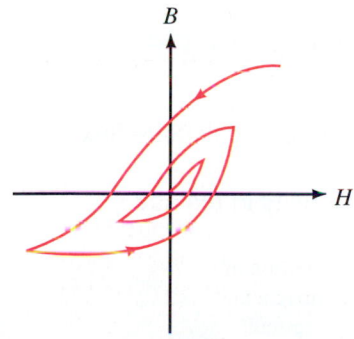

Figure 12–38   Demagnetization by successively shrinking hysteresis loop.

12.15   Measuring Magnetic Fields

One way to measure magnetic field strength is to use the *Hall effect* (after E. H. Hall). The basic idea is illustrated in Figure 12–39. When a strip of semiconductor material such as indium arsenide is placed in a magnetic field, a small voltage, called the Hall voltage, V_H, appears across opposite edges. For a fixed current *I*, V_H is proportional to magnetic field strength *B*. Instruments using this principle are known as *Hall-effect gaussmeters*. To measure a magnetic field with such a meter, insert its probe into the field perpendicular to the field (Figure 12–40). The meter indicates flux density directly.

Figure 12–39   The Hall effect.

Figure 12–40   Magnetic field measurement.

Problems

12.3   Flux and Flux Density

1. Refer to Figure 12–41:
   a. Which area, A_1 or A_2, do you use to calculate flux density?
   b. If $\Phi = 28$ mWb, what is flux density in teslas?

Figure 12–41

2. For Figure 12–41, if $\Phi = 250\ \mu$Wb, $A_1 = 1.25$ in^2, and $A_2 = 2.0$ in^2, what is the flux density in the English system of units?

3. The toroid of Figure 12–42 has a circular cross section and $\Phi = 628\ \mu\text{Wb}$. If $r_1 = 8$ cm and $r_2 = 12$ cm, what is the flux density in teslas?

Figure 12–42

4. If r_1 of Figure 12–42 is 3.5 inches and r_2 is 4.5 inches, what is the flux density in the English system of units if $\Phi = 628\ \mu\text{Wb}$?

12.5   Air Gaps, Fringing, and Laminated Cores

5. If the section of core in Figure 12–43 is 0.025 m by 0.04 m, has a stacking factor of 0.85, and $B = 1.45$ T, what is Φ in Webers?

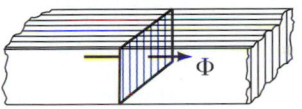

Figure 12–43

6. If the core of Figure 12–43 is 3 cm by 2 cm, has a stacking factor of 0.9, and $B = 8 \times 10^3$ gauss, what is Φ in maxwells?

12.6   Series Elements and Parallel Elements

7. For the iron core of Figure 12–44, flux density $B_2 = 0.6$ T. Compute B_1 and B_3.

$A_1 = 0.02\ \text{m}^2$

$A_3 = 0.01\ \text{m}^2$

$A_2 = 0.015\ \text{m}^2$
$B_2 = 0.6$ T

Figure 12–44

$A_1 = 0.02\ \text{m}^2$

$A_3 = 0.016\ \text{m}^2$

$A_2 = 0.01\ \text{m}^2$

Figure 12–45

8. For the section of iron core of Figure 12–45, if $\Phi_1 = 12$ mWb and $\Phi_3 = 2$ mWb, what is B_2?

9. For the section of iron core of Figure 12–45, if $B_1 = 0.8$ T and $B_2 = 0.6$ T, what is B_3?

(a) $\mathfrak{R} = \dfrac{\ell}{\mu A}$

(b) $B = \mu H$

Figure 12–46 $\mathcal{F} = NI$.

12.8   Magnetic Field Intensity and Magnetization Curves

10. Consider again Figure 12–42. If $I = 10$ A, $N = 40$ turns, $r_1 = 5$ cm, and $r_2 = 7$ cm, what is H in ampere-turns per meter?

11. Figure 12–46 shows the two electric circuit equivalents for magnetic circuits. Show that μ in $\mathfrak{R} = \ell/\mu A$ is the same as μ in $B = \mu H$.

12.9   Ampere's Circuital Law

12. Let H_1 and ℓ_1 be the magnetizing force and path length respectively, where flux Φ_1 exists in Figure 12–47 and similarly for Φ_2 and Φ_3. Write Ampere's law around each of the windows.

Figure 12–47

13. Assume that a coil N_2 carrying current I_2 is added on leg 3 of the core shown in Figure 12–47 and that it produces flux directed upward. Assume, however, that the net flux in leg 3 is still downward. Write the Ampere's law equations for this case.

14. Repeat Problem 13 if the net flux in leg 3 is upward.

12.10   Series Magnetic Circuits: Given Φ, Find NI

15. Find the current I in Figure 12–48 if $\Phi = 0.16$ mWb.

16. Let everything be the same as in Problem 15 except that the cast steel portion is replaced with laminated sheet steel with a stacking factor of 0.85.

17. A gap of 0.5 mm is cut in the cast steel portion of the core in Figure 12–48. Find the current for $\Phi = 0.128$ mWb. Neglect fringing.

18. Two gaps, each 1 mm, are cut in the circuit of Figure 12–48, one in the cast steel portion and the other in the cast iron portion. Determine current for $\Phi = 0.128$ mWb. Neglect fringing.

Cast steel   Cast iron

$\ell_{steel} = 0.14$ m $\ell_{iron} = 0.06$ m

$A = 3.2 \times 10^{-4}$ m^2
$N = 300$ turns

Figure 12–48

19. For the circuit of Figure 12–49, $\Phi = 30$ μWb and $N = 2000$ turns. Neglecting fringing, find current I.

$\ell_{iron} = 3$ cm
$\ell_{steel} = 8$ cm

Cast iron

$\ell_g = 2$ mm

A (everywhere) = 0.5 cm²

Figure 12–49

$\ell_1 = 2$ in $A_1 = 1$ in²

Cast iron

$\ell_2 = 3.5$ in

$\ell_g = 0.2$ in

$\ell_3 = 5.8$ in

$\ell_4 = 7.5$ in

$N = 600$

Sheet steel

Cast steel

Area of all sections (except A_1) = 2 in²

Figure 12–50

20. For the circuit of Figure 12–50, $\Phi = 25{,}000$ lines. The stacking factor for the sheet steel portion is 0.95. Find current I.

21. A second coil of 450 turns with $I_2 = 4$ amps is wound on the cast steel portion of Figure 12–50. Its flux is in opposition to the flux produced by the original coil. The resulting flux is 35 000 lines in the counterclockwise direction. Find the current I_1.

12.11   Series-Parallel Magnetic Circuits

22. For Figure 12–51, if $\Phi_g = 80$ μWb, find I.

Cast steel

Φ_1 •a Φ_3

x

•d

y ḃ

Φ_2

100 turns

•c

$\ell_g = \ell_{xy} = 0.001$ m
$\ell_{abc} = 0.14$ m
$\ell_{cda} = 0.16$ m
$\ell_{ax} = \ell_{cy} = 0.039$ m
$A = 4$ cm² everywhere

Figure 12–51

23. If the circuit of Figure 12–51 has no gap and $\Phi_3 = 0.2$ mWb, find I.

12.12   Series Magnetic Circuits: Given *NI*, Find Φ

24. A cast steel magnetic circuit with $N = 2500$ turns, $I = 200$ mA, and a cross-sectional area of 0.02 m² has an air gap of 0.00254 m. Assuming 90% of the mmf appears across the gap, estimate the flux in the core.

25. If $NI = 644$ At for the cast steel core of Figure 12–52, find the flux, Φ.

Diameter = 2 cm

Radius = 6 cm

Figure 12–52   Toroidal core.

26. A gap $\ell = 0.004$ m is cut in the core of Figure 12–52. Everything else remains the same. Find the flux, Φ.

12.13   Force Due to an Electromagnet

27. For the relay of Figure 12–34, if the pole face is 2 cm by 2.5 cm and a force of 2 pounds is required to close the gap, what flux (in webers) is needed?

28. For the solenoid of Figure 12–28, $\Phi = 4 \times 10^{-4}$ Wb. Find the force of attraction on the plunger in newtons and in pounds.

Inductance and Inductors

CHAPTER PREVIEW

In its simplest form, an *inductor* is a coil of wire (Figure 13–1)(a). Its effect on circuit operation is to oppose changes in current. The property of an inductor that causes it to oppose changes in current is termed its *inductance*. Inductance is of great importance in circuits with time-varying currents such as ac circuits or dc circuits under transient conditions.

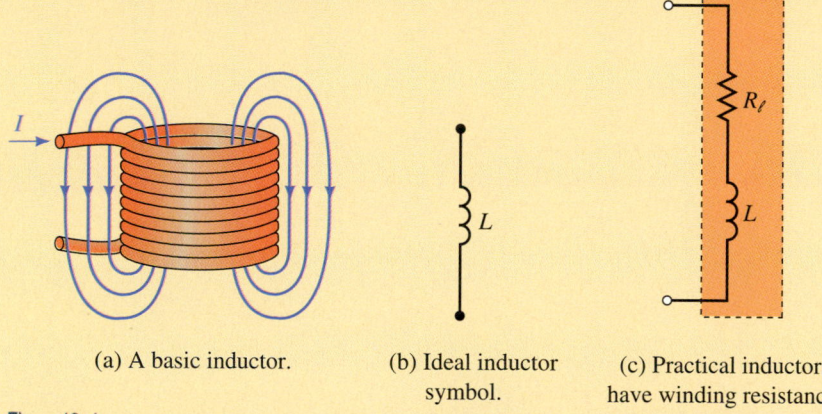

(a) A basic inductor. (b) Ideal inductor symbol. (c) Practical inductors have winding resistance.

Figure 13–1   Inductance is due to the magnetic field created by an electric current.

Many types of inductors are made. (The chapter opener photo shows a few. As indicated, a variety of coil shapes and core materials are used.) Some inductors, for example, have coils wound on iron cores, some on plastic forms, some on circular rods, some on toroidal rings, while others have free-standing, self-supporting windings that require no core at all. Depending on their use, inductors may be referred to as *chokes* or *reactors*.

On circuit diagrams and in equations, inductance is represented by the letter *L*. Its circuit symbol is a coil as shown in Figure 13–1(b). The unit of inductance is the *henry*.

Ideally, inductors have only inductance. However, since inductors are made of wire, they also have resistance, as shown in Figure 13–1(c). Although this resistance is usually small, it must sometimes be included in the analysis of a circuit. Initially, however, we will ignore resistance and treat inductors as ideal circuit elements that possess only inductance.

Commercial inductors are made in a variety of shapes and sizes. The top row includes some transformers. *(Courtesy Goguen Industries, Liverpool, New York, USA)*

OBJECTIVES

After studying this chapter, you will be able to

- describe what an inductor is and what its effect on circuit operation is,

- explain Faraday's law and Lenz's law,

- compute induced voltage using Faraday's law,

- define inductance,

- compute voltage across an inductance,

- compute inductance for series and parallel configurations,

- compute inductor voltages and currents for steady state dc excitation,

- compute energy stored in an inductance,

- describe common inductor problems and how to test for them.

Inductors are used in many places. In radios, they are part of the tuning circuit that you adjust when you select a station. In fluorescent lamps, they are part of the ballast circuit that limits current when the lamp is turned on; in power systems, they are part of the protection circuitry used to control short-circuit currents during fault conditions.

Since inductance depends on induced voltage, we begin with a look at electromagnetic induction.

in process
l e a r n i n g
check I

The Discovery of Electromagnetic Induction

*M*ost of our ideas concerning inductance and induced voltages are due to Michael Faraday (recall Chapter 12) and Joseph Henry (1797–1878). Working independently (Faraday in England and Henry in the USA), they discovered, almost simultaneously, the fundamental laws governing electromagnetic induction.

While experimenting with magnetic fields, Faraday developed the transformer. He wound two coils on an iron ring and energized one of them from a battery. As he closed the switch energizing the first coil, Faraday noticed that a momentary voltage was induced in the second coil, and when he opened the switch, he found that a momentary voltage was again induced but with opposite polarity. When the current was steady, no voltage was produced at all.

Faraday explained this effect in terms of his magnetic lines of flux concept. When current was first turned on, he visualized the lines as springing outward into space; when it was turned off, he visualized the lines as collapsing inward. He then visualized that voltage was produced by these lines as they cut across circuit conductors. Companion experiments showed that voltage was also produced when a magnet was passed through a coil or when a conductor was moved through a magnetic field. Again, he visualized these voltages in terms of flux cutting a conductor.

Working independently in the United States, Henry discovered essentially the same results. In fact, Henry's work preceded Faraday's by a few months, but because he did not publish them first, credit was given to Faraday. However, Henry is credited with the discovery of self-induction, and in honor of his work the unit of inductance was named the henry.

13.1 Electromagnetic Induction

We begin by examining Faraday's and Henry's results. Consider Figure 13–2. In (a), a magnet is moved through a coil of wire, and this action induces a voltage in the coil. When the magnet is thrust into the coil, the meter deflects upscale; when it is withdrawn, the meter deflects downscale, indicating that polarity has changed. The voltage magnitude is proportional to how fast the magnet is moved. In (b), when the conductor is moved through the field, voltage is induced. If the conductor is moved to the right, its far end is positive; if it is moved to the left, the polarity reverses and its far end becomes negative. Again, the voltage magnitude is proportional to how fast the wire is moved. In (c), voltage is induced in coil 2 due to the magnetic field created by the current in coil 1. At the instant the switch is closed, the meter kicks upscale; at the instant it is opened, the meter kicks downscale. In (d) voltage is induced in a coil by its own current. At the instant the switch is closed, the top end of the coil becomes positive, while at the instant it is opened, the polarity reverses and the top end becomes negative.

(a) Motional emf.

(b) Motional emf.

(c) Mutually induced voltage.

(d) Self-induced voltage.

Figure 13–2   Principle of electromagnetic induction. Voltage is created as long as the flux linking a circuit is changing.

Faraday's Law

Based on these observations, Faraday concluded *that voltage is induced in a circuit whenever the flux linking* (i.e., passing through) *the circuit is changing and that the magnitude of the voltage is proportional to the rate of change of the flux linkages*. This result, known as *Faraday's law*, is also sometimes stated in terms of the rate of cutting flux lines. We look at this viewpoint in Chapter 15.

Lenz's Law

Heinrich Lenz (a Russian physicist, 1804–1865) determined a companion result. He showed that *the polarity of the induced voltage is such as to oppose the cause producing it*. This result, known as *Lenz's law*, can be used to determine the polarity of induced voltages by examining the buildup and collapse of magnetic fields. (We do this in Chapter 24.)

13.2 Induced Voltage and Induction

We now apply Faraday's concept to the inductor (Figure 13–3). First, consider (a). Since the current is constant, the magnetic field is constant and hence no voltage is induced. Thus, inductance has no effect on steady, unchanging, dc current.

Now consider (b). If you try to increase the current as in (b), you find that a voltage develops with polarity as shown. As you can see, this voltage opposes current buildup. Since its magnitude is proportional to the rate of change of flux, and since flux is proportional to current, the faster the current increases, the larger the opposing voltage.

Now consider (c). If you try to decrease current, the polarity of the voltage reverses, that is, the collapsing field produces a voltage that tries to keep the current going. Again, the faster the rate of change of current, the larger the induced voltage.

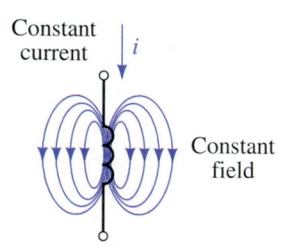

(a) Steady dc. Field is constant.
Therefore, induced voltage is zero.

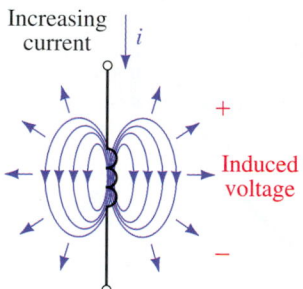

(b) Increasing current causes expanding field. The expanding field induces a voltage that opposes the current build-up.

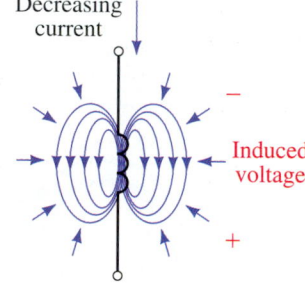

(c) Decreasing current causes collapsing magnetic field. The collapsing field induces a voltage that opposes the current decrease.

Figure 13–3   Effect of self-induced voltage.

Counter EMF

Because the induced voltage in Figure 13–3 tries to counter changes in current, it is called a *counter emf* or *back voltage*. Note carefully, however, that this voltage does not oppose current, it opposes only changes in current. It also does not prevent the current from changing; it only prevents it from changing abruptly. The result is that current in an inductor changes gradually and smoothly from one value to another as indicated in Figure 13–4(b). The effect of inductance is thus similar to the effect of inertia in a mechanical system. The flywheel used on an engine for example prevents abrupt changes in engine speed but does not prevent the engine from gradually changing from one speed to another.

(a) Current cannot jump from one value to another like this.

(b) Current must change smoothly with no abrupt jumps.

Figure 13–4   Current in an inductance.

Iron-Core and Air-Core Inductors

As Faraday discovered, the voltage induced in a coil depends on flux linkages, and flux linkages depend on core materials. Coils with ferromagnetic cores (called *iron-core coils*) have their flux almost entirely confined to their cores, while coils wound on nonferromagnetic materials do not. (The latter are sometimes called *air-core coils* because all nonmagnetic core materials have the same permeability as air and thus behave magnetically the same as air.)

First, consider the iron-core case, in Figure 13–5. Ideally, all flux lines are confined to the core and hence pass through (link) all windings. The product of flux times the number of turns that it passes through is defined as the *flux linkage* of the coil. For Figure 13–5, Φ lines pass through N turns yielding a flux linkage of NΦ. By Faraday's law, the induced voltage is thus equal to the rate of change of NΦ. Since N is constant, this becomes

Figure 13–5   When flux Φ passes through all N turns, the flux linking the coil is NΦ.

$$e = N \times \text{the rate of change of } \Phi \qquad (13\text{–}1)$$

In calculus notation,

$$e = N\frac{d\Phi}{dt} \quad \text{(volts, V)} \qquad (13\text{–}2)$$

In the SI system, Φ is in webers, t in seconds, and e in volts.

■ **EXAMPLE 13–1**   If the flux through a 200-turn coil changes steadily from 1 Wb to 4 Wb in one second, what is the voltage induced?

Solution   The flux changes by 3 Wb in one second. Thus, its rate of change is 3 Wb/s.

$$e = N \times \text{rate of change of flux}$$
$$= (200 \text{ turns})(3 \text{ Wb/s}) = 600 \text{ volts}$$

notes

...*tion 13–2*

is sometimes shown with a minus sign. However, the minus sign is unnecessary. In circuit theory, we use Equation 13–2 to determine the magnitude of the induced voltage and Lenz's law to determine its polarity.

Now consider an air-core inductor (Figure 13–6). Since not all lines pass through all windings, it is difficult to determine flux linkages as above. However, (since no ferromagnetic material is present) flux is directly proportional to current. In this case, then, since induced voltage is proportional to the rate of change of flux, and since flux is proportional to current, induced voltage will be proportional to the rate of change of current. Let the constant of proportionality be L. Thus,

$$e = L \times \text{rate of change of current} \qquad (13\text{–}3)$$

In calculus notation, this can be written as

$$e = L\frac{di}{dt} \quad \text{(volts, V)} \qquad (13\text{–}4)$$

L is called the *self-inductance* of the coil and in the SI system, its unit is the henry. (This is discussed in Section 13.3.)

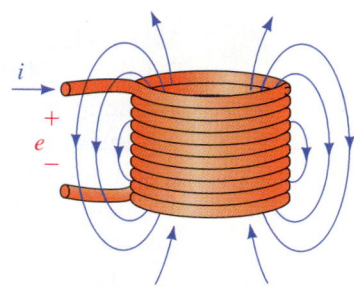

Figure 13–6   Not all flux links all turns. Nonetheless, the flux linking the coil is proportional to current.

We now have two equations for coil voltage. Equation 13–4 is the more useful form for this chapter, while Equation 13–2 is the more useful form for the circuits of Chapter 24. We look at Equation 13–4 in the next section.

in process
l e a r n i n g
check I

1. Which of the currents shown in Figure 13–7 cannot be the current in an inductor? Why?

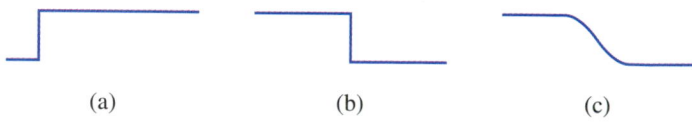

(a)   (b)   (c)

Figure 13–7

2. Compute the flux linkage for the coil of Figure 13–5, given $\Phi = 500$ mWb and $N = 1200$ turns.

3. If the flux Φ of Question 2 changes steadily from 500 mWb to 525 mWb in 1 s, what is the voltage induced in the coil?

4. If the flux Φ of Question 2 changes steadily from 500 mWb to 475 mWb in 100 ms, what is the voltage induced?

13.3 Self-Inductance

In the preceding section, we showed that the voltage induced in a coil is $e = L\,di/dt$, where L is the self-inductance of the coil (usually referred to simply as inductance) and di/dt is the rate of change of its current. In the SI system, L is measured in henries. As can be seen from Equation 13–4, it is the ratio of voltage induced in a coil to the rate of change of current producing it. From this, we get the definition of the henry. By definition, *the inductance of a coil is one henry if the voltage created by its changing current is one volt when its current changes at the rate of one ampere per second.*

Figure 13–8   Voltage-current convention for inductance. The plus sign for voltage goes at the tail of the current arrow.

In practice, the voltage across an inductance is denoted by v_L rather than by e. Thus,

$$v_L = L\frac{di}{dt} \quad \text{(V)} \tag{13–5}$$

Voltage and current references are as shown in Figure 13–8.

■ EXAMPLE 13–2   If the current through a 5-mH inductance changes at the rate of 1000 A/s, what is the voltage induced?

Solution

$$v_L = L \times \text{rate of change of current}$$
$$= (5 \times 10^{-3}\ \text{H})(1000\ \text{A/s}) = 5\ \text{volts}$$

---

practice problems 1

1. The voltage across an inductance is 250 V when its current changes at the rate of 10 mA/μs. What is L?

2. If the voltage across a 2-mH inductance is 50 volts, how fast is the current changing?

Inductance Formulas

Inductance for some simple shapes can be determined using the principles of Chapter 12. For example, the approximate inductance of the coil of Figure 13–9 can be shown to be

$$L = \frac{\mu N^2 A}{\ell} \quad \text{(H)} \tag{13–6}$$

where ℓ is in meters, A is in square meters, N is the number of turns, and μ is the permeability of the core. (Details can be found in many physics books.)

Figure 13–9   Inductance formula for a single-layer coil.

$$L = \frac{\mu N^2 A}{\ell}\ \text{H}$$

■ **EXAMPLE 13-3**   A 0.15 m long air-core coil has a radius of 0.006 m and 120 turns. Compute its inductance.

Solution

$$A = \pi r^2 = 1.131 \times 10^{-4} \, \text{m}^2$$

$$\mu = \mu_0 = 4\pi \times 10^{-7}$$

Thus,

$$L = 4\pi \times 10^{-7} \, (120)^2 \, (1.131 \times 10^{-4})/0.15 = 13.6 \, \mu\text{H}$$

The accuracy of Equation 13–6 breaks down for small ℓ/d ratios. (If ℓ/d is greater than 10, the error is less than 4%.) Improved formulas may be found in design handbooks, such as the *Radio Amateur's Handbook* published by the American Radio Relay League (ARRL).

To provide greater inductance in smaller spaces, iron cores are sometimes used. Unless the core flux is kept below saturation, however, permeability varies and inductance is not constant. To get around this problem, an air gap may be used (Figure 13–10). If the gap is wide enough to dominate, coil inductance is approximately

$$L = \frac{\mu_0 N^2 A_g}{\ell_g} \quad \text{(H)} \tag{13-7}$$

where μ_0 is the permeability of air, A_g is the area of the air gap, and ℓ_g is its length. (The derivation of this equation is left as an exercise; see end-of-chapter Problem 11.)

Figure 13-10   Iron-core coil with an air gap to control saturation. The reason for the laminated core is discussed in Chapter 24.

■ **EXAMPLE 13-4**   The inductor of Figure 13–10 has 1000 turns, a 5-mm gap, and a cross-sectional area at the gap of $5 \times 10^{-4} \, \text{m}^2$. What is its inductance?

Solution

$$L = (4\pi \times 10^{-7})(1000)^2(5 \times 10^{-4})/(5 \times 10^{-3}) = 0.126 \, \text{H}$$

practical notes

1. *Since inductance is due to a conductor's magnetic field, it depends on the same factors that the magnetic field depends on. The stronger the field for a given current, the greater the inductance. Thus, a coil of many turns will have more inductance than a coil of a few turns (L is proportional to N²) and a coil wound on a magnetic core will have greater inductance than a coil wound on a nonmagnetic form.*

2. *However, if a coil is wound on a magnetic core, the core's permeability μ may change with flux density. Since flux density depends on current and inductance depends on permeability, inductance in this case depends on current, i.e., it is not constant, but is a non-linear function of current. For example, the inductor of Figure 13–11 has a nonlinear inductance due to core saturation. All inductors encountered in this book are assumed to be linear, i.e., of constant value.*

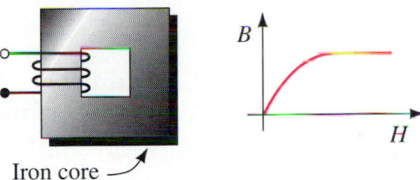

Iron core

Figure 13–11   This coil does not have a fixed inductance since its flux is not proportional to its current.

in process
l e a r n i n g
check *II*

1. The voltage across an inductance whose current changes uniformly by 10 mA in 4 μs is 70 volts. What is its inductance?

2. If you triple the number of turns in the inductor of Figure 13–10, but everything else remains the same, by what factor does the inductance increase?

13.4 Computing Induced Voltage

Earlier, we determined that the voltage across an inductance is given by $v_L = L\,di/dt$, where the voltage and current references are shown in Figure 13–8. Note that the polarity of v_L depends on whether the current is increasing or decreasing. For example, if the current is increasing, di/dt is positive and so v_L is positive, while if the current is decreasing, di/dt is negative and v_L must be negative.

To compute voltage, we need to determine di/dt. In general, this requires calculus. However, since di/dt is slope, you can determine voltage without calculus for currents that can be described by straight lines, as in Figure 13–12. For any Δt segment, slope $= \Delta i/\Delta t$, where Δi is the amount that the current changes during time interval Δt.

■ **EXAMPLE 13–5**   Figure 13–12 is the current through a 10-mH inductance. Determine voltage v_L and sketch it.

Figure 13–12

Solution   Break the problem into intervals over which the slope is constant, determine the slope for each segment, then compute voltage using $v_L = L \times$ slope for that interval:

0 to 1 ms: Slope = 0. Thus, $v_L = 0$ V.
1 ms to 2 ms: Slope = $\Delta i/\Delta t = 4$ A/$(1 \times 10^{-3}$ s) = 4×10^3 A/s. Thus, $v_L = L\Delta i/\Delta t = (0.010 \text{ H})(4 \times 10^3 \text{ A/s}) = 40$ V.
2 ms to 4 ms: Slope = $\Delta i/\Delta t = -8$ A/$(2 \times 10^{-3}$ s) = -4×10^3 A/s. Thus, $v_L = L\Delta i/\Delta t = (0.010 \text{ H})(-4 \times 10^3 \text{ A/s}) = -40$ V.
4 ms to 5 ms: Slope = 0. Thus, $v_L = 0$ V.
5 ms to 6 ms: Same slope as from 1 ms to 2 ms. Thus, $v_L = 40$ V.

The voltage waveform is shown in Figure 13–13.

Figure 13–13

**practice
problems
II**

1. Figure 13–14 shows the current through a 5-H inductance. Determine voltage v_L and sketch it.

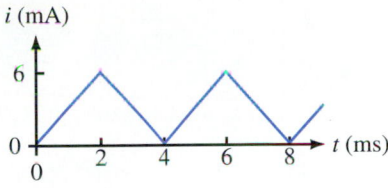

Figure 13–14

2. If the current of Figure 13–12 is applied to an unknown inductance and the voltage change from 1 ms to 2 ms is 28 volts, what is L?

**in process
learning
check III**

1. An inductance L_1 of 50 mH is in series with an inductance L_2 of 35 mH. If the voltage across L_1 at some instant is 125 volts, what is the voltage across L_2 at that instant? Hint: Since the same current passes through both inductances, the rate of change of current is the same for both.

2. Current through a 5-H inductance changes linearly from 10 A to 12 A in 0.5 s. Suppose now the current changes linearly from 2 mA to 6 mA in 1 ms. Although the currents are significantly different, the induced voltage is the same in both cases. Why? Compute the voltage.

13.5 Inductances in Series and Parallel

For inductances in series or parallel, the equivalent inductance is found by using the same rules that you used for resistance. (To show this is left as an exercise for the student.) For the series case (Figure 13–15) the total inductance is the sum of the individual inductances:

$$L_T = L_1 + L_2 + L_3 + \cdots + L_N \qquad (13\text{–}8)$$

For the parallel case (Figure 13–16),

$$\frac{1}{L_T} = \frac{1}{L_1} + \frac{1}{L_2} + \frac{1}{L_3} + \cdots + \frac{1}{L_N} \qquad (13\text{–}9)$$

For two inductances, Equation 13–9 reduces to

$$L_T = \frac{L_1 L_2}{L_1 + L_2} \qquad (13\text{–}10)$$

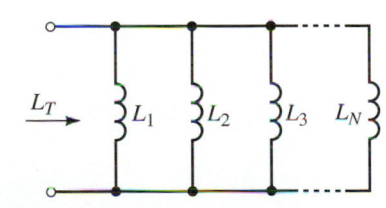

Figure 13–15 $L_T = L_1 + L_2 + \cdots + L_N$.

Figure 13–16 $1/L_T = 1/L_1 + 1/L_2 + \cdots + 1/L_N$

■ **EXAMPLE 13–6**   Find L_T for the circuit of Figure 13–17.

Figure 13–17

Solution   The parallel combination of L_2 and L_3 is

$$L_{eq} = \frac{L_2 L_3}{L_2 + L_3} = \frac{6 \times 2}{6 + 2} = 1.5 \text{ H}$$

This is in series with L_1 and L_4. Thus, $L_T = 2.5 + 1.5 + 11 = 15$ H.

practice
problems
III

1. For Figure 13–18, $L_T = 2.25$ H. Determine L_x.

Figure 13–18

2. For Figure 13–15, the current is the same in each inductance and $v_1 = L_1 di/dt$, $v_2 = L_2 di/dt$, and so on. Apply KVL and show that $L_T = L_1 + L_2 + L_3 + \cdots + L_N$.

13.6 Practical Considerations

Core Types

The type of core used in an inductor depends to a great extent on its intended use and frequency range. (Although you have not studied frequency yet, you can get a feel for frequency by noting that the electrical power system operates at low frequency [60 cycles per second, called 60 hertz], while radio and TV systems operate at high frequency [hundreds of megahertz].) Inductors used in audio or power supply applications generally have iron cores (because they need large inductance values), while inductors for radio-frequency circuits generally use air or ferrite cores. (Ferrite is a mixture of iron oxide in a ceramic binder. It has characteristics that make it suitable for high-frequency work.) Iron cannot be used, however, since it has large power losses at high frequencies (for reasons discussed in Chapter 17).

Variable Inductors

Inductors can be made so that their inductance is variable. In one approach, inductance is varied by changing coil spacing with a screwdriver adjustment. In another approach, a threaded ferrite slug is moved in or out of the coil's winding. Figure 13–19 shows an inductor of this type with its slug removed for viewing.

Figure 13–19   A variable inductor with its core removed. The threaded ferrite slug is screwed in or out of the coil to vary its inductance.

Circuit Symbols

Figure 13–20 shows inductor symbols. Iron-core inductors are identified by double solid lines, while dashed double lines denote a ferrite core. (Air-core inductors have no core symbol.) An arrow indicates a variable inductor.

Stray Capacitance

Because the turns of an inductor are separated from each other by insulation, a small amount of capacitance exists from winding to winding. This capacitance is called *stray* or *parasitic capacitance*. Although this capacitance is distributed from turn to turn, its effect can be approximated by lumping it as in Figure 13–21(a). The effect of stray capacitance depends on frequency. At low frequencies, it can usually be neglected; at high frequencies, it may have to be taken into account. Some coils are wound in multiple sections as in Figure 13–21(b) to reduce stray capacitance.

(a) Iron-core.

(b) Ferrite-core.

(c) Variable.

(d) Air-core.

Figure 13–20   Circuit symbols for inductors.

(a) C_w is stray capacitance.

(b) Separating coil sections
to reduce stray capacitance.
Shown is a 2.4 mH rf (radio
frequency) choke wound on
a powdered iron core.

Figure 13–21   Inductors have stray capacitance. While negligible at low frequencies, it may be important at high frequencies.

Stray Inductance

Because inductance is due entirely to the magnetic effects of electric current, all current-carrying conductors have inductance. This means that leads on circuit components such as resistors, capacitors, transistors, and so on, all have inductance, as do traces on printed circuit boards and wires in cables. We call this inductance *"stray inductance."* Fortunately, in most cases, the stray inductance is so small that it can be neglected.

practical

note

*A*lthough stray inductance is small, it is not always negligible. In general, stray inductance will not be a problem for short wires at low to moderate frequencies. However, even a short piece of wire can be a problem at high frequencies, or a long piece of wire at low frequencies. For example, the inductance of just a few centimeters of conductor in a high-speed logic system may be non-negligible because the current through it changes at such a high rate.

13.7 Inductance and Steady State DC

$v_L = 0$

Constant dc current

(a) Since the field is constant, induced voltage is zero.

$v_L = 0$

(b) Equivalent of inductor is a short circuit.

Figure 13–22 Inductance looks like a short circuit in steady state dc.

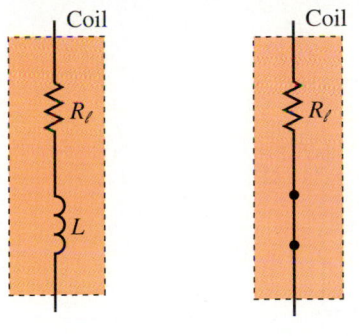

Coil   Coil

R_ℓ R_ℓ

L

(a) Coil.   (b) DC equivalent.

Figure 13–23 Steady state dc equivalent of a coil.

We now look at inductive circuits with constant dc current. Consider Figure 13–22. The voltage across an inductance with constant dc current is zero because the rate of change of current is zero. This is indicated in (a). Since the inductor has current through it but no voltage across it, it looks like a short circuit, (b). This is true in general, that is, *an ideal inductor looks like a short circuit in steady state dc.* (This should not be surprising since it is just a piece of wire to dc.) For a nonideal inductor, its dc equivalent is its coil resistance (Figure 13–23).

■ **EXAMPLE 13–7**   In Figure 13–24(a), the resistance of the coil is 14.4 Ω. What is *I*?

6 Ω

I

120 V

9 Ω

R_ℓ

L

Coil

$R_\ell = 14.4\ \Omega$

(a)

Thévenin equivalent

R_{Th}

3.6 Ω

I

14.4 Ω

E_{Th} 72 V

Coil equivalent

(b)

Figure 13–24

Solution   For the circuit external to the coil,

$$E_{Th} = (9/15)(120) = 72\ \text{V}$$

$$R_{Th} = 6\Omega \| 9\Omega = 3.6\ \Omega$$

Now replace the coil by its dc equivalent circuit as in (b). Thus,

$$I = E_{Th}/R_T = 72/(3.6 + 14.4) = 4\ \text{A}$$

■ **EXAMPLE 13–8**   The resistance of coil 1 in Figure 13–25(a) is 30 Ω and that of coil 2 is 15 Ω. Find the voltage across the capacitor assuming steady state dc.

(a)

(b)

Figure 13–25

Solution   Replace each coil inductance with a short circuit and the capacitor with an open circuit. As you can see from (b), the voltage across C is the same as the voltage across R_{ℓ_2}. Thus,

$$V_C = [R_{\ell_2}/(R_{\ell_1} + R_{\ell_2})]E = (15/45)(60) = 20 \text{ V}$$

practice problem IV

For Figure 13–26, find I, V_{C_1}, and V_{C_2} in the steady state.

Figure 13–26

13.8 Energy Stored by an Inductance

When power flows into an inductor, energy is stored in its magnetic field. When the field collapses, this energy is returned to the circuit. For an ideal inductor, $R_\ell = 0$ ohm and hence no power is dissipated; thus, an ideal inductor has zero power loss.

To determine the energy stored by an ideal inductor, consider Figure 13–27. Power to the inductor is given by $p = v_L i$ watts, where $v_L = L di/dt$. By summing this power (see ∫), the energy is found to be

$$W = \frac{1}{2}Li^2 \quad \text{(J)} \tag{13–11}$$

where i is the instantaneous value of current. When current reaches its steady state value I, $W = 1/2 LI^2$ J. This energy remains stored in the field as long as the current continues. When the current goes to zero, the field collapses and the energy is returned to the circuit.

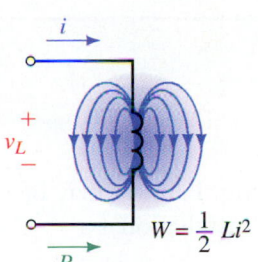

$$W = \frac{1}{2}Li^2$$

Figure 13–27   Energy is stored in the magnetic field of an inductance.

■ **EXAMPLE 13–9**   The coil of Figure 13–28(a) has a resistance of 15 Ω. When the current reaches its steady state value, the energy stored is 12 J. What is the inductance of the coil?

$W = 12$ J

(a)

(b)

Figure 13–28

Solution

$$I = 100 \text{ V}/25 \ \Omega = 4 \text{ A}$$
$$W = 1/2 LI^2 \quad \text{J}$$
$$12 \text{ J} = 1/2 L(4 \text{ A})^2$$

Thus,

$$L = 2(12)/4^2 = 1.5 \text{ H}$$

Deriving Equation 13–11

The power to the inductor in Figure 13–27 is given by $p = v_L i$, where $v_L = Ldi/dt$. Therefore, $p = Lidi/dt$. However, $p = dW/dt$. Integrating yields

$$W = \int_0^t p\,dt = \int_0^t Li\frac{di}{dt}\,dt = L\int_0^i i\,di = \frac{1}{2}Li^2$$

13.9   Inductor Troubleshooting Hints

Inductors may fail by either opening or shorting. Failures may be caused by misuse, defects in manufacturing, or faulty installation.

Open Coil

Opens can be the result of poor solder joints or broken connections. First, make a visual inspection. If nothing wrong is found, disconnect the inductor and check it with an ohmmeter. An open-circuited coil has infinite resistance.

Shorts

Shorts can occur between windings or between the coil and its core (for an iron-core unit). A short may result in excessive current and overheating. Again, check visually. Look for burned insulation, discolored components, an acrid odor, and other evidence of overheating. An ohmmeter can be used to check for shorts between windings and the core. However, checking coil resistance for shorted turns is often of little value, especially if only a few turns are shorted. This is because the shorting of a few windings will not change the overall resistance enough to be measurable. Sometimes the only conclusive test is to substitute a known good inductor for the suspected one.

Problems

13.2   Induced Voltage and Inductance

1. If the flux linking a 75-turn coil changes at the rate of 3 Wb/s, what is the voltage across the coil?

2. If 80 volts is induced when the flux linking a coil changes at a uniform rate from 3.5 mWb to 4.5 mWb in 0.5 ms, how many turns does the coil have?

3. Flux changing at a uniform rate for 1 ms induces 60 V in a coil. What is the induced voltage if the same flux change takes place in 0.01 s?

13.3   Self-Inductance

4. The current in a 0.4-H inductance is changing at the rate of 200 A/s. What is the voltage across it?

5. The current in a 75-mH inductance changes uniformly by 200 μA in 0.1 ms. What is the voltage across it?

6. The voltage across an inductance is 25 volts when the current changes at 5 A/s. What is L?

7. The voltage induced when current changes uniformly from 3 amps to 5 amps in a 10-H inductance is 180 volts. How long did it take for the current to change from 3 to 5 amps?

8. Current changing at a uniform rate for 1 ms induces 45 V in a coil. What is the induced voltage if the same current change takes place in 100 μs?

9. Compute the inductance of the air-core coil of Figure 13–29, given $\ell = 20$ cm, $N = 200$ turns, and $d = 2$ cm.

10. The iron-core inductor of Figure 13–30 has 2000 turns, a cross-section of 1.5×1.2 inches, and an air gap of 0.2 inch. Compute its inductance.

11. The iron-core inductor of Figure 13–30 has a high-permeability core. Therefore, by Ampere's law, $NI \approx H_g\ell_g$. Because the air gap dominates, saturation does not occur and the core flux is proportional to the current, i.e., the flux linkage equals LI. In addition, since all flux passes through the coil, the flux linkage equals $N\Phi$. By equating the two values of flux linkage and using ideas from Chapter 12, show that the inductance of the coil is

$$L = \frac{\mu_0 N^2 A_g}{\ell_g}$$

13.4   Computing Induced Voltage

12. Figure 13–31 shows the current in a 0.75-H inductance. Determine v_L and plot its waveform.

13. Figure 13–32 shows the current in an inductance. If the voltage from 0 to 2 ms is 100 volts, what is L?

14. Why is Figure 13–33 not a valid inductor current? Sketch the voltage across L to show why.

Figure 13–29

Figure 13–30

Figure 13–31

Figure 13–32

Figure 13–33

v_L (V)

Figure 13–34

15. Figure 13–34 shows the graph of the voltage across an inductance. The current changes from 4 A to 5 A during the time interval from 4 s to 5 s.
  a. What is L?
  b. Determine the current waveform and plot it.
  c. What is the current at $t = 10$ s?

16. If the current in a 25-H inductance is $i_L = 20e^{-12t}$ mA, what is v_L? (This is a calculus problem.)

13.5   Inductances in Series and Parallel

17. Determine L_T for the circuits of Figure 13–35.

(a)

(b)

(c)

Figure 13–35

18. Determine L_T for the circuits of Figure 13–36.

(a)

(b)

Figure 13–36

Figure 13–37

19. For Figure 13–37, determine L_x.

20. For the circuits of Figure 13–38, determine L_3 and L_4.

(a)   (b)

Figure 13–38

21. A 6-H and a 4-H inductance are connected in parallel. After a third inductance is added, $L_T = 4$ H. What is the value of the third inductance and how was it connected?

22. Inductances of 2 H, 4 H, and 9 H are connected in a circuit. If $L_T = 3.6$ H, how are the inductors connected?

Figure 13–39

23. Inductances of 8 H, 12 H, and 1.2 H are connected in a circuit. If $L_T = 6$ H, how are the inductors connected?

24. For inductors in parallel (Figure 13–39), the same voltage appears across each. Thus, $v = L_1 di_1/dt$, $v = L_2 di_2/dt$, etc. Apply KCL and show that $1/L_T = 1/L_1 + 1/L_2 + \cdots + 1/L_N$.

25. Using the result of Problem 24, show that for two inductors in parallel,

$$L_T = \frac{L_1 L_2}{L_1 + L_2}$$

13.7 Inductance and Steady State DC

26. For each of the circuits of Figure 13–40, the voltages and currents have reached their final (steady state) values. Solve for the quantities indicated.

(a)

(b)

Figure 13–40

13.8 Energy Stored by an Inductance

27. Find the energy stored in the inductor of Figure 13–41.

28. In Figure 13–42, $L_1 = 2L_2$. The total energy stored is $W_T = 75$ J. Find L_1 and L_2.

Figure 13–41

Figure 13–42

13.9 Inductor Troubleshooting Hints

29. In Figure 13–43, an inductance meter measures 7 H. What is the likely fault?

Figure 13–43

30. Referring to Figure 13–44, an inductance meter measures 8 mH. What is the likely fault?

Figure 13–44

14

Inductive Transients

CHAPTER PREVIEW

In Chapter 11, you learned that transients occur in capacitive circuits because capacitor voltage cannot change instantaneously. In this chapter, you will learn that transients occur in inductive circuits because inductor current cannot change instantaneously. Although the details differ, you will find that many of the basic ideas are the same.

Inductive transients result when circuits containing inductance are disturbed. More so than capacitive transients, inductive transients are potentially destructive and dangerous. For example, when you break the current in an inductive circuit, an extremely large and damaging voltage may result.

In this chapter, we study basic *RL* transients. We look at transients during current buildup and decay and learn how to calculate the voltages and currents that result.

The sudden collapse of the magnetic field induces a large voltage in the coil. This voltage is an example of an inductive transient.

OBJECTIVES

After studying this chapter, you will be able to

* explain why transients occur in *RL* circuits,

* explain why an inductor looks like an open circuit when first energized,

* compute time constants for *RL* circuits,

* compute voltage and current transients in *RL* circuits during the current buildup phase,

* compute voltage and current transients in *RL* circuits during the current decay phase,

* solve moderately complex *RL* transient problems using circuit simplification techniques,

* solve *RL* transient problems using PSpice.

Inductance, the Dual of Capacitance

*I*nductance is the dual of capacitance. This means that the effect that inductance has on circuit operation is identical with that of capacitance if you interchange the term current for voltage, open circuit for short circuit, and so on. For example, for simple dc transients, current in an RL circuit has the same form as voltage in an RC circuit: they both rise to their final value exponentially according to $1 - e^{-t/\tau}$. Similarly, voltage across inductance decays in the same manner as current through capacitance, i.e., according to $e^{-t/\tau}$.

Duality applies to steady state and initial condition representations as well. To steady state dc, for example, a capacitor looks like an open circuit, while an inductor looks like a short circuit. Similarly, the dual of a capacitor that looks like a short circuit at the instant of switching is an inductor that looks like an open circuit. Finally, the dual of a capacitor that has an initial condition of V_0 volts is an inductance with an initial condition of I_0 amps. Duality carries over to ac analysis as well.

The principle of duality is helpful in circuit analysis as it lets you transfer the principles and concepts learned in one area directly into another. You will find, for example, that many of the ideas learned in Chapter 11 reappear here in their dual form.

14.1 Introduction

As you saw in Chapter 11, when a circuit containing capacitance is disturbed, voltages and currents do not change to their new values immediately, but instead pass through a transitional phase as the circuit capacitance charges or discharges. The voltages and currents during this transitional interval are called *transients*. In a dual fashion, transients occur when circuits containing inductances are disturbed. In this case, however, transients occur because current in inductance cannot change instantaneously.

To get at the idea, consider Figure 14–1. In (a), we see a purely resistive circuit. At the instant the switch is closed, current jumps from 0 to E/R as required by Ohm's law. Thus, no transitional phase occurs because cur-

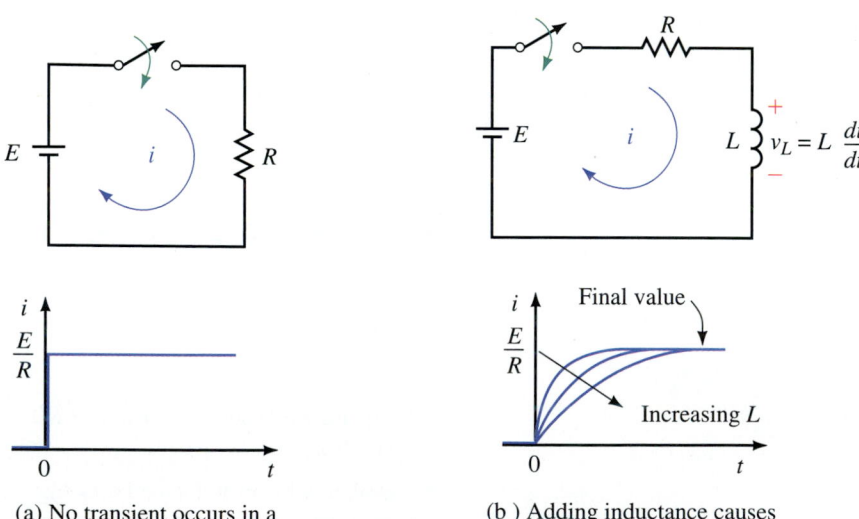

(a) No transient occurs in a purely resistive circuit

(b) Adding inductance causes a transient to appear. *R* is held constant here.

Figure 14–1   Transient due to inductance. For fixed resistance, the larger the inductance, the longer the transient lasts.

notes

T*he* *continuity* *statement has a sound mathematical basis. Recall, induced voltage is proportional to the rate of change of current. In calculus notation,*

$$v_L = L \frac{di}{dt}$$

This means that the faster that current changes, the larger the induced voltage. If current could change from one value to another instantaneously as in Figure 14–1(a), its rate of change (i.e., di/dt) would be infinite and hence the induced voltage would be infinite. But infinite voltage is not possible. Thus, inductor current cannot change instantaneously.

rent reaches its final value immediately. Now consider (b). Here, we have added inductance. At the instant the switch is closed, a counter emf appears across the inductance. This voltage attempts to stop the current from changing and consequently slows its rise. Current thus does not jump to E/R immediately as in (a), but instead climbs gradually and smoothly as in (b). The larger the inductance, the longer the transition takes.

Continuity of Current

As Figure 14–1(b) illustrates, *current through an inductance cannot change instantaneously, i.e., it cannot jump abruptly from one value to another; but must be continuous at all values of time.* This observation is known as the statement of *continuity of current for inductance.* You will find this statement of great value when analyzing circuits containing inductance. We will use it many times in what follows.

Inductor Voltage

Now consider inductor voltage. When the switch is open in Figure 14–2(a), the current in the circuit and voltage across L are both zero. Now close the switch. Immediately after the switch is closed, the current is still zero, (since it cannot change instantaneously). Since $v_R = Ri$, the voltage across R is also zero and thus the full source voltage appears across L as shown in (b). The inductor voltage therefore jumps from 0 V just before the switch is closed to E volts just after. It then decays to zero, since, as we saw in Chapter 13, the voltage across inductance is zero for steady state dc. This is indicated in (c).

(a) Circuit with switch open. Current $i = 0$.

(b) Circuit just after the switch has been closed. Current is still equal to zero. Thus, $v_L = E$.

(c) Voltage across L

Figure 14–2   Voltage across inductance *L*.

Open-Circuit Equivalent of an Inductance

Consider again Figure 14–2(b). Note that just after the switch is closed, the inductor has voltage across it but no current through it. It therefore momentarily looks like an open circuit. This is indicated in Figure 14–3. This observation is true in general, that is, *an inductor with zero initial current looks like an open circuit at the instant of switching.* (Later, we extend this statement to include inductors with nonzero initial currents.)

Figure 14–3   Inductor with zero initial current looks like an open circuit at the instant the switch is closed.

Initial Condition Circuits

Voltages and currents in circuits just after switching must sometimes be calculated. These can be determined with the aid of the open-circuit equivalent. By replacing inductances with open circuits, you can see what a circuit looks like just after switching. Such a circuit is called an *initial condition circuit*.

■ **EXAMPLE 14–1**   A coil and two resistors are connected to a 20-V source as in Figure 14–4(a). Determine source current i and inductor voltage v_L at the instant the switch is closed.

(a) Original circuit   (b) Initial condition network

Figure 14–4

Solution

Replace the inductance with an open circuit. This yields the network shown in (b). Thus $i = E/R_T = 20 \text{ V}/10 \ \Omega = 2$ A and the voltage across R_2 is $V_2 = (2 \text{ A})(4 \ \Omega) = 8$ V. Since $v_L = V_2$, $v_L = 8$ volts as well.

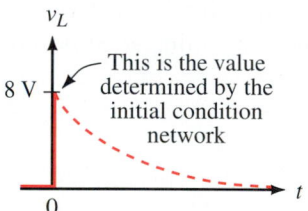

Figure 14–5   The initial condition network yields only the value at $t = 0^+$ s.

Initial condition networks yield voltages and currents only at the instant of switching, i.e., at $t = 0^+$ s. Thus, the value of 8 V calculated in Example 14–1 is only a momentary value as illustrated in Figure 14–5. Sometimes such an initial value is all that you need. In other cases, you need the complete solution. This is considered next, in Section 14.2.

practice problem I

Determine all voltages and currents in the circuit of Figure 14–6 just after the switch is closed.

Figure 14–6

14.2　Current Buildup Transients

Current

We will now develop equations to describe voltages and current during energization. Consider Figure 14–7. KVL yields

$$v_L + v_R = E \qquad (14\text{–}1)$$

Substituting $v_L = L\,di/dt$ and $v_R = Ri$ into Equation 14–1 yields

$$L\frac{di}{dt} + Ri = E \qquad (14\text{–}2)$$

Equation 14–2 can be solved using basic calculus in a manner similar to what we did for RC circuits in Chapter 11. The result is

$$i = \frac{E}{R}\,(1 - e^{-Rt/L}) \quad (A) \qquad (14\text{–}3)$$

where R is in ohms, L is in henries, and t is in seconds. Equation 14–3 describes current buildup. Values of current at any point in time for the circuit of Figure 14–7 can thus be found by direct substitution as we illustrate next.

Figure 14–7　For this circuit, $v_L + v_R = E$.

■ **EXAMPLE 14–2**　For the circuit of Figure 14–7, suppose $E = 50$ V, $R = 10\ \Omega$, and $L = 2$ H:

    a.　Determine the expression for i.

    b.　Compute and tabulate values of i at $t = 0^+$, 0.2, 0.4, 0.6, 0.8, and 1.0 s.

    c.　Using these values, plot the current.

Solution

    a.　Substituting the values into Equation 14–3 yields

$$i = \frac{E}{R}(1 - e^{-Rt/L}) = \frac{50\ \text{V}}{10\ \Omega}\,(1 - e^{-10t/2}) = 5\,(1 - e^{-5t})\ \text{amps}$$

    b.　At $t = 0^+$ s, $i = 5(1 - e^{-5t}) = 5(1 - e^0) = 5(1 - 1) = 0$ A.
        At $t = 0.2$ s, $i = 5(1 - e^{-5(0.2)}) = 5(1 - e^{-1}) = 3.16$ A.
        At $t = 0.4$ s, $i = 5(1 - e^{-5(0.4)}) = 5(1 - e^{-2}) = 4.32$ A.

Continuing in this manner, you get Table 14–1.

c. Values are plotted in Figure 14–8. Note that this curve looks exactly like the curves we determined intuitively in Figure 14–1(b).

TABLE 14–1

TIME	CURRENT
0	0
0.2	3.16
0.4	4.32
0.6	4.75
0.8	4.91
1.0	4.97

Figure 14–8   Current buildup transient.

Circuit Voltages

With i known, circuit voltages can be determined. Consider voltage v_R. Since $v_R = Ri$, when you multiply R times Equation 14–3, you get

$$v_R = E(1 - e^{-Rt/L}) \quad (V) \tag{14–4}$$

Note that v_R has exactly the same shape as the current. Now consider v_L. Voltage v_L can be found by subtracting v_R from E as per Equation 14–1:

$$v_L = E - v_R = E - E(1 - e^{-Rt/L}) = E - E + Ee^{-Rt/L}$$

Thus,

$$v_L = Ee^{-Rt/L} \tag{14–5}$$

An examination of Equation 14–5 shows that v_L has an initial value of E and then decays exponentially to zero. This agrees with our earlier observation in Figure 14–2(c).

■ **EXAMPLE 14–3**   Repeat Example 14–2 for voltage v_L.

Solution

a. From equation 14–5,

$$v_L = Ee^{-Rt/L} = 50e^{-5t} \text{ volts}$$

b. At $t = 0^+$ s, $v_L = 50e^{-5t} = 50e^0 = 50(1) = 50$ V.
At $t = 0.2$ s, $v_L = 50e^{-5(0.2)} = 50e^{-1} = 18.4$ V.
At $t = 0.4$ s, $v_L = 50e^{-5(0.4)} = 50e^{-2} = 6.77$ V.

Continuing in this manner, you get Table 14–2.

c. The waveform is shown in Figure 14–9.

TABLE 14–2

TIME (S)	VOLTAGE (V)
0	50.0
0.2	18.4
0.4	6.77
0.6	2.49
0.8	0.916
1.0	0.337

Figure 14–9   Inductor voltage transient.

practice problem II

For the circuit of Figure 14–7, assume that $E = 80$ V, $R = 5$ kΩ, and $L = 2.5$ mH:

a. Determine expressions for i, v_L, and v_R.
b. Compute and tabulate values at $t = 0^+$, 0.5, 1.0, 1.5, 2.0, and 2.5 μs.
c. At each point in time, does $v_L + v_R = E$?
d. Plot i, v_L, and v_R using the values computed in (b).

Time Constant

The ratio L/R in Equations 14–3 to 14–5 is the time constant of the circuit. That is,

$$\tau = \frac{L}{R} \quad \text{(s)} \tag{14–6}$$

τ has units of seconds. (This is left as an exercise for the student.) In terms of τ, equations 14–3, 14–4, and 14–5 may be written as

$$i = \frac{E}{R}(1 - e^{-t/\tau}) \quad \text{(A)} \tag{14–7}$$

$$v_L = Ee^{-t/\tau} \quad \text{(V)} \tag{14–8}$$

$$v_R = E(1 - e^{-t/\tau}) \quad \text{(V)} \tag{14–9}$$

Curves are plotted in Figure 14–10 versus time constant. As expected, transitions take approximately 5τ; thus, *for all practical purposes, inductive transients last five time constants.*

Figure 14–10   Universal curves for *RL* circuit.

■ **EXAMPLE 14–4**   In a circuit where $L = 2$ mH, transients lasts 50 μs. What is R?

Solution   Transients last five time constants. Thus, $\tau = 50$ μs/5 = 10 μs. Now $\tau = L/R$. Therefore, $R = L/\tau = 2$ mH/10 μs = 200 Ω.

■ **EXAMPLE 14–5**   For an RL circuit, $i = 40(1 - e^{-5t})$ A and $v_L = 100e^{-5t}$ V.

   a. What are E and τ?
   b. What is R?
   c. Determine L.

Solution

   a. From Equation 14–8, $v_L = Ee^{-t/\tau} = 100e^{-5t}$. Therefore, $E = 100$ V and $\tau = \dfrac{1}{5} = 0.2$s.

   b. From Equation 14–7,

$$i = \frac{E}{R}\,(1 - e^{-t/\tau}) = 40(1 - e^{-5t}).$$

Therefore, $E/R = 40$ A and $R = E/40$ A = 100 V/40 A = 2.5 Ω.
   c. $\tau = L/R$. Therefore, $L = R\tau = (2.5)(0.2) = 0.5$ H.

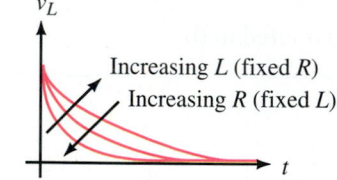

Figure 14–11   Effect of R and L on transient duration.

The effect of inductance and resistance on transient duration is shown in Figure 14–11. The larger the inductance, the longer the transient for a given resistance. Resistance has the opposite effect: for a fixed inductance, the larger the resistance, the shorter the transient. [This is not hard to understand. As R increases, the circuit looks more and more resistive. If you get to a point where inductance is negligible compared with resistance, the circuit looks purely resistive, as in Figure 14–1(a), and no transient occurs.]

in process
learning
check I

1. For the circuit of Figure 14–12, the switch is closed at $t = 0$ s.
   a. Determine expressions for v_L and i.
   b. Compute v_L and i at $t = 0^+$, 10 μs, 20 μs, 30 μs, 40 μs, and 50 μs.
   c. Plot curves for v_L and i.

Figure 14–12

2. For a certain RL circuit, transients last 25 s. If $L = 10$ H and steady state current is 2 A, what is E?

3. An RL circuit has $E = 50$ V and $R = 10\ \Omega$. The switch is closed at $t = 0$ s. What is the current at the end of 1.5 time constants?

14.3 Interrupting Current in an Inductive Circuit

We now look at what happens when inductor current is interrupted. Consider Figure 14-13. At the instant the switch is opened, the field begins to collapse, which induces a voltage in the coil. If inductance is large and current is high, a great deal of energy is released in a very short time, creating a huge voltage that may damage equipment and create a shock hazard. (This induced voltage is referred to as an *inductive kick*.) For example, abruptly breaking the current through a large inductor (such as a motor or generator field coil) can create voltage spikes up to several thousand volts, a value large enough to draw long arcs as indicated in Figure 14–13. Even moderate sized inductances in electronic systems can create enough voltage to cause damage if protective circuitry is not used.

Figure 14–13   Sudden collapse of the magnetic field causes a large induced voltage in the coil. The switch arcs over due to the large voltage. Voltages of several thousand volts may occur.

The dynamics of the switch flashover are not hard to understand. When the field collapses, the voltage across the coil rises rapidly. Part of this voltage appears across the switch. As the switch voltage rises, it quickly exceeds the breakdown strength of air, causing a flashover between its contacts. Once struck, the arc is easily maintained, as it creates ionized gases that provide a relatively low resistance path for conduction. As the contacts spread apart, the arc elongates and eventually disappears as the coil energy is dissipated and coil voltage drops below that required to sustain the arc.

There are several important points to note here. 1) Flashovers, as in Figure 14–13, are generally undesirable. However, they can be controlled through proper engineering design. (One way is to use a discharge resistor, as in the next example; another way is to use a diode, as you will see in your electronics course.) 2) On the other hand, the large

Figure 14–14　Discharge resistor R_2 helps limit the size of the induced voltage when the switch is opened.

voltages created by breaking inductive currents have their uses. One is in the ignition system of automobiles, where current in the primary winding of a transformer coil is interrupted at the appropriate time by a control circuit to create the spark needed to fire the engine. 3) It is not possible to rigorously analyze the circuit of Figure 14–13 because the resistance of the arc changes as the switch opens. However, the main ideas can be established by studying circuits using fixed resistors as we see next.

The Basic Ideas

We begin with the circuit of Figure 14–14. Assume the switch is closed and the circuit is in steady state. Since the inductance looks like a short circuit [Figure 14–15(a)], its current is $i_L = 120\ \text{V}/30\ \Omega = 4\ \text{A}$.

(a) Circuit just before the switch is opened

(b) Circuit just after SW is opened. Since v_L is negative, its actual polarity is opposite to that shown

Figure 14–15　Circuit of Figure 14–14 just before and after the switch is opened. Coil voltage jumps from 0 V to –2520 V for this example.

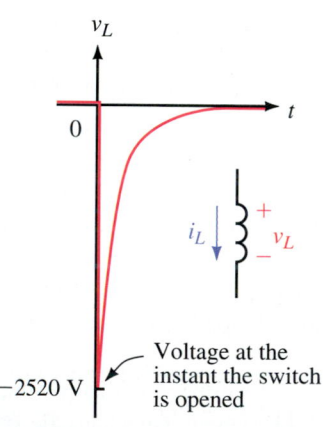

Voltage at the instant the switch is opened

Figure 14–16　Voltage spike for the circuit of Figure 14–14. This voltage is more than 20 times larger than the source voltage.

Now open the switch. Just prior to opening the switch, $i_L = 4\ \text{A}$; therefore, just after opening the switch, it must still be 4 A. As indicated in (b), this 4 A passes through resistances R_1 and R_2, creating voltages $v_{R_1} = 4\ \text{A} \times 30\ \Omega = 120\ \text{V}$ and $v_{R_2} = 4\ \text{A} \times 600\ \Omega = 2400\ \text{V}$ with the polarity shown. From KVL, $v_L + v_{R_1} + v_{R_2} = 0$. Therefore at the instant the switch is opened,

$$v_L = -(v_{R_1} + v_{R_2}) = -2520 \text{ volts}$$

appears across the coil, yielding a negative voltage spike as in Figure 14–16. Note that this spike is more than 20 times larger than the source voltage. As we see in the next section, the size of this spike depends on the ratio of R_2 to R_1; the larger the ratio, the larger the voltage.

　Consider again Figure 14–15. Note that current i_2 changes abruptly from 0.2 A just prior to switching to –4 A just after. This is permissible, however, since i_2 does not pass through the inductor and only currents through inductance must be continuous.

practice
problem
III

Figure 14–16 shows the voltage across the coil of Figure 14–14. Make a similar sketch for the voltage across the switch and across resistor R_2. Hint: Use KVL to find v_{switch} and v_{R_2}.

The Inductor at Switching

Figure 14–17 shows the current through L of Figure 14–15. Because the current is the same just after switching as it is just before, it is constant over the interval from $t = 0^-$ s to $t = 0^+$ s. Since this is true in general, we see that *an inductance with an initial current looks like a current source at the instant of switching*. Its value is the value of the current at the instant of switching. This is shown in Figure 14–18.

Figure 14–17 Inductor current for the circuit of Figure 14–14.

(a) Just before switching   (b) Just after switching

Figure 14–18 Inductor carrying current looks like a current source at the instant of switching.

14–4   De-Energizing Transients

We now look at equations for the voltages and currents described in the previous section. As we go through this material, you should focus on the basic principles involved, rather than just the resulting equations. You will probably forget formulas, but if you understand principles, you should be able to reason your way through many problems using just the basic current and voltage relationships.

Consider Figure 14–19(a). Let the initial current in the inductor be denoted as I_0 amps. Now open the switch as in (b). KVL yields $v_L + v_{R_1} + v_{R_2} = 0$. Substituting $v_L = L\,di/dt$, $v_{R_1} = R_1 i$, and $v_{R_2} = R_2 i$ yields $L\,di/dt + (R_1 + R_2)i = 0$. Now using calculus, it can be shown that

$$i = I_0 e^{-t/\tau'} \quad \text{(A)} \tag{14–10}$$

where

$$\tau' = \frac{L}{R_T} = \frac{L}{R_1 + R_2} \quad \text{(s)} \tag{14–11}$$

is the time constant of the discharge circuit. If the circuit is in steady state before the switch is opened, initial current $I_0 = E/R_1$ and Equation 14–10 becomes

$$i = \frac{E}{R_1} e^{-t/\tau'} \quad \text{(A)} \tag{14–12}$$

(a) Original circuit

(b) Decay circuit

Figure 14–19 Circuit for studying decay transients.

■ **EXAMPLE 14–6**   For Figure 14–19, assume the current has reached steady state with the switch closed. Suppose that $E = 120$ V, $R_1 = 30$ Ω, $R_2 = 600$ Ω, and $L = 126$ mH:

a. Determine I_0.
b. Determine the decay time constant.
c. Determine the equation for the current decay.
d. Compute the current i at $t = 0^+$ s and $t = 0.5$ ms.

Solution

a. Consider Figure 14–19(a). Since the inductor looks like a short circuit to dc, $I_0 = E/R_1 = 4$ A.

b. Consider Figure 14–19(b). $\tau' = L/(R_1 + R_2) = 126$ mH/630 Ω = 0.2 ms.

c. $i = I_0 e^{-t/\tau'} = 4e^{-t/0.2 \text{ ms}}$ A.

d. At $t = 0^+$ s, $i = 4e^{-0} = 4$ A.

   At $t = 0.5$ ms, $i = 4e^{-0.5 \text{ ms}/0.2 \text{ ms}} = 4e^{-2.5} = 0.328$ A.

Now consider voltage v_L. It can be shown to be

$$v_L = V_0 e^{-t/\tau'} \tag{14–13}$$

where V_0 is the voltage across L just after the switch is opened. Letting $i = I_0$ in Figure 14–19(b), you can see that $V_0 = -I_0(R_1 + R_2) = -I_0 R_T$. Thus Equation 14–13 can be written as

$$v_L = -I_0 R_T e^{-t/\tau'} \tag{14–14}$$

Finally, if the current has reached steady state before the switch is opened, $I_0 = E/R_1$, and Equation 14–14 becomes

$$v_L = -E\left(1 + \frac{R_2}{R_1}\right)e^{-t/\tau'} \tag{14–15}$$

Note that v_L starts at V_0 volts (which is negative) and decays to zero as shown in Figure 14–20.

Now consider the resistor voltages. Each is the product of resistance times current. Thus,

$$v_{R_1} = R_1 I_0 e^{-t/\tau'} \tag{14–16}$$

and

$$v_{R_2} = R_2 I_0 e^{-t/\tau'} \tag{14–17}$$

If current has reached steady state before switching, these become

$$v_{R_1} = E e^{-t/\tau'} \tag{14–18}$$

and

$$v_{R_2} = \frac{R_2}{R_1} E e^{-t/\tau'} \tag{14–19}$$

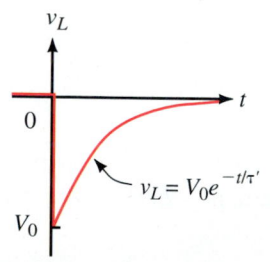

Figure 14–20   Inductor voltage during de-energization V_0 is negative.

Substituting the values of Example 14–6 into these equations, we get for the circuit of Figure 14–19 $v_L = -2520e^{-t/0.2 \text{ ms}}$ V, $v_{R_1} = 120e^{-t/0.2 \text{ ms}}$ V and $v_{R_2} = 2400e^{-t/0.2 \text{ ms}}$ V. These can also be written as $v_L = -2520e^{-5000t}$ V and so on if desired.

14.5 More Complex Circuits

The equations that we have developed so far apply only to circuits of the form of Figures 14–7 or 14–19. Fortunately, many circuits can be reduced to these forms using circuit reduction techniques such as series and parallel combinations, source conversions, Thévenin's theorem, and so on.

■ **EXAMPLE 14–7** Determine i_L for the circuit of Figure 14–21(a) if $L = 5$ H.

(a) Circuit

(b) Thévenin equivalent

Figure 14–21

Solution The circuit can be reduced to its Thévenin equivalent (b) as you saw in Chapter 11 (Section 11.5). For this circuit, $\tau = L/R_{\text{Th}} = 5$ H$/200$ Ω $= 25$ ms. Now apply Equation 14–7. Thus,

$$i_L = \frac{E_{\text{Th}}}{R_{\text{Th}}}(1 - e^{-t/\tau}) = \frac{40}{200}(1 - e^{-t/25\text{ms}}) = 0.2\,(1 - e^{-40t})\quad \text{(A)}$$

■ **EXAMPLE 14–8**   For the circuit of Example 14–7, at what time does current reach 0.12 amps?

Solution

$$i_L = 0.2(1 - e^{-40t})   \text{(A)}$$

Thus,

$$0.12 = 0.2(1 - e^{-40t})   \text{(Figure 14–22)}$$

$$0.6 = 1 - e^{-40t}$$

$$e^{-40t} = 0.4$$

Taking the natural log of both sides,

$$\ln e^{-40t} = \ln 0.4$$

$$-40t = -0.916$$

$$t = 22.9 \text{ ms}$$

Figure 14–22

practice problems IV

1. For the circuit of Figure 14–21, let $E = 120$ V, $R_1 = 600$ Ω, $R_2 = 3$ kΩ, $R_3 = 2$ kΩ, $R_4 = 100$ Ω, and $L = 0.25$ H:
   a. Determine i_L and sketch it.
   b. Determine v_L and sketch it.

2. Let everything be as in Problem 1 except L. If $i_L = 0.12$ A at $t = 20$ ms, what is L?

A Note About Time Scales

Until now, we have considered energization and de-energization phases separately. When both occur in the same problem, we must clearly define what we mean by time. One way to handle this problem (as we did with RC circuits) is to define $t = 0$ s as the beginning of the first phase and solve for voltages and currents in the usual manner, then shift the time axis to the beginning of the second phase, redefine $t = 0$ s and then solve the second part. This is illustrated in Example 14–9. Note that only the first time scale is shown explicitly on the graph.

■ **EXAMPLE 14–9**

Refer to the circuit of Figure 14–23:

a. Close the switch and determine equations for i_L and v_L.

b. After the circuit has reached steady state, open the switch and determine equations for i_L and v_L during the decay phase.

c. Sketch i_L and v_L.

Figure 14–23

Solution

a. Thévenize the circuit to the left of L. As indicated in Figure 14–24(a), $R_{\text{Th}} = 60\|30 + 80 = 100\ \Omega$. From (b), $E_{\text{Th}} = V_2$. Voltage V_2 can be found by the voltage divider rule as

$$\frac{60\ \Omega}{60\ \Omega + 30\ \Omega} \times 300\ \text{V} = 200\ \text{V}$$

(a)   (b)

Figure 14–24

The Thévenin equivalent circuit is shown in Figure 14–25. $\tau = L/R_{\text{Th}} = 50$ ms. Thus,

$$i_L = \frac{E_{\text{Th}}}{R_{\text{Th}}}\,(1 - e^{-t/\tau}) = \frac{200}{100}\,(1 - e^{-t/50\ \text{ms}}) = 2\,(1 - e^{-20t})\quad \text{A}$$

$$v_L = E_{\text{Th}}e^{-t/\tau} = 200\,e^{-20t}\quad \text{V}$$

In both equations, t is measured from the instant the switch is closed. Current rises to a steady state value of 2 A.

(a) Thévenin equivalent
of Figure 14–23

(b)

Figure 14–25 Circuit and current for the buildup phase.

b. The current has an initial value of 2 A when the switch is
opened, as shown in Figure 14–25(b). It then decays to zero
through a resistance of $60 + 80 = 140\ \Omega$ as shown in Figure
14–26. Thus, $\tau' = 5\text{H}/140\ \Omega = 35.7$ ms. If $t = 0$ s is redefined as
the instant the switch is opened, the equation for the decay
current is

$$i_L = I_0 e^{-t/\tau'} = 2e^{-t/35.7\text{ms}} = 2e^{-28t}\quad\text{A}$$

(a) Decay circuit

(b) As it looks at $t = 0^+$.
KVL yields $v_L = -280$ V

Figure 14–26 The circuit of Figure 14–23 as it looks during the decay phase.

Now consider voltage. As indicated in Figure 14–26(b), the voltage
across L just after the switch is open is $V_0 = -280$ V. Thus

$$v_L = V_0 e^{-t/\tau'} = -280e^{-28t}\quad\text{V}$$

c. The waveforms are shown in Figure 14–27. The decay transient
is shorter than the buildup transient because the circuit
resistance is larger and hence the decay time constant is smaller.

Figure 14–27

The basic principles that we have developed in this chapter permit us to solve problems that do not correspond exactly to the circuits of Figure 14–7 and 14–19. This is illustrated in the following example. Note that the decay circuit of Figure 14–28(b) differs slightly from the decay circuit of Figure 14–19, but the same principles apply.

■ **EXAMPLE 14–10**   The circuit of Figure 14–28(a) is in steady state with the switch open. At $t = 0$ s, the switch is closed.

a. Sketch the circuit as it looks after the switch is closed and determine τ'.
b. Determine current i_L at $t = 0^+$ s.
c. Determine the expression for i_L.
d. Determine v_L at $t = 0^+$ s.
e. Determine the expression for v_L.
f. How long does the transient last?
g. Sketch i_L and v_L.

(a) Steady state current with the
switch open is $\dfrac{100\text{ V}}{50\ \Omega} = 2$ A

(b) Decay circuit
$\tau' = \dfrac{L}{R_2} = 2.5$ ms

Figure 14–28

Solution

a. When you close the switch, you short out E and R_1, leaving the decay circuit of (b). Thus $\tau' = L/R_2 = 100\text{ mH}/40\ \Omega = 2.5$ ms.

b. In steady state with the switch open, $i_L = I_0 = 100/50 = 2$ A. This is the current just before the switch is closed. Therefore, just after the switch is closed, i_L will still be 2 A.

c. i_L decays from 2 A to 0. From Equation 14–10, $i_L = I_0 e^{-t/\tau'} = 2e^{-t/2.5\text{ ms}} = 2e^{-400t}$ A.

d. KVL yields $v_L = -v_{R_2} = -R_2 I_0 = -(40\ \Omega)\,(2\text{A}) = -80$ V. Thus, $V_0 = -80$ V.

e. v_L decays from -80 V to 0. Thus, $v_L = V_0 e^{-t/\tau'} = -80e^{-400t}$ V.

f. Transients last $5\tau' = 5(2.5\text{ ms}) = 12.5$ ms.

g. See Figure 14–29.

Figure 14–29

14.6 *RL* Transients Using PSpice

A few new PSpice ideas are needed. First, consider inductors. Inductors are specified in PSpice in a manner similar to capacitance. To specify a 40-mH inductor designated L3 between nodes 4 and 5 with an initial current of 4 mA, for example, use the statement

$$\text{L3}\qquad 4\qquad 5\qquad 40\text{mH}\qquad \text{IC} = 4\text{mA}$$

where IC stands for *Initial Condition*. When the initial current is zero, omit the "IC = " entry.

An Example

Consider Figure 14–30. The switch is closed at $t = 0$. The effect of closing the switch is to apply a 40-V step across R and L. The easiest way to do this is to use a dc source and a .TRAN statement with the UIC (use initial

Figure 14-30   PSpice example.

condition) keyword as we did in Chapter 11. This causes the dc source to be applied as a step voltage rather than as a steady state dc voltage. The *Input File* is shown below and waveforms are shown in Figure 14–31.

```
Input File
* Figure 14-30: RL Circuit
Vs 1 0 40V
R  1 2 80
L  2 0 8H
.TRAN 25ms 500ms UIC
.PROBE
.END
```

Figure 14–31   Inductor voltage and current for the circuit of Figure 14–30.

Another Example

Figure 14–32 shows another example. The switch is closed at $t = 0$ s, then opened 35 μs later. To model this, we can use a PSpice voltage controlled switch (VSWITCH), Appendix A, Table A-2, controlled by a pulse that closes it for 35 μs, then opens and leaves it open. To use the switch, specify its location in the circuit and the node voltage that controls it. In this example, the switch is located between nodes 1 and 2 and its opening and closing is controlled by a pulse source between nodes 1 and 0. When the pulse is on, the switch is closed, applying 100 V to the circuit; when the pulse is off, the switch is open. To define the switch, use

```
S1    1    2    1    0    SW1
.MODEL SW1 VSWITCH
```

Figure 14–32   PSpice example.

Here, S1 designates the switch and the .MODEL statement defines that it is a voltage-controlled switch. (The designation "SW1" links S1 to its model definition.) The complete input file is shown below. Waveforms are shown in Figure 14–33. (If you have forgotten how PULSE works, review Section 11.8.)

```
Input File
* Figure 14-32: Transient Analysis
Vs 1 0 Pulse(0 100V 0 1ns 1ns 35us)
S1 1 2 1 0  SW1
R1 2 0 10kohm
R2 2 3 5kohm
L  3 0 10mH
.Model SW1 VSWITCH
.TRAN 4us 50us
.PROBE
.END
```

Figure 14–33   Inductor voltage and current for the circuit of Figure 14–32.

Problems

14.1   Introduction

1. Consider Figure 14–34. How much current does the source supply at the instant the switch is closed?

Figure 14–34

2. Repeat Problem 1 if L_2 is replaced with an uncharged capacitor.

3. Repeat Problem 1 if L_2 remains in the circuit but L_1 is replaced with an uncharged capacitor.

14.2   Current Buildup Transients

4. a. If $i_L = 8(1-e^{-500t})$ A, what is the current at $t = 6$ ms?
   b. If $v_L = 125e^{-500t}$ V, what is the voltage v_L at $t = 5$ ms?

5. The switch of Figure 14–35 is closed at $t = 0$ s.
   a. How long is it until current reaches its steady value?
   b. Determine the equations for i_L and v_L.
   c. Compute values for i_L and v_L at intervals of one time constant from $t = 0$ to $5\,\tau$.
   d. Sketch i_L and v_L. Label the axis in τ and in seconds.

6. Repeat Problem 5 for the circuit of Figure 14–36 with $L = 4$ H.

Figure 14–35

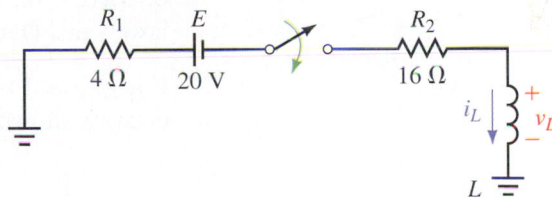

Figure 14–36

7. For the circuit of Figure 14–1(b), the voltage across the inductance at the instant the switch is closed is 80 V, the final steady state current is 4 A, and the transient lasts 0.5 s. Determine E, R, and L.

8. For an RL circuit, $i_L = 20(1-e^{-t/\tau})$ mA and $v_L = 40e^{-t/\tau}$ V. If the transient lasts 0.625 ms, what are E, R, and L?

9. For Figure 14–1(b), if $v_L = 40e^{-2000t}$ V and the steady state current is 10 mA, what are E, R, and L?

14.4 De-Energizing Transients

10. For Figure 14–37, $E = 80$ V, $R_1 = 200$ Ω, $R_2 = 300$ Ω, and $L = 0.5$ H.

   a. When the switch is closed, how long does it take for i_L to reach steady state?

   b. When the switch is opened, how long does it take for i_L to reach steady state?

   c. After the circuit has reached steady state with the switch closed, it is opened. Determine equations for i_L and v_L.

Figure 14–37

11. For Figure 14–37, $R_1 = 20$ Ω, $R_2 = 230$ Ω, and $L = 0.5$ H, and the inductor current has reached a steady value of 5 A with the switch closed. At $t = 0$ s, the switch is opened.

   a. What is the decay time constant?

   b. Determine equations for i_L and v_L.

   c. Compute values for i_L and v_L at intervals of one time constant from $t = 0$ to 5 τ.

   d. Sketch i_L and v_L. Label the axis in τ and in seconds.

12. For Figure 14–37, the inductor voltage at the instant the switch is closed is 150 V and $i_L = 0$ A. After the circuit has reached steady state, the switch is opened. At the instant the switch is opened, $i_L = 3$ A and v_L jumps to -750 V. The decay transient lasts 5 ms. Determine E, R_1, R_2, and L.

13. For Figure 14–37, $L = 20$ H. The current during buildup and decay is shown in Figure 14–38. Determine R_1 and R_2.

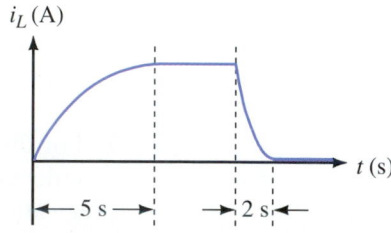

Figure 14–38

14. For Figure 14–19, when the switch is moved to energization, $i_L = 2(1 - e^{-10t})$. Now open the switch and redefine $t = 0$ s as the instant the switch is opened. For this case, $v_L = -400\ e^{-25t}$. Determine E, R_1, R_2, and L.

14.5   More Complex Circuits

15. Refer to Figure 14–39:

   a. What is the energizing circuit time constant?

   b. Close the switch and determine the equation for i_L and v_L during current buildup.

   c. What is the voltage across the inductor and the current through it at $t = 20$ μs?

$$L = 0.36 \text{ H}$$

Figure 14–39

16. For Figure 14–39, the circuit has reached steady state with the switch closed. Now suppose the switch is opened:

   a. Determine the de-energizing circuit time constant.

   b. Determine the equations for i_L and v_L.

   c. What is the voltage across the inductor and current through it at $t = 20$ μs?

17. a. Repeat Problem 15, Parts (a) and (b) for the circuit of Figure 14–40.

   b. What are i_L and v_L at $t = 25$ ms?

Figure 14–40

18. Repeat Problem 16 for the circuit of Figure 11–40, except find v_L and i_L at $t = 13.8$ ms.

19. An unknown circuit containing dc sources and resistors has an open-circuit voltage of 45 volts. When its output terminals are shorted, the short-circuit current is 0.15 A. A switch, resistor, and inductance are connected (Figure 14–41). Determine the inductor current and voltage 2.5 ms after the switch is closed.

Figure 14–41

14.6   *RL* Transients Using PSpice

20. We solved the circuit of Figure 14–21(a) by reducing it to its Thévenin equivalent. Using PSpice, analyze the circuit in its original form and plot the inductor current. Check a few points on the curve by computing values according to the solution of Example 14–7 and compare to values obtained from the .PROBE plot using the screen cursor.

21. Use PSpice to solve for v_L and i_L for the circuit of Figure 14–36. Use the screen cursor to obtain a few values from the .PROBE plot and compare them to the solution of Problem 6.

22. For the circuit of Figure 14–39, the switch is closed for 100 μs, then opened and left open. Use PSpice to compute and plot i_L and v_L. Check a few points on the curves against the solutions of Problems 15 and 16 and compare to values obtained from the .PROBE plot using the screen cursor.

PART IV

Foundation ac Concepts

Although we have concentrated to this point on dc, the majority of the world's electrical energy is produced as ac (alternating current). In ac systems, voltages alternate in polarity and currents alternate in direction, rather than being fixed and unchanging as with dc systems.

Alternating current is important to us for a number of reasons. All commercial electrical power systems are ac. The signals transmitted by radio and TV stations arc ac and many of the signals internal to electronic products are ac. In fact, nearly every electrical and electronic device that we use in our daily lives operates from or involves the use of ac in some way.

We begin Part IV of this book with a look at fundamental ac concepts. We examine ways to generate ac voltages, methods used to represent ac voltages and currents, relationships between ac quantities in resistive, inductive, and capacitive circuits, and, finally, the meaning and representation of power in ac systems. This sets the stage for succeeding chapters which deal with ac circuit analysis techniques, including ac versions of the various methods that you have used for dc circuits throughout previous chapters of this book.

15

AC Fundamentals

CHAPTER PREVIEW

In ac systems, voltage polarities and current directions are not fixed as they are in dc systems; instead, voltage polarities continually cycle between positive and negative, and currents continually alternate in direction, passing first one way, then the other through a circuit. Such currents are called *alternating currents* or *ac*. By convention, alternating currents are called *ac currents* and alternating voltages are called *ac voltages*.

The variation of an ac voltage or current versus time is called its *waveform*. Since waveforms vary with time, they are designated by lowercase letters $v(t)$, $i(t)$, $e(t)$, and so on, rather than by uppercase letters V, I, and E as for dc. Often we drop the functional notation and simply use v, i, and e.

While many waveforms are important to us, the most fundamental is the *sine wave* (also called *sinusoidal ac*). In fact, the sine wave is of such importance that many people associate the term ac with sinusoidal, even though ac refers to any quantity that alternates with time.

In this chapter, we look at basic ac principles, including the generation of ac voltages and ways to represent and manipulate ac quantities. These ideas are then used throughout the remainder of the book to develop methods of analysis for ac circuits.

OBJECTIVES

After studying this chapter, you will be able to

- explain how ac voltages and currents differ from dc,

- draw waveforms for ac voltage and currents and explain what they mean,

- explain the voltage polarity and current direction conventions used for ac,

- describe the basic ac generator and explain how ac voltage is generated,

- define and compute frequency, period, amplitude, and peak-to-peak values,

- compute instantaneous sinusoidal voltage or current at any instant in time,

- define the relationships between ω, T, and f for a sine wave,

- define and compute phase differences between waveforms,

- use phasors to represent sinusoidal voltages and currents,

- determine phase relationships between waveforms using phasors,

- define and compute average values for time-varying waveforms,

- define and compute effective values for time-varying waveforms.

15.1 Introduction

So far we have considered only dc circuits. In dc circuits, voltage polarities and current directions do not change. For example, the top end of the battery of Figure 15–1(a) is always positive and current is always in the clockwise direction. In contrast, ac voltage polarities and current directions constantly vary with time.

(a)

(b) Voltage and current versus time

Figure 15–1   In a dc circuit, voltage polarities and current directions do not change.

Sinusoidal AC Voltage

To illustrate, consider the voltage at the wall outlet in your home. Called a *sine wave* or *sinusoidal ac waveform* (for reasons discussed in Section 15.5), this voltage has the shape shown in Figure 15–2. One complete variation is referred to as a *cycle*. During the first half-cycle, the voltage is positive, while during the second half-cycle, it is negative. Sinusoidal ac consists of a continuous stream of such cycles as illustrated in (b). Since the waveform repeats itself at regular intervals, it is called a *periodic* waveform.

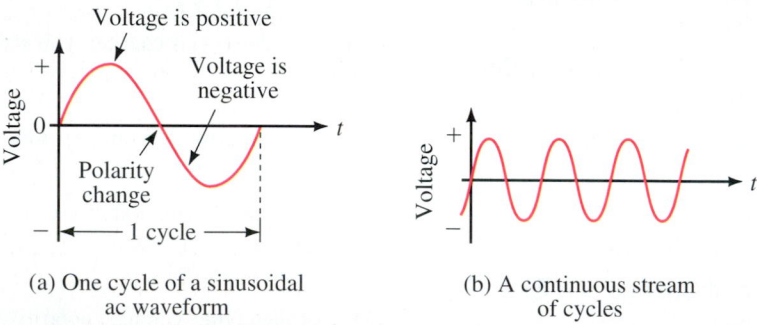

(a) One cycle of a sinusoidal ac waveform

(b) A continuous stream of cycles

Figure 15–2   Sinusoidal ac waveforms. Values above the axis are positive, while values below are negative.

Symbol for an AC Voltage Source

The symbol for a sinusoidal voltage source is shown in Figure 15–3. Note that a lowercase *e* is used to represent voltage rather than *E*, since it is a function of time. Polarity marks are also shown although, since the polarity of the source varies, their meaning has yet to be established.

Figure 15–3   Symbol for a sinusoidal voltage source. Lowercase letter indicates that the voltage varies with time.

Sinusoidal AC Current

Figure 15–4 shows a resistor connected to an ac source. During the first half-cycle, the voltage is positive; therefore, the current is in the clockwise direction. During the second half-cycle, the voltage polarity reverses; therefore, the current is in the counterclockwise direction. Since current is proportional to voltage, its shape is also sinusoidal (Figure 15–5). Note that both voltage and current alternate, voltage in polarity and current in direction.

(a)   (b)

Figure 15–4   Current direction reverses when the source polarity reverses.

Figure 15–5   Current has the same waveshape as voltage.

15.2 Generating ac Voltages

We now look at how ac voltages are generated. Recall from Chapter 13, when a conductor is moved through a magnetic field, a voltage is induced whose magnitude is proportional to the rate at which flux lines are cut and whose polarity depends on the direction in which the conductor is moved through the field. Figure 15–6 shows an ac generator based on this principle. As the coil rotates through the field, voltage is induced in coil sides AA' and BB'. Slip rings connect to the coil and *brushes* bear on these rings to connect the coil to the load.

Coil voltage depends on the coil's position. In (a), the coil is in the 0° position, and at this instant, conductors AA' and BB' are moving parallel to the magnetic field. Since they are cutting no flux, the induced voltage and hence current are zero.

As the coil leaves the 0° position, conductors AA' and BB' start to cut across flux lines and voltage starts to build. Voltage reaches a maximum when the coil arrives at its 90° position (b), at which time sides AA' and BB' cut perpendicularly across the field. Note the polarity of the voltage. Using the principles of Chapter 13, it can be shown that ring A is positive with respect to ring B. Thus, current has the direction indicated.

Rotation

(a) 0° Position: Coil sides move parallel to flux lines. Since no flux is being cut, induced voltage is zero.

(b) 90° Position: Coil end *A* is positive with respect to *B*. Current direction is out of slip ring *A*.

(c) 180° Position: Coil again cutting no flux. Induced voltage is zero.

(d) 270° Position: Voltage polarity reverses, therefore, current direction reverses.

Figure 15–6 Generation of an ac voltage. The 0° position of the coil is defined as in (a) where the coil sides move parallel to the flux lines.

As the coil passes 90°, the voltage starts to decrease, reaching zero when the coil arrives at its 180° position (c). Since the induced voltage is zero, the current is also zero. At this point, the coil has gone through half a revolution.

As the coil passes through 180° and begins its second half revolution, the positions of conductors *AA'* and *BB'* are interchanged relative to their positions at 0°; thus, the polarity of the induced voltage reverses. The voltage now increases, reaching its maximum negative value at 270° as indicated in (d). Since *A* is now negative, the current is opposite in direction to what it was during the first half-cycle.

As the coil passes 270°, the voltage begins to decrease, reaching zero at 360°. Since this is the same as the 0° position (a), the cycle starts over again.

Figure 15–7 shows the waveform produced by this generator. (Although we won't prove it, this waveform is sinusoidal.) Since the coil rotates continuously, it is a repetitive, periodic waveform.

Rotation

(a) Coil Position

(b)

Figure 15–7   Simplified representation of Figure 15–6 showing generation of ac voltage. (Adapted from Meade, *Foundations of Electronics,* © 1991, Delmar Publishers Inc., Albany, New York)

practical

notes

*I*n practice, the coil of Figure 15–6 consists of many turns wound on an iron core. The coil, core, and slip rings rotate as a unit.

In Figure 15–6, the magnetic field is fixed and the coil rotates. While small generators are built this way, large ac generators usually have the opposite construction, that is, their coils are fixed and the magnetic field is rotated instead. In addition, large ac generators are usually made as three-phase machines with three sets of coils instead of one. This is covered in Chapter 23. However, although its details are oversimplified, the generator of Figure 15–6 gives a true picture of the voltage produced by a real ac generator.

Time Scales

The horizontal axis of Figure 15–7 is scaled in degrees. Often we need it scaled in time. The length of time required to generate one cycle depends on the velocity of rotation. To illustrate, assume that the coil rotates at 600 rpm (revolutions per minute). Six hundred revolutions in one minute equals 600 rev/60 s = 10 revolutions in one second. At ten revolutions per second, the time for one revolution is one tenth of a second, i.e., 100 ms. Since one cycle is 100 ms, a half-cycle is 50 ms, a quarter-cycle is 25 ms, and so on. Figure 15–8 shows the waveform rescaled in time.

Instantaneous Value

As Figure 15–8 shows, the coil voltage changes from instant to instant. The value of voltage at any point on the waveform is referred to as its *instantaneous value.* This is illustrated in Figure 15–9. Figure 15–9(a) shows

Figure 15–8   Cycle scaled in time. At 600 rpm, cycle length is 100 ms.

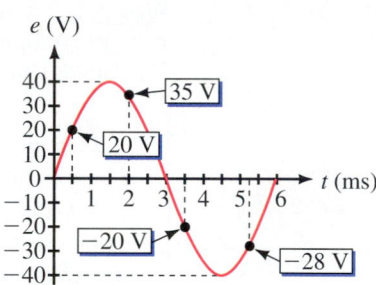

(a) Sinusoidal voltage

(b) Values scaled from the photograph

Figure 15–9   Instantaneous values.

a photograph of an actual waveform, and (b) shows it redrawn, with values scaled from the photo. Note that the voltage has a peak value of 100 volts and a cycle time of 6 ms. From the graph, we see that at $t = 0$ ms, the voltage is zero. At $t = 0.5$ ms, it is 20 V. At $t = 2$ ms, it is 35 V. At $t = 3.5$ ms, it is −20 V, and so on.

Electronic Signal Generators

Alternating current waveforms may also be created electronically using signal generators. In fact, with signal generators, you are not limited to sinusoidal ac. The general-purpose lab signal generator of Figure 15–10, for example, can produce a variety of variable-frequency waveforms, including sinusoidal, square wave, triangular, and so on. Waveforms such as these are commonly used to test electronic gear. You will use a signal generator similar to this in many of your electronics lab courses.

(a) A typical signal generator

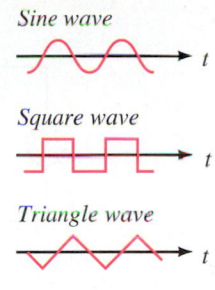

(b) Typical waveforms

Figure 15–10   Electronic signal generators produce variable-frequency waveforms of different shapes.

15.3 Voltage and Current Conventions for ac

In Section 15.1, we looked briefly at voltage polarities and current directions. At that time, we used separate diagrams for each half-cycle (Figure 15–4). However, this is unnecessary; one diagram and one set of references is all that is required. This is illustrated in Figure 15–11. First, we choose reference polarities for the source and a reference direction for the current. We then we use the convention that, *when e has a positive value, its actual polarity is the same as the reference polarity, and when e has a negative value, its actual polarity is opposite to that of the reference.* For current, we use the convention that *when i has a positive value, its actual direction is the same as the reference arrow, and when i has a negative value, its actual direction is opposite to that of the reference.*

To illustrate, consider Figure 15–12. At time t_1, e has a value of 10 volts. This means that at this instant, the voltage of the source is 10 V and its top end is positive with respect to its bottom end. This is indicated in (b). With a voltage of 10 V and a resistance of 5 Ω, the instantaneous value of current is $i = e/R = 10$ V/5 Ω = 2 A. Since i is positive, the current is in the direction of the reference arrow.

Now consider time t_2. Here, $e = -10$ V. This means that source voltage is again 10 V, but now its top end is negative with respect to its

(a) Reference for voltage
and current.

(b) During the first half-cycle,
voltage polarity and current
direction are as shown in (a).
During the second half-cycle,
voltage polarity and current
direction are opposite to that
shown in (a). This is why we
designate them as negative.

Figure 15–11   Voltage polarity and current references for ac.

(a)

(b) Voltage and current at
time t_1. $e = 10$ V and $i = 2$ A.
Thus i is in the direction of
the reference arrow.

(c) Voltage and current at
time t_2. $e = -10$ V and $i = -2$ A.
Thus i is opposite in direction to
the reference arrow.

Figure 15–12   Illustrating ac voltage and current convention.

bottom end. Again applying Ohm's law, you get $i = e/R = -10$ V/5 $\Omega = -2$ A. Since i is negative, current is actually opposite in direction to the reference arrow. This is indicated in (c).

The above concept is valid for any ac signal, regardless of waveshape.

■ **EXAMPLE 15–1**   Figure 15–13 shows one cycle of a triangular voltage wave. Determine the current and its direction at $t = 0, 1, 2, 3, 4, 5, 6, 7, 8, 9, 10, 11,$ and 12 μs.

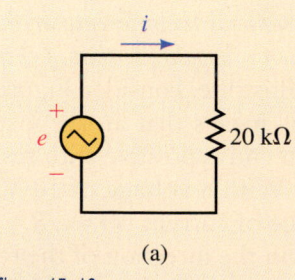

(a)

(b)

Figure 15–13

Solution Apply Ohm's law at each point in time. At $t = 0$ μs, $e = 0$ V, so $i = e/R = 0$ V/20 kΩ = 0 mA. At $t = 1$ μs, $e = 30$ V. Thus, $i = e/R = 30$ V/20 kΩ = 1.5 mA. At $t = 2$ μs, $e = 60$ V. Thus, $i = e/R = 60$ V/20 kΩ = 3 mA. Continuing in this manner, you get the values shown in Table 15–1.

TABLE 15–1 Values for Example 15–1

t (μS)	e (V)	i (mA)	DIRECTION
0	0	0	
1	30	1.5	
2	60	3.0	
3	90	4.5	→
4	60	3.0	
5	30	1.5	
6	0	0	
7	–30	–1.5	
8	–60	–3.0	
9	–90	–4.5	←
10	–60	–3.0	
11	–30	–1.5	
12	0	0	

practice problems 1

1. Let the source voltage of Figure 15–11 be the waveform of Figure 15–9. If $R = 2.5$ kΩ, determine the current at $t = 0, 0.5, 1, 1.5, 3, 4.5,$ and 5.25 ms.

2. For Figure 15–13, if $R = 180$ Ω, determine the current at $t = 1.5, 3, 7.5,$ and 9 μs.

15.4 Frequency, Period, Amplitude, and Peak Value

Periodic waveforms, regardless of their waveshape, may be described by a group of attributes such as frequency, period, amplitude, peak value, and so on.

Frequency

The number of cycles per second of a waveform is defined as its *frequency*. In Figure 15–14(a), one cycle occurs in one second, in (b), two cycles occur in one second, and in (c), sixty cycles occur in one second. Thus the frequency of (a) is one cycle per second, that of (b) is two cycles per second, and that of (c) is sixty cycles per second. Frequency is denoted by the lowercase letter *f*. In the SI system, its unit is the *hertz* (Hz, named in honor of pioneer researcher Heinrich Hertz, 1857–1894). By definition,

$$1 \text{ Hz} = 1 \text{ cycle per second} \qquad (15–1)$$

Thus, the examples depicted in Figure 15–14 represent 1 Hz, 2 Hz, and 60 Hz respectively.

(a) 1 cycle per second = 1 Hz

(b) 2 cycles per second = 2 Hz

(c) 60 cycles per second = 60 Hz

Figure 15–14   Concept of frequency.

The range of frequencies is immense. Power line frquencies, for example, are 60 Hz in North America and 50 Hz in many other parts of the world. Audible sound frequencies range from about 20 Hz to about 20 kHz. Radiowave frequencies extend from 3 kHz (the VLF or very low frequency end) to 300 GHz (the EHF or extremely high frequency end). Within this range, the standard AM radio band occupies from 550 kHz to 1.6 MHz, while the FM band extends from 88 MHz to 108 MHz. TV transmissions occupy several bands in the 54-MHz to 890-MHz range. Above 300 GHz are optical and X-ray frequencies.

Period

The *period*, *T*, of a waveform, (Figure 15–15) is the duration of one cycle. It is the inverse of frequency. To illustrate, consider again Figure 15–14. In (a), the frequency is 1 cycle per second; thus, the duration of each cycle is $T = 1$ s. In (b), the frequency is two cycles per second; thus, the duration of each cycle is $T = 1/2$ s, and so on. In general,

$$T = \frac{1}{f} \quad \text{(s)} \tag{15–2}$$

and

$$f = \frac{1}{T} \quad \text{(Hz)} \tag{15–3}$$

Figure 15–15   Definition of period. *T* has units of seconds.

■ EXAMPLE 15–2

a. What is the period of a 50-Hz voltage?

b. What is the period of a 1-MHz current?

Solution

(a) $T = \dfrac{1}{f} = \dfrac{1}{50 \text{ Hz}} = 20 \text{ ms}$

(b) $T = \dfrac{1}{f} = \dfrac{1}{1 \times 10^6 \text{ Hz}} = 1 \text{ }\mu\text{s}$

■ **EXAMPLE 15–3** Figure 15–16 shows an oscilloscope trace of a square wave. Each horizontal division represents 50 µs. Determine the frequency.

Figure 15–16 The concepts of frequency and period apply to nonsinusoidal waveforms.

Solution Since the wave repeats itself every 200 µs, its period is 200 µs and

$$f = \frac{1}{200 \times 10^{-6} \text{ s}} = 5 \text{ kHz}$$

The period of a waveform can be measured between any two corresponding points (Figure 15–17). Often it is measured between zero points because they are easy to establish on an oscilloscope trace.

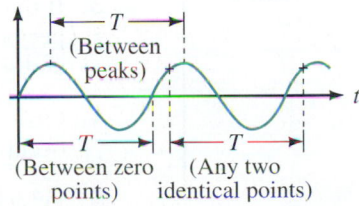

Figure 15–17 Period may be measured between any two corresponding points.

■ **EXAMPLE 15–4**   Determine the period and frequency of the waveform of Figure 15–18.

Figure 15–18

Solution   Time interval T_1 does not represent a period as it is not measured between corresponding points. Interval T_2, however, is. Thus, $T = 10$ ms and

$$f = \frac{1}{T} = \frac{1}{10 \times 10^{-3}\,\text{s}} = 100\,\text{Hz}$$

(a)

(b)

Figure 15–19   Definitions.

Peak Amplitude and Peak-to-Peak Value

The *peak amplitude* of a sine wave is the distance from its average to its peak. Thus, the *peak* amplitude of the voltage in Figures 15–19(a) and (b) is E_m.

   Peak-to-peak voltage is also indicated in Figure 15–19(a). It is measured between minimum and maximum peaks. Peak-to-peak voltages are denoted $E_{p\text{-}p}$ or $V_{p\text{-}p}$ in this book. (Some authors use $V_{pk\text{-}pk}$ or the like.) Similarly, peak-to-peak currents are denoted as $I_{p\text{-}p}$. To illustrate, consider again Figure 15–9. The amplitude of this voltage is $E_m = 40$ V, and its peak-to-peak voltage is $E_{p\text{-}p} = 80$ V.

Peak Value

The *peak value* of a voltage or current is its maximum value with respect to zero. Consider Figure 15–19(b). Here, a sine wave rides on top of a dc value, yielding a peak that is the sum of the dc voltage and the ac waveform amplitude. For the case indicated, the peak voltage is $E + E_m$.

in process
l e a r n i n g
check I

1. What is the period of the commercial ac power system voltage in North America?

2. If you double the rotational speed of an ac generator, what happens to the frequency and period of the waveform?

3. If the generator of Figure 15–6 rotates at 3000 rpm, what is the period and frequency of the resulting voltage? Sketch four cycles and scale the horizontal axis in units of time.

4. On Figure 15–9, at what values of time do $e = 10$ V and $e = -35$ V occur on the waveform?

5. Which of the waveform pairs of Figure 15–20 are valid combinations? Why?

(a) Circuit   (b)   (c)   (d)

Figure 15–20   Which waveform pairs are valid?

6. For the waveform in Figure 15–21, determine the frequency.

Figure 15–21

7. Two waveforms have periods of $T_1 = 10$ ms and $T_2 = 30$ ms respectively. Which has the higher frequency? Compute the frequencies of both waveforms.

8. Two sources have frequencies f_1 and f_2 respectively. If $f_2 = 20f_1$, and T_2 is 1 μs, what is f_1?

9. Consider Figure 15–22. What is the frequency of the waveform?

Figure 15–22

10. For Figure 15–11, if $f = 20$ Hz, what is the current direction at $t = 12$ ms, 37 ms, and 60 ms? Hint: Sketch the waveform and scale the horizontal axis in ms. The answers should be apparent.

11. A 10-Hz sinusoidal current has a value of 5 amps at $t = 25$ ms. What is its value at $t = 75$ ms? See hint in Problem 10.

15.5 Angular and Graphic Relationships for Sine Waves

The Basic Sine Wave Equation

Consider again the generator of Figure 15–6, redrawn in end view as Figure 15–23. The voltage produced by this generator is

$$e = E_m \sin \alpha \quad \text{(V)} \tag{15–4}$$

where E_m is the maximum coil voltage and α is the instantaneous angular position of the coil. (For a given generator and rotational velocity, E_m is constant.) Note that $\alpha = 0°$ represents the horizontal position of the coil and that one complete cycle corresponds to 360°. Equation 15–4 states that the voltage at any point on the sine wave may be found by multiplying E_m times the sine of the angle at that point.

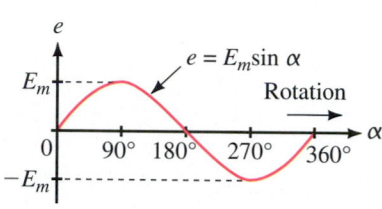

(a) End view of ac generator (b) Voltage waveform

Figure 15–23   Voltage versus angle.

■ **EXAMPLE 15–5**   If the amplitude of the waveform of Figure 15–23(b) is $E_m = 100$ V, determine the coil voltage at 30° and 330°.

Solution   At $\alpha = 30°$, $e = E_m \sin \alpha = 100 \sin 30° = 50$ V. At 330°, $e = 100 \sin 330° = -50$ V. These are shown on the graph of Figure 15–24.

Figure 15–24

practice problem II

Table 15–2 is a tabulation of voltage versus angle computed from $e = 100 \sin \alpha$. Use your calculator to verify each value, then plot the result on graph paper. The resulting waveshape should look like Figure 15–24.

TABLE 15–2 Data for Plotting $e = 100 \sin \alpha$

ANGLE α	VOLTAGE e
0	0
30	50
60	86.6
90	100
120	86.6
150	50
180	0
210	−50
240	−86.6
270	−100
300	−86.6
330	−50
360	0

Angular Velocity, ω

The rate at which the generator coil rotates is called its *angular velocity*. If the coil rotates through an angle of 30° in one second, for example, its angular velocity is 30° per second. Angular velocity is denoted by the Greek letter ω (omega). For the case cited, $\omega = 30°/s$. (Normally angular velocity is expressed in radians per second instead of degrees per second. We will make this change shortly.) When you know the angular velocity of a coil and the length of time that it has rotated, you can compute the angle through which it has turned. For example, a coil rotating at 30°/s rotates through an angle of 30° in one second, 60° in two seconds, 90° in three seconds, and so on. In general,

$$\alpha = \omega t \tag{15–5}$$

Expressions for t and ω can now be found. They are

$$t = \frac{\alpha}{\omega} \quad \text{(s)} \tag{15–6}$$

$$\omega = \frac{\alpha}{t} \tag{15–7}$$

■ **EXAMPLE 15–6**   If the coil of Figure 15–23 rotates at $\omega = 300°/s$, how long does it take to complete one revolution?

Solution   One revolution is 360°. Thus,

$$t = \frac{\alpha}{\omega} = \frac{360 \text{ degrees}}{300 \dfrac{\text{degrees}}{\text{s}}} = 1.2 \text{ s}$$

Since this is one period, we should use the symbol *T*. Thus, *T* = 1.2 s, as in Figure 15–25.

Figure 15–25

*practice
problem
III*

If the coil of Figure 15–23 rotates at 3600 rpm, determine its angular velocity, ω, in degrees per second.

(a) A full circle may be expressed as 360° or 2π radians

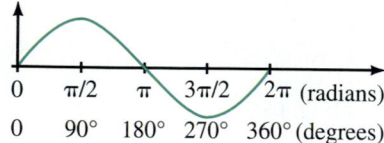

(b) Cycle length is 2π, half-cycle is π, quarter cycle is π/2 radians, etc.

Figure 15–26 Radian measure.

Radian Measure

In practice, ω is usually expressed in radians per second, where radians and degrees are related by the identity

$$2\pi \text{ radians} = 360° \qquad (15\text{–}8)$$

One radian therefore equals $360°/2\pi = 57.296°$. A full circle, as shown in Figure 15–26(a), can be designated as either 360° or 2π radians. Likewise, the cycle length of a sinusoid, shown in Figure 15–26(b), can be stated as either 360° or 2π radians; a half-cycle can be given as π radians, and so on.

To convert from degrees to radians, multiply by $\pi/180$, while to convert from radians to degrees, multiply by $180/\pi$.

$$\alpha_{\text{radians}} = \frac{\pi}{180°} \times \alpha_{\text{degrees}} \qquad (15\text{–}9)$$

$$\alpha_{\text{degrees}} = \frac{180°}{\pi} \times \alpha_{\text{radians}} \qquad (15\text{–}10)$$

Table 15–3 shows selected angles in both measures.

TABLE 15–3 Selected Angles
in Degrees and Radians

DEGREES	RADIANS
30	π/6
45	π/4
60	π/3
90	π/2
180	π
270	3π/2
360	2π

(a)

(b)

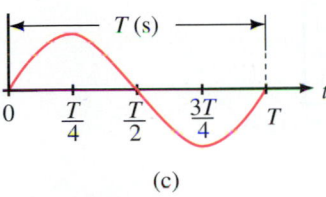

(c)

Figure 15–27   Various horizontal scales. All are independent of frequency.

■ **EXAMPLE 15–7**

   a. Convert 315° to radians.

   b. Convert 5π/4 radians to degrees.

Solution

   a. $\alpha_{radians} = (\pi/180°)(315°) = 5.5$ rad

   b. $\alpha_{degrees} = (180°/\pi)(5\pi/4) = 225°$

Graphing Sine Waves

A sinusoidal waveform can be graphed with its horizontal axis scaled in degrees, radians, or time. When scaled in degrees or radians, one cycle is always 360° or 2π radians (Figure 15–27); when scaled in time, it is frequency dependent, since the length of a cycle depends on the coil's velocity of rotation. However, if scaled in terms of period T instead of in seconds, the waveform is also frequency independent, since one cycle is always T, as shown in Figure 15–27(c).

When graphing a sine wave, you don't actually need many points to get a good sketch: Values every 45° (one eighth of a cycle) are generally adequate. Table 15–4 shows corresponding values for sin α at this spacing. Example 15–8 shows how to use this table.

TABLE 15–4 Values for Rapid Sketching

α (deg)	α (RAD)	t (T)	VALUE OF SIN α
0	0	0	0.0
45	π/4	T/8	0.707
90	π/2	T/4	1.0
135	3π/4	3T/8	0.707
180	π	T/2	0.0
225	5π/4	5T/8	−0.707
270	3π/2	3T/4	−1.0
315	7π/4	7T/8	−0.707
360	2π	T	0.0

■ **EXAMPLE 15–8**   Sketch the waveform for a 25-kHz sinusoidal current that has an amplitude of 4 mA. Scale the axis in seconds.

Solution   The easiest approach is to use $T = 1/f$, then scale the graph accordingly. For this waveform, $T = 1/25$ kHz = 40 μs. Proceed as follows.

   1. Mark the end of the cycle as 40 μs, the half-cycle point as 20 μs, the quarter-cycle point as 10 μs, and so on (Figure 15–28).

   2. The peak value (i.e., 4 mA) occurs at the quarter-cycle point, which is 10 μs on the waveform. Likewise, −4 mA occurs at 30 μs.

3. To find values at other points, you can either relate them to angular measure or use Table 15–3. For example, 5 μs is one eighth of a cycle, or 45°. Thus, $i = 4 \sin 45°$ mA = 2.83 mA. Alternately, from Table 15–3, at $T/8$, $i = (4 \text{ mA})(0.707) = 2.83$ mA. As many points as you need can be computed and plotted in this manner.

4. Values at arbitrary angles can also be located easily if desired. For instance, if you want a value at 30°, the required value is $i = 4 \sin 30°$ mA = 2.0 mA. To locate this point, note that 30° is one twelfth of a cycle or $T/12 = (40 \text{ μs})/12 = 3.33$ μs. The point is shown on Figure 15–28.

Figure 15–28

15.6 Voltages and Currents as Functions of Time

Relationship between ω, T, and f

Earlier you learned that the cycle length of a sine wave is $\alpha = 2\pi$ when measured in radians and $t = T$ when measured in periods (Figure 15–27). You also know that $\alpha = \omega t$. Substituting 2π and T into this equation yields $2\pi = \omega T$. Transposing,

$$\omega T = 2\pi \quad \text{(rad)} \qquad (15\text{–}11)$$

Thus,

$$\omega = \frac{2\pi}{T} \quad \text{(rad/s)} \qquad (15\text{–}12)$$

Recall, $f = 1/T$ Hz. Substituting this into Equation 15–12 you get

$$\omega = 2\pi f \quad \text{(rad/s)} \qquad (15\text{–}13)$$

■ **EXAMPLE 15–9** In some parts of the world, the power system frequency is 60 Hz; in other parts, it is 50 Hz. Determine ω for each.

Solution For 60 Hz, $\omega = 2\pi f = 2\pi(60) = 377$ rad/s. For 50 Hz, $\omega = 2\pi f = 2\pi(50) = 314.2$ rad/s.

practice problems
IV

1. If $\omega = 240$ rad/s, how many cycles occur in 27 s?

2. If 56 000 cycles occur in 3.5 s, what is ω?

Sinusoidal Voltages and Currents as Functions of Time

Recall from Equation 15–4, $e = E_m \sin \alpha$, and from Equation 15–5, $\alpha = \omega t$. Combining these equations yields

$$e = E_m \sin \omega t \qquad (15–14a)$$

Similarly,

$$v = V_m \sin \omega t \qquad (15–14b)$$

$$i = I_m \sin \omega t \qquad (15–14c)$$

■ EXAMPLE 15–10   A 100-Hz sinusoidal voltage source has an amplitude of 150 volts. Write the equation for e.

Solution $\omega = 2\pi f = 2\pi(100) = 628$ rad/s and $E_m = 150$ V. Thus, $e = E_m \sin \omega t = 150 \sin 628t$ V.

Equations 15–14 may be used to compute voltages or currents at any instant in time. Usually, ω is in radians per second, and thus ωt is in radians. You can work directly in radians or you can convert to degrees. For example, suppose you want to know the voltage at $t = 1.25$ ms for $e = 150 \sin 628t$ V.

Working in Rads.   With your calculator in the RAD mode, $e = 150 \sin(628)(1.25 \times 10^{-3}) = 150 \sin 0.785$ rad $= 106$ V.

Working in Degrees.   0.785 rad $= 45°$. Thus, $e = 150 \sin 45° = 106$ V as before.

■ EXAMPLE 15–11   For $v = 170 \sin 2450t$, determine v at $t = 3.65$ ms and show the point on the v waveform.

Solution $\omega = 2450$ rad/s. Therefore $\omega t = (2450)(3.65 \times 10^{-3}) = 8.943$ rad $= 512.4°$. Thus, $v = 170 \sin 512.4° = 78.8$ V. Alternatively, $v = 170 \sin 8.943$ rad $= 78.8$ V. The point is plotted on the waveform in Figure 15–29.

Figure 15–29

A sinusoidal current has a peak amplitude of 10 amps and a period of 120 ms.

a. Determine its equation as a function of time using Equation 15–14c.

b. Using this equation, compute a table of values at 10-ms intervals and plot one cycle of the waveform scaled in seconds.

c. Sketch one cycle of the waveform using the procedure of Example 15–8. (Note how much less work this is.)

Determining when a Particular Value Occurs

Sometimes you need to know when a particular value of voltage or current occurs. Given $v = V_m \sin \alpha$. Rewrite this as $\sin \alpha = v/V_m$. Now compute the angle at which the desired value occurs using the inverse function

$$\alpha = \sin^{-1} \frac{v}{V_m} \qquad (15\text{–}15)$$

then determine the time from

$$t = \alpha/\omega$$

■ **EXAMPLE 15–12**   A sinusoidal current has an amplitude of 10 A and a period of 0.120 s. Determine the times at which

a. $i = 5.0$ A,
b. $i = -5$ A.

Solution

a. Consider Figure 15–30. As you can see, there are two points on the waveform where $i = 5$ A. Let these be denoted t_1 and t_2 respectively. First, determine ω:

$$\omega = \frac{2\pi}{T} = \frac{2\pi}{0.120 \text{ s}} = 52.36 \text{ rad/s}$$

Thus, $i = 10 \sin 52.36t$ A. Now, find the angle α_1 at which $i = 5$ A:

$$\alpha_1 = \sin^{-1} \frac{i}{I_m} = \sin^{-1} \frac{5 \text{ A}}{10 \text{ A}} = \sin^{-1} 0.5 = 30° = 0.5236 \text{ rad}$$

Thus, $t_1 = \alpha_1/\omega = (0.5236 \text{ rad})/(52.36 \text{ rad/s}) = 0.01$ s $= 10$ ms. This is indicated in Figure 15–30. Now consider t_2. Note that t_2 is the same distance back from the half-cycle point as t_1 is in from the beginning of the cycle. Thus, $t_2 = 60$ ms $- 10$ ms $= 50$ ms.

b. Similarly, t_3 (the first point at which $i = -5$ A occurs) is 10 ms past midpoint, while t_4 is 10 ms back from the end of the cycle. Thus, $t_3 = 70$ ms and $t_4 = 110$ ms.

Figure 15–30

practice
problem
VI

Given $v = 10 \sin 52.36t$, determine both occurrences of $v = -8.66$ V.

Voltages and Currents with Phase Shifts

A sine wave that does not pass through zero at $t = 0$ s as in Figure 15–30 has a *phase shift*. Waveforms may be shifted to the left or to the right (see Figure 15–31). For a waveform shifted left as in (a),

$$v = V_m \sin(\omega t + \theta) \tag{15–16a}$$

while, for a waveform shifted right as in (b),

$$v = V_m \sin(\omega t - \theta) \tag{15–16b}$$

(a) $v = V_m \sin(\omega t + \theta)$ (b) $v = V_m \sin(\omega t - \theta)$

Figure 15–31   Waveforms with phase shifts. Similar equations hold for currents.

■ **EXAMPLE 15–13**   Demonstrate that $v = 20 \sin(\omega t - 60°)$, where $\omega = \pi/6$ rad/s (i.e, $= 30°/s$), yields the shifted waveform shown in Figure 15–32.

Figure 15–32

Solution

1. Since ωt and 60° are both angles, $(\omega t - 60°)$ is also an angle. Let us define it as β. Then $v = 20 \sin \beta$, which means that the shifted wave is also sinusoidal.

2. Consider $v = \sin(\omega t - 60°)$. At $t = 0$ s, $v = 20 \sin(0 - 60°) = 20 \sin(-60°) = -17.3$ V as indicated in Figure 15–32.

3. Since $\omega = 30°/s$, it takes 2 s for ωt to reach 60°. Thus, at $t = 2$ s, $v = 20 \sin(60° - 60°) = 0$ V, and the waveform passes through zero at $t = 2$ s.

Summary: Since $v = \sin(\omega t - 60°)$ is a sine wave and since it passes through zero at $t = 2$ s, where $\omega t = 60°$, it represents the shifted wave shown in Figure 15–32.

■ EXAMPLE 15–14

a. Determine the equation for the waveform of Figure 15–33(a), given $f = 60$ Hz.

b. Repeat (a) for Figure 15–33(b).

Solution

a. $I_m = 2$ A and $\omega = 2\pi(60) = 377$ rad/s. This waveform corresponds to Figure 15–31(b). Therefore,

$$i = 2 \sin(\omega t - \theta) = 2 \sin(377t - 120°) \text{ A}$$

b. This waveform matches neither Figure 15–31(a) nor (b). However, if you extend the waveform back 90° from its peak as in (c), you obtain an intercept of −40° and it now corresponds to Figure 15–31(a). Thus,

$$i = 2 \sin(377t + 40°) \text{ A}$$

(a) (b) (c)

Figure 15–33

In the above examples, ωt is in radians and θ is in degrees. This yields mixed units (which is permissible and common in symbolic notation). However, when performing computations, you must convert both angles to the same units, either degrees or radians.

1. Given $i = 2 \sin(377t + 60°)$, compute the current at $t = 3$ ms by converting $377t + 60°$ to radians. Repeat by converting $377t + 60°$ to degrees.

2. Sketch each of the following:
   a. $v = 10 \sin(\omega t + 20°)$ V.   **b.** $i = 80 \sin(\omega t - 50°)$ A.
   c. $i = 50 \sin(\omega t + 120°)$ A.   **d.** $v = 5 \sin(\omega t + 180°)$ V.

Probably the easiest way to deal with shifted waveforms is to use phasors. We introduce the idea next.

15.7 Introduction to Phasors

A *phasor* is a rotating line whose projection on a vertical axis can be used to represent sinusoidally varying quantities. To get at the idea, consider the red line of length V_m shown in Figure 15–34(a). (It is the phasor.) The vertical projection of this line (indicated in dotted red) is $V_m \sin \alpha$. Now, assume that the phasor rotates at angular velocity of ω rad/s in the counterclockwise direction. Then, $\alpha = \omega t$, and its vertical projection is $V_m \sin \omega t$. If we designate this projection (height) as v, we get $v = V_m \sin \omega t$, which is the familiar sinusoidal voltage equation.

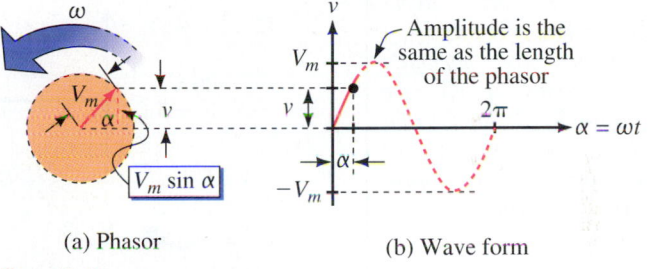

(a) Phasor   (b) Wave form

Figure 15–34   Rotating line and resulting sine wave.

If you plot a graph of v versus α, you get the sine wave of Figure 15–34(b). Figure 15–35 illustrates the process. It shows snapshots of the phasor and the evolving waveform at various instants of time for a phasor of magnitude $V_m = 100$ V rotating at $\omega = 30°/s$. For example, consider $t = 0, 1, 2,$ and 3s:

1. At $t = 0$ s, $\alpha = 0$, the phasor is at its 0° position, and its vertical projection is $v = V_m \sin \omega t = 100 \sin 0° = 0$ V. The point is at the origin.

2. At $t = 1$ s, the phasor has rotated 30° and its vertical projection is $v = 100 \sin 30° = 50$ V. This point is plotted at $\alpha = 30°$ on the horizontal axis.

3. At $t = 2$ s, $\alpha = 60°$ and $v = 100 \sin 60° = 87$ V, which is plotted at $\alpha = 60°$. Similarly, at $t = 3$ s, $\alpha = 90°$, and $v = 100$ V. Continuing in this manner, the complete waveform is evolved.

From the foregoing, we conclude that *a sinusoidal waveform can be created by plotting the vertical projection of a phasor that rotates in the counterclockwise direction at constant angular velocity* ω. *If the phasor has a length of* V_m, *the waveform represents voltage; if the phasor has a length of* I_m, *it represents current.* Note carefully: **Phasors apply only to sinusoidal waveforms.**

practical

notes

1. *Although we have indicated phasor rotation in Figure 15–35 by a series of "snapshots," this is too cumbersome; in practice, we show only the phasor at its $t = 0$ s (reference) position and imply rotation rather than show it explicitly.*

2. *Although we are using maximum values (E_m and I_m) here, phasors are normally drawn in terms of effective values (considered in Section 15.9). For the moment, we will continue to use maximum values. We make the change in Chapter 16.*

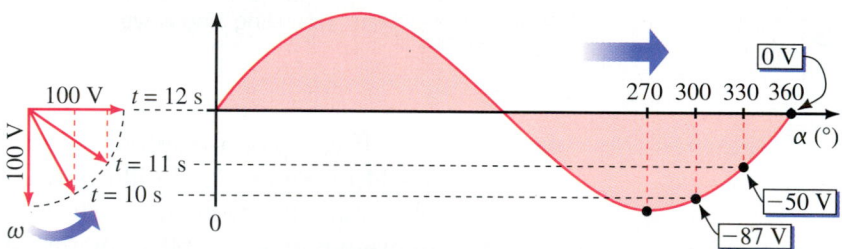

Figure 15–35   Evolution of a sine wave.

■ EXAMPLE 15–15 Draw the phasor and waveform for current $i = 25 \sin \omega t$ mA for $f = 100$ Hz.

Solution The phasor has a length of 25 mA and is drawn at zero degrees as indicated in Figure 15–36. Since $f = 100$ Hz, the period is $T = 1/f = 10$ ms.

Figure 15–36

Shifted Sine Waves

Phasors may be used to represent shifted waveforms, $v = V_m\sin(\omega t \pm \theta)$ or $i = I_m\sin(\omega t \pm \theta)$ as indicated in Figure 15–37. Angle θ is the position of the phasor at $t = 0$ s.

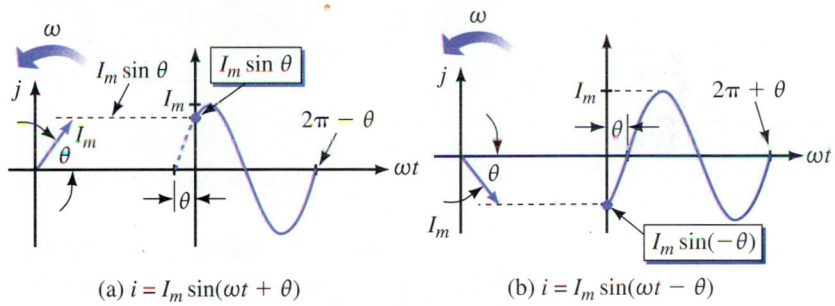

(a) $i = I_m \sin(\omega t + \theta)$ (b) $i = I_m \sin(\omega t - \theta)$

Figure 15–37   Representing shifted waveforms by phasors.

■ **EXAMPLE 15–16**   Consider $v = 20 \sin(\omega t - 60°)$, where $\omega = \pi/6$ rad/s (i.e., 30°/s). Show that the phasor of Figure 15–38(a) represents this waveform.

(a) Phasor   (b) $v = 20 \sin(\omega t - 60°)$, $\omega = 30°/s$

Figure 15–38

Solution   The phasor has length 20 V and at time $t = 0$ is at $-60°$ as indicated in (a). Using simple geometry, its value at $t = 0$ s can be seen to be $20 \sin(-60°)$. Now, as the phasor rotates, it generates a sinusoidal waveform, oscillating between ±20 V as indicated in (b). Note that the zero crossover point occurs at $t = 2$ s, since it takes 2 seconds for the phasor to rotate from $-60°$ to $0°$ at 30 degrees per second. Now compare the waveform of (b) to the waveform of Figure 15–32, Example 15–13. They are identical. Thus, the phasor of (a) represents the shifted waveform $v = 20 \sin(\omega t - 60°)$.

■ **EXAMPLE 15–17**   With the aid of a phasor, sketch the waveform for $v = V_m \sin(\omega t + 90°)$.

Solution   Place the phasor at $90°$ as in Figure 15–39(a). Note that the resultant waveform (b) is a cosine waveform, i.e., $v = V_m \cos \omega t$. From this, we conclude that

$$\sin(\omega t + 90°) = \cos \omega t$$

(a) Phasor at $90°$ position   (b) Waveform can also be described as a cosine wave

Figure 15–39   Demonstrating that $\sin(\omega t + 90°) = \cos \omega t$.

With the aid of phasors, show that

a. $\sin(\omega t - 90°) = -\cos \omega t$,

b. $\sin(\omega t \pm 180°) = -\sin \omega t$,

c. $-\sin(\omega t + 45°) = \sin(\omega t - 135°)$

Phase Difference

Phase difference refers to the angular displacement between different waveforms of the same frequency. Consider Figure 15–40. If the angular displacement is $0°$ as in (a), the waveforms are said to be *in phase*; otherwise, they are *out of phase*. When describing a phase difference, select one waveform as reference. Other waveforms then lead, lag, or are in phase with this reference. For example, in (b), for reasons to be discussed in the next paragraph, the current waveform is said to lead the voltage waveform, while in (c) the current waveform is said to lag.

The terms *lead* and *lag* can be understood in terms of phasors. If you observe phasors rotating as in Figure 15–41(a), the one that you see passing first is leading and the other is lagging. By definition, *the waveform generated by the leading phasor leads the waveform generated by the lagging phasor and vice versa*. In Figure 15–41, phasor I_m leads phasor V_m; thus current $i(t)$ leads voltage $v(t)$.

(a) In phase   (b) Current leads   (c) Current lags

Figure 15–40   Illustrating the phase concept. In these examples, voltage is selected as reference.

(a) I_m leads V_m (b) Therefore, $i(t)$ leads $v(t)$

Figure 15–41   Defining lead and lag.

practical

notes

*I*f you have trouble determining which waveform leads and which lags, make a quick sketch of their phasors, and the answer will be apparent. Note also that the terms lead and lag are relative. In Figure 15–41, we said that current leads voltage; you can just as correctly say that voltage lags current.

■ **EXAMPLE 15–18**   Voltage and current are out of phase by 40°, and voltage lags. Using current as the reference, sketch the phasor diagram and the corresponding waveforms.

Solution   Since current is the reference, place its phasor in the 0° position and the voltage phasor at –40°. Figure 15–42 shows the phasors and corresponding waveforms.

Figure 15–42

■ **EXAMPLE 15–19**   Given $v = 20 \sin(\omega t + 30°)$ and $i = 18 \sin(\omega t - 40°)$, draw the phasor diagram, determine phase relationships, and sketch the waveforms.

Solution The phasors are shown in Figure 15–43(a). From these, you can see that v leads i by 70°. The waveforms are shown in (b).

(a)

(b)

Figure 15–43

■ **EXAMPLE 15–20** Voltage and current waveforms as viewed on an oscilloscope are shown in Figure 15–44. Each major division on the horizontal scale is 20 μs and on the vertical scale is 20 V for voltage and 5 A for current. Voltage leads. Prepare a phasor diagram and, using current as the reference, determine equations for voltage and current.

(a)

(b)

Figure 15–44

Solution From the photograph, $T = 120$ μs and the waveform displacement is 20 μs. Now, 20 μs represents $(20/120) = 1/6$ of a cycle, or 60°. The voltage and current amplitudes, as scaled from the photograph, are 60 V and 10 A respectively. Selecting current as the reference, phasors are drawn as in (b), with voltage leading by 60°. Angular frequency ω is $2\pi/T = 2\pi/120 \times 10^{-6}$s $= 52.36 \times 10^{3}$ rad/s. Thus, $i = 10 \sin 52.36 \times 10^{3}t$ A, and $v = 80 \sin(52.36 \times 10^{3}t + 60°)$ V.

Sometimes voltages and currents are expressed in terms of cos ωt rather than sin ωt. As Example 15–17 shows, a cosine wave is a sine wave shifted by +90°, or alternatively, a sine wave is a cosine wave shifted by −90°. To convert from cosine to sine, replace "cos" with "sin," and add 90° to the phase angle; to convert from sine to cosine, replace "sin" with "cos" and subtract 90° from the phase angle. Thus,

$$\cos(\omega t + \theta) = \sin(\omega t + \theta + 90°) \tag{15–17a}$$

$$\sin(\omega t + \theta) = \cos(\omega t + \theta - 90°) \tag{15–17b}$$

To illustrate, consider $\cos(\omega t + 30°)$. From Equation 15–17a, $\cos(\omega t + 30°) = \sin(\omega t + 30° + 90°) = \sin(\omega t + 120°)$. Figure 15–45 illustrates this relationship graphically. The red phasor in (a) generates cos ωt as was shown in Example 15–17. Therefore, the green phasor generates a waveform that leads it by 30°, namely $\cos(\omega t + 30°)$. For (b), the red phasor generates sin ωt, and the green phasor generates a waveform that leads it by 120°, i.e., $\sin(\omega t + 120°)$. Since the green phasor is the same in both cases, you can see that $\cos(\omega t + 30°) = \sin(\omega t + 120°)$.

Figure 15–45   Illustrating that $\cos(\omega t + 30°) = \sin(\omega t + 120°)$.

■ **EXAMPLE 15–21**   Determine the phase angle between $v = 30 \cos(\omega t + 20°)$ and $i = 25 \sin(\omega t + 70°)$.

Solution   Since cos ωt is represented by a phasor in the 90° position, $\cos(\omega t + 20°)$ is represented by a phasor at 110°, i.e., by the red phasor of Figure 15–46(a). The phasor for $\sin(\omega t + 70°)$ is straightforward. As you can see, v leads i by 40°. Waveforms are shown in (b).

Figure 15–46

The importance of phasors to ac circuit analysis cannot be overstated—you will find that they are one of your main tools for representing ideas and for solving problems in later chapters. We will leave them for the moment, but pick them up again in Chapter 16.

in process
l e a r n i n g
check *II*

(a) f = 40 Hz

(b) T = 100 ms

(c) f = 100 Hz

(d) f = 50 kHz

Figure 15–47

1. If $i = 15 \sin \alpha$ mA, compute the current at $\alpha = 0°$, 45°, 90°, 135°, 180°, 225°, 270°, 315°, and 360°.

2. Convert the following angles to radians:
   a. 20°.   b. 50°.
   c. 120°.   d. 250°.

3. If a coil rotates at $\omega = \pi/60$ radians per millisecond, how many degrees does it rotate through in 10 ms? In 40 ms? In 150 ms?

4. A current has an amplitude of 50 mA and $\omega = 0.2\pi$ rad/s. Sketch the waveform with the horizontal axis scaled in
   a. degrees,   b. radians,   c. seconds.

5. If 2400 cycles of a waveform occur in 10 ms, what is ω in radians per second?

6. A sinusoidal current has a period of 40 ms and an amplitude of 8 A. Write its equation in the form of $i = I_m \sin \omega t$, with numerical values for I_m and ω.

7. A current $i = I_m \sin \omega t$ has a period of 90 ms. If $i = 3$ A at $t = 7.5$ ms, what is its equation?

8. Write equations for each of the waveforms in Figure 15–47.

9. Given $v = 10 \sin \omega t$, where $f = 50$ Hz, find all occurrences of
   a. $i = 8$ A between $t = 0$ and $t = 40$ ms,
   b. $i = -5$ A between $t = 0$ and $t = 40$ ms.

10. Sketch the following waveforms with the horizontal axis scaled in degrees:
    a. $v_1 = 80 \sin(\omega t + 45°)$ V.   b. $v_2 = 40 \sin(\omega t - 80°)$ V.
    c. $i_1 = 10 \cos \omega t$ mA.   d. $i_2 = 5 \cos(\omega t - 20°)$ A.

11. Given $\omega = \pi/3$ rad/s, determine when voltage first crosses through 0 for
    a. $v_1 = 80 \sin(\omega t + 45°)$ V,   b. $v_2 = 40 \sin(\omega t - 80°)$ V.

12. Consider the voltages of Question 10:
    a. Sketch phasors for v_1 and v_2
    b. What is the phase difference between v_1 and v_2?
    c. Determine which voltage leads and which lags.

13. Repeat Question 12 for the currents of Question 10.

15.8 AC Waveforms and Average Value

While we can describe ac quantities in terms of frequency, period, instantaneous value, etc., we do not yet have any way to give a meaningful value to an ac current or voltage in the same sense that we can say of a car battery that it has a voltage of 12 volts. This is because ac quantities constantly change and thus there is no one single numerical value that truly represents a waveform over its complete cycle. For this reason, ac quantities are generally described by a group of characteristics, including instantaneous, peak, average, and effective values. The first two of these we have already seen. In this section, we look at average values; in Section 15.9, we consider effective values.

Average Values

Many quantities are measured by their average, for instance, test and examination scores. To find the average of a set of marks for example, you add them, then divide by the number of items summed. For waveforms, the process is conceptually the same. For example, to find the average of a waveform, you can sum the instantaneous values over a full cycle, then divide by the number of points used. The trouble with this approach is that waveforms do not consist of discrete values.

Average in Terms of the Area Under a Curve

An approach more suitable for use with waveforms is to find the area under the curve, then divide by the baseline of the curve. To get at the idea, we can use an analogy. Consider again the technique of computing the average for a set of numbers. Assume that you earn marks of 80, 60, 60, 95, and 75 on a group of tests. Your average mark is therefore

$$\text{Average} = (80 + 60 + 60 + 95 + 75)/5 = 74$$

An alternate way to view these marks is graphically as in Figure 15–48. The area under this curve can be computed as

$$\text{area} = (80 \times 1) + (60 \times 2) + (95 \times 1) + (75 \times 1)$$

Now divide this by the length of the base, namely 5. Thus,

$$\frac{(80 \times 1) + (60 \times 2) + (95 \times 1) + (75 \times 1)}{5} = 74$$

which is exactly the answer obtained above. That is,

$$\text{average} = \frac{\text{area under curve}}{\text{length of base}} \qquad (15\text{–}18)$$

This result is true in general. Thus, *to find the average value of a waveform, divide the area under the waveform by the length of its base. Areas above the axis are counted as positive, while areas below the axis are counted as negative.* This approach is valid regardless of waveshape.

Average values are also called *dc values*, because dc meters indicate average values rather than instantaneous values. Thus, if you measure a non-dc quantity with a dc meter, the meter will read the average of the waveform, i.e., the value calculated according to Equation 15–18.

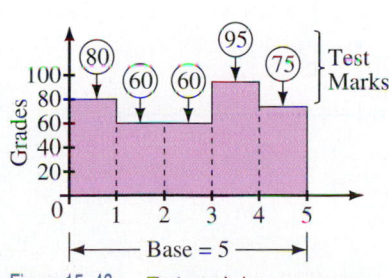

Figure 15–48   Determining average by area.

■ EXAMPLE 15–22

a. Compute the average for the current waveform of Figure 15–49.

b. If the negative portion of Figure 15–49 is −3 A instead of −1.5 A, what is the average?

c. If the current is measured by a dc ammeter, what will the ammeter indicate?

Figure 15–49

Solution

a. The waveform repeats itself after 7 ms. Thus, $T = 7$ ms and the average is

$$I_{avg} = \frac{(2 \text{ A} \times 3 \text{ ms}) - (1.5 \text{ A} \times 4 \text{ ms})}{7 \text{ ms}} = \frac{6 - 6}{7} = 0 \text{ A}$$

b. $I_{avg} = \dfrac{(2 \text{ A} \times 3 \text{ ms}) - (3 \text{ A} \times 4 \text{ ms})}{7 \text{ ms}} = \dfrac{-6 \text{ A}}{7} = -0.857 \text{ A}$

c. A dc ammeter measuring (a) will indicate zero, while for (b) it will indicate −0.857 A.

■ EXAMPLE 15–23   Compute the average value for the waveforms of Figures 15–50(a) and (c). Sketch the averages for each.

(a)   (b)

Figure 15–50

Figure 15–50

Solution   For the waveform of (a), $T = 6$ s. Thus,

$$V_{avg} = \frac{(10\ V \times 2\ s) + (20\ V \times 1\ s) + (30\ V \times 2\ s) + (0\ V \times 1\ s)}{6\ s} = \frac{100\ V\text{-}s}{6\ s} = 16.7\ V$$

The average is shown as (b). A dc voltmeter would indicate 16.7 V. For the waveform of (c), $T = 8$ s and

$$I_{avg} = \frac{\frac{1}{2}(40\ mA \times 3\ s) - (20\ mA \times 2\ s) - (40\ mA \times 2\ s)}{8\ s} = \frac{-60}{8}\ mA = -7.5\ mA$$

In this case, a dc ammeter would indicate −7.5 mA.

practice problem IX

Determine the averages for Figures 15–51(a) and (b).

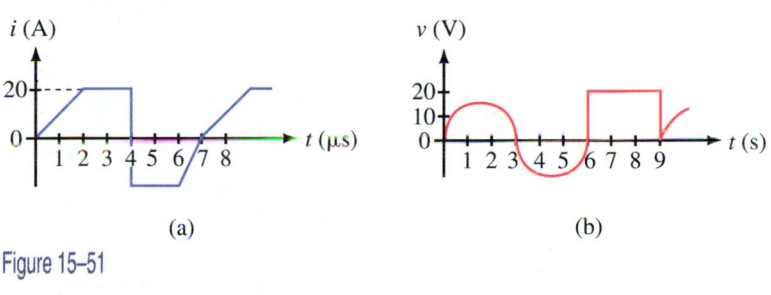

Figure 15–51

Sine Wave Averages

Due to symmetry, the average of a sinusoidal wave over a full cycle is zero, independent of frequency and phase angle. Thus, the average of sin ωt, sin($\omega t \pm \theta$), sin $2\omega t$, cos ωt, cos($\omega t \pm \theta$), cos $2\omega t$, and so on are each zero. The average of half a sine wave, however, is not zero. Consider Figure 15–52. The area under the half-cycle may be found using calculus as

$$\text{area} = \int_0^\pi I_m \sin\alpha\,d\alpha = -I_m \cos\alpha \Big]_0^\pi = 2I_m \qquad (15\text{–}19)$$

Similarly, the area under a half-cycle of voltage is $2V_m$. (If you haven't studied calculus, you can approximate this area using the numerical methods as described later in this section.)

Figure 15–52   Area under a half-cycle.

Figure 15–53   Full-wave average.

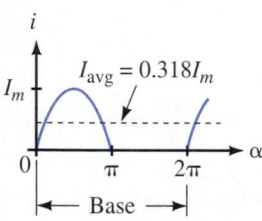

Figure 15–54   Half-wave average.

Two cases are important; full-wave average and half-wave average. The full-wave case is illustrated in Figure 15–53. The area from 0 to 2π is $2(2I_m)$ and the base is 2π. Thus, the average is

$$I_{avg} = \frac{2(2I_m)}{2\pi} = \frac{2I_m}{\pi} = 0.637I_m$$

For the half-wave case (Figure 15–54),

$$I_{avg} = \frac{2I_m}{2\pi} = \frac{I_m}{\pi} = 0.318I_m$$

The corresponding expressions for voltage are

$$V_{avg} = 0.637V_m \quad \text{(full-wave)}$$
$$V_{avg} = 0.318V_m \quad \text{(half-wave)}$$

Numerical Methods

If the area under a curve cannot be computed exactly, it can be approximated. One method is to approximate the curve by straight line segments as in Figure 15–55. (If the straight lines closely fit the curve, the accuracy is very good.) Each element of area is a trapezoid (b) whose area is its average height times its base. Thus, $A_1 = \frac{1}{2}(y_0+y_1)\Delta x$, $A_2 = \frac{1}{2}(y_1+y_2)\Delta x$, etc. Summing areas and combining terms yields

$$\text{area} = \left(\frac{y_0}{2} + y_1 + y_2 + \cdots + y_{k-1} + \frac{y_k}{2}\right)\Delta x \qquad (15\text{–}20)$$

This result is known as the *trapezoidal rule*. Example 15–24 illustrates its use.

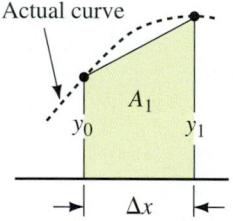

(a) Area under curve   (b) Element of area

Figure 15–55   Calculating area using the trapezoidal rule.

■ **EXAMPLE 15–24**   Approximate the area under $y = \sin(\omega t - 30°)$, Figure 15–56. Use an increment size of $\pi/6$ rad, i.e., 30°.

Solution   Points on the curve $\sin(\omega t - 30°)$ have been computed by calculator and plotted as Figure 15–56. Substituting these values into equation 15–20 yields

$$\text{area} = \left(\frac{1}{2}(-0.5) + 0 + 0.5 + 0.866 + 1.0 + 0.866 + 0.5 + \frac{1}{2}(0)\right)\left(\frac{\pi}{6}\right) = 1.823$$

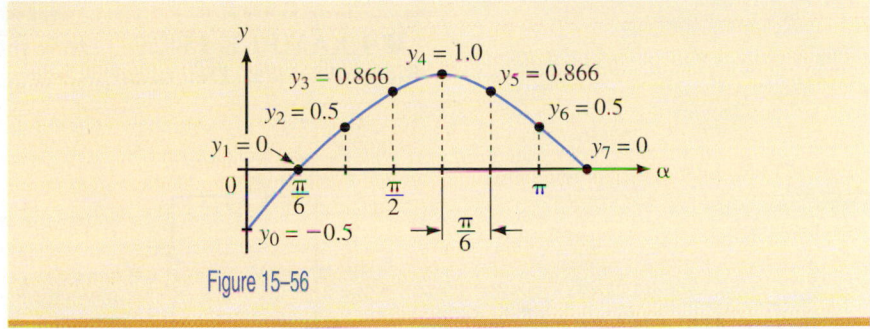

Figure 15–56

The exact area (found using calculus) is 1.866; thus, the above value is in error by 2.3%. If you decrease the increment to $\pi/12$, the answer is 1.855, an error of only 0.59%.

practice problems
X

1. Repeat Example 15–24 using an increment size of $\pi/12$ rad.

2. Approximate the area under $v = 50 \sin(\omega t + 30°)$ from $\omega t = 0°$ to $\omega t = 210°$. Use an increment size of $\pi/12$ rad.

Superimposed ac and dc

Sometimes ac and dc are found in the same circuit. For example, amplifiers are powered by dc but the signals they amplify are ac. Figure 15–57 shows a simple circuit with combined ac and dc.

Figure 15–58(c) shows superimposed ac and dc. Since we know that the average of a sine wave is zero, the average value of the combined waveform will be its dc component, E. However, peak voltages depend on both components as illustrated in (c). Note for the case illustrated that although the waveform varies sinusoidally, it does not alternate in polarity since it never changes polarity to become negative.

Figure 15–57

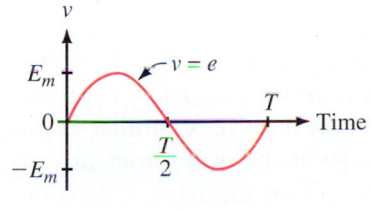

(a) AC alone. $E = 0$ V.
$V_{avg} = 0$ V.

(b) DC alone. $e = 0$.
$V_{avg} = E$ V.

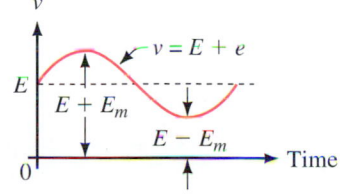

(c) Superimposed ac and dc.
$V_{avg} = E$ V.

Figure 15–58   Superimposed dc and ac.

■ **EXAMPLE 15–25**   Draw the voltage waveform for the circuit of Figure 15–59(a). Determine average, peak, and minimum voltages.

(a)   (b)

Figure 15–59

Solution   The waveform consists of a 10-V dc value with 15 V ac riding on top of it. The average is the dc value, $V_{avg} = 10$ V. The peak voltage is $10 + 15 = 25$ V, while the minimum voltage is $10 - 15 = -5$ V. This waveform alternates in polarity, although not symmetrically (as is the case when there is no dc component).

practice
problem
XI

Repeat Example 15–25 if the dc source of Figure 15–59 is $E = -5$ V.

15.9 Effective Values

While instantaneous, peak, and average values provide useful informa- tion about a waveform, none of them truly represents the ability of the waveform to do useful work. In this section, we look at a representation that does. It is called the waveform's *effective value*. The concept of effec- tive value is an important one; in practice, most ac voltages and currents are expressed as effective values. *Effective values are also called rms values* for reasons discussed shortly.

What Is an Effective Value?

An effective value is an equivalent dc value: it tells you how many volts or amps of dc that a time-varying waveform is equal to in terms of its ability to produce average power. Effective values depend on the waveform. A familiar exam- ple of such a value is the value of the voltage at the wall outlet in your home. In North America its value is 120 Vac. This means that the sinu- soidal voltage at the wall outlets of your home is capable of producing the same average power as 120 volts of steady dc.

Effective Values for Sine Waves

The effective value of a waveform can be determined using the circuits of Figure 15–60. Consider a sinusoidally varying current, $i(t)$. By definition, the effective value of i is that value of dc current that produces the same average power. Consider (b). Let the dc source be adjusted until its average power is the same as the average power in (a). The resulting dc current is then the effective value of the current of (a). To determine this value, determine the average power for both cases, then equate them.

First, consider the dc case. Since current is constant, power is constant, and average power is

$$P_{avg} = P = I^2R \qquad (15\text{--}21)$$

Now consider the ac case. Power to the resistor at any value of time is $p(t) = i^2R$, where i is the instantaneous value of current. A sketch of $p(t)$ is shown in Figure 15–60(a), obtained by squaring values of current at various points along the axis, then multiplying by R. Average power is the average of $p(t)$. Since $i = I_m \sin \omega t$,

$$p(t) = i^2R \qquad (15\text{--}22)$$

$$= (I_m \sin \omega t)^2 R = I^2_m R \sin^2 \omega t$$

$$= I^2_m R \left[\frac{1}{2}(1 - \cos 2\omega t) \right]$$

Thus,

$$p(t) = \frac{I_m^2 R}{2} - \frac{I_m^2 R}{2} \cos 2\omega t \qquad (15\text{--}23)$$

where we have used the trigonometric identity $\sin^2 \omega t = \frac{1}{2}(1 - \cos 2\omega t)$, from the mathematics tables inside the front cover to expand $\sin^2 \omega t$. To get the average of $p(t)$, note that the average of $\cos 2\omega t$ is zero and thus the last term of Equation 15–23 drops off leaving

$$P_{avg} = \text{average of } p(t) = \frac{I_m^2 R}{2} \qquad (15\text{--}24)$$

Now equate P_{avg} to the dc case (Equation 15–21) and then cancel R. Thus,

$$I^2 = \frac{I_m^2}{2}$$

Now take the square root of both sides. Thus,

$$I = \sqrt{\frac{I_m^2}{2}} = \frac{I_m}{\sqrt{2}} = 0.707 I_m$$

$p(t) = i^2R$. Therefore, $p(t)$ varies cyclically.

(a) AC circuit

$P = I^2R$. Therefore, P is constant.

(b) DC circuit

Figure 15–60   Determining the effective value of sinusoidal ac.

Current I is the value that we are looking for; it is the effective value of current i. To emphasize that it is an effective value, we will initially use subscripted notation I_{eff}. Thus,

$$I_{\text{eff}} = \frac{I_m}{\sqrt{2}} = 0.707I_m \qquad (15\text{--}25)$$

Effective values for voltage are found in the same way:

$$E_{\text{eff}} = \frac{E_m}{\sqrt{2}} = 0.707E_m \qquad (15\text{--}26a)$$

$$V_{\text{eff}} = \frac{V_m}{\sqrt{2}} = 0.707V_m \qquad (15\text{--}26b)$$

As you can see, *effective values depend only on magnitude.*

note

B*ecause ac currents alternate in direction, you might expect average power to be zero, with power during the negative half-cycle being equal and opposite to power during the positive half-cycle and hence cancelling. However, as Equation 15–24 shows, current is squared, and hence power is never negative. This is consistent with the idea that insofar as power dissipation is concerned, the direction of current through a resistor does not matter (Figure 15–61).*

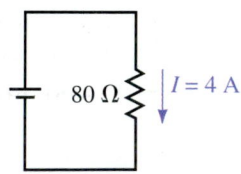

(a) $P = (4)^2(80) = 1280$ W

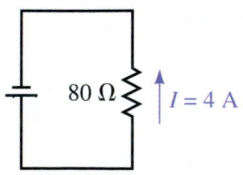

(b) $P = (4)^2(80) = 1280$ W

Figure 15–61 Power is the same for both current directions.

■ **EXAMPLE 15–26**   Determine the effective values of

a. $i = 10 \sin \omega t$ A,   b. $i = 50 \sin(\omega t + 20°)$ mA,
c. $v = 100 \cos 2\omega t$ V

Solution   Since effective values depend only on magnitude,

a. $I_{\text{eff}} = (0.707)(10$ A$) = 7.07$ A,
b. $I_{\text{eff}} = (0.707)(50$ mA$) = 35.35$ mA,
c. $V_{\text{eff}} = (0.707)(100$ V$) = 70.7$ V.

To obtain peak values from effective values, rewrite Equations 15–25 and 15–26. Thus,

$$I_m = \sqrt{2}I_{\text{eff}} = 1.414I_{\text{eff}} \qquad (15\text{--}27)$$

$$E_m = \sqrt{2}E_{\text{eff}} = 1.414E_{\text{eff}} \qquad (15\text{--}28a)$$

$$V_m = \sqrt{2}V_{\text{eff}} = 1.414V_{\text{eff}} \qquad (15\text{--}28b)$$

It is important to note that these relationships hold only for sinusoidal waveforms. However, the concept of effective value applies to all waveforms, as we soon see.

Consider again the ac voltage at the wall outlet in your home. Since $E_{\text{eff}} = 120$ V, $E_m = (\sqrt{2})(120$ V$) = 170$ V. This means that a sinusoidal voltage alternating between ±170 V produces the same average power as 120 V of steady dc (Figure 15–62).

Figure 15–62   120 V of steady dc is capable of producing the same average power as 170 V of sinusoidal ac.

General Equation for Effective Values

The $\sqrt{2}$ relationship holds only for sinusoidal waveforms. For other waveforms, you need a more general formula. Using calculus, it can be shown that

$$I_{\text{eff}} = \sqrt{\frac{1}{T}\int_{0}^{T} i^2\, dt} \qquad (15\text{–}29)$$

with a similar equation for voltage. This equation can be used to compute effective values for any waveform, In addition, it leads to a graphic approach to finding effective values. In Equation 15–29, the integral of i^2 represents the area under the i^2 waveform. Thus,

$$I_{\text{eff}} = \sqrt{\frac{\text{area under the } i^2 \text{ curve}}{\text{base}}} \qquad (15\text{–}30)$$

To compute effective values using this equation, do the following:

Step 1: Square the current (or voltage) curve.
Step 2: Find the area under the squared curve.
Step 3: Divide the area by the length of the curve.
Step 4: Find the square root of the value from Step 3.

This process is easily carried out for rectangular-shaped waveforms since the area under their squared curves is easy to compute. For other waveforms, you may have to approximate the area using numerical methods. For the special case of superimposed ac and dc (Figure 15–58), Equation 15–29 leads to the following formula:

$$I_{\text{eff}} = \sqrt{I^2_{\text{dc}} + I^2_{\text{ac}}} \qquad (15\text{–}31)$$

where I_{dc} is the dc current value, I_{ac} is the effective value of the ac current, and I_{eff} is the effective value of the combined ac and dc currents. Equations 15–30 and 15–31 also hold for voltage when V is substituted for I.

RMS Values

Consider again Equation 15–30. To use this equation, we compute the root of the mean square to obtain the effective value. For this reason, effective values are called *root mean square* or *rms* values and the terms effective and rms are synonymous. Since, in practice, ac quantities are almost always expressed as rms values, we shall assume from here on that, unless otherwise noted, *all voltages and currents are rms values.*

■ **EXAMPLE 15–27** One cycle of a voltage waveform is shown in Figure 15–63(a). Determine its effective value.

(a) Voltage waveform   (b) Squared waveform

Figure 15–63

Solution Square the voltage waveform and plot it as in (b). Apply Equation 15–30:

$$V_{eff} = \sqrt{\frac{(400 \times 4) + (900 \times 2) + (100 \times 2)}{10}} = \sqrt{\frac{3600}{10}} = 19.0 \text{ V}$$

Since effective values are dc values, the waveform of Figure 15–63(a) is equivalent to 19.0 V of steady dc.

■ **EXAMPLE 15–28** Determine the effective value of the waveform of Figure 15–64(a).

(a)   (b)

Figure 15–64

Solution Square the curve, then apply Equation 15–30. Thus,

$$I_{eff} = \sqrt{\frac{(9 \times 3) + (1 \times 2) + (4 \times 3)}{8}} = \sqrt{\frac{41}{8}} = 2.264 \text{ A}$$

■ **EXAMPLE 15–29**   Compute the effective value of the waveform of Figure 15–59(b).

Solution   Use Equation 15–31 (with I replaced by V). First, compute the rms value of the ac component. $V_{ac} = 0.707 \times 15 = 10.61$ V. Now substitute this into Equation 15–31. Thus,

$$V_{rms} = \sqrt{I_{dc}^2 + I_{ac}^2} = \sqrt{(10)^2 + (10.61)^2} = 14.58 \text{ V}$$

practice problems XII

1. Determine the effective value of the current of Figure 15–49.

2. Repeat for the voltage graphed in Figure 15–50(a).

One Final Note

The subscripts *eff* and *rms* are not used in practice. Once the concept is familiar, we drop them.

15.10 Rate of Change of a Sine Wave

Several important circuit effects depend on the rate of change of sinusoidal quantities. The rate of change of a quantity is the slope (written d/dt in the calculus) of its waveform versus time. Consider the waveform of Figure 15–65. As indicated, the slope is maximum positive at the beginning of the cycle, zero at both its peaks, maximum negative at the half-cycle crossover point, and maximum positive at the end of the cycle. This slope is plotted in Figure 15–66. Note that the original waveform and its slope are 90° out of phase. Thus, if sinusoidal waveform *A* is taken as reference, its slope *B* leads it by 90°, whereas, if the slope *B* is taken as reference, *A* lags it by 90°. Thus, if *A* is a sine wave, *B* is a cosine wave, and so on. (This relationship can be proven rigorously using calculus since the derivative of a sine wave is a cosine wave.) The result shown in Figure 15–66 is important to us in Chapter 16.

Figure 15–65   Rates of change of a sinusoidal waveform. Slope is plotted in Figure 15–66.

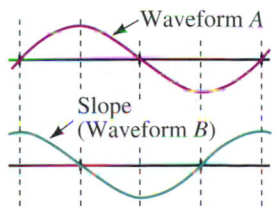

Figure 15–66   Showing the 90° shift.

15.11 AC Voltage and Current Measurement

Two of the most commonly used instruments for measuring ac quantities are the multimeter and the oscilloscope. Multimeters read voltage, current, and sometimes frequency. Oscilloscopes show waveshape and period and permit determination of frequency, phase difference, and so on.

Meters for Voltage and Current Measurement

There are two basic classes of ac meters: one measures rms correctly for sinusoidal waveforms only (called "average responding" instruments); the other measures rms correctly regardless of waveform (called "true rms" meters). Most common meters are average responding meters.

Average Responding Meters.   Average responding meters use a rectifier circuit to convert incoming ac to dc. They then respond to the average value of the rectified input, which, as shown in Figure 15–53, is $0.637V_m$ for a "full-wave" rectified sine wave. However, the rms value of a sine wave is $0.707V_m$. Thus, the scale of such a meter is modified by the factor $0.707V_m/0.637V_m = 1.11$ so that it indicates rms values directly. Other meters use a "half-wave" circuit, which yields the waveform of Figure 15–54 for a sine wave input. In this case, its average is $0.318V_m$, yielding a scale factor of $0.707V_m/0.318V_m = 2.22$. Figure 15–67 shows a typical average responding DMM; in addition to voltage and current, it can also measure frequency.

Figure 15–67   A DMM. In addition to voltage and current measurements, this instrument can also measure frequency.

True RMS Measurement.   If you need to measure the rms value of a nonsinusoidal waveform, you will need a true rms meter. A true rms meter indicates true rms voltages and currents regardless of waveform. For example, for the waveform of Figure 15–62, any ac meter will correctly read 120 V (since it is a sine wave), but for the waveform of 15–59(b), only a true rms meter will correctly read 14.58 V (the rms value that we calculated earlier, in Example 15–29). True rms instruments tend to be more expensive than standard meters.

Oscilloscopes

Oscilloscopes (frequently referred to as 'scopes, Figure 15–68) are used for time domain measurement, i.e., waveshape, frequency, period, phase difference, and so on. Usually, you scale values from the screen, although some higher-priced models can compute and display them for you on a digital readout.

Figure 15–68   An oscilloscope may be used for waveform analysis.

Oscilloscopes measure voltage. To measure current, you need a current-to-voltage converter. One type of converter is a clip-on device, known as a *current gun* that clamps over the current-carrying conductor and monitors its magnetic field. (It works only with ac.) The varying magnetic field induces a voltage which is then displayed on the screen. With such a device, you can monitor current waveshapes and make current-related measurements. Alternately, you can place a small resistor in the current path, measure voltage across it with the oscilloscope, then use Ohm's law to determine the current.

A Final Note

AC meters measure voltage and current only over a limited frequency range, typically from 50 Hz to a few kHz, although others are available that work up to the 100-kHz range. Note, however, that accuracy may be affected by frequency. (Check the manual.) Oscilloscopes, on the other hand, can measure very high frequencies; even moderately priced oscilloscopes work at frequencies up to hundreds of MHz.

Problems

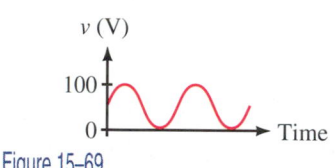

Figure 15–69

15.1   Introduction

1. The voltage waveform of Figure 15–69 is sinusoidal, yet it is not ac. Why not?

2. Sketch two cycles of a sinusoidal ac voltage and identify the points where the voltage polarity changes from plus to minus and from minus to plus.

15.2   Generating ac Voltages

3. The waveform of Figure 15–8 is created by a 600-rpm generator. If the speed of the generator changes so that its cycle time is 50 ms, what is its new speed?

4. a. What do we mean by instantaneous value?
   b. For Figure 15–9(a), find the instantaneous voltages at $t =$ 1, 1.5, 2.5, 4, 4.5, and 5.5 ms from the photo.

5. The concept of instantaneous value applies regardless of waveform. For Figure 15–70, determine instantaneous voltages at $t = 0$, 1, 2, 3, 4, 5, 6, 7, and 8 ms.

Figure 15–70

15.3   Voltage and Current Conventions for ac

6. For Figure 15–71, what is I when the switch is in position 1? When in position 2?

7. The voltage of Figure 15–70 is applied to the circuit of Figure 15–72. Determine the current at $t = 0$, 1, 2, 3, 4, 5, 6, 7, and 8 ms. What is the significance of the plus and minus signs?

15.4   Frequency, Period, Amplitude, and Peak Value

8. For each of the following, determine the period:
   a. $f = 100$ Hz.   b. $f = 40$ kHz.   c. $f = 200$ MHz.

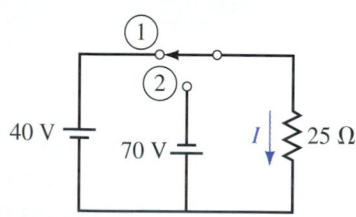

Figure 15–71

9. For each of the following, determine the frequency:
   a. $T = 0.5$ s.   b. $T = 100$ ms.   c. $5T = 80$ μs.

10. For a certain waveform, $625T = 12.5$ ms. What is the waveform's period and frequency?

Figure 15–72

11. A square wave with a frequency of 847 Hz goes through how many cycles in 2 minutes and 57 seconds?

Figure 15–73

Figure 15–74

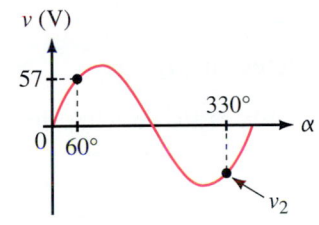

Figure 15–75

12. What is the period and frequency of the waveform of Figure 15–70?

13. For the waveform of Figure 15–73, determine
a. period,   b. frequency,   c. peak-to-peak value.

15.5   Angular and Graphic Relationships for Sine Waves

14. Refer to the current graph of Figure 15–74:
a. Determine the equation for i.
b. Determine current at all points marked.

15. A sinusoidal voltage has a value of 50 V at $\alpha = 150°$. What is V_m?

16. Convert the following angles from radians to degrees:
a. $\pi/12$,   b. $\pi/1.5$,   c. $3\pi/2$
d. 1.43,   e. 17,   f. 32π.

17. Convert the following angles from degrees to radians:
a. 10°,   b. 25°,   c. 80°,
d. 150°,   e. 350°,   f. 620°.

18. A 50-kHz sine wave has an amplitude of 150 V. Sketch the waveform with its axis scaled in microseconds.

19. If the period of the waveform in Figure 15–74 is 180 ms, compute current at $t = 30$, 75, 140, and 315 ms.

20. A sinusoidal waveform has a period of 60 μs and $V_m = 80$ V. What is its voltage at 4 μs?

21. A 20-kHz sine wave has a value of 50 volts at $t = 5$ μs. Determine V_m and sketch the waveform.

22. For the waveform of Figure 15–75, determine v_2.

15.6   Voltages and Currents as Functions of Time

23. Calculate ω in radians per second for each of the following:
a. $T = 100$ ns,   b. $f = 30$ Hz,   c. 100 cycles in 4 s,
d. period = 20 ms,   e. 5 periods in 20 ms.

24. For each of the following values of ω, compute f and T:
a. 100 rad/s,   b. 40 rad in 20 ms,   c. 34×10^3 rad/s.

25. Determine equations for sine waves with the following:
a. $V_m = 170$ V, $f = 60$ Hz.   b. $I_m = 40$ μA, $T = 10$ ms.
c. $T = 120$ μs, $v = 10$ V at $t = 12$ μs.

26. Determine f and T for each of the following:
a. $v = 75 \sin 200\pi t$.   b. $i = -8 \sin 300t$

27. A sine wave has a peak-to-peak voltage of 40 V and $T = 50$ ms. Determine its equation.

28. For $v = 100 \sin(100\pi/3)t$, sketch the waveform, scale the axis in time, and determine all occurrences of $v = 50$ V during the first cycle.

29. Repeat Problem 28 for $v = -50$ V.

30. Sketch the following waveforms with the horizontal axis scaled in degrees, radians, and seconds:
   a. $v = 100 \sin 200\pi t$ V.
   b. $i = 90 \sin \omega t$ mA, $T = 80$ μs.

31. Sketch the following waveforms with the horizontal axis scaled in degrees and seconds:
   a. $v = 100 \sin(232.7t + 40°)$ V.
   b. $i = 20 \sin(\omega t - 60°)$ mA, $f = 200$ Hz.

32. Given $v = 5 \sin(\omega t + 45°)$. If $\omega = 20\pi$ rad/s, what is v at $t = 20$, 75, and 90 ms?

33. For the waveform of Figure 15–76, determine i_2.

34. Given $v = 30 \sin(\omega t - 45°)$ where $\omega = 40\pi$ rad/s. At what time does v reach 0 V?

15.7   Introduction to Phasors

35. For each of the phasors of Figure 15–77, determine the equation for $v(t)$ or $i(t)$ as applicable, and sketch the waveform.

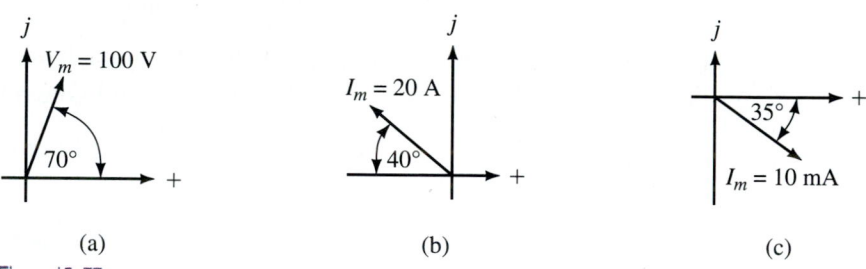

(a)   (b)   (c)

Figure 15–77

36. With the aid of phasors, sketch the waveforms for each of the following pairs and determine the phase difference and which waveform leads:
   a. $v = 100 \sin \omega t$
      $i = 80 \sin(\omega t + 20°)$
   b. $v_1 = 200 \sin(\omega t - 30°)$
      $v_2 = 150 \sin(\omega t - 30°)$
   c. $i_1 = 40 \sin(\omega t + 30°)$
      $i_2 = 50 \sin(\omega t - 20°)$
   d. $v = 100 \sin(\omega t + 140°)$
      $i = 80 \sin(\omega t - 160°)$
   e. $i = 40 \sin(\omega t + 80°)$
      $v = -30 \sin(\omega t - 70°)$
   f. $v = 20 \cos(\omega t + 10°)$
      $i = 15 \sin(\omega t - 10°)$
   g. $v = 20 \cos(\omega t + 10°)$
      $i = 15 \sin(\omega t + 120°)$

i

I_m

$i_1 = 15.6$ A

$350°$

0

$180°$

ωt

i_2

$I_m \sin(-60°)$

Figure 15–76

37. For the waveforms in Figure 15–78, determine the phase differences.

(a)   (b)

Figure 15–78

15.8   AC Waveforms and Average Value

38. What is the average value of each of the following over an integral number of cycles?

a. $i = 5 \sin \omega t$.   b. $i = 40 \cos \omega t$.

c. $v = 400 \sin(\omega t + 30°)$.   d. $v = 20 \cos 2\omega t$.

39. Compute I_{avg} for the waveforms of Figure 15–79.

(a)

(b)

(c)

Figure 15–79

Figure 15–80

Figure 15–81

40. By summing elemental areas, show that the area under the curve in Figure 15–55 is given by Equation 15–20.

41. Using Equation 15–20, compute the area under the half-cycle of Figure 15–52 using increments of $\pi/12$ rad.

42. For the waveform of Figure 15–80, compute I_m.

43. For the circuit of Figure 15–81, $e = 25 \sin \omega t$ V and period $T = 120$ ms.

a. Sketch voltage $v(t)$ with the axis scaled in milliseconds.

b. Determine the peak and minimum voltages.

c. Compute v at $t = 10, 20, 70,$ and 100 ms.

d. Determine V_{avg}.

15.9   Effective Values

44. What is the effective voltage of a 12-V battery?

45. Determine the effective values of each of the following:

a. $v = 100 \sin \omega t$ V.   b. $i = 8 \sin 377t$ A.

c. $v = 40 \sin(\omega t + 40°)$ V.   d. $i = 120 \cos \omega t$ mA.

46. For a sine wave, $V_{eff} = 9$ V. What is its amplitude?

47. Compute the rms values for Figures 15–79(a), and 15–80. For Figure 15–80, $I_m = 30$ A.

48. Compute the rms value of the superimposed dc and ac voltage $v(t)$ of Figure 15–81, given $e = 25 \sin \omega t$.

49. Compute the rms value of the waveform of Figure 15–50(c). To handle the triangular portion, use Equation 15–20. Use a time interval $\Delta t = 1$s.

50. Repeat Problem 49, using calculus to handle the triangular portion.

R, L, and C Elements and the Impedance Concept

CHAPTER PREVIEW

While it is possible to analyze ac circuits in the time domain using voltages and currents expressed as functions of time as in Chapter 15, this is not a very practical approach. A more practical approach is to represent voltages and currents as phasors and analyze circuits in the phasor domain using complex algebra. With this approach, ac circuit analysis is handled much like dc circuit analysis, and all basic relationships and theorems—Ohm's law, Kirchhoff's laws, mesh and nodal analysis, superposition and so on—apply. The major difference is that ac quantities are complex rather than real as with dc. While this complicates computational details, it does not alter basic circuit principles. This is the approach used in practice. The basic ideas are developed in this chapter.

$$Z = \frac{V}{I}$$

Impedance represents the opposition that circuit elements present to current in the phasor domain.

OBJECTIVES

After studying this chapter, you will be able to

- express complex numbers in rectangular and polar forms,

- represent ac voltage and current phasors as complex numbers,

- represent voltage and current sources in transformed form,

- add and subtract currents and voltages using phasors,

- compute inductive and capacitive reactance,

- determine voltages and currents in simple ac circuits,

- explain the impedance concept,

- determine impedance for *R, L,* and *C* circuit elements,

- determine voltages and currents in simple ac circuits using the impedance concept,

- use PSpice to solve simple ac circuit problems.

16.1 Complex Number Review

A *complex number* is a number of the form $\mathbf{c} = a + jb$, where *a* and *b* are real numbers and $j = \sqrt{-1}$. The number *a* is called the *real* part of **c** and *b* is called its *imaginary* part. (In circuit theory, *j* is used to denote the imaginary component rather than *i* to avoid confusion with current *i*.)

Geometrical Representation

Complex numbers may be represented geometrically, either in *rectangular form* or in *polar form* as points on a two-dimensional plane called the *complex plane* (Figure 16–1). The complex number $\mathbf{C} = 6 + j8$, for example, represents a point whose coordinate on the real axis is 6 and whose coordinate on the imaginary axis is 8. This form of representation is called the *rectangular form.*

Complex numbers may also be represented in *polar form* by magnitude and angle. Thus, $\mathbf{C} = 10\angle 53.13°$ (Figure 16–2) is a complex number with magnitude 10 and angle 53.13°. This magnitude and angle representation is just an alternate way of specifying the location of the point represented by $\mathbf{C} = a + jb.$

Conversion between Rectangular and Polar Forms

To convert between forms, note from Figure 16–3 that

$$\mathbf{C} = a + jb \quad \text{(rectangular form)} \tag{16–1}$$

$$\mathbf{C} = C\angle\theta \quad \text{(polar form)} \tag{16–2}$$

where *C* is the magnitude of **C.** From the geometry of the triangle,

$$a = C\cos\theta \tag{16–3a}$$

$$b = C\sin\theta \tag{16–3b}$$

where

$$C = \sqrt{a^2 + b^2} \tag{16–4a}$$

Figure 16–1   A complex number in rectangular form.

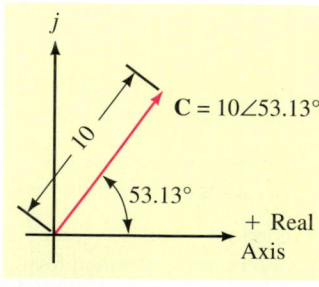

Figure 16–2   A complex number in polar form.

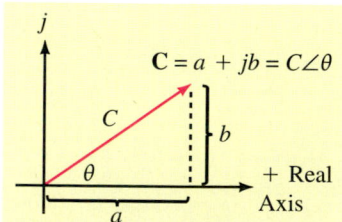

Figure 16–3   Polar and rectangular equivalence.

and

$$\theta = \tan^{-1} \frac{b}{a} \qquad (16\text{--}4b)$$

Equations 16–3 and 16–4 permit conversion between forms. When using Equation 16–4b, however, be careful when the number to be converted is in the second or third quadrant, as the angle obtained is the supplementary angle rather than the actual angle in these two quadrants. This is illustrated in Example 16–1 for the complex number **W**.

■ **EXAMPLE 16–1**   Determine rectangular and polar forms for the complex numbers **C, D, V,** and **W** of Figure 16–4(a).

(a) Complex numbers

(b) In polar form, **C** = 5∠36.87°

(c) In polar form, **D** = 5.66∠ − 45°

(d) In polar form, **W** = 5.66∠135°

Figure 16–4

Solution   *Point C:* Real part = 4; imaginary part = 3. Thus, **C** = 4 + j3. In polar form, $C = \sqrt{4^2 + 3^2} = 5$ and $\theta_C = \tan^{-1}(3/4) = 36.87°$. Thus, **C** = 5∠36.87° as indicated in (b).

Point D: In rectangular form, **D** = 4 − j4. Thus, $D = \sqrt{4^2 + 4^2} = 5.66$ and $\theta_D = \tan^{-1}(-4/4) = -45°$. Therefore, **D** = 5.66 ∠−45°, as shown in (c).

Point V: In rectangular form, **V** = −j2. In polar form, **V** = 2∠−90°.

Point W: In rectangular form, **W** = −4 + j4. Thus, $W = \sqrt{4^2 + 4^2} = 5.66$ and $\tan^{-1}(-4/4) = -45°$. Inspection of Figure 16–4(d) shows, however, that this 45° angle is the supplementary angle. The actual angle (measured from the positive horizontal axis) is 135°. Thus, **W** = 5.66∠135°.

Powers of *j*

Powers of *j* are frequently required in calculations. Here are some useful powers:

$$j^2 = (\sqrt{-1})(\sqrt{-1}) = -1$$
$$j^3 = j^2 j = -j$$
$$j^4 = j^2 j^2 = (-1)(-1) = 1 \tag{16-5}$$
$$(-j)j = 1$$
$$\frac{1}{j} = \frac{1}{j} \times \frac{j}{j} = \frac{j}{j^2} = -j$$

Addition and Subtraction of Complex Numbers

Addition and subtraction of complex numbers can be performed analytically or graphically. Analytic addition and subtraction is most easily done in rectangular form, while graphical addition and subtraction is best done in polar form. For analytic addition, add real and imaginary parts separately. Similarly for subtraction. For graphical addition, add vectorially as in Figure 16–5(a); for subtraction, change the sign of the subtrahend, then add, as in Figure 16–5(b).

■ **EXAMPLE 16–2**   Given **A** = 2 + *j*1 and **B** = 1 + *j*3. Determine their sum and difference analytically and graphically.

Solution

$\mathbf{A} + \mathbf{B} = (2 + j1) + (1 + j3) = (2 + 1) + j(1 + 3) = 3 + j4.$
$\mathbf{A} - \mathbf{B} = (2 + j1) - (1 + j3) = (2 - 1) + j(1 - 3) = 1 - j2.$

Graphical addition and subtraction are shown in Figure 16–5.

(a)   (b)

Figure 16–5

Multiplication and Division of Complex Numbers

These operations are usually performed in polar form. For multiplication, multiply magnitudes and add angles algebraically. For division, divide the magnitude of the denominator into the magnitude of the numerator, then subtract algebraically the angle of the denominator from that of the numerator. Thus, given $\mathbf{A} = A\angle\theta_A$ and $\mathbf{B} = B\angle\theta_B$,

$$\mathbf{A} \cdot \mathbf{B} = AB \underline{/\theta_A + \theta_B} \qquad (16\text{–}6)$$

$$\mathbf{A}/\mathbf{B} = A/B \underline{/\theta_A - \theta_B} \qquad (16\text{–}7)$$

■ **EXAMPLE 16–3**   Given $\mathbf{A} = 3\angle 35°$ and $\mathbf{B} = 2\angle{-20°}$, determine the product $\mathbf{A} \cdot \mathbf{B}$ and the quotient $\mathbf{A}/\mathbf{B}$.

Solution

$$\mathbf{A} \cdot \mathbf{B} = (3\angle 35°)(2\angle{-20°}) = (3)(2)\underline{/35° - 20°} = 6\angle 15°$$

$$\frac{\mathbf{A}}{\mathbf{B}} = \frac{(3\angle 35°)}{(2\angle{-20°})} = \frac{3}{2}\underline{/35° - (-20°)} = 1.5\angle 55°$$

■ **EXAMPLE 16–4**   For computations involving purely real, purely imaginary, or small integer numbers, it is sometimes easier to multiply directly in rectangular form than it is to convert to polar. Compute the following directly:

a. $(-j3)(2 + j4)$.

b. $(2 + j3)(1 + j5)$.

Solution

a. $(-j3)(2 + j4) = (-j3)(2) + (-j3)(j4) = -j6 - j^2 12 = 12 - j6$

b. $(2 + j3)(1 + j5) = (2)(1) + (2)(j5) + (j3)(1) + (j3)(j5)$
$$= 2 + j10 + j3 + j^2 15 = 2 + j13 - 15 = -13 + j13$$

practice problems
I

1. Polar numbers with the same angle can be added or subtracted directly without conversion to rectangular form. For example, the sum of $6\angle 36.87°$ and $4\angle 36.87°$ is $6\angle 36.87° + 4\angle 36.87° = 10\angle 36.87°$, while the difference is $6\angle 36.87° - 4\angle 36.87° = 2\angle 36.87°$. By means of sketches, indicate why this procedure is valid.

2. To compare methods of multiplication with small integer values, convert the numbers of Example 16–4 to polar form, multiply them, then convert the answers back to rectangular form.

Reciprocals

The reciprocal of a complex number $\mathbf{C} = C\angle\theta$ is

$$\frac{1}{\mathbf{C}\angle\theta} = \frac{1}{C}\angle{-\theta} \qquad (16\text{–}8)$$

Thus,

$$\frac{1}{20\angle 30°} = 0.05\angle{-30°}$$

Complex Conjugates

The *conjugate* of a complex number is a complex number with the same real part but the opposite imaginary part. Thus, the conjugate of $\mathbf{C} = C\angle\theta = a + jb$ is $\mathbf{C^*} = C\angle{-\theta} = a - jb$. For example, if $\mathbf{C} = 3 + j4 = 5\angle 53.13°$, then $\mathbf{C^*} = 3 - j4 = 5\angle{-53.13°}$.

Calculators for ac Analysis

Analysis of ac circuits involves a great deal of complex number computation. To speed up the process, you will need a calculator that works easily with complex numbers. At the very least, you will need direct rectangular to polar ($\rightarrow$P) and polar to rectangular ($\rightarrow$R) conversion capabilities. There are many such calculators on the market. You should acquire one and learn to use it proficiently.

■ **EXAMPLE 16–5**   The following illustrates the type of calculations that you will encounter. Use your calculator to reduce the following to rectangular form:

$$(6 + j5) + \frac{(3 - j4)\,(10\angle 40°)}{6 + 30\angle 53.13°}$$

Solution

$$\text{answer} = (6 + j5) + \frac{(5\angle{-53.13})\,(10\angle 40)}{6 + (18 + j24)}$$

$$= (6 + j5) + \frac{(5\angle{-53.13})\,(10\angle 40)}{24 + j24}$$

$$= (6 + j5) + \frac{(5\angle{-53.13})\,(10\angle 40)}{33.94\angle 45}$$

$$= (6 + j5) + 1.473\angle{-58.13} = (6 + j5) + (0.778 - j1.251)$$

$$= 6.778 + j3.749$$

Not all calculators require the intermediate conversions shown. The calculator of Figure 16–6, for instance, can perform any calculation in either rectangular or polar form directly, with no conversion between forms required. (The above problem was solved on this calculator without writing down a single intermediate result.) However, it is a programmable calculator and not all colleges permit the use of programmable calculators in introductory courses. Fortunately, there are several inexpensive, nonprogrammable calculators that work similarly. One of these is the Sharp EL-546G. It and similar calculators can save you many hours of problem solving time.

Figure 16–6   Powerful calculators make working with complex numbers easy. The calculator shown displays results in standard mathematical notation.

16.2 Complex Numbers in ac Analysis

Representing ac Voltages and Currents by Complex Numbers

As you learned in Chapter 15, ac voltages and currents can be represented as phasors. Since phasors have magnitude and angle, they can be viewed as complex numbers. To get at the idea, consider the voltage source of Figure 16–7(a). Its phasor equivalent (b) has magnitude E_m and angle θ. It therefore can be viewed as the complex number

$$\mathbf{E} = E_m \angle \theta \qquad (16\text{–}9)$$

From this point of view, the sinusoidal voltage $e(t) = 200\ \sin(\omega t + 40°)$ of Figure 16–8(a) can be represented by its phasor equivalent, $\mathbf{E} = 200\ \text{V}\angle 40°$, as in (c).

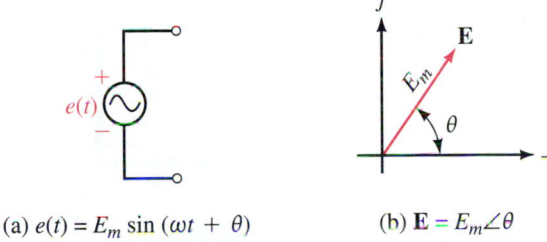

(a) $e(t) = E_m \sin(\omega t + \theta)$ (b) $\mathbf{E} = E_m\angle\theta$

Figure 16–7   Representation of a sinusoidal source voltage as a complex number.

  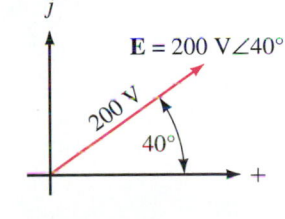

(a) $e = 200 \sin(\omega t + 40°)$ V   (b) Waveform   (c) Phasor equivalent of voltage
$e = 200 \sin(\omega t + 40°)$ V

Figure 16–8   Transforming a source voltage.

$E = 200 \text{ V} \angle 40°$

Transformed source

Figure 16–9 Direct transformation of the source.

We can take advantage of this equivalence. *Rather than show a source as a time-varying voltage e(t) that we subsequently convert to a phasor, we can represent the source by its phasor equivalent right from the start.* This viewpoint is illustrated in Figure 16–9. Since $\mathbf{E} = 200 \text{ V} \angle 40°$, this representation retains all the original information of Figure 16–8 with the exception of the sinusoidal time variation. However, since sine waves and phasors are equivalent, we can easily restore the time information if needed.

The idea illustrated in Figure 16–9 is of fundamental importance to circuit theory. *By replacing the time function e(t) with its phasor equivalent* **E**, *we have transformed the source from the time domain to the phasor domain.* (This idea applies to any voltage or current in an ac circuit. All steady state ac circuit analysis is based on this concept, although phasor voltages and currents are usually shown as rms values rather than the peak amplitudes. We will make this change shortly.) The value of this approach is illustrated next.

Summing ac Voltages and Currents

Sinusoidal quantites must sometimes be added or subtracted as in Figure 16–10. Here, we want the sum of e_1 and e_2, where $e_1 = 10 \sin \omega t$ and $e_2 = 15 \sin(\omega t + 60°)$. The sum of e_1 and e_2 can be found by adding waveforms point by point as in (b). For example, at $\omega t = 0°$, $e_1 = 10 \sin 0° = 0$ and $e_2 = 15 \sin(0° + 60°) = 13 \text{ V}$, and their sum is 13 V. Similarly, at $\omega t = 90°$, $e_1 = 10 \sin 90° = 10 \text{ V}$ and $e_2 = 15 \sin(90° + 60°) = 15 \sin 150° = 7.5$, and their sum is 17.5 V. Continuing in this manner, the sum of $e_1 + e_2$ (the green waveform) is obtained.

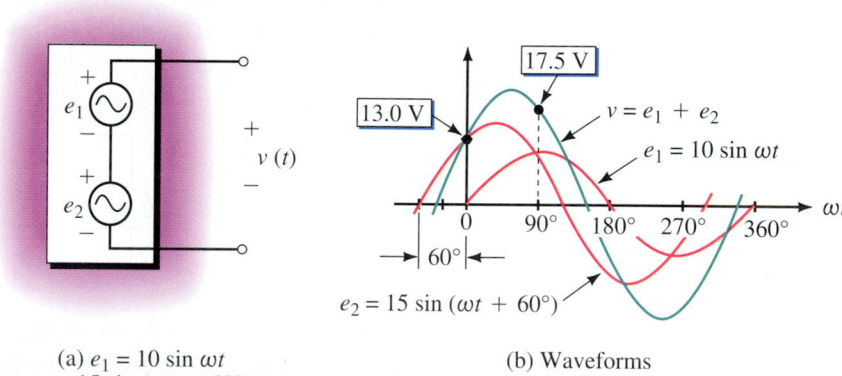

(a) $e_1 = 10 \sin \omega t$
    $e_2 = 15 \sin (\omega t + 60°)$

(b) Waveforms

Figure 16–10 Summing waveforms e_1 and e_2 point by point. While this process works, it is tedious.

As you can see, the process is tedious and provides no analytic expression for the resulting voltage. A better way is to transform the sources and use complex number addition. This is shown in Figure 16–11. Here, we have replaced voltages e_1 and e_2 with their phasor equivalents, $\mathbf{E}_1$ and $\mathbf{E}_2$, and v with its phasor equivalent, $\mathbf{V}$. Since $v = e_1 + e_2$, replacing v, e_1, and e_2 with their phasor equivalents yields $\mathbf{V} = \mathbf{E}_1 + \mathbf{E}_2$.

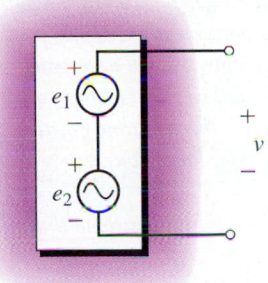

(a) Original network.
$v(t) = e_1(t) + e_2(t)$

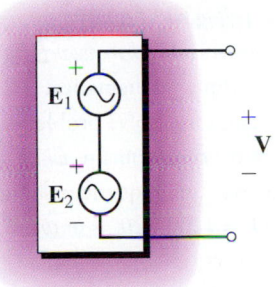

(b) Transformed circuit.
$V = E_1 + E_2$

Figure 16–11   Transformed circuit. This is one of the key ideas of sinusoidal circuit analysis.

Now $\mathbf{V}$ can be found by adding $\mathbf{E}_1$ and $\mathbf{E}_2$ as complex numbers. Once $\mathbf{V}$ is known, its corresponding time equation and companion waveform can be determined.

■ **EXAMPLE 16–6**   Given $e_1 = 10 \sin \omega t$ V and $e_2 = 15 \sin(\omega t + 60°)$ V as before, determine v and sketch it.

Solution $e_1 = 10 \sin \omega t$ V. Thus, $\mathbf{E}_1 = 10$ V$\angle 0°$.

$e_2 = 15 \sin(\omega t + 60°)$ V. Thus, $\mathbf{E}_2 = 15$ V$\angle 60°$.

Transformed sources are shown in Figure 16–12(a) and phasors in (b).

$$\mathbf{V} = \mathbf{E}_1 + \mathbf{E}_2 = 10\angle 0° + 15\angle 60° = (10 + j0) + (7.5 + j13)$$

$$= (17.5 + j13) = 21.8 \text{ V}\angle 36.6°$$

$$v = 21.8 \sin(\omega t + 36.6°) \text{ V}$$

Waveforms are shown in (c).

(a) Phasor summation

(b) Phasors   (c) Waveforms

Figure 16–12

practice problem 11

Verify by direct substitution that $v = 21.8 \sin(\omega t + 36.6°)$ V, as in Figure 16–12, is the sum of e_1 and e_2. To do this, compute e_1 and e_2 at a point, add them, then compare the sum to $21.8 \sin(\omega t + 36.6°)$ V computed at the same point. Perform this computation at $\omega t = 30°$ intervals over the complete cycle to satisfy yourself that the result is true everywhere. (For example, at $\omega t = 0°$, $v = 21.8 \sin(\omega t + 36.6°) = 21.8 \sin(36.6°) = 13$ V, as we saw earlier in Figure 16–10.)

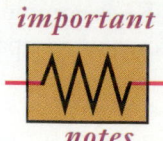

important

notes

1. *To this point, we have used peak values such as V_m and I_m to represent the magnitudes of phasor voltages and currents, as this has been most convenient for our purposes. In practice, however, rms values are used instead. Accordingly, we will now change to rms. Thus, from here on, the phasor $V = 120\ V\ \angle 0°$ will be taken to mean a voltage of 120 volts rms at an angle of $0°$. If you need to convert this to a time function, first multiply the rms value by $\sqrt{2}$, then follow the usual procedure. Thus, $v = \sqrt{2}\ (120)\ sin\ \omega t = 170\ sin\ \omega t$.*

2. *To add or subtract sinusoidal voltages or currents, follow the three steps outlined in Example 16–6. That is,*

   • *convert sine waves to phasors and express them in complex number form,*

   • *add or subtract the complex numbers,*

   • *convert back to time functions if desired.*

3. *Although we use phasors to represent sinusoidal waveforms, it should be noted that sine waves and phasors are not the same thing. Sinusoidal voltages and currents are real— they are the actual quantities that you measure with meters and whose waveforms you see on oscilloscopes. Phasors, on the other hand, are simply mathematical abstractions that we use to help visualize relationships and solve problems.*

4. *Quantities expressed as time functions are said to be in the time domain, while quantities expressed as phasors are said to be in the phasor (or frequency) domain. Thus, $e = 170\ sin\ \omega t\ V$ is in the time domain, while $V = 120\ V\ \angle 0°$ is in the phasor domain.*

■ **EXAMPLE 16–7** Express the voltages and currents of Figure 16–13 in both the time and the phasor domains.

(a)   (b)

Figure 16–13

Solution

a. Time domain: $v = 100\ sin(\omega t + 80°)$ volts.
   Phasor domain: $V = (0.707)\ (100\ V\ \angle 80°) = 70.7\ V\angle 80°$.

b. Time domain: $i = 40\ sin(\omega t - 25°)$ mA.
   Phasor Domain: $I = (0.707)(40\ mA\angle -25°) = 28.3\ mA\angle -25°$.

■ **EXAMPLE 16–8**   If $i_1 = 14.14 \sin(\omega t - 55°)$ A and $i_2 = 4 \sin(\omega t + 15°)$ A, determine their sum, i. Work with rms values.

Solution

$\mathbf{I}_1 = (0.707)(14.14 \text{ A})\angle{-55°} = 10 \text{ A}\angle{-55°}$

$\mathbf{I}_2 = (0.707)(4 \text{ A})\angle{15°} = 2.828 \text{ A}\angle{15°}$

$\mathbf{I} = \mathbf{I}_1 + \mathbf{I}_2 = 10 \text{ A}\angle{-55°} + 2.828 \text{ A}\angle{15°} = (5.74 \text{ A} - j8.19 \text{ A}) + (2.73 \text{ A} + j0.732 \text{ A})$

$= 8.47 \text{ A} - j7.46 \text{ A} = 11.3 \text{ A}\angle{-41.4}$

$i(t) = \sqrt{2}(11.3) \sin(\omega t - 41.4°) = 16 \sin(\omega t - 41.4°)$ A

practice
problems
III

1. Convert the following to time functions and sketch waveforms. Values are rms.

   a. $\mathbf{E} = 500 \text{ mV}\angle{-20°}$.   b. $\mathbf{I} = 80 \text{ A}\angle{40°}$.

2. For the circuit of Figure 16–14, determine voltage e_1.

Figure 16–14

in process
learning
check I

1. Convert the following to polar form:
   a. $j6$.   b. $-j4$.   c. $3 + j3$.   d. $4 - j6$.
   e. $-5 + j8$.   f. $1 - j2$.   g. $-2 - j3$.

2. Convert the following to rectangular form:
   a. $4\angle{90°}$.   b. $3\angle{0°}$.   c. $2\angle{-90°}$.   d. $5\angle{40°}$
   e. $6\angle{120°}$.   f. $2.5\angle{-20°}$.   g. $1.75\angle{-160°}$.

3. If $-\mathbf{C} = 12\angle{-140°}$, what is $\mathbf{C}$?

4. Given: $\mathbf{C}_1 = 36 + j4$ and $\mathbf{C}_2 = 52 - j11$. Determine $\mathbf{C}_1 + \mathbf{C}_2$ and $\mathbf{C}_1 - \mathbf{C}_2$.

5. Given: $\mathbf{C}_1 = 24\angle{25°}$ and $\mathbf{C}_2 = 12\angle{-125°}$. Determine $\mathbf{C}_1 \cdot \mathbf{C}_2$ and $\mathbf{C}_1/\mathbf{C}_2$.

6. Compute the following and express answers in rectangular form:

   a. $\dfrac{6 + j4}{10\angle{20°}} + (14 + j2)$.   b. $(1 + j6) + \left[2 + \dfrac{(12\angle{0°})(14 + j2)}{6 - (10\angle{20°})(2\angle{-10°})}\right]$

7. For Figure 16–15, determine i_T where $i_1 = 10 \sin \omega t$, $i_2 = 20 \sin(\omega t - 90°)$, and $i_3 = 5 \sin(\omega t + 90°)$.

Figure 16–15

16.3 *R, L, and C* Circuits with Sinusoidal Excitation

R, *L*, and *C* circuit elements each have quite different electrical properties. Resistance, for example, opposes current, while inductance opposes changes in current and capacitance opposes changes in voltage. These differences result in quite different voltage-current relationships as you saw earlier. We now investigate these relationships for the case of sinusoidal ac. Sine waves have several important characteristics that you will discover from this investigation:

1. When a circuit consisting of linear circuit elements *R*, *L*, and *C* is connected to a sinusoidal source, all currents and voltages in the circuit will be sinusoidal.

2. These sine waves have the same frequency as the source and differ from it only in terms of their magnitudes and phase angles.

16.4 Resistance and Sinusoidal ac

For a purely resistive circuit, current is directly proportional to voltage. Current variations therefore follow voltage variations, reaching their peak when voltage reaches its peak, changing direction when voltage changes polarity, and so on (Figure 16–16). From this, we conclude that *for a purely resistive circuit, current and voltage are in phase*. Since voltage and current waveforms coincide, their phasors also coincide (Figure 16–17).

(a) Source voltage is a sine wave. Therefore, v_R is a sine wave.

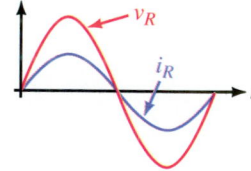

(b) $i_R = V_R/R$. Therefore i_R is a sine wave.

Figure 16–16   Ohm's law applies for ac circuits. Note that current and voltage are in phase.

Figure 16–17   Phasor relationship for a resistive circuit.

The relationship illustrated in Figure 16–16 may be stated mathematically as

$$i_R = \frac{v_R}{R} = \frac{V_m \sin \omega t}{R} = \frac{V_m}{R} \sin \omega t = I_m \sin \omega t \qquad (16\text{–}10)$$

where

$$I_m = V_m / R \qquad (16\text{–}11)$$

Transposing,

$$V_m = I_m R \qquad (16\text{–}12)$$

The in-phase relationship is true regardless of reference. Thus, if $v_R = V_m \sin(\omega t + \theta)$, then $i_R = I_m \sin(\omega t + \theta)$.

■ **EXAMPLE 16–9**   For the circuit of Figure 16–16(a), if $R = 5\ \Omega$ and $i_R = 12 \sin (\omega t - 18)$ A, determine v_R.

Solution $v_R = R i_R = 5 \times 12 \sin(\omega t - 18°) = 60 \sin(\omega t - 18°)$ V. The waveforms are shown in Figure 16–18.

Figure 16–18

practice
problems
IV

1. If $v_R = 150 \cos \omega t$ V and $R = 25$ kΩ, determine i_R and sketch both waveforms.

2. If $v_R = 100 \sin(\omega t + 30°)$ V and $R = 0.2$ MΩ, determine i_R and sketch both waveforms.

16.5 Inductance and Sinusoidal ac

Phase Lag in an Inductive Circuit

As you saw in Chapter 13, for an ideal inductor, voltage $v_L + 2$ is proportional to the rate of change of current (Figure 16–19). Because of this relationship, voltage and current are not in phase as they are for a resistive circuit. To see why, consider Figure 16–20. Here, current i_L is a sine wave; thus, v_L will be a cosine wave (since the rate of change of a sine wave is a cosine wave as you saw in Chapter 15, Section 15.10). Phasors are shown in Figure 16–21. As you can see, *for a purely inductive circuit, current lags voltage by 90°* or alternatively, voltage leads current by 90°.

Figure 16–19 Voltage v_L is proportional to the rate of change of current i_L.

Figure 16–20 Current lags voltage by 90°.

Figure 16–21 Phasors for the waveforms of Figure 16–20 showing the 90° lag of current I_L.

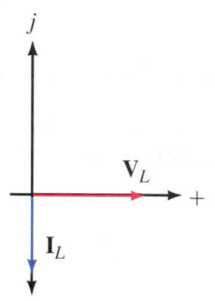

(a) Current I_L always lags voltage V_L by 90°

Although we have shown that current lags voltage by 90° for the case of Figure 16–20, this relationship is true in general, that is, current lags voltage by 90° regardless of the choice of reference. This is illustrated in Figure 16–22. Here, V_L is at 0° and I_L at −90°. Thus, voltage v_L will be a sine wave and current i_L a negative cosine wave, i.e., $i_L = -I_m\cos \omega t$. Since i_L is a negative cosine wave, it can also be expressed as $i_L = I_m\sin(\omega t - 90°)$. The waveforms are shown in (b).

Since current always lags voltage by 90° for a pure inductance, you can, if you know the phase of the voltage, determine the phase of the current, and vice versa. Thus, if v_L is known, i_L must lag it by 90°, while if i_L is known, v_L must lead it by 90°.

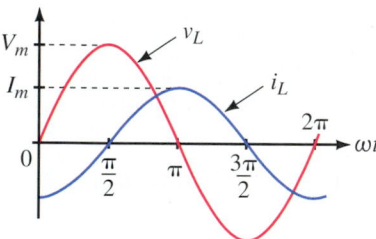

Figure 16–22 Phasors and waveforms when V_L is selected as reference.

Inductive Reactance

We now need to determine the relationship between voltage and current amplitudes. Consider again the sine wave of current $i_L = I_m\sin \omega t$ of Figure 16–20. Applying Faraday's law yields

$$v_L = L\frac{di_L}{dt} = L\frac{d}{dt}(I_m\sin \omega t) = \omega L I_m\cos \omega t$$

$$= V_m\cos \omega t = V_m\sin(\omega t + 90°) \qquad (16\text{–}13)$$

where

$$V_m = \omega L I_m \qquad (16\text{–}14)$$

From this, we see that the ratio V_m to I_m is

$$\frac{V_m}{I_m} = \omega L \qquad (16\text{–}15)$$

This ratio is defined as *inductive reactance* and is given the symbol X_L. Since the ratio of volts to amps is ohms, reactance has units of ohms. Thus,

$$X_L = \frac{V_m}{I_m} \quad (\Omega) \qquad (16\text{–}16)$$

Combining Equations 16–15 and 16–16 yields

$$X_L = \omega L \quad (\Omega) \qquad (16\text{–}17)$$

where ω is in radians per second and L is in henries. *Reactance X_L represents the opposition that inductance presents to current for the sinusoidal ac case.*

We now have everything that we need to solve simple inductive circuits with sinusoidal excitation, that is, we know that current lags voltage by 90° and that their amplitudes are related by

$$I_m = \frac{V_m}{X_L} \tag{16–18}$$

and

$$V_m = I_m X_L \tag{16–19}$$

■ **EXAMPLE 16–10**   The voltage across a 0.2-H inductance is $v_L = 100 \sin(400t + 70°)$. Determine i_L and sketch it.

Solution $\omega = 400$ rad/s. Therefore, $X_L = \omega L = (400)(0.2) = 80\ \Omega$.

$$I_m = \frac{V_m}{X_L} = \frac{100\text{ V}}{80\ \Omega} = 1.25\text{ A}$$

The current lags the voltage by 90°. Therefore $i_L = 1.25 \sin(400t - 20°)$ as indicated in Figure 16–23.

Figure 16–23

■ **EXAMPLE 16–11**   The current through a 0.01-H inductance is $i_L = 20 \sin(\omega t - 50°)$ A and $f = 60$ Hz. Determine v_L.

Solution

$$\omega = 2\pi f = 2\pi(60) = 377\text{ rad/s}$$
$$X_L = \omega L = (377)(0.01) = 3.77\ \Omega$$
$$V_m = I_m X_L = (20\text{ A})(3.77\ \Omega) = 75.4\text{ V}$$

Voltage leads current by 90°. Thus, $v_L = 75.4 \sin(377t + 40°)$ V as shown in Figure 16–24.

(a)   (b)

Figure 16–24

practice
problems
V

1. Two inductances are connected in series (Figure 16–25). If $e = 100 \sin \omega t$ and $f = 10$ kHz, determine the current. Sketch voltage and current waveforms.

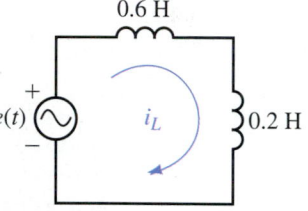

Figure 16–25

2. The voltage across a 0.5-H inductance is $v_L = 100 \sin(2400t + 60°)$. Determine i_L and sketch voltage and current waveforms.

Variation of Inductive Reactance with Frequency

Since $X_L = \omega L = 2\pi f L$, inductive reactance is directly proportional to frequency (Figure 16–26). Thus, if frequency is doubled, reactance doubles, while if frequency is halved, reactance halves, and so on. In addition, X_L is directly proportional to inductance. Thus, if inductance is doubled, X_L is doubled, and so on. Note also that at $f = 0$, $X_L = 0$. This means that inductance looks like a short circuit to dc. (We already concluded this earlier in Chapter 13.)

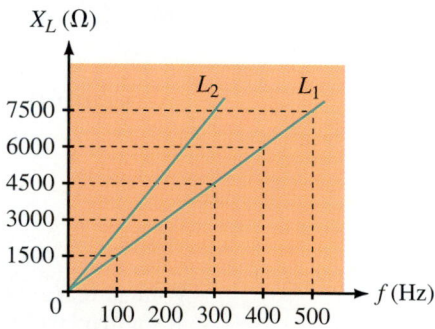

Figure 16–26   Variation of X_L with frequency. Note that $L_2 > L_1$.

A circuit has 50 ohms inductive reactance. If both the inductance and the frequency are doubled, what is the new X_L?

16.6 Capacitance and Sinusoidal ac

Phase Lead in a Capacitive Circuit

As you learned in Chapter 10, capacitor current is proportional to the rate of change of capacitor voltage. Consider Figure 16–27. Here, v_C is a sine wave; thus, i_C will be a cosine wave (since the rate of change of a sine wave is a cosine wave). This is demonstrated in (b) and (c). Phasors are shown in (d). As these indicate, *for a purely capacitive circuit, current leads voltage by 90°*, or alternatively, voltage lags current by 90°. This relationship is true under all circumstances. Thus, if the voltage is known, the current must lead by 90°; and if the current is known, the voltage must lag by 90°. For example, if $\mathbf{I}_C$ is at 60° as in (e), $\mathbf{V}_C$ must be at −30°.

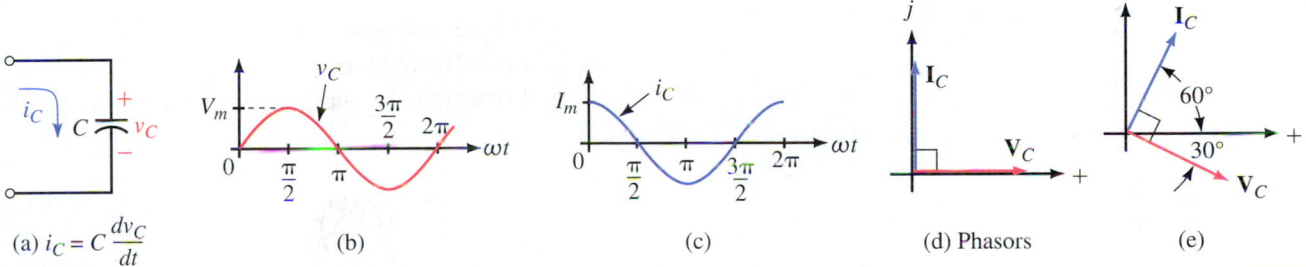

(a) $i_C = C\dfrac{dv_C}{dt}$ (b)   (c)   (d) Phasors   (e)

Figure 16–27   Current $\mathbf{I}_C$ always leads voltage $\mathbf{V}_c$ by 90°.

1. The current source of Figure 16–28 is a sine wave. Sketch capacitor voltage v_C.

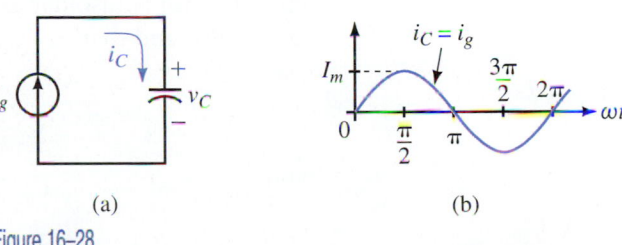

(a)   (b)

Figure 16–28

2. Refer to the circuit of Figure 16–29:
  a. Sketch the phasors.
  b. Sketch capacitor current i_C.

(a)   (b)

Figure 16–29

Capacitive Reactance

Now consider the relationship between V_m and I_m. This can be determined from $i_C = C dv_C/dt$. Assume $v_C = V_m \sin \omega t$. Then,

$$i_C = C\frac{dv_C}{dt} = C\frac{d}{dt}(V_m\sin \omega t) = \omega C V_m \cos \omega t$$

$$= I_m\cos \omega t = I_m\sin (\omega t + 90°) \qquad (16\text{–}20)$$

where

$$I_m = \omega C V_m \qquad (16\text{–}21)$$

The ratio of V_m to I_m is defined as *capacitive reactance* and is given the symbol X_C. Thus,

$$X_C = \frac{V_m}{I_m} \quad (\Omega)$$

From Equation 16–21,

$$X_C = \frac{1}{\omega C} \quad (\Omega) \qquad (16\text{–}22)$$

where ω is in radians per second and C is in farads. *Reactance X_C represents the opposition that capacitance presents to current for the sinusoidal ac case.*

We now have everything that we need to solve simple capacitive circuits with sinusoidal excitation, i.e., we know that current leads voltage by 90° and that

$$I_m = \frac{V_m}{X_C} \qquad (16\text{–}23)$$

and

$$V_m = I_m X_C \qquad (16\text{–}24)$$

■ **EXAMPLE 16–12**   The voltage across a 10-μF capacitance is v_C = 100 sin(ωt – 40°) V and f = 1000 Hz. Determine i_C and sketch its waveform.

Solution

$$\omega = 2\pi f = 2\pi(1000 \text{ Hz}) = 6283 \text{ rad/s}$$

$$X_C = \frac{1}{\omega C} = \frac{1}{(6283)(10 \times 10^{-6})} = 15.92 \ \Omega$$

$$I_m = \frac{V_m}{X_C} = \frac{100 \text{ V}}{15.92 \ \Omega} = 6.28 \text{ A}$$

Since current leads voltage by 90°, i_C = 6.28 sin(6283t + 50°) A as indicated in Figure 16–30.

(a)   (b)

Figure 16–30   Phasors not to scale with waveform.

■ **EXAMPLE 16–13**   The current through a 0.1-μF capacitance is i_C = 5 sin(1000t + 120°) mA. Determine v_C.

Solution

$$X_C = \frac{1}{\omega C} = \frac{1}{(1000 \text{ rad/s}) (0.1 \times 10^{-6} \text{ F})} = 10 \text{ k}\Omega$$

Thus, $V_m = I_m X_C$ = (5 mA)(10 kΩ) = 50 V. Since voltage lags current by 90°, v_C = 50 sin(1000t + 30°). Waveforms and phasors are shown in Figure 16–31.

(a)   (b)

Figure 16–31   Phasors not to scale with waveform.

practice
problem
VIII

Two capacitances are connected in parallel (Figure 16–32). If $e = 100 \sin \omega t$ V and $f = 10$ Hz, determine the source current. Sketch current and voltage.

Figure 16–32

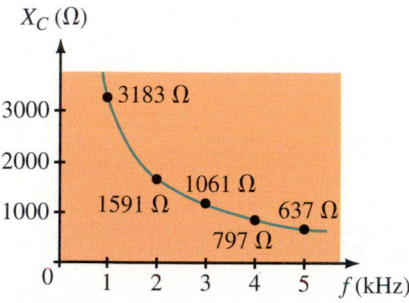

Figure 16–33 X_C varies inversely with frequency. Values shown are for $C = 0.05$ μF.

Variation of Capacitive Reactance with Frequency

Since $X_C = 1/\omega C = 1/2\pi f C$, the opposition that capacitance presents varies inversely with frequency. This means that the higher the frequency, the lower the reactance, and vice versa (Figure 16–33). At $f = 0$ (i.e., dc), capacitive reactance is infinite. This means that a capacitance looks like an open circuit to dc. (We already concluded this earlier in Chapter 10.) Note that X_C is also inversely proportional to capacitance. Thus, if capacitance is doubled, X_C is halved, and so on.

in process
learning
check II

1. For a pure resistance, $v_R = 100 \sin(\omega t + 30°)$ V. If $R = 2$ Ω, what is the expression for i_R?

2. For a pure inductance, $v_L = 100 \sin(\omega t + 30°)$ V. If $X_L = 2$ Ω, what is the expression for i_L?

3. For a pure capacitance, $v_C = 100 \sin(\omega t + 30°)$ V. If $X_C = 2$ Ω, what is the expression for i_C?

4. If $f = 100$ Hz and $X_L = 400$ Ω, what is L?

5. If $f = 100$ Hz and $X_C = 400$ Ω, what is C?

6. For each of the phasor sets of Figure 16–34, identify whether the circuit is resistive, inductive, or capacitive. Justify your answers.

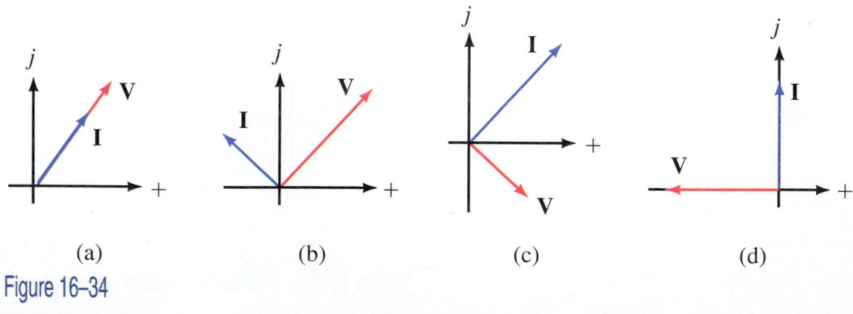

Figure 16–34

16.7 The Impedance Concept

In Sections 16.5 and 16.6, we handled magnitude and phase analysis separately. However, this is not the way to do it in practice. In practice, we represent circuit elements by their *impedance*, and determine magnitude and phase relationships in one step. Before we do this, however, we need to learn how to represent circuit elements as impedances.

$$Z = \frac{V}{I} \text{ ohms}$$

Figure 16–35   Impedance concept.

Impedance

The opposition that a circuit element presents to current in the phasor domain is defined as its *impedance*. The impedance of the element of Figure 16–35, for example, is the ratio of its voltage phasor to its current phasor. Impedance is denoted by the boldface, uppercase letter **Z**. Thus,

$$\mathbf{Z} = \frac{\mathbf{V}}{\mathbf{I}} \quad \text{(ohms)} \qquad (16\text{–}25)$$

(This equation is sometimes referred to as Ohm's law for ac circuits.)

Since phasor voltages and currents are complex, **Z** is also complex. That is,

$$\mathbf{Z} = \frac{\mathbf{V}}{\mathbf{I}} = \frac{V}{I}\angle\theta \qquad (16\text{–}26)$$

where V and I are the rms magnitudes of **V** and **I** respectively, and θ is the angle between them. From Equation 16–26,

$$\mathbf{Z} = Z\angle\theta \qquad (16\text{–}27)$$

where $Z = V/I$. Since $V = 0.707V_m$ and $I = 0.707I_m$, Z can also be expressed as V_m/I_m. Once the impedance of a circuit is known, the current and voltage can be determined using

$$\mathbf{I} = \frac{\mathbf{V}}{\mathbf{Z}} \qquad (16\text{–}28)$$

and

$$\mathbf{V} = \mathbf{IZ} \qquad (16\text{–}29)$$

Note: Although **Z** is a complex number, it is not a phasor since it does not represent a sinusoidally varying quantity.

Let us now determine impedance for the basic circuit elements R, L, and C.

Resistance: For a pure resistance, voltage and current are in phase. Assuming a 0° angle for voltage (we can assume any angle we want because we are interested only in the angle between **V** and **I**), we can write $\mathbf{V}_R = V_R\angle 0°$ and $\mathbf{I} = I\angle 0°$. The impedance of a pure resistance (Figure 16–36) is therefore

$$\mathbf{Z}_R = \frac{\mathbf{V}_R}{\mathbf{I}} = \frac{V_R\angle 0°}{I\angle 0°} = \frac{V_R}{I}\angle 0° = R\angle 0° = R$$

(a) Voltage
and current

$Z_R = R \ \Omega$

(b) Impedance

Figure 16–36 Impedance of a pure resistance.

Thus,

$$\mathbf{Z}_R = R \qquad (16\text{–}30)$$

This agrees with what we know about resistive circuits, i.e., that the ratio of voltage to current is R, and that the angle between them is 0°.

Inductance.   For a pure inductance, current lags voltage by 90°. Assuming a 0° angle for voltage (again we can assume any reference we want), we can write $\mathbf{V}_L = V_L \angle 0°$ and $\mathbf{I} = I\angle -90°$. The impedance of a pure inductance (Figure 16–37) is therefore

$$\mathbf{Z}_L = \frac{\mathbf{V}_L}{\mathbf{I}} = \frac{V_L \angle 0°}{I\angle -90°} = \frac{V_L}{I}\angle 90° = \omega L\angle 90° = j\omega L$$

where we have used the fact that $V_L/I_L = \omega L$. Thus,

$$\mathbf{Z}_L = j\omega L = jX_L \qquad (16\text{–}31)$$

since ωL is equal to X_L.

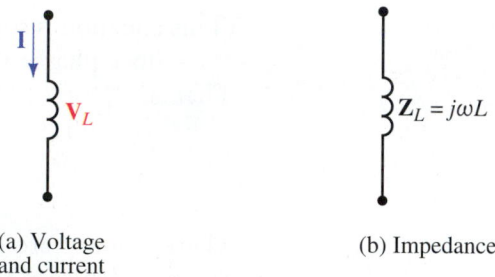

(a) Voltage
and current

(b) Impedance

Figure 16–37   Impedance of a pure inductance.

■ **EXAMPLE 16–14**   Consider again the circuit of Example 16–10. Given $v_L = (100 \sin 400t + 70°)$ and $L = 0.2$ H, determine i_L using the impedance concept.

Solution   See Figure 16–38.

$$\mathbf{V}_L = 70.7 \text{ V}\angle 70° \quad \text{and} \quad \omega = 400 \text{ rad/s}$$

$$\mathbf{Z}_L = j\omega L = j(400)(0.2) = j80 \ \Omega$$

$$\mathbf{I}_L = \frac{\mathbf{V}_L}{\mathbf{Z}_L} = \frac{70.7\angle 70°}{j\,80} = \frac{70.7\angle 70°}{80\angle 90°} = 0.884 \text{ A}\angle -20°$$

In the time domain, $i_L = \sqrt{2}(0.884) \sin(400t - 20°) = 1.25 \sin(400t - 20°)$ A as before.

$\mathbf{Z}_L = j80 \ \Omega$ $\mathbf{V}_L = 70.7 \text{ V}\angle 0°$

Figure 16–38

$$\mathbf{Z}_C = -j\frac{1}{\omega C}$$

Figure 16–39 Impedance of a pure capacitance.

Capacitance. For a pure capacitance, current leads voltage by 90°. Its impedance (Figure 16–39) is therefore

$$\mathbf{Z}_C = \frac{\mathbf{V}_C}{\mathbf{I}} = \frac{V_C \angle 0°}{I \angle 90°} = \frac{V_C}{I} \angle -90° = \frac{1}{\omega C} \angle -90° = -j\frac{1}{\omega C} \quad \text{(ohms)}$$

Thus,

$$\mathbf{Z}_C = -j\frac{1}{\omega C} = -jX_C \quad \text{(ohms)} \tag{16–32}$$

since $1/\omega C$ is equal to X_C.

■ **EXAMPLE 16–15**   Given $v_C = 100 \sin(\omega t - 40°)$, $f = 1000$ Hz, and $C = 10\ \mu F$, determine i_C.

Solution

$$\omega = 2\pi f = 2\pi(1000 \text{ Hz}) = 6283 \text{ rads/s}$$

$$\mathbf{V}_C = 70.7 \text{ V} \angle -40°$$

$$\mathbf{Z}_C = -j\frac{1}{\omega C} = -j\left(\frac{1}{6283 \times 10 \times 10^{-6}}\right) = -j15.92\ \Omega.$$

$$\mathbf{I}_C = \frac{\mathbf{V}_C}{\mathbf{Z}_C} = \frac{70.7 \angle -40°}{-j15.92} = \frac{70.7 \angle -40°}{15.92 \angle -90°} = 4.442 \text{ A} \angle 50°$$

In the time domain, $i_C = \sqrt{2}(4.442) \sin(6283t + 50°) = 6.28 \sin(6283t + 50°)$ A, which agrees with our previous solution, in Example 16–12.

$$\mathbf{Z}_C = -j15.92\ \Omega \qquad \mathbf{V}_C = 70.7 \text{ V} \angle -40°$$

Figure 16–40

---

important

notes

In the above examples, we changed from peak to rms by dividing by $\sqrt{2}$ at the beginning of the problem, then changed back to peak at the end by multiplying by $\sqrt{2}$. Although this is a bit clumsy (since you can work directly with peak values all the way through), we did it for a reason—to get used to the idea of working with rms values. In later chapters (and in practice), you will use rms values almost exclusively when you work with sinusoidal ac.

---

*practice
problems
IX*

1. If $\mathbf{I}_L = 5 \text{ mA} \angle -60°$, $L = 2$ mH, and $f = 10$ kHz, what is $\mathbf{V}_L$?

2. A capacitor has a reactance of 50 Ω at 1200 Hz. If $v_C = 80 \sin 800t$ V, what is i_C?

A Final Note

The real power of the impedance method becomes apparent when you consider complex circuits with elements in series, parallel, and so on. This we do later, beginning in Chapter 18. Before we do this, however, there are some ideas on power that you need to know. These are considered in Chapter 17.

16.8 Computer Analysis of ac Circuits

Analysis of ac circuits requires some new ideas. In this section, we introduce the .AC and the .PRINT AC commands. We also use the .PROBE command (which was introduced in Chapter 11). All are summarized in Appendix A.

An ac Example

To begin, let us calculate current at 1 kHz for the circuit of Figure 16–41. A PSpice solution follows. (Explanations of new ideas follow the input file.)

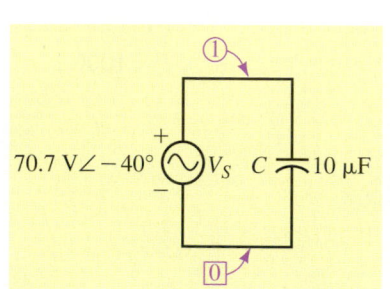

Figure 16–41   Circuit for PSpice example.

```
Figure 16-41: Capacitance circuit (f = 1000 Hz)
Vs      1     0      AC      70.7V     -40deg
C       1     0      10uF
.AC     LIN    1   1000Hz 1000Hz
.PRINT AC    IM(C)   IP(C)
.OPTIONS NOPAGE
.END
```

Analysis

Line 2: Specifies that the voltage source V_S between nodes 1 and 0 is ac and that it has a magnitude of 70.7 volts and a phase angle of –40°.

Line 4: This sets the frequency to 1000 Hz as described in Appendix A.

Line 5: Prints results to an output file. Here, I specifies current, M specifies magnitude, and P phase. Thus, IM(C) denotes the magnitude of the current through capacitor C, while IP(C) denotes its phase.

Key in and run this program. An abbreviated output file follows. Note that the magnitude of capacitor current is 4.442 A and its angle is 50 degrees. Thus, $\mathbf{I} = 4.442$ A $\angle 50°$ as we determined earlier in Example 16–15.

```
Output:
****        AC ANALYSIS
FREQ              IM(C)              IP(C)
1.000E + 03     4.442E + 00     5.000E + 01
```

Another Example

The next example uses the .AC statement to compute values over a range of frequencies and the .PROBE statement to plot results.

■ **EXAMPLE 16–16**   Compute and plot the capacitive reactance for the circuit of Figure 16–41 over the range 1 Hz to 200 Hz.

Solution   PSpice has no command to compute reactance; however, we can calculate the voltage and current over the desired frequency range, then use .PROBE to plot their ratio. This gives reactance. The input file is shown below. Note that we have changed the .AC statement to calculate values at 50 linearly spaced points from 1 Hz to 200 Hz. The .PRINT command has been replaced with the .PROBE command so that you can view results on the screen.

```
Figure 16-41: Computing Data for a Reactance Plot
Vs    1    0    AC    70.7V    -40deg
C     1    0    10uF
.AC    LIN    50    1Hz    200Hz
.PROBE
.END
```

After the program is run, a set of *xy* axes appears on the screen. To display reactance, do the following:

1. Move the cursor to *X_axis* and press the Enter key.
2. Move the cursor to *Linear* and press the Enter key.
3. Move the cursor to *Exit* and press the Enter key.
4. Move cursor to *Add_trace* and press the Enter key.
5. Key in *V(1)/I(C)* and press the Enter key. The display shown as Figure 16–42 should appear on the screen. Compare this to Figure 16–33.

Figure 16–42   Reactance for a 10-μF capacitor versus frequency.

Modify Example 16–16 to plot both capacitor current and reactance on the same graph. To do this, follow steps 1, 2, and 3 as above, then follow the procedure in Appendix A, Displaying Multiple Traces.

practice
problem
X

Problems

16.1 Complex Number Review

1. Convert each of the following to polar form:
   a. $5 + j12$.   b. $9 - j6$.   c. $-8 + j15$.   d. $-10 - j4$.

2. Convert each of the following to rectangular form:
   a. $6\angle 30°$.   b. $14\angle 90°$.   c. $16\angle 0°$.   d. $6\angle 150°$.
   e. $20\angle -140°$.   f. $-12\angle 30°$.   g. $-15\angle -150°$.

3. Plot each of the following on the complex plane:
   a. $4 + j6$.   b. $j4$.   c. $6\angle -90°$.   d. $10\angle 135°$.

4. Simplify the following using powers of j:
   a. $j(1 - j1)$.   b. $(-j)(2 + j5)$.   c. $j[j(1 + j6)]$.
   d. $(j4)(-j2 + 4)$.   e. $(2 + j3)(3 - j4)$.

5. Add or subtract as indicated. Express your answer in rectangular form.
   a. $(4 + j8) + (3 - j2)$.   b. $(4 + j8) - (3–j2)$.
   c. $(4.1 - j7.6) + 12\angle 20°$.   d. $2.9\angle 25° - 7.3\angle -5°$.
   e. $9.2\angle -120° - (2.6 + j4.1)$.

6. Multiply or divide as indicated. Express your answer in polar form.
   a. $(37 + j9.8)(3.6–j12.3)$.   b. $(41.9\angle -80°)(16 + j2)$.
   c. $\dfrac{42 + j18.6}{19.1 - j4.8}$ d. $\dfrac{42.6 + j187.5}{11.2\ 38°}$

7. Reduce each of the following to polar form:
   a. $15 - j6 - \left[\dfrac{18\angle 40° + (12 + j8)}{11 + j11}\right]$.

   b. $\dfrac{21\angle 20° - j41}{36\angle 0° + (1 + j12) - 11\angle 40°}$.

   c. $\dfrac{18\angle 40° - 18\angle -40°}{7 + j12} -$

   $\dfrac{16 + j17 + 21\angle -60°}{4}$.

16.2 Complex Numbers in ac Analysis

8. In the manner of Figure 16–9, represent each of the following

a. $e = 100 \sin(\omega t + 30°)$ V.   b. $e = 15 \sin(\omega t - 20°)$ V.

c. $e = 50 \sin(\omega t + 90°)$ V.   d. $e = 50 \cos \omega t$ V.

e. $e = -40 \sin(\omega t + 120°)$ V.

9. Determine the sinusoidal equivalent for each of the transformed sources of Figure 16–43.

(a)   (b)

Figure 16–43

10. Given: $e_1 = 10 \sin(\omega t + 30°)$ V and $e_2 = 15 \sin(\omega t - 20°)$ V. Determine their sum $v = e_1 + e_2$ in the manner of Example 16–6:

a. Convert e_1 and e_2 to phasor form.

b. Determine $\mathbf{V} = \mathbf{E}_1 + \mathbf{E}_2$.

c. Convert $\mathbf{V}$ to the time domain.

d. Sketch e_1, e_2, and v as per Figure 16–12.

11. Repeat Problem 10 for $v = e_1 - e_2$.

Note: For the remaining problems and throughout the remainder of the book, express phasor quantities as rms values rather than as peak values.

12. Express the voltages and currents of Figure 16–44 as time domain and phasor domain quantities.

13. For Figure 16–45, $i_T = 50 \sin(\omega t + 60°)$ A and $i_2 = 20 \sin(\omega t - 30°)$ A.

a. Determine phasors $\mathbf{I}_T$ and $\mathbf{I}_2$.

b. Determine $\mathbf{I}_1$.

c. From (b), determine i_1.

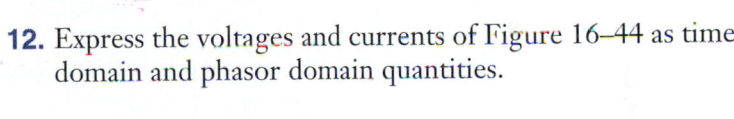

Figure 16–45

14. For Figure 16–15, $i_T = 38.08 \sin(\omega t - 21.8°)$, $i_1 = 35.36 \sin \omega t$, and $i_2 = 28.28 \sin(\omega t - 90°)$ A. Determine i_3.

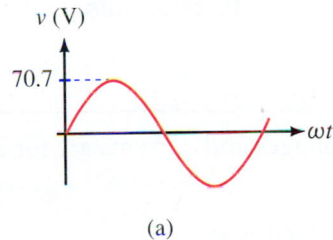

v (V)

70.7

ωt

(a)

v (V)

90

150°

ωt

(b)

i (mA)

100

ωt

(c)

v (V)

10

ωt

200°

(d)

Figure 16–44

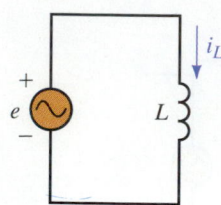

Figure 16–46

16.5 Inductance and Sinusoidal ac

15. For a certain inductance, $V_m = 10$ V and $I_m = 5$ A. For each of the following, determine the missing quantity:

a. $v_L = 10 \sin(\omega t + 60°)$, $i_L = $ _____.

b. $v_L = 10 \sin(\omega t - 15°)$, $i_L = $ _____.

c. $i_L = 5 \cos(\omega t - 60°)$, $v_L = $ _____.

d. $i_L = 5 \sin(\omega t + 10°)$, $v_L = $ _____.

16. What is the reactance of a 0.5-H inductance at

a. 60 Hz,   b. 1000 Hz,   c. 500 rad/s?

17. For Figure 16–46, $e = 100 \sin \omega t$ and $L = 0.5$ H. Determine i_L at

a. 60 Hz,   b. 1000 Hz,   c. 500 rad/s.

16.6 Capacitance and Sinusoidal ac

18. Repeat Problem 15 if the given voltages and currents are for a capacitor.

19. What is the reactance of a 5-μF capacitor at

a. 60 Hz,   b. 1000 Hz,   c. 500 rad/s?

20. For Figure 16–47, $e = 100 \sin \omega t$ and $C = 5$ μF. Determine i_C at

a. 60 Hz,   b. 1000 Hz,   c. 500 rad/s.

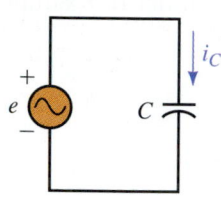

Figure 16–47

16.7 The Impedance Concept

21. Determine the impedance of each circuit element of Figure 16–48.

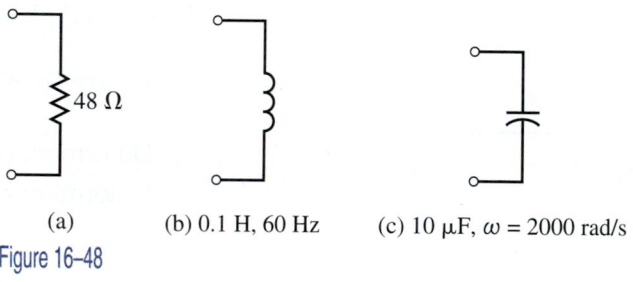

(a)   (b) 0.1 H, 60 Hz   (c) 10 μF, $\omega = 2000$ rad/s

Figure 16–48

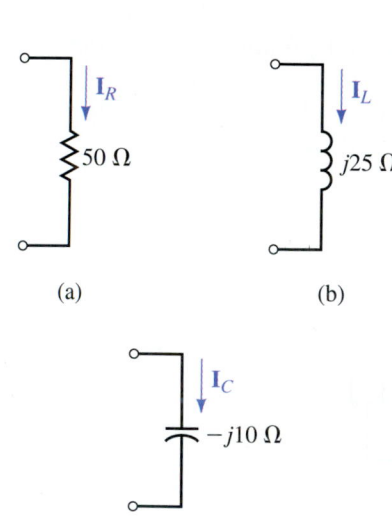

(a)   (b)

(c)

Figure 16–49

22. If $\mathbf{E} = 100$ V∠0° is applied across each of the circuit elements of Figure 16–49:

a. Determine each current in phasor form.

b. Express each current in time domain form.

23. If the current through each circuit element of Figure 16–49 is 0.5 A∠0°:

a. Determine each voltage in phasor form.

b. Express each voltage in time domain form.

$I_L = 2 \text{ A}\angle -90°$

100 V∠0°   E

(a) $L = 0.2$ H.
Determine f.

$I_C = 0.4 \text{ A}\angle 90°$

100 V∠0°

(b) $f = 100$ Hz.
Determine C.

Figure 16–50

24. For each of the following, determine the impedance of the circuit element and state whether it is resistive, inductive, or capacitive.
   a. **V** = 40 V∠30°, **I** = 4 A∠−60°.
   b. **V** = 60 V∠−30°, **I** = 4 A∠60°.
   c. **V** = 140 V∠−30°, **I** = 14 mA∠−120°.
   d. **V** = 240 V∠−30°, **I** = 4 A∠−30°.

25. For each circuit of Figure 16–50, determine the unknown.

16.8 Computer Analysis of AC Circuits

26. In Figure 16–50(a), if f = 50 Hz, use PSpice to solve for **I**$_L$ (magnitude and angle). See note below.

27. Use PSpice to compute and plot current and reactance for Figure 16–50(a) from f = 1 Hz to 200 Hz.

Note: PSpice does not permit source/inductor loops. To get around this, add a very small resistor in series, for example, R = 0.00001 Ω.

17

Power in AC Circuits

CHAPTER PREVIEW

Average power to a resistive circuit is equal to the product of effective voltage times effective current. This power is termed *active power* or *real power* and is the power that does useful work such as light a lamp, power an electric heater, and so on.

For ac circuits that contain reactive elements, (i.e., inductance or capacitance), a second component of power also exists. This component, termed *reactive power*, represents energy that oscillates back and forth throughout the system. For example, during the buildup of current in an inductance, energy flows from the power source to the inductance to create its magnetic field. When the magnetic field collapses, this energy is returned to the circuit. This movement of energy in and out of the inductance constitutes a flow of power. However, since it flows first in one direction, then in the other, it contributes nothing to the average flow of power from the source to the load. For this reason, reactive power is sometimes referred to as *wattless power*.

For a circuit that contains resistive as well as reactive elements, some energy is dissipated while the remainder is shuttled back and forth as described above; thus, both active and reactive components of power are present. This combination of real and reactive power is termed *apparent power*.

In this chapter, we look at all three components of power. New ideas that emerge include the concept of *power factor*, the *power triangle*, the measurement of power in ac circuits, and the concept of *effective resistance*.

Electronics play an important part in the monitoring and control of vehicles.

OBJECTIVES

After studying this chapter, you will be able to

- explain what is meant by active, reactive, and apparent power,

- compute the active power to a load,

- compute the reactive power to a load,

- compute the apparent power to a load,

- construct and use the power triangle to analyze power to complex loads,

- compute power factor,

- explain why equipment is rated in VA instead of watts,

- measure power in single-phase circuits,

- describe why effective resistance differs from geometric resistance,

- describe energy relations in ac circuits.

$$p = vi$$

Figure 17–1 Voltage, current, and power references. When *p* is positive, power is in the direction of the reference arrow.

17.1 Instantaneous Power

At any given instant, the power to a load is equal to the product of voltage times current (Figure 17–1). This means that if voltage and current vary with time, so will power. This time-varying power is referred to as *instantaneous power* and is given the symbol *p(t)* or just *p*. Thus,

$$p = vi \quad \text{(watts)} \tag{17–1}$$

Consider the case of sinusoidal ac. Since voltage and current are positive at various times during their cycle and negative at others, instantaneous power will also be positive at some times and negative at others. This is illustrated in Figure 17–2, where we have multiplied voltage times current point by point to get the power waveform. For example, from $t = 0$ s to $t = t_1$, v and i are both positive; therefore, power is positive. At $t = t_1$, $v = 0$ V and thus $p = 0$ W. From t_1 to t_2, i is positive and v is negative; therefore, p is negative. From t_2 to t_3, both v and i are negative; therefore power is positive, and so on. As discussed in Chapter 4, a positive value means that power transfer is in the direction of the reference arrow, i.e., from the source to the load, while a negative value means that it is in the opposite direction. Thus, during positive parts of the power cycle, power flows from the source to the load, while during negative parts, it flows from the load back to the source.

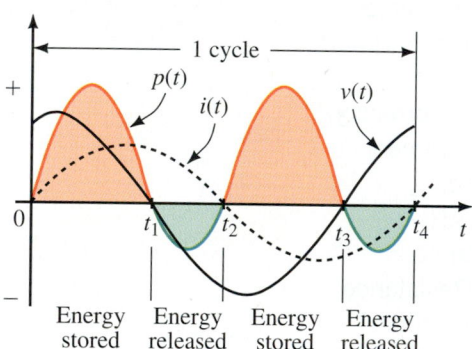

Figure 17–2 Instantaneous power in an ac circuit. Positive *p* represents power to the load; negative *p* represents power returned from the load.

Active Power

Since *p* is constantly changing, a more useful measure of power is its average, *P*. If *P* is positive, then, on average, more power flows to the load than is returned from it. If *P* is zero, all power sent to the load is returned.

The difference between the power supplied to the load and the power returned is the power consumed by the load. Thus, if *P* has a positive average value, it represents the power that is really dissipated by the load. For this reason, *P* is called *real power*. In modern terminology, real power is also called *active power*. Thus, *active power is the average value of the instantaneous power, and the terms real power, active power, and average power mean the same thing.* In this book, we will use them interchangeably.

Reactive Power

Consider again Figure 17–2. During the intervals that p is negative, power is being returned from the load. (This can only happen if the load contains reactive elements: L or C.) The power that flows into the load then back out is called *reactive power*. Since it first flows one way then the other, *its average value is zero*; thus, reactive power contributes nothing to the average power to the load.

　　Although reactive power does no useful work, it cannot be ignored. Extra current is required to create reactive power and this current must be supplied by the source; this also means that conductors, circuit breakers, switches, transformers, and other equipment must be made physically larger to handle the extra current. This increases the cost of a system.

　　At this point, it should be noted that real power and reactive power do not exist as separate entities. Rather, they are components of the power waveform shown in Figure 17–2. However, as you will see, we are able to conceptually separate them for purposes of analysis.

17.2　Power to a Resistive Load

First consider power to a purely resistive load (Figure 17–3). Here, current is in phase with voltage. Assume $i = I_m \sin \omega t$ and $v = V_m \sin \omega t$. Since $p = vi$,

$$p = (V_m \sin \omega t)(I_m \sin \omega t) = V_m I_m \sin^2 \omega t$$

Load

(a)

(b)

Figure 17–3　Power to a purely resistive load. All energy is dissipated by the load.

Therefore,

$$p = \frac{V_m I_m}{2}(1 - \cos 2\omega t) \tag{17–2}$$

where we have used the trigonometric relationship $\sin^2 \omega t = \frac{1}{2}(1 - \cos 2\omega t)$.

A sketch of p versus time is shown in (b). Note that p has a peak value of $V_m I_m$, and that the waveform cycles between zero and this peak, but never goes negative. This means that power flows only from the source to the load. Since none is ever returned, all power delivered by the source is absorbed by the load. We therefore conclude that *power to a pure resistance consists of real or active power only*. Note also that the frequency of the power waveform is double that of the voltage and current waveforms. (This is confirmed by the 2ω in Equation 17–2.)

Average Power

Inspection of the power waveform of Figure 17–3 shows that its average value lies half way between zero and $V_m I_m$. That is,

$$P = V_m I_m / 2$$

(You can also get the same result by averaging Equation 17–2 as we did in Chapter 15.) Since V (the magnitude of the rms value of voltage) is $V_m / \sqrt{2}$ and I (the magnitude of the rms value of current) is $I_m / \sqrt{2}$, this can be written as $P = VI$. Thus, average power to a purely resistive load is

$$P = VI \quad \text{(watts)} \tag{17–3}$$

Equation 17–3 is the basic power relationship for resistive circuits. Alternate forms are obtained by substituting $V = IR$ and $I = V/R$ into Equation 17–3. They are

$$P = I^2 R \quad \text{(watts)} \tag{17–4}$$

$$= V^2/R \quad \text{(watts)} \tag{17–5}$$

These are identical with the dc power equations for resistive circuits.

■ **EXAMPLE 17–1**   Compute the power dissipated by each resistor of Figure 17–4.

$I = 2$ A

E

$V_1 = 40$ V P_1

$V_2 = 60$ V P_2

Figure 17–4

Solution

$$P_1 = V_1 I = (40 \text{ V})(2 \text{ A}) = 80 \text{ W}$$

$$P_2 = V_2 I = (60 \text{ V})(2 \text{ A}) = 120 \text{ W}$$

17.3 Power to an Inductive Load

For a purely inductive load as in Figure 17–5(a), current lags voltage by 90°. If we select current as reference, $i = I_m \sin \omega t$ and $v = V_m \sin(\omega t + 90°)$. A sketch of p versus time (obtained by multiplying v times i) then looks as shown in (b). Note that during the first quarter-cycle, p is positive and hence power flows to the inductance, while during the second quarter-cycle, p is negative and all power transferred to the inductance during the first quarter-cycle flows back out. Similarly for the third and fourth quarter-cycles. Thus, *the average power to an inductance over a full cycle is zero i.e., there are no power losses associated with a pure inductance.* Consequently, $P_L = 0$ W and the only power flowing in the circuit is reactive power. This is true in general, that is, *the power that flows into and out of a pure inductance is reactive power only.*

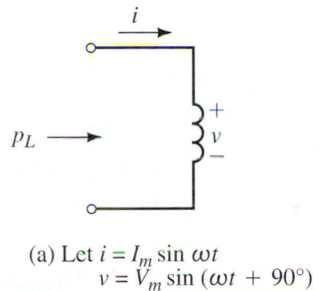

(a) Let $i = I_m \sin \omega t$
$v = V_m \sin (\omega t + 90°)$

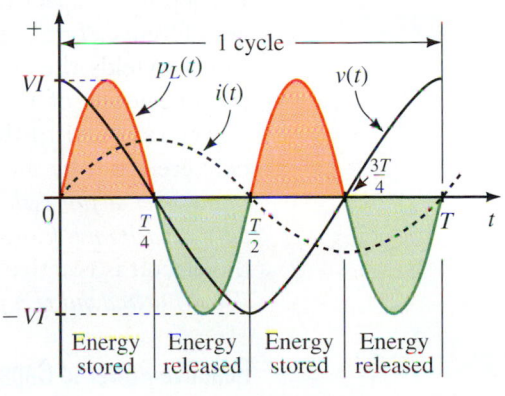

(b)

Figure 17–5   Power to a purely inductive load. Energy stored during each quarter-cycle is returned during the next quarter-cycle.

Reactive Power to Inductance

To determine this power, consider again equation 17–1.
With $v = V_m\sin(\omega t + 90°)$ and $i = I_m\sin \omega t$, $p_L = vi$ becomes

$$p_L = V_m I_m \sin(\omega t + 90°)\sin \omega t$$

After some trigonometric manipulation, this reduces to

$$p_L = VI \sin 2\omega t \qquad\qquad (17\text{--}6)$$

where V and I are the magnitudes of the rms values of the voltage and current respectively.

The product VI in Equation 17–6 is defined as *reactive power* and is given the symbol Q_L. Because it represents "power" that alternately flows into, then out of the inductance, Q_L contributes nothing to the average power to the load and as noted earlier, is sometimes referred to as watt-less power. As you will soon see, however, reactive power is of major concern in the operation of electrical power systems.

Since Q_L is the product of voltage times current, its unit is the volt-amp (VA). To indicate that Q_L represents reactive volt-amps, an "R" is tagged onto the end of VA to yield a new unit, the *VAR (volt-amps reactive)*. Thus,

$$Q_L = VI \quad (\text{VAR}) \qquad\qquad (17\text{--}7)$$

Substituting $V = IX_L$ and $I = V/X_L$ yields the following alternate forms:

$$Q_L = I^2 X_L = \frac{V^2}{X_L} \quad (\text{VAR}) \qquad\qquad (17\text{--}8)$$

By convention, Q_L is taken to be positive. Thus, if $I = 4$ A and $X_L = 2\ \Omega$, $Q_L = (4\ \text{A})^2(2\ \Omega) = +32$ VAR. Note that the VAR (like the watt) is a scalar quantity with magnitude only and no angle.

17.4 Power to a Capacitive Load

For a purely capacitive load, current leads voltage by 90°. Taking current as reference, $i = I_m\sin \omega t$ and $v = V_m\sin(\omega t - 90°)$. Multiplication of v times i yields the power curve of Figure 17–6. Note that negative and positive loops of the power wave are identical; thus, over a cycle, the power returned to the circuit by the capacitance is exactly equal to that delivered to it by the source. This means that *the average power to a capacitance over a full cycle is zero, i.e., there are no power losses associated with a pure capacitance.* Consequently, $P_C = 0$ W and the only power flowing in the circuit is reactive power. This is true in general, that is, *the power that flows into and out of a pure capacitance is reactive power only.*

Reactive Power to Capacitance

This reactive power is given by

$$p_C = vi = V_m I_m \sin \omega t \sin(\omega t - 90°)$$

(a) Let $i = I_m \sin \omega t$
$v = V_m \sin(\omega t - 90°)$

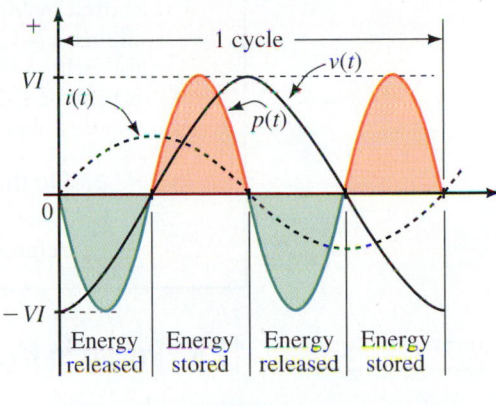

(b)

Figure 17–6   Power to a purely capacitive load.

which reduces to

$$p_L = -VI \sin 2\omega t \qquad (17\text{–}9)$$

where V and I are the magnitudes of the rms values of the voltage and current respectively. Now define the product VI as Q_C. This product represents reactive power. That is,

$$Q_C = VI \quad \text{(VAR)} \qquad (17\text{–}10)$$

Since $V = IX_C$ and $I = V/X_C$, Q_C can also be expressed as

$$Q_C = I^2 X_C = \frac{V^2}{X_C} \quad \text{(VAR)} \qquad (17\text{–}11)$$

By convention, reactive power to capacitance is defined as negative. Thus, if $I = 4$ A and $X_C = 2\ \Omega$, then $I^2 X_C = (4\ \text{A})^2 (2\ \Omega) = 32$ VAR. We can either explicitly show the minus sign as $Q_C = -32$ VAR or imply it by stating that Q represents capacitive vars, i.e. $Q = 32$ VAR (cap.).

**practice
problems
1**

1. If the power at some instant in Figure 17–1 is $p = -27$ W, in what direction is the power at that instant?

2. For a purely resistive load, v and i are in phase. Given $v = 10 \sin \omega t$ V and $i = 5 \sin \omega t$ A. Using graph paper, carefully plot v and i at 30° intervals. Now multiply the values of v and i at these points and plot the power. (The result should look like Figure 17–3(b).

   a. From the graph, determine the peak and average values for power.

b. Note that $V_{rms} = 0.707V_m$ and $I_{rms} = 0.707I_m$. Determine V_{rms} and I_{rms}, then substitute them into Equation 17–3. Compare this to the results determined from the graph in (a).

3. For a purely inductive load, i lags v by 90°. Assume $v = 10 \sin \omega t$ V and $i = 5 \sin(\omega t - 90°)$ A. Using graph paper, carefully plot v and i at 30° intervals. Now multiply the values of v and i at these points and plot the power. (When you compare this graph to Figure 17–5, you will see that the v, i, and p waveforms are shifted by 90° because we chose v as reference instead of i as in Figure 17–5.) During the analysis of Figure 17–5, we made the following observations: 1) All power supplied to the inductance is returned. 2) The peak value of the power waveform is $Q_L = VI$, where V and I are the rms values of v and i.

    a. Do these same observations hold true here?

    b. Does this mean that it does not matter what we use for reference as long as the 90° relationship holds?

17.5 Power in More Complex Circuits

The relationships described above were developed using the load of Figure 17–1. However, they hold true for every element in a circuit, no matter how complex the circuit or how its elements are interconnected.

    Total power to a circuit is often of interest. Total real power P_T is found by summing real power to all circuit elements, while total reactive power Q_T is found by summing reactive power to all circuit elements, taking into account that inductive Q is positive and capacitive Q is negative.

    It is sometimes convenient to show power to circuit elements symbolically as illustrated in the next example.

■ **EXAMPLE 17–2**   For the RL circuit of Figure 17–7(a), $I = 5$ A. Determine P and Q.

(a)

(b) Symbolic representation

Figure 17–7   From the terminals, P and Q are the same for both (a) and (b).

Solution

$$P = I^2R = (5 \text{ A})^2(3\ \Omega) = 75 \text{ W}$$

$$Q_L = I^2X_L = (5 \text{ A})^2(4\ \Omega) = 100 \text{ VAR (ind.)}$$

These can be represented symbolically as in Figure 17–7(b).

■ **EXAMPLE 17–3**   For the *RC* circuit of Figure 17–8(a), determine *P* and *Q*.

(a)   (b)

Figure 17–8   From the terminals, *P* and *Q* are the same for both (a) and (b).

Solution

$$P = V^2/R = (40 \text{ V})^2/(20 \text{ }\Omega) = 80 \text{ W}$$

$$Q_C = V^2/X_C = (40 \text{ V})^2/(80 \text{ }\Omega) = 20 \text{ VAR (cap.)}$$

These can be represented symbolically as in Figure 17–8(b).

■ **EXAMPLE 17–4**   A coil with resistance $R = 4 \text{ }\Omega$ and inductive reactance $X_L = 6 \text{ }\Omega$ is connected in series with a capacitor and resistor as shown in Figure 17–9. Current $I = 10 \text{ A}$ (rms). Determine P_T and Q_T.

(a) Circuit   (b) $P = 1200 \text{ W}$;
                                 $Q = 600 - 900 = -300 \text{ VAR}$

(c) Resultant

Figure 17–9

Solution

$$P_1 = I^2 R_1 = (10 \text{ A})^2 (4 \text{ }\Omega) = 400 \text{ W}$$

$$Q_L = I^2 X_L = (10 \text{ A})^2 (6 \text{ }\Omega) = 600 \text{ VAR (ind.)}$$

$$Q_C = I^2 X_C = (10 \text{ A})^2 (9 \text{ }\Omega) = 900 \text{ VAR (cap.)}$$

$$P_2 = I^2 R_2 = (10 \text{ A})^2 (8 \text{ }\Omega) = 800 \text{ W}$$

This is represented symbolically in Figure 17–9(b). $Q_T = Q_L - Q_C =$ 600 VAR – 900 VAR = –300 VAR This can also be expressed as $Q_T =$ 300 VAR (cap.) as indicated in (c).

practice
problem
11

For the circuit of Figure 17–10, $P_T = 1.9$ kW and $Q_T = 900$ VAR (ind.). Determine P_2 and Q_2.

Figure 17–10

As you can see from these examples, in terms of determining total P and Q, it does not matter how circuits are connected. Elements can be connected in series, parallel, series-parallel, and so on, and P_T is still found by summing the real power to individual elements, while Q_T is found by summing (algebraically) their reactive power.

Real circuits have circuit components (such as coils) that are not ideal. As illustrated in Example 17–4, these are handled by treating them as combinations of idealized elements. In addition, real circuits and systems may contain devices such as motors that have a real power component and a reactive power component. These are handled in the same fashion as any other real/reactive power consuming device.

17.6 Apparent Power

When a load has voltage V across it and current I through it as in Figure 17–11, the power that appears to flow to it is VI. However, if the load contains both resistance and reactance, this product represents neither real power nor reactive power. Since it appears to represent power, it is called *apparent power*. Apparent power is given the symbol S and has units of volt-amperes (VA). Thus,

$$S = VI \quad \text{(VA)} \tag{17–12}$$

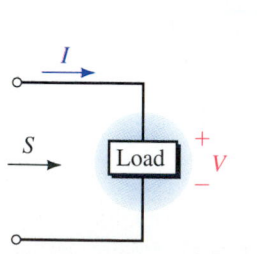

Figure 17–11   Apparent power $S = VI$ (volt-amperes: VA).

where V and I are the magnitudes of the rms voltage and current respectively. For heavy power apparatus, VI is too small a unit and apparent power is then expressed in kilovolt-amperes (kVA), i.e.,

$$S = \frac{VI}{1000} \quad \text{(kVA)} \tag{17–13}$$

Since $V = IZ$ and $I = V/Z$, S can also be written as

$$S = I^2Z = V^2/Z \quad \text{(VA)} \tag{17–14}$$

Figure 17–12   Power apparatus is rated in apparent power. The transformer shown is a 167-kVA unit. *(Courtesy Carte International Ltd)*

The concept of apparent power also applies to every component or element in a circuit or system.

Equipment Ratings in VA

One of the places where you will encounter apparent power is in the rating of electrical apparatus. For example, the transformer of Figure 17–12 (a type used by electric utilities) has a rating of 167 kVA. At generating stations, transformers as large as several hundred megavolt-amps are used. At the other end of the scale, transformers of only a few volt-amps are used in electronic gear.

■ **EXAMPLE 17–5**   To a very good approximation, for large power transformers, apparent power out equals apparent power in, i.e., $S_{out} = S_{in}$, as in Figure 17–13(a). If the transformer is rated 2400 V on its input side and 120 V on its output side, what are its input and output currents when operated at $S = 48$ kVA?

Solution

$$I_{in} = S/V_{in} = (48\ 000\ \text{VA})/2400\ \text{V} = 20\ \text{A}$$

$$I_{out} = S/V_{out} = (48\ 000\ \text{VA})/120\ \text{V} = 400\ \text{A}$$

(a) $S_{out} \approx S_{in}$ since losses are small

(b) $S_{in} = (2400)I_{in}$
$S_{out} = (120)I_{out}$

Figure 17–13 When losses are small, apparent power out and apparent power in are approximately equal.

17.7 The Relationship Between P, Q, and S

Until now, we have treated real, reactive, and apparent power separately. However, they are related by a very simple relationship through the power triangle.

The Power Triangle

Consider the series circuit of Figure 17–14(a). Let the current through the circuit be $\mathbf{I} = I\angle0°$, with phasor representation (b). The voltages across the resistor and inductance are $\mathbf{V}_R$ and $\mathbf{V}_L$ respectively. As noted in Chapter 16, $\mathbf{V}_R$ is in phase with $\mathbf{I}$, while $\mathbf{V}_L$ leads it by 90°. Kirchhoff's voltage law applies for ac voltages in phasor form. Thus, $\mathbf{V} = \mathbf{V}_R + \mathbf{V}_L$ as indicated in (c).

(a)

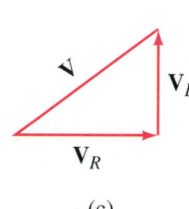

(b)

The voltage triangle of (c) may be redrawn as in Figure 17–15(a) with magnitudes of V_R and V_L replaced by IR and IX_L respectively. Now multiply all quantities by I. This yields sides of I^2R, I^2X_L, and hypotenuse VI as indicated in (b). Note that these represent P, Q, and S respectively as indicated in (c). This is the triangle that we want. It is called the *power triangle*. From the geometry of this triangle, you can see that

$$S = \sqrt{P^2 + Q_L^2} \qquad (17–15)$$

Alternatively, the relationship between P, Q, and S may be expressed as a complex number:

$$\mathbf{S} = P + jQ_L \qquad (17–16a)$$

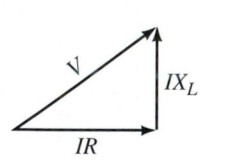

(c)

Figure 17–14 Steps in the development of the power triangle.

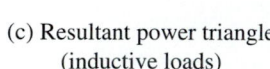

(a) Magnitudes only shown

(b) Multiplied by I

(c) Resultant power triangle (inductive loads)

Figure 17–15 Steps in the development of the power triangle (continued).

or

$$\mathbf{S} = S\angle\theta \qquad (17\text{–}16b)$$

If the circuit is capacitive instead of inductive, Equation 17–16a becomes

$$\mathbf{S} = P - jQ_C \qquad (17\text{–}17)$$

The power triangle in this case has a negative imaginary part as indicated in Figure 17–16.

The power relationships may be written in generalized forms as

$$\mathbf{S} = \mathbf{P} + \mathbf{Q} \qquad (17\text{–}18)$$

and

$$\mathbf{S} = \mathbf{VI^*} \qquad (17\text{–}19)$$

where $\mathbf{P} = P\angle 0°$, $\mathbf{Q}_L = jQ_L$, $\mathbf{Q}_C = -jQ_C$, and $\mathbf{I^*}$ is the conjugate of current $\mathbf{I}$. These relationships hold true for all networks regardless of what they contain or how they are configured. For example, as noted in Section 17.5, in terms of determining P and Q, it does not matter whether a circuit is connected in series or parallel, or whether it contains resistors, inductors, or capacitors. Circuits may also contain motors, transformers, transmission lines, ac sources, and so on and may be connected in any configuration possible, and these relationships still apply.

When solving problems involving power, remember that P values can be added to get P_T, and Q values to get Q_T (where Q is positive for inductive circuits and negative for capacitive). However, apparent power values cannot be added to get S_T, i.e., $S_T \neq S_1 + S_2 + \ldots + S_N$. Instead, determine P_T and Q_T, then use the power triangle to obtain S_T.

Figure 17–16   Power triangle for capacitive loads.

■ **EXAMPLE 17–6**   The *P* and *Q* values for a circuit are shown in Figure 17–17(a).

   a. Determine the power triangle.

   b. Determine the magnitude of the current supplied by the source.

(a)

(b)

Figure 17–17

Solution

a. $P_T = 700 + 800 + 80 + 120 = 1700$ W

$Q_T = 1300 - 600 - 100 - 1200 = -600$ VAR $= 600$ VAR (cap.)

$\mathbf{S}_T = P_T + jQ_T = 1700 - j600 = 1803\angle-19.4°$ VA

The power triangle is as shown.

b. $I = S_T/E = 1803$ VA$/120$ V $= 15.0$ A

Active and Reactive Power Equations

An examination of the power triangle of Figures 17–15 and 17–16 shows that P and Q may be expressed respectively as

$$P = VI \cos \theta = S \cos \theta \quad \text{(W)} \tag{17–20}$$

and

$$Q = VI \sin \theta = S \sin \theta \quad \text{(VAR)} \tag{17–21}$$

where V and I are the magnitudes of the rms values of the voltage and current respectively and θ is the angle between them. P is always positive, while Q is positive for inductive circuits and negative for capacitive circuits. Thus, if $V = 120$ volts, $I = 50$ A, and $\theta = 30°$, $P = (120)(50)\cos 30° = 5196$ W and $Q = (120)(50)\sin 30° = 3000$ VAR.

17.8 Power Factor

The quantity $\cos \theta$ in Equation 17–20 is defined as *power factor* and is given the symbol F_p. Thus,

$$F_p = \cos \theta \tag{17–22}$$

From Equation 17–20, we see that F_p may be computed as the ratio of real power to apparent power. Thus,

$$\cos \theta = P/S \tag{17–23}$$

Power factor is expressed as a number or as a percent. From Equation 17–23, it is apparent that power factor cannot exceed 1.0 (or 100% if expressed in percent).

The *power factor angle* θ is of interest. It can be found as

$$\theta = \cos^{-1}(P/S) \tag{17–24}$$

Angle θ is the angle between voltage and current. For a pure resistance, therefore, $\theta = 0°$. For a pure inductance, $\theta = 90°$; for a pure capacitance, $\theta = -90°$. For a circuit containing both resistance and inductance, θ will be somewhere between $0°$ and $90°$; for a circuit containing both resistance and capacitance, θ will be somewhere between $0°$ and $-90°$.

Unity, Lagging, and Leading Power Factor

As indicated by Equation 17–23, a load's power factor shows how much of its apparent power is actually real power. For example, for a purely resistive circuit, $\theta = 0°$ and $F_p = \cos 0° = 1.0$. Therefore, $P = VI$ (watts) and all the load's apparent power is real power. This case ($F_p = 1$) is referred to as *unity* power factor.

For a load containing only resistance and inductance, the load current lags voltage. The power factor in this case is described as *lagging*. On the other hand, for a load containing only resistance and capacitance, current leads voltage and the power factor is described as *leading*. Thus, *an inductive circuit has a lagging power factor, while a capacitive circuit has a leading power factor.*

A load with a very poor power factor can draw excessive current. This is discussed next.

Why Equipment Is Rated in VA

We now examine why electrical apparatus is rated in VA instead of watts. Consider Figure 17–18. Assume that the generator is rated at 600 V, 120 kVA. This means that it is capable of supplying $I = 120$ kVA/600 V = 200 A. In (a), the generator is supplying a purely resistive load with 120 kW. Since $S = P$ for a purely resistive load, $S = 120$ kVA and the generator is supplying its rated kVA. In (b), the generator is supplying a load with $P = 120$ kW as before, but $Q = 160$ kVAR. Its apparent power is therefore $S = 200$ kVA, which means that the generator current is $I = 200$ kVA/600 V = 333.3 A. Even though it is supplying the same power as in (a), the generator is now greatly overloaded, and damage may result as indicated in (b).

This example illustrates clearly that rating a load or device in terms of power is a poor choice, as its current-carrying capability can be greatly exceeded (even though its power rating is not). Thus, *the size of electrical apparatus (generators, interconnecting wires, transformers, etc.) required to supply a load is governed, not by the load's power requirements, but rather by its VA requirements.*

(a) $S = 120$ kVA

(b) $S = \sqrt{(120)^2 + (160)^2} = 200$ kVA
    The generator is overloaded.

Figure 17–18   Illustrating why electrical apparatus is rated in VA instead of watts. Both loads dissipate 120 kW but the current rating of generator (b) is exceeded because of the power factor of its load.

Power Factor Correction

The problem shown in Figure 17–18 can be alleviated by cancelling some or all of the reactive component of power by adding reactance of the opposite type to the circuit. This is referred to as *power factor correction*.

■ **EXAMPLE 17–7**   For the circuit of Figure 17–18(b), a capacitance with $Q_C = 160$ kVAR is added in parallel with the load as in Figure 17–19(a). Determine current I.

Solution $Q_T = 160$ kVAR − 160 kVAR = 0. Therefore, $\mathbf{S}_T = 120$ kW + $j0$ kVAR. Thus, $S_T = 120$ kVA, and $I = 120$ kVA/600 V = 200 A as

(a) $Q_C = 160$ kVAR

(b) Corrected load looks purely resistive

Figure 17–19   Power factor correction. The parallel capacitor greatly reduces source current.

in Figure 17–18(a). The generator is no longer overloaded. Note that the corrected circuit looks purely resistive, as in (b).

As a residential customer, you pay your electrical utility bill based solely on the electrical energy that you use. The power company is not concerned about your power factor because all residential power factors are basically the same and the cost of supplying reactive power to homes is simply factored into the tariff. On the other hand, industrial customers have widely different power factors and the utility may have to take their power factors into account.

To illustrate, assume that the loads of Figures 17–18(a) and (b) are two small industrial plants. If the utility based its charge solely on power, both customers would pay the same amount. However, it costs the utility more to supply customer (b) since larger conductors, larger transformers, larger switchgear, and so on are required to handle the larger current. For this reason, industrial customers may pay a penalty if their power factor drops below a prescribed value.

■ **EXAMPLE 17–8**   An industrial client is charged a penalty if the plant power factor drops below 0.85. The plant loads are as shown in Figure 17–20. The frequency is 60 Hz.

   a. Determine P_T and Q_T.
   b. Determine what value of capacitance (in microfarads) is required to bring the power factor up to 0.85.

Solution   a. The powers are as follows:

Lights: $P = 12$ kW, $Q = 0$ kVAR

Furnace: $P = I^2R = (150)^2(2.4) = 54$ kW

$Q = I^2X = (150)^2(3.2) = 72$ kVAR (ind.)

Motor: $\theta_m = \cos^{-1}(0.8) = 36.9°$. Thus, from the motor power triangle,

$Q_m = P_m \tan \theta_m = 80 \tan 36.9° = 60$ kVAR (ind.)

Figure 17–20

(a)

(b) Power triangle for motor.

Total: $P_T = 12 \text{ kW} + 54 \text{ kW} + 80 \text{ kW} = 146 \text{ kW}$

$Q_T = 0 + 72 \text{ kVAR} + 60 \text{ kVAR} = 132 \text{ kVAR}$

b. The power triangle for the plant is shown in Figure 17–21(a). However, we must correct the power factor to 0.85. Thus we need $\theta' = \cos^{-1}(0.85) = 31.8°$, where θ' is the power factor angle of the corrected load as indicated in Figure 17–21(b). The maximum reactive power that we can tolerate is thus $Q'_T = P_T \tan \theta' = 146 \tan 31.8° = 90.5 \text{ kVAR}$.

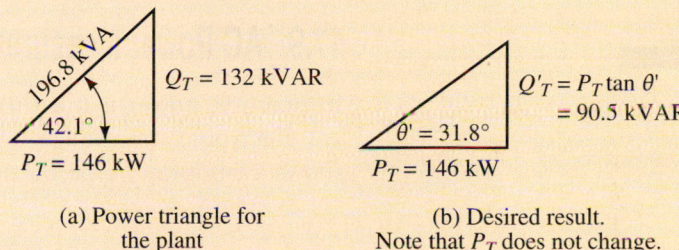

(a) Power triangle for the plant

(b) Desired result. Note that P_T does not change.

Figure 17–21

Now consider Figure 17–22. $Q'_T = Q_C + 132 \text{ kVAR}$, where $Q'_T = 90.5 \text{ kVAR}$. Therefore, $Q_C = -41.5 \text{ kVAR} = 41.5 \text{ kVAR (cap.)}$. But $Q_C = V^2/X_C$. Therefore, $X_C = V^2/Q_C = (600)^2/41.5 \text{ kVAR} = 8.67 \text{ Ω}$. But $X_C = 1/\omega C$. Thus a capacitor of

$$C = \frac{1}{\omega X_C} = \frac{1}{(2\pi)(60)(8.67)} = 306 \text{ μF}$$

will provide the required correction.

Figure 17–22

Due to plant expansion, 102 kW of purely resistive load is added to the plant of Figure 17–20. Determine whether power factor correction is needed to correct the expanded plant to 0.85 F_p, and if so, how much.

In practice, almost all loads (industrial, residential, and commercial) are inductive due to the presence of motors, fluorescent lamp ballasts, and the like. Consequently, you will likely never run into capacitive loads that need power factor correcting.

1. Sketch the power triangle for Figure 17–9(c). Using this triangle, determine the magnitude of the input voltage E.

2. For Figure 17–10, assume a source of $E = 240$ volts, $P_2 = 700$ W, and $Q_2 = 650$ VAR (cap.). What is the magnitude of the source current I?

3. What is the power factor of each of the circuits of Figure 17– 7, 17–8, and 17–9? Indicate whether they are leading or lagging.

4. Consider the circuit of Figure 17–18(b). If $P = 100$ kW and $Q_L = 80$ kVAR, is the source overloaded, assuming it is capable of handling a 120-kVA load?

17.9 AC Power Measurement

In practice, power is measured in ac circuits using either a *wattmeter* or an electronic transducer-based measurement system. Figure 17–23 shows a laboratory type wattmeter.

Figure 17–23   A laboratory wattmeter.

The Wattmeter

The heart of this wattmeter is an electrodynamometer movement (Figure 17–24). It consists of a set of stationary coils (called current coils: CC) and a movable coil (called the voltage or potential coil: PC). A pointer is attached to the potential coil. The current coils are made from large-diameter wire to keep their impedance low. The potential coil circuit uses a large series resistance to keep its impedance high so as not to load the circuit.

Figure 17–25 shows how the wattmeter is connected in a circuit. As indicated, its current coil is placed in series with the load, while its potential coil is placed across the load. The magnetic field created by the current coil is proportional to load current, while the magnetic field created by the potential coil is proportional to load voltage. Instantaneous torque on the pointer coil is proportional to the product of these two magnetic fields, that is, to the product of v_L times i_L (i.e., instantaneous power). Because of its inertia, the meter movement cannot follow the rapidly varying torque and thus takes up a position corresponding to average torque, i.e., to average power. The scale of the meter is calibrated directly in watts.

Figure 17–24   Electrodynamometer movement used in a wattmeter. The green arrows indicate the forces due to the interaction of the magnetic fields. The spiral spring counters the torque due to these forces. The rotation of the pivoted coil is proportional to average power.

Figure 17–25   Measuring power with a wattmeter. Here, the wattmeter is represented by its schematic circuit.

Note the polarity markings on the coils. These must be connected properly. For example, in Figure 17–25, the meter is connected so that load current enters the ± terminal of the current coil (CC), while the potential coil (PC) has its ± terminal connected to the higher-potential end of the load. This is the standard single-phase connection. Three-phase connections, as discussed in Chapter 23, may be different.

When power is to be measured in a low power factor circuit, a *low power factor wattmeter* must be used. This is because, for low power factor loads, currents can be very high, even though the power is low. Thus, you can easily exceed the current rating of a standard wattmeter and damage it, even though the indication on the meter is small.

■ **EXAMPLE 17–9**   For the circuit of Figure 17–25, what does the wattmeter indicate if

  a. $V_L = 100 \text{ V} \angle 0°$ and $I_L = 15 \text{ A} \angle 60°$,
  b. $V_L = 100 \text{ V} \angle 10°$ and $I_L = 15 \text{ A} \angle 30°$?

Solution   A wattmeter reads average power, i.e., $W = V_L I_L \cos \theta_L$

  a. $\theta_L = 60°$. Thus, $W = (100)(15)\cos 60° = 750 \text{ W}$,
  b. $\theta_L = 10° - 30° = -20°$. Thus, $W = (100)(15)\cos(-20°) = 1410 \text{ W}$.

Note: For (b), since $\cos(-20°) = \cos(+20°)$, it does not matter whether we include the minus sign.

■ **EXAMPLE 17–10**   For Figure 17–26, determine the wattmeter reading.

Figure 17–26   A wattmeter measures only active power. Note the simplified representation of the wattmeter.

Solution   A wattmeter reads only power. Thus, it indicates 600 W.

practice problem IV

Determine the wattmeter reading for Figure 17–27.

Figure 17–27

Measuring VARs

The basic varmeter is the same as the basic wattmeter except that it requires a circuit to shift PC coil voltage by 90°. With this change, the meter responds to sin θ rather than to cos θ, i.e., to VARs rather than watts. Varmeters are most commonly used in three-phase circuits (Chapter 23). In this case, a set of transformers is usually used to obtain the needed 90° phase shift. If you major in power technology, you will study this circuit later.

17.10 Effective Resistance

Up to now, we have assumed that resistance is constant, independent of frequency. However, this is not entirely true. For a number of reasons, the resistance of a circuit to ac is greater than its resistance to dc. While this effect is small at low frequencies, it is very pronounced at high frequencies. When working with ac therefore, you must be sure to use ac resistance, which is known as *effective resistance*.

Before looking at why ac resistance is greater than dc resistance, we need to reexamine the concept of resistance itself. Recall from Chapter 3 that resistance was originally defined as opposition to current, that is, $R = V/I$. (This is ohmic resistance.) Building on this, you learned in Chapter 4 that $P = I^2 R$. It is this latter viewpoint that allows us to give meaning to ac resistance. That is, we define ac or effective resistance as

$$R_{\text{eff}} = \frac{P}{I^2} \qquad (17\text{–}25)$$

where P is dissipated power (as determined by a wattmeter). From this, you can see that anything that affects dissipated power affects resistance.

DC and AC Resistance

For dc and low-frequency ac, both definitions for R, i.e., $R = V/I$ and $R = P/I^2$ yield the same value. However, as frequency increases, other factors cause an increase in resistance. We will now consider some of these.

Eddy Currents and Hysteresis

The magnetic field surrounding a coil or other circuit carrying ac current varies with time and thus induces voltages in nearby conductive material such as metal equipment cabinets, transformer cores, and so on. The resulting currents (called *eddy currents* because they flow in circular patterns like eddies in a brook) are unwanted and create power losses called *eddy current losses*. Since additional power must be supplied to the coil to make up for these losses, it looks to the source as if the coil's resistance has increased.

If ferromagnetic material is also present, an additional power loss occurs due to hysteresis effects caused by the magnetic field alternately magnetizing the material in one direction, then the other. Hysteresis and eddy current losses are important even at low frequencies, such as the 60-Hz power system frequency. This is discussed in Chapter 24.

Skin Effect

Magnetically induced voltages created inside a conductor by its own changing magnetic field force electrons to the periphery of the conductor (Figure 17–28), resulting in a nonuniform distribution of current, with current density greatest near the periphery and smallest in the center. This phenomenon is known as *skin effect*. Because the center of the wire carries little current, its cross-sectional area has effectively been reduced, thus increasing resistance. While skin effect is generally negligible at power line frequencies (except for conductors larger than several hundred thousand circular mils), it is so pronounced at microwave frequencies that the center of a wire carries almost no current. For this reason, hollow conductors are often used instead of solid wires, as shown in Figure 17–28(c).

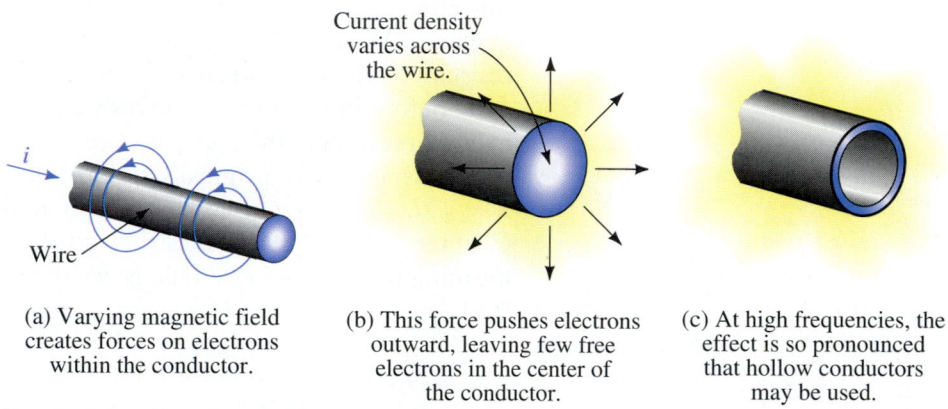

(a) Varying magnetic field creates forces on electrons within the conductor.

(b) This force pushes electrons outward, leaving few free electrons in the center of the conductor.

(c) At high frequencies, the effect is so pronounced that hollow conductors may be used.

Figure 17–28   Skin effect in ac circuits.

Radiation Resistance

At high frequencies some of the energy supplied to a circuit escapes as radiated energy. For example, a radio transmitter supplies power to an antenna, where it is converted into radio waves and radiated into space. The resistance effect here is known as *radiation resistance*. This resistance is much higher than simple dc resistance. For example, a TV transmitting antenna may have a resistance of a fraction of an ohm to dc but several hundred ohms effective resistance at its operating frequency.

A Final Note

Many of the effects noted above will be treated in detail in your various electronics courses. We will not pursue them further here.

17.11   Energy Relationships for AC

Recall, power and energy are related by the equation $p = dw/dt$. Thus, energy can be found by integration as

$$W = \int p\, dt = \int vi\, dt \qquad (17\text{–}26)$$

Inductance   For an inductance, $v = Ldi/dt$. Substituting this into Equation 17–26, cancelling dt, and rearranging terms yields

$$W_L = \int \left(L\frac{di}{dt} \right) i\, dt = L \int i\, di \qquad (17\text{–}27)$$

Recall from Figure 17–5(b), energy flows into an inductor during time interval 0 to $T/4$ and is released during time interval $T/4$ to $T/2$. The process then repeats itself. The energy stored (and subsequently released) can thus be found by integrating power from $t = 0$ to $t = T/4$. Current at $t = 0$ is 0 and current at $t = T/4$ is I_m. Using these as our limits of integration, we find

$$W_L = L \int_0^{I_m} i\, di = \frac{1}{2}LI_m^2 = LI^2 \quad \text{(J)} \qquad (17\text{–}28)$$

where we have used $I = I_m/\sqrt{2}$ to express energy in terms of effective current.

Capacitance.   For a capacitance, $i = Cdv/dt$. Substituting this into Equation 17–26 yields

$$W_C = \int v \left(C\frac{dv}{dt} \right) dt = C \int v\, dv \qquad (17\text{–}29)$$

Consider Figure 17–6(b). Energy stored can be found by integrating power from $T/4$ to $T/2$. The corresponding limits for voltage are 0 to V_m. Thus,

$$W_C = C \int_0^{V_m} v\, dv = \frac{1}{2}CV_m^2 = CV^2 \quad \text{(J)} \qquad (17\text{–}30)$$

where we have used $V = V_m/\sqrt{2}$. You will use these relationships later.

Problems

Figure 17–29

Figure 17–30

17.1–17.4   Instantaneous Power to Loads

1. What is real power? What is reactive power? Which power, real or reactive, has an average value of zero?

2. Instantaneous power (Equation 17–2) is given by

$$p = \frac{V_m I_m}{2} (1 - \cos 2\omega t)$$

Show that this has an average value of $P = VI$ (watts), where V and I are rms values.

3. Refer to the circuit of Figure 17–29:
   a. Determine the power to each resistor.
   b. What is total power output by the source?

4. For the circuit of Figure 17–30, determine Q_L.

5. For the circuit of Figure 17–31, $f = 100$ Hz. Find
   a. R,
                   b. X_C,
                   c. C.

5 A
250 W
150 VAR

Figure 17–31

6. For the circuit of Figure 17–32, $f = 10$ Hz. Find
   a. P,
                   b. X_L,
                   c. L.

P
40 VAR
$8\ \Omega$
$+$ 60 V $-$
L

Figure 17–32

7. For Figure 17–33, $\omega = 10$ rad/s. Determine
   a. R_T,
              b. R_2,
               c. X_C,
               d. L_{eq}.

$I = 6$ A R_1 $j20$ $-jX_C$
$10\ \Omega$
720 W
432 VAR (ind)
R_3
$4\ \Omega$
R_2

6 A
720 W 432 VAR
(ind.)

R_T L_{eq}

(a)   (b)   (c)

Figure 17–33

8. For Figure 17–34, determine the total P and Q.

P
Q
1152 W
633.6 VAR
(cap.)
2.4 A
R_2 40 Ω
$+$ 600 V $-$
R_1 20 Ω
X_C 50 Ω
X_L 40 Ω

Figure 17–34

Figure 17–35

Figure 17–37

17.7 The Relationship Between *P*, *Q*, and *S*

9. a. For the circuit of Figure 17–7, draw the power triangle and determine the apparent power.

    b. Repeat for Figure 17–8.

10. For the circuit of Figure 17–35, what is the source current?

11. For Figure 17–36, the generator supplies 30 A. What is *R*?

Figure 17–36

12. Suppose $\mathbf{V} = 100\ V\angle 60°$ and $\mathbf{I} = 10A\angle 40°$:

    a. What is θ, the angle between **V** and **I**?

    b. Determine *P* from $P = VI \cos\theta$.

    c. Determine *Q* from $Q = VI \sin\theta$.

    d. Sketch the power triangle and from it, determine **S**.

    e. Show that $\mathbf{S} = \mathbf{VI}^*$ gives the same answer as (d).

13. For Figure 17–37, $S_{gen} = 4835$ VA. What is *R*?

14. Refer to the circuit of Figure 17–17:

    a. Determine the apparent power for each box.

    b. Show that the sum of the apparent powers for the boxes does not equal the actual apparent power, $S_T = 1803$ VA.

17.8 Power Factor

15. Refer to the circuit of Figure 17–38:

    a. Determine P_T, Q_T, and S_T.

    b. Determine whether the fuse will blow.

Figure 17–38

Figure 17–39

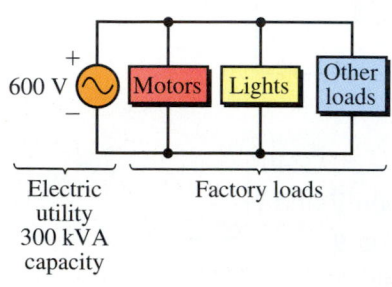

Electric
utility
300 kVA
capacity

Factory loads

(a)

(b) Factory power triangle

Figure 17–40

16. A motor with an efficiency of 87% supplies 10 hp to a load (Figure 10–39). Its power factor is 0.65 (lag).
   a. What is the power input to the motor?
   b. What is the reactive power to the motor?
   c. Draw the motor power triangle. What is the apparent power to the motor?

17. To correct the circuit power factor of Figure 17–39 to unity, a power factor correction capacitor is added.
   a. Show where the capacitor is connected.
   b. Determine its value in microfarads.

18. Consider Figure 17–20. The motor is replaced with a new unit requiring $\mathbf{S}_m = (120 + j35)$ kVA. Everything else remains the same. Find the following:
   a. P_T.   b. Q_T.   c. S_T.
   d. Sketch where a power factor correcting capacitor should be placed.
   e. Determine how much kVAR capacitive correction is needed to correct to unity F_p.

19. A small electrical utility has a 600-V, 300-kVA capacity. It supplies a factory (Figure 17–40) with the power triangle shown in (b). This fully loads the utility. If a power factor correcting capacitor corrects the load to unity power factor, how much more power can the utility sell to other customers?

17.9 AC Power Measurement

20. a. Why does the wattmeter of Figure 17–41 indicate only 1200 watts?
   b. Where would the wattmeter have to be placed to measure power delivered by the source? Sketch the modified circuit.
   c. What would the wattmeter indicate in (b)?

Figure 17–41

PART

V

Impedance Networks

As you have already observed, the impedance of an inductor or a capacitor is dependent upon the frequency of the signal applied to the element. When capacitors and inductors are combined with resistors and voltage or current sources, the circuit will behave in a predictable manner for all frequencies.

The final part of this textbook examines how circuits consisting of various combinations of impedances and sources behave under specific conditions. In particular, we find that all the laws, rules, and theorems developed previously apply to even the most complicated impedance network.

Ohm's law and Kirchhoff's voltage and current laws are easily modified to give the framework for developing methods of network analysis. Just as in dc circuits, Thévenin's and Norton's theorems will allow us to simplify a complicated circuit to a single source and corresponding impedance.

The theorems and methods of analysis are applied to numerous types of circuits which are commonly encountered throughout electrical and electronics technology. Resonant circuits and filter circuits are commonly used to restrict the range of output frequencies for a given range of input frequencies.

The study of three-phase systems and transformers is particularly useful for anyone interested in commercial power distribution. These topics deal with practical applications and the drawbacks of using various types of circuits.

Finally, we examine how a circuit reacts to nonsinusoidal alternating voltages. This topic involves complex signals which are processed by impedance networks resulting in outputs which are often dramatically different from the input.

18

AC Series-Parallel Circuits

Chapter Preview

In this chapter we examine how simple circuits containing resistors, inductors, and capacitors behave when subjected to sinusoidal voltages and currents. Principally, we find that the rules and laws which were developed for dc circuits will apply equally well for ac circuits. The major difference between solving dc and ac circuits is that analysis of ac circuits requires using vector algebra.

In order to proceed successfully, it is suggested that the student spend time reviewing the important topics covered in dc analysis. These include Ohm's law, the voltage divider rule, Kirchhoff's voltage law, Kirchhoff's current law, and the current divider rule.

You will also find that a brief review of vector algebra will make your understanding of this chapter more productive. In particular, you should be able to add and subtract any number of vector quantities.

KEY TERMS

AC Parallel Circuits
AC Series Circuits
Current Divider Rule
Frequency Effects
Impedance
Kirchhoff's Current Law
Kirchhoff's Voltage Law
Voltage Divider Rule

(Courtesy of Martin Marietta Corp.)

OBJECTIVES

After studying this chapter, you will be able to

- apply Ohm's law to analyze simple series circuits,

- apply the voltage divider rule to determine the voltage across any element in a series circuit,

- apply Kirchhoff's voltage law to verify that the summation of voltages around a closed loop is equal to zero,

- apply Kirchhoff's current law to verify that the summation of currents entering a node is equal to the summation of currents leaving the same node,

- determine unknown voltage, current, and power for any series/parallel circuit,

- determine the series or parallel equivalent of any network consisting of a combination of resistors, inductors, and capacitors.

Heinrich Rudolph Hertz

Heinrich Hertz was born in Hamburg, Germany, on February 22, 1857. He is known mainly for his research into the transmission of electromagnetic waves.

Hertz began his career as an assistant to Hermann von Helmholtz in the Berlin Institute physics laboratory. In 1885, he was appointed Professor of Physics at Karlsruhe Polytechnic, where he did much to verify James Clerk Maxwell's theories of electromagnetic waves.

In one of his experiments, Hertz discharged an induction coil with a rectangular loop of wire having a very small gap. When the coil discharged, a spark jumped across the gap. He then placed a second, identical coil close to the first, but with no electrical connection. When the spark jumped across the gap of the first coil, a smaller spark was also induced across the second coil. Today, more elaborate antennas use similar principles to transmit radio signals over vast distances. Through further research, Hertz was able to prove that electromagnetic waves have many of the characteristics of light: they have the same speed as light; they travel in straight lines; they can be reflected and refracted; and they can be polarized.

Hertz's experiments ultimately led to the development of radio communication by such electrical engineers as Guglielmo Marconi and Reginald Fessenden.

Heinrich Hertz died at the age of 36 on January 1, 1894.

18.1 Ohm's Law for AC Circuits

This section is a brief review of the relationship between voltage and current for resistors, inductors, and capacitors. Unlike Chapter 16, all phasors are given as rms rather than as peak values. As you saw in Chapter 17, this approach simplifies the calculation of power.

Resistors

In Chapter 16, we saw that when a resistor is subjected to a sinusoidal voltage as shown in Figure 18–1, the resulting current will also be sinusoidal and in phase with the voltage.

The sinusoidal voltage $v = V_m \sin(\omega t + \theta)$ may be written in phasor form as $\mathbf{V} = V\angle\theta$. Whereas the sinusoidal expression gives the instantaneous value of voltage for a waveform having an amplitude of V_m (volts peak), the phasor form has a magnitude which is the effective (or rms) value. The relationship between the magnitude of the phasor and the peak of the sinusoidal voltage is given as

$$V_m = \sqrt{2}V$$

Because the resistance vector may be expressed as $\mathbf{Z}_R = R\angle 0°$, we evaluate the current phasor as follows:

$$\mathbf{I} = \frac{\mathbf{V}}{\mathbf{Z}_R} = \frac{V\angle\theta}{R\angle 0°} = \frac{V}{R}\angle\theta = I\angle\theta$$

If we wish to convert the current from phasor form to its sinusoidal equivalent in the time domain, we would have $i = I_m \sin(\omega t + \theta)$. Again, the relationship between the magnitude of the sinusoidal and the phasor equivalent is given as

note

*A*lthough currents and voltages may be shown in either time domain (as sinusoidal quantities) or in phasor domain (as vectors), resistance and reactance are never shown as sinusoidal quantities. The reason for this is that whereas currents and voltages vary as functions of time, resistance and reactance do not.

(a)

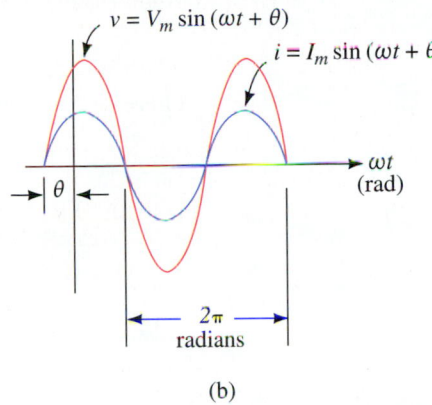

(b)

Figure 18–1   Sinusoidal voltage and current for a resistor.

$$I_m = \sqrt{2}I$$

The voltage and current phasors may be shown on a phasor diagram as in Figure 18–2.

Because one phasor is a current and the other is a voltage, the relative lengths of these phasors are purely arbitrary. Regardless of the angle θ, we see that the voltage across and the current through a resistor will always be in phase.

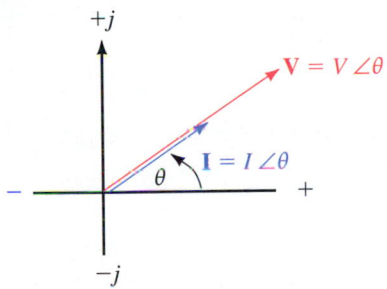

Figure 18–2   Voltage and current phasors for a resistor.

■ **EXAMPLE 18–1**   Refer to the resistor shown in Figure 18–3:

a. Find the sinusoidal current i using phasors.
b. Sketch the sinusoidal waveforms for v and i.
c. Sketch the phasor diagram of **V** and **I**.

Figure 18–3

Solution:

a. The phasor form of the voltage is determined as follows:

$$v = 72 \sin \omega t \Leftrightarrow \mathbf{V} = 50.9 \text{ V} \angle 0°$$

From Ohm's law, the current phasor is determined to be

$$I = \frac{V}{Z_R} = \frac{50.9 \text{ V} \angle 0°}{18 \text{ } \Omega \angle 0°} = 2.83 \text{ A} \angle 0°$$

which results in the sinusoidal current waveform having an amplitude of

$$I_m = (\sqrt{2})(2.83 \text{ A}) = 4.0 \text{ A}$$

Therefore, the current i will be written as

$$i = 4 \sin \omega t \quad \text{(A)}$$

b. The voltage and current waveforms are shown in Figure 18–4.

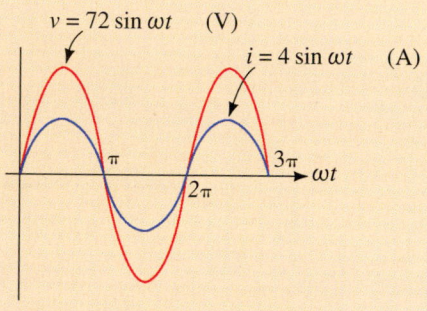

Figure 18–4

c. Figure 18–5 shows the voltage and current phasors.

Figure 18–5

■ **EXAMPLE 18–2**   Refer to the resistor of Figure 18–6:

a. Use phasor algebra to find the sinusoidal voltage, v.
b. Sketch the sinusoidal waveforms for v and i.
c. Sketch a phasor diagram showing **V** and **I**.

Solution

a. The sinusoidal current has a phasor form as follows:

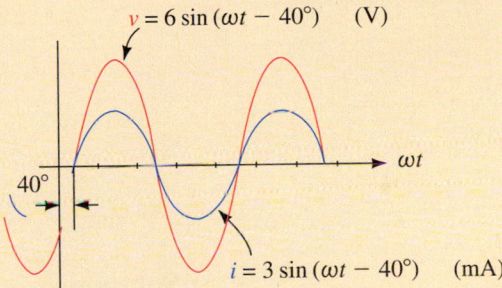

$$i = 3 \sin(\omega t - 40°) \quad \text{(mA)}$$

2 kΩ v

Figure 18–6

$$i = 3 \times 10^{-3}\sin(\omega t - 40°) \Leftrightarrow \mathbf{I} = 2.12 \text{ mA}\angle{-40°}$$

From Ohm's law, the voltage across the 2-kΩ resistor is determined as the phasor product

$$\mathbf{V} = \mathbf{I}\mathbf{Z}_R$$
$$= (2.12 \text{ mA}\angle{-40°})(2 \text{ k}\Omega\angle 0°)$$
$$= 4.24 \text{ V}\angle{-40°}$$

The amplitude of the sinusoidal voltage is

$$V_m = (\sqrt{2})(4.24 \text{ V}) = 6.0 \text{ V}$$

The voltage may now be written as

$$v = 6.0 \sin(\omega t - 40°) \quad \text{(V)}$$

b. Figure 18–7 shows the sinusoidal waveforms for v and i.

$$v = 6 \sin(\omega t - 40°) \quad \text{(V)}$$

ωt

40°

$$i = 3 \sin(\omega t - 40°) \quad \text{(mA)}$$

Figure 18–7

c. The corresponding phasors for the voltage and current are shown in Figure 18–8.

$+j$

40° $\mathbf{I} = 2.12 \text{ mA}\angle{-40°}$

$+$

$-j$ $\mathbf{V} = 4.24 \text{ V}\angle{-40°}$

Figure 18–8

Inductors

When an inductor is subjected to a sinusoidal current, a sinusoidal voltage is induced across the inductor such that the voltage across the inductor leads the current waveform by exactly 90°. If we know the reactance of an inductor, then from Ohm's law the current in the inductor may be expressed in phasor form as

$$\mathbf{I} = \frac{\mathbf{V}}{\mathbf{Z}_L} = \frac{V\angle\theta}{X_L\angle 90°} = \frac{V}{X_L}\angle(\theta - 90°)$$

In vector form, the reactance of the inductor is given as

$$\mathbf{Z}_L = X_L\angle 90°$$

where $X_L = \omega L = 2\pi f L$.

■ **EXAMPLE 18–3**   Consider the inductor shown in Figure 18–9:

a. Determine the sinusoidal expression for the current i using phasors.
b. Sketch the sinusoidal waveforms for v and i.
c. Sketch the phasor diagram showing **V** and **I**.

$$V = 1.05 \sin(\omega t + 120°) \quad (V)$$

Figure 18–9

Solution:

a. The phasor form of the voltage is determined as follows:

$$v = 1.05 \sin(\omega t + 120°) \Leftrightarrow \mathbf{V} = 0.742 \text{ V}\angle 120°$$

From Ohm's law, the current phasor is determined to be

$$\mathbf{I} = \frac{\mathbf{V}}{\mathbf{Z}_L} = \frac{0.742 \text{ V}\angle 120°}{25 \text{ }\Omega\angle 90°} = 29.7 \text{ mA}\angle 30°$$

The amplitude of the sinusoidal current is

$$I_m = (\sqrt{2})(29.7 \text{ mA}) = 42 \text{ mA}$$

The current i is now written as

$$i = 0.042 \sin(\omega t + 30°) \quad (\text{amp})$$

b. Figure 18–10 shows the sinusoidal waveforms of the voltage and current.

Figure 18–10   Sinusoidal voltage and current for an inductor.

c. The voltage and current phasors are shown in Figure 18–11.

Figure 18–11   Voltage and current phasors for an inductor.

Capacitors

When a capacitor is subjected to a sinusoidal voltage, a sinusoidal current results. The current through the capacitor leads the voltage by exactly 90°. If we know the reactance of a capacitor, then from Ohm's law the current in the capacitor expressed in phasor form is

$$\mathbf{I} = \frac{\mathbf{V}}{\mathbf{Z}_C} = \frac{V\angle\theta}{X_L\angle{-90°}} = \frac{V}{X_L}\angle(\theta + 90°)$$

In vector form, the reactance of the capacitor is given as

$$\mathbf{Z}_C = X_C\angle{-90°}$$

where

$$X_C = \frac{1}{\omega C} = \frac{1}{2\pi f C}$$

■ **EXAMPLE 18–4**   Consider the capacitor of Figure 18–12.

a. Find the voltage v across the capacitor.

b. Sketch the sinusoidal waveforms for v and i.

c. Sketch the phasor diagram showing **V** and **I**.

$$i = 2.4 \sin(\omega t + 62°) \quad \text{(mA)}$$

$$X_C \qquad 1.2 \text{ k}\Omega \quad V$$

Figure 18–12

Solution

a. Converting the sinusoidal current into its equivalent phasor form gives

$$i = 2.4 \times 10^{-3}\sin(\omega t + 62°) \Leftrightarrow \mathbf{I} = 1.70 \text{ mA}\angle 62°$$

From Ohm's law, the phasor voltage across the capacitor must be

$$\mathbf{V} = \mathbf{I}\mathbf{Z}_C$$

$$= (1.70 \text{ mA}\angle 62°)(1.2 \text{ k}\Omega\angle -90°)$$

$$= 2.04 \text{ V}\angle -28°$$

The amplitude of the sinusoidal voltage is

$$V_m = (\sqrt{2})(2.04 \text{ V}) = 2.88 \text{ V}$$

The voltage v is now written as

$$v = 2.88 \sin(\omega t - 28°) \quad \text{(V)}$$

b. Figure 18–13 shows the waveforms for v and i.

$$v = 2.88 \sin(\omega t - 28°) \qquad i = 2.4 \times 10^{-3} \sin(\omega t + 62°)$$

Figure 18–13   Sinusoidal voltage and current for a capacitor.

c. The corresponding phasor diagram for $\mathbf{V}$ and $\mathbf{I}$ is shown in Figure 18–14.

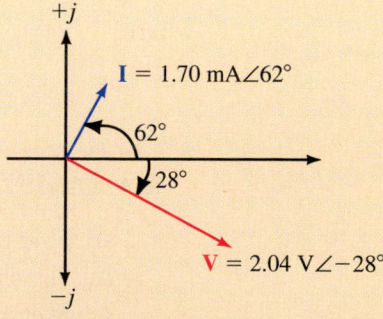

$$\mathbf{I} = 1.70 \text{ mA}\angle 62°$$

$$\mathbf{V} = 2.04 \text{ V}\angle -28°$$

Figure 18–14   Voltage and current phasors for a capacitor.

The relationships between voltage and current, as illustrated in the previous three examples, will always hold for resistors, inductors, and capacitors.

in process
learning
check I

1. What is the phase relationship between current and voltage for a resistor?

2. What is the phase relationship between current and voltage for a capacitor?

3. What is the phase relationship between current and voltage for an inductor?

practice
problem
I

A voltage source, $\mathbf{E} = 10 \text{ V} \angle 30°$, is applied to an inductive impedance of 50 Ω.

   a. Solve for the phasor current, $\mathbf{I}$.
   b. Sketch the phasor diagram for $\mathbf{E}$ and $\mathbf{I}$.
   c. Write the sinusoidal expressions for e and i.
   d. Sketch the sinusoidal expressions for e and i.

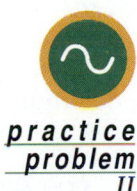

practice
problem
II

A voltage source, $\mathbf{E} = 10 \text{ V} \angle 30°$, is applied to a capacitive impedance of 20 Ω.

   a. Solve for the phasor current, $\mathbf{I}$.
   b. Sketch the phasor diagram for $\mathbf{E}$ and $\mathbf{I}$.
   c. Write the sinusoidal expressions for e and i.
   d. Sketch the sinusoidal expressions for e and i.

18.2 AC Series Circuits

When we examined dc circuits we saw that the current everywhere in a series circuit is always constant. This same principle applies when we have series elements with an ac source. Further, we had seen that the total resistance of a dc series circuit consisting of N resistors was determined as the summation

$$R_T = R_1 + R_2 + \cdots + R_N$$

When working with ac circuits we no longer work with only resistance but also with capacitive and inductive reactance. *Impedance is a term used to collectively determine how the resistance, capacitance, and inductance "impede" the current in a circuit.* The symbol for impedance is the letter Z and the unit is the ohm (Ω). Because impedance may be made up of any combination of resistances and reactances, it is written as a vector quantity $\mathbf{Z}$, where

$$\mathbf{Z} = Z \angle \theta \quad (\Omega)$$

Figure 18–15

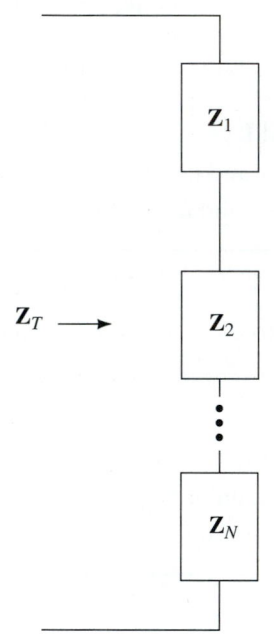

Figure 18–16

Each impedance may be represented as a vector on the complex plane, such that the length of the vector is representative of the magnitude of the impedance. The diagram showing one or more impedances is referred to as an *impedance diagram*.

Resistance **R** is a vector having a magnitude of R along the positive real axis. Inductive reactance $\mathbf{Z}_L$ is a vector having a magnitude of X_L along the positive imaginary axis, while the capacitive reactance $\mathbf{Z}_C$ is a vector having a magnitude of X_C along the negative imaginary axis. Mathematically, each of the vector impedances is written as follows:

$$\mathbf{Z}_R = R\angle 0° = R + j0 = R$$
$$\mathbf{Z}_L = X_L\angle 90° = 0 + jX_L = jX_L$$
$$\mathbf{Z}_C = X_C\angle 90° = 0 - jX_C = -jX_C$$

An impedance diagram showing each of the above impedances is shown in Figure 18–15.

All impedance vectors will appear in either the first or the fourth quadrants, since the resistance vector is always positive.

For a series ac circuit consisting of N impedances, as shown in Figure 18–16, the total impedance of the circuit will be found as the vector sum

$$\mathbf{Z}_T = \mathbf{Z}_1 + \mathbf{Z}_2 + \cdots + \mathbf{Z}_N \qquad (18\text{–}1)$$

Consider the branch of Figure 18–17.

By applying Equation 18–1, we may determine the total impedance of the circuit as

$$\mathbf{Z}_T = (3\ \Omega + j0) + (0 + j4\ \Omega) = 3\ \Omega + j4\ \Omega$$
$$= 5\ \Omega\angle 53.13°$$

The above quantities are shown on an impedance diagram as in Figure 18–18.

From Figure 18–18 we see that the total impedance of the series elements consists of a real component and an imaginary component. The corresponding total impedance vector may be written in either polar or rectangular form.

Figure 18–17

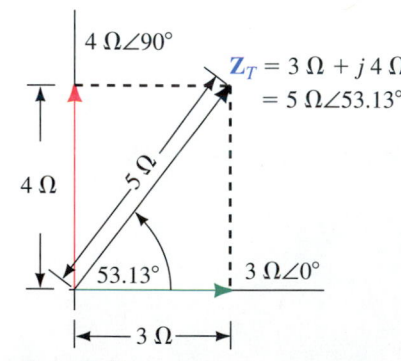

Figure 18–18

The rectangular form of an impedance is written as

$$\mathbf{Z} = R \pm jX$$

If we are given the polar form of the impedance, then we may determine the equivalent rectangular expression from

$$R = Z \cos \theta$$

and

$$X = Z \sin \theta$$

In the rectangular representation for impedance, the resistance term, R, is the total of all resistance looking into the network. The reactance term, X, is the difference between the total capacitive and inductive reactances. The sign for the imaginary term will be positive if the inductive reactance is greater than the capacitive reactance. In such a case, the impedance vector will appear in the first quadrant of the impedance diagram and is referred to as being an *inductive* impedance. If the capacitive reactance is larger, then the sign for the imaginary term will be negative. In such a case, the impedance vector will appear in the fourth quadrant of the impedance diagram and the impedance is said to be *capacitive*.

The polar form of any impedance will be written in the form

$$\mathbf{Z} = Z\angle\theta$$

The value Z is the magnitude (in ohms) of the impedance vector $\mathbf{Z}$ and is determined as follows:

$$Z = \sqrt{R^2 + X^2} \quad (\Omega) \tag{18-2}$$

The corresponding angle of the impedance vector is determined as

$$\theta = \pm\tan^{-1}\left(\frac{X}{R}\right) \tag{18-3}$$

Whenever a capacitor and an inductor having equal reactances are placed in series, as shown in Figure 18–19, the equivalent circuit of the two components is a short circuit since the inductive reactance will be exactly balanced by the capacitive reactance.

Figure 18–19

Any ac circuit having a total impedance with only a real component, is referred to as a *resistive* circuit. In such a case, the impedance vector $\mathbf{Z}_T$ will be located along the positive real axis of the impedance diagram and the angle of the vector will be 0°. The condition under which series reactances are equal is referred to as "series resonance" and is examined in greater detail in a later chapter.

If the impedance $\mathbf{Z}$ is written in polar form, then the angle θ will be positive for an inductive impedance and negative for a capacitive impedance. In the event that the circuit is purely reactive, the resulting angle θ will be either +90° (inductive) or −90° (capacitive). If we reexamine the impedance diagram of Figure 18–18, we conclude that the original circuit is inductive.

■ **EXAMPLE 18–5**   Consider the network of Figure 18–20.

Figure 18–20

a. Find $\mathbf{Z}_T$.

b. Sketch the impedance diagram for the network and indicate whether the total impedance of the circuit is inductive, capacitive, or resistive.

c. Use Ohm's law to determine $\mathbf{I}$, $\mathbf{V}_R$, and $\mathbf{V}_C$.

Solution

a. The total impedance is the vector sum

$$\mathbf{Z}_T = 25\ \Omega + j200\ \Omega + (-j225\ \Omega)$$

$$= 25\ \Omega - j25\ \Omega$$

$$= 35.36\ \Omega\angle{-45°}$$

b. The corresponding impedance diagram is shown in Figure 18–21.

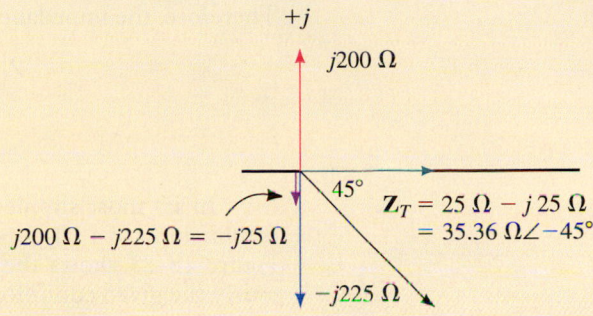

Figure 18–21

Because the total impedance has a negative reactance term ($-j25\ \Omega$), $\mathbf{Z}_T$ is capacitive.

$$\text{c.} \qquad \mathbf{I} = \frac{10\ \text{V}\angle 0°}{35.36\ \Omega\angle -45°} = 282.8\ \text{mA}\angle 45°$$

$$\mathbf{V}_R = (282.8\ \text{mA}\angle 45°)(25\ \Omega\angle 0°) = 7.071\ \text{V}\angle 45°$$

$$\mathbf{V}_C = (282.8\ \text{mA}\angle 45°)(225\ \Omega\angle -90°) = 63.64\ \text{V}\angle -45°$$

Notice that the magnitude of the voltage across the capacitor is many times larger than the source voltage applied to the circuit. This example illustrates that the voltages across reactive elements must be calculated to ensure that maximum ratings for the components are not exceeded.

■ **EXAMPLE 18–6**   Determine the impedance $\mathbf{Z}$ which must be within the indicated block of Figure 18–22 if the total impedance of the network is $13\ \Omega\angle 22.62°$.

Figure 18–22

Solution   Converting the total impedance from polar to rectangular form, we get

$$\mathbf{Z}_T = 13\ \Omega\angle 22.62° \Leftrightarrow 12\ \Omega + j5\ \Omega$$

Now, we know that the total impedance is determined from the summation of the individual impedance vectors, namely

$$\mathbf{Z}_T = 2\ \Omega + j10\ \Omega + \mathbf{Z} = 12\ \Omega + j5\ \Omega$$

Therefore, the impedance **Z** is found as

$$\mathbf{Z} = 12\ \Omega + j5\ \Omega - (2\ \Omega + j10\ \Omega)$$

$$= 10\ \Omega - j5\ \Omega$$

$$= 11.18\ \Omega\angle-26.57°$$

In its most simple form, the impedance **Z** will consist of a series combination of a 10-Ω resistor and a capacitor having a reactance of 5-Ω. Figure 18–23 shows the elements which may be contained within **Z** to satisfy the given conditions.

$$\mathbf{Z} = 10\ \Omega - j\,5\Omega$$
$$= 11.18\ \Omega\ \angle-26.57°$$

Figure 18–23

■ **EXAMPLE 18–7**   Find the total impedance for the network of Figure 18–24. Sketch the impedance diagram showing $\mathbf{Z}_1, \mathbf{Z}_2$, and $\mathbf{Z}_T$.

Figure 18–24

Solution:

$$\mathbf{Z}_T = \mathbf{Z}_1 + \mathbf{Z}_2$$

$$= (50\ \Omega + j90\ \Omega) + (30\ \Omega - j30\ \Omega)$$

$$= (80\ \Omega + j60\ \Omega) = 100\ \Omega\angle36.87°$$

The polar forms of the vectors $\mathbf{Z}_1$ and $\mathbf{Z}_2$ are as follows:

$$\mathbf{Z}_1 = 50\ \Omega + j90\ \Omega = 102.96\ \Omega\angle60.95°$$

$$\mathbf{Z}_2 = 30\ \Omega - j30\ \Omega = 42.43\ \Omega\angle-45°$$

The resulting impedance diagram is shown in Figure 18–25.

Figure 18–25

Figure 18–26

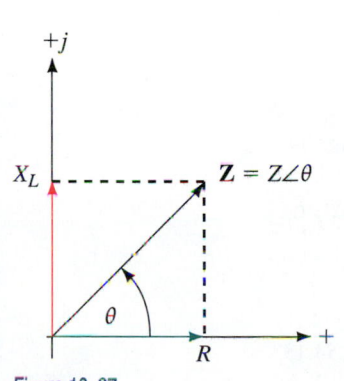

Figure 18–27

The phase angle θ for the impedance vector $\mathbf{Z} = Z\angle\theta$ provides the phase angle between the voltage $\mathbf{V}$ across $\mathbf{Z}$ and the current $\mathbf{I}$ through the impedance. For an inductive impedance the voltage will lead the current by θ. If the impedance is capacitive, then the voltage will lag the current by an amount equal to the magnitude of θ.

The phase angle θ is also useful for determining the power dissipated by the circuit. In the simple series circuit shown in Figure 18–26, we know that only the resistor will dissipate power.

The power dissipated by the resistor may be determined as follows:

$$P = V_R I = \frac{V_R^2}{R} = I^2 R \qquad (18\text{–}4)$$

Notice that Equation 18–4 uses only the *magnitudes* of the voltage, current, and impedance vectors. *Power is never determined by using phasor products.*

Ohm's law provides the magnitude of the current phasor as

$$I = \frac{V}{Z}$$

Substituting this expression into Equation 18–4, we obtain the expression for power as

$$P = \frac{V^2}{Z^2} R = \frac{V^2}{Z}\left(\frac{R}{Z}\right) \qquad (18\text{–}5)$$

From the impedance diagram of Figure 18–27, we see that

$$\cos\theta = \frac{R}{Z}$$

The previous chapter had defined the power factor as $F_p = \cos\theta$, where θ is the angle between the voltage and current phasors. We now see that for a series circuit, the power factor of the circuit is determined from the magnitudes of resistance and total impedance.

$$F_p = \cos\theta = \frac{R}{Z} \qquad (18\text{–}6)$$

The power factor, F_p, is said to be *leading* if the current leads the voltage (capacitive circuit) and *lagging* if the current lags the voltage (inductive circuit).

Now substituting the expression for the power factor into Equation 18–5, we express power delivered to the circuit as

$$P = VI\cos\theta$$

Since $V = IZ$, power may be expressed as

$$P = VI\cos\theta = I^2 Z\cos\theta = \frac{V^2}{Z}\cos\theta \qquad (18\text{–}7)$$

■ **EXAMPLE 18–8**   Refer to the circuit of Figure 18–28.

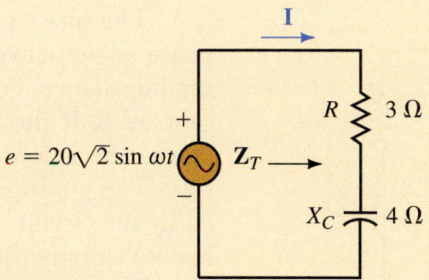

Figure 18–28

a. Find the impedance $\mathbf{Z}_T$.
b. Calculate the power factor of the circuit.
c. Determine $\mathbf{I}$.
d. Sketch the phasor diagram for $\mathbf{E}$ and $\mathbf{I}$.
e. Find the power delivered to the circuit by the voltage source.
f. Calculate the power dissipated by both the resistor and the capacitor.

Solution

a. $\mathbf{Z}_T = 3\,\Omega - j4\,\Omega = 5\,\Omega\angle{-53.13°}$

b. $F_p = \cos\theta = \dfrac{3\,\Omega}{5\,\Omega} = 0.6$ (leading)

c. The phasor form of the applied voltage is

$$\mathbf{E} = \frac{(\sqrt{2})(20\text{ V})}{\sqrt{2}}\angle 0° = 20\text{ V}\angle 0°$$

which gives a current of

$$\mathbf{I} = \frac{20\text{ V}\angle 0°}{5\,\Omega\angle{-53.13°}} = 4.0\text{ A}\angle 53.13°$$

d. The phasor diagram is shown in Figure 18–29.

Figure 18–29

From this phasor diagram, we see that the current phasor for the capacitive circuit leads the voltage phasor by 53.13°.

e. The power delivered to the circuit by the voltage source is

$$P = (20 \text{ V})(4 \text{ A}) \cos 53.13° = 48.0 \text{ W}$$

f. The power dissipated by the resistor and capacitor will be

$$P_R = (4 \text{ A})^2 (3 \text{ }\Omega)\cos 0° = 48 \text{ W}$$

$$P_C = (4 \text{ A})^2 (4 \text{ }\Omega)\cos 90° = 0 \text{ W} \quad \text{(as expected!)}$$

Notice that the power factor used in determining the power dissipated by each of the elements is the power factor for that element and not the total power factor for the circuit.

As expected, the summation of powers dissipated by the resistor and capacitor is equal to the total power delivered by the voltage source.

practice problem III

A circuit consists of a voltage source $\mathbf{E} = 50 \text{ V}\angle 25°$ in series with $L = 20$ mH, $C = 50$ μH, and $R = 25$ Ω. The circuit operates at an angular frequency of 2 krad/s.

a. Determine the current phasor, **I**.

b. Solve for the power factor of the circuit.

c. Calculate the power dissipated by the circuit and verify that this is equal to the power delivered by the source.

d. Use Ohm's law to find $\mathbf{V}_R$, $\mathbf{V}_L$, and $\mathbf{V}_C$.

18.3 Kirchhoff's Voltage Law and the Voltage Divider Rule

When a voltage is applied to impedances in series, as shown in Figure 18–30, Ohm's law may be used to determine the voltage across any impedance as

$$V_x = \mathbf{I}\mathbf{Z}_x$$

Figure 18–30

The current in the circuit is

$$I = \frac{E}{Z_T}$$

Now, by substitution we arrive at the voltage divider rule for any series combination of elements as

$$V_x = \frac{Z_x}{Z_T} E \qquad (18-8)$$

Equation 18–8 is very similar to the equation for the voltage divider rule in dc circuits. The fundamental differences in solving ac circuits are that we use impedances rather than resistances and that the voltages found are phasors. Because the voltage divider rule involves solving products and quotients of phasors, we generally use the polar form rather than the rectangular form of phasors.

Kirchhoff's voltage law must apply for all circuits whether they are dc or ac circuits. However, because ac circuits have voltages expressed as either sinusoidal or phasor form, Kirchhoff's voltage law for ac circuits may be stated as follows:

The phasor sum of voltage drops and voltage rises around a closed loop is equal to zero.

When adding phasor voltages, we find that the summation is generally done more easily in rectangular form rather than the polar form.

■ **EXAMPLE 18–9**   Consider the circuit of Figure 18–31.

Figure 18–31

a. Find Z_L.

b. Determine the voltages V_R and V_L using the voltage divider rule.

c. Verify Kirchhoff's voltage law around the closed loop.

Solution

a. $Z_T = 5 \text{ k}\Omega + j12 \text{ k}\Omega = 13 \text{ k}\Omega\angle 67.38°$

b. $V_R = \left(\dfrac{5 \text{ k}\Omega\angle 0°}{13 \text{ k}\Omega\angle 67.38°} \right)(26 \text{ V}\angle 0°) = 10 \text{ V}\angle -67.38°$

$V_L = \left(\dfrac{12 \text{ k}\Omega\angle 90°}{13 \text{ k}\Omega\angle 67.38°} \right)(26 \text{ V}\angle 0°) = 24 \text{ V}\angle 22.62°$

c. Kirchhoff's voltage law around the closed loop will give

$$26\text{ V}\angle 0° - 10\text{ V}\angle -67.38° - 24\text{ V}\angle 22.62° = 0$$

$$(26 + j0) - (3.846 - j9.231) - (22.154 + j9.231) = 0$$

$$(26 - 3.846 - 22.154) + j(0 + 9.231 - 9.231) = 0$$

$$0 + j0 = 0$$

■ **EXAMPLE 18–10**   Consider the circuit of Figure 18–32:

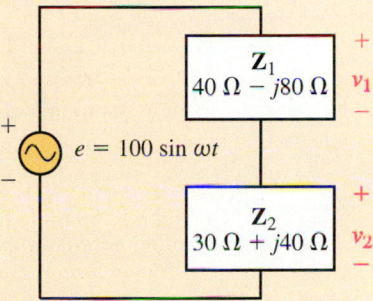

Figure 18–32

a. Calculate the sinusoidal voltages v_1 and v_2 using phasors and Kirchhoff's voltage law.

b. Sketch the phasor diagram showing $\mathbf{E}$, $\mathbf{V}_1$, and $\mathbf{V}_2$.

c. Sketch the sinusoidal waveforms of e, v_1, and v_2.

Solution

a. The phasor form of the voltage source is determined as

$$e = 100 \sin \omega t \Leftrightarrow \mathbf{E} = 70.71\angle \text{V } 0°$$

Applying VDR, we get

$$\mathbf{V}_1 = \left(\frac{40\ \Omega - j80\ \Omega}{(40\ \Omega - j80\ \Omega) + (30\ \Omega + j40\ \Omega)}\right)(70.71\text{ V}\angle 0°)$$

$$= \left(\frac{89.44\ \Omega\angle -63.43°}{80.62\ \Omega\angle -29.74°}\right)(70.71\text{ V}\angle 0°)$$

$$= 78.45\text{ V}\angle -33.69°$$

and

$$\mathbf{V}_2 = \left(\frac{30\ \Omega + j40\ \Omega}{(40\ \Omega - j80\ \Omega) + (30\ \Omega + j40\ \Omega)}\right)(70.71\text{ V}\angle 0°)$$

$$= \left(\frac{50.00\ \Omega\angle 53.13°}{80.62\ \Omega\angle -29.74°}\right)(70.71\text{ V}\angle 0°)$$

$$= 43.85\text{ V}\angle 82.87°$$

The sinusoidal voltages are determined to be

$$v_1 = (\sqrt{2})\,(78.45)\sin(\omega t - 33.69°)$$
$$= 110.9\,\sin(\omega t - 33.69°)$$

and

$$v_2 = (\sqrt{2})(43.85)\sin(\omega t + 82.87°)$$
$$= 62.01\,\sin(\omega t + 82.87°)$$

b. The phasor diagram is shown in Figure 18–33.

Figure 18–33

c. The corresponding sinusoidal voltages are shown in Figure 18–34.

Figure 18–34

in process
l e a r n i n g
check II

1. Express Kirchhoff's voltage law as it applies to ac circuits.

2. What is the fundamental difference between how Kirchhoff's voltage law is used in ac circuits as compared with dc circuits?

practice problem IV

A circuit consists of a voltage source $\mathbf{E} = 50\ \text{V}\angle 25°$ in series with $L = 20\ \text{mH}$, $C = 50\ \mu\text{H}$, and $R = 25\ \Omega$. The circuit operates at an angular frequency of 2 krad/s.

a. Use the voltage divider rule to determine the voltage across each element in the circuit.

b. Verify that Kirchhoff's voltage law applies for the circuit.

18.4 AC Parallel Circuits

The *admittance* $\mathbf{Y}$ of any impedance is defined as a vector quantity which is the reciprocal of the impedance $\mathbf{Z}$.

Mathematically, admittance is expressed as

$$\mathbf{Y}_T = \frac{1}{\mathbf{Z}_T} = \frac{1}{Z_T \angle \theta} = \left(\frac{1}{Z_T}\right)\angle -\theta = Y_T \angle -\theta \quad \text{(S)} \tag{18–9}$$

where the unit of admittance is the siemens (S).

In particular, we have seen that the admittance of a resistor R is called *conductance* and is given the symbol $\mathbf{Y}_R$. If we consider resistance as a vector quantity, then the corresponding vector form of the conductance is

$$\mathbf{Y}_R = \frac{1}{R\angle 0°} = \frac{1}{R}\angle 0° = G\angle 0° = G + j0 \quad \text{(S)} \tag{18–10}$$

If we determine the admittance of a purely reactive component X, the resultant admittance is called the *susceptance* of the component and is assigned the symbol B. The unit for susceptance is the siemens (S). In order to distinguish between inductive susceptance and capacitive susceptance, we use the subscripts L and C respectively. Since reactances are represented as vector quantities, we also express susceptances in vector form as follows:

$$\mathbf{Y}_L = \frac{1}{X_L \angle 90°} = \frac{1}{X_L}\angle -90° = B_L\angle -90° = 0 - jB_L \quad \text{(S)} \tag{18–11}$$

$$\mathbf{Y}_C = \frac{1}{X_C \angle -90°} = \frac{1}{X_C}\angle 90° = B_C\angle 90° = 0 + jB_C \quad \text{(S)} \tag{18–12}$$

In a manner similar to impedances, admittances may be represented on the complex plane in an *admittance diagram* as shown in Figure 18–35.

The lengths of the various vectors are proportional to the magnitudes of the corresponding admittances. The conductance vector $\mathbf{G}$ is shown on the positive real axis, whereas the inductive and capacitive susceptance vectors $\mathbf{Y}_L$ and $\mathbf{Y}_C$ are shown on the negative and positive imaginary axes respectively.

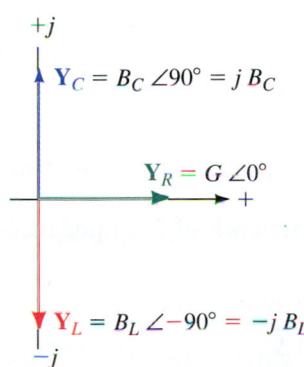

Figure 18–35 Admittance diagram showing conductance ($\mathbf{Y}_R$) and susceptance ($\mathbf{Y}_L$ and $\mathbf{Y}_C$).

■ **EXAMPLE 18–11**   Determine the admittances of the following impedances. Sketch the corresponding admittance diagram.

a. $R = 10\ \Omega$.   **b.** $X_L = 20\ \Omega$.   **c.** $X_C = 40\ \Omega$.

Solutions

a. $$\mathbf{Y}_R = \frac{1}{R} = \frac{1}{10\ \Omega\angle 0°} = 100\ \text{mS}\angle 0°$$

b. $$\mathbf{Y}_L = \frac{1}{X_L} = \frac{1}{20\ \Omega\angle 90°} = 50\ \text{mS}\angle -90°$$

c. $$\mathbf{Y}_C = \frac{1}{X_C} = \frac{1}{40\ \Omega\angle -90°} = 25\ \text{mS}\angle 90°$$

The admittance diagram is shown in Figure 18–36.

Figure 18–36

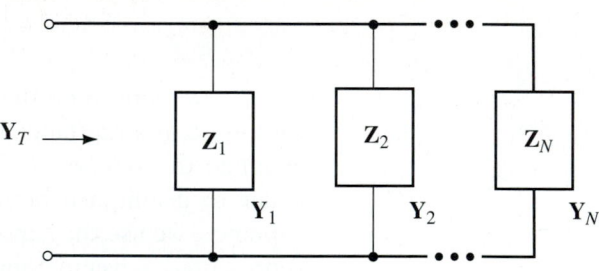

Figure 18–37

For any network of N admittances as shown in Figure 18–37, the total admittance is the vector sum of the admittances of the network. Mathematically, the total admittance of a network is given as

$$\mathbf{Y}_T = \mathbf{Y}_1 + \mathbf{Y}_2 + \cdots + \mathbf{Y}_N \quad (S) \tag{18–13}$$

The resultant impedance of a parallel network of N impedances is determined to be

$$\mathbf{Z}_T = \frac{1}{\mathbf{Y}_T} = \frac{1}{\mathbf{Y}_1 + \mathbf{Y}_2 + \cdots + \mathbf{Y}_N}$$

$$\mathbf{Z}_T = \frac{1}{\dfrac{1}{\mathbf{Z}_1} + \dfrac{1}{\mathbf{Z}_2} + \cdots + \dfrac{1}{\mathbf{Z}_N}} \quad (\Omega) \tag{18–14}$$

■ **EXAMPLE 18–12**   Find the equivalent admittance and impedance of the network of Figure 18–38. Sketch the admittance diagram.

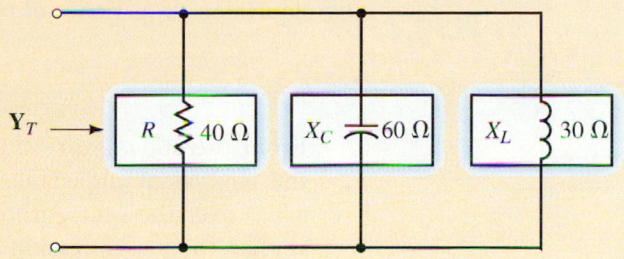

Figure 18–38

Solution   The admittances of the various parallel elements are

$$\mathbf{Y_1} = \frac{1}{40 \ \Omega \angle 0°} = 25.0 \ \text{mS} \angle 0° = 25.0 \ \text{mS} + j0$$

$$\mathbf{Y_2} = \frac{1}{60 \ \Omega \angle -90°} = 16.\overline{6} \ \text{mS} \angle 90° = 0 + j16.\overline{6} \ \text{mS}$$

$$\mathbf{Y_3} = \frac{1}{30 \ \Omega \angle 90°} = 33.\overline{3} \ \text{mS} \angle -90° = 0 - j33.\overline{3} \ \text{mS}$$

The total admittance is determined as

$$\mathbf{Y_T} = \mathbf{Y_1} + \mathbf{Y_2} + \mathbf{Y_3}$$

$$= 25.0 \ \text{mS} + j16.\overline{6} \ \text{mS} + (-j33.\overline{3} \ \text{mS})$$

$$= 25.0 \ \text{mS} - j16.\overline{6} \ \text{mS}$$

$$= 30.05 \ \text{mS} \angle -33.69°$$

This results in a total impedance for the network of

$$\mathbf{Z_T} = \frac{1}{\mathbf{Y_T}}$$

$$= \frac{1}{30.05 \ \text{mS} \angle -33.69°}$$

$$= 33.28 \ \Omega \angle 33.69°$$

The admittance diagram is shown in Figure 18–39.

Figure 18–39

Two Impedances in Parallel

By applying Equation 18–14 for two impedances, we determine the equivalent impedance of two impedances as

$$\mathbf{Z}_T = \frac{\mathbf{Z}_1 \mathbf{Z}_2}{\mathbf{Z}_1 + \mathbf{Z}_2} \quad (\Omega) \tag{18–15}$$

From the above expression, we see that for two impedances in parallel, the equivalent impedance is determined as the product of the impedances over the sum. Although the expression for two impedances is very similar to the expression for two resistors in parallel, the difference is that the calculation of impedance involves the use of complex algebra.

■ **EXAMPLE 18–13** Find the total impedance for the network shown in Figure 18–40.

Figure 18–40

Solution

$$\mathbf{Z}_T = \frac{(200\ \Omega\angle-90°)(250\ \Omega\angle90°)}{-j200\ \Omega + j250\ \Omega}$$

$$= \frac{50\ \text{k}\Omega\angle0°}{50\angle90°} = 1\ \text{k}\Omega\angle-90°$$

The previous example illustrates that unlike total parallel resistance, the total impedance of a combination of parallel reactances may be much larger that either of the individual impedances. Indeed, if we are given a parallel combination of equal inductive and capacitive reactances, the total impedance of the combination is equal to infinity (namely an open circuit). Consider the network of Figure 18–41.

The total impedance $\mathbf{Z}_T$ is found as

$$\mathbf{Z}_T = \frac{(X_L\angle90°)(X_C\angle-90°)}{jX_L - jX_C} = \frac{X^2\angle0°}{0\angle0°} = \infty\angle0°$$

Because the denominator of the above expression is equal to zero, the magnitude of the total impedance will be undefined ($Z = \infty$). Although the magnitude is undefined, we see that the algebra yields a phase angle $\theta = 0°$, which indicates that the vector lies on the positive real axis of the impedance diagram.

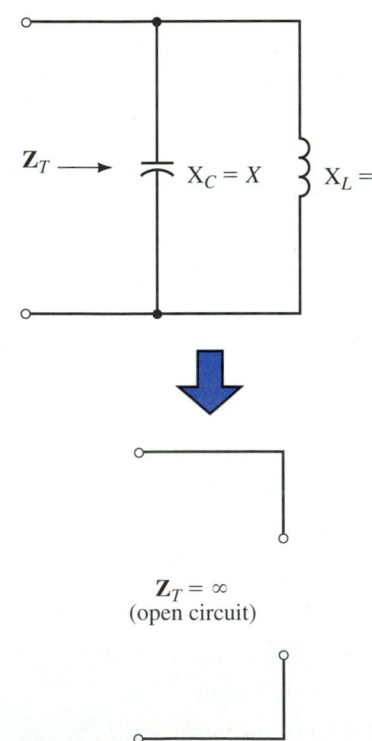

Figure 18–41

Whenever a capacitor and an inductor having equal reactances are placed in parallel, the equivalent circuit of the two components is an open circuit.

The principle of equal parallel reactances will be studied in a later chapter dealing with "resonance."

Three Impedances in Parallel

Equation 18–14 may be solved for three impedances to give the equivalent impedance as

$$Z_T = \frac{Z_1 Z_2 Z_3}{Z_1 Z_2 + Z_1 Z_3 + Z_2 Z_3} \quad (\Omega) \qquad (18\text{–}16)$$

although this is less than the general equation.

■ **EXAMPLE 18–14**   Find the equivalent impedance of the network of Figure 18–42.

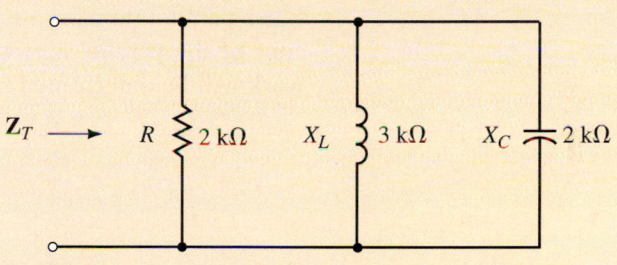

Figure 18–42

Solution

$$Z_T = \frac{(2\ k\Omega\angle 0°)(3\ k\Omega\angle 90°)(2\ k\Omega\angle -90°)}{(2\ k\Omega\angle 0°)(3\ k\Omega\angle 90°) + (2\ k\Omega\angle 0°)(2\ k\Omega\angle -90°) + (3\ k\Omega\angle 90°)(2\ k\Omega\angle -90°)}$$

$$= \frac{12 \times 10^9\ \Omega\angle 0°}{6 \times 10^6 \angle 90° + 4 \times 10^6 \angle -90° + 6 \times 10^6 \angle 0°}$$

$$= \frac{12 \times 10^9\ \Omega\angle 0°}{6 \times 10^6 + j2 \times 10^6} = \frac{12 \times 10^9\ \Omega\angle 0°}{6.325 \times 10^6 \angle 18.43°}$$

$$= 1.90\ k\Omega\angle -18.43°$$

And so the equivalent impedance of the network is

$$Z_T = 1.80\ k\Omega - j0.6\ k\Omega$$

---

practice
problem
V

A circuit consists of a current source, $i = 0.030 \sin 500t$, in parallel with $L = 20$ mH, $C = 50$ μF, and $R = 25$ Ω.

  a. Determine the voltage **V** across the circuit.

  b. Solve for the power factor of the circuit.

  c. Calculate the power dissipated by the circuit and verify that this is equal to the power delivered by the source.

  d. Use Ohm's law to find the phasor quantities, $\mathbf{I}_R$, $\mathbf{I}_L$, and $\mathbf{I}_C$.

A circuit consists of a 2.5-A$_{rms}$ current source connected in parallel with a resistor, an inductor, and a capacitor. The resistor has a value of 10 Ω and dissipates 40 W of power.

a. Calculate the values of X_L and X_C if $X_L = 3X_C$.

b. Determine the magnitudes of current through the inductor and the capacitor.

18.5 Kirchhoff's Current Law and the Current Divider Rule

The current divider rule for ac circuits has the same form as for dc circuits with the notable exception that currents are expressed as phasors. For a parallel network as shown in Figure 18–43, where the total impedance of the parallel network is $\mathbf{Z}_T$, the current in any branch of the network will be determined to be

$$\mathbf{I}_x = \frac{\mathbf{Z}_T}{\mathbf{Z}_x}\mathbf{I} \qquad (18\text{–}17)$$

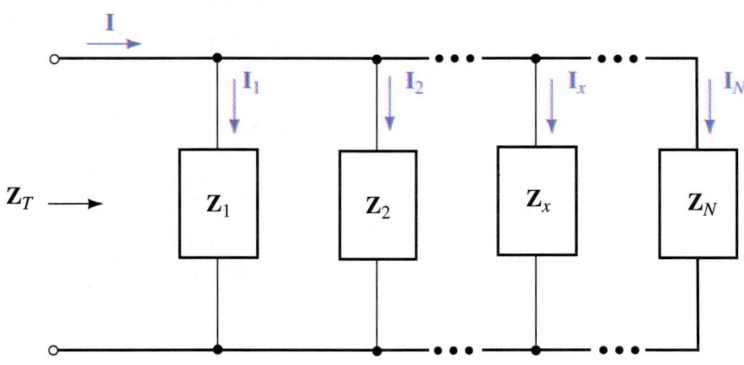

Figure 18–43

For two branches in parallel the current in either branch is determined as

$$\mathbf{I}_1 = \frac{\mathbf{Z}_2}{\mathbf{Z}_1 + \mathbf{Z}_2}\mathbf{I} \qquad (18\text{–}18)$$

Also, as one would expect, Kirchhoff's current law must apply to any node within an ac circuit. For such circuits, KCL may be stated as follows:

The summation of current phasors entering and leaving a node is equal to zero.

■ **EXAMPLE 18–15**   Calculate the current in each of the branches in the network of Figure 18–44.

$\mathbf{I} = 2\ \text{A}\angle 0°$

$\mathbf{I}_1$ $\mathbf{I}_2$

$X_L \gtrless 200\ \Omega$ $X_C \rightleftharpoons 250\ \Omega$

Figure 18–44

Solution

$$\mathbf{I}_1 = \left(\frac{250\ \Omega\angle -90°}{j200\ \Omega - j250\ \Omega} \right)(2\ \text{A}\angle 0°)$$

$$= \left(\frac{250\ \Omega\angle -90°}{50\ \Omega\angle -90°} \right)(2\ \text{A}\angle 0°) = 10\ \text{A}\angle 0°$$

and

$$\mathbf{I}_2 = \left(\frac{200\ \Omega\angle 90°}{j200\ \Omega - j250\ \Omega} \right)(2\ \text{A}\angle 0°)$$

$$= \left(\frac{200\ \Omega\angle 90°}{50\ \Omega\angle -90°} \right)(2\ \text{A}\angle 0°) = 8\ \text{A}\angle 180°$$

The above results illustrate that the currents in parallel reactive components may be significantly larger than the applied current. If the current through the component exceeds the maximum current rating of the element, severe damage may occur.

■ **EXAMPLE 18–16**   Refer to the circuit of Figure 18–45:

$\mathbf{I}_T$

a

$\mathbf{I}_1$ $\mathbf{I}_1$ $\mathbf{I}_1$

$\mathbf{E} = 5\ \text{V}\angle 0°$ $\mathbf{Z}_T \longrightarrow$ $R \gtrless 20\ \text{k}\Omega$ $X_L \gtrless 1\ \text{k}\Omega$ $X_C \rightleftharpoons 1\ \text{k}\Omega$

Figure 18–45

a. Find the total impedance, $\mathbf{Z}_T$.

b. Determine the supply current, $\mathbf{I}_T$.

c. Calculate $\mathbf{I}_1$, $\mathbf{I}_2$, and $\mathbf{I}_3$ using the current divider rule.

d. Verify Kirchhoff's current law at node a.

Solution

a. Because the inductive and capacitive reactances are in parallel and have the same value, we may replace the combination by an open circuit. Consequently, only the resistor R needs to be considered. As a result

$$\mathbf{Z}_T = 20 \text{ k}\Omega\angle 0°$$

b.
$$\mathbf{I}_T = \frac{5 \text{ V}\angle 0°}{20 \text{ k}\Omega\angle 0°} = 250 \ \mu\text{A}\angle 0°$$

c.
$$\mathbf{I}_1 = \left(\frac{20 \text{ k}\Omega\angle 0°}{20 \text{ k}\Omega\angle 0°} \right)(250 \ \mu\text{A}\angle 0°) = 250 \ \mu\text{A}\angle 0°$$

$$\mathbf{I}_2 = \left(\frac{20 \text{ k}\Omega\angle 0°}{1 \text{ k}\Omega\angle 90°} \right)(250 \ \mu\text{A}\angle 0°) = 5.0 \text{ mA}\angle -90°$$

$$\mathbf{I}_3 = \left(\frac{20 \text{ k}\Omega\angle 0°}{1 \text{ k}\Omega\angle -90°} \right)(250 \ \mu\text{A}\angle 0°) = 5.0 \text{ mA}\angle 90°$$

d. Notice that the currents through the inductor and capacitor are 180° out of phase. By adding the current phasors in rectangular form, we have

$$I_T = 250 \ \mu\text{A} - j5.0 \text{ A} + j5.0 \text{ A} = 250 \ \mu\text{A} + j0 = 250 \ \mu\text{A}\angle 0°$$

The above result verifies Kirchhoff's current law at the node.

in process
learning
check III

1. Express Kirchhoff's current law as it applies to ac circuits.

2. What is the fundamental difference between how Kirchhoff's current law is applied to ac circuits as compared with dc circuits?

practice
problem
VII

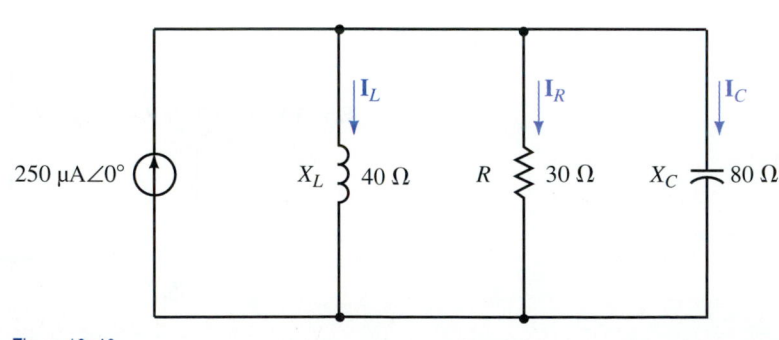

Figure 18–46

a. Use the current divider rule to determine current through each branch in the circuit of Figure 18–46.

b. Verify that Kirchhoff's current law applies to the circuit of Figure 18–46.

18.6 Series-Parallel Circuits

We may now apply the analysis techniques of series and parallel circuits in solving more complicated circuits. As in dc circuits, the analysis of such circuits is simplified by starting with easily recognized combinations. If necessary, the original circuit may be redrawn to make further simplification more apparent. Regardless of the complexity of the circuits, we find that the fundamental rules and laws of circuit analysis must apply in all cases.

Consider the network of Figure 18–47.

We see that the impedances Z_2 and Z_3 are in series. The branch containing this combination is then seen to be in parallel with the impedance Z_1.

The total impedance of the network is expressed as

$$Z_T = Z_1 \,\|\, (Z_2 + Z_3)$$

Solving for Z_T gives the following:

$$Z_T = (2\ \Omega - j8\ \Omega) \,\|\, (2\ \Omega - j5\ \Omega + 6\ \Omega + j7\ \Omega)$$

$$= (2\ \Omega - j8\ \Omega) \,\|\, (8\ \Omega + j2\ \Omega)$$

$$= \frac{(2\ \Omega - j8\ \Omega)(8\ \Omega + j2\ \Omega)}{2\ \Omega - j8\ \Omega + 8\ \Omega + j2\ \Omega}$$

$$= \frac{(8.246\ \Omega\angle{-75.96°})(8.246\ \Omega\angle14.04°)}{11.66\ \Omega\angle{-30.96°}}$$

$$= 5.832\ \Omega\angle{-30.96°} = 5.0\ \Omega - j3.0\ \Omega$$

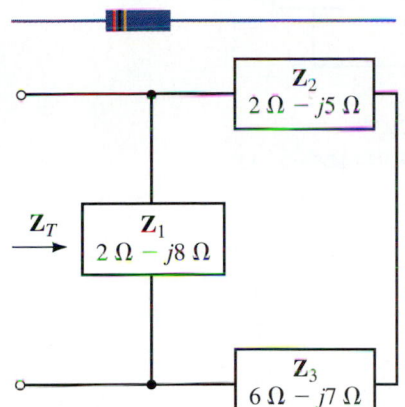

Figure 18–47

■ **EXAMPLE 18–17**   Determine the total impedance of the network of Figure 18–48. Express the impedance in both polar form and rectangular form.

Figure 18–48

Solution   After redrawing and labelling the given circuit, we have the circuit shown in Figure 18–49.

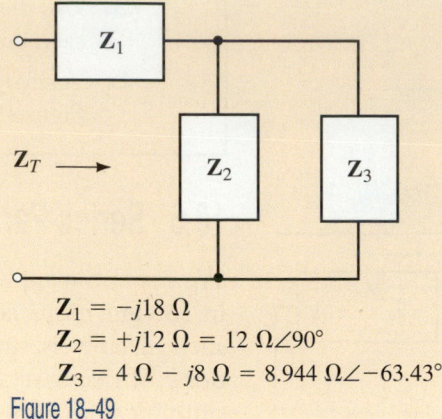

$$\mathbf{Z}_1 = -j18\ \Omega$$
$$\mathbf{Z}_2 = +j12\ \Omega = 12\ \Omega\angle90°$$
$$\mathbf{Z}_3 = 4\ \Omega - j8\ \Omega = 8.944\ \Omega\angle-63.43°$$

Figure 18–49

The total impedance is given as

$$\mathbf{Z}_T = \mathbf{Z}_1 + \mathbf{Z}_2 \,\|\, \mathbf{Z}_3$$

where

$$\mathbf{Z}_1 = -j18\ \Omega = 18\Omega\angle-90°$$

$$\mathbf{Z}_2 = j12\ \Omega = 12\Omega\angle90°$$

$$\mathbf{Z}_3 = 4\ \Omega - j8\ \Omega = 8.944\ \Omega\angle-63.43°$$

We determine the total impedance as

$$\mathbf{Z}_T = -j18\ \Omega + \left[\frac{(12\ \Omega\angle90°)(8.944\ \Omega\angle-63.43°)}{j12\ \Omega + 4\ \Omega - j8\ \Omega}\right]$$

$$= -j18\ \Omega + \left(\frac{107.33\ \Omega\angle26.57°}{5.657\angle45°}\right)$$

$$= -j18\ \Omega + 18.97\ \Omega\angle-18.43°$$

$$= -j18\ \Omega + 18\ \Omega - j6\ \Omega$$

$$= 18\ \Omega - j24\ \Omega = 30\ \Omega\angle-53.13°$$

■ **EXAMPLE 18–18**   Consider the circuit of Figure 18–50:

Figure 18–50

a. Find $\mathbf{Z}_T$.

b. Determine the currents $\mathbf{I}_1$, $\mathbf{I}_2$, and $\mathbf{I}_3$.

c. Calculate the total power provided by the voltage source.

d. Determine the powers P_1, P_2, and P_3 dissipated by each of the impedances. Verify that the power delivered to the circuit is the same as the power dissipated by the impedances.

Solution

a. The total impedance is determined by the combination

$$\mathbf{Z}_T = \mathbf{Z}_1 + \mathbf{Z}_2 \parallel \mathbf{Z}_3$$

For the parallel combination we have

$$\mathbf{Z}_2 \parallel \mathbf{Z}_3 = \frac{(1\ \text{k}\Omega + j2\ \text{k}\Omega)\,(-j2\ \text{k}\Omega)}{1\ \text{k}\Omega + j2\ \text{k}\Omega - j2\ \text{k}\Omega}$$

$$= \frac{(2.236\ \text{k}\Omega \angle 63.43°)\,(2\ \text{k}\Omega \angle -90°)}{1\ \text{k}\Omega \angle 0°}$$

$$= 4.472\ \text{k}\Omega \angle -26.57° = 4.0\ \text{k}\Omega - j2.0\ \text{k}\Omega$$

And so the total impedance is

$$\mathbf{Z}_T = 5\ \text{k}\Omega - j2\ \text{k}\Omega = 5.385\ \text{k}\Omega \angle -21.80°$$

$$\mathbf{I}_1 = \frac{50\ \text{V} \angle 0°}{5.385\ \text{k}\Omega \angle -21.80°}$$

$$= 9.285\ \text{mA} \angle 21.80°$$

Applying the current divider rule, we get

$$\mathbf{I}_2 = \frac{(2\ \text{k}\Omega \angle -90°)(9.285\ \text{mA} \angle 21.80°)}{1\ \text{k}\Omega + j2\ \text{k}\Omega - j2\ \text{k}\Omega}$$

$$= 18.57\ \text{mA} \angle -68.20°$$

and

$$\mathbf{I}_3 = \frac{(1\ \text{k}\Omega + j2\ \text{k}\Omega)(9.285\ \text{mA} \angle 21.80°)}{1\ \text{k}\Omega + j2\ \text{k}\Omega - j2\ \text{k}\Omega}$$

$$= \frac{(2.236\ \text{k}\Omega \angle 63.43°)(9.285\ \text{mA} \angle 21.80°)}{1\ \text{k}\Omega \angle 0°}$$

$$= 20.761\ \text{mA} \angle 85.23°$$

c.
$$P_T = (50\ \text{V})(9.285\ \text{mA})\cos 21.80°$$

$$= 431.0\ \text{mW}$$

d. Because only the resistors will dissipate power, we may use $P = I^2 R$:

$$P_1 = (9.285\ \text{mA})^2(1\ \text{k}\Omega) = 86.2\ \text{mW}$$

$$P_2 = (18.57\ \text{mA})^2(1\ \text{k}\Omega) = 344.8\ \text{mW}$$

Alternatively, the power dissipated by $\mathbf{Z}_2$ may have been determined as $P = I^2 Z \cos \theta$:

$$P_2 = (18.57\ \text{mA})^2(2.236\ \text{k}\Omega)\cos 63.43° = 344.8\ \text{mW}$$

Since $\mathbf{Z}_3$ is purely capacitive, it will not dissipate any power:

$$P_3 = 0$$

By combining these powers, the total power dissipated is found:

$$P_T = 86.2 \text{ mW} + 344.8 \text{ mW} + 0 = 431.0 \text{ mW} \quad \text{(checks!)}$$

*practice
problem
VIII*

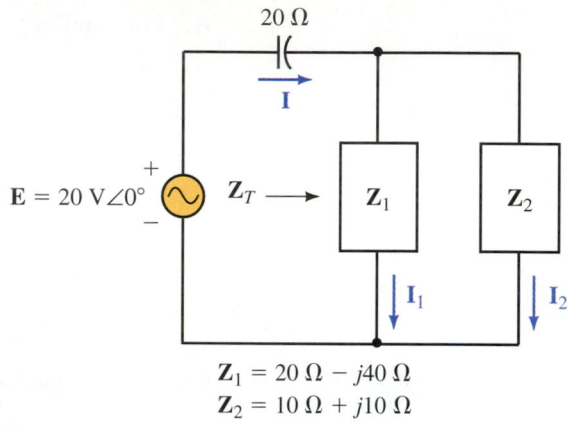

$$\mathbf{Z}_1 = 20 \ \Omega - j40 \ \Omega$$
$$\mathbf{Z}_2 = 10 \ \Omega + j10 \ \Omega$$

Figure 18–51

Refer to the circuit of Figure 18–51:

a. Calculate the total impedance, $\mathbf{Z}_T$.
b. Find the current $\mathbf{I}$.
c. Use the current divider rule to find $\mathbf{I}_1$ and $\mathbf{I}_2$.
d. Determine the power factor for each impedance, $\mathbf{Z}_1$ and $\mathbf{Z}_2$.
e. Determine the power factor for the circuit.
f. Verify that the total power dissipated by impedances $\mathbf{Z}_1$ and $\mathbf{Z}_2$ is equal to the power delivered by the voltage source.

18.7 Frequency Effects

As we have already seen, the reactance of inductors and capacitors depends on frequency. Consequently, the total impedance of any network having reactive elements is also frequency dependent. Any such circuit would need to be analyzed separately at each frequency of interest. We will examine several fairly simple combinations of resistors, capacitors, and inductors to see how the various circuits operate at different frequencies. Some of the more important combinations will be examined in greater detail in later chapters which deal with resonance and filters.

RC Circuits

As the name implies, *RC* circuits consist of a resistor and a capacitor. The components of an *RC* circuit may be connected either in series or in parallel as shown in Figure 18–52.

(a) Series *RC* circuit   (b) Parallel *RC* circuit

Figure 18–52

Figure 18–53

Consider the *RC* series circuit of Figure 18–53. If we recall that the capacitive reactance, X_C, is given as

$$X_C = \frac{1}{\omega C} = \frac{1}{2\pi f C}$$

The total impedance of the circuit is a vector quantity expressed as

$$\mathbf{Z}_T = R - j\frac{1}{\omega C} = R + \frac{1}{j\omega C}$$

$$\mathbf{Z}_T = \frac{1 + j\omega RC}{j\omega C} \qquad (18\text{--}19)$$

If we define the *corner frequency* for an *RC* circuit as

$$\omega_c = \frac{1}{RC} = \frac{1}{\tau} \quad (\text{rad/s}) \qquad (18\text{--}20)$$

or equivalently as

$$f_c = \frac{1}{2\pi RC} \quad (\text{Hz}) \qquad (18\text{--}21)$$

then several important points become evident.

For $\omega \le \omega_c/10$ (or $f \le f_c/10$) Equation 18–19 can be expressed as

$$\mathbf{Z}_T \simeq \frac{1 + j0}{j\omega C} = \frac{1}{j\omega C}$$

and for $\omega \ge 10\omega_c$, the expression of (18–19) can be simplified as

$$\mathbf{Z}_T \simeq \frac{0 + j\omega RC}{j\omega C} = R$$

Solving for the magnitude of the impedance at several angular frequencies, we have the results shown in Table 18–1.

If the magnitude of the impedance $\mathbf{Z}_T$ is plotted as a function of angular frequency ω, we get the graph of Figure 18–54. Notice that the abscissa and ordinate of the graph are not scaled linearly, but rather logarithmically. This allows for the the display of results over a wide range of frequencies.

TABLE 18–1

ANGULAR FREQUENCY, ω (RAD/S)	X_C (Ω)	Z_T (Ω)
0	∞	∞
1	1 M	1 M
10	100 k	100 k
100	10 k	10.05 k
200	5 k	5.099 k
500	2 k	2.236 k
1000	1 k	1.414 k
2000	500	1118
5000	200	1019
10 k	100	1005
100 k	10	1000

Figure 18–54   Impedance versus angular frequency for the network of Figure 18–53.

Figure 18–55

The graph illustrates that the reactance of a capacitor is very high (effectively an open circuit) at low frequencies. Consequently, the impedance of the series circuit will also be very high at low frequencies. Secondly, we notice that as the frequency increases, the reactance decreases. Therefore, as the frequency gets higher, the capacitive reactance has a diminished effect in the circuit. At very high frequencies (typically for $\omega \geq 10\omega_c$), the impedance of the circuit will effectively be $R = 1$ kΩ.

Consider the parallel RC circuit of Figure 18–55.

The total impedance, $\mathbf{Z}_T$, of the circuit is determined as

$$\mathbf{Z}_T = \frac{\mathbf{Z}_R \mathbf{Z}_C}{\mathbf{Z}_R + \mathbf{Z}_C}$$

$$= \frac{R\left(\dfrac{1}{j\omega C}\right)}{R + \dfrac{1}{j\omega C}}$$

$$= \frac{\dfrac{R}{j\omega C}}{\dfrac{1 + j\omega RC}{j\omega C}}$$

which may be simplified as

$$\mathbf{Z}_T = \frac{R}{1 + j\omega RC} \qquad (18\text{–}22)$$

As before, we have the corner frequency given by Equation 18–20. Now, by examining the expression of (18–22) for $\omega \leq \omega_c/10$, we have the following result:

$$\mathbf{Z}_T \simeq \frac{R}{1 + j0} = R$$

TABLE 18–2

ANGULAR FREQUENCY, ω (RAD/S)	X_c (Ω)	Z_T (Ω)
0	∞	1000
1	1 M	1000
10	100 k	1000
100	10 k	995
200	5 k	981
500	2 k	894
1 k	1 k	707
2 k	500	447
5 k	200	196
10 k	100	99.5
100 k	10	10

For $\omega \geq 10\omega_c$, we have

$$\mathbf{Z}_T \simeq \frac{R}{0 + j\omega RC} = \frac{1}{j\omega C}$$

If we solve for the impedance of the circuit in Figure 18–55 at various angular frequencies, we obtain the results of Table 18–2.

Plotting the magnitude of the impedance $\mathbf{Z}_T$ as a function of angular frequency ω, we get the graph of Figure 18–56. Notice that the abscissa and ordinate of the graph are again scaled logarithmically, allowing for the display of results over a wide range of frequencies.

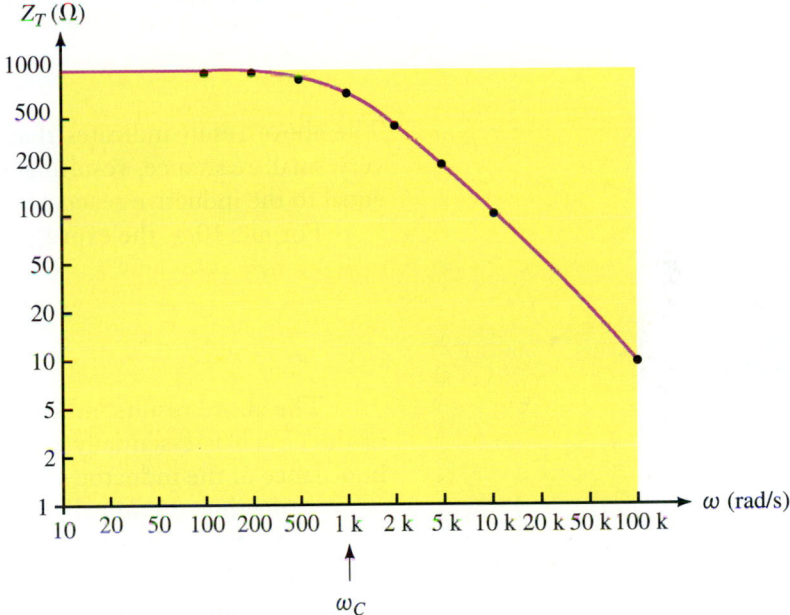

Figure 18–56   Impedance versus angular frequency for the network of Figure 18–55.

The previous results indicate that at dc ($f = 0$ Hz) the capacitor, which behaves as an open circuit, will result in a circuit impedance of $R = 1$ kΩ. As the frequency increases, the capacitor reactance approaches 0 Ω, resulting in a corresponding decrease in circuit impedance.

RL Circuits

RL circuits may be analyzed in a manner similar to the analysis of *RC* circuits. Consider the parallel *RL* circuit of Figure 18–57.

The total impedance of the parallel circuit is found as follows:

$$\mathbf{Z}_T = \frac{\mathbf{Z}_R \mathbf{Z}_L}{\mathbf{Z}_R + \mathbf{Z}_L}$$

$$= \frac{R(j\omega L)}{R + j\omega L}$$

$$\mathbf{Z}_T = \frac{j\omega L}{1 + j\omega\dfrac{L}{R}} \qquad (18\text{–}23)$$

Figure 18–57

If we define the *corner frequency* for an *RL* circuit as

$$\omega_c = \frac{R}{L} = \frac{1}{\tau} \quad \text{(rad/s)} \tag{18–24}$$

or equivalently as

$$f_c = \frac{1}{2\pi\frac{R}{L}} \quad \text{(Hz)} \tag{18–25}$$

then several important points become evident.

For $\omega \le \omega_c/10$ (or $f \le f_c/10$) Equation 18–23 can be expressed as

$$\mathbf{Z}_T \simeq \frac{j\omega L}{1 + j0} = j\omega L$$

The above result indicates that for low frequencies, the inductor has a very small reactance, resulting in a total impedance which is essentially equal to the inductive reactance.

For $\omega \ge 10\omega_c$, the expression of (18–23) can be simplified as

$$\mathbf{Z}_T \simeq \frac{j\omega L}{0 + j\omega\frac{L}{R}} = R$$

The above results indicate that for high frequencies, the impedance of the circuit is essentially equal to the resistance, due to the very high impedance of the inductor.

Evaluating the impedance at several angular frequencies, we have the results of Table 18–3.

When the magnitude of the impedance $\mathbf{Z}_T$ is plotted as a function of angular frequency ω, we get the graph of Figure 18–58.

TABLE 18–3

ANGULAR FREQUENCY, ω (RAD/S)	X_L (Ω)	Z_T (Ω)
0	0	0
1	1	1
10	10	10
100	100	99.5
200	200	196
500	500	447
1 k	1 k	707
2 k	2 k	894
5 k	5 k	981
10 k	10 k	995
100 k	100 k	1000

Figure 18–58   Impedance versus angular frequency for the network of Figure 18–57.

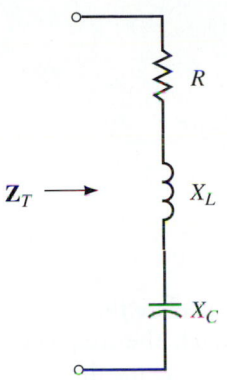

Figure 18–59

RLC Circuits

When numerous capacitive and inductive components are combined with resistors in series-parallel circuits, the total impedance $\mathbf{Z}_T$ of the circuit may rise and fall several times over the full range of frequencies. The analysis of such complex circuits is outside the scope of this textbook. However, for illustrative purposes we examine the simple series *RLC* circuit of Figure 18–59.

The impedance $\mathbf{Z}_T$ at any frequency will be determined as

$$\mathbf{Z}_T = R + jX_L - jX_C$$
$$= R + j(X_L - X_C)$$

At very low frequencies, the inductor will appear as a very low impedance (effectively a short circuit), while the capacitor will appear as a very high impedance (effectively an open circuit). Because the capacitive reactance will be much larger than the inductive reactance, the circuit will have a very large capacitive reactance. This results in a very high circuit impedance, $\mathbf{Z}_T$.

As the frequency increases, the inductive reactance increases, while the capacitive reactance decreases. At some frequency, f_0, the inductor and the capacitor will have the same magnitude of reactance. At this frequency, the reactances cancel, resulting in a circuit impedance which is equal to the resistance value.

As the frequency increases still further, the inductive reactance becomes larger than the capacitive reactance. The circuit becomes inductive and the magnitude of the total impedance of the circuit again rises. Figure 18–60 shows how the impedance of a series *RLC* circuit varies with frequency.

The complete analysis of the series *RLC* circuit and the parallel *RLC* circuit is left until we examine the principle of *resonance* in a later chapter.

Figure 18–60

1. For a series network consisting of a resistor and a capacitor, what will be the impedance of the network at a frequency of 0 Hz (dc)? What will be the impedance of the network as the frequency approaches infinity?

2. For a parallel network consisting of a resistor and an inductor, what will be the impedance of the network at a frequency of 0 Hz (dc)? What will be the impedance of the network as the frequency approaches infinity?

Figure 18–61

Given the series RC network of Figure 18–61, calculate the corner frequency in hertz and in radians per second. Sketch the frequency response of Z_T (magnitude) versus angular frequency ω for the network. Show the magnitude Z_T at $\omega_c/10$, ω_c, and $10\omega_c$.

Figure 18–62

Figure 18–63

18.8 Applications

As we have seen, we may determine the impedance of any ac circuit as a vector $\mathbf{Z} = R \pm jX$. This means that any ac circuit may now be simplified as a series circuit having a resistance and a reactance, as shown in Figure 18–62.

Additionally, an ac circuit may be represented as an equivalent parallel circuit consisting of a single resistor and a single reactance as shown in Figure 18–63. *Any equivalent circuit will be valid only at the given frequency of operation.*

We will now examine the technique used to convert any series impedance into its parallel equivalent. Suppose that the two circuits of Figure 18–62 and Figure 18–63 are exactly equivalent at some frequency. These circuits can be equivalent only if both circuits have the same total impedance, $\mathbf{Z}_T$, and the same total admittance, $\mathbf{Y}_T$.

From the circuit of Figure 18–62, the total impedance may be written as

$$\mathbf{Z}_T = R_S \pm jX_S$$

Therefore, the total admittance of the circuit is

$$\mathbf{Y}_T = \frac{1}{\mathbf{Z}_T} = \frac{1}{R_S \pm jX_S}$$

Multiplying the numerator and denominator by the complex conjugate, we obtain the following:

$$\mathbf{Y}_T = \frac{R_S \mp jX_S}{(R_S \pm jX_S)(R_S \mp jX_S)}$$

$$= \frac{R_S \mp jX_S}{R_S^2 + X_S^2}$$

$$\mathbf{Y}_T = \frac{R_S}{R_S^2 + X_S^2} \mp j\frac{X_S}{R_S^2 + X_S^2} \qquad (18\text{–}26)$$

Now, from the circuit of Figure 18–63, the total admittance of the parallel circuit may be found from the parallel combination of R_P and X_P as

$$\mathbf{Y}_T = \frac{1}{R_P} + \frac{1}{\pm jX_P}$$

which gives

$$\mathbf{Y}_T = \frac{1}{R_P} \mp j\frac{1}{X_P} \qquad (18\text{–}27)$$

Two vectors can only be equal if both the real components are equal and the imaginary components are equal. Therefore the circuits of Figure 18–62 and Figure 18–63 can only be equivalent if the following conditions are met:

$$R_P = \frac{R_S^{\,2} + X_S^{\,2}}{R_S} \qquad (18\text{–}28)$$

and

$$X_P = \frac{R_S^{\,2} + X_S^{\,2}}{X_S} \qquad (18\text{–}29)$$

In a similar manner, we use impedances to arrive at the following conversion from a parallel circuit to an equivalent series circuit:

$$R_S = \frac{R_P X_P^{\,2}}{R_P^{\,2} + X_P^{\,2}} \qquad (18\text{–}30)$$

and

$$X_S = \frac{R_P^{\,2} X_P}{R_P^{\,2} + X_P^{\,2}} \qquad (18\text{–}31)$$

■ EXAMPLE 18–19 A circuit has a total impedance of $\mathbf{Z}_T = 10\ \Omega + j50\ \Omega$. Sketch the equivalent series and parallel circuits.

Solution   The series circuit will be an inductive circuit having $R_S = 10\ \Omega$ and $X_{LS} = 50\ \Omega$.

The equivalent parallel circuit will also be an inductive circuit having the following values:

$$R_P = \frac{(10\ \Omega)^2 + (50\ \Omega)^2}{10\ \Omega} = 260\ \Omega$$

$$X_{LP} = \frac{(10\ \Omega)^2 + (50\ \Omega)^2}{50\ \Omega} = 52\ \Omega$$

The equivalent series and parallel circuits are shown in Figure 18–64.

Figure 18–64

■ **EXAMPLE 18–20** A circuit has a total admittance of $\mathbf{Y}_T = 0.559\ \text{mS}\angle 63.43°$. Sketch the equivalent series and parallel circuits.

Solution   Because the admittance is written in polar form, we first convert to the rectangular form of the admittance.

$$G_P = (0.559\ \text{mS}) \cos 63.43° = 0.250\ \text{mS} \quad \Leftrightarrow \quad R_P = 4.0\ \text{k}\Omega$$

$$B_{CP} = (0.559\ \text{mS}) \sin 63.43° = 0.500\ \text{mS} \quad \Leftrightarrow \quad X_{CP} = 2.0\ \text{k}\Omega$$

The equivalent series circuit is found as

$$R_S = \frac{(4\ \text{k}\Omega)(2\ \text{k}\Omega)^2}{(4\ \text{k}\Omega)^2 + (2\ \text{k}\Omega)^2} = 0.8\ \text{k}\Omega$$

and

$$X_{CS} = \frac{(4\ \text{k}\Omega)^2(2\ \text{k}\Omega)}{(4\ \text{k}\Omega)^2 + (2\ \text{k}\Omega)^2} = 1.6\ \text{k}\Omega$$

The equivalent circuits are shown in Figure 18–65.

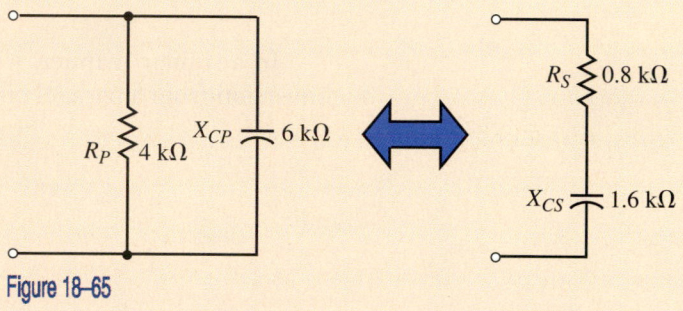

Figure 18–65

■ **EXAMPLE 18–21**   Refer to the circuit of Figure 18–66.

Figure 18–66

a. Find $\mathbf{Z}_T$.
b. Sketch the equivalent series circuit.
c. Determine $\mathbf{I}_T$.

Solution

a. The circuit consists of two parallel networks in series. We apply Equations 18–30 and 18–31 to arrive at equivalent series elements for each of the parallel networks as follows:

$$R_{S_1} = \frac{(20 \text{ k}\Omega)(10 \text{ k}\Omega)^2}{(20 \text{ k}\Omega)^2 + (10 \text{ k}\Omega)^2} = 4 \text{ k}\Omega$$

$$X_{CS} = \frac{(20 \text{ k}\Omega)^2(10 \text{ k}\Omega)}{(20 \text{ k}\Omega)^2 + (10 \text{ k}\Omega)^2} = 8 \text{ k}\Omega$$

and

$$R_{S_2} = \frac{(30 \text{ k}\Omega)(10 \text{ k}\Omega)^2}{(30 \text{ k}\Omega)^2 + (10 \text{ k}\Omega)^2} = 3 \text{ k}\Omega$$

$$X_{LS} = \frac{(30 \text{ k}\Omega)^2(10 \text{ k}\Omega)}{(30 \text{ k}\Omega)^2 + (10 \text{ k}\Omega)^2} = 9 \text{ k}\Omega$$

The equivalent circuits are shown in Figure 18–67.

Figure 18–67

The total impedance of the circuit is found to be

$$\mathbf{Z}_T = (4 \text{ k}\Omega - j8 \text{ k}\Omega) + (3 \text{ k}\Omega + j9 \text{ k}\Omega) = 7 \text{ k}\Omega + j1 \text{ k}\Omega = 7.071 \text{ k}\Omega\angle 8.13°$$

b. Figure 18–68 shows the equivalent series circuit.

Figure 18–68

c.
$$\mathbf{I}_T = \frac{200 \text{ V}\angle 0°}{7.071 \text{ k}\Omega\angle 8.13°} = 28.28 \text{ mA}\angle -8.13°$$

An inductor of 10 mH has a series resistance of 5 Ω.
- **a.** Determine the parallel equivalent of the inductor at a frequency of 1 kHz. Sketch the equivalent showing the values of L_P (in henries) and R_P.
- **b.** Determine the parallel equivalent of the inductor at a frequency of 1 MHz. Sketch the equivalent showing the values of L_P (in henries) and R_P.
- **c.** If the frequency were increased still further, predict what would happen to the values of L_P and R_P.

A network has an impedance of $\mathbf{Z}_T = 50\ k\Omega\angle75°$ at a frequency of 5 kHz.

- **a.** Determine the most simple equivalent series circuit (L and R).
- **b.** Determine the most simple equivalent parallel circuit.

18.9 Circuit Analysis Using Computers

In this section we will use the PROBE postprocessor of PSpice to show how the impedance of an *RC* circuit circuit changes as a function of frequency. The PROBE output will provide a graphical result which is very similar to the frequency responses determined in previous sections of this chapter.

■ **EXAMPLE 18–22**   Refer to the network of Figure 18–69:

R_1

1 kΩ

$\mathbf{Z}_T \longrightarrow$ C ⊥ 1 μF

Figure 18–69

Write the input file to use the PROBE processor of PSpice to obtain a graphical display of the circuit impedance as a function of frequency up to 500 Hz.

Solution   In order to determine the input impedance of the circuit, we will use the relationship $Z = E/I$. The given circuit is placed into an acceptable form as shown in Figure 18–70.

Figure 18–70

Notice that we have added a voltage source of 10 V∠0° into the circuit to provide the necessary current. The choice of this value over any other value is purely arbitrary, since the impedance does not depend upon either the magnitude or the phase angle of the voltage source.

In the input file, we enter the ac voltage source using the following syntax:

Vxxx n+   n−   AC   mag   [phase]

where

    mag is the magnitude of the ac voltage, and
    [phase] is the optional phase angle in degrees.

The program will calculate voltages and currents for a range of frequencies if we use the control statement:

.AC   [type]   num   start   end

where

    [type] indicates whether the sweep is to be linear or logarithmic,
    num is the number of points to be used in the sweep. The number of points will be equal to the number of intervals plus 1,
    start is the frequency (in hertz) at which the calculation begins,
    end is the frequency at which the calculation ends.

Although we would like the impedance to be shown for the frequency range between 0 Hz and 500 Hz, the calculation will involve division by zero. For this reason, we let the program determine values from 50 Hz to 500 Hz. The program will evaluate ac currents and voltages over 100 equal intervals (101 points). If we had wanted frequency displayed on a logarithmic axis, we would use the specify either OCT or DEC corresponding to whether the sweep is by octave or decade. These terms will be defined in a later chapter.

The PSpice input file is shown below:

```
Input File:
Example 18-22: Frequency response of RC circuit,
Figure 18-70
Vs 1 0 ac 10V 0
R1 1 2 1kohm
C  2 0 1uF
.AC LIN 101 50  500
.PROBE
.OPTIONS NOPAGE
.END
```

Once the PROBE postprocessor has been installed, the screen will display a blank graph, with the frequency indicated along the abscissa (X-axis). Since the frequency is shown in a logarithmic scale, we first change the scale to a linear scale as follows. Select the *X_axis* menu item. Then select the *Linear* menu item, followed by the *Exit* item. Now, we obtain the impedance display by using *Add_trace* and keying in the following command:

V(1)/I(R1)

Figure 18–71 shows the resulting output display.

Figure 18–71   PSpice display.

practice
problem
XI

Write a PSpice input file to use the PROBE postprocessor to obtain a graph of input impedance Z_T versus frequency f (up to 500 Hz) for the circuit of Figure 18–55.

Problems

18.1   Ohm's Law for AC Circuits

1. Refer to the resistors shown in Figure 18–72.

(a)   (b)

Figure 18–72

   a. Find the sinusoidal current i using phasors.
   b. Sketch the sinusoidal waveforms for v and i.
   c. Sketch the phasor diagram for $\mathbf{V}$ and $\mathbf{I}$.

2. Refer to the resistors shown in Figure 18–73.

(a)

(b)

Figure 18–73

a. Find the sinusoidal voltage *v* using phasors.
b. Sketch the sinusoidal waveforms for *v* and *i*.
c. Sketch the phasor diagram for **V** and **I**.

3. Consider the elements shown in Figure 18–74.

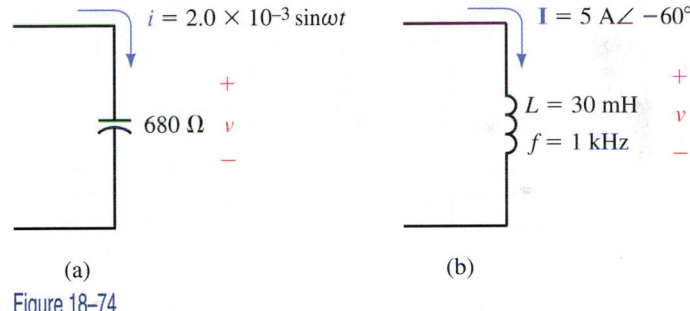

(a)

(b)

Figure 18–74

a. Find the sinusoidal voltage *v* using phasors.
b. Sketch the sinusoidal waveforms for *v* and *i*.
c. Sketch the phasor diagram for **V** and **I**.

4. Consider the elements shown in Figure 18–75.

(a)

(b)

Figure 18–75

a. Find the sinusoidal current *i* using phasors.
b. Sketch the sinusoidal waveforms for *v* and *i*.
c. Sketch the phasor diagram for **V** and **I**.

5. Consider the elements shown in Figure 18–76.

(a)   (b)

Figure 18–76

   a. Find the sinusoidal voltage v using phasors.
   b. Sketch the sinusoidal waveforms for v and i.
   c. Sketch the phasor diagram for **V** and **I**.

6. Consider the elements shown in Figure 18–77.

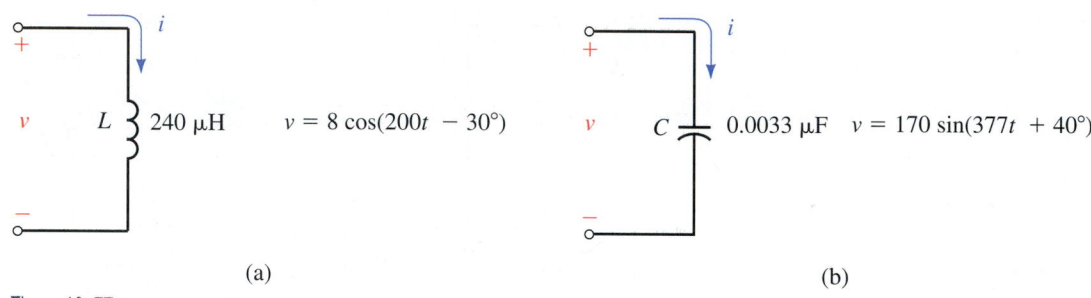

(a)   (b)

Figure 18–77

   a. Find the sinusoidal current i using phasors.
   b. Sketch the sinusoidal waveforms for v and i.
   c. Sketch the phasor diagram for **V** and **I**.

18.2   AC Series Circuits

7. Find the total impedance of each of the networks shown in Figure 18–78.

(a)   (b)

Figure 18–78

8. Repeat Problem 7 for the networks of Figure 18–79.

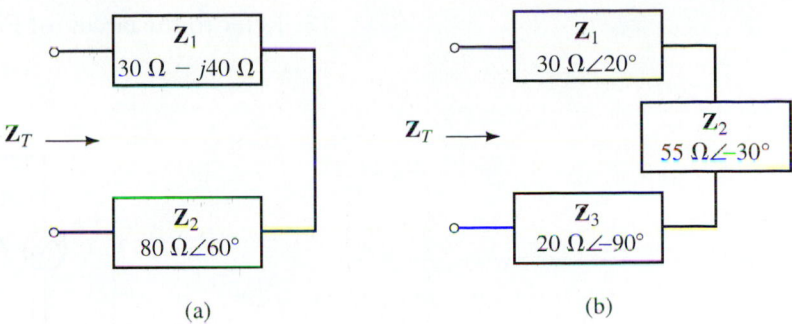

(a)   (b)

Figure 18–79

9. Refer to the networks of Figure 18–80.

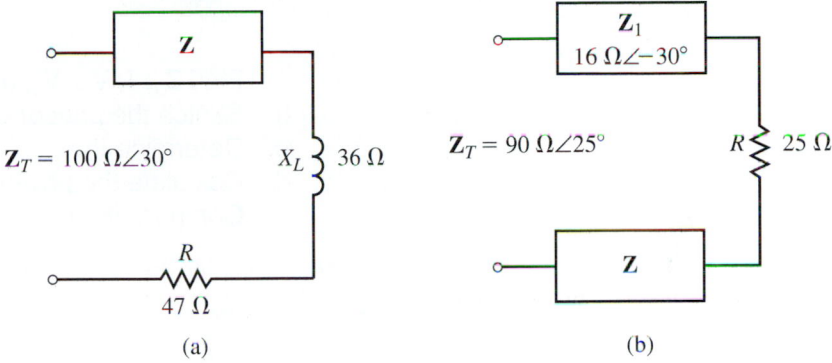

(a)   (b)

Figure 18–80

    a. Determine the series impedance **Z** which will result in the given total impedance, $\mathbf{Z}_T$, for each of the networks. Express your answers in rectangular and polar form.

    b. Sketch an impedance diagram showing $\mathbf{Z}_T$ and **Z**.

10. A circuit consisting of two elements has a total impedance of $\mathbf{Z}_T = 2 \text{ k}\Omega\angle15°$ at a frequency of 18 kHz. Determine the values in ohms, henries, or farads of the unknown elements.

11. Given that the circuit of Figure 18–81 is to operate at a frequency of 1 kHz, what series components R and L (in henries) or C (in farads) must be in the indicated block to result in a total circuit impedance of $\mathbf{Z}_T = 50 \ \Omega\angle60°$?

Figure 18–81

12. Repeat Problem 11 for a frequency of 2 kHz.

13. Refer to the circuit of Figure 18–82.

Figure 18–82

a. Find $\mathbf{Z}_T$, $\mathbf{I}$, $\mathbf{V}_R$, $\mathbf{V}_L$, and $\mathbf{V}_C$.
b. Sketch the phasor diagram showing $\mathbf{I}$, $\mathbf{V}_R$, $\mathbf{V}_L$, and $\mathbf{V}_C$.
c. Determine the power dissipated by the resistor.
d. Calculate the power delivered by the voltage source. Compare the result to (c).

14. Consider the circuit of Figure 18–83.

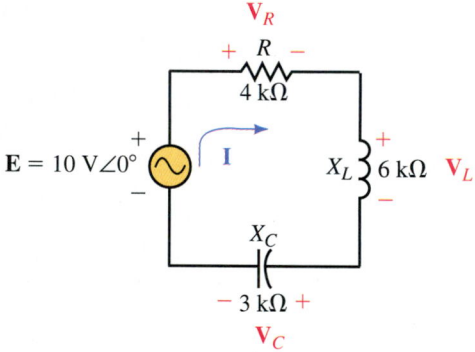

Figure 18–83

a. Find $\mathbf{Z}_T$, $\mathbf{I}$, $\mathbf{V}_R$, $\mathbf{V}_L$, and $\mathbf{V}_C$.
b. Sketch the phasor diagram showing $\mathbf{I}$, $\mathbf{V}_R$, $\mathbf{V}_L$, and $\mathbf{V}_C$.
c. Write the sinusoidal expressions for the current, i, and the voltages e, v_R, v_C, and v_L.
d. Sketch the sinusoidal current and voltages found in (c).
e. Determine the power dissipated by the resistor.
f. Calculate the power delivered by the voltage source. Compare the result to (e).

15. Refer to the circuit of Figure 18–84.
a. Determine the circuit impedance, $\mathbf{Z}_T$.
b. Use phasors to solve for i, v_R, v_C, and v_L.
c. Sketch the phasor diagram showing $\mathbf{I}$, $\mathbf{V}_R$, $\mathbf{V}_L$, and $\mathbf{V}_C$.

v_C
$+$ X_C $-$
47 Ω
i
$e = 24 \sin(\omega t)$ $\mathbf{Z}_T \longrightarrow$ R $\gtrless$ 36 Ω v_R
X_L
$-$ 20 Ω $+$
v_L

Figure 18–84

   d. Sketch the sinusoidal expressions for the current and voltages found in (b).
   e. Determine the power dissipated by the resistor.
   f. Calculate the power delivered by the voltage source. Compare the result to (e).

16. Refer to the circuit of Figure 18–85.

$P = 200$ mW
R
i 100 Ω
$e = 10 \sin(\omega t)$ X_C
X_L
50 Ω

Figure 18–85

   a. Determine the value of the capacitor reactance, X_C, needed so that the resistor in the circuit of Figure 18–85 dissipates a power of 200 mW.
   b. Using the value of X_C from (a), determine the sinusoidal expression for the current i in the circuit.

18.3   Kirchhoff's Voltage Law and the Voltage Divider Rule

17. a. Suppose a voltage of 10 V∠0° is applied across each of the networks in Figure 18–78. Use the voltage divider rule to find the voltage appearing across each element.
    b. Verify Kirchhoff's voltage law for each network.

18. a. Suppose a voltage of 240 V∠30° is applied across each of the networks in Figure 18–79. Use the voltage divider rule to find the voltage appearing across each impedance.
    b. Verify Kirchhoff's voltage law for each network.

19. Let the circuit of Figure 18–86 be given.
   a. Find the voltages $\mathbf{V}_C$ and $\mathbf{V}_L$.
   b. Determine the value of R.

Figure 18–86

Figure 18–87

20. Refer to the circuit of Figure 18–87.
   a. Find the voltages $\mathbf{V}_R$ and $\mathbf{V}_L$.
   b. Determine the value of X_C.

21. Refer to the circuit of Figure 18–88:
   a. Find the voltage across $\mathbf{X}_C$.
   b. Use Kirchhoff's voltage law to find the voltage across the unknown impedance.
   c. Calculate the value of the unknown impedance $\mathbf{Z}$.
   d. Determine the power dissipated by the circuit.

Figure 18–88

Figure 18–89

22. Given that the circuit of Figure 18–89 has a current with a magnitude of 2.0 A and dissipates a total power of 500 W:
   a. Calculate the value of the unknown impedance $\mathbf{Z}$. (Hint: Two solutions are possible.)
   b. Calculate the phase angle θ of the current $\mathbf{I}$.
   c. Find the voltages $\mathbf{V}_R$, $\mathbf{V}_L$, and $\mathbf{V}_Z$.

18.4   AC Parallel Circuits

23. Determine the input impedance, $\mathbf{Z}_T$, for each of the networks of Figure 18–90.

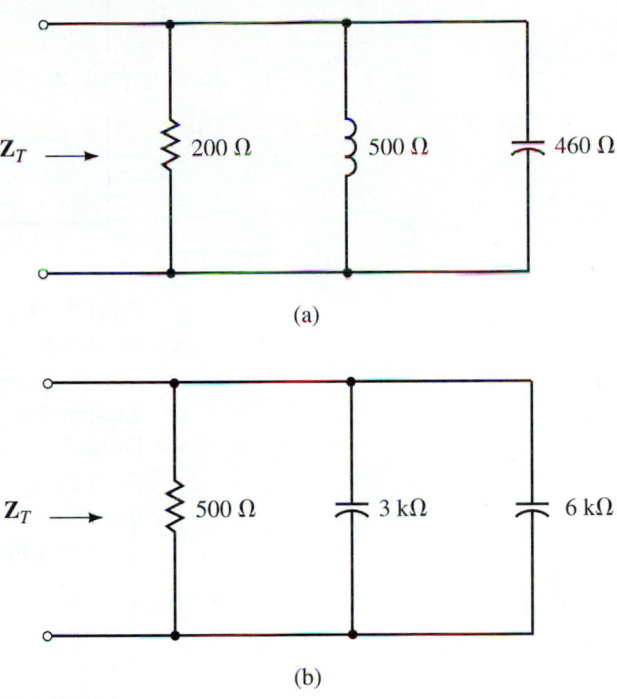

(a)

(b)

Figure 18–90

24. Repeat Problem 23 for Figure 18–91.

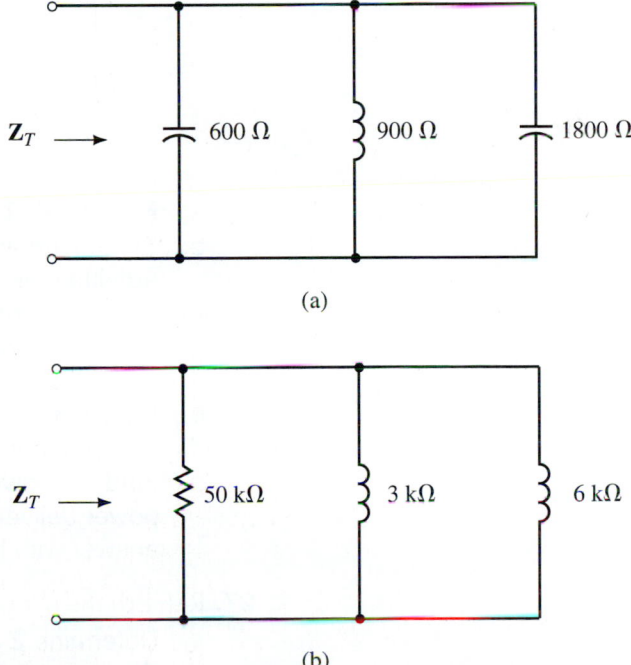

(a)

(b)

Figure 18–91

25. Look at the circuit of Figure 18–92.

Figure 18-92

a. Find $\mathbf{Z}_T$, $\mathbf{I}_T$, $\mathbf{I}_1$, $\mathbf{I}_2$, and $\mathbf{I}_3$..
b. Sketch the admittance diagram showing each of the admittances.
c. Sketch the phasor diagram showing $\mathbf{E}$, $\mathbf{I}_T$, $\mathbf{I}_1$, $\mathbf{I}_2$, and $\mathbf{I}_3$.
d. Determine the power dissipated by the resistor.
e. Find the power factor of the circuit and calculate the power delivered by the voltage source. Compare the answer with the result obtained in (d).

26. Refer to the circuit of Figure 18–93.

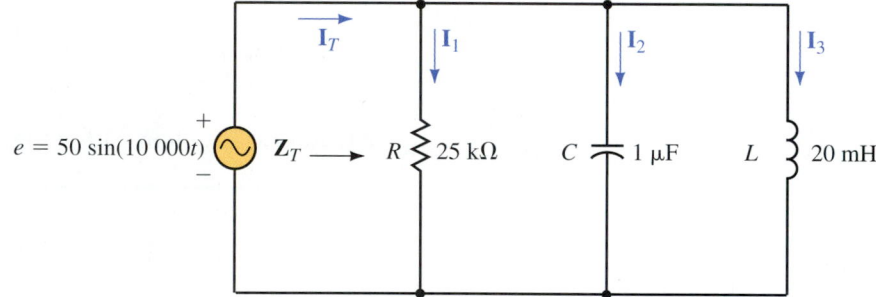

Figure 18-93

a. Find $\mathbf{Z}_T$, $\mathbf{I}_T$, $\mathbf{I}_1$, $\mathbf{I}_2$, and $\mathbf{I}_3$.
b. Sketch the admittance diagram for each of the admittances.
c. Sketch the phasor diagram showing $\mathbf{E}$, $\mathbf{I}_T$, $\mathbf{I}_1$, $\mathbf{I}_2$, and $\mathbf{I}_3$.
d. Determine the expressions for the sinusoidal currents i_T, i_1, i_2, and i_3.
e. Sketch the sinusoidal voltage e and current i_T.
f. Determine the power dissipated by the resistor.
g. Find the power factor of the circuit and calculate the power delivered by the voltage source. Compare the answer with the result obtained in (f).

27. Refer to the network of Figure 18–94.
a. Determine $\mathbf{Z}_T$.
b. Given the indicated current, use Ohm's law to find the voltage, $\mathbf{V}$, across the network.

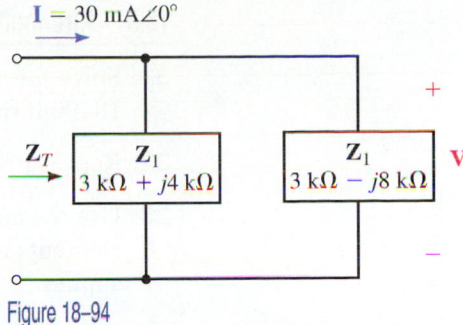

Figure 18–94

28. Consider the network of Figure 18–95.

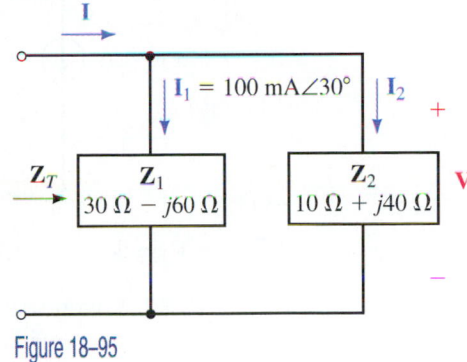

Figure 18–95

a.   Determine $\mathbf{Z}_T$.
b.   Given the indicated current, use Ohm's law to find the voltage, $\mathbf{V}$, across the network.
c.   Solve for $\mathbf{I}_2$ and $\mathbf{I}$.

29. Determine the impedance, $\mathbf{Z}_2$, which will result in the total impedance shown in Figure 18–96.

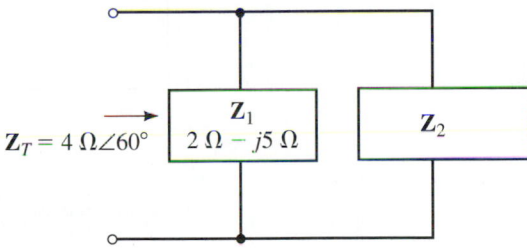

Figure 18–96

30. Determine the impedance, $\mathbf{Z}_2$, which will result in the total impedance shown in Figure 18–97.

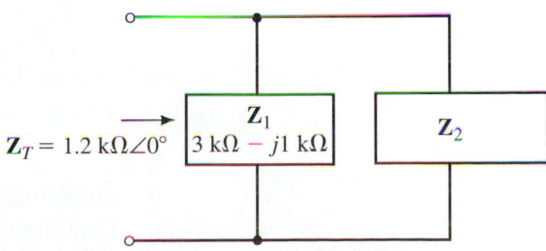

Figure 18–97

18.5   Kirchhoff's Current Law and the Current Divider Rule

31. Solve for the current in each element of the networks in Figure 18–90 if the current applied to each network is 10 mA∠–30°.

32. Repeat Problem 31 for Figure 18–91.

33. Use the current divider rule to find the current in each of the elements in Figure 18–98. Verify that Kirchhoff's current law applies.

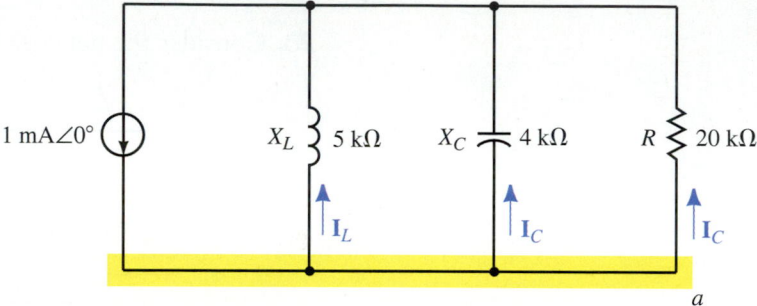

Figure 18–98

34. Given that $I_R = 4$ A∠30° in the circuit of Figure 18–99, find the currents I, I_C, and I_L. Verify that Kirchhoff's current law appplies for this circuit.

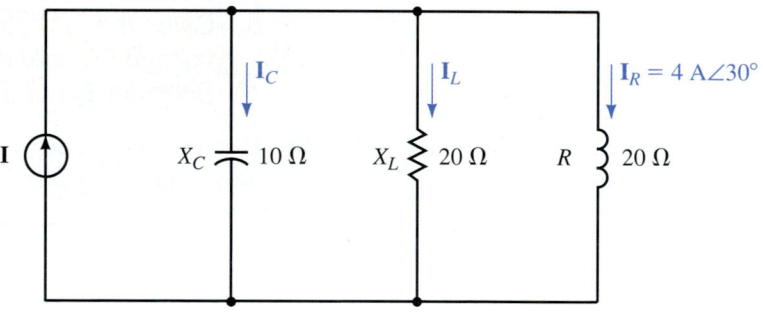

Figure 18–99

35. Suppose that the circuit of Figure 18–100 has a current I with a magnitude of 8 A:

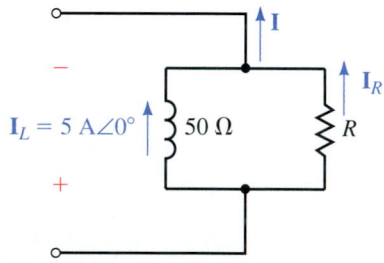

Figure 18–100

a. Determine the current, I_R, through the resistor.
b. Calculate the value of resistance, R.
c. What is the phase angle of the current I?

Figure 18–101

36. Assume that the circuit of Figure 18–101 has a current **I** with a magnitude of 3 A:

    a. Determine the current, I_R, through the resistor.
    b. Calculate the value of capacitive reactance, X_C.
    c. What is the phase angle of the current **I**?

18.6 Series-Parallel Circuits

37. Refer to the circuit of Figure 18–102.

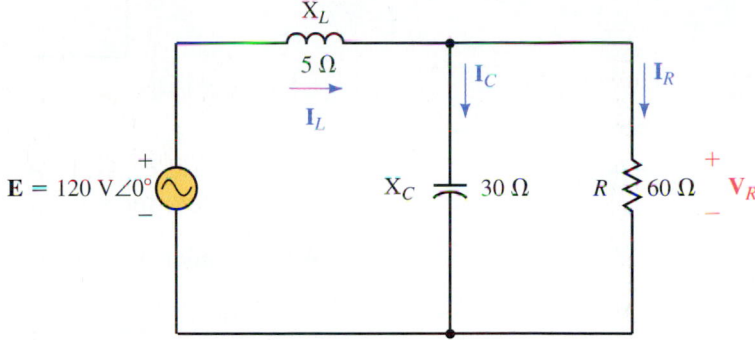

Figure 18–102

    a. Find Z_T, I_L, I_C, and I_R.
    b. Sketch the phasor diagram showing **E**, I_L, I_C, and I_R.
    c. Calculate the power dissipated by the resistor.
    d. Use the circuit power factor to calculate the power delivered by the voltage source. Compare the answer with the result obtained in (c).

38. Refer to the circuit of Figure 18–103.

Figure 18–103

    a. Find Z_T, I_1, I_2, and I_3.
    b. Sketch the phasor diagram showing **E**, I_1, I_2, and I_3.
    c. Calculate the power dissipated by each of the resistors.
    d. Use the circuit power factor to calculate the power delivered by the voltage source. Compare the answer with the result obtained in (c).

39. Refer to the circuit of Figure 18–104.

Figure 18–104

a. Find $\mathbf{Z}_T$, $\mathbf{I}_T$, $\mathbf{I}_1$, and $\mathbf{I}_2$.
b. Determine the voltage $\mathbf{V}_{ab}$.

40. Consider the circuit of Figure 18–105.

Figure 18–105

a. Find $\mathbf{Z}_T$, $\mathbf{I}_T$, $\mathbf{I}_1$, and $\mathbf{I}_2$.
b. Determine the voltage $\mathbf{V}$.

41. Refer to the circuit of Figure 18–106:

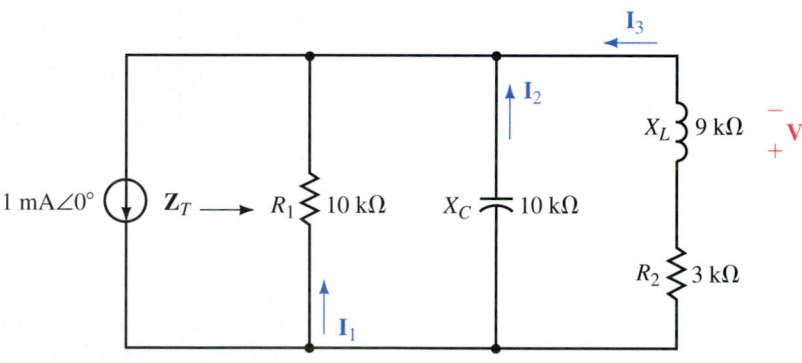

Figure 18–106

a. Find $\mathbf{Z}_T$, $\mathbf{I}_1$, $\mathbf{I}_2$, and $\mathbf{I}_3$.
b. Determine the voltage $\mathbf{V}$.

42. Refer to the circuit of Figure 18–107:

Figure 18–107

    a. Find $\mathbf{Z}_T$, $\mathbf{I}_1$, $\mathbf{I}_2$, and $\mathbf{I}_3$.
    b. Determine the voltage $\mathbf{V}$.

18.7 Frequency Effects

43. A 50-kΩ resistor is placed in series with a 0.01-μF capacitor. Determine the corner frequency, ω_C (in rad/s), and sketch the frequency response (Z_T vs. ω) of the network.

44. A 2-mH inductor is placed in parallel with a 2-kΩ resistor. Determine the corner frequency, ω_c (in rad/s), and sketch the frequency response (Z_T vs. ω) of the network.

45. A 100-kΩ resistor is placed in parallel with a 0.47-μF capacitor. Determine the corner frequency, f_C (in Hz), and sketch the frequency response (Z_T vs. f) of the network.

46. A 2.7-kΩ resistor is placed in parallel with a 20-mH inductor. Determine the corner frequency, f_C (in Hz), and sketch the frequency response (Z_T vs. f) of the network.

18.8 Applications

47. Convert each each of the networks of Figure 18–108 into an equivalent series network consisting of two elements.

48. Convert each each of the networks of Figure 18–108 into an equivalent parallel network consisting of two elements.

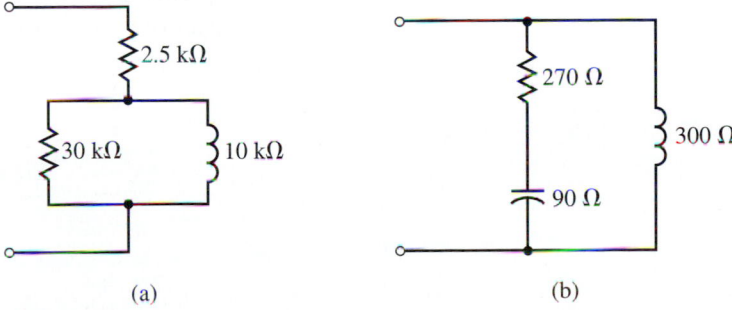

(a)   (b)

Figure 18–108

49. Show that the networks of Figure 18–109 have the same input impedance at frequencies of 1 krad/s and 10 krad/s. (It can be shown that these networks are equivalent at all frequencies.)

(a)   (b)

Figure 18–109

50. Show that the networks of Figure 18–110 have the same input impedance at frequencies of 5 rad/s and 10 rad/s.

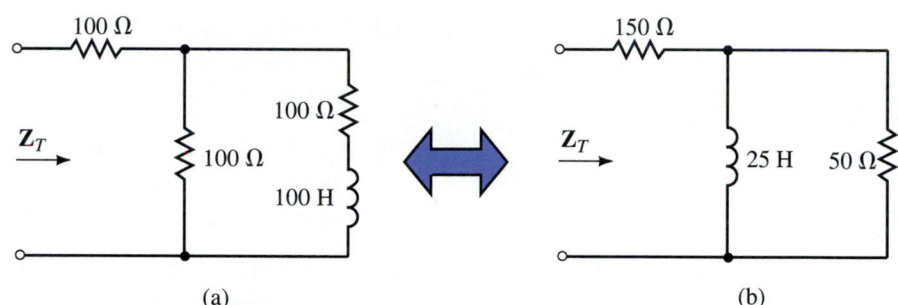

(a)   (b)

Figure 18–110

18.9   Circuit Analysis Using Computers

51. Write the PSpice input file for the *RC* network of Problem 43. Use the PROBE processor of to obtain a graphical display of the circuit impedance as a function of frequency. Let the frequency go from 1 Hz to 1 kHz. Use a linear sweep.

52. Write the PSpice input file for the *RL* network of Problem 44. Use the PROBE processor of to obtain a graphical display of the circuit impedance as a function of frequency. Let the frequency go from 1 Hz to 100 kHz. Use a logarithmic sweep, and let the number of points per decade be 51.

Methods of AC Analysis

CHAPTER PREVIEW

To this point, we have examined only circuits having a single ac source. In this chapter, we continue our study by analyzing multisource circuits and bridge networks. You will find that most of the techniques used in analyzing ac circuits parallel those of dc circuit analysis. Consequently, a review of Chapter 8 will help you understand the topics of this chapter.

Near the end of the chapter, we examine how computer techniques are used to analyze even the most complex ac circuits. It must be emphasized that although computer techniques are much simpler than using a pencil and calculator, there is virtually no knowledge to be gained by mindlessly entering data into a computer. We use the computer merely as a tool to verify our results and to provide a greater dimension to the analysis of circuits.

KEY TERMS

Balanced Bridges
Controlling Elements
Delta-Wye Conversion
Dependent Sources
Hay Bridge
Maxwell Bridge
Mesh Analysis
Nodal Analysis
Schering Bridge
Source Conversion

CAD software lets designers create complex printed circuit boards and multichip models. (*Courtesy of Harris Electronic Design Automation, Inc.*)

OBJECTIVES

After studying this chapter, you will be able to

- convert an ac voltage source into its equivalent current source, and conversely, convert a current source into an equivalent voltage source,

- solve for the current or voltage in a circuit having either a depen-

- dent current source or a dependent voltage source,

- set up simultaneous linear equations to solve an ac circuit using mesh analysis,

- use complex determinants to find the solutions for a given set of linear equations,

- set up simultaneous linear equations to solve an ac circuit using nodal analysis,

- perform delta-to-wye and wye-to-delta conversions for circuits having reactive elements,

- solve for the balanced condition in a given ac bridge circuit. In particular, you will examine the Maxwell, Hay, and Schering bridges,

- use PSpice to calculate current and voltage in an ac circuit.

19.1 Dependent Sources

The voltage and current sources we have worked with up to now have been *independent sources,* meaning that the voltage or current of the supply was not in any way dependent upon any voltage or current elsewhere in the circuit. In many amplifier circuits, particularly those involving transistors, it is possible to explain the operation of the circuits by replacing the device with an equivalent electronic model. These models often use voltage and current sources which have values dependent upon some internal voltage or current. Such sources are called *dependent sources.* Figure 19–1 compares the symbols for both independent and dependent sources.

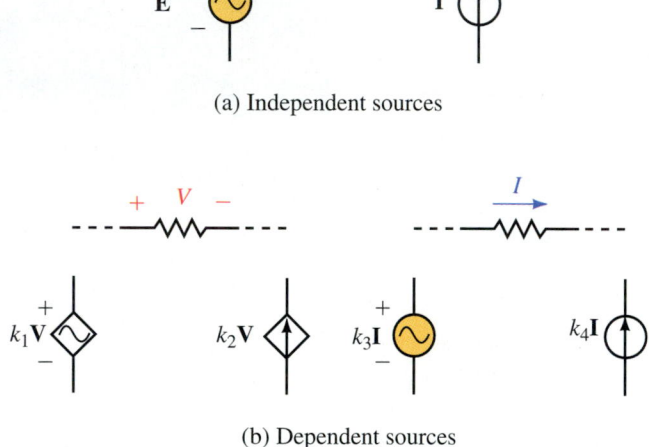

(a) Independent sources

(b) Dependent sources

Figure 19–1

Although the diamond is now most often used for representing dependent sources, many articles and textbooks still use a circle. In this textbook we use both forms of the dependent source to familiarize the student with the various notations. The dependent source has a magnitude and phase angle determined by voltage or current at some internal element multiplied by a constant, k. The magnitude of the constant is determined by parameters within the particular model, and the units of the constant correspond to the required quantities in the equation.

■ **EXAMPLE 19–1**   Refer to the resistor shown in Figure 19–2. Determine the voltage V_R across the resistor given that the following voltages are applied:

   a. $V = 0$ V.
   b. $V = 5$ V$\angle 30°$.
   c. $V = 3$ V$\angle -150°$.

Solution   Notice that the dependent source of this example has a constant, g_m, called the *transconductance.* Here, $g_m = 4$ mS.

Given: $g_m = 4$ mS

Figure 19–2

a. $\mathbf{I} = (4\text{ mS})(0\text{ V}) = 0$

$\mathbf{V}_R = 0$ V

b. $\mathbf{I} = (4\text{ mS})(5\text{ V}\angle 30°) = 20\text{ mA}\angle 30°$

$\mathbf{V}_R = (20\text{ mA}\angle 30°)(2\text{ k}\Omega) = 40\text{ V}\angle 30°$

c. $\mathbf{I} = (4\text{ mS})(3\text{ V}\angle{-}150°) = 12\text{ mA}\angle 150°$

$\mathbf{V}_R = (12\text{ mA}\angle{-}150°)(2\text{ k}\Omega) = 24\text{ V}\angle{-}150°$

practice
problem
1

The circuit of Figure 19–3 represents a simplified model of a transistor amplifier.

$h_{fe} = 120$

Figure 19–3

Determine the voltage $\mathbf{V}_R$ for each of the following voltages:

a. $\mathbf{V} = 10\text{ mV}\angle 0°$.

b. $\mathbf{V} = 2\text{ mV}\angle 180°$.

c. $\mathbf{V} = 0.03\text{ V}\angle 90°$.

19.2 Source Conversion

When working with dc circuits, the analysis of a circuit is often simplified by replacing the source (whether a voltage source or a current source) with its equivalent. The conversion of any ac source is similar to the method used in dc circuit analysis.

A voltage source **E** in series with an impedance **Z** is equivalent to a current source **I** having the same impedance **Z** in parallel. Figure 19–4 shows the equivalent sources.

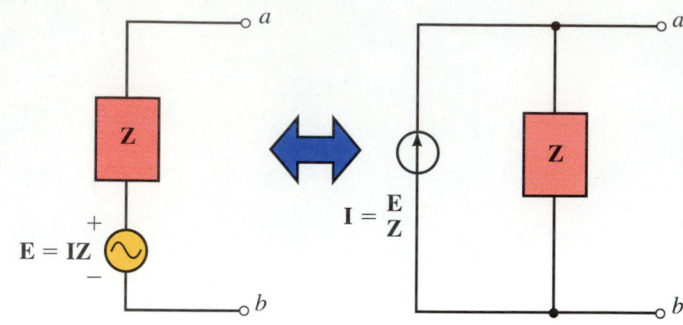

Figure 19–4

From Ohm's law, we perform the source conversion as follows:

$$\mathbf{I} = \frac{\mathbf{E}}{\mathbf{Z}}$$

and

$$\mathbf{E} = \mathbf{IZ}$$

It is important to realize that the two circuits of Figure 19–4 are equivalent between points *a* and *b*. This means that any network connected to points *a* and *b* will behave exactly the same regardless of which type of source is used. However, the voltages or currents within the sources will seldom be the same. In order to determine the current through or the voltage across the source impedance, the circuit must be returned to its original state.

■ **EXAMPLE 19–2**   Convert the voltage source of Figure 19–5 into an equivalent current source.

Figure 19–5

Solution

$$\mathbf{Z}_T = 3 \ \Omega + j4 \ \Omega = 5 \ \Omega \angle 53.13°$$

$$\mathbf{I} = \frac{10 \ \text{V} \angle 0°}{5 \ \Omega \angle 53.13°} = 2 \ \text{A} \angle -53.13°$$

The equivalent current source is shown in Figure 19–6.

Figure 19–6

■ **EXAMPLE 19–3**   Convert the current source of Figure 19–7 into an equivalent voltage source.

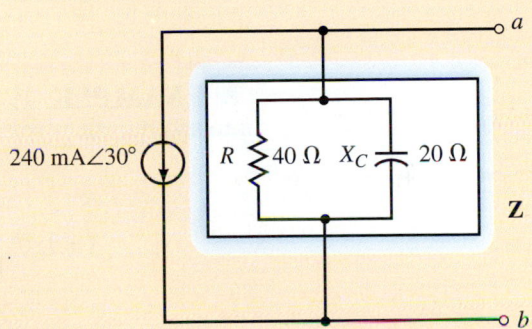

Figure 19–7

Solution   The impedance of the parallel combination is determined to be

$$\mathbf{Z} = \frac{(40 \ \Omega \ 0°)(20 \ \Omega \ \angle -90°)}{40 \ \Omega - j20 \ \Omega}$$

$$= \frac{800 \ \Omega \ \angle -90°}{44.72 \angle -26.57°}$$

$$= 17.89 \ \Omega \angle -63.43° = 8 \ \Omega - j16 \ \Omega$$

and so

$$\mathbf{E} = (240 \ \text{mA} \angle 30°)(17.89 \ \Omega \angle -63.43°)$$

$$= 4.294 \ \text{V} \angle -33.43°$$

The resulting equivalent circuit is shown in Figure 19–8.

Figure 19–8

It is possible to use the same procedure to convert a dependent source into its equivalent provided that the controlling element is external to the circuit in which the source appears. *If the controlling element is in the same circuit as the dependent source, this procedure may not be used.*

■ **EXAMPLE 19–4** Convert the current source of Figure 19–9 into an equivalent voltage source.

Figure 19–9

Solution In the circuit of Figure 19–9 the controlling element, R_1, is in a separate circuit. Therefore, the current source is converted into an equivalent voltage source as follows:

$$\mathbf{E} = (100\mathbf{I}_1)(\mathbf{Z})$$

$$= (100\mathbf{I}_1)(25 \text{ k}\Omega\angle 0°)$$

$$= (2.5 \times 10^6 \ \Omega)\mathbf{I}_1$$

The resulting voltage source is shown in Figure 19–10. Notice that the equivalent voltage source is dependent on the current, **I**, just as the original current source.

Figure 19–10

practice
problem
II

Convert the voltage sources of Figure 19–11 into equivalent current sources.

Figure 19–11

in process
learning
check I

Imagine a 40-mA∠0° current source in parallel with an impedance, **Z**. Determine the equivalent voltage source for each of the following impedances:

a. **Z** = 25 kΩ∠30°.

b. **Z** = 100 Ω ∠–90°.

c. **Z** = 20 kΩ – j16 kΩ.

19.3 Mesh (Loop) Analysis

Mesh analysis allows us to determine each loop current within a circuit, regardless of the number of sources within the circuit. The following steps provide a format which simplifies the process of using mesh analysis:

1. Convert all sinusoidal expressions into equivalent phasor notation. If necessary, convert current sources into equivalent voltage sources.

2. Redraw the given circuit, simplifying the given impedances wherever possible and labelling the impedances ($\mathbf{Z}_1$, $\mathbf{Z}_2$, etc.).

3. Arbitrarily assign clockwise loop currents to each interior closed loop within a circuit. Show the polarities of all impedances using the assumed current directions. If an impedance is common to two loops, it may be thought to have two simultaneous currents. Although in fact two currents will not occur simultaneously, this maneuver makes the algebraic calculations fairly simple. The actual current through a common impedance is the vector sum of the individual loop currents.

4. Apply Kirchhoff's voltage law to each closed loop in the circuit, writing each equation as follows:

$$\Sigma\,(\mathbf{ZI}) = \Sigma\,\mathbf{E}$$

If the current directions are originally assigned in a clockwise direction, then the resulting linear equations may be simplified to the following format:

$$\text{Loop 1: } +(\Sigma\mathbf{Z}_1)\mathbf{I}_1 - (\Sigma\mathbf{Z}_{1\text{-}2})\mathbf{I}_2 - \cdots -(\Sigma\mathbf{Z}_{1\text{-}n})\mathbf{I}_n = (\Sigma\mathbf{E}_1)$$

$$\text{Loop 2: } -(\Sigma\mathbf{Z}_{2\text{-}1})\mathbf{I}_1 + (\Sigma\mathbf{Z}_2)\mathbf{I}_2 - \cdots -(\Sigma_{2\text{-}n})\mathbf{I}_n = (\Sigma\mathbf{E}_2)$$

$$\text{Loop } n\text{: } -(\Sigma\mathbf{Z}_{n-1})\mathbf{I}_1 - (\Sigma\mathbf{Z}_{n-2})\mathbf{I}_2 - \cdots +(\Sigma\mathbf{Z}_n)\mathbf{I}_n = (\Sigma\mathbf{E}_n)$$

In the above format, $\Sigma\mathbf{Z}_x$ is the summation of all impedances around loop x. The sign in front of all loop impedances will be positive.

$\Sigma\mathbf{Z}_{x\text{-}y}$ is the summation of impedances which are common between loop x and loop y. If there are no common impedances between two loops, this term is simply set to zero. All common impedance terms in the linear equations are given a negative sign.

$\Sigma\mathbf{E}_x$ is the summation of voltage rises in the direction of the assumed current $\mathbf{I}_x$. If a voltage source has a polarity such that the loop current passes from plus to minus in the direction of the assumed current, then the voltage is given a negative sign.

5. Solve the resulting simultaneous linear equations using substitution or determinants. If necessary, refer to Appendix B for a review of how to solve simultaneous linear equations.

■ **EXAMPLE 19–5**   Solve for the loop equations in the circuit of Figure 19–12.

Figure 19-12

Solution   **Step 1:** The current source is first converted into an equivalent voltage source as shown in Figure 19–13.

$$E_2 = (1.25 \text{ A}\angle{-90°})(4 \text{ }\Omega\angle 90°)$$
$$= 5 \text{ V}\angle 0°)$$

Figure 19-13

Steps 2 and 3: Next the circuit is redrawn as shown in Figure 19–14. The impedances have been simplified and the loop currents are drawn in a clockwise direction.

$$Z_1 = 3 \text{ }\Omega - j3 \text{ }\Omega$$
$$Z_2 = 1 \text{ }\Omega + j0$$
$$Z_3 = 2 \text{ }\Omega + j4 \text{ }\Omega$$

Figure 19-14

Step 4: The loop equations are written as

$$\text{Loop 1: } (\mathbf{Z}_1 + \mathbf{Z}_2)\mathbf{I}_1 - (\mathbf{Z}_2)\mathbf{I}_2 = -\mathbf{E}_1$$

$$\text{Loop 2: } -(\mathbf{Z}_2)\mathbf{I}_1 + (\mathbf{Z}_2 + \mathbf{Z}_3)\mathbf{I}_2 = -\mathbf{E}_2$$

The solution for the currents may be found by using determinants:

$$\mathbf{I}_1 = \frac{\begin{vmatrix} -\mathbf{E}_1 & -\mathbf{Z}_2 \\ -\mathbf{E}_2 & \mathbf{Z}_2 + \mathbf{Z}_3 \end{vmatrix}}{\begin{vmatrix} \mathbf{Z}_1 + \mathbf{Z}_2 & -\mathbf{Z}_2 \\ -\mathbf{Z}_2 & \mathbf{Z}_2 + \mathbf{Z}_3 \end{vmatrix}}$$

$$= \frac{(-\mathbf{E}_1)(\mathbf{Z}_2 + \mathbf{Z}_3) - \mathbf{E}_2\mathbf{Z}_2}{(\mathbf{Z}_1 + \mathbf{Z}_2)(\mathbf{Z}_2 + \mathbf{Z}_3) - \mathbf{Z}_2\mathbf{Z}_2}$$

$$= \frac{(-\mathbf{E}_1)(\mathbf{Z}_2 + \mathbf{Z}_3) - \mathbf{E}_2\mathbf{Z}_2}{\mathbf{Z}_1\mathbf{Z}_2 + \mathbf{Z}_1\mathbf{Z}_3 + \mathbf{Z}_2\mathbf{Z}_3}$$

and

$$\mathbf{I}_2 = \frac{\begin{vmatrix} \mathbf{Z}_1 + \mathbf{Z}_2 & -\mathbf{E}_1 \\ -\mathbf{Z}_2 & -\mathbf{E}_2 \end{vmatrix}}{\begin{vmatrix} \mathbf{Z}_1 + \mathbf{Z}_2 & -\mathbf{Z}_2 \\ -\mathbf{Z}_2 & \mathbf{Z}_2 + \mathbf{Z}_3 \end{vmatrix}}$$

$$= \frac{-\mathbf{E}_2(\mathbf{Z}_1 + \mathbf{Z}_2) - \mathbf{E}_1\mathbf{Z}_2}{(\mathbf{Z}_1 + \mathbf{Z}_2)(\mathbf{Z}_2 + \mathbf{Z}_3) - \mathbf{Z}_2\mathbf{Z}_2}$$

$$= \frac{-\mathbf{E}_2(\mathbf{Z}_1 + \mathbf{Z}_2) - \mathbf{E}_1\mathbf{Z}_2}{\mathbf{Z}_1\mathbf{Z}_2 + \mathbf{Z}_1\mathbf{Z}_3 + \mathbf{Z}_2\mathbf{Z}_3}$$

Solving these equations using the actual values of impedances and voltages, we have

$$\mathbf{I}_1 = \frac{-(5)(3 + j4) - (5)(1)}{(3 - j3)(1) + (3 - j3)(2 + j4) + (1)(2 + j4)}$$

$$= \frac{(-15 - j20) - 5}{(3 - j3) + (6 + j6 - j^2 12) + (2 + j4)}$$

$$= \frac{-20 - j20}{23 + j7}$$

$$= \frac{28.28\angle -135°}{24.04\angle 16.93°}$$

$$= 1.176\,\text{A}\angle -151.93°$$

and

$$\mathbf{I}_2 = \frac{-(5)(4 - j3) - (5)(1)}{23 + j7}$$

$$= \frac{(-20 + j15) - (5)}{23 + j7}$$

$$= \frac{-25 + j15}{23 + j7}$$

$$= \frac{29.15\angle 149.04°}{24.04\angle 16.93°}$$

$$= 1.213\,\text{A}\angle 132.11°$$

■ **EXAMPLE 19–6**   Given the circuit of Figure 19–15, write the loop equations and solve for the loop currents. Determine the voltage, **V**.

Figure 19–15

Solution   **Step 1:** Converting the current source into an equivalent voltage source gives us the circuit of of Figure 19–16.

Figure 19–16

Steps 2 and 3: After simplifying the impedances and assigning clockwise loop currents, we have the circuit of Figure 19–17.

$$\mathbf{Z}_1 = 3\ \Omega - jZ\ \Omega$$
$$\mathbf{Z}_2 = 0 + j4\ \Omega$$
$$\mathbf{Z}_3 = 6\ \Omega + j4\ \Omega$$

Figure 19–17

Step 4: The loop equations for the circuit of Figure 19–17 are as follows:

Loop 1: $(\mathbf{Z}_1 + \mathbf{Z}_2)\mathbf{I}_1 - (\mathbf{Z}_2)\mathbf{I}_2 = \mathbf{E}_1$

Loop 2: $-(\mathbf{Z}_2)\mathbf{I}_1 + (\mathbf{Z}_2 + \mathbf{Z}_3)\mathbf{I}_2 = -\mathbf{E}_2$

which, after substituting the impedance values into the expressions, become

Loop 1: $(3\ \Omega + j2\ \Omega) - (j4\ \Omega) = 3\ \text{V}\angle40°$

Loop 2: $-(j4\ \Omega) + (6\ \Omega + j4\ \Omega) = -12\ \text{V}\angle0°$

Step 5: Solve for the currents using determinants, where the elements of the determinants are expressed as vectors.

Since the determinant of the denominator is common to both terms we find this value first.

$$\mathbf{D} = \begin{vmatrix} 3+j2 & -j4 \\ -j4 & 6+j4 \end{vmatrix}$$

$$= (3+j2)(6+j4) - (-j4)(-j4)$$

$$= 18 + j12 + j12 - 8 + 16$$

$$= 26 + j24 = 35.38\angle 42.71°$$

Now, the currents are found as

$$\mathbf{I}_1 = \frac{\begin{vmatrix} 3\angle 40° & -j4 \\ -12\angle 0° & 6+j4 \end{vmatrix}}{\mathbf{D}}$$

$$= \frac{(3\angle 40°)(7.211\angle 33.69°) - (12\angle 0°)(4\angle 90°)}{35.38\angle 42.71°}$$

$$= \frac{21.633\angle 73.69° - 48\angle 90°}{35.38\angle 42.71°}$$

$$= \frac{27.91\angle -77.43°}{35.38\angle 42.71°}$$

$$= 0.7887\ \text{A}\angle -120.14°$$

and

$$\mathbf{I}_2 = \frac{\begin{vmatrix} 3+j2 & 3\angle 40° \\ -j4 & -12\angle 0° \end{vmatrix}}{\mathbf{D}}$$

$$= \frac{(3.606\angle 33.69°)(-12\angle 0°) + (3\angle 40°)(4\angle 90°)}{35.38\angle 42.71°}$$

$$= \frac{-43.27\angle 33.69° + 12\angle 130°}{35.38\angle 42.71°}$$

$$= \frac{46.15\angle -161.29°}{35.38\angle 42.71°}$$

$$= 1.304\ \text{A}\angle 156.00°$$

The current through the 4-Ω inductive reactance is

$$\mathbf{I} = \mathbf{I}_1 - \mathbf{I}_2$$

$$= (0.7887\ \text{A}\angle -120.14°) - (1.304\ \text{A}\angle 156.00°)$$

$$= (-0.3960\ \text{A} - j0.6821\ \text{A}) - (-1.1913\ \text{A} + j0.5304\ \text{A})$$

$$= 0.795\ \text{A} - j1.213\ \text{A} = 1.45\ \text{A}\angle -56.75°$$

The voltage is now easily found from Ohm's law as

$$\mathbf{V} = \mathbf{I}\mathbf{Z}_L$$

$$= (1.45\ \text{A}\angle -56.75°)(4\ \Omega\angle 90°) = 5.80\ \text{V}\angle 33.25°$$

■ **EXAMPLE 19–7**   Given the circuit of of Figure 19–18, write the loop equations. Do not solve them.

Figure 19–18

Solution   The circuit is redrawn in Figure 19–19, showing loop currents and impedances together with the appropriate voltage polarities.

Figure 19–19

The loop equations may now be written as

$$\text{Loop 1: } (\mathbf{Z}_1 + \mathbf{Z}_3)\mathbf{I}_1 - (\mathbf{Z}_3)\mathbf{I}_2 - (\mathbf{Z}_1)\mathbf{I}_3 = \mathbf{E}_1$$

$$\text{Loop 2: } -(\mathbf{Z}_3)\mathbf{I}_1 + (\mathbf{Z}_2 + \mathbf{Z}_3 + \mathbf{Z}_4)\mathbf{I}_2 - (\mathbf{Z}_4)\mathbf{I}_3 = 0$$

$$\text{Loop 3: } -(\mathbf{Z}_1)\mathbf{I}_1 - (\mathbf{Z}_4)\mathbf{I}_2 + (\mathbf{Z}_1 + \mathbf{Z}_4)\mathbf{I}_3 = \mathbf{E}_2$$

The determinant for the coefficients of the loop equations will be as follows:

$$\mathbf{D} = \begin{vmatrix} \mathbf{Z}_1 + \mathbf{Z}_3 & -\mathbf{Z}_3 & -\mathbf{Z}_1 \\ -\mathbf{Z}_3 & \mathbf{Z}_2 + \mathbf{Z}_3 + \mathbf{Z}_4 & -\mathbf{Z}_4 \\ -\mathbf{Z}_1 & -\mathbf{Z}_4 & \mathbf{Z}_1 \mathbf{Z}_4 \end{vmatrix}$$

Notice that each of the terms along the principal diagonal is positive and that all other terms in the determinant are negative. In addition, we see that the determinant is symmetrical about the principal diagonal. These characteristics of the determinant may be used to detect whether an error has occurred in writing the loop equations for any number of simultaneous linear equations. However, these conditions apply only if the loop currents are assigned in the clockwise direction.

practice problem III

Given the circuit of Figure 19–20, write the mesh equations and solve for the loop currents. Use the results to determine the current **I**.

Figure 19–20

in process learning check II

Briefly list the steps followed in using mesh analysis to solve for the loop currents of a circuit.

19.4 Nodal Analysis

Nodal analysis allows us to calculate all node voltages with respect to an arbitrary reference point in a circuit. The following steps provide a simple format to apply nodal analysis.

1. Convert all sinusoidal expressions into equivalent phasor notation. If necessary, convert voltage sources into equivalent current sources.

2. Redraw the given circuit, simplifying the given impedances wherever possible and relabelling the impedances as admittances (Y_1, Y_2, etc.).

3. Select and label an appropriate reference node. Arbitrarily assign subscripted voltages (V_1, V_2, etc.) to each of the remaining n nodes within the circuit.

4. Indicate assumed current directions through all admittances in the circuit. If an admittance is common to two nodes, it is considered in each of the two node equations.

5. Apply Kirchhoff's current law to each of the n nodes in the circuit, writing each equation as follows:

$$\Sigma(\mathbf{Y}\mathbf{V}) = \Sigma\mathbf{I}_{sources}$$

The resulting linear equations may be simplified to the following format:

$$\text{Node 1: } +(\Sigma \mathbf{Y}_1)\mathbf{V}_1 - (\Sigma \mathbf{Y}_{1\text{-}2})\mathbf{V}_2 - \cdots -(\Sigma \mathbf{Y}_{1\text{-}n})\mathbf{V}_n = (\Sigma \mathbf{I}_1)$$

$$\text{Node 2: } -(\Sigma \mathbf{Y}_{2\text{-}1})\mathbf{I}_1 + (\Sigma \mathbf{Y}_2)\mathbf{I}_2 - \cdots -(\Sigma \mathbf{Y}_{2\text{-}n})\mathbf{I}_n = (\Sigma \mathbf{I}_2)$$

$$\text{Node } n\text{: } -(\Sigma \mathbf{Y}_{n\text{-}1})\mathbf{I}_1 - (\Sigma \mathbf{Y}_{n\text{-}2})\mathbf{I}_2 - \cdots + (\Sigma \mathbf{Y}_n)\mathbf{I}_n = (\Sigma \mathbf{I}_n)$$

In the above format, $\Sigma \mathbf{Y}_x$ is the summation of all admittances connected to node x. The sign in front of all node admittances will be positive.

$\Sigma \mathbf{Y}_{x\text{-}y}$ is the summation of common admittances between node x and node y. If there are no common admittances between two nodes, this term is simply set to zero. All common admittance terms in the linear equations will have a negative sign.

$\Sigma \mathbf{I}_x$ is the summation of current sources entering node x. If a current source leaves the node, the current is given a negative sign.

6. Solve the resulting simultaneous linear equations using substitution or determinants.

■ **EXAMPLE 19–8**   Given the circuit of Figure 19–21, write the nodal equations and solve for the node voltages.

Figure 19–21

Solution   The circuit is redrawn in Figure 19–22, showing the nodes and a simplified representation of the admittances.

The nodal equations are written as

$$\text{Node 1: } (\mathbf{Y}_1 + \mathbf{Y}_2)\mathbf{V}_1 - (\mathbf{Y}_2)\mathbf{V}_2 = \mathbf{I}_1$$

$$\text{Node 2: } -(\mathbf{Y}_2)\mathbf{V}_1 + (\mathbf{Y}_2 + \mathbf{Y}_3)\mathbf{V}_2 = -\mathbf{I}_2$$

Using determinants, the following expressions for nodal voltages are obtained:

$$Y_1 = 0.5 \text{ S}$$
$$Y_2 = -j\, 0.5 \text{ S}$$
$$Y_3 = +j\, 0.25 \text{ S}$$

Figure 19–22

$$\mathbf{V}_1 = \frac{\begin{vmatrix} \mathbf{I}_1 & -\mathbf{Y}_2 \\ -\mathbf{I}_2 & \mathbf{Y}_2 + \mathbf{Y}_3 \end{vmatrix}}{\begin{vmatrix} \mathbf{Y}_1 + \mathbf{Y}_2 & -\mathbf{Y}_2 \\ -\mathbf{Y}_2 & \mathbf{Y}_2 + \mathbf{Y}_3 \end{vmatrix}}$$

$$= \frac{\mathbf{I}_1(\mathbf{Y}_2 + \mathbf{Y}_3) - \mathbf{I}_2\mathbf{Y}_2}{\mathbf{Y}_1\mathbf{Y}_2 + \mathbf{Y}_1\mathbf{Y}_3 + \mathbf{Y}_2\mathbf{Y}_3}$$

and

$$\mathbf{V}_2 = \frac{\begin{vmatrix} \mathbf{Y}_1 + \mathbf{Y}_2 & \mathbf{I}_1 \\ -\mathbf{Y}_2 & -\mathbf{I}_2 \end{vmatrix}}{\begin{vmatrix} \mathbf{Y}_1 + \mathbf{Y}_2 & -\mathbf{Y}_2 \\ -\mathbf{Y}_2 & \mathbf{Y}_2 + \mathbf{Y}_3 \end{vmatrix}}$$

$$= \frac{-\mathbf{I}_2(\mathbf{Y}_1 + \mathbf{Y}_2) + \mathbf{I}_1\mathbf{Y}_2}{\mathbf{Y}_1\mathbf{Y}_2 + \mathbf{Y}_1\mathbf{Y}_3 + \mathbf{Y}_2\mathbf{Y}_3}$$

Substituting the appropriate values into the above expressions we find the nodal voltages as

$$\mathbf{V}_1 = \frac{1(-j0.5 + j0.25) - (2)(-j0.5)}{(0.5)(-j0.5) + (0.5)(j0.25) + (-j0.5)(j0.25)}$$

$$= \frac{j0.75}{0.125 - j0.125}$$

$$= \frac{0.75\angle 90°}{0.1768\angle -45°}$$

$$= 4.243 \text{ V}\angle 135°$$

and

$$\mathbf{V}_2 = \frac{-2(0.5 - j0.5) + (1)(-j0.5)}{(0.5)(-j0.5) + (0.5)(j0.25) + (-j0.5)(j0.25)}$$

$$= \frac{-1 + j0.5}{0.125 - j0.125}$$

$$= \frac{1.118\angle 153.43°}{0.1768\angle -45°}$$

$$= 6.324 \text{ V}\angle 198.43°$$

$$= 6.324 \text{ V}\angle -161.57°$$

■ **EXAMPLE 19–9** Use nodal analysis to determine the voltage **V** for the circuit of Figure 19–23. Compare the results to those obtained when the circuit was analyzed using mesh analysis in Example 19–6.

Figure 19–23

Solution **Step 1:** Convert the voltage source into an equivalent current source as illustrated in Figure 19–24.

$$\mathbf{I}_2 = \frac{\mathbf{E}_2}{6\ \Omega} = \frac{12\ V\angle 0°}{6\ \Omega} = 2\ A\angle 0°$$

Figure 19–24

Steps 2, 3, and 4: The reference node is selected to be at the bottom of the circuit and the admittances are simplified as shown in Figure 19–25.

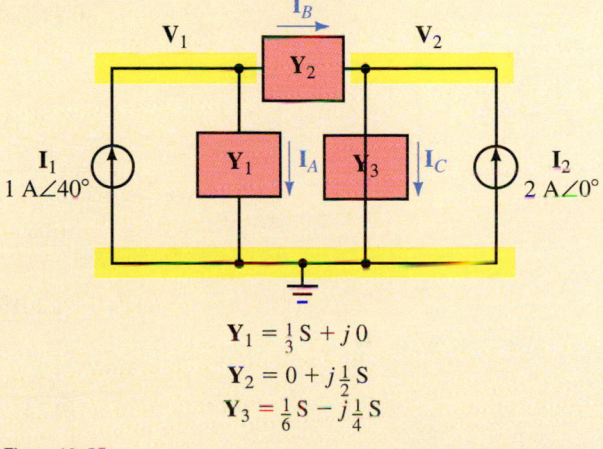

$$\mathbf{Y}_1 = \tfrac{1}{3}S + j\,0$$
$$\mathbf{Y}_2 = 0 + j\tfrac{1}{2}S$$
$$\mathbf{Y}_3 = \tfrac{1}{6}S - j\tfrac{1}{4}S$$

Figure 19–25

Step 5: Applying Kirchhoff's current law to each node, we have the following:

$$\text{Node 1: } \mathbf{I}_A + \mathbf{I}_B = \mathbf{I}_1$$

$$\mathbf{Y}_1 \, \mathbf{V}_1 + \mathbf{Y}_2(\mathbf{V}_1 - \mathbf{V}_2) = \mathbf{I}_1$$

$$(\mathbf{Y}_1 + \mathbf{Y}_2)\mathbf{V}_1 - \mathbf{Y}_2\mathbf{V}_2 = \mathbf{I}_1$$

$$\text{Node 2: } \mathbf{I}_C = \mathbf{I}_B + \mathbf{I}_2$$

$$\mathbf{Y}_3\mathbf{V}_2 = \mathbf{Y}_2(\mathbf{V}_1 - \mathbf{V}_2) + \mathbf{I}_2$$

$$-\mathbf{Y}_2\mathbf{V}_1 + (\mathbf{Y}^2 + \mathbf{Y}_3)\mathbf{V}_2 = \mathbf{I}_2$$

After substituting the appropriate values into the above equations, we have the following simultaneous linear equations:

$$(0.3333 \text{ S} + j0.5 \text{ S})\mathbf{V}_1 - (j0.5 \text{ S})\mathbf{V}_2 = 1 \text{ A}\angle 40°$$

$$-(j0.5 \text{ S})\mathbf{V}_1 + (0.16667 \text{ S} + j0.25 \text{ S})\mathbf{V}_2 = 2 \text{ A}\angle 0°$$

Step 6: We begin by solving for the determinant for the denominator:

$$\mathbf{D} = \begin{vmatrix} 0.3333 + j0.5 & -j0.5 \\ -j0.5 \ 0.1667 & +j0.25 \end{vmatrix}$$

$$= (0.3333 + j0.5)(0.1667 + j0.25) - (j0.5)(j0.5)$$

$$= 0.0556 + j0.0833 + j0.0833 - 0.125 + 0.25$$

$$= 0.1806 + j0.1667 = 0.2457\angle 42.71°$$

Next we solve the node voltages as

$$\mathbf{V}_1 = \frac{\begin{vmatrix} 1\angle 40° & -j0.5 \\ 2\angle 0° & 0.1667 + j0.25 \end{vmatrix}}{\mathbf{D}}$$

$$= \frac{(1\angle 40°)(0.3005\angle 56.31°) - (2\angle 0°)(0.5\angle -90°)}{\mathbf{D}}$$

$$= \frac{0.3005\angle 96.31° - 1.0\angle -90°}{0.2457\angle 42.71°}$$

$$= \frac{1.299\angle 91.46°}{0.2457\angle 42.71°} = 5.29 \text{ V}\angle 48.75°$$

and

$$\mathbf{V}_2 = \frac{\begin{vmatrix} 0.3333 + j0.5 & 1\angle 40° \\ -j0.5 & 2\angle 0° \end{vmatrix}}{\mathbf{D}}$$

$$= \frac{(0.6009\angle 56.31°)(2\angle 0°) - (0.5\angle -90°)(1\angle 40°)}{\mathbf{D}}$$

$$= \frac{1.2019\angle 56.31° - 0.5\angle -50°}{0.2457\angle 42.71°}$$

$$= \frac{1.426\ 75\angle 98°}{0.2457\angle 42.71°} = 5.80 \text{ V}\angle 33.27°$$

Examining the circuit of Figure 19–23, we see that the voltage **V** is the same as the node voltage $\mathbf{V}_2$. Therefore $\mathbf{V} = 5.80 \text{ V}\angle 33.27°$, which is the same result obtained in Example 19–6. (The slight difference in phase angle is the result of rounding error.)

■ **EXAMPLE 19–10**   Given the circuit of Figure 19–26, write the nodal equations expressing all coefficients in rectangular form. Do not solve the equations.

Figure 19–26

Solution   As in the previous example, we need to first convert the voltage source into its equivalent current source. The current source will be a phasor $\mathbf{I}_4$, where

$$\mathbf{I}_4 = \frac{\mathbf{V}_4}{\mathbf{X}_C} = \frac{2\text{ V}\angle 0°}{2\text{ }\Omega\angle -90°} = 1.0\text{ A}\angle 90°$$

Figure 19–27 shows the circuit as it appears after the source conversion. Notice that the direction of the current source is downward to correspond with the polarity of the voltage source $\mathbf{V}_4$.

$$\mathbf{I}_4 = \frac{2\text{ V}\angle 0°}{2\text{ }\Omega\angle -90°} = 1\text{ A}\angle 90°$$

Figure 19–27

Now, by labelling the nodes and admittances, the circuit may be simplified as shown in Figure 19–28.

$$\mathbf{Y}_1 = 0 - j\tfrac{1}{2}\,\text{S} \qquad \mathbf{Y}_3 = \tfrac{1}{2}\,\text{S} + j\,0 \qquad \mathbf{Y}_5 = 1\,\text{S} - j\tfrac{1}{2}\,\text{S}$$
$$\mathbf{Y}_2 = 0 + j\tfrac{1}{2}\,\text{S} \qquad \mathbf{Y}_4 = 0 - j\tfrac{1}{2}\,\text{S}$$

Figure 19–28

The admittances of Figure 19–28 are as follows:

$$\mathbf{Y}_1 = 0 - j0.5\ \text{S}$$
$$\mathbf{Y}_2 = 0 + j0.5\ \text{S}$$
$$\mathbf{Y}_3 = 0.5\ \text{S} + j0$$
$$\mathbf{Y}_4 = 0 - j0.5\ \text{S}$$
$$\mathbf{Y}_5 = 1.0\ \text{S} - j0.5\ \text{S}$$

Using the assigned admittances, the nodal equations are written as follows:

$$\text{Node 1: } (\mathbf{Y}_1 + \mathbf{Y}_2)\mathbf{V}_1 - (\mathbf{Y}_2)\mathbf{V}_2 - (0)\mathbf{V}_3 = -\mathbf{I}_1 + \mathbf{I}_2$$
$$\text{Node 2: } -(\mathbf{Y}_2)\mathbf{V}_1 + (\mathbf{Y}_2 + \mathbf{Y}_3 + \mathbf{Y}_5)\mathbf{V}_2 - (\mathbf{Y}_3)\mathbf{V}_3 = -\mathbf{I}_2 - \mathbf{I}_3$$
$$\text{Node 3: } -(0)\mathbf{V}_1 - (\mathbf{Y}_3)\mathbf{V}_2 + (\mathbf{Y}_3 + \mathbf{Y}_4)\mathbf{V}_3 = \mathbf{I}_3 - \mathbf{I}_4$$

By substituting the rectangular form of the admittances and current into the above linear equations, the equations are rewritten as

$$\text{Node 1: } (-j0.5 + j0.5)\mathbf{V}_1 - (j0.5)\mathbf{V}_2 - (0)\mathbf{V}_3 = -1 + j2$$
$$\text{Node 2: } -(j0.5)\mathbf{V}_1 - (j0.5 + 0.5 + 1 - j0.5)\mathbf{V}_2 - (0.5)\mathbf{V}_3 = -j2 - j2$$
$$\text{Node 3: } -(0)\mathbf{V}_1 - (0.5)\mathbf{V}_2 + (0.5 - j0.5)\mathbf{V}_3 = j2 - j1$$

Finally, the nodal equations are simplified as follows:

$$\text{Node 1: } (0)\mathbf{V}_1 - (j0.5)\mathbf{V}_2 - (0)\mathbf{V}_3 = -1 + j2$$
$$\text{Node 2: } -(j0.5)\mathbf{V}_1 - (1.5)\mathbf{V}_2 - (0.5)\mathbf{V}_3 = -j4$$
$$\text{Node 3: } -(0)\mathbf{V}_1 - (0.5)\mathbf{V}_2 + (0.5 - j0.5)\mathbf{V}_3 = j1$$

Given the circuit of Figure 19–29, use nodal analysis to find the voltages V_1 and V_2. Use your results to find the current I.

Figure 19–29

Briefly list the steps followed in using nodal analysis to solve for the node voltages of a circuit.

19.5   Delta-to-Wye and Wye-to-Delta Conversions

In Chapter 8, we derived the relationships showing the equivalence of "delta" (or "pi") connected resistance to resistors connected in a "wye" (or "tee") configuration.

In a similar manner, *impedances* connected in a Δ configuration are equivalent to a unique Y configuration. Figure 9–30 shows the equivalent circuits.

A Δ configuration is converted to a Y equivalent by using the following:

$$\mathbf{Z}_1 = \frac{\mathbf{Z}_b \mathbf{Z}_c}{\mathbf{Z}_a + \mathbf{Z}_b + \mathbf{Z}_c} \tag{19–1}$$

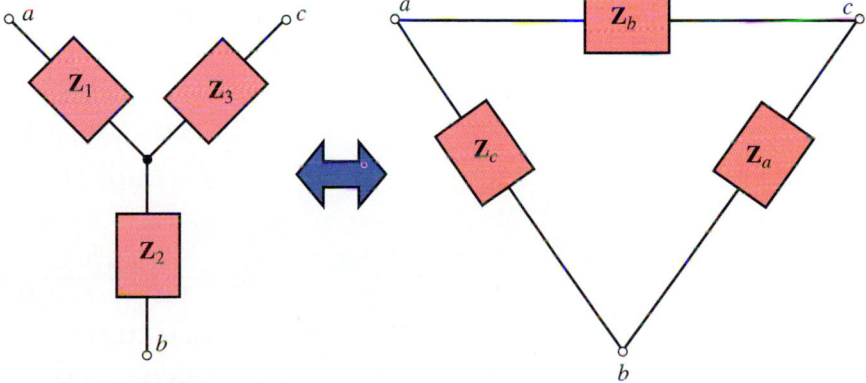

Figure 19–30   Delta-wye equivalence.

$$\mathbf{Z}_2 = \frac{\mathbf{Z}_a \mathbf{Z}_c}{\mathbf{Z}_a + \mathbf{Z}_b + \mathbf{Z}_c} \qquad (19\text{-}2)$$

$$\mathbf{Z}_3 = \frac{\mathbf{Z}_a \mathbf{Z}_b}{\mathbf{Z}_a + \mathbf{Z}_b + \mathbf{Z}_c} \qquad (19\text{-}3)$$

The above conversion indicates that the impedance in any arm of a Y circuit is determined by taking the product of the two adjacent Δ impedances at this arm and dividing by the summation of the Δ impedances.

If the impedances of the sides of the Δ network are all equal (magnitude and phase angle), the equivalent Y network will have identical impedances, where each impedance is determined as

$$\mathbf{Z}_Y = \frac{\mathbf{Z}_\Delta}{3} \qquad (19\text{-}4)$$

A Y configuration is converted to a Δ equivalent by using the following:

$$\mathbf{Z}_a = \frac{\mathbf{Z}_1 \mathbf{Z}_2 + \mathbf{Z}_1 \mathbf{Z}_3 + \mathbf{Z}_2 \mathbf{Z}_3}{\mathbf{Z}_1} \qquad (19\text{-}5)$$

$$\mathbf{Z}_b = \frac{\mathbf{Z}_1 \mathbf{Z}_2 + \mathbf{Z}_1 \mathbf{Z}_3 + \mathbf{Z}_2 \mathbf{Z}_3}{\mathbf{Z}_2} \qquad (19\text{-}6)$$

$$\mathbf{Z}_c = \frac{\mathbf{Z}_1 \mathbf{Z}_2 + \mathbf{Z}_1 \mathbf{Z}_3 + \mathbf{Z}_2 \mathbf{Z}_3}{\mathbf{Z}_3} \qquad (19\text{-}7)$$

Any impedance in a "Δ" is determined by summing the possible two-impedance product combinations of the "Y" and then dividing by the impedance found in the opposite branch of the "Y."

If the arms of a "Y" have identical impedances, the equivalent "Δ" will have impedances given as

$$\mathbf{Z}_\Delta = 3\mathbf{Z}_Y \qquad (19\text{-}8)$$

■ **EXAMPLE 19–11**   Determine the Y equivalent of the Δ network shown in Figure 19–31.

Solution

$$\mathbf{Z}_1 = \frac{(3\ \Omega)(-j6\ \Omega)}{3\ \Omega - j6\ \Omega + j9\ \Omega} = \frac{-j18\ \Omega}{3 + j3} = \frac{18\ \Omega\angle{-90°}}{4.242\angle{45°}}$$

$$= 4.242\ \Omega\angle{-135°}$$

$$= -3.0\ \Omega - j3.0\ \Omega$$

$$\mathbf{Z}_2 = \frac{(3\ \Omega)(j9\ \Omega)}{3\ \Omega - j6\ \Omega + j9\ \Omega} = \frac{j27\ \Omega}{3 + j3} = \frac{27\ \Omega\angle{90°}}{4.242\angle{45°}}$$

$$= 6.364\ \Omega\angle{45°}$$

$$= 4.5\ \Omega + j4.5\ \Omega$$

$$\mathbf{Z}_3 = \frac{(j9\ \Omega)(-j6\ \Omega)}{3\ \Omega - j6\ \Omega + j9\ \Omega} = \frac{54\ \Omega}{3 + j3} = \frac{54\ \Omega \angle 0°}{4.242 \angle 45°}$$

$$= 12.73\ \Omega \angle -45°$$

$$= 9.0\ \Omega - j9.0\Omega$$

Figure 19–31

In the above solution, we see that the given Δ network has an equivalent Y network with one arm having a negative resistance. This result indicates that although the Δ circuit has an equivalent Y circuit, the Y circuit cannot actually be constructed from real components since *negative resistors* do not exist (although some active components may demonstrate negative resistance characteristics). If the given conversion is used to simplify a circuit we would treat the impedance $\mathbf{Z}_1 = -3\ \Omega - j3\ \Omega$ as if the resistance actually were a negative value. Figure 19–32 shows the equivalent Y circuit.

Figure 19–32

■ **EXAMPLE 19–12**   Find the total impedance of the network in Figure 19–33.

Solution   If we take a moment to examine this network, we see that the circuit contains both a Δ and a Y. In calculating the total impedance, the solution is exactly the same regardless of whether we choose to convert the Y to a Δ or the Δ to a Y.

We will convert the Y shown in Figure 19–34 to a Δ.

Figure 19–33

Figure 19–34

$$\mathbf{Z}_a = \frac{\mathbf{Z}_1\mathbf{Z}_2 + \mathbf{Z}_1\mathbf{Z}_3 + \mathbf{Z}_2\mathbf{Z}_3}{\mathbf{Z}_1}$$

$$= \frac{(3\ \Omega)(j3\ \Omega) + (3\ \Omega)(-j3\ \Omega) + (j3\ \Omega)(-j3\ \Omega)}{3\ \Omega}$$

$$= \frac{-j^2 9\ \Omega}{3} = 3\ \Omega$$

$$\mathbf{Z}_b = \frac{\mathbf{Z}_1\mathbf{Z}_2 + \mathbf{Z}_1\mathbf{Z}_3 + \mathbf{Z}_2\mathbf{Z}_3}{\mathbf{Z}_2} = \frac{9\ \Omega}{-j3} = j3\ \Omega$$

$$\mathbf{Z}_c = \frac{\mathbf{Z}_1\mathbf{Z}_2 + \mathbf{Z}_1\mathbf{Z}_3 + \mathbf{Z}_2\mathbf{Z}_3}{\mathbf{Z}_3} = \frac{9\ \Omega}{j3} = -j3\ \Omega$$

Now, substituting the equivalent Δ into the original network, we have the revised network of Figure 19–35.

Figure 19–35

The network of Figure 19–35 shows that the corresponding sides of the Δ are parallel. Because the inductor and the capacitor in the left side of the Δ have the same values, we may replace the parallel combination of these two components with an open circuit. The resulting impedance of the network is now easily determined as

$$Z_T = 3\ \Omega \| 6\ \Omega + (j3\ \Omega) \| (-j6\ \Omega) = 2\ \Omega + j6\ \Omega$$

practice
problem
V

Find the equivalent impedance, Z_T, between the indicated terminals of Figure 19–33 by converting the Δ into an equivalent Y. Compare your result to that obtained in Example 19–12.

in process
learning
check IV

A Δ network consists of a resistor, inductor, and capacitor, each having an impedance of 150 Ω. Determine the corrresponding Y network.

19.6 Bridge Networks

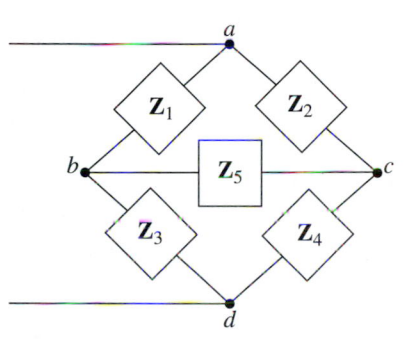

Figure 19–36

Bridge circuits, similar to the network of Figure 19–36, are used extensively in electronics to measure the values of unknown components.

Recall from Chapter 8 that any bridge circuit is said to be balanced when the current through the branch between the two arms is zero. In a practical circuit, component values of very precise resistors are adjusted until the current through the central element (usually a sensitive galvanometer) is exactly equal to zero. For ac circuits, the condition of a balanced bridge occurs when the impedance vectors of the various arms satisfy the following condition:

$$\frac{Z_1}{Z_3} = \frac{Z_2}{Z_4} \tag{19–9}$$

When a balanced bridge occurs in a circuit, the equivalent impedance of the bridge network is easily determined by removing the central impedance and replacing it by either an open or a short circuit. The resulting impedance of the bridge circuit is then found as either of the following:

$$Z_T = Z_1 \| Z_2 + Z_3 \| Z_4$$

or

$$Z_T = (Z_1 + Z_3) \| (Z_2 + Z_4)$$

If, on the other hand, the bridge is not balanced, then the total impedance must be determined by performing a Δ-to-Y conversion. Alternatively, the circuit may be analyzed by using either mesh analysis or nodal analysis.

■ **EXAMPLE 19–13** Consider the balanced bridge circuit of Figure 19–37.

$$\mathbf{Z}_1 = 30 \text{ k}\Omega\angle{-20°}$$
$$\mathbf{Z}_2 = 10 \text{ k}\Omega\angle{0°}$$
$$\mathbf{Z}_3 = 100 \text{ }\Omega\angle{0°}$$

Figure 19–37

a. Calculate the unknown impedance, $\mathbf{Z}_x$.

b. Determine the values of L_x and R_x if the circuit operates at a frequency of 1 kHz.

Solution

a. The expression for the unknown impedance is determined from Equation 19–9 as

$$\mathbf{Z}_x = \frac{\mathbf{Z}_2 \mathbf{Z}_3}{\mathbf{Z}_1}$$

$$= \frac{(10 \text{ k}\Omega)(100 \text{ }\Omega)}{30 \text{ k}\Omega\angle{-20°}}$$

$$= 33.3 \text{ }\Omega\angle{20°}$$

$$= 31.3 + j11.4 \text{ }\Omega$$

b. From the above result, we have

$$R_x = 31.3 \text{ }\Omega$$

and

$$L_x = \frac{X_L}{2\pi f} = \frac{11.4 \text{ }\Omega}{2\pi(1000 \text{ Hz})} = 1.81 \text{ mH}$$

We will now consider various forms of bridge circuits which are used in electronic circuits to determine the values of unknown inductors and capacitors. As in resistor bridges, the circuits use variable resistors together with very sensitive galvanometer movements to ensure a balanced condition for the bridge. However, rather than using a dc source to provide current in the circuit, the bridge circuits use ac sources operating at a known frequency (usually 1 kHz). Once the bridge is balanced, the value of unknown inductance or capacitance may be easily determined by obtaining the reading directly from the instrument. Most instruments using bridge circuitry will incorporate several different bridges to enable the measurement of various types of unknown impedances.

Maxwell Bridge

The Maxwell bridge, shown in Figure 19–38, is used to determine the inductance and series resistance of an inductor having a relatively large series resistance (in comparison with $X_L = \omega L$).

Figure 19–38   Maxwell bridge.

Resistors R_1 and R_3 are adjusted to provide the balanced condition (when the current through the galvanometer is zero: $I_G = 0$).

When the bridge is balanced, we know that the following condition must apply:

$$\frac{\mathbf{Z}_1}{\mathbf{Z}_2} = \frac{\mathbf{Z}_3}{\mathbf{Z}_4}$$

If we write the impedances using the rectangular forms, we obtain

$$\frac{\left[\dfrac{(R_1)\left(-j\dfrac{1}{\omega C}\right)}{R_1 - j\dfrac{1}{\omega C}}\right]}{R_2} = \frac{R_3}{R_x + j\omega L_x}$$

$$\frac{\left(-j\dfrac{R_1}{\omega C}\right)}{\left(\dfrac{\omega R_1 C - j1}{\omega C}\right)} = \frac{R_2 R_3}{R_x + j\omega L_x}$$

$$\frac{-jR_1}{\omega C R_1} - j = \frac{R_2 R_3}{R_x + j\omega L_x}$$

$$(-jR_1)(R_x + j\omega L_x) = R_2 R_3(\omega C R_1 - j)$$

$$\omega L_x R_1 - jR_1 R_x = \omega R_1 R_2 R_3 C - jR_2 R_3$$

Now, since two complex numbers can be equal only if their real parts are equal and if their imaginary parts are equal, we must have the following:

$$\omega L_x R_1 = \omega R_1 R_2 R_3 C$$

and

$$R_1 R_x = R_2 R_3$$

Simplifying these expressions, we get the following equations for a Maxwell bridge:

$$L_x = R_2 R_3 C \qquad (19\text{–}10)$$

and

$$R_x = \frac{R_2 R_3}{R_1} \qquad (19\text{–}11)$$

■ **EXAMPLE 19–14**

Figure 19–39

a. Determine the values of R_1 and R_3 so that the bridge of Figure 19–39 is balanced.

b. Calculate the current **I** when the bridge is balanced.

Solution

a. Rewriting Equations 19–10 and 19–11 and solving for the unknowns, we have

$$R_3 = \frac{L_x}{R_2 C} = \frac{16 \text{ mH}}{(10 \text{ k}\Omega)(0.01 \text{ }\mu\text{F})} = 160 \text{ }\Omega$$

and

$$R_1 = \frac{R_2 R_3}{R_x} = \frac{(10 \text{ k}\Omega)(160 \text{ }\Omega)}{50 \text{ }\Omega} = 32 \text{ k}\Omega$$

b. The total impedance is found as

$$\mathbf{Z}_T = (\mathbf{Z}_C \| \mathbf{R}_1 \| \mathbf{R}_2) + [\mathbf{R}_3 \| (\mathbf{R}_x + \mathbf{Z}_{Lx})]$$

$$\mathbf{Z}_T = (-j15.915 \text{ k}\Omega) \| 32 \text{ k}\Omega \| 10 \text{ k}\Omega + [160 \text{ }\Omega \| (50 \text{ }\Omega + j100.5 \text{ }\Omega)]$$

$$= 6.87 \text{ k}\Omega \angle -25.6° + 77.2 \text{ }\Omega \angle 38.0°$$

$$= 6.91 \text{ k}\Omega \angle -25.0°$$

The resulting circuit current is

$$\mathbf{I} = \frac{10 \text{ V}\angle0°}{6.91 \text{ k}\Omega\angle-25°} = 1.45 \text{ mA}\angle25.0°$$

Hay Bridge

In order to measure the inductance and series resistance of an inductor having a small series resistance, a Hay bridge may be used. The circuit of a Hay bridge is shown in Figure 19–40.

By applying a method similar to that used to determine the values of the unknown inductance and resistance of the Maxwell bridge, it may be shown that the following equations for the Hay bridge apply:

$$L_x = \frac{R_2 R_3 C}{\omega^2 R_1^2 C^2 + 1} \qquad (19\text{--}12)$$

and

$$R_x = \frac{\omega^2 R_1 R_2 R_3 C^2}{\omega^2 R_1^2 C^2 + 1} \qquad (19\text{--}13)$$

Schering Bridge

The Schering bridge, shown in Figure 19–41, is a circuit used to determine the value of unknown capacitance.

By solving for the balanced bridge condition, we may arrive at the following equations for the unknown quantities of the circuit:

$$C_x = \frac{R_1 C_3}{R_2} \qquad (19\text{--}14)$$

$$R_x = \frac{C_1 R_2}{C_3} \qquad (19\text{--}15)$$

Figure 19–40   Hay bridge.

Figure 19–41   Schering bridge.

■ **EXAMPLE 19–15**　Determine the values of C_1 and C_3 which will result in a balanced bridge for the circuit of Figure 19–42.

Figure 19–42

Solution　Rewriting Equations 19–14 and 19–15, we solve for the unknown capacitances as

$$C_3 = \frac{R_2 C_x}{R_1} = \frac{(1\ \mu F)(10\ k\Omega)}{5\ M\Omega} = 0.002\ \mu F$$

and

$$C_1 = \frac{C_3 R_x}{R_2} = \frac{(0.002\ \mu F)(200\ \Omega)}{10\ k\Omega} = 40\ pF$$

*practice
problem
VI*

Determine the values of R_1 and R_3 so that the bridge of Figure 19–43 is balanced.

Figure 19–43

*in process
learning
check V*

If the Schering bridge of Figure 19–41 is used to measure an unknown capacitance C_x in parallel with a resistance R_x, write the expression for C_x and R_x in terms of the known quantities C_1, R_1, R_2, C_3, and ω.

19.7 Circuit Analysis Using Computers

As you have already seen, the PSpice program is particularly useful for solving complicated dc circuits. The same is true for ac circuits, with the added advantage that PSpice allows us to obtain a response which varies with frequency. In some of the examples of this chapter, we examined circuits which had as many as three simultaneous linear equations. Although it is possible to use linear algebra to solve for the unknown variables, such solutions are very lengthy and prone to error. The PSpice program is able to analyze complicated circuits easily and without error.

■ **EXAMPLE 19–16**   Write the PSpice input file to determine the currents through each element of the circuit in Figure 19–18.

Solution   Since the original circuit has all impedances given in ohms, we must first convert the reactances into corresponding values of inductance (in henries) and capacitance (in farads). In order to do this, we assume that the circuit is operating at some frequency, say $f = 1$ kHz. Figure 19–44 illustrates the resulting circuit, with corresponding component values and designations which are in an acceptable format for the PSpice analysis.

Figure 19–44

In order to print the required output information, we enter the following line in the input file:

.PRINT   AC   VM(1)   VP(1)   VM(2)   VP(2)   ...etc.

The output will provide the magnitudes, VM(1) and VM(2), and the phase angles, VP(1) and VP(2), for the voltages occurring at nodes 1 and 2. The format for finding the ac current through an element is similar.

The input file and the relevent portion of the output file are given as follows:

```
Input File:
Example 19-16: ac analysis, Figure 19-44
Vs1 1 0 ac 3V 0
Vs2 3 0 ac 6V 0
R1  2 0 1ohm
R2 1 3 2ohms
L1 1 2 0.3183mH
C1 2 3 53.05uF
.AC LIN 1 1kHz 1kHz
.PRINT AC VM(1) VP(1) VM(2) VP(2)
.PRINT AC VM(3) VP(3) IM(L1) IP(L1)
.PRINT AC IM(C1) IP(C1) IM(R1) IP(R1)
.PRINT AC IM(R2) IP(R2)
.OPTIONS NOPAGE
.END
Output File:
FREQ VM(1) VP(1) VM(2) VP(2) 1.000E+03 3.000E+00
 0.000E+00 4.931E-01 9.946E+01


FREQ VM(3) VP(3) IM(L1) IP(L1)
1.000E+03 6.000E+00 0.000E+00 1.560E+00 -9.897E+01


FREQ IM(C1) IP(C1) IM(R1) IP(R1)
1.000E+03 2.033E+00 -9.457E+01 4.931E-01 9.946E+01


FREQ IM(R2) IP(R2)
1.000E+03 1.500E+00 1.800E+02
```

practice
problem
VII

Write the PSpice input file to evaluate the node voltages for the circuit of Figure 19–23. Assume that the circuit operates at an angular frequency of $\omega = 1000$ rad/s ($f = 159.15$ Hz).

Problems

19.1 Dependent Sources

1. Refer to the circuits of Figure 19–45.

   Find **V** when the controlling current, **I**, is the following:
   a. 20 μA∠0°.
   b. 50 μA∠−180°.
   c. 60 μA∠60°.

Figure 19–45

2. Consider the circuits of Figure 19–46.

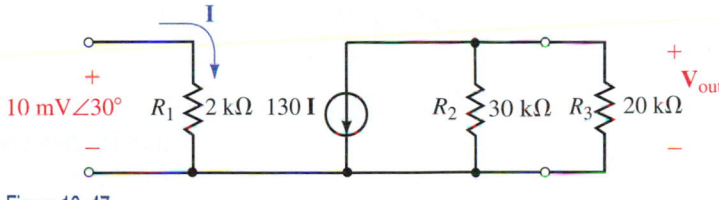

Figure 19–46

Find **I** when the controlling voltage, **V**, is the following:

a. 30 mV∠0°.

b. 60 mV∠−180°.

c. 100 mV∠−30°.

3. Find the output voltage, **V**_{out}, for the circuit of Figure 19– 47.

Figure 19–47

4. Repeat Problem 3 for the circuit of Figure 19–48.

Figure 19–48

19.2 Source Conversion

5. Given the circuits of Figure 19–49, convert each of the current sources into an equivalent voltage source. Use the resulting circuit to find V_L.

(a)   (b)

Figure 19–49

6. Convert each voltage source of Figure 19–50 into an equivalent current source.

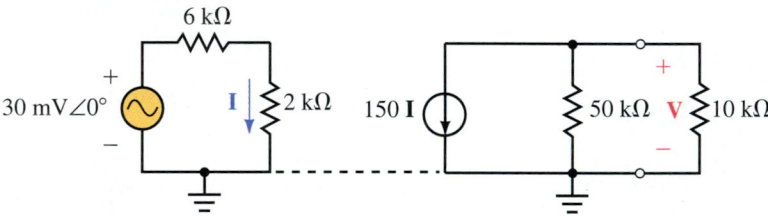

(a)   (b)

Figure 19–50

7. Refer to the circuit of Figure 19–51.

Figure 19–51

  a. Solve for the voltage, **V**.
  b. Convert the current source into an equivalent voltage source and again solve for **V**. Compare to the result obtained in (a).

8. Refer to the circuit of Figure 19–52.

$\mathbf{I} = 3\ \mu A\angle 0°$

Figure 19–52

  a. Solve for the voltage, $\mathbf{V}_L$.
  b. Convert the current source into an equivalent voltage source and again solve for $\mathbf{V}_L$.
  c. If $\mathbf{I} = 5\ \mu A\angle 90°$, what is $\mathbf{V}_L$?

19.3 Mesh (Loop) Analysis

9. Consider the circuit of Figure 19–53.

Figure 19–53

a. Write the mesh equations for the circuit.
b. Solve for the loop currents.
c. Determine the current **I** through the 4-Ω resistor.

10. Refer to the circuit of Figure 19–54.

Figure 19–54

a. Write the mesh equations for the circuit.
b. Solve for the loop currents.
c. Determine the current through the 25-Ω inductor.

11. Refer to the circuit of Figure 19–55.

Figure 19–55

a. Simplify the circuit and write the mesh equations.
b. Solve for the loop currents.
c. Determine the voltage **V** across the 15-Ω capacitor.

12. Consider the circuit of Figure 19–56.

Figure 19–56

a. Simplify the circuit and write the mesh equations.
b. Solve for the loop currents.
c. Determine the voltage **V** across the 2-Ω resistor.

13. Use mesh analysis to find the current **I** and the voltage **V** in the circuit of Figure 19–57.

Figure 19–57

14. Repeat Problem 13 for the circuit of Figure 19–58.

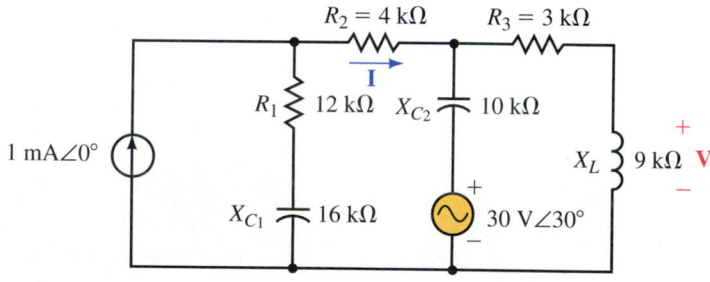

Figure 19–58

19.4 Nodal Analysis

15. Consider the circuit of Figure 19–59.

Figure 19–59

a. Write the nodal equations.
b. Solve for the node voltages.
c. Determine the current **I** through the 4-Ω capacitor.

16. Refer to the circuit of Figure 19–60.

Figure 19–60

a. Write the nodal equations.
b. Solve for the node currents.
c. Determine the voltage **V** across the 3-Ω capacitor.

17. a. Simplify the circuit of Figure 19–55, and write the nodal equations.
  b. Solve for the node voltages.
  c. Determine the voltage across the 15-Ω capacitor.

18. a. Simplify the circuit of Figure 19–56, and write the nodal equations.
  b. Solve for the node voltages.
  c. Determine the current through the 2-Ω resistor.

19. Use nodal analysis to determine the node voltages in the circuit of Figure 19–57. Use the results to find the current **I** and the voltage **V**. Compare your answers to those obtained using mesh analysis in Problem 13.

20. Use nodal analysis to determine the node voltages in the circuit of Figure 19–58. Use the results to find the current **I** and the voltage **V**. Compare your answers to those obtained using mesh analysis in Problem 14.

19.5 Delta-to-Wye and Wye-to-Delta Conversions

21. Convert each of the Δ networks of Figure 19–61 into an equivalent Y network.

(a)   (b)

Figure 19–61

22. Convert each of the Y networks of Figure 19–62 into an equivalent Δ network.

23. Using Δ→Y or Y→Δ conversion, calculate **I** for the circuit of Figure 19–63.

(a)

(b)

Figure 19–62

Figure 19–63

24. Using Δ→Y or Y→Δ conversion, calculate **I** for the circuit of Figure 19–64.

25. Refer to the circuit of Figure 19–65:
a. Determine the equivalent impedance, **Z**$_T$, of the circuit.
b. Find the currents **I** and **I**$_1$.

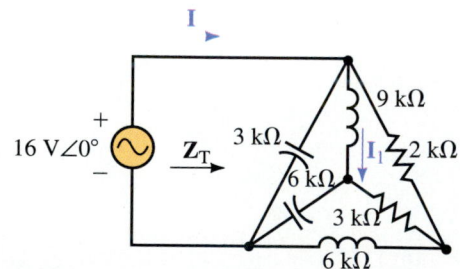

Figure 19–64   Figure 19–65

26. Refer to the circuit of Figure 19–66:

Figure 19–66

a. Determine the equivalent impedance, $\mathbf{Z}_T$, of the circuit.
b. Find the voltages $\mathbf{V}$ and $\mathbf{V}_1$.

19.6 Bridge Networks

27. Consider the balanced bridge circuit of Figure 19–67:

Figure 19–67

a. Determine the value of the unknown impedance.
b. Solve for the current $\mathbf{I}$.

28. Consider the balanced bridge circuit of Figure 19–68:

Figure 19–68

a. Determine the value of the unknown impedance.
b. Solve for the current **I**.

29. Show that the bridge circuit of Figure 19–69 is balanced.

Figure 19–69

30. Show that the bridge circuit of Figure 19–70 is balanced.

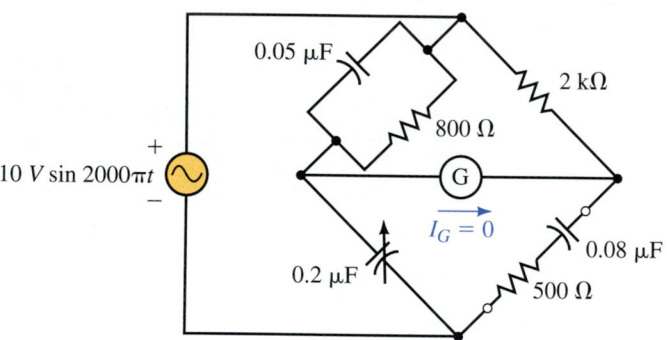

Figure 19–70

31. Derive Equations 19–14 and 19–15 for the balanced Schering bridge.

32. Derive Equations 19–12 and 19–13 for the balanced Hay bridge.

33. Determine the values of the unknown resistors which will result in a balanced bridge for the circuit of Figure 19–71.

Figure 19–71

34. Determine the values of the unknown capacitors which will result in a balanced bridge for the circuit of Figure 19–72.

Figure 19–72

19.7 Circuit Analysis Using Computers

35. Write the PSpice input file to determine the node voltages and the currents through each element of the circuit in Figure 19–21. Assume that the circuit operates at an angular frequency $\omega = 2$ krad/s.

36. Write the PSpice input file to determine the currents through each element of the circuit in Figure 19–29. Assume that the circuit operates at angular frequency, $\omega = 1$ krad/s.

AC Network Theorems

CHAPTER PREVIEW

In this chapter we apply the superposition, Thévenin, Norton, and maximum power transfer theorems in the analysis of ac circuits. Although the Millman and reciprocity theorems apply to ac circuits as well as to dc circuits, they are omitted since the applications are virtually identical with those used in analyzing dc circuits.

Many of the techniques used in this chapter are similar to those used in Chapter 9, and as a result, most students will find a brief review of dc theorems useful.

This chapter examines the application of the network theorems by considering both independent and dependent sources. In order to show the distinctions between the methods used in analyzing the various types of sources, the sections are labelled according to the types of sources involved.

An understanding of dependent sources is particularly useful when working with transistor circuits and operational amplifiers. Sections 20.2 and 20.5 are intended to provide the background for analyzing the operation of feedback amplifiers. Your instructor may find that these topics are best left until you cover this topic in a course dealing with such amplifiers. Consequently, the omission of sections 20.2 and 20.5 will not in any way detract from the continuity of the important ideas presented in this chapter.

KEY TERMS

Absolute Maximum Power
Maximum Power Transfer
Norton's Theorem
Relative Maximum Power
Superposition Theorem

Telstar 401. *(Courtesy of Martin Marietta Corp.)*

OBJECTIVES

After studying this chapter you will be able to

- apply the superposition theorem to determine the voltage across or current through any component in a given circuit,

- determine the Thévenin equivalent of circuits having independent and/or dependent sources,

- determine the Norton equivalent of circuits having independent and/or dependent sources,

- apply the maximum power transfer theorem to determine the load impedance for which maximum power is transferred to the load from a given circuit,

- use PSpice to find the Thévenin and Norton equivalents of circuits having either independent or dependent sources.

William Bradford Shockley

Shockley was born the son of a mining engineer in London, England on February 13, 1910. After graduating from the California Institute of Technology and Massachusetts Institute of Technology, Shockley joined Bell Telephone Laboratories.

With his co-workers, John Bardeen and Walter Brattain, Shockley developed an improved solid-state rectifier using a germanium crystal which had been injected with minute amounts of impurities. Unlike vacuum tubes, the resulting diodes were able to operate at much lower voltages without the need for inefficient heater elements.

In 1948, Shockley combined three layers of germanium to produce a device which was able to not only rectify a signal but to amplify it. Thus was developed the first transistor. Since its humble beginning, the transistor has been improved and decreased in size to the point where now a circuit containing thousands of transistors can easily fit into an area not much bigger than the head of a pin.

The advent of the transistor has permitted the construction of elaborate spacecraft, unprecedented communication, and new forms of energy generation.

Shockley, Bardeen, and Brattain received the 1956 Nobel Prize in Physics for their discovery of the transistor.

20.1 Superposition Theorem—Independent Sources

The superposition theorem states the following:

The voltage across (or current through) an element is determined by summing the voltage (or current) due to each independent source.

In order to apply this theorem all sources, other than the one being considered, are eliminated. As in dc circuits, this is done by replacing current sources with open circuits and by replacing voltage sources with short circuits. The process is repeated until the effects due to all sources have been determined.

Although we generally work with circuits having all sources at the same frequency, occasionally a circuit may operate at more than one frequency at a time. This is particularly true in diode and transistor circuits which use a dc source to set a "bias" (or operating) point and an ac source to provide the signal to be conditioned or amplified. In such cases, the resulting voltages or currents are still determined by applying the superposition theorem. The topic of how to solve circuits operating at several different frequencies simultaneously is covered in Chapter 25.

note

A*s in dc circuits, the superposition theorem can be applied only to voltage and current; it cannot be used to solve for the total power dissipated by an element. This is because power is not a linear quantity, but rather follows a square-law relationship ($P = V^2/R = I^2R$).*

■ **EXAMPLE 20–1** Determine the current **I** in Figure 20–1 by using the superposition theorem.

Figure 20–1

Solution   i) **Current due to the 5 V∠0° voltage source:** Eliminating the current source, we obtain the circuit shown in Figure 20–2.

Current source is replaced with an open circuit.

Figure 20–2

Applying Ohm's law, we have

$$\mathbf{I}_{(1)} = \frac{5\ V\angle0°}{4 - j2\ \Omega} = \frac{5\ V\angle0°}{4.472\Omega\angle{-}26.57°}$$

$$= 1.118\ A\angle26.57°$$

ii) **Current due to the 2 A∠0° voltage source:** Eliminating the current source, we obtain the circuit shown in Figure 20–3.

Voltage source is replaced with a short circuit.

Figure 20–3

The current $\mathbf{I}_{(2)}$ due to this source is determined by applying the current divider rule:

$$\mathbf{I}_{(2)} = -(2\ A\angle0°)\frac{4\ \Omega\angle0°}{4\ \Omega - j2\ \Omega}$$

$$= \frac{8\ V\angle180°}{4.472\ \Omega\angle{-}26.57°}$$

$$= 1.789\ A\angle{-}153.43°$$

In the previous calculation the current $\mathbf{I}_{(2)}$ is initially given a negative sign since the source would actually result in current in a direction opposite to the assumed reference direction. The negative sign is eliminated by adding (or subtracting) 180° from the phase angle of the solution.

The total current is determined as the summation of currents $\mathbf{I}_{(1)}$ and $\mathbf{I}_{(2)}$:

$$\mathbf{I} = \mathbf{I}_{(1)} + \mathbf{I}_{(2)}$$
$$= 1.118 \text{ A} \angle 26.57° + 1.789 \text{ A} \angle -153.43°$$
$$= (1.0 \text{ A} + j0.5 \text{ A}) + (-1.6 \text{ A} - j0.8 \text{ A})$$
$$= -0.6 \text{ A} - j0.3 \text{ A}$$
$$= 0.671 \text{ A} \angle -153.43°$$

■ **EXAMPLE 20–2**   Consider the circuit of 20–4:

Figure 20–4

Find the following:

a. $\mathbf{V}_R$ and $\mathbf{V}_C$ using the superposition theorem.
b. Power dissipated by the circuit.
c. Power delivered to the circuit by each of the sources.

Solution

a. The superposition theorem may be employed as follows:

i) **Voltages due to the current source:** Eliminating the voltage source, we obtain the circuit shown in Figure 20–5.

The impedance "seen" by the current source will be the parallel combination of $\mathbf{R}\|\mathbf{Z}_C$.

$$\mathbf{Z}_1 = \frac{(20 \text{ }\Omega)(-j15 \text{ }\Omega)}{20 \text{ }\Omega - j15 \text{ }\Omega} = \frac{300 \text{ }\Omega\angle -90°}{25 \text{ }\Omega\angle -36.87°} = 12 \text{ }\Omega\angle -53.13°$$

The voltage $\mathbf{V}_{R(1)}$ is the same as the voltage across the capacitor, $\mathbf{V}_{C(1)}$. Hence,

Figure 20–5

$$\mathbf{V}_{R(1)} = \mathbf{V}_{C(1)}$$

$$= (2 \text{ A}\angle 0°)(12 \text{ }\Omega\angle -53.13°)$$

$$= 24 \text{ V}\angle -53.13°$$

ii) **Voltages due to the voltage source:** Eliminating the current source, we have the circuit shown in Figure 20–6.

Figure 20–6

The voltages $\mathbf{V}_{R(2)}$ and $\mathbf{V}_{C(2)}$ are determined by applying the voltage divider rule,

$$\mathbf{V}_{R(2)} = \frac{20 \text{ }\Omega\angle 0°}{20 \text{ }\Omega - j15 \text{ }\Omega}(20 \text{ V}\angle 0°)$$

$$= \frac{400 \text{ V}\angle 0°}{25 \text{ V}\angle -36.87°} = 16 \text{ V}\angle -36.87°$$

and

$$\mathbf{V}_{C(2)} = \frac{-15 \text{ }\Omega\angle -90°}{20 \text{ }\Omega - j15 \text{ }\Omega}(20 \text{ V}\angle 0°)$$

$$= \frac{300 \text{ V}\angle 90°}{25 \text{ V}\angle -36.87°} = 12 \text{ V}\angle 126.87°$$

Notice that $\mathbf{V}_{C(2)}$ is assigned to be negative relative to the originally assumed polarity.

By applying superposition, we get

$$\mathbf{V}_R = \mathbf{V}_{R(1)} + \mathbf{V}_{R(2)}$$
$$= 24\ \text{V}\angle{-53.13°} + 16\ \text{V}\angle{36.87°}$$
$$= (14.4\ \text{V} - j19.2\ \text{V}) + (12.8\ \text{V} + j9.6\ \text{V})$$
$$= 27.2\ \text{V} - j9.6\ \text{V}$$
$$= 28.84\ \text{V}\angle{-19.44°}$$

and

$$\mathbf{V}_C = \mathbf{V}_{C(1)} + \mathbf{V}_{C(2)}$$
$$= 24\ \text{V}\angle{-53.13°} + 12\ \text{V}\angle{126.87°}$$
$$= (14.4\ \text{V} - j19.2\ \text{V}) + (-7.2\ \text{V} + j9.6\ \text{V})$$
$$= 7.2\ \text{V} - j9.6\ \text{V}$$
$$= 12\ \text{V}\angle{-53.13°}$$

b. Since only the resistor will dissipate power, the total power dissipated by the circuit is found as

$$P_T = \frac{(28.84\ \text{V})^2}{20\ \Omega} = 41.60\ \text{W}$$

c. The power delivered to the circuit by the current source is

$$P_1 = V_1 I \cos \theta_1$$

where $\mathbf{V}_1$ is the voltage across the current source and θ_1 is the phase angle between $\mathbf{V}_1$ and $\mathbf{I}$.

Since the voltage across the current source is the same as the voltage across the capacitor, the power delivered by the current source is

$$P_1 = (12\ \text{V})(2\ \text{A}) \cos 53.13° = 14.4\ \text{W}$$

The power delivered to the circuit by the voltage source is similarly determined as

$$P_2 = E I_2 \cos \theta_2$$

where $\mathbf{I}_2$ is the current through the voltage source and θ_2 is the phase angle between $\mathbf{E}$ and $\mathbf{I}_2$.

$$P_2 = (20\ \text{V})\left(\frac{28.84\ \text{V}}{20\ \Omega}\right) \cos 19.44° = 27.2\ \text{W}$$

As expected, the total power delivered to the circuit must be the summation

$$P_T = P_1 + P_2 = 41.6\ \text{W}$$

practice
problem
I

Use superposition to find **V** and **I** for the circuit of Figure 20–7.

Figure 20–7

in process
learning
check I

A 20-Ω resistor is in a circuit having three sinusoidal sources. After analyzing the circuit, it is found that the current through the resistor due to each of the sources is as follows:

$$I_1 = 1.5 \text{ A} \angle 20°$$

$$I_2 = 1.0 \text{ A} \angle 110°$$

$$I_3 = 2.0 \text{ A} \angle 0°$$

a. Use superposition to calculate the resultant current through the resistor.
b. Calculate the power dissipated by the resistor.
c. Show that the power dissipated by the resistor cannot be found by applying superposition, namely, $P_T \neq I_1^2 R + I_2^2 R + I_3^2 R$.

20.2 Superposition Theorem—Dependent Sources

Chapter 19 introduced the concept of dependent sources. We now examine ac circuits which are powered by dependent sources. In order to analyze circuits having dependent sources, it is first necessary to determine whether the dependent source is conditional upon a controlling element in its own circuit or whether the controlling element is located in some other circuit.

If the controlling element is external to the circuit under consideration, the method of analysis is the same as for an independent source. However, if the controlling element is in the same circuit, the analysis follows a slightly different stategy. The next two examples show the techniques used to analyze circuits having dependent sources.

■ **EXAMPLE 20–3**   Consider the circuit of Figure 20–8.

a. Determine the general expression for **V**.
b. Calculate **V** if **I** = 1.0 A $\angle 0°$.
c. Calculate **V** if **I** = 0.3 A $\angle 90°$.

Figure 20–8

Solution   a. Since the current source in the circuit is dependent on current through an element which is located outside of the circuit of interest, the circuit may be analyzed in the same manner as for independent sources.

i) **Voltage due to the voltage source:** Eliminating the current source, we obtain the circuit shown in Figure 20–9.

Figure 20–9

$$\mathbf{V}_{(1)} = \frac{8\ \Omega}{10\ \Omega}\ (6\ \text{V}\angle 0°) = 4.8\ \text{V}\angle 0°$$

ii) **Voltage due to the current source:** Eliminating the voltage source, we have the circuit shown in Figure 20–10.

Figure 20–10

$$\mathbf{Z}_T = 2\ \Omega \| 8\ \Omega = 1.6\ \Omega\angle 0°$$

$$\mathbf{V}_{(2)} = \mathbf{V}_{Z_T} = -(5\mathbf{I})(1.6\ \Omega\angle 0°) = -8.0\ \Omega\mathbf{I}$$

From superposition, the general expression for voltage is determined to be

$$\mathbf{V} = \mathbf{V}_{(1)} + \mathbf{V}_{(2)}$$

$$= 4.8\ \text{V}\angle 0° - 8.0\ \Omega\mathbf{I}$$

b.   If $\mathbf{I} = 1.0 \text{ A}\angle 0°$,

$$\mathbf{V} = 4.8 \text{ V}\angle 0° - (8.0 \ \Omega)(1.0 \text{ A}\angle 0°) = -3.2 \text{ V}$$

$$= 3.2 \text{ V}\angle 180°$$

c.   If $\mathbf{I} = 0.3 \text{ A}\angle 90°$,

$$\mathbf{V} = 4.8 \text{ V}\angle 0° - (8.0 \ \Omega)(0.3 \text{ A}\angle 90°) = 4.8 \text{ V} - j2.4 \text{ V}$$

$$= 5.367 \text{ V}\angle -26.57°$$

■ **EXAMPLE 20–4**   Given the circuit of Figure 20–11, calculate the voltage across the 40-Ω resistor.

Figure 20–11

Solution   In the circuit of Figure 20–11, the dependent source is controlled by an element located in the circuit. Unlike the sources in the previous examples, the dependent source cannot be eliminated from the circuit since doing so would contradict Kirchhoff's voltage law and/or Kirchhoff's current law.

The circuit must be analyzed by considering all effects simultaneously.

Applying Kirchhoff's current law, we have

$$\mathbf{I}_1 + \mathbf{I}_2 = 2 \text{ A}\angle 0°$$

and from Kirchhoff's voltage law, we have

$$(10 \ \Omega) \mathbf{I}_1 = \mathbf{V} + 0.2 \ \mathbf{V} = 1.2 \ \mathbf{V}$$

$$\mathbf{I}_1 = 0.12 \ \mathbf{V}$$

Finally,

$$\mathbf{I}_2 = \frac{\mathbf{V}}{40 \ \Omega} = 0.025 \ \mathbf{V}$$

Combining the above expressions, we have

$$0.12 \ \mathbf{V} + 0.025 \ \mathbf{V} = 2.0 \text{ A}\angle 0°$$

$$0.145 \ \mathbf{V} = 2.0 \text{ A}\angle 0°$$

$$\mathbf{V} = 13.79 \text{ V}\angle 0°$$

practice problem II

Determine the voltage **V** in the circuit of Figure 20–12.

Figure 20–12

20.3 Thévenin's Theorem—Independent Sources

Thévenin's theorem is a method which converts any linear bilateral ac circuit into a single ac voltage source in series with an equivalent impedance as shown in Figure 20–13.

The resulting two-terminal network will be equivalent when it is connected to any external branch or component. If the original circuit contains reactive elements, the Thévenin equivalent circuit will be valid only at the frequency at which the reactances were determined. The following method may be used to determine the Thévenin equivalent of an ac circuit having either independent sources or sources which are dependent upon voltage or current in some other circuit. The outlined method may not be used in circuits having dependent sources controlled by voltage or current in the same circuit.

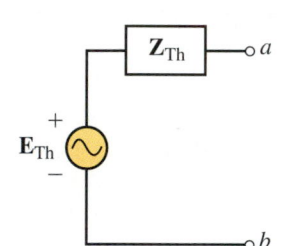

Figure 20–13   Thévenin equivalent circuit.

1. Remove the branch across which the Thévenin equivalent circuit is to be found. Label the resulting two terminals. Although any designation will do, we will use the notations *a* and *b*.

2. Set all sources to zero. As in dc circuits, this is achieved by replacing voltage sources with short circuits and current sources with open circuits.

3. Determine the Thévenin equivalent impedance, Z_{Th} by calculating the impedance seen between the open terminals *a* and *b*. Occasionally it may be necessary to redraw the circuit to simplify this process.

4. Replace the sources removed in Step 3 and determine the open-circuit voltage across the terminals *a* and *b*. If any of the sources are expressed in sinusoidal form, it is first necessary to convert these sources into an equivalent phasor form. For circuits having more than one source, it may be necessary to apply the superposition theorem to calculate the open-circuit voltage. Since all voltages will be phasors, the resultant is found by using vector algebra. The open-circuit voltage is the Thévenin voltage, E_{Th}.

5. Sketch the resulting Thévenin equivalent circuit by including that portion of the circuit removed in Step 1.

■ **EXAMPLE 20–5**   Find the Thévenin equivalent circuit external to $\mathbf{Z}_L$ for the circuit of Figure 20–14.

Figure 20–14

Solution   **Steps 1 and 2:** Removing the load impedance $\mathbf{Z}_L$ and setting the voltage source to zero, we have the circuit of Figure 20–15.

Voltage source is replaced with a short circuit.

Figure 20–15

Step 3: The Thévenin impedance between terminals a and b is found as

$$\mathbf{Z}_{\text{Th}} = \mathbf{R} \| (\mathbf{Z}_L + \mathbf{Z}_C)$$

$$= \frac{(40\ \Omega \angle 0°)(20\ \Omega \angle 90°)}{40\ \Omega + j20\ \Omega}$$

$$= \frac{800\ \Omega \angle 90°}{44.72\ \Omega \angle 26.57°}$$

$$= 17.89\ \Omega \angle 63.43°$$

$$= 8\ \Omega + j16\ \Omega$$

Step 4: The Thévenin voltage is found by using the voltage divider rule as shown in the circuit of Figure 20–16.

Figure 20–16

$$\mathbf{E}_{Th} = \mathbf{V}_{ab} = \frac{40\ \Omega\angle 0^\circ}{40\ \Omega + j80\ \Omega - j60\ \Omega}(20\text{V}\angle 0^\circ)$$

$$= \frac{800\ \text{V}\angle 0^\circ}{44.72\ \Omega\angle 26.57^\circ}$$

$$= 17.89\ \text{V}\angle -26.57^\circ$$

Step 5: The resultant Thévenin equivalent circuit is shown in Figure 20–17.

Figure 20–17

■ **EXAMPLE 20–6**   Determine the Thévenin equivalent circuit external to $\mathbf{Z}_L$ in the circuit in Figure 20–18.

Figure 20–18

Solution   **Step 1:** Removing the branch containing $\mathbf{Z}_L$ and labelling the terminals a and b, we have the circuit of Figure 20–19.

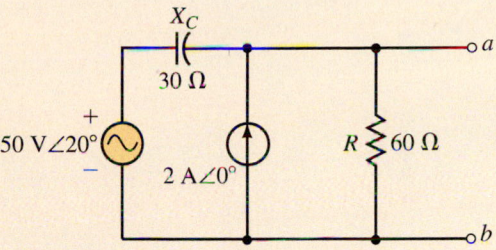

Figure 20–19

Step 2: After setting the voltage and current sources to zero, we have the circuit of Figure 20–20.

Voltage source replaced with a short circuit.

Current source replaced with an open circuit.

Figure 20–20

Step 3: The Thévenin impedance is determined as

$$\mathbf{Z}_{Th} = \mathbf{Z}_C \| \mathbf{Z}_R$$

$$= \frac{(30\ \Omega\angle{-90°})(60\ \Omega\angle{0°})}{60\ \Omega - j30\ \Omega}$$

$$= \frac{1800\ \Omega\angle{-90°}}{67.08\ \Omega\angle{-26.57°}}$$

$$= 26.83\ \Omega\angle{-63.43°}$$

Step 4: Because the given network consists of two independent sources, we consider the individual effects of each upon the open-circuit voltage. The total effect is then easily determined by applying the superposition theorem. Reinserting only the voltage source into the original circuit, as shown in Figure 20–21, allows us to find the open-circuit voltage, $\mathbf{V}_{ab}(1)$, by applying the voltage divider rule:

$$\mathbf{V}_{ab}(1) = \frac{60\ \Omega}{60\ \Omega - j30\ \Omega}(50\ V\angle{20°})$$

$$= \frac{3000\ V\angle{20°}}{67.08\angle{-26.57°}}$$

$$= 44.72\ V\angle{46.57°}$$

Figure 20–21

Now, considering only the current source as shown in Figure 20– 22, we determine $\mathbf{V}_{ab}(2)$ by Ohm's law:

Figure 20–22

$$\mathbf{V}_{ab}(2) = (2\ A\angle 0°) \frac{(30\ \Omega\angle{-90°})(60\ \Omega\angle 0°)}{60\ \Omega - j30\ \Omega}$$

$$= (2\ A\angle 0°)(26.83\ \Omega\angle{-63.43°})$$

$$= 53.67\ V\angle{-63.43°}$$

From the superposition theorem, the Thévenin voltage is determined as

$$\mathbf{E}_{\text{Th}} = \mathbf{V}_{ab}(1) + \mathbf{V}_{ab}(2)$$

$$= 44.72\ V\angle 46.57° + 53.67\ V\angle{-63.43°}$$

$$= (30.74\ V + j32.48\ V) + (24.00\ V - j48.00\ V)$$

$$= (54.74\ V - j15.52\ V) = 56.90\ V\angle{-15.83°}$$

Step 5: The resulting Thévenin equivalent circuit is shown in Figure 20–23.

Figure 20–23

Refer to the circuit shown in Figure 20–24 of Practice Problem III. List the steps that you would use to find the Thévenin equivalent circuit.

practice
problem
III

Figure 20–24

a. Find the Thévenin equivalent circuit external to the inductor in the circuit in Figure 20–24. (Notice that the voltage source is shown as sinusoidal.)

b. Use the Thévenin equivalent circuit to find the phasor output voltage, $\mathbf{V}_L$.

c. Convert the answer of (b) into the equivalent sinusoidal voltage.

20.4 Norton's Theorem—Independent Sources

Norton's theorem converts any linear bilateral network into an equivalent circuit consisting of a single current source and a parallel impedance as shown in Figure 20–25.

Although Norton's equivalent circuit may be determined by first finding the Thévenin equivalent circuit and then performing a source conversion, we generally use the more direct method outlined below. The steps to find the Norton equivalent circuit are as follows:

Figure 20–25   Norton equivalent circuit.

1. Remove the branch across which the Norton equivalent circuit is to be found. Label the resulting two terminals *a* and *b*.

2. Set all sources to zero.

3. Determine the Norton equivalent impedance, $\mathbf{Z}_N$, by calculating the impedance seen between the open terminals *a* and *b*.

NOTE: *Since the previous steps are identical with those followed for finding the Thévenin equivalent circuit, we conclude that the Norton impedance must be the same as the Thévenin impedance.*

4. Replace the sources removed in Step 3 and determine the current that would occur between terminals *a* and *b* if these terminals were shorted. Any voltages and currents that are

given in sinusoidal notation must first be expressed in equivalent phasor notation. If the circuit has more than one source it may be necessary to apply the superposition theorem to calculate the total short-circuit current. Since all currents will be in phasor form, any addition must be done using vector algebra. The resulting current is the Norton current, I_N.

5. Sketch the resulting Norton equivalent circuit by inserting that portion of the circuit removed in Step 1.

As mentioned previously, it is possible to find the Norton equivalent circuit from the Thévenin equivalent by simply performing a source conversion. We have already determined that both the Thévenin and Norton impedances are determined in the same way. Consequently, the impedances must be equivalent, and so we have

$$\mathbf{Z}_N = \mathbf{Z}_{Th} \tag{20–1}$$

Now, applying Ohm's law, we determine the Norton current source from the Thévenin voltage and impedance, namely,

$$\mathbf{I}_N = \frac{\mathbf{E}_{Th}}{\mathbf{Z}_{Th}} \tag{20–2}$$

Figure 20–26 shows the equivalent circuits.

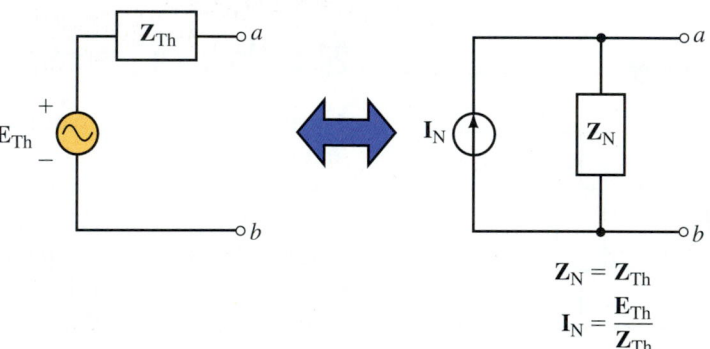

$$\mathbf{Z}_N = \mathbf{Z}_{Th}$$
$$\mathbf{I}_N = \frac{\mathbf{E}_{Th}}{\mathbf{Z}_{Th}}$$

Figure 20–26

■ **EXAMPLE 20–7**   Given the circuit of Figure 20–27, find the Norton equivalent.

Solution   **Steps 1 and 2:** By removing the load impedance, $\mathbf{Z}_L$, and setting the voltage source to zero, we have the network of Figure 20–28.

Step 3: The Norton impedance may now be determined by evaluating the impedance between terminals a and b. Hence, we have

Figure 20–27

Figure 20–28

$$\mathbf{Z_N} = \frac{(40\ \Omega\angle 0°)(20\ \Omega\angle 90°)}{40\ \Omega + j20\ \Omega}$$

$$= \frac{800\ \Omega\angle 90°}{44.72\angle 26.57°}$$

$$= 17.89\ \Omega\angle 63.43°$$

$$= 8\ \Omega + j16\ \Omega$$

Step 4: Reinserting the voltage source, as in Figure 20–29, we find the Norton current by calculating the current between the shorted terminals, *a* and *b*.

Figure 20–29

Because the resistor, $R = 40\ \Omega$ is shorted, the current is determined by the impedances X_L and X_C as

$$\mathbf{I}_N = \mathbf{I}_{ab} = \frac{20 \text{ V}\angle 0°}{j80 \ \Omega - j60 \ \Omega}$$

$$= \frac{20 \text{ V}\angle 0°}{20 \ \Omega\angle -90°}$$

$$= 1.00 \text{ A}\angle -90°$$

Step 5: The resultant Norton equivalent circuit is shown in Figure 20–30.

$$\mathbf{Z}_N = 8 \ \Omega + j16 \ \Omega = 17.89 \ \Omega\angle 63.43°$$

Figure 20–30

■ **EXAMPLE 20–8**   Find the Norton equivalent circuit external to R_L in the circuit of Figure 20–31. Use the equivalent circuit to calculate the current $\mathbf{I}_L$ when $R_L = 0 \ \Omega$, 400 Ω, and 2 kΩ.

Figure 20–31

Solution   **Steps 1 and 2:** Removing the load resistor and setting the sources to zero, we obtain the network shown in Figure 20–32.

Figure 20–32

Step 3: The Norton impedance is determined as

$$\mathbf{Z}_N = \frac{(400\ \Omega \angle 90°)(400\ \Omega - j400\ \Omega)}{j400\ \Omega + 400\ \Omega - j400\ \Omega}$$

$$= \frac{(400\ \Omega \angle 90°)(565.69\ \Omega \angle -45°)}{400\ \Omega \angle 0°}$$

$$= 565.69\ \Omega \angle -45°$$

Step 4: Because the network consists of two sources, we determine the effects due to each source separately and then apply superposition to evaluate the Norton current source.

Reinserting the voltage source into the original network, we see from Figure 20–33 that the short-circuit current between the terminals *a* and *b* is easily found by using Ohm's law.

Notice that inductor is short circuited.

Figure 20–33

$$\mathbf{I}_{ab}(1) = \frac{50\ V \angle 45°}{400\ \Omega - j400\ \Omega}$$

$$= \frac{50\ V \angle 45°}{565.69\ \Omega \angle -45°}$$

$$= 88.4\ mA \angle 90°$$

Since short circuiting the current source effectively removes all impedances, as illustrated in Figure 20–34, the short-circuit current between the terminals *a* and *b* is given as follows:

$$\mathbf{I}_{ab}(2) = -100\ mA \angle 0°$$

$$= 100\ mA \angle 180°$$

Now applying the superposition theorem, the Norton current is determined as the summation

$$\mathbf{I}_N = \mathbf{I}_{ab}(1) + \mathbf{I}_{ab}(2)$$

$$= 88.4\ mA \angle 90° + 100\ mA \angle 180°$$

$$= -100\ mA + j88.4\ mA$$

$$= 133.5\ mA \angle 138.52°$$

These components are
short circuited.

Figure 20–34

Step 5: The resulting Norton equivalent circuit is shown in Figure
20–35.

$\mathbf{Z}_N = 565.7\ \Omega\angle45°$

$\mathbf{I}_N$
$133.5\ \text{mA}\angle138.52°$

$\mathbf{Z}_N$

R_L
$0\rightarrow 2\ \text{k}\Omega$

$\mathbf{I}_L$

Figure 20–35

From the above circuit, we express the current through the load,
$\mathbf{I}_L$, as

$$\mathbf{I}_L = \frac{\mathbf{Z}_N}{R_L + \mathbf{Z}_N}\mathbf{I}_N$$

Let $R_L = 0\ \Omega$,

$$\mathbf{I}_L = \mathbf{I}_N = 133.5\ \text{mA}\angle138.52°$$

Let $R_L = 400\ \Omega$,

$$\mathbf{I}_L = \frac{\mathbf{Z}_N}{R_L + \mathbf{Z}_N}\mathbf{I}_N$$

$$= \frac{(565.7\ \Omega\angle45°)(133.5\ \text{mA}\angle138.52°)}{400\ \Omega + 400\ \Omega + j400\ \Omega}$$

$$= \frac{75.24\ \text{V}\angle183.52°}{894.43\ \Omega\angle26.57°}$$

$$= 84.12\ \text{mA}\angle156.95°$$

Let $R_L = 2$ kΩ,

$$\mathbf{I}_L = \frac{\mathbf{Z}_N}{R_L + \mathbf{Z}_N}\mathbf{I}_N$$

$$= \frac{(565.7\ \Omega\angle 45°)(133.5\ \text{mA}\angle 138.52°)}{2000\ \Omega + 400\ \Omega + j400\ \Omega}$$

$$= \frac{75.24\ \text{V}\angle 183.52°}{2433.1\ \Omega\angle 9.46°}$$

$$= 30.92\ \text{mA}\angle 174.06°$$

in process
learning
check III

Refer to the circuit shown in Figure 20–36 (Practice Problem IV). List the steps that you would use to find the Norton equivalent circuit.

practice
problem
IV

Find the Norton equivalent circuit external to R_L in the circuit of Figure 20–36. Use the equivalent circuit to find the current $\mathbf{I}_L$.

Figure 20–36

20.5 Thévenin's and Norton's Theorem for Dependent Sources

If a circuit contains a dependent source which is controlled by an element outside the circuit of interest, the methods outlined in Sections 20.2 and 20.3 may be used to find either the Thévenin or Norton equivalent circuit.

■ **EXAMPLE 20–9** Given the circuit of Figure 20–37, find the Thévenin equivalent circuit external to R_L. If the voltage applied to the resistor R_1 is 10 mV, use the Thévenin equivalent circuit to calculate the minimum and maximum voltage across R_L.

Figure 20–37

Solution

Step 1: Removing the load resistor from the circuit and labelling the remaining terminals a and b, we have the circuit shown in Figure 20–38.

Figure 20–38

Steps 2 and 3: The Thévenin resistance is found by open circuiting the current source and calculating the impedance observed between the terminals a and b. Since the circuit is purely resistive, we have

$$R_{\text{Th}} = 20 \text{ k}\Omega \| 5 \text{ k}\Omega$$

$$= \frac{(20 \text{ k}\Omega)(5 \text{ k}\Omega)}{20 \text{ k}\Omega + 5 \text{ k}\Omega}$$

$$= 4 \text{ k}\Omega$$

Step 4: The open-circuit voltage between the terminals is found to be

$$\mathbf{V}_{ab} = -(100\mathbf{I})(4 \text{ k}\Omega)$$

$$= -(4 \times 10^5 \text{ }\Omega)\mathbf{I}$$

As expected, the Thévenin voltage source is dependent upon the current $\mathbf{I}$.

Step 5: Because the Thévenin voltage is a dependent voltage source we may use the appropriate symbol when sketching the equivalent circuit, as shown in Figure 20–39.

Figure 20–39

Using the given conditions, we have

$$I = \frac{10 \text{ mV}}{1 \text{ k}\Omega} = 10 \text{ }\mu\text{A}$$

The voltage across the load is now determined as follows:

$R_L = 1$ kΩ: $V_{ab} = -\dfrac{1 \text{ k}\Omega}{1 \text{ k}\Omega + 4 \text{ k}\Omega}(4 \times 10^5 \text{ }\Omega)(10 \text{ }\mu\text{A})$

$$= -0.8 \text{ V}$$

$R_L = 4$ kΩ: $V_{ab} = -\dfrac{4 \text{ k}\Omega}{4 \text{ k}\Omega + 4 \text{ k}\Omega}(4 \times 10^5 \text{ }\Omega)(10 \text{ }\mu\text{A})$

$$= -2.0 \text{ V}$$

For an applied voltage of 10 mV, the voltage across the load resistance will vary between 0.8 V and 2.0 V as R_L is adjusted between 1 kΩ and 4 kΩ.

If a circuit contains one or more dependent sources which are controlled by an element in the circuit being analyzed, all previous methods fail to provide equivalent circuits which correctly model the circuit's behavior. In order to determine the Thévenin or Norton equivalent circuit of a circuit having a dependent source controlled by a local voltage or current, the following steps must be taken:

1. Remove the branch across which the Norton equivalent circuit is to be found. Label the resulting two terminals *a* and *b*.

2. Calculate the open-circuit voltage (Thévenin voltage) across the two terminals *a* and *b*. Because the circuit contains a dependent source controlled by an element in the circuit, the dependent source may not be set to zero. Its effects must be considered together with the effects of any independent source(s).

3. Determine the short-circuit current (Norton current) that would occur between the terminals. Once again, the dependent source may not be set to zero, but rather must have its effects considered concurrently with the effects of any independent source(s).

4. Determine the Thévenin or Norton impedance by applying Equations 20–1 and 20–2 as follows:

$$\mathbf{Z}_N = \mathbf{Z}_{Th} = \frac{\mathbf{E}_{Th}}{\mathbf{I}_N} \qquad (20\text{–}3)$$

5. Sketch the Thévenin or Norton equivalent circuit, as shown previously in Figure 20–26. Ensure that the portion of the network that was removed in Step 1 is reinserted as part of the equivalent circuit.

■ **EXAMPLE 20–10** For the circuit of Figure 20–40, find the Norton equivalent circuit external to the load resistor, R_L.

Figure 20–40

Solution

Step 1: After removing the load resistor from the circuit, we have the network shown in Figure 20–41.

Figure 20–41

Step 2: The presence of the locally controlled dependent source makes the analysis of this circuit fairly involved. We know, however, know that the basic laws of circuit analysis must be followed for all circuits, and so we use Ohm's law and Kirchhoff's laws as a starting point.

The current **I** is easily determined from Ohm's law as

$$\mathbf{I} = \frac{100 \text{ mV}}{20 \text{ k}\Omega} = 5 \ \mu A$$

and so the dependent current source must have a magnitude of

$$10\mathbf{I} = 50 \ \mu A$$

From Kirchhoff's voltage law, we have

$$V_{R_2} + V_{R_3} = 100 \text{ mV}$$

Applying Kirchhoff's current law at node a as illustrated in Figure 20–42 yields the following expression:

$$I_2 = 50 \ \mu A + I_3$$

Now, combining the latter two expressions, we have

Figure 20–42

$$(20 \text{ k}\Omega)I_2 + (20 \text{ k}\Omega)I_3 = 100 \text{ mV}$$

$$(20 \text{ k}\Omega)(I_3 + 50 \text{ } \mu\text{A}) + (20 \text{ k}\Omega)I_3 = 100 \text{ mV}$$

$$(40 \text{ k}\Omega)I_3 + 1000 \text{ mV} = 100 \text{ mV}$$

$$I_3 = \frac{-900 \text{ mV}}{40 \text{ k}\Omega} = -22.5 \text{ } \mu\text{A}$$

And so the open-circuit voltage is found to be

$$V_{ab} = I_3 R_3 = (-22.5 \text{ } \mu\text{A})(20 \text{ k}\Omega) = -0.45 \text{ V}$$

Step 3: The short-circuit current between terminals *a* and *b* is determined by examining the network of Figure 20–43.

Figure 20–43

From Figure 20–43, we see that the current **I** is once again **I** = 5 μA. Consequently, the dependent source has a value of

$$10\mathbf{I} = 50 \text{ } \mu\text{A}$$

Note: Although in this example the dependent source has the same value for both the open-circuited and the short-circuited terminals, we may not assume that this situation will occur in all circuits.

We know that the voltage across a short circuit is equal to zero and so the voltage across R_2 must be $V_2 = 100$ mV. Applying Ohm's law, the current through R_2 is found to be

$$I_2 = \frac{100 \text{ mV}}{20 \text{ k}\Omega} = 5 \text{ } \mu\text{A}$$

Once again, we isolate node *a* as illustrated in Figure 20–44.

Figure 20–44

From Kirchhoff's current law, we write the nodal equation as

$$I_2 = 50\ \mu A + I_{ab}$$

which allows us to calculate the short-circuit current as

$$I_{ab} = I_2 - 50\ \mu A$$
$$= 5\mu A - 50\ \mu A$$
$$= -45\ \mu A$$

Step 4: By applying Equation 20–3, we are able to determine the Norton impedance as

$$Z_N = \frac{E_{Th}}{I_N} = \frac{V_{ab}}{I_{ab}}$$

$$= \frac{-0.45\ V}{-45\ \mu A} = 10\ k\Omega$$

Step 5: Combining the impedance of Step 4 and the current of Step 3 gives the Norton equivalent circuit shown in Figure 20–45.

Figure 20–45

practice
problem
V

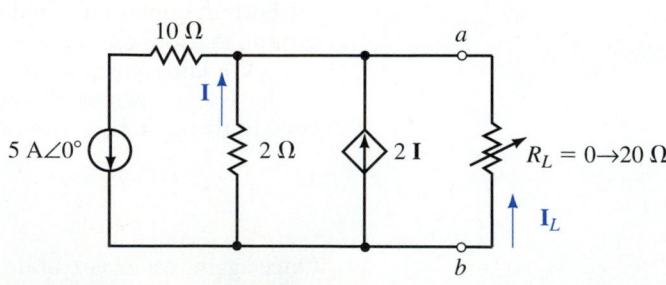

Figure 20–46

   a. Find the Thévenin equivalent circuit external to R_L in the circuit of Figure 20–46.
   b. Determine the current I_L when $R_L = 0$ and when $R_L = 20\ \Omega$.

If a given circuit has more than one independent source, it is necessary to determine the open-circuit voltage and short-circuit current due to each dependent source while simultaneously considering the effects of the independent source. The following example illustrates the principle.

■ **EXAMPLE 20–11**   Find the Thévenin equivalent circuit external to the load resistor, R_L, for the circuit of Figure 20–47.

Figure 20–47

Solution

Step 1: Removing the load resistor and labelling the resulting terminals a and b, we obtain the circuit shown in Figure 20–48.

Figure 20–48

Step 2: In order to find the open-circuit voltage, V_{ab} of Figure 20–48, we need to isolate the effects due to each independent source. The resultant is then determined by applying the superposition theorem.

   i) **Open-circuit voltage due to the voltage source:** Removing the current source, we have the network of Figure 20–49.

The controlling element (R_1) has a voltage, **V**, determined as

$$\mathbf{V} = V_{R_1} = \frac{6\ \text{k}\Omega}{6\ \text{k}\Omega + 4\ \text{k}\Omega}(20\ \text{V}) = 12.0\ \text{V}$$

Figure 20–49

The voltage across R_2 is

$$V_{R_2} = E - V_{R_1} = 20\ \text{V} - 12\ \text{V} = 8.0\ \text{V}$$

Now, applying Kirchhoff's voltage law, we determine the voltage $V_{ab}(1)$ as

$$V_{ab}(1) = -2\ \mathbf{V} + V_{R_3} + V_{R_2}$$
$$= -2(12\ \text{V}) + 0 + 8\ \text{V}$$
$$= -16.0\ \text{V}$$

    ii) **Open-circuit voltage due to the current source:** After removing the voltage source from the circuit of Figure 20–48, we have the network of Figure 20–50.

Figure 20–50

The controlling element has a voltage, $\mathbf{V}$, determined as

$$\mathbf{V} = -V_{R_2} = -(5\ \text{mA})\left[\dfrac{(6\ \text{k}\Omega)(4\ \text{k}\Omega)}{6\ \text{k}\Omega + 4\ \text{k}\Omega}\right]$$
$$= -(5\ \text{mA})(2.4\ \text{k}\Omega)$$
$$= -12.0\ \text{V}$$

Applying Kirchhoff's voltage law to the open-circuited terminals, we have

$$V_{ab}(2) = -2\ \mathbf{V} + V_{R_3} + V_{R_2}$$
$$= -2(-12\ \text{V}) + 0 + 12\ \text{V}$$
$$= +36.0\ \text{V}$$

Now, by superposition, we find the open-circuit (Thévenin) voltage to be

$$V_{ab} = V_{ab}(1) + V_{ab}(2)$$
$$= -16\text{ V} + 36\text{ V}$$
$$= +20.0\text{ V}$$

Step 3: In order to determine the short-circuit current of the circuit, we begin by redrawing the circuit as illustrated in Figure 20–51.

Figure 20–51

Once again, we will need to consider the effects due to each of the dependent sources. The combined result will be found by applying the superposition theorem.

i) **Short-circuit current due to the voltage source:** Because the analysis of the network involves both a dependent and an independent source, we need to simultaneously apply several of the basic laws of circuit analysis to correctly arrive at a solution.

We begin by removing the current source as shown in the network of Figure 20–52.

Figure 20–52

By examining the circuit and applying Kirchhoff's voltage law on the right-hand loop, we write the following expressions:

$$V_{R_2} = V_{R_3} + 2\text{ V}$$

The voltages may be rewritten as follows:

$$20\text{ V} - \mathbf{V} = (1\text{ k}\Omega)I_{ab}(1) + 2\text{ V}$$

By simplifying this expression, we have

$$I_{ab}(1) = \frac{20\text{ V} - 3\text{ V}}{1\text{ k}\Omega} \tag{1}$$

Next, we apply Kirchhoff's current law at the indicated node, which has been expanded in Figure 20–53.

$$I_1 = I_2 + I_{ab}(1)$$

Figure 20–53

and so we may write

$$I_{ab}(1) = I_1 - I_2 \qquad (2)$$

But from Ohm's law, we have the following:

$$I_1 = \frac{V_{R_1}}{R_1} = \frac{\mathbf{V}}{6 \text{ k}\Omega}$$

and

$$I_2 = \frac{V_{R_2}}{R_2} = \frac{20 \text{ V} - \mathbf{V}}{4 \text{ k}\Omega}$$

Therefore, Equation 2 may be rewritten as

$$I_{ab}(1) = \frac{\mathbf{V}}{6 \text{ k}\Omega} - \frac{20 \text{ V} - \mathbf{V}}{4 \text{ k}\Omega} \qquad (3)$$

Now, if we combine Equations 1 and 3, we may solve for the controlling voltage, **V,** as follows:

$$\frac{\mathbf{V}}{6 \text{ k}\Omega} - \frac{20 \text{ V} - \mathbf{V}}{4 \text{ k}\Omega} = \frac{20 \text{ V} - 3 \text{ V}}{1 \text{ k}\Omega}$$

$$\left(3 + \frac{1}{6} + \frac{1}{4}\right)\mathbf{V} = 20 \text{ V} + 5 \text{ V}$$

$$\mathbf{V} = \frac{25 \text{ V}}{3.417} = 7.3171 \text{ V}$$

Substituting this result into Equation 3, we get

$$I_{ab}(1) = \frac{20 \text{ V} - 3(7.3171 \text{ V})}{1 \text{ k}\Omega}$$

$$= -1.9512 \text{ mA}$$

ii) Short-circuit current due to the current source: Redrawing the network to examine the effects of the current source, we have the network of Figure 20–54.

Figure 20–54

By applying Kirchhoff's voltage law on the outside loop, we write the following expressions:

$$V_{R_2} = V_{R_3} + 2\,\mathbf{V}$$

$$-\mathbf{V} = [I_{ab}(2)](1\text{ k}\Omega) + 2\,\mathbf{V}$$

which may be simplified as

$$I_{ab}(2) = \frac{-3\,\mathbf{V}}{1\text{ k}\Omega} \tag{4}$$

Now applying Kirchhoff's current law at the indicated node, which has been expanded as shown in Figure 20–55, we write the following equations:

Figure 20–55

$$I_1 + I_2 + I_{ab}(2) = 5\text{ mA}$$

$$\frac{-\mathbf{V}}{6\text{ k}\Omega} + \frac{-\mathbf{V}}{4\text{ k}\Omega} + I_{ab}(2) = 5\text{ mA}$$

$$I_{ab}(2) = 5\text{ mA} + \frac{\mathbf{V}}{6\text{ k}\Omega} + \frac{\mathbf{V}}{4\text{ k}\Omega} \tag{5}$$

Combining Equations 4 and 5, we solve for the controlling voltage, **V**, as follows:

$$5\text{ mA} + \frac{\mathbf{V}}{6\text{ k}\Omega} + \frac{\mathbf{V}}{4\text{ k}\Omega} = -\frac{3\,\mathbf{V}}{1\text{ k}\Omega}$$

$$\left(\frac{3}{1\text{ k}\Omega} + \frac{1}{6\text{ k}\Omega} + \frac{1}{4\text{ k}\Omega}\right)\mathbf{V} = -5\text{ mA}$$

$$\mathbf{V} = -1.4634\text{ V}$$

Substituting this result into Equation 4, we get

$$I_{ab}(2) = \frac{-3(-1.4634\ V)}{1\ k\Omega}$$

$$= 4.3902\ mA$$

Finally, the superposition theorem allows us to determine the short-circuit current as

$$I_{ab} = I_{ab}(1) + I_{ab}(2)$$

$$= -1.9512\ mA + 4.3902\ mA$$

$$= 2.4390\ mA$$

Step 4: The Thévenin impedance is determined from Equation 20–3 as

$$Z_{Th} = \frac{V_{ab}}{I_{ab}} = \frac{20\ V}{2.4390\ mA}$$

$$= 8.200\ k\Omega$$

Step 5: Combining the results of Step 2 and Step 4 gives the Thévenin circuit as illustrated in Figure 20–56.

Figure 20–56

practice problem VI

Find the Thévenin equivalent of the circuit in Figure 20–47 if the dependent source has a magnitude of 4 **V** rather than 2 **V**.

20.6 Maximum Power Transfer Theorem

The maximum power transfer theorem is used to determine the value of load impedance required so that the load receives the maximum amount of power from the circuit. Consider the Thévenin equivalent circuit shown in Figure 20–57.

For any load impedance $\mathbf{Z}_L$ consisting of a resistance and a reactance such that $\mathbf{Z}_L = R_L \pm jX_L$, the power dissipated by the load will be determined as follows:

Figure 20–57

$$P_L = I^2 R_L$$

$$I = \frac{E_{Th}}{\sqrt{(R_{Th} + R_L)^2 + (X_{Th} \pm X_L)^2}}$$

$$R_L = \frac{E_{Th}^2 R_L}{(R_{Th} + R_L)^2 + (X_{Th} \pm X_L)^2}$$

Consider only the reactance portion, $\mathbf{X}_L$, of the load impedance for the moment and neglect the effect of the load resistance. We see that the power dissipated by the load will be maximum when the denominator is kept to a minimum. If the load were to have a reactance such that $\mathbf{X}_L = -\mathbf{X}_{Th}$, then the power delivered to the load would be given as

$$P_L = \frac{E_{Th}^2 R_L}{(R_{Th} + R_L)^2} \tag{20–4}$$

We recognize that this is the same expression for power as that determined for the Thévenin equivalent of dc circuits in Chapter 9. Recall that maximum power was delivered to the load when

$$R_L = R_{Th}$$

For ac circuits, the maximum power transfer theorem states the following:

Maximum power will be delivered to a load whenever the load has an impedance which is equal to the complex conjugate of the Thévenin (or Norton) impedance of the equivalent circuit.

The maximum power delivered to the load may be calculated by using Equation 20–4, which is simplified as follows:

$$P_{max} = \frac{E_{Th}^2}{4R_{Th}} \tag{20–5}$$

For a Norton equivalent circuit, the maximum power delivered to a load is determined by substituting $\mathbf{E}_{Th} = I_N \mathbf{Z}_N$ into the above expression as follows:

$$P_{max} = \frac{I_N^2 Z_N^2}{4R_N} \tag{20–6}$$

■ **EXAMPLE 20–12**   Determine the load impedance $\mathbf{Z}_L$ which will allow maximum power to be delivered to the load in the circuit of Figure 20–58. Find the maximum power.

Figure 20–58

Solution   Expressing the Thévenin impedance in its rectangular form, we have

$$\mathbf{Z}_{\text{Th}} = 500\ \Omega\angle 60° = 250\ \Omega + j433\ \Omega$$

In order to deliver maximum power to the load, the load impedance must be the complex conjugate of the Thévenin impedance. Hence,

$$\mathbf{Z}_L = 250\ \Omega - j433\ \Omega = 500\ \Omega\angle{-60°}$$

The power delivered to the load is now easily determined by applying Equation 20–5:

$$P_{\text{max}} = \frac{(20\ \text{V})^2}{4(250\ \Omega)} = 400\ \text{mW}$$

practice problem VII

Given the circuit of Figure 20–58, determine the power dissipated by the load if the load impedance is equal to the Thévenin impedance, $\mathbf{Z}_L = 500\ \Omega\angle 60°$. Compare your answer to that obtained in Example 20–12.

Occasionally it is not possible to adjust the reactance portion of a load. In such cases, a **relative maximum** power will be delivered to the load when the load resistance has a value determined as

$$R_L = \sqrt{R_{Tb}^2 + (X_{Tb} \pm X_L)^2} \tag{20–7}$$

If the reactance of the Thévenin impedance is of the same type (capacitive or inductive) as the reactance in the load, then the reactances are added.

If one reactance is capacitive and the other is inductive, however, then the reactances are subtracted.

To determine the power delivered to the load in such cases, the power will need to be calculated by finding either the voltage across the load or the current through the load. Equations 20–5 and 20–6 will no longer apply, since these equations were based on the premise that the load impedance is the complex conjugate of the Thévenin impedance.

■ **EXAMPLE 20–13**   For the circuit of Figure 20–59, determine the value of the load resistor, R_L, such that maximum power will be delivered to the load.

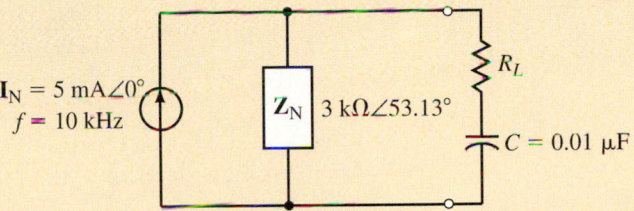

Figure 20–59

Solution   Notice that the load impedance consists of a resistor in series with a capacitance of 0.010 μF. Since the capacitive reactance is determined by the frequency, it is quite likely that the maximum power for this circuit may only be a relative maximum, rather than the absolute maximum. For the **absolute maximum power** to be delivered to the load, the load impedance would need to be

$$\mathbf{Z}_L = 3 \text{ k}\Omega\angle{-53.13°} = 1.80 \text{ k}\Omega - j2.40 \text{ k}\Omega$$

The reactance of the capacitor at a frequency of 10 kHz is determined to be

$$X_C = \frac{1}{2\pi(10 \text{ kHz})(0.010 \text{ }\mu\text{F})} = 1.592 \text{ k}\Omega$$

Because the capacitive reactance is not equal to the inductive reactance of the Norton impedance, the circuit will not deliver the absolute maximum power to the load. However, **relative maximum power** will be delivered to the load when

$$R_L = \sqrt{R_{Tb}^2 + (X_{Tb} - X_L^2)}$$
$$= \sqrt{(1.800 \text{ k}\Omega)^2 + (2.4 \text{ k}\Omega - 1.592 \text{ k}\Omega)^2}$$
$$= 1.973 \text{ k}\Omega$$

Figure 20–60 shows the circuit with all impedance values.

Figure 20–60

The load current will be

$$I_L = \frac{\mathbf{Z}_N}{\mathbf{Z}_N + \mathbf{Z}_L} \mathbf{I}_N$$

$$= \frac{1.80 \text{ k}\Omega + j2.40 \text{ k}\Omega}{(1.80 \text{ k}\Omega + j2.40 \text{ k}\Omega) + (1.973 \text{ k}\Omega - j1.592 \text{ k}\Omega)} (5 \text{ mA}\angle 0°)$$

$$= \frac{3 \text{ k}\Omega\angle 53.13°}{3.773 \text{ k}\Omega + j0.808 \text{ k}\Omega} (5 \text{ mA}\angle 0°)$$

$$= \frac{15.0 \text{ V}\angle 53.13°}{3.859 \text{ k}\Omega\angle 12.09°} = 3.887 \text{ mA}\angle 41.04°$$

We now determine the power delivered to the load for the given conditions as

$$P_L = I_L^2 R_L$$

$$= (3.887 \text{ mA})^2 (1.973 \text{ k}\Omega) = 29.82 \text{ mW}$$

If we had applied Equation 20–6, we would have found the absolute maximum power to be

$$P_{max} = \frac{(5 \text{ mA})^2 (3.0 \text{ k}\Omega)^2}{4(1.8 \text{ k}\Omega)} = 31.25 \text{ mW}$$

practice problem VIII

Refer to the Norton equivalent circuit of Figure 20–61:

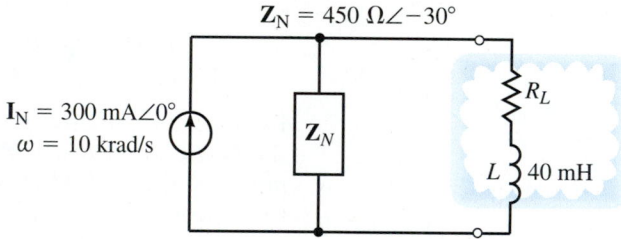

Figure 20–61

a. Find the value of load resistance, R_L, such that the load receives maximum power.

b. Determine the maximum power received by the load for the given conditions.

in process learning check IV

Refer to the Thévenin equivalent circuit of Figure 20–62.

Figure 20–62

a. Determine the value of the unknown load resistance, R_L, which will result in a relative maximum power at an angular frequency of 1 krad/s.
b. Solve for the power dissipated by the load at $\omega_1 = 1$ krad/s.
c. Assuming that the Thévenin impedance remains constant at all frequencies, at what angular frequency, ω_2, will the circuit provide absolute maximum power?
d. Solve for the power dissipated by the load at ω_2.

20.7　Circuit Analysis Using Computers

As demonstrated in Chapter 9, the PSpice program is easily able to determine the equivalent circuit between specified terminals of a dc circuit. The method for determining the Thévenin equivalent and Norton equivalent ac circuits is essentially the same as that for dc circuits, with the obvious exception that the voltages and currents are entered as ac quantities. The program is particularly useful in helping to find the Thévenin equivalent circuit of a circuit which has dependent sources. As we have already seen, the work in analyzing such a circuit manually is lengthy and very time-consuming. The following statements are used by PSpice to define dependent sources.

Voltage-controlled dependent voltage source:

Exxx　　n+　　n−　　cn+　　cn−　　gain

Current-controlled current source:

Fxxx n+　　n−　　cn+　　cn−　　gain

Voltage-controlled current source:

Gxxx　　n+　　n−　　cn+　　cn−　　gain

Current-controlled voltage source:

Hxxx　　n+　　n−　　Vyyy　　gain

where

> **Nxxx** is the the unique name of the dependent source.
>
> **n+** is the node to which the positive terminal of the source is connected.
>
> **n−** is the node to which the negative terminal of the source is connected.
>
> **cn+** is the node to which the positive terminal of the controlling source is connected.
>
> **cn−** is the node to which the negative terminal of the controlling source is connected.
>
> **Vyyy** is the voltage source through which the controlling current passes. (Since the original circuit may not have a voltage source in the controlling branch, this voltage is generally set to zero.)
>
> **gain** is the ratio between the dependent source and the controlling source.

■ **EXAMPLE 20–14**   Write the PSpice input file to find the Thévenin equivalent circuit of the circuit in Figure 20–47.

Solution   The original circuit is redrawn and relabelled to conform to the PSpice format as shown in Figure 20–63. Notice that the load impedance, R_L, has been removed.

Figure 20–63

The corresponding input file is given below.

```
Input File:
Example 20-14: Thévenin Equivalent - Dep. E source,
Figure 20-63
Vs   1   0   AC   20V
Is   0   2   AC   5mA
R1   1   2   6kohms
R2   2   0   4kohms
R3   2   3   1kohms
Edep 3   4   1   2   2
.AC  LIN 1   1kH 1kH
.PRINT AC VM(4)  VP(4)
.TF  V(4,0)  Vs
.OPTIONS NOPAGE
.END
```

The pertinent data from the output file are as follows:

```
Output File:
V(4,0)/Vs =-8.000E-01
INPUT RESISTANCE AT Vs = 1.000E+04
OUTPUT RESISTANCE AT V(4,0) = 8.200E+03

FREQ      VM(4)      VP(4)
1.000E+03  2.000E+01  0.000E+00
```

From the above data, we determine the Thévenin voltage and impedance to be

$$E_{Th} = 20 \text{ V}$$

$$R_{Th} = 8.2 \text{ k}\Omega$$

The resulting circuit is shown in Figure 20–64.

Figure 20–64

■ **EXAMPLE 20–15**   Write the PSpice input file to find the Norton equivalent circuit external to the load resistor of the circuit in Figure 20–40.

Solution   The original circuit is redrawn and relabelled to conform to the PSpice format as shown in Figure 20–65.

Figure 20–65

The corresponding input file is given below.

```
Input File:
Example 20-15: Thévenin Equivalent - Dep. I source,
Figure 20-65
Vs  1  0  AC  100mV
Vref 2  0  DC  0
```

```
R1   1   2   20kohms
R2   1   3   20kohms
R3   3   0   20kohms
F1   3   0   Vref   10
.AC  LIN 1  1kH 1kH
.PRINT AC VM(3)   VP(3)
.TF  V(3,0)   Vs
.OPTIONS NOPAGE
.END
```

Output File:
```
V(3,0)/Vs = -4.500E+00
INPUT RESISTANCE AT Vs = 3.077E+03
OUTPUT RESISTANCE AT V(3,0) = 1.000E+04


FREQ          VM(3)          VP(3)
1.000E+03   4.500E-01   1.800E+02
```

The above results show that the open-circuit impedance is $R_{Th} = 10$ kΩ and the open-circuit voltage is given as $V(3) = -0.45$ V. These results give us the Thévenin equivalent circuit shown in Figure 20–66(a) and the corresponding Norton equivalent circuit shown in Figure 20–66(b).

(a)   (b)

Figure 20–66

The above results are consistent with those obtained in Example 20–10.

practice
problem
IX

Write the PSpice input file to find the Norton equivalent circuit external to the load resistor in the circuit of Figure 20–31. Assume that the circuit operates at a frequency of 20 kHz. Compare your results to those obtained in Example 20–8.

Problems

Figure 20–67

20.1   Superposition Theorem—Independent Sources

1. Use superposition to determine the current in the indicated branch of the circuit in Figure 20–67.

2. Repeat Problem 1 for the circuit of Figure 20–68.

Figure 20–68

3. Consider the circuit of Figure 20–69.

Figure 20–69

a. Use superposition to determine the indicated voltage, **V**.
b. Show that the power dissipated by the indicated resistor cannot be determined by superposition.

4. Repeat Problem 3 for the circuit of Figure 20–70.

Figure 20–70

20.2   Superposition Theorem—Dependent Sources

5. Refer to the circuit of Figure 20–71.

$g = 4.0 \text{ mS}$

Figure 20–71

 a. Use superposition to find $\mathbf{V}_L$.
 b. If the magnitude of the applied voltage, $\mathbf{V}$, is increased to 200 mV, solve for the resulting $\mathbf{V}_L$.

6. Consider the circuit of Figure 20–72.

$r = 2 \text{ k}\Omega$

Figure 20–72

 a. Use superposition to find $\mathbf{V}_L$.
 b. If the magnitude of the applied current, $\mathbf{I}$, is decreased to 2 mA, solve for the resulting $\mathbf{V}_L$.

7. Use superposition to find $\mathbf{V}_L$ in the circuit of Figure 20–73.

Figure 20–73

8. Use superposition to find $\mathbf{V}_L$ in the circuit of Figure 20–74.

Figure 20–74

20.3   Thévenin's Theorem—Independent Sources

9. Find the Thévenin equivalent circuit external to the load impedance of Figure 20–67.

10. Refer to the circuit of Figure 20–75.

Figure 20–75

a. Find the Thévenin equivalent circuit external to indicated load.
b. Determine the power dissipated by the load.

11. Refer to the circuit of Figure 20–76.

Figure 20–76

a. Find the Thévenin equivalent circuit external to indicated load at a frequency of 5 kHz.
b. Determine the power dissipated by the load if $Z_L = 100\ \Omega\angle30°$.

12. Repeat Problem 11 for a frequency of 1 kHz.

13. Consider the circuit of Figure 20–77.

Figure 20–77

a. Find the Thévenin equivalent circuit external to the indicated load.
b. Determine the power dissipated by the load if $Z_L = 20\ \Omega\angle{-60}°$

14. Repeat Problem 13 if a 10-Ω resistor is placed in series with the voltage source.

20.4   Norton's Theorem—Independent Sources

15. Find the Norton equivalent circuit external to the load impedance of Figure 20–68.

16. a. Using the outlined procedure, find the Norton equivalent circuit external to the indicated load impedance in Figure 20–75.
b. Determine the current through the indicated load.
c. Find the power dissipated by the load.

17. a. Using the outlined procedure, find the Norton equivalent circuit external to the indicated load impedance (located between terminals *a* and *b*) in Figure 20–69.
b. Determine the current through the indicated load.
c. Find the power dissipated by the load.

18. Repeat Problem 17 for the circuit of Figure 20–70.

19. Suppose that the circuit of Figure 20–76 operates at a frequency of 2 kHz.
a. Find the Norton equivalent circuit external to load impedance.
b. If a 30-Ω load resistor is connected between terminals *a* and *b,* find the current through the load.

20. Repeat Problem 19 for a frequency of 8 kHz.

20.5   Thévenin's and Norton's Theorem for Dependent Sources

21. a. Find the Thévenin equivalent circuit external to the load impedance in Figure 20–71.
b. Solve for the current through R_L.
c. Determine the power dissipated by R_L.

22. a. Find the Norton equivalent circuit external to the load impedance in Figure 20–72.
b. Solve for the current through R_L.
c. Determine the power dissipated by R_L.

23. Find the Thévenin and Norton equivalent circuits external to the load impedance of Figure 20–73.

24. Find the Thévenin equivalent circuit external to the load impedance of Figure 20–74.

20.6   Maximum Power Transfer Theorem

25. Refer to the circuit of Figure 20–78.

Figure 20–78

a. Determine the load impedance, Z_L, needed to ensure that the load receives maximum power.
b. Find the maximum power to the load.

26. Repeat Problem 25 for the circuit of Figure 20–79.

27. Repeat Problem 25 for the circuit of Figure 20–80.

Figure 20–79

Figure 20–80

30 Ω

30 Ω

a

$10\ V\angle{-}40°$

Z_L

b

40 Ω

20 Ω

$0.2\ A\angle{-}70°$

Figure 20–81

28. Repeat Problem 25 for the circuit of Figure 20–81.

29. What load impedance is required for the circuit of Figure 20–70 to ensure that the load receives maximum power from the circuit?

30. Determine the load impedance required for the circuit of Figure 20–77 to ensure that the load receives maximum power from the circuit.

31. a. Determine the required load impedance, Z_L, for the circuit of Figure 20–76 to deliver maximum power to the load at a frequency of 5 kHz.
    b. If the load impedance contains a resistor and a 1-μF capacitor, determine the value of the resistor to result in a relative maximum power transfer.
    c. Solve for the power delivered to the load in (b).

32. a. Determine the required load impedance, Z_L, for the circuit of Figure 20–76 to deliver maximum power to the load at a frequency of 1 kHz.
    b. If the load impedance contains a resistor and a 1-μF capacitor, determine the value of the resistor to result in a relative maximum power transfer.
    c. Solve for the power delivered to the load in (b).

20.7 Circuit Analysis Using Computers

33. Write the PSpice input file to find the Thévenin equivalent circuit external to R_L of the circuit in Figure 20–69.

34. Write the PSpice input file to find the Norton equivalent circuit external to the load impedance of the circuit in Figure 20–70.

35. Write the PSpice input file to find the Thévenin equivalent circuit external to the load impedance of the circuit in Figure 20–73.

36. Write the PSpice input file to find the Norton equivalent circuit external to the load impedance of the circuit in Figure 20–74.

Resonance

CHAPTER PREVIEW

In this chapter, we build upon the knowledge obtained in previous chapters to observe how resonant circuits are able to pass a desired range of frequencies from a signal source to a load. In its most simple form, the **resonant circuit** consists of an inductor and a capacitor together with a voltage or current source. Although the circuit is simple, it is one of the most important circuits used in electronics. As an example, the resonant circuit, in one of its many forms, allows us to select a desired radio or television signal from the vast number of signals that are around us at any time.

Whereas there are various configurations of resonant circuits, they all have several common characteristics. Resonant electronic circuits contain at least one inductor and one capacitor and have a bell-shaped response curve centered at some resonant frequency, f_r, as illustrated in Figure 21–1.

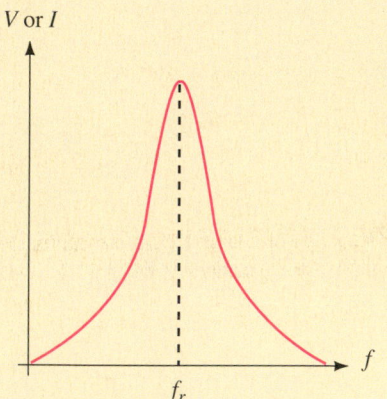

Figure 21–1 Response curve of a resonant circuit.

KEY TERMS
Bandwidth
Damped Oscillations
Half-Power Frequencies
Quality Factor
Parallel Resonance
Selectivity Curve
Series Resonance

The response curve of Figure 21–1 indicates that current or voltage will be at a maximum at the resonant frequency, f_r. Varying the frequency in either direction results in a reduction of the voltage or current.

If we were to apply variable-frequency sinusoidal signals to a circuit consisting of an inductor and capacitor, we would find that maximum energy will transfer back and forth between the two elements at the resonant frequency. In an ideal LC circuit (one containing no resistance), these oscillations would continue unabated even if the signal source were turned off. However, in the practical situation, all circuits have some resistance. As a result, the stored energy will eventually be dissipated by the resistance, resulting in **damped oscillations.** In a manner similar to pushing a child on a swing, the oscillations will continue indefinitely if a small amount of energy is applied to the circuit at exactly the right moment. This phenomenon illustrates the basis of how oscillator circuits operate and therefore provides us with another application of the resonant circuit.

In this chapter, we examine in detail the two main types of resonant circuits: the **series resonant circuit** and the **parallel resonant circuit.**

Microwave communication permits the simultaneous transmission of thousands of telephone conversations. *(Courtesy of Harris Corporation—Farinon Division.)*

OBJECTIVES

After studying this chapter, you will be able to

- determine the resonant frequency and bandwidth of a simple series or parallel circuit,

- determine the voltages, currents, and power of elements in a resonant circuit,

- sketch the impedance, current, and power response curves of a series resonant circuit,

- find the quality factor, Q, of a resonant circuit and use Q to determine the bandwidth for a given set of conditions,

- explain the dependence of bandwidth on the L/C ratio and on R for both a series and a parallel resonant circuit,

- design a resonant circuit for a given set of parameters,

- convert a series RL network into an equivalent parallel network for a given frequency.

Edwin Howard Armstrong—Radio Reception

*E*dwin Armstrong was born in New York City on December 18, 1890. As a young man, he was keenly interested in experiments involving radio transmission and reception.

After earning a degree in electrical engineering at Columbia University, Armstrong used his theoretical background to explain and improve the operation of the triode vacuum tube, which had been invented by Lee de Forest. Edwin Armstrong was able to improve the sensitivity of receivers by using feedback to amplify a signal many times. By increasing the amount of signal feedback, Armstrong also designed and patented a circuit which used the vacuum tube as an oscillator.

Armstrong is best known for conceiving the concept of superheterodyning, in which a high frequency is lowered to a more usable intermediate frequency. Superheterodyning is still used in modern AM and FM receivers and in numerous other electronic circuits such as radar and communication equipment.

Edwin Armstrong was the inventor of FM transmission, which led to greatly improved fidelity in radio transmission.

Although Armstrong was a brilliant engineer, he was an uncompromising person who was involved in numerous lawsuits with Lee de Forest and the communications giant, RCA.

After spending nearly two million dollars in legal battles, Edwin Armstrong jumped to his death from his thirteenth-floor apartment window on January 31, 1954.

21.1 Series Resonance

A simple series resonant circuit is constructed by combining an ac source with an inductor, a capacitor, and optionally, a resistor as shown in Figure 21–2a. By combining the generator resistance, R_G, with the series resistance, R_S, and the resistance of the inductor coil, R_{coil}, the circuit may be simplified as illustrated in Figure 21–2b.

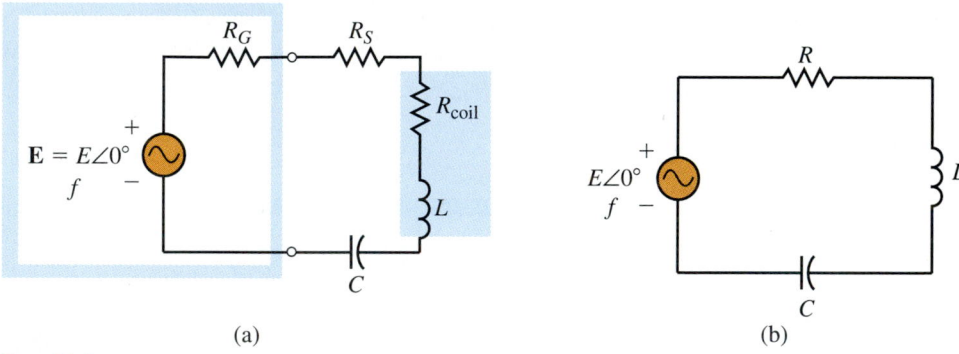

(a)   (b)

Figure 21–2

In this circuit, the total resistance is expressed as

$$R = R_G + R_S + R_{coil} \tag{21–1}$$

Because the circuit of Figure 21–2 is a series circuit, we calculate the total impedance as follows:

$$\mathbf{Z}_T = R + jX_L - jX_C$$
$$= R + j(X_L - X_C)$$

Resonance occurs when the reactance of the circuit is effectively eliminated, resulting in a total impedance that is purely resistive. We know that the reactances of the inductor and capacitor are given as follows:

$$X_L = \omega L = \pi f L \tag{21-2}$$

$$X_C = \frac{1}{\omega C} = \frac{1}{2\pi f C} \tag{21-3}$$

Examining Equation 21–1, we see that by setting the reactances of the capacitor and inductor equal to one another, the total impedance, $\mathbf{Z}_T$, is purely resistive since the inductive reactance which is on the positive j axis cancels the capacitive reactance on the negative j axis. The total impedance of the series circuit at resonance is equal to the total circuit resistance, R. Hence, at resonance,

$$Z_T = R \tag{21-4}$$

By letting the reactances be equal we are able to determine the series resonance frequency, ω_s (in radians per second) as follows:

$$\omega L = \frac{1}{\omega C}$$

$$\omega^2 = \frac{1}{LC}$$

$$\omega_s = \frac{1}{\sqrt{LC}} \quad \text{(rad/s)} \tag{21-5}$$

Since the calculation of the angular frequency, ω, in radians per second is usually easier than solving for frequency, f, in hertz, we generally express our resonant frequencies in the more simple form. Further calculations of voltage and current will usually be much easier by using ω rather than f. If, however, it becomes necessary to determine a frequency in hertz, recall that the relationship between ω and f is as follows:

$$\omega = 2\pi f \quad \text{(rad/s)} \tag{21-6}$$

Equation 21–6 is inserted into Equation 21–5 to give the resonant frequency as

$$f_s = \frac{1}{2\pi\sqrt{LC}} \quad \text{(Hz)} \tag{21-7}$$

where

f_s is the resonant frequency in hertz (Hz),
L is the inductance in henries (H),
C is the capacitance in farads (F).

The subscript s in the above equations indicates that the frequency determined is the series resonant frequency.

At resonance, the total current in the circuit is determined from Ohm's law as

$$\mathbf{I} = \frac{\mathbf{E}}{\mathbf{Z}_T} = \frac{E\angle 0°}{R\angle 0°} = \frac{E}{R}\angle 0° \tag{21-8}$$

+j (Imaginary)

V_L

I $V_R = E$

+(Real)

V_C

−j

Figure 21-3

By again applying Ohm's law, we find the voltage across each of the elements in the circuit as follows:

$$V_R = IR\angle 0° \qquad (21\text{–}9)$$

$$V_L = IX_L\angle 90° \qquad (21\text{–}10)$$

$$V_C = IX_C\angle -90° \qquad (21\text{–}11)$$

The phasor form of the voltages and current is shown in Figure 21–3.

Notice that since the inductive and capacitive reactances have the same magnitude, the voltages across the elements must have the same magnitude but be 180° out of phase.

We determine the average power dissipated by the resistor and the reactive powers of the inductor and capacitor as follows:

$$P_R = I^2 R \quad (\text{W})$$

$$Q_L = I^2 X_L \quad (\text{VAR})$$

$$Q_C = I^2 X_C \quad (\text{VAR})$$

These powers are illustrated graphically in Figure 21–4.

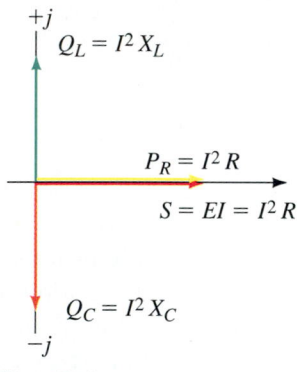

$+j$

$Q_L = I^2 X_L$

$P_R = I^2 R$

$S = EI = I^2 R$

$Q_C = I^2 X_C$

$-j$

Figure 21–4

21.2 Quality Factor, Q

For any resonant circuit, we define the *quality factor*, Q, as the ratio of reactive power to average power, namely,

$$Q = \frac{\text{reactive power}}{\text{average power}} \qquad (21\text{–}12)$$

Because the reactive power of the inductor is equal to the reactive power of the capacitor at resonance, we may express Q in terms of either reactive power. Consequently, the above expression is written as follows:

$$Q_s = \frac{I^2 X_L}{I^2 R}$$

and so we have

$$Q_s = \frac{X_L}{R} = \frac{\omega L}{R} \qquad (21\text{–}13)$$

Quite often, the inductor coil of a given circuit will have a Q expressed in terms of its reactance and internal resistance, as follows:

$$Q_{coil} = \frac{X_L}{R_{coil}}$$

If an inductor with a specified Q_{coil} is included in a circuit, it is necessary to include its effects in the overall calculation of the total circuit Q.

We now examine how the Q of a circuit is used in determining other quantities of the circuit. By multiplying both the numerator and denominator of Equation 21–13 by the current, I, we have the following:

$$Q_s = \frac{IX_L}{IR} = \frac{V_L}{E} \tag{21–14}$$

Now, since the magnitude of the voltage across the capacitor is equal to the magnitude of the voltage across the inductor at resonance, we see that the voltages across the inductor and capacitor are related to the Q by the following expression:

$$V_C = V_L = Q_s E \quad \text{at resonance} \tag{21–15}$$

Note: Since the Q of a resonant circuit is generally significantly larger than 1, we see that the voltage across reactive elements can be many times greater than the applied source voltage. Therefore, it is always necessary to ensure that the reactive elements used in a resonant circuit are able to handle the expected voltages and currents.

■ **EXAMPLE 21–1**   Find the indicated quantities for the circuit of Figure 21–5.

Figure 21–5

a. Resonant frequency expressed as ω(rad/s) and f(Hz).
b. Total impedance at resonance.
c. Current at resonance.
d. $\mathbf{V}_L$ and $\mathbf{V}_C$.
e. Reactive powers, Q_C and Q_L.
f. Quality factor of the circuit, Q_s.

Solution

a.
$$\omega_s = \frac{1}{\sqrt{LC}}$$

$$= \frac{1}{\sqrt{(10\text{ mH})(1\ \mu\text{F})}}$$

$$= 10\ 000\text{ rad/s}$$

$$f_s = \frac{\omega}{2\pi} = 1592\text{ Hz}$$

b.
$$X_L = \omega L = (10\ 000\text{ rad/s})(10\text{ mH}) = 100\ \Omega$$

$$R_{\text{coil}} = \frac{X_L}{Q_{\text{coil}}} = \frac{100\ \Omega}{50} = 2.00\ \Omega$$

$$R_T = R + R_{\text{coil}} = 10.0\ \Omega$$

$$\mathbf{Z}_T = 10\ \Omega\angle 0°$$

c.
$$\mathbf{I} = 1.0\text{ A}\angle 0°$$

d.
$$\mathbf{V}_L = (100\ \Omega\angle 90°)(1.0\text{ A}\angle 0°) = 100\text{ V}\angle 90°$$

$$\mathbf{V}_C = (100\ \Omega\angle -90°)(1.0\text{ A}\angle 0°) = 100\text{ V}\angle -90°$$

Notice that the voltage across the reactive elements is ten times greater than the applied signal voltage.

e. Although we use the symbol Q to designate both reactive power and the quality factor, the context of the question generally provides us with a clue as to which meaning to use.

$$Q_L = (1.0\text{ A})^2(100\ \Omega) = 100\text{ VAR}$$

$$Q_C = (1.0\text{ A})^2(100\ \Omega) = 100\text{ VAR}$$

f.
$$Q_s = \frac{Q_L}{P} = \frac{100\text{ VAR}}{10\text{ W}} = 10$$

practice
problem
1

Consider the circuit of Figure 21–6:

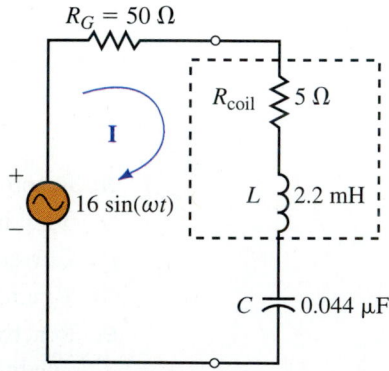

Figure 21–6

a. Find the resonant frequency expressed as ω(rad/s) and f(Hz).

b. Determine the total impedance at resonance.

c. Solve for $\mathbf{I}$, $\mathbf{V}_L$, and $\mathbf{V}_C$ at resonance.

d. Calculate reactive powers Q_C and Q_L at resonance.

e. Find the quality factor, Q_s, of the circuit.

21.3 Impedance of a Series Resonant Circuit

In this section, we examine how the impedance of a series resonant circuit varies as a function of frequency. Because the impedances of inductors and capacitors are dependent upon frequency, the total impedance of a series resonant circuit must similarly vary with frequency. For algebraic simplicity, we use frequency expressed as ω in radians per second. If it becomes necessary to express the frequency in hertz, the conversion of Equation 21–6 is used.

The total impedance of a simple series resonant circuit is written as

$$\mathbf{Z}_T = R + j\omega L - j\frac{1}{\omega C}$$

$$= R + j\left(\frac{\omega^2 LC - 1}{\omega C}\right)$$

The magnitude and phase angle of the impedance vector, $\mathbf{Z}_T$, is expressed as follows:

$$Z_T = \sqrt{R^2 + \left(\frac{\omega^2 LC - 1}{\omega C}\right)^2} \tag{21–16}$$

$$\theta = \tan^{-1}\left(\frac{\omega^2 LC - 1}{\omega RC}\right) \tag{21–17}$$

Examining these equations for various values of frequency, we note that the following conditions will apply:

When $\omega = \omega$:

$$Z_T = R$$

and

$$\theta = \tan^{-1} 0 = 0°$$

This result is consistent with the results obtained in the previous section.

When $\omega < \omega$:

As we decrease ω from resonance, Z_T will get larger until $\omega = 0$. At this point, the magnitude of the impedance will be undefined, corresponding to an open circuit. As one might expect, the large impedance occurs because the capacitor behaves like an open circuit at dc.

The angle θ will occur between of 0° and −90° since the numerator of the argument of the arctangent function will always be negative, corresponding to an angle in the fourth quadrant. Because the angle of the impedance has a negative sign, we conclude that the impedance must appear capacitive in this region.

When $\omega > \omega_s$:

As ω is made larger than resonance, the impedance Z_T will again increase due to the increasing reactance of the inductor.

For these values of ω, the angle θ will always be within 0° and +90° because the both the numerator and the denominator of the arctangent function are positive. Because the angle of $\mathbf{Z}_T$ occurs in the first quadrant, the impedance must be inductive.

Sketching the magnitude and phase angle of the impedance $\mathbf{Z}_T$ as a function of angular frequency, we have the curves shown in Figure 21–7.

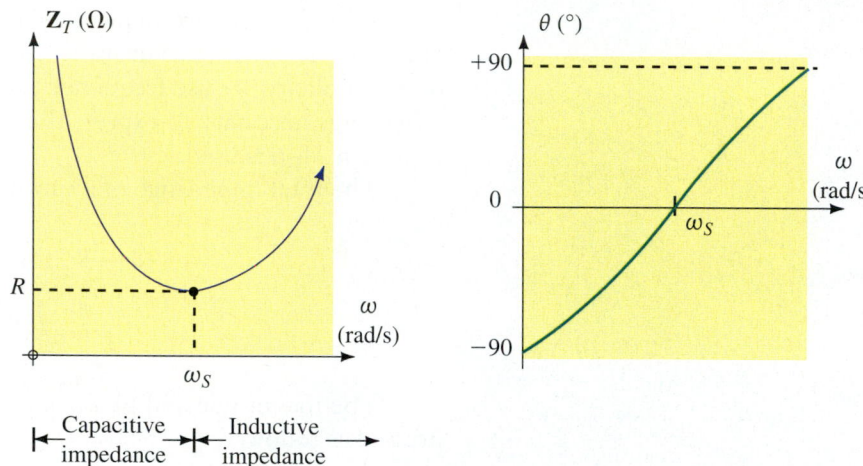

Figure 21–7   Impedance (magnitude and phase angle) versus angular frequency for a series resonant circuit.

21.4   Power, Bandwidth, and Selectivity of a Series Resonant Circuit

Due to the changing impedance of the circuit, we conclude that if a constant-amplitude voltage is applied to the series resonant circuit, the current and power of the circuit will not be constant at all frequencies. In this section, we examine how current and power are affected by changing the frequency of the voltage source.

Applying Ohm's law gives the magnitude of the current at resonance as follows:

$$I_{\max} = \frac{E}{R} \tag{21–18}$$

For all other frequencies, the magnitude of the current will be less than $I_{\max}$ because the impedance is greater than at resonance. Indeed, when the frequency is zero (dc), the current will be zero since the capacitor is effectively an open circuit. On the other hand, at increasingly higher frequencies, the inductor begins to approximate an open circuit, once again causing the current in the circuit to approach zero. The current response curve for a typical series resonant circuit is shown in Figure 21–8.

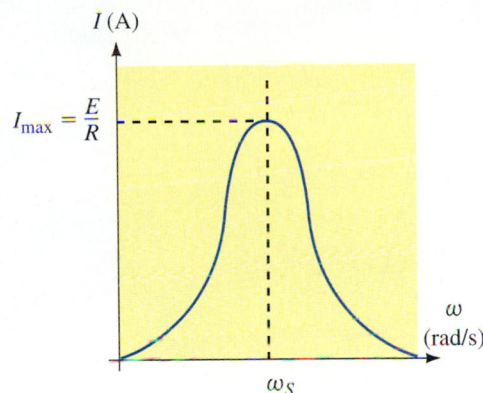

Figure 21–8   Current versus angular frequency for a series resonant circuit.

The total power dissipated by the circuit at any frequency is given as

$$P = I^2 R \qquad (21\text{–}19)$$

Since the current is maximum at resonance, it follows that the power must similarly be maximum at resonance. The maximum power dissipated by the series resonant circuit is therefore given as

$$P_{max} = I_{max}{}^2 R = \frac{E^2}{R} \qquad (21\text{–}20)$$

The power response of a series resonant circuit has a bell-shaped curve called the *selectivity curve*, which is similar to the current response. Figure 21–9 illustrates the typical selectivity curve.

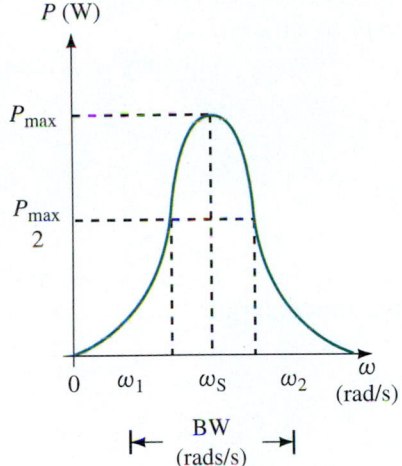

Figure 21–9   Selectivity curve.

Examining Figure 21–9, we see that only frequencies around ω_s will permit significant amounts of power to be dissipated by the circuit. We define the *bandwidth*, BW, of the resonant circuit to be the difference between the frequencies at which the circuit delivers half of the maximum power. The frequencies ω_1 and ω_2 are called the *half-power frequencies*, the *cutoff frequencies*, or the *band frequencies*.

If the bandwidth of a circuit is kept very narrow, the circuit is said to have a *high selectivity*, since it is highly selective to signals occuring within a very narrow range of frequencies. On the other hand, if the bandwidth of a circuit is large, the circuit is said to have a *low selectivity*.

The elements of a series resonant circuit determine not only the the frequency at which the circuit is resonant, but also the shape (and hence the bandwidth) of the power response curve. Consider a circuit in which the resistance, R, and the resonant frequency, ω_s, are held constant. We find that by increasing the ratio of L/C, the sides of the power response curve become steeper. This in turn results in a decrease in the bandwidth. Inversely, decreasing the ratio of L/C causes the sides of the curve become more gradual, resulting in an increased bandwidth. These characteristics are illustrated in Figure 21–10.

If, on the other hand, L and C are kept constant we find that the bandwidth will decrease as R is decreased and will increase as R is increased. Figure 21–11 shows how the shape of the selectivity curve is de-

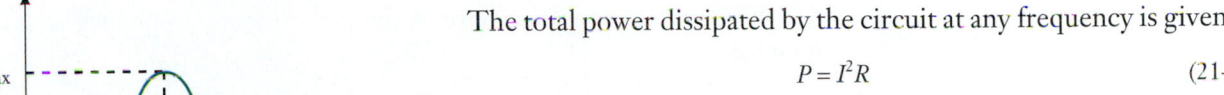

notes

W*e see from Figure 21–9 that the selectivity curve is not perfectly symmetrical on both sides of the resonant frequency. As a result, ω_s is not exactly centered between the half-power frequencies. However, as Q increases, we find that the resonant frequency approaches the midpoint between ω_1 and ω_2. In general, if $Q > 10$, then we assume that the resonant frequency is at the midpoint of half-power frequencies.*

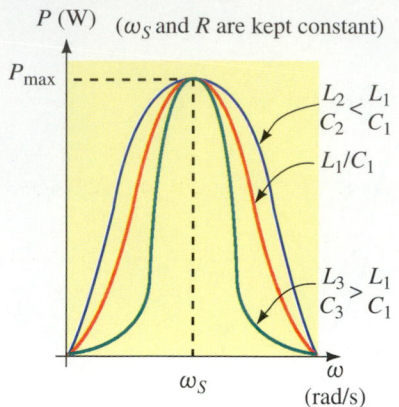

P (W)   (ω_S and R are kept constant)

P_{max}

$\dfrac{L_2}{C_2} < \dfrac{L_1}{C_1}$

L_1/C_1

$\dfrac{L_3}{C_3} > \dfrac{L_1}{C_1}$

ω_S

ω (rad/s)

Figure 21–10

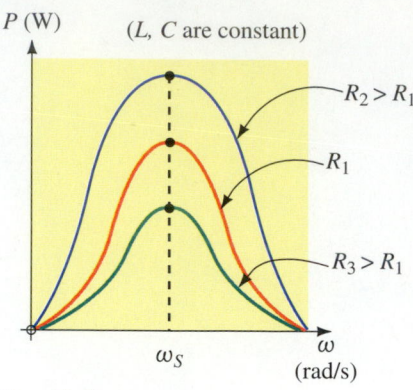

P (W)   (L, C are constant)

$R_2 > R_1$

R_1

$R_3 > R_1$

ω_S

ω (rad/s)

Figure 21–11

pendent upon the value of resistance. A series circuit has the highest selectivity if the resistance of the circuit is kept to a minimum.

For the series resonant circuit the power at any frequency is determined as

$$P = I^2 R$$

$$= \left(\frac{E}{Z_T} \right)^2 R$$

By substituting Equation 21–16 into the above expression, we arrive at the general expression for power as a function of frequency, ω:

$$P = \frac{E^2 R}{R^2 + \left(\dfrac{\omega^2 LC - 1}{\omega C} \right)^2} \qquad (21\text{–}21)$$

At the half-power frequencies, the power must be

$$P_{hpf} = \frac{E^2}{2R} \qquad (21\text{–}22)$$

Since the maximum current in the circuit is given as $I_{max} = E/R$, we see that by manipulating the above expression, the magnitude of current at the half-power frequencies is

$$I_{hpf} = \sqrt{\frac{P_{hpf}}{R}} = \sqrt{\frac{E^2}{2R^2}} = \sqrt{\frac{I_{max}^{\,2}}{2}}$$

$$I_{hpf} = \frac{I_{max}}{\sqrt{2}} \qquad (21\text{–}23)$$

The cutoff frequencies are found by evaluating the frequencies at which the power dissipated by the circuit is half of the maximum power. Combining Equations 21–21 and 21–22, we see that at the half-power frequencies we have the following:

$$\frac{E^2}{2R} = \frac{E^2 R}{R^2 + \left(\dfrac{\omega^2 LC - 1}{\omega C} \right)^2}$$

$$2R^2 = R^2 + \left(\frac{\omega^2 LC - 1}{\omega C} \right)^2$$

$$\frac{\omega^2 LC - 1}{\omega C} = \pm R$$

$$\omega^2 LC - 1 = \pm \omega RC \quad \text{(at half-power)} \tag{21-24}$$

From the selectivity curve for a series circuit, we see that the two half-power points occur on both sides of the resonant angular frequency, ω_s.

When $\omega < \omega_s$, the term $\omega^2 LC$ must be less than 1. In this case the solution is determined as follows:

$$\omega^2 LC - 1 = -\omega RC$$

$$\omega^2 LC + \omega RC - 1 = 0$$

The solution of this quadratic equation is found as

$$\omega_1 = \frac{-RC + \sqrt{(RC)^2 + 4LC}}{2LC}$$

or

$$\omega_1 = \frac{-R}{2L} + \sqrt{\frac{R^2}{4L^2} + \frac{1}{LC}} \tag{21-25}$$

In a similar manner, for $\omega > \omega_s$, the half-power frequency is

$$\omega_2 = \frac{R}{2L} + \sqrt{\frac{R^2}{4L^2} + \frac{1}{LC}} \tag{21-26}$$

Taking the difference between Equations 21–26 and 21–25, we find the bandwidth of the circuit as

$$\text{BW} = \omega_2 - \omega_1$$

$$= \frac{R}{2L} + \sqrt{\frac{R^2}{4L^2} + \frac{1}{LC}} - \left(-\frac{R}{2L} + \sqrt{\frac{R^2}{4L^2} + \frac{1}{LC}} \right)$$

which gives

$$\text{BW} = \frac{R}{L} \quad \text{(rad/s)} \tag{21-27}$$

If the above expression is multiplied by ω_s/ω_s we obtain

$$\text{BW} = \frac{\omega_s R}{\omega_s L}$$

and since $Q_s = \omega_s L/R$ we further simplify the bandwidth as

$$\text{BW} = \frac{\omega_s}{Q_s} \quad \text{(rad/s)} \tag{21-28}$$

Because the bandwidth may alternately be expressed in hertz, the above expression is equivalent to having

$$\text{BW} = \frac{f_s}{Q_s} \quad \text{(Hz)} \tag{21-29}$$

■ **EXAMPLE 21–2** Refer to the circuit of Figure 21–12.

Figure 21–12

a. Determine the maximum power dissipated by the circuit.
b. Use the results obtained from Example 21–1 to determine the bandwidth of the resonant circuit and to arrive at the approximate half-power frequencies, ω_1 and ω_2.
c. Calculate the actual half-power frequencies, ω_1 and ω_2, from the given component values.
d. Solve for the circuit current, **I**, and power dissipated at the lower half-power frequency, ω_1, found in Part (c).

Solution:

a.
$$P_{\text{max}} = \frac{E^2}{R} = 10.0 \text{ W}$$

b. From Example 21–1, we had the following circuit characteristics:

$$Q_s = 10, \quad \omega_s = 10 \text{ krad/s}$$

The bandwidth of the circuit is determined to be

$$BW = \omega_s Q_s = 1.0 \text{ krad/s}$$

If the resonant frequency were centered in the bandwidth, then the half-power frequencies occur at approximately

$$\omega_1 = 9.50 \text{ krad/s}$$

and

$$\omega_2 = 10.50 \text{ krad/s}$$

c.
$$\omega_1 = -\frac{R}{2L} + \sqrt{\frac{R^2}{4L^2} + \frac{1}{LC}}$$

$$= -\frac{10 \text{ } \Omega}{(2)(10 \text{ mH})} + \sqrt{\frac{(10 \text{ } \Omega)^2}{(4)(10 \text{ mH})^2} + \frac{1}{(10 \text{ mH})(1 \text{ } \mu\text{F})}}$$

$$= -500 + 10\,012.49 = 9512.49 \text{ rad/s}$$

$$\omega_2 = \frac{R}{2L} + \sqrt{\frac{R^2}{4L^2} + \frac{1}{LC}}$$

$$= 500 + 10\,012.49 = 10\,512.49 \text{ rad/s}$$

Notice that the actual half-power frequencies are very nearly equal to the approximate values. For this reason, if $Q \geq 10$, it is often sufficient to calculate the cutoff frequencies by using the easier approach of Part (b).

d. At $\omega_1 = 9.51249$ krad/s, the reactances are as follows:

$$X_L = \omega L = (9.51249 \text{ krad/s})(10 \text{ mH}) = 95.12 \ \Omega$$

$$X_C = \frac{1}{\omega C} = \frac{1}{(9.51249 \text{ krad/s})(1 \ \mu\text{F})} = 105.12 \ \Omega$$

The current is now determined to be

$$\mathbf{I} = \frac{10 \text{ V} \angle 0°}{10 \ \Omega + j95.12 \ \Omega - j105.12 \Omega}$$

$$= \frac{10 \text{ V} \angle 0°}{14.14 \ \Omega \angle -45°}$$

$$= 0.707 \text{ A} \angle 45°$$

and the power is given as

$$P = I^2 R = (0.707 \text{ A})^2 (10 \ \Omega) = 5.0 \text{ W}$$

As expected, we see that the power at the frequency ω_1 is indeed equal to half of the power dissipated by the circuit at resonance.

■ **EXAMPLE 21–3**   Refer to the circuit of Figure 21–13.

Figure 21–13

a. Calculate the values of R_L and C for the circuit to have a resonant frequency of 200 kHz and a bandwidth of 16 kHz.

b. Use the designed component values to determine the power dissipated by the circuit at resonance.

c. Solve for $v_{out}(t)$ at resonance.

Solution

a. Because the circuit is at resonance, we must have the following conditions:

$$Q_s = \frac{f_s}{\text{BW}}$$

$$= \frac{200 \text{ kHz}}{16 \text{ kHz}}$$

$$= 12.5 \text{ kHz}$$

$$X_L = 2\pi f L$$

$$= 2\pi (200 \text{ kHz})(200 \text{ μH})$$

$$= 251.3 \text{ Ω}$$

$$R = R_L + R_{\text{coil}} = \frac{X_L}{Q_s}$$

$$= 20.1 \text{ Ω}$$

and so R_L must be

$$R_L = 20.1 \text{ Ω} - 5 \text{ Ω} = 15.1 \text{ Ω}$$

Since $X_C = X_L$, we determine the capacitance as

$$C = \frac{1}{2\pi f X_C}$$

$$= \frac{1}{2\pi (200 \text{ kHz})(251.3 \text{ Ω})}$$

$$= 3.17 \text{ nF } (\equiv 0.00317 \text{ μF})$$

b. The power at resonance is found from Equation 21–20 as

$$P_{\text{max}} = \frac{E^2}{R} = \frac{\left(\dfrac{30 \text{ V}}{\sqrt{2}}\right)^2}{20.1 \text{ Ω}}$$

$$= 22.4 \text{ W}$$

c. We see from the circuit of Figure 21–13 that the voltage $v_{\text{out}}(t)$ may be determined by applying the voltage divider rule to the circuit. However, we must first convert the source voltage from time domain into phasor domain as follows:

$$e(t) = 30 \sin \omega t \Leftrightarrow \mathbf{E} = 21.21 \text{ V} \angle 0°$$

Now, applying the voltage divider rule to the circuit, we have

$$\mathbf{V}_{\text{out}} = \frac{(R_1 + j\omega L)}{\mathbf{R}} \mathbf{E}$$

$$= \frac{(5 \text{ Ω} + j251.3 \text{ Ω})}{20.1 \text{ Ω}\angle 0°} 21.21 \text{ V} \angle 0°$$

$$= (251.4 \text{ Ω}\angle 88.86°)(1.056 \text{ A}\angle 0°)$$

$$= 265.5 \text{ V}\angle 88.86°$$

which in time domain is given as

$$v_{\text{out}}(t) = 375 \sin(\omega t + 88.86°) \quad \text{(V)}$$

Refer to the circuit of Figure 21–14.

Figure 21–14

a. Determine the maximum power dissipated by the circuit.

b. Use the results obtained from Practice Problem I to determine the bandwidth of the resonant circuit. Solve for the approximate values of the half-power frequencies, ω_1 and ω_2.

c. Calculate the actual half-power frequencies, ω_1 and ω_2, from the given component values. Compare your results to those obtained in Part (b). Briefly explain why there is a discrepancy between the results.

d. Solve for the circuit current, **I**, and power dissipated at the lower half-power frequency, ω_1, found in Part (c).

Refer to the series resonant circuit of Figure 21–15.

f_S = 600 kHz   220 pF
BW = 10 kHz

Figure 21–15

Suppose the circuit has a resonant frequency of 600 kHz and a bandwidth of 10 kHz:

a. Determine the value of inductor L in henries.

b. Calculate the value of resistor R in ohms.

c. Find **I**, $\mathbf{V}_L$, and power, P, at resonance.

d. Find the appoximate values of the half-power frequencies, f_1 and f_2.

e. Using the results of Part (d), determine the current in the circuit at the lower half-power frequency, f_1, and show that the power dissipated by the resistor at this frequency is half the power dissipated at the resonant frequency.

Consider the series resonant circuit of Figure 21–16:

$R = 20 \ \Omega$

$100 \ \text{mV} \angle 0°$

L 20 mH

C

$0.5 \ \mu\text{F}$

Figure 21–16

a. Solve for the resonant frequency of the circuit, ω_s, and calculate the power dissipated by the circuit at resonance.

b. Determine Q, BW, and the half-power frequencies, ω_1 and ω_2, in radians per second.

c. Sketch the selectivity curve of the circuit, showing P (in watts) versus ω (in radians per second).

d. Repeat Parts (a) through (c) if the value of resistance is reduced to 10 Ω.

e. Explain briefly how selectivity depends upon the value of resistance in a series resonant circuit.

21.5 Series-to-Parallel *RL* and *RC* Conversion

As we have already seen, an inductor will always have some series resistance due to the length of wire used the coil winding. Even though the resistance of the wire is generally small in comparison with the reactances in the circuit, this resistance may occasionally contribute tremendously to the overall circuit response of a parallel resonant circuit. We begin by converting the series *RL* network as shown in Figure 21–17 into an equivalent parallel *RL* network. It must be emphasized, however, that *the equivalence is only valid at a single frequency, ω.*

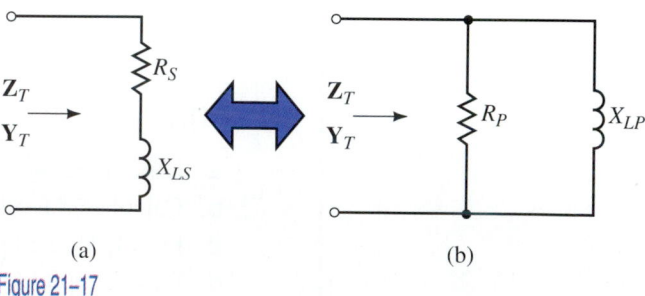

$\mathbf{Z}_T$
$\mathbf{Y}_T$
R_S
X_{LS}

$\mathbf{Z}_T$
$\mathbf{Y}_T$
R_P
X_{LP}

(a)   (b)

Figure 21–17

The networks of Figure 21–17 can be equivalent only if they each have the same input impedance, $\mathbf{Z}_T$ (and also the same input admittance, $\mathbf{Y}_T$).

The input impedance of the series network of Figure 21–17(a) is given as

$$\mathbf{Z}_T = R_S + jX_{LS}$$

which gives the input admittance as

$$\mathbf{Y}_T = \frac{1}{\mathbf{Z}_T} = \frac{1}{R_S + jX_{LS}}$$

Multiplying numerator and denominator by the complex conjugate, we have

$$Y_T = \frac{R_S - jX_{LS}}{(R_S + jX_{LS})(R_S - jX_{LS})}$$

$$= \frac{R_S - jX_{LS}}{R_S^2 + X_{LS}^2}$$

$$= \frac{R_S}{R_S^2 + X_{LS}^2} - j\frac{X_{LS}}{R_S^2 + X_{LS}^2} \qquad (21\text{–}30)$$

From Figure 21–17(b), we see that the input admittance of the parallel network must be

$$\mathbf{Y}_T = G_P - jB_{LP}$$

which may also be written as

$$\mathbf{Y}_T = \frac{1}{R_P} - j\frac{1}{X_{LP}} \qquad (21\text{–}31)$$

The admittances of Equations 21–30 and 21–31 can only be equal if the real and the imaginary components are equal. As a result, we see that for a given frequency, the following equations enable us to convert a series *RL* network into its equivalent parallel network:

$$R_P = \frac{R_S^2 + X_{LS}^2}{R_S} \qquad (21\text{–}32)$$

$$X_{LP} = \frac{R_S^2 + X_{LS}^2}{X_{LS}} \qquad (21\text{–}33)$$

If we were given a parallel *RL* network, it is possible to show that the conversion to an equivalent series network is accomplished by applying the following equations:

$$R_S = \frac{R_P X_{LP}^2}{R_P^2 + X_{LP}^2} \qquad (21\text{–}34)$$

$$X_{LS} = \frac{R_P^2 X_{LP}}{R_P^2 + X_{LP}^2} \qquad (21\text{–}35)$$

The derivation of the above equations is left as an exercise for the student.

Equations 21–32 to 21–35 may be simplified by using the quality factor of the coil. Multiplying Equation 21–32 by R_S/R_S and then using Equation 21–13, we have

$$R_P = R_S \frac{R_S^2 + X_{LS}^2}{R_S^2}$$

$$R_P = R_S(1 + Q^2) \tag{21-36}$$

Equation 21–33 is also simplified as

$$X_{LP} = X_{LS} \frac{R_S^{\ 2} + X_{LS}^{\ 2}}{X_{LS}^{\ 2}}$$

$$X_{LP} = X_{LS}\left(1 + \frac{1}{Q^2}\right) \tag{21-37}$$

The quality factor of the resulting parallel network must be the same as for the original series network because the reactive and the average powers must be the same. Using the parallel elements, the quality factor is expressed as

$$Q = \frac{X_{LS}}{R_S}$$

$$= \frac{\left(\dfrac{R_P^{\ 2} X_{LP}}{R_P^{\ 2} + X_{LP}^{\ 2}}\right)}{\left(\dfrac{R_P X_{LP}^{\ 2}}{R_P^{\ 2} + X_{LP}^{\ 2}}\right)}$$

$$= \frac{R_P^{\ 2} X_{LP}}{R_P X_{LP}^{\ 2}}$$

$$Q = \frac{R_P}{X_{LP}} \tag{21-38}$$

■ **EXAMPLE 21–4**   For the series network of Figure 21–18, find the Q of the coil at $\omega = 1000$ rad/s and convert the series *RL* network into its equivalent parallel network. Repeat the above steps for $\omega = 10$ krad/s.

Figure 21–18

Solution   For $\omega = 1000$ rad/s,

$$X_L = \omega L = 20\ \Omega$$

$$Q = \frac{X_{LS}}{R_S} = 2.0$$

$$R_P = R_S(1 + Q^2) = 50\ \Omega$$

$$X_{LP} = X_{LS}\left(1 + \frac{1}{Q^2}\right) = 25\ \Omega$$

The resulting parallel network for $\omega = 1000$ rad/s is shown in Figure 21–19.

Figure 21–19

For $\omega = 10$ krad/s,

$$X_L = \omega L = 200 \ \Omega$$

$$Q = \frac{X_{LS}}{R_S} = 20$$

$$R_P = R_S(1 + Q^2) = 4010 \ \Omega$$

$$X_{LP} = X_{LS}\left(1 + \frac{1}{Q^2}\right) = 200.5 \ \Omega$$

The resulting parallel network for $\omega = 10$ krad/s is shown in Figure 21–20.

Figure 21–20

■ **EXAMPLE 21–5**   Find the *Q* of each of the networks of Figure 21–21 and determine the series equivalent of each.

Figure 21–21

Solution   For the network of Figure 21–21(a),

$$Q = \frac{R_P}{X_{LP}} = \frac{10 \text{ k}\Omega}{250 \ \Omega} = 40$$

$$R_S = \frac{R_P}{1 + Q^2} = \frac{10 \text{ k}\Omega}{1 + 40^2} = 6.25 \ \Omega$$

$$X_{LS} = QR_S = (40)(6.25 \ \Omega) = 250 \ \Omega$$

$$L = \frac{X_L}{\omega} = \frac{250 \ \Omega}{2500 \text{ rad/s}} = 0.1 \text{ H}$$

For the network of Figure 21–21(b),

$$Q = \frac{R_P}{X_{LP}} = \frac{2700 \ \Omega}{900 \ \Omega} = 3$$

$$R_S = \frac{R_P}{1 + Q^2} = \frac{2700 \ \Omega}{1 + 3^2} = 270 \ \Omega$$

$$X_{LS} = QR_S = (3)(270 \ \Omega) = 810 \ \Omega$$

$$L = \frac{X_L}{\omega} = \frac{810 \ \Omega}{2000 \ \text{rad/s}} = 0.405 \ \text{H}$$

The resulting equivalent series networks are shown in Figure 21– 22.

(a)   (b)

Figure 21–22

practice
problem
III

Refer to the networks of Figure 21–23.

100 Ω   50 mH   2 kΩ   0.4 H

(a)   (b)

Figure 21–23

a. Find the quality factors, Q, of the networks at $\omega_1 = 5$ krad/s.

b. Use the Q to find the equivalent parallel networks (resistance and reactance) at an angular frequency of $\omega_1 = 5000$ rad/s.

c. Repeat Parts (a) and (b) for an angular frequency of $\omega_2 = 25$ krad/s.

The previous examples illustrate two important points which are valid if the Q of the network is large ($Q \geq 10$).

1. The resistance of the parallel network is approximately Q^2 larger than the resistance of the series network.

2. The inductive reactances of the series and parallel networks are approximately equal. Hence

$$R_P \cong Q^2 R_S \quad (Q \geq 10) \tag{21–39}$$

$$X_{LP} \cong X_{LS} \quad (Q \geq 10) \tag{21–40}$$

Although we have performed conversions between series and parallel *RL* circuits, it is easily shown that if the reactive element is a capacitor, the conversions apply equally well. In all cases, the equations are simply changed by replacing the terms X_{LS} and X_{LP} with X_{CS} and X_{CP} respectively. The *Q* of the network is determined by the ratios

$$Q = \frac{X_{CS}}{R_S} = \frac{R_P}{X_{CP}} \qquad (21\text{--}41)$$

practice
problem
_____*IV*

Consider the networks of Figure 21–24:

(a)   (b)

Figure 21–24

a. Find the *Q* of each network at a frequency of $f_1 = 1$ kHz.

b. Determine the series equivalent of the network in Figure 21–24(a) and the parallel equivalent of the network in Figure 21–24(b).

c. Repeat Parts (a) and (b) for a frequency of $f_2 = 200$ kHz.

in process
l e a r n i n g
check III

Refer to the networks of Figure 21–25:

(a)   (b)

Figure 21–25

a. Determine the resistance, R_P, for each network.

b. Find the equivalent series network by using the quality factor for the given networks.

21.6 Parallel Resonance

A simple parallel resonant circuit is illustrated in Figure 21– 26. The parallel resonant circuit is best analyzed using a constant-current source, unlike the series resonant circuit which used a constant-voltage source.

Figure 21-26   Simple parallel resonant circuit.

Figure 21-27

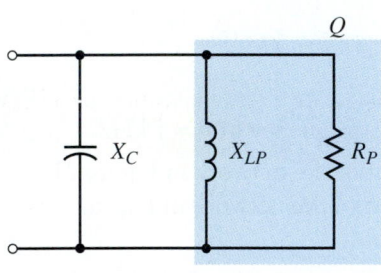

Figure 21-28

Consider the *LC* "tank" circuit shown in Figure 21–27. The tank circuit consists of a capacitor in parallel with an inductor. Due to its high *Q* and frequency response, the tank circuit is used extensively in communications equipment such as AM, FM, and television transmitters and receivers.

The circuit of Figure 21–27 is not exactly a parallel resonant circuit, since the resistance of the coil is in series with the inductance. In order to determine the frequency at which the circuit is purely resistive, we must first convert the series combination of resistance and inductance into an equivalent parallel network. The resulting circuit is shown in Figure 21–28.

At resonance, the capacitive and inductive reactances of the circuit of Figure 21–28 are equal. As we have observed previously, placing equal inductive and capacitive reactances in parallel effectively results in an open circuit at the given frequency. The input impedance of the above network at resonance is therefore purely resisitive and given as $Z_T = R_P$. We determine the resonant frequency of a tank circuit by first letting the reactances of the equivalent parallel circuit be equal:

$$X_C = X_{LP}$$

Now, using the component values of the tank circuit, we have

$$X_C = \frac{(R_{coil})^2 + X_L^2}{X_{LS}}$$

$$\frac{1}{\omega C} = \frac{(R_{coil})^2 + (\omega L)^2}{\omega L}$$

$$\frac{L}{C} = (R_{coil})^2 + (\omega L)^2$$

which may be further reduced to

$$\omega = \sqrt{\frac{1}{LC} - \frac{R^2}{L^2}}$$

Factoring $\sqrt{LC}$ from the denominator, we express the parallel resonant frequency as

$$\omega_P = \frac{1}{\sqrt{LC}} \sqrt{1 - \frac{(R_{coil})^2 C}{L}} \qquad (21\text{–}42)$$

Notice that if $R_{coil}^2 \ll L/C$, then the term under the radical is approximately equal to 1.

Consequently, if $L/C \geq 100 R_{coil}$, the parallel resonant frequency may be simplified as

$$\omega_P \cong \frac{1}{\sqrt{LC}} \qquad \text{(for } L/C \geq 100 R_{coil}\text{)} \qquad (21\text{--}43)$$

Recall that the quality factor, Q, of a circuit is defined as the ratio of reactive power to average power for a circuit at resonance. If we consider the parallel resonant circuit of Figure 21–29, we make several important observations.

Figure 21–29

The inductor and capacitor reactances cancel, resulting in a circuit voltage simply determined by Ohm's law as

$$\mathbf{V} = \mathbf{I}R = IR\angle 0°$$

The frequency response of the impedance of the parallel circuit is shown in Figure 21–30.

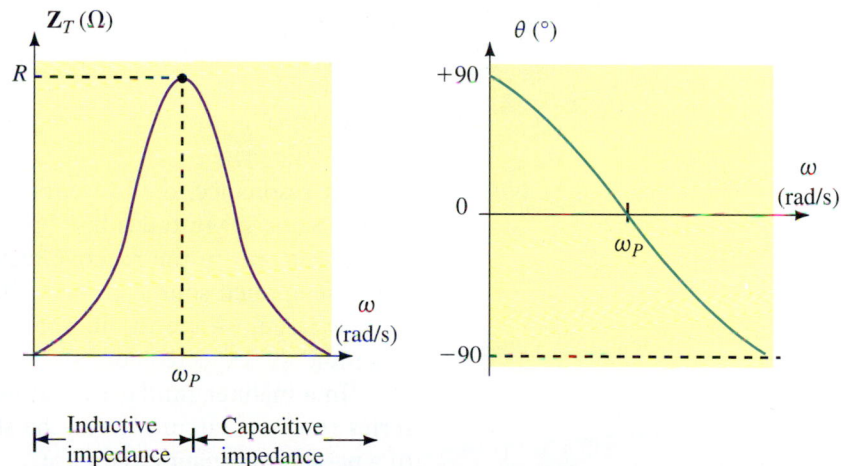

Figure 21–30   Impedance (magnitude and phase angle) versus angular frequency for a parallel resonant circuit.

Notice that the impedance of the entire circuit is maximum at resonance and minimum at the boundary conditions ($\omega = 0$ rad/s and $\omega \rightarrow \infty$). This result is exactly opposite to that observed in series resonant circuits which have minimum impedance at resonance. We also see that for parallel circuits, the impedance will appear inductive for frequencies less than the resonant frequency, ω_P. Inversely, the impedance is capacitive for frequencies greater than ω_P.

The Q of the parallel circuit is determined from the definition as

$$Q_P = \frac{\text{reactive power}}{\text{average power}}$$

$$= \frac{V^2/X_L}{V^2/R}$$

$$Q_P = \frac{R_L}{X_L} = \frac{R}{X_C} \tag{21-44}$$

This is precisely the same result as that obtained when we converted an RL series network into its equivalent parallel network. If the resistance of the coil is the only resistance within a circuit, then the circuit Q will be equal to the Q of the coil. However, if the circuit has other sources of resistance, then the additional resistance will reduce the circuit Q.

For a parallel RLC resonant circuit, the currents in the various elements are found from Ohm's law as follows:

$$\mathbf{I}_R = \frac{\mathbf{V}}{\mathbf{R}} = \mathbf{I} \tag{21-45}$$

$$\mathbf{I}_L = \frac{\mathbf{V}}{\mathbf{X}_L}$$

$$= \frac{V}{R/Q_P} \angle{-90°}$$

$$= Q_P I \angle{-90°} \tag{21-46}$$

$$\mathbf{I}_C = \frac{\mathbf{V}}{\mathbf{X}_C}$$

$$= \frac{V}{R/Q_P} \angle{90°}$$

$$= Q_P I \angle{90°} \tag{21-47}$$

At resonance, the currents through the inductor and the capacitor are the same magnitudes but 180° out of phase. Notice that the magnitude of current in the reactive elements at resonance is Q times greater than the applied source current. Because the Q of a parallel circuit may be very large, we see the importance of choosing elements that are able to handle the expected currents.

In a manner similar to that used in determining the bandwidth of a series resonant circuit, it may be shown that the half-power frequencies of a parallel resonant circuit are

$$\omega_1 = \frac{1}{2RC} + \sqrt{\frac{1}{4R^2C^2} + \frac{1}{LC}} \quad \text{(rad/s)} \tag{21-48}$$

$$\omega_2 = \frac{1}{2RC} + \sqrt{\frac{1}{4R^2C^2} + \frac{1}{LC}} \quad \text{(rad/s)} \tag{21-49}$$

The bandwidth is therefore

$$\text{BW} = \omega_2 - \omega_1 = \frac{1}{RC} \quad \text{(rad/s)} \tag{21-50}$$

If $Q \geq 10$, then the selectivity curve is very nearly symmetrical around ω_P, resulting in half-power frequencies which are located at $\omega_P \pm BW/2$.

Multiplying Equation 21–50 by ω_P / ω_P results in the following:

$$BW = \frac{\omega_P}{R(\omega_P C)} = \frac{X_C}{R}\omega_P$$

$$BW = \frac{\omega_P}{Q_P} \quad \text{(rad/s)} \tag{21–51}$$

Notice that Equation 21–51 is the same for both series and parallel resonant circuits.

■ **EXAMPLE 21–6**   Consider the circuit shown in Figure 21–31.

Figure 21–31

a. Determine the resonant frequencies, ω_r(rad/s) and f_r(Hz) of the tank circuit.

b. Find the Q of the circuit at resonance.

c. Calculate the voltage across the circuit at resonance.

d. Solve for currents through the inductor and the resistor at resonance.

e. Determine the bandwidth of the circuit in both radians per second and hertz.

f. Sketch the voltage response of the circuit, showing the voltage at the half-power frequencies.

g. Sketch the selectivity curve of the circuit showing P(watts) versus ω(rad/s).

Solution

a. $\omega_P = \dfrac{1}{\sqrt{LC}} = \dfrac{1}{\sqrt{(16\text{ mH})(0.4\ \mu\text{F})}} = 12.5\text{ krad/s}$

$f_P = \dfrac{\omega}{2\pi} = \dfrac{12.5\text{ krad/s}}{2\pi} = 1989\text{ Hz}$

b. $Q_P = \dfrac{R_P}{\omega L} = \dfrac{500\ \Omega}{(12.5\text{ krad/s})(16\text{ mH})} = \dfrac{500\ \Omega}{200\ \Omega} = 2.5$

c. At resonance, $\mathbf{V}_C = \mathbf{V}_L = \mathbf{V}_R$, and so

$\mathbf{V} = \mathbf{I}R = (3.6\text{ mA}\angle 0°)(500\ \Omega\angle 0°) = 1.8\text{ V}\angle 0°$

d.
$$I_L = \frac{\mathbf{V}_L}{\mathbf{X}_L} = \frac{1.8\ \mathrm{V}\angle 0°}{200\ \Omega\angle 90°} = 9.0\ \mathrm{mA}\angle -90°$$

$$\mathbf{I}_R = \mathbf{I} = 3.6\ \mathrm{mA}\angle 0°$$

e.
$$\mathrm{BW(rad/s)} = \frac{\omega_P}{Q_P} = \frac{12.5\ \mathrm{krad/s}}{2.5} = 5\ \mathrm{krad/s}$$

$$\mathrm{BW(Hz)} = \frac{\mathrm{BW(rad/s)}}{2\pi} = \frac{5\ \mathrm{krad/s}}{2\pi} = 795.8\ \mathrm{Hz}$$

f. The half-power frequencies are calculated from Equations 21–48 and 21–49 since the Q of the circuit is less than 10.

$$\omega_1 = -\frac{1}{2RC} + \sqrt{\frac{1}{4R^2C^2} + \frac{1}{LC}}$$

$$= -\frac{1}{0.0004} + \sqrt{\frac{1}{1.6 \times 10^{-7}} + \frac{1}{6.4 \times 10^{-9}}}$$

$$= -2500 + 12\ 748$$

$$= 10\ 248\ \mathrm{krad/s}$$

$$\omega_2 = \frac{1}{2RC} + \sqrt{\frac{1}{4R^2C^2} + \frac{1}{LC}}$$

$$= \frac{1}{1.0004} + \sqrt{\frac{1}{1.6 \times 10^{-7}} + \frac{1}{6.4 \times 10^{-9}}}$$

$$= 2500 + 12\ 748$$

$$= 15.248\ \mathrm{krad/s}$$

The resulting voltage response curve is illustrated in Figure 21–32.

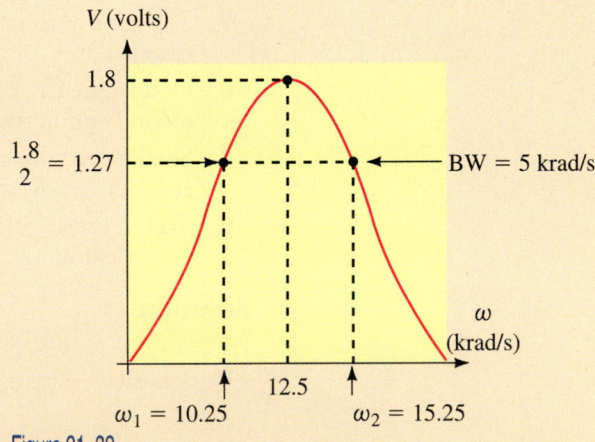

Figure 21–32

g. The power dissipated by the circuit at resonance is

$$P = \frac{V^2}{R} = \frac{(1.8\ \mathrm{V})^2}{500\ \Omega} = 6.48\ \mathrm{mW}$$

The selectivity curve is now easily sketched as shown in Figure 21–33.

Figure 21–33

■ **EXAMPLE 21–7**   Consider the circuit of Figure 21–34.

Figure 21–34

a. Calculate the resonant frequency, ω_r, of the tank circuit.
b. Find the Q of the coil at resonance.
c. Sketch the equivalent parallel circuit.
d. Determine the Q of the entire circuit at resonance.
e. Solve for the voltage across the capacitor at resonance.
f. Find the bandwidth of the circuit in radians per second.
g. Sketch the voltage response of the circuit showing the voltage at the half-power frequencies.

Solution

a. Since the ratio $L/C = 1000 \geq 100R$, we use the approximation:

$$\omega_r = \frac{1}{\sqrt{LC}} = \frac{1}{\sqrt{(1 \text{ mH})(1 \text{ nF})}} = 1 \text{ Mrad/s}$$

b.

$$Q_{coil} = \frac{\omega L}{R_{coil}} = \frac{(1 \text{ Mrad/s})(1 \text{ mH})}{10 \ \Omega} = 100$$

c. $$R_P \cong Q^2 R_{\text{coil}} = (100)^2 (10\ \Omega) = 100\ \text{k}\Omega$$

$$X_{LP} \cong X_{LS} = \omega L = (1\ \text{Mrad/s})\ (1\ \text{mH}) = 1\ \text{k}\Omega$$

The circuit of Figure 21–35 shows the circuit with the parallel equivalent of the inductor.

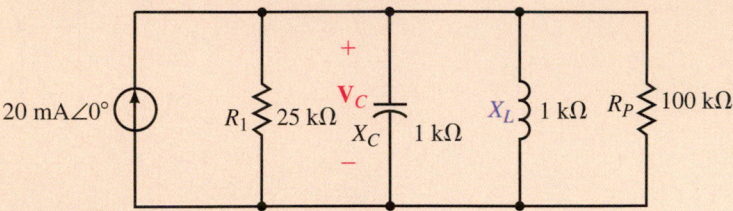

Figure 21–35

We see that the previous circuit may be further simplified by combining the parallel resistances:

$$R_{\text{eq}} = R_1 \| R_P = \frac{(25\ \text{k}\Omega)(100\ \text{k}\Omega)}{25\ \text{k}\Omega + 100\ \text{k}\Omega} = 20\ \text{k}\Omega$$

The simplified equivalent circuit is shown in Figure 21–36.

Figure 21–36

d. $$Q = \frac{R_{\text{eq}}}{X_L} = \frac{20\ \text{k}\Omega}{1\ \text{k}\Omega} = 20$$

e. At resonance,

$$\mathbf{V}_C = \mathbf{I} R_{\text{eq}} = (20\ \text{mA}\angle 0°)\ (20\ \text{k}\Omega) = 400\ \text{V}\angle 0°$$

f. $$\text{BW} = \frac{\omega_r}{Q} = \frac{1\ \text{Mrad/s}}{20} = 50\ \text{krad/s}$$

g. The voltage response curve is shown in Figure 21–37. Since $Q \geq 10$, the half-power frequencies will occur at the following angular frequencies:

$$\omega_1 \cong \omega_r - \frac{\text{BW}}{2} = 1.0\ \text{Mrad/s} - \frac{50\ \text{krad/s}}{2} = 0.975\ \text{Mrad/s}$$

and

$$\omega_2 \cong \omega_r + \frac{\text{BW}}{2} = 1.0\ \text{Mrad/s} + \frac{50\ \text{krad/s}}{2} = 1.025\ \text{Mrad/s}$$

Figure 21–37

■ **EXAMPLE 21–8** Determine the values of R_1 and C for the resonant tank circuit of Figure 21–38 so that the given conditions are met. Solve for the current, $\mathbf{I}_L$, through the inductor.

Figure 21–38

$L = 10$ mH, $R_{coil} = 30 \ \Omega$

$f_P = 58$ kHz

BW $= 1$ kHz

Solution

$$Q = \frac{f_P}{\text{BW(Hz)}} = \frac{580 \text{ kHz}}{10 \text{ kHz}} = 58$$

Now, because the frequency expressed in radians per second is more useful than hertz, we convert f_P to ω_P:

$$\omega_P = 2\pi f_P = (2\pi)(58 \text{ kHz}) = 364.4 \text{ krad/s}$$

The capacitance is determined from Equation 21–43 as

$$C = \frac{1}{\omega_P^2 L} = \frac{1}{(364.4 \text{ krad/s})^2 (10 \text{ mH})} = 753 \text{ pF}$$

Solving for the Q of the coil permits us to easily convert the series RL network into its equivalent parallel network.

$$Q_{\text{coil}} = \frac{\omega_L}{R_{\text{coil}}}$$

$$= \frac{(364.4 \text{ krad/s})(10 \text{ mH})}{30 \ \Omega}$$

$$= \frac{3.644 \text{ k}\Omega}{30 \ \Omega} = 121.5$$

$$R_P \cong Q_{\text{coil}}^2 \ R_S = (121.5)^2(30 \ \Omega) = 443 \text{ k}\Omega$$

$$X_{LP} \cong X_{LS} = 3644 \ \Omega$$

The resulting equivalent parallel circuit is shown in Figure 21–39.

Figure 21–39

The quality factor, Q, is used to determine the total resistance of the circuit as

$$R = QX_C = (58)(3.644 \text{ k}\Omega) = 211 \text{ k}\Omega$$

But

$$\frac{1}{R} = \frac{1}{R_1} + \frac{1}{R_P}$$

$$\frac{1}{R_1} = \frac{1}{R} - \frac{1}{R_P} = \frac{1}{211 \text{ k}\Omega} - \frac{1}{443 \text{ k}\Omega} = 2.47 \ \mu\text{S}$$

And so

$$R_1 = 405 \text{ k}\Omega$$

The voltage across the circuit is determined to be

$$\mathbf{V} = \mathbf{I}R = (10 \ \mu\text{A}\angle0°)(211 \text{ k}\Omega) = 2.11 \text{ V}\angle0°$$

and the current through the inductor is

$$\mathbf{I}_L = \frac{\mathbf{V}}{R_{\text{coil}} + jX_L}$$

$$= \frac{2.11 \text{ V}\angle0°}{30 + j3644 \ \Omega} = \frac{2.11 \text{ V}\angle0°}{3644\angle89.95°} = 579 \ \mu\text{A}\angle-89.95°$$

practice
problem
V

Refer to the circuit of Figure 21–40:

Figure 21-40

a. Determine the resonant frequency and express it in radians per second and in hertz.

b. Calculate the quality factor of the circuit.

c. Solve for the bandwidth.

d. Determine the voltage **V** at resonance.

in process
learning
check *IV*

Refer to the parallel resonant circuit of Figure 21–41:

$f_P = 800$ kHz
BW $= 25$ kHz

Figure 21-41

Suppose the circuit has a resonant frequency of 800 kHz and a bandwidth of 25 kHz.

a. Determine the value of the inductor, L, in henries.

b. Calculate the value of the resistance, R, in ohms.

c. Find **V**, $\mathbf{I}_L$, and power, P, at resonance.

d. Find the appoximate values of the half-power frequencies, f_1 and f_2.

e. Determine the voltage across the circuit at the lower half-power frequency, f_1, and show that the power dissipated by the resistor at this frequency is half the power dissipated at the resonant frequency.

21.7 Circuit Analysis Using Computers

PSpice is particularly useful in examining the operation of resonant circuits. The ability of the software to provide a visual display of the frequency response is used to evaluate the resonant frequency, maximum current, and bandwidth of a circuit. The Q of the given circuit is then easily determined.

■ **EXAMPLE 21–9**   Write the PSpice input file to find the frequency response of the current for the circuit shown in Figure 21–13.

Solution   The original circuit is redrawn and labelled to conform to the PSpice format as shown in Figure 21–42.

Figure 21–42

We have determined that the resonant frequency for the circuit is $\omega_s =$ 10 krad/s ($f_s = 1.59$ kHz). Therefore, we will perform an ac analysis of the circuit from 1 kHz to 10 kHz. The program will perform analysis at 1001 points per decade. (Fewer points will result in a faster output with less resolution). The PSpice input file is shown below.

```
Input File:
Example 21-9: Series Resonance, Figure 21-42
Vs   1   0   AC   10V 0
R    1   2   10ohms
L    2   3   10mH
C    3   0   1uF
.AC   DEC   1001   1kH   10kH
.PROBE
.END
```

Once we are in the PROBE postprocessor, we select a trace by using the *Add_trace* menu item. The current is selected by simply entering *I(R)*. In order to obtain the cursors we use the *Cursor* menu item. Cursor C1 is moved with the arrow keys on the keyboard and cursor C2 is moved by holding down the *Shift* key while pressing the appropriate arrow key. Additional circuit variables are easily displayed by using the *Add_trace* menu item. For example, the power dissipated by the circuit is determined as the product $V(1,2) * I(R)$. By displaying this response, we would find the half-power frequencies and consequently the bandwidth. Figure 21–43 shows a typical output for the given circuit.

Figure 21–43

By using the cursors, we see that the bandwidth of the circuit is 159.8 Hz (1004 rad/s). This corresponds well to the theoretical value predicted by Example 21–2.

■ **EXAMPLE 21–10**   Write the PSpice input file to find the frequency response of the current for the circuit shown in Figure 21–34.

Solution   The original circuit is redrawn and labelled to conform to the PSpice format as shown in Figure 21–44.

Figure 21–44

From previous calculations, we know that the resonant frequency will occur at $\omega = 1$ Mrad/s (159 kHz). In this example we select to have the ac analysis done between 100 kHz and 300 kHz using a linear sweep with 1001 points. The input file is given below:

```
Input File:
Example 21-10: Parallel resonant circuit, Figure
21-44
Is    1    0   AC   20mA   0
R1    1    0   25kohms
C     1    0   1nF
Rcoil 1    2   10ohms
L     2    0   1mH
.AC LIN 1001 100kH 300kH
.PROBE
.END
```

The resulting output using the PROBE postprocessor is shown in Figure 21–45. The bandwidth found by using the cursors is 8.0 kHz, which corresponds to the theoretical bandwidth of 50 krad/s.

Figure 21–45

practice
problem
VI

Write the PSpice input file to obtain the frequency response of voltage, V versus f, for the circuit of Figure 21–40. Use cursors to determine the approximate values of the half-power frequencies and the bandwidth of the circuit. Compare the results to those obtained in Practice Problem V.

Problems

21.1   Series Resonance

1. Consider the circuit of Figure 21–46:

Figure 21–46

a. Determine the resonant frequency of the circuit in both radians per second and hertz.
b. Calculate the current, **I**, at resonance.
c. Solve for the voltages **V**$_R$, **V**$_L$, and **V**$_C$. (Notice that the voltage **V**$_L$ includes the voltage dropped across the internal resistance of the coil.)
d. Determine the power (in watts) dissipated by the inductor. (Hint: The power will not be zero.)

2. Refer to the circuit of Figure 21–47:

Figure 21–47

a. Determine the resonant frequency of the circuit in both radians per second and hertz.
b. Calculate the phasor current, **I**.
c. Determine the power dissipated by the circuit at resonance.
d. Calculate the phasor voltages, **V**$_L$ and **V**$_R$.
e. Write the sinusoidal form of the voltages v_L and v_R.

3. Consider the circuit of Figure 21–48:

Figure 21–48

a. Determine the values of R and C such that the circuit has a resonant frequency of 25 kHz and an rms current of 25 mA at resonance.
b. Calculate the power dissipated by the circuit at resonance.
c. Determine the phasor voltages, $\mathbf{V}_C$, $\mathbf{V}_L$, and $\mathbf{V}_R$.
d. Write the sinusoidal expressions for the voltages v_C, v_L, and v_R.

4. Refer to the circuit of Figure 21–49.

Figure 21–49

a. Determine the capacitance required in order that the circuit have a resonant frequency of 100 kHz.
b. Solve for the phasor quantities $\mathbf{I}$, $\mathbf{V}_L$, and $\mathbf{V}_R$.
c. Find the sinusoidal expressions for i, v_L, and v_R.
d. Determine the power dissipated by each element in the circuit.

21.2 Quality Factor, Q

5. Refer to the circuit of Figure 21–50:

Figure 21–50

a. Determine the resonant frequency expressed as ω(rad/s) and f(Hz).
b. Calculate the total impedance, $\mathbf{Z}_T$, at resonance.
c. Solve for current $\mathbf{I}$ at resonance.
d. Solve for $\mathbf{V}_R$, $\mathbf{V}_L$, and $\mathbf{V}_C$ at resonance.
e. Calculate the power dissipated by the circuit and evaluate the reactive powers, Q_C and Q_L.
f. Find the quality factor, Q_S, of the circuit.

6. Suppose that the circuit of Figure 21–51 has a resonant frequency of $f_S = 25$ kHz and a quality factor of $Q_S = 10$:

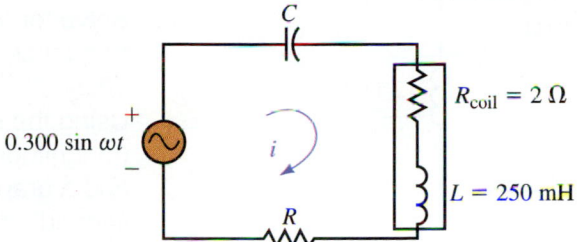

$0.300 \sin \omega t$

$R_{\text{coil}} = 2\ \Omega$

$L = 250$ mH

Figure 21–51

a. Determine the values of R and C.
b. Solve for the quality factor of inductor, Q_{coil}.
c. Find $\mathbf{Z}_T$, $\mathbf{I}$, $\mathbf{V}_C$, and $\mathbf{V}_R$ at resonance.
d. Solve for the sinusoidal expression of current i at resonance.
e. Calcuate the sinusoidal expressions v_C and v_R at resonance.
f. Calculate the power dissipated by the circuit and determine the reactive powers, Q_C and Q_L.

7. Refer to the circuit of Figure 21–52:

R

$2.5\ \text{V}\angle 0°$

R_{coil} $3.6\ \Omega$

V_L

L 5 mH

C

Figure 21–52

a. Design the circuit to have a resonant frequency of $\omega = 50$ krad/s and a quality factor $Q_S = 25$.
b. Calculate the power dissipated by the circuit at the resonant frequency.
c. Determine the voltage, $\mathbf{V}_L$, across the inductor at resonance.

$C = 0.01\ \mu F$

Figure 21–53

Figure 21–54

8. Consider the circuit of Figure 21–53:
   a. Design the circuit to have a resonant frequency of $\omega =$ 400 krad/s and a quality factor $Q_S = 10$.
   b. Calculate the power dissipated by the circuit at the resonant frequency.
   c. Determine the voltage, $\mathbf{V}_L$, across the inductor at resonance.

21.3   Impedance of a Series Resonant Circuit

9. Refer to the series resonant circuit of Figure 21–54.
   a. Determine the resonant frequency, ω_S.
   b. Solve for the input impedance, $\mathbf{Z}_T = Z\angle\theta$, of the circuit at frequencies of $0.1\omega_S$, $0.2\omega_S$, $0.5\omega_S$, ω_S, $2\omega_S$, $5\omega_S$, and $10\omega_S$.
   c. Using the results from b), sketch a graph of Z (magnitude in ohms) versus ω (in radians per second) and a graph of θ (in degrees) versus ω (in radians per second). If possible, use log-log graph paper for the former and semilog graph paper for the latter.
   d. Using your results from (b), determine the magnitude of current at each of the given frequencies.
   e. Use the results from (d) to plot a graph of I (magnitude in amps) versus ω (in radians per second) on log-log graph paper.

10. Repeat Problem 9 if the 10-Ω resistor is replaced with a 50-Ω resistor.

21.4   Power, Bandwidth, and Selectivity of a Series Resonant Circuit

11. Refer to the circuit of Figure 21–55.
   a. Find ω_S, Q, and BW (in radians per second).
   b. Calculate the maximum power dissipated by the circuit.
   c. From the results obtained in (a) solve for the approximate half-power frequencies, ω_1 and ω_2.
   d. Calculate the actual half-power frequencies, ω_1 and ω_2 using the component values and the appropriate equations.
   e. Are the results obtained in (c) and (d) comparable? Explain.
   f. Solve for the circuit current, $\mathbf{I}$, and power dissipated at the lower half-power frequency, ω_1, determined in (d).

Figure 21–55

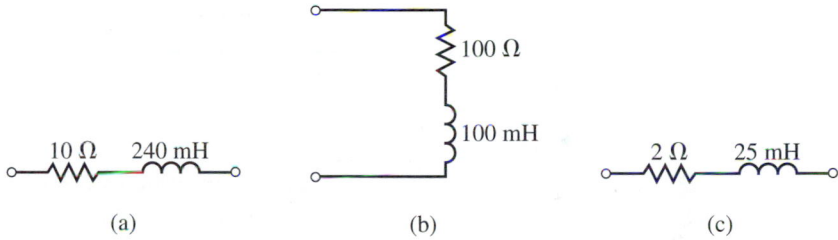

Figure 21–56   Figure 21–57

12. Repeat Problem 11 for the circuit of Figure 21–56.

13. Consider the circuit of Figure 21–57.
   a. Calculate the values of R and C for the circuit to have a resonant frequency of 200 kHz and a bandwidth of 16 kHz.
   b. Use the designed component values to determine the power dissipated by the circuit at resonance.
   c. Solve for v_{out} at resonance.

14. Repeat Problem 13 if the resonant frequency is to be 580 kHz and the bandwidth is 10 kHz.

21.5 Series-to-Parallel *RL* and *RC* Conversion

15. Refer to the series networks of Figure 21–58.

Figure 21–59

Figure 21–58

   a. Find the Q of each network at $\omega = 1000$ rad/s.
   b. Convert each series RL network into an equivalent parallel network, having R_P and X_{LP} in ohms.
   c. Repeat (a) and (b) for $\omega = 10$ krad/s.

16. Consider the series networks of Figure 21–59.
   a. Find the Q of each coil at $\omega = 20$ krad/s.
   b. Convert each series RL network into an equivalent parallel network consisting of R_P and X_{LP} in ohms.
   c. Repeat (a) and (b) for $\omega = 100$ krad/s.

17. For the series networks of Figure 21–60, find the Q and convert each network into its parallel equivalent.

Figure 21–60

18. Derive Equations 21–34 and 21–35, which enable us to convert a parallel *RL* network into its series equivalent. (Hint: Begin by determining the expression for the input impedance of the parallel network.)

19. Find the *Q* of each of the networks of Figure 21–61 and determine the series equivalent of each. Express all answers in ohms.

(a)

Figure 21–61

(b)   (c)

20. Repeat Problem 19 for networks of Figure 21–62.

(a)

Figure 21–62

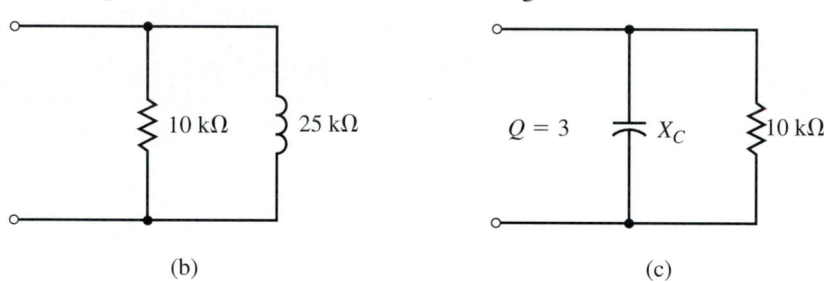

(b)   (c)

21. Determine the values of L_S and L_P in henries, given that the networks of Figure 21–63 are equivalent at a frequency of 250 krad/s.

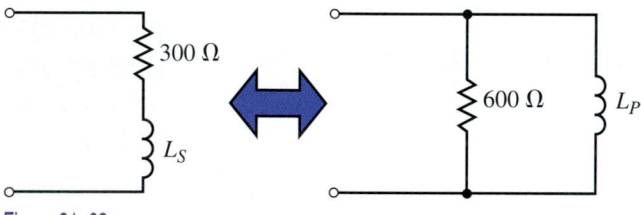

Figure 21–63

22. Determine the values of C_S and C_P in farads, given that the networks of Figure 21–64 are equivalent at a frequency of 48 krad/s.

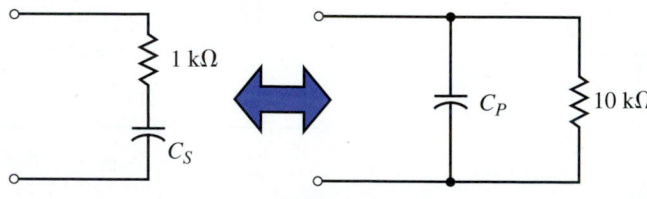

Figure 21–64

21.6   Parallel Resonance

23. Consider the circuit of Figure 21–65.

2 mA∠0° R 20 kΩ C 12.5 nF L 200 mH   **V**

Figure 21–65

a. Determine the resonant frequency, ω_P, in radians per second.
b. Solve for the input impedance, $\mathbf{Z}_T = Z\angle\theta$, of the circuit at frequencies of $0.1\omega_P$, $0.2\omega_P$, $0.5\omega_P$, ω_P, $2\omega_P$, $5\omega_P$, and $10\omega_P$.
c. Using the results obtained in (b), sketch graphs of Z (magnitude in ohms) versus ω (in radians per second) and θ (in degrees) versus ω. If possible, use log-log graph paper for the former and semilog for the latter.
d. Using the results from (b), determine the voltage **V** at each of the indicated frequencies.
e. Sketch a graph of magnitude V versus ω on log-log graph paper.

24. Repeat Problem 23 if the 20-kΩ resistor is replaced with a 40-kΩ resistor.

25. Refer to the circuit shown in Figure 21–66.

100 mA∠0° $\mathbf{I}_R$ $\mathbf{I}_C$ $\mathbf{I}_L$ 133 Ω   +
60 kΩ   25 nF   **V**   25 mH   –

Figure 21–66

a. Determine the resonant frequencies, ω_P(rad/s) and f_P(Hz).
b. Find the Q of the circuit.
c. Calculate **V**, $\mathbf{I}_R$, $\mathbf{I}_L$, and $\mathbf{I}_C$ at resonance.
d. Determine the power dissipated by the circuit at resonance.
e. Solve for the bandwidth of the circuit in both radians per second and hertz.
f. Sketch the voltage response of the circuit, showing the voltage at the half-power frequencies.

26. Repeat Problem 25 for the circuit of Figure 21–67.

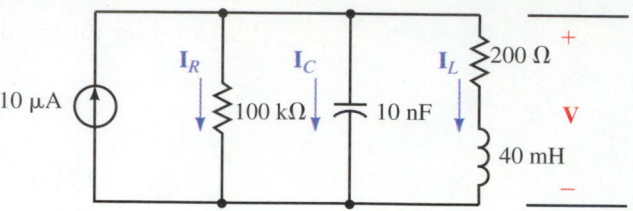

Figure 21–67

27. Determine the values of R_1 and C for the resonant tank circuit of Figure 21–68 so that the given conditions are met. Solve for current $\mathbf{I}_L$ through the inductor.

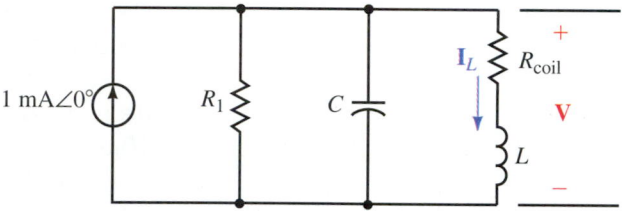

Figure 21–68

$L = 25$ mH, $R_{\text{coil}} = 100$ Ω
$f_P = 50$ kHz
BW = 10 kHz

28. Determine the values of R_1 and C for the resonant circuit of Figure 21–68 so that the given conditions are met. Solve for the voltage, $\mathbf{V}$, across the circuit.

$L = 50$ mH, $R_{\text{coil}} = 50$ Ω
$\omega_P = 100$ krad/s
BW = 10 krad/s

21.7 Circuit Analysis Using Computers

29. Write the PSpice input file to find the frequency response of the current for the circuit shown in Figure 21–54. Use the PROBE postprocessor to display the response of current versus frequency.

30. Write the PSpice input file to find the frequency response of the current for the circuit shown in Figure 21–67. Use the PROBE postprocessor to display the response of voltage versus frequency.

31. Write the PSpice input file for the circuit of Figure 21–55. The output should show the voltage (magnitude and phase angle) across all elements at resonance.

32. Write the PSpice input file for the circuit of Figure 21–66. The output should show the current (magnitude and phase angle) through all elements at resonance.

Filters and the Bode Plot

CHAPTER PREVIEW

In the previous chapter we examined how *LRC* resonant circuits react to changes in frequency. In this chapter we will continue to study how changes in frequency affect the behavior of other simple circuits. We will analyze simple low-pass, high-pass, band-pass, and band-reject filter circuits. The analysis will compare the amplitude and phase shift of the output signal with the input signal.

As their names imply, low-pass and high-pass filter circuits are able to pass low frequencies and high frequencies while blocking other frequency components. A good understanding of these filters provides a basis for understanding why circuits such as amplifiers and oscilloscopes are not able to pass all signals from their input to their output.

Band-reject filters and band-pass filters are circuits which are similar to resonant circuits with the exception that these circuits are able to reject or pass certain frequencies without the need for *LC* combinations.

The analysis of all filters may be simplified by plotting the output/input voltage relationship on a semilogarithmic graph called a *Bode plot*.

Washington National GPS River Approach

3000'
1800'
1200'
900'
DIFFERENTIAL
GPS
CORRECTION
TRANSMITTER

The Gobal Positioning System (GPS) can be used to fly instrument approaches to virtually any airport in the world. *(Courtesy of Honeywell Inc.)*

OBJECTIVES

After studying this chapter you will be able to

- evaluate the power gain and voltage gain of a given system,

- express power gain and voltage gain in *decibels,*

- express power levels in dBm and voltage levels in dBV and use these levels to determine power gain and voltage gain,

- identify and design simple (first-order) *RL* and *RC* low-pass and high-pass filters and explain the principles of operation of each type of filter,

- write the standard form of a transfer function for a given filter. The circuits which are studied will include band-pass and band-stop as well as low- and high-pass circuits,

- compute τ_c and use the time constant to determine the cutoff frequency(ies) in both radians per second and hertz for the transfer function of any first-order filter,

- sketch the Bode plot showing the frequency response of voltage gain and phase shift of any first-order filter,

- use PSpice to verify the operation of any first-order filter circuit.

Alexander Graham Bell

*A*lexander Graham Bell was born in Edinburgh, Scotland on March 3, 1847. As a young man, Bell followed his father and grandfather in research dealing with the deaf.

In 1873, Bell was appointed professor of vocal physiology at Boston University. His research was primarily involved in converting sound waves into electrical fluctuations. With the encouragement of Joseph Henry, who had done a great deal of work with inductors, Bell eventually developed the telephone.

In his now-famous accident in which Bell spilled acid on himself, Bell uttered the words "Watson, please come here. I want you". Watson, who was on another floor, ran to Bell's assistance.

Although others had worked on the principle of the telephone, Alexander Graham Bell was awarded the patent for the telephone in 1876.

The telephone was a simple device which passed a current through carbon powder. The density of the carbon powder was determined by air fluctuations due to the sound of a person's voice. When the carbon was compressed, resistivity would decrease, allowing more current.

Bell's name has been adopted for the decibel, which is the unit used to describe sound intensities and power gain.

Although the invention of the telephone made Bell wealthy, he continued experimenting in electronics, air conditioning, and animal breeding.

Bell died at the age of 75 in Baddeck, Nova Scotia on August 2, 1922.

22.1 The Decibel

In electronics, we often wish to consider the effects of a circuit without examining the actual operation of the circuit itself. This *black-box* approach is a common technique used to simplify transistor circuits and for depicting integrated circuits which may contain hundreds or even thousands of elements. Consider the system shown in Figure 22–1.

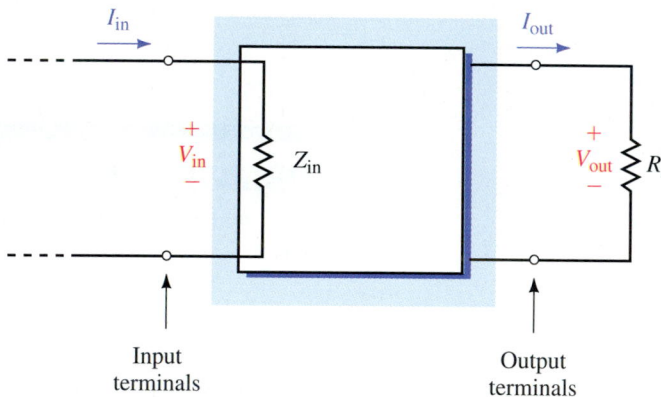

Figure 22–1

Although the circuit within the box may contain many elements, any source connected to the input terminals will effectively see only the input impedance, Z_{in}. Similarly, any load impedance, R_L, connected to the output terminals will have voltage and current determined by certain parameters in the circuit. These parameters usually result in an output at the load which is easily predicted for certain conditions.

We now define several terms which are used to analyze any system having two input terminals and two output terminals.

The *power gain*, A_p, is defined as the ratio of output signal power to the input signal power:

$$A_P = \frac{P_{out}}{P_{in}} \qquad (22\text{--}1)$$

We must emphasize that the total output power delivered to any load can never exceed the total input power to a circuit. When we refer to the power gain of a system, we are interested in only the power contained in the ac signal, so we neglect any power due to dc. In many circuits, the ac power will be significantly less than the dc power. However, ac power gains in the order of tens of thousands are quite possible.

The *voltage gain*, A_v, is defined as the ratio of output signal voltage to the input signal voltage:

$$A_v = \frac{V_{out}}{V_{in}} \qquad (22\text{--}2)$$

As mentioned, the power gain of a system may be a very large. For other applications the output power may be much smaller than the input power, resulting in a loss, or *attenuation*. Any circuit in which the output signal power is greater than the input signal power is referred to as an *amplifier*. Inversely, any circuit in which the output signal power is less than the input signal power is referred to as an *attenuator*.

The ratios expressing power gain or voltage gain may be either very large or very small, making it inconvenient to express the power gain as a simple ratio of two numbers. The *bel*, which is a logarithmic unit named after Alexander Graham Bell, was selected to represent a ten-fold increase or decrease in power. Stated mathematically, the power gain in bels is given as

$$A_p(\text{bels}) = \log_{10} \frac{P_{out}}{P_{in}}$$

Because the bel is an awkwardly large unit, the *decibel* (dB), which is one tenth of a bel, has been adopted as a more acceptable unit for describing the logarithmic change in power levels. One bel contains 10 decibels and so the power gain in decibels is given as

$$A_p(\text{dB}) = 10 \log_{10} \frac{P_{out}}{P_{in}} \qquad (22\text{--}3)$$

If the power level in the system increases from the input to the output, then the gain in dB will be positive. Inversely, if the power at the output is less than the power at the input, then the power gain will be negative. Notice that if the input and output have the same power levels, then the power gain will be 0 dB, since log 1 = 0.

■ **EXAMPLE 22–1**   An amplifier has the indicated input and output power levels. Determine the power gain both as a ratio and in dB for each of the conditions:

a. $P_{in} = 1$ mW, $P_{out} = 100$ W.
b. $P_{in} = 4$ μW, $P_{out} = 2$ μW.

c. $P_{in} = 6$ mW, $P_{out} = 12$ mW.

d. $P_{in} = 25$ mW, $P_{out} = 2.5$ mW.

Solution

a. $$A_p = \frac{P_{out}}{P_{in}} = \frac{100 \text{ W}}{1 \text{ mW}} = 100\ 000$$

$$A_p(dB) = 10 \log_{10}(100\ 000) = (10)(5) = 50 \text{ dB}$$

b. $$A_p = \frac{P_{out}}{P_{in}} = \frac{2 \ \mu W}{4 \ \mu W} = 0.5$$

$$A_p(dB) = 10 \log_{10}(0.5) = (10)(-0.30) = -3.0 \text{ dB}$$

c. $$A_p = \frac{P_{out}}{P_{in}} = \frac{12 \text{ mW}}{6 \text{ mW}} = 2$$

$$A_p(dB) = 10 \log_{10}2 = (10)(.30) = 3.0 \text{ dB}$$

d. $$A_p = \frac{P_{out}}{P_{in}} = \frac{2.5 \text{ mW}}{25 \text{ mW}} = 0.10$$

$$A_p(dB) = 10 \log_{10}0.10 = (10)(-1) = -10 \text{ dB}$$

The above example illustrates that if the power is increased or decreased by a factor of two, the resultant power gain will be either +3 dB or −3 dB, respectively. You will recall that in the previous chapter a similar reference was made when the half-power frequencies of a resonant circuit were referred to as the 3-dB-down frequencies.

The voltage gain of a system may also be expressed in dB. In order to derive the expression for voltage gain, we first assume that the input resistance and the load resistance are the same value. Then, using the definition of power gain as given by Equation 22–3, we have the following:

$$A_p(dB) = 10 \log_{10}\frac{P_{out}}{P_{in}}$$

$$= 10 \log_{10}\frac{V_{out}^{2}/R}{V_{in}^{2}/R}$$

$$= 10 \log_{10}\left(\frac{V_{out}}{V_{in}}\right)^{2}$$

which gives

$$A_p(dB) = 20 \log_{10}\frac{V_{out}}{V_{in}}$$

Because the above expression represents a decibel equivalent of the voltage gain, we simply write the voltage gain in dB as follows:

$$A_v(dB) = 20 \log_{10}\frac{V_{out}}{V_{in}} \qquad (22\text{–}4)$$

■ **EXAMPLE 22–2** The amplifier circuit of Figure 22–2 has the given conditions. Calculate the voltage gain and power gain in dB.

$Z_{in} = 10\ k\Omega$

$R_L = 600\ \Omega$

$V_{in} = 20\ mV_{rms}$

$V_{out} = 500\ mV_{rms}$

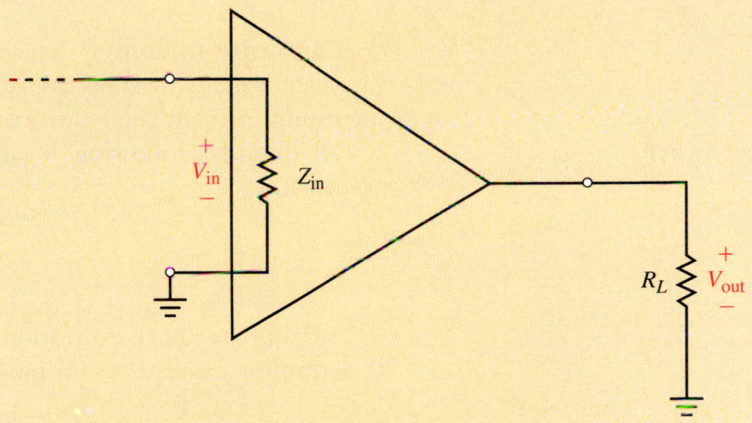

Figure 22–2

Solution   The voltage gain of the amplifier is

$$A_v = 20\ \log_{10} \frac{V_{out}}{V_{in}}$$

$$= 20\ \log_{10} \frac{500\ mV}{20\ mV} = 20\ \log_{10} 25 = 28.0\ dB$$

The signal power available at the input of the amplifier is

$$P_{in} = \frac{V_{in}^{2}}{Z_{in}} = \frac{(20\ mV)^2}{10\ k\Omega} = 0.040\ \mu W$$

The signal at the output of the amplifier has a power of

$$P_{out} = \frac{V_{out}^{2}}{R_L} = \frac{(500\ mV)^2}{600\ \Omega} = 416.7\ \mu W$$

The power gain of the amplifier is

$$A_p = 10\ \log_{10} \frac{P_{out}}{P_{in}}$$

$$= 10\ \log_{10} \frac{416.7\ \mu W}{0.040\ \mu W}$$

$$= 10\ \log_{10}(10\ 417) = 40.2\ dB$$

Calculate the voltage gain and power gain (in dB) for the amplifier of Figure 22–2, given the following conditions:

$$Z_{in} = 2 \text{ k}\Omega$$

$$R_L = 5 \text{ }\Omega$$

$$V_{in} = 16 \text{ }\mu V_{rms}$$

$$V_{out} = 32 \text{ mV}_{rms}$$

In order to convert a gain from decibels into a simple ratio of power or voltage, it is necessary to perform the inverse operation of the logarithm; namely, solve the unknown quantity by using the exponential. Recall that the following logarithmic and exponential operations are equivalent:

$$y = \log_b x$$

$$x = b^y$$

Using the above expressions, Equations 22–3 and 22–4 may be used to determine expressions for power and voltage gains as follows:

$$\frac{A_p(\text{dB})}{10} = \log_{10} \frac{P_{out}}{P_{in}}$$

$$\frac{P_{out}}{P_{in}} = 10^{A_p(\text{dB})/10} \tag{22–5}$$

$$\frac{A_v(\text{dB})}{20} = \log_{10} \frac{V_{out}}{V_{in}}$$

$$\frac{V_{out}}{V_{in}} = 10^{A_v(\text{dB})/20} \tag{22–6}$$

■ **EXAMPLE 22–3**   Convert the following from decibels to ratios:

a. $A_p = 25$ dB.

b. $A_p = -6$ dB.

c. $A_v = 10$ dB.

d. $A_v = -6$ dB.

Solution

a.
$$A_p = \frac{P_{out}}{P_{in}} = 10^{A_p(\text{dB})/10}$$

$$= 10^{25/10} = 316$$

b.
$$A_p = 10^{A_p(\text{dB})/10}$$

$$= 10^{-6/10} = 0.251$$

c.
$$A_v = \frac{V_{out}}{V_{in}} = 10^{A_v(dB)/20}$$
$$= 10^{10/20} = 3.16$$

d.
$$A_v = 10^{A_v(dB)/20}$$
$$= 10^{-6/20} = 0.501$$

Applications of Decibels

Decibels were originally intended as a measure of changes in acoustical levels. The human ear is not a linear instrument; rather, it responds to sounds in a logarithmic fashion. Because of this peculiar phenomenon, a ten-fold increase in sound intensity results in a perceived doubling of sound. This means that if we wish to double the sound heard from a 10-W power amplifier, we must increase the output power to 100 W.

The minimum sound level which may be detected by the human ear is called the *threshold of hearing* and is usually taken to be $I_0 = 1 \times 10^{-12}$ W/m^2. Table 22–1 lists approximate sound intensities of several common sounds. The decibel levels are determined from the expression

$$\beta(dB) = 10 \log_{10} \frac{I}{I_0}$$

Some electronic circuits operate with very small power levels. These power levels may be referenced to some arbitrary level and then expressed in decibels in a manner similar to how sound intensities are

TABLE 22–1 Intensity Levels of Common Sounds

SOUND	INTENSITY LEVEL (dB)	INTENSITY (W/m^2)
threshold of hearing, I_0	0	10^{-12}
virtual silence	10	10^{-11}
quiet room	20	10^{-10}
watch ticking a 1 m	30	10^{-9}
quiet street	40	10^{-8}
quiet conversation	50	10^{-7}
quiet motor at 1 m	60	10^{-6}
busy traffic	70	10^{-5}
door slamming	80	10^{-4}
busy office room	90	10^{-3}
jackhammer	100	10^{-2}
motorcycle	110	10^{-1}
loud indoor rock concert	120	1
threshold of pain	130	10

represented. For instance, power levels may be referenced to a standard power of 1 mW. In such cases the power level is expressed in dBm and is determined as

$$P_{\text{dBm}} = 10 \log_{10} \frac{P}{1 \text{ mW}} \qquad (22\text{–}7)$$

If a power level is referenced to a standard of 1 W, then we have

$$P_{\text{dBW}} = 10 \log_{10} \frac{P}{1 \text{ W}} \qquad (22\text{–}8)$$

■ **EXAMPLE 22–4** Express the following powers in dBm and in dBW.

 a. $P_1 = 0.35 \ \mu\text{W}$.
 b. $P_2 = 20 \ \text{mW}$.
 c. $P_3 = 1000 \ \text{W}$.
 d. $P_4 = 1 \ \text{pW}$.

Solution

 a. $\quad P_1(\text{dBm}) = 10 \log_{10} \dfrac{0.35 \ \mu\text{W}}{1 \text{ mW}} = -34.6 \text{ dBm}$

 $\quad P_1(\text{dBW}) = 10 \log_{10} \dfrac{0.35 \ \mu\text{W}}{1 \text{ W}} = -64.6 \text{ dBW}$

 b. $\quad P_2(\text{dBm}) = 10 \log_{10} \dfrac{20 \ \text{mW}}{1 \text{ mW}} = 13.0 \text{ dBm}$

 $\quad P_2(\text{dBW}) = 10 \log_{10} \dfrac{20 \ \text{mW}}{1 \text{ W}} = -17.0 \text{ dBW}$

 c. $\quad P_3(\text{dBm}) = 10 \log_{10} \dfrac{1000 \ \text{W}}{1 \text{ mW}} = 60 \text{ dBm}$

 $\quad P_3(\text{dBW}) = 10 \log_{10} \dfrac{1000 \ \text{W}}{1 \text{ W}} = 30 \text{ dBW}$

 d. $\quad P_4(\text{dBm}) = 10 \log_{10} \dfrac{1 \ \text{pW}}{1 \text{ mW}} = -90 \text{ dBm}$

 $\quad P_4(\text{dBW}) = 10 \log_{10} \dfrac{1 \ \text{pW}}{1 \text{ W}} = -120 \text{ dBW}$

Many voltmeters have a separate scale calibrated in decibels. In such cases, the voltage expressed in dBV uses 1 V_{rms} as the reference voltage. In general, any voltage reading may be expressed in dBV as follows:

$$V_{\text{dBV}} = 20 \log_{10} \frac{V_{\text{out}}}{1 \text{ V}} \qquad (22\text{–}9)$$

practice
problem
II

Consider the resistors of Figure 22–3:

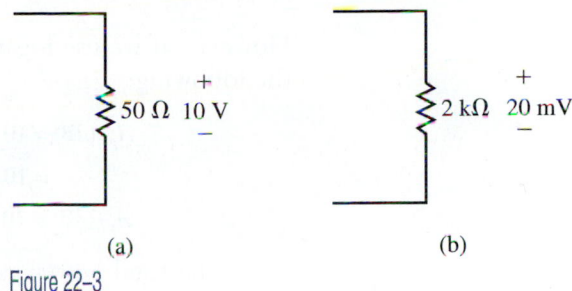

50 Ω 10 V

2 kΩ 20 mV

(a)

(b)

Figure 22-3

a. Determine the power levels in dBm and in dBW.
b. Express the voltages in dBV.

22.2 Multistage Systems

Quite often a system consists of several stages. In order to find the total voltage gain or power gain of the system, we would need to solve for the product of the individual gains. The use of decibels makes the solution of a multistage system easy to find. If the gain of each stage is given in decibels, then the resultant gain is simply determined as the summation of the individual gains.

Consider the system of Figure 22–4, which represents a three-stage system.

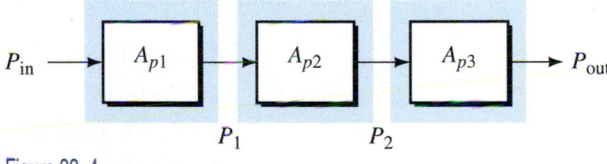

Figure 22-4

The power at the output of each stage is determined as follows:

$$P_1 = A_{p1} \, P_{in}$$
$$P_2 = A_{p2} \, P_1$$
$$P_{out} = A_{p3} \, P_2$$

The total power gain of the system is found as

$$A_{pT} = \frac{P_{out}}{P_{in}} = \frac{A_{p3} \, P_2}{P_{in}}$$

$$= \frac{A_{p3} \, (A_{p2} \, P_1)}{P_{in}}$$

$$= \frac{A_{p3} \, A_{p2} \, (A_{p1} \, P_i)}{P_{in}}$$

$$= A_{p1} \, A_{p2} \, A_{p3}$$

In general, for *n* stages the total power gain is found as the product:

$$A_{pT} = A_{p1}\, A_{p2} \cdots A_{pn} \tag{22–10}$$

However, if we use logarithms to solve for the gain in decibels, we have the following:

$$A_{pT}(\text{dB}) = 10\log_{10}A_{pT}$$
$$= 10\log_{10}(A_{p1}\,A_{p2}\cdots A_{pn})$$
$$A_{pT}(\text{dB}) = 10\log_{10}A_{p1} + 10\log_{10}A_{p2} + \cdots + 10\log_{10}A_{pn}$$

The total power gain in decibels for n stages is determined as the summation of the individual decibel power gains:

$$A_p(\text{dB}) = A_{p1}(\text{dB}) + A_{p2}(\text{dB}) + \cdots + A_{pn}(\text{dB}) \tag{22–11}$$

The advantage of using decibels in solving for power gains and power levels is illustrated in the following example.

■ **EXAMPLE 22–5**   The circuit of Figure 22–5 represents the first three stages of a typical AM or FM receiver.

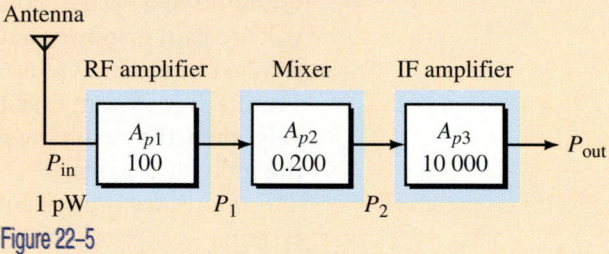

Figure 22–5

Find the following quantities:

a. $A_{p1}(\text{dB})$, $A_{p2}(\text{dB})$, and $A_{p3}(\text{dB})$.
b. $A_{pT}(\text{dB})$.
c. P_1, P_2, and P_{out}.
d. $P_{\text{in}}(\text{dBm})$, $P_1(\text{dBm})$, $P_2(\text{dBm})$, and $P_{\text{out}}(\text{dBm})$.

Solution

a. $A_{p1}(\text{dB}) = 10\log_{10}A_{p1} = 10\log_{10}100 = 20\text{ dB}$

$A_{p2}(\text{dB}) = \log_{10}A_{p2} = 10\log_{10}0.2 = -7.0\text{ dB}$

$A_{p3}(\text{dB}) = 10\log_{10}A_{p3} = 10\log_{10}(10\,000) = 40\text{ dB}$

b. $A_{pT}(\text{dB}) = A_{p1}(\text{dB}) + A_{p2}(\text{dB}) + A_{p3}(\text{dB})$
$$= 20\text{ dB} - 7.0\text{ dB} + 4.0\text{ dB}$$
$$= 53.0\text{ dB}$$

c. $P_1 = A_{p1}P_{\text{in}} = (100)(1\text{ pW}) = 100\text{ pW}$

$P_2 = A_{p2}P_1 = \dfrac{100\text{ pW}}{5} = 20\text{ pW}$

$P_{\text{out}} = A_{p3}P_2 = (10\,000)(20\text{ pW}) = 0.20\ \mu\text{W}$

d. $P_{in}(dBm) = 10 \log_{10} \dfrac{P_{in}}{1 \text{ mW}} = 10 \log_{10} \dfrac{1 \text{ pW}}{1 \text{ mW}} = -90 \text{ dBm}$

$P_1(dBm) = 10 \log_{10} \dfrac{P_1}{1 \text{ mW}} = 10 \log_{10} \dfrac{100 \text{ pW}}{1 \text{ mW}} = -70 \text{ dBm}$

$P_2(dBm) = 10 \log_{10} \dfrac{P_2}{1 \text{ mW}} = 10 \log_{10} \dfrac{20 \text{ pW}}{1 \text{ mW}} = -77.0 \text{ dBm}$

$P_{out}(dBm) = 10 \log_{10} \dfrac{P_{out}}{1 \text{ mW}} = 10 \log_{10} \dfrac{0.20 \text{ μW}}{1 \text{ mW}} = -37.0 \text{ dBm}$

Notice that the power level (in dBm) at the output of any stage is easily determined as the sum of the input power level (in dBm) and the gain of the stage (in dB). It is for this reason that many communication circuits express power levels in decibels rather than in watts.

Calculate the power level at the output of each of the stages in Figure 22–6.

Figure 22–6

1. Given that an amplifier has a power gain of 25 dB, calculate the output power level (in dBm) for the following input characteristics:

   a. $V_{in} = 10 \text{ mV}_{rms}$, $Z_{in} = 50 \text{ Ω}$.
   b. $V_{in} = 10 \text{ mV}_{rms}$, $Z_{in} = 1 \text{ kΩ}$.
   c. $V_{in} = 400 \text{ μV}_{rms}$, $Z_{in} = 200 \text{ Ω}$.

2. Given amplifiers with the following output characteristics, determine the output voltage (in volts rms).

   a. $P_{out} = 8.0 \text{ dBm}$, $R_L = 50 \text{ Ω}$.
   b. $P_{out} = -16.0 \text{ dBm}$, $R_L = 2 \text{ kΩ}$.
   c. $P_{out} = -16.0 \text{ dBm}$, $R_L = 5 \text{ kΩ}$.

22.3   Simple *RC* and *RL* Transfer Functions

Electronic circuits usually operate in a highly predictable fashion. If a certain signal is applied at the input of a system, the output will be determined by the physical characteristics of the circuit. The frequency of the incoming signal is one of the many physical conditions which determine the relationship between a given input signal and the resulting output. Although outside the scope of this text, other conditions that may deter-

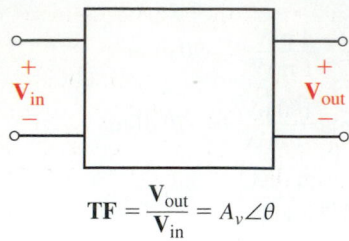

$$\mathbf{TF} = \frac{\mathbf{V}_{out}}{\mathbf{V}_{in}} = A_v \angle \theta$$

Figure 22–7

mine the relationship between the input and output signals of a given circuit are temperature, light, radiation, etc.

For any system subjected to a sinusoidal input voltage, as shown in Figure 22–7, we define the *transfer function* as the ratio of the output voltage phasor to the input voltage phasor for any frequency ω (in radians per second).

$$\mathbf{TF}(\omega) = \frac{\mathbf{V}_{out}}{\mathbf{V}_{in}} = A_v \angle \theta \qquad (22\text{–}12)$$

Notice that the definition of transfer function is almost the same as voltage gain. The difference is that the transfer function takes into account both *amplitude* and *phase shift* of the voltages, whereas voltage gain compared only the amplitudes.

From Equation 22–12 we see that the amplitude of the transfer function is in fact the voltage gain. The phase angle, θ, represents the phase shift between the input and output voltage phasors. The angle θ will be positive if the output leads the input and negative if the output lags the input waveform.

If the elements within the block of Figure 22–7 are resistors, then the output and the input voltages will always be in phase. Also, since resistors have the same value at all frequencies, the voltage gain will remain constant at all frequencies. (This text does not take into account resistance variations due to very high frequencies.) The resulting circuit is called an *attenuator* since resistance within the block will dissipate some power, thereby reducing (or attenuating) the signal as it passes through the circuit.

If the elements within the block are combinations of resistors, inductors, and capacitors, then the output voltage and phase will depend on frequency since the impedances of inductors and capacitors are frequency dependent. Figure 22–8 illustrates an example of how voltage gain and phase shift of a circuit may change as a function of frequency.

In order to examine the operation of the circuit over a wide range of frequencies, the abscissa (horizontal axis) is usually shown as a logarithmic scale. The ordinate (vertical axis) is usually shown as a linear

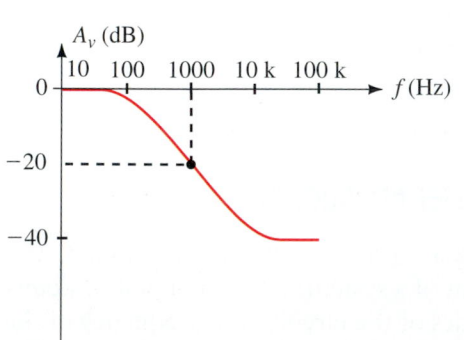

(a) Typical frequency response of voltage gain.

(b) Typical frequency response of phase shift.

Figure 22–8   Frequency response of a circuit.

scale in decibels. Such graphs are said to be semilogarithmic since one scale is logarithmic and the other is linear.

By examining the frequency response of a circuit, we are able to determine at a glance the voltage gain (in dB) and phase shift (in degrees) for any sinusoidal input at a given frequency. For example, at a frequency of 1000 Hz, the voltage gain is −20 dB and the phase shift is 45°. This means that the output signal is one tenth as large as the input signal and the output leads the input by 45°.

From the frequency response of Figure 22–8, we see that the circuit which corresponds to this response is able to pass low-frequency signals, while at the same time partially attenuating high-frequency signals. Any circuit which passes a particular range of frequencies while blocking others is referred to as a *filter* circuit. Filter circuits are usually named according to their function although certain filters are named after their inventors. The filter having the response of Figure 22–8 is referred to as a stepfilter, since the voltage gain occurs between two limits (steps). Other types of filters which are used regularly in electrical and electronic circuits include *low-pass, high-pass, band-pass,* and *band-reject filters.*

Although the design of filters is a topic in itself, we will examine a few of the more common types of filters which are used extensively.

Sketching Transfer Functions

Transfer functions are easily sketched on semilogarithmic graph paper by following a few basic steps. Consider a transfer function given as

$$\mathbf{TF} = \frac{1}{1 + j0.01\omega}$$

Several important characteristics are evident by examining the transfer function above. The magnitude of the transfer function (voltage gain) will be less than 1 for all frequencies other than zero hertz (dc). As the frequency increases, the voltage gain will decrease. For this reason, the circuit which has the above transfer function is called a *low-pass filter.* Finally, we see that the output voltage will lag the input voltage at all frequencies other than at zero hertz.

We define the *cutoff frequency,* ω_c (also called the *break frequency*), as the frequency at which the imaginary component of $1 + j0.01\omega$ is equal to the real component. This results in a voltage gain which is equal to

$$A_v = \frac{1}{\sqrt{2}} = 0.7071$$

From the transfer function, we see that the cutoff frequency occurs when the angular frequency $\omega = \omega_c = 1/0.01 = 100$ rad/s. At the cutoff frequency, the voltage gain (in dB) is $A_v = -3.0$ dB and the phase shift is $\theta = -45°$. To determine the cutoff frequency in hertz, we simply perform the following operation:

$$f_c = \frac{\omega_c}{2\pi}$$

The voltage gain (in decibels) at any angular frequency is found from the transfer function as

$$A_v(\text{dB}) = -20 \log(\sqrt{1 + (0.01\omega)^2})$$

or

$$A_v(\text{dB}) = -10 \log[1 + (0.01\omega)^2]$$

and the phase shift is given as

$$\theta = -\arctan(0.01\omega)$$

The frequency response of the transfer function may be sketched by determining the voltage gain and phase shift at various angular frequencies around the cutoff frequency. Table 22–2 provides the voltage gain and phase shift for frequencies between $\omega = 0.01\omega_c$ and $\omega = 100\omega_c$.

TABLE 22–2   Frequency Response of a Low-Pass Filter

ω (rad/s)	TF(ω)	A_v (V/V)	A_v (db)	θ (°)
1	1.00∠−0.6°	1.000	0.00	−0.6°
5	0.999∠−2.9°	0.999	0.00	−2.9°
10	0.995∠−5.7°	0.995	−0.04	−5.7°
20	0.981∠−11.3°	0.981	−0.17	−11.3°
50	0.894∠−26.6°	0.894	−0.97	−26.6°
100	0.707∠−45°	0.707	−3.01	−45.0°
200	0.447∠−63.4°	0.447	−6.99	−63.4°
500	0.196∠−78.9°	0.196	−14.2	−78.9°
1000	0.100∠−84.3°	0.100	−20.0	−84.3°
2000	0.050∠−87.1°	0.050	−26.0	−87.1°
10k	0.010∠−89.4°	0.010	−40.0	−89.4°

The data of Table 22–2 are plotted in Figure 22–9.

Figure 22–9   (a) Frequency response of voltage gain.

Figure 22–9   (b) Frequency response of phase shift.

While the previous method can be used to provide a frequency response for any circuit, it is tedious and very time-consuming. A better method is to approximate the results as follows:

For $\omega \ll \omega_c$ ($\omega \le 0.1\omega_c$),

$$A_v \simeq 0 \text{ dB}$$
$$\theta \simeq 0$$

For $\omega \gg$ ($\omega \ge 10\omega_c$),

$$A_v \simeq -20 \log(0.01\omega)$$
$$\theta \simeq -90°$$

The above information is plotted on semilogarithmic graphs using straight lines to approximate the actual frequency response. Such a graph is called a *Bode plot*. The actual response will approach the straight lines for frequencies much less than and much greater than the cutoff frequency. For this reason, the straight lines are also called the *asymptotes* of the transfer function. Figure 22–10 shows the Bode plot of the transfer function for a low-pass filter.

Figure 22–10   Comparison of Bode plot and actual frequency response.

The Bode plot illustrates several very important points. Because the abscissa of the Bode plot is not measured linearly, the slope of the asymptotes cannot be expressed in the usual terms of rise over run. A *decade* is defined as the ten-fold increase or decrease in frequency. For instance, increasing a frequency from 10 Hz to 100 Hz represents a decade. Similarly, a ten-fold change in angular frequency is also a decade.

An alternate way of describing a frequency change is by using the term *octave*. An octave is defined as a twofold increase or decrease in frequency. Therefore, a frequency of 400 Hz, is one octave higher than a frequency of 200 Hz, while 800 Hz is two octaves higher than 200 Hz.

Both voltage gain and the phase shift have changes which are easily expressed in terms of a decade. The asymptotic response for voltage gain shows that the voltage gain will be approximately 0 dB for frequencies less than ω_c. For frequencies above ω_c, the voltage gain changes at a rate of -20 dB for each decade that the frequency increases (-20 dB/decade). A change of -20 dB/decade is equivalent to having a change of -6 dB/octave.

The asymptotic response for phase shift illustrates that for frequencies less that $0.1\omega_c$ the phase shift will be approximately 0°. For frequencies above $10\omega_c$ the phase shift will be essentially constant at $-90°$. In the region between $0.1\omega_c$ and $10\omega_c$, the phase shift may be approximated to be $-45°$/decade although it actually follows a curved line in this region.

All transfer functions of first-order filters (the only types of filters covered in this text) can be written as

$$\mathbf{TF} = \frac{(j\omega\tau_{Z_1})(1+j\omega\tau_{Z_2})\cdots(1+j\omega\tau_{Z_N})}{(j\omega\tau_{P_1})(1+j\omega\tau_{P_2})\cdots(1+j\omega\tau_{P_M})}$$

where each of the cutoff frequencies (in radians per second) is found as

$$\omega = \frac{1}{\tau}$$

Since voltage gain is expressed in dB, the Bode plot of any transfer function is determined from the summation of the effects due to the various terms in the transfer function.

■ **EXAMPLE 22–6** Sketch the Bode plot of the following transfer function:

$$\mathbf{TF} = \frac{j0.01\omega}{1+j0.1\omega}$$

Solution The voltage gain in dB is given as

$$A_v(\text{dB}) = 20\log(0.01\omega) - 20\log(\sqrt{1+(0.1\omega)^2})$$

and the phase shift is given as

$$\theta = 90° - \arctan(0.1\omega)$$

The individual terms of the voltage gain are shown in Figure 22–11(a) and the individual terms of the phase shift are shown in Figure 22–11(b).

Figure 22-11

The combined results are shown in Figure 22–12. The total voltage gain at each frequency is determined by summing the gain due to each term of the transfer function. Similarly, the total phase shift at each frequency is the summation of phase shifts due to each of the two terms.

Figure 22-12   (*Continues*)

Figure 22–12   *(Continued)*

From the Bode plot of Figure 22–12, we conclude that the filter having this response is a high-pass filter with a cutoff frequency of 10 rad/s. The high-frequency gain of this filter is −20 dB.

practice problem IV

Sketch the Bode plot for each of the transfer functions given.

$$\text{a. } \mathbf{TF} = 1 + j0.02\omega$$

$$\text{b. } \mathbf{TF} = \frac{100}{1 + j0.005\omega}$$

$$\text{c. } \mathbf{TF} = \frac{1 + j0.02\omega}{20(1 + j0.002\omega)}$$

Writing Transfer Functions

The transfer function of any circuit is found by following a few simple steps. As we have already seen, a properly written transfer function allows us to easily calculate the cutoff frequencies and quickly sketch the corresponding Bode plot for the circuit. The steps are as follows:

1. Determine the boundary conditions for the given circuit by solving for the voltage gain when the frequency is zero (dc) and when the frequency approaches infinity. The boundary conditions are found by using the following approximations:

    At $\omega = 0$,   • inductors are short circuits,
                       • capacitors are open circuits.

    As $\omega \to \infty$,   • inductors are open circuits,
                       • capacitors are short circuits.

    By using the above approximations, all capacitors and inductors are easily removed from the circuit. The resulting voltage gain is determined for each boundary condition by simply applying the voltage divider rule.

2. Use the voltage divider rule to write the general expression for the transfer function in terms of the frequency, ω. In order to simplify the algebra, all capacitive and inductive reactance vectors are written as follows:

$$\mathbf{X}_C = \frac{1}{j\omega C}$$

and

$$\mathbf{X}_L = j\omega L$$

3. Simplify the resulting transfer function so that it is in the following format:

$$\mathbf{TF} = \frac{(j\omega\tau_{Z_1})(1 + j\omega\tau_{Z_2}) \cdots (1 + j\omega\tau_{Z_N})}{(j\omega\tau_{P_1})(1 + j\omega\tau_{P_2}) \cdots (1 + j\omega\tau_{P_M})}$$

Once the function is in this format, it is good practice to verify the boundary conditions found in Step 1. The boundary conditions are determined algebraically by first letting $\omega = 0$ and then solving for the resulting dc voltage gain. Next, we let $\omega \rightarrow \infty$. The various $(1 + j\omega\tau)$ terms of the transfer function may now be approximated as simply $j\omega\tau$, since the imaginary terms will be much larger (≥ 10) than the real components. The resultant gain will give the high-frequency gain.

4. Determine the break frequency(ies) as $\omega = 1/\tau$ (in radians per second) where the time constants will be expressed as either $\tau = RC$ or $\tau = L/R$.

5. Sketch the Bode plot by separately considering the effects of each term in the transfer function.

6. Approximate the actual response of the circuit from the Bode plot. The actual response will be a smooth, continuous curve which follows the asymptotic curve but which usually has a 3-dB difference at the cutoff frequency(ies). This approximation will not apply if two cutoff frequencies are separated by less than a decade.

These steps will now be used in analyzing several important types of filters.

22.4 The Low-Pass Filter

The *RC* Low-Pass Filter

The circuit of Figure 22–13 is referred to as a low-pass *RC* filter circuit since it permits low-frequency signals to pass from the input to the output while attenuating high frequency signals.

At low frequencies, the capacitor has a very large reactance. Consequently, at low frequencies the capacitor is essentially an open circuit resulting in the voltage across the capacitor, $\mathbf{V}_{out}$, to be essentially equal to the applied voltage, $\mathbf{V}_{in}$.

Figure 22–13   *RC* low-pass filter.

At high frequencies, the capacitor has a very small reactance, which essentially short circuits the output terminals. The voltage at the output will therefore approach zero as the frequency increases. Although we are able to easily predict what happens at the two extremes of frequency, called the *boundary conditions*, we do not yet know what occurs between the two extremes.

The circuit of Figure 22–13 is easily analyzed by applying the voltage divider rule. Namely,

$$\mathbf{V}_{out} = \frac{\mathbf{X}_C}{\mathbf{R} + \mathbf{X}_C} \mathbf{V}_{in}$$

In order to simplify the algebra, the reactance of a capacitor is expressed as follows:

$$\mathbf{X}_C = -j\frac{1}{\omega C} = -j\frac{j}{j\omega C} = \frac{1}{j\omega C} \qquad (22\text{–}13)$$

The transfer function for the circuit of Figure 22–13 is now evaluated as follows:

$$\mathbf{TF}(\omega) = \frac{\mathbf{V}_{out}}{\mathbf{V}_{in}} = \frac{\dfrac{1}{j\omega C}}{R + \dfrac{1}{j\omega C}} = \frac{\dfrac{1}{j\omega C}}{\dfrac{1 + j\omega RC}{j\omega C}}$$

$$= \frac{1}{1 + j\omega RC}$$

If we define the *cutoff frequency*, ω_c, as the frequency at which the output power is equal to half of the maximum output power (3 dB down from the maximum). This frequency occurs when the output voltage has an amplitude which is 0.7071 of the input voltage. The cutoff frequency occurs when

$$\omega_c = \frac{1}{\tau} = \frac{1}{RC} \qquad (22\text{–}14)$$

Then the transfer function may be simply written as

$$\mathbf{TF}(\omega) = \frac{1}{1 + j\dfrac{\omega}{\omega_c}} \qquad (22\text{–}15)$$

The above transfer function results in the Bode plot shown in Figure 22–14.

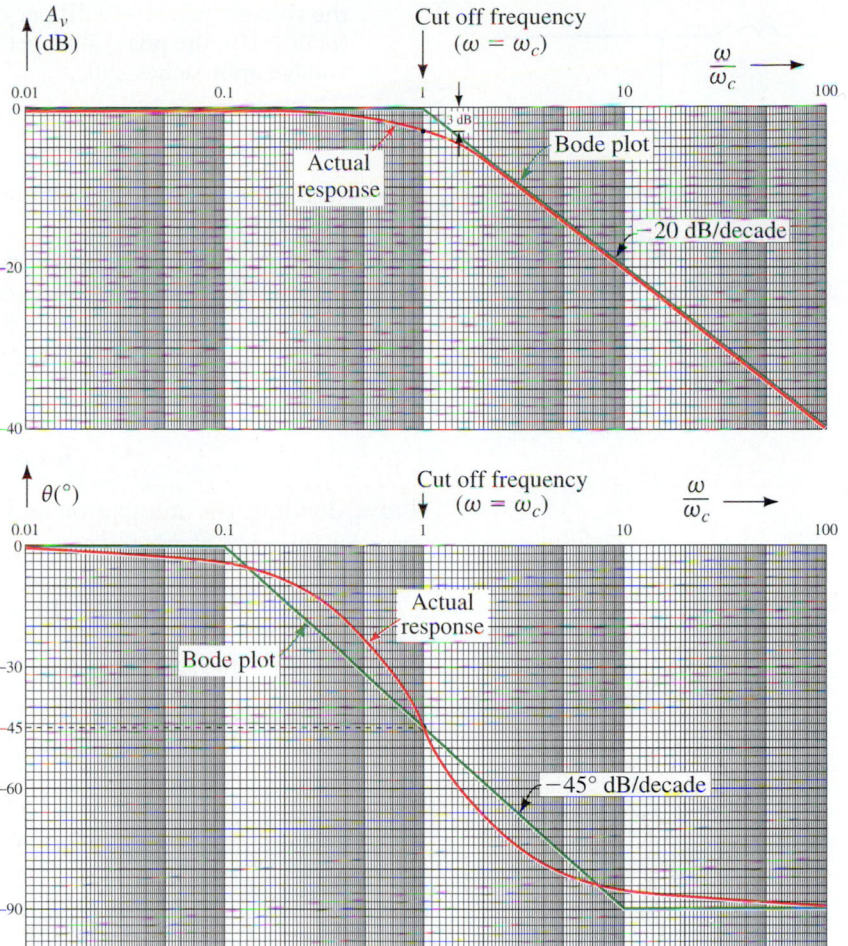

Figure 22-14   Normalized frequency response for an *RC* low-pass filter. (a) Voltage gain response. (b) Phase shift response.

Notice that the abscissas (horizontal axes) of the graphs in Figure 22–14 are shown as a ratio of ω/ω_c. Such a graph is called a *normalized plot* and eliminates the need to determine the actual cutoff frequency, ω_c. The normalized plot will have the same values for all low-pass *RC* filters. The actual frequency response of the *RC* low-pass filter can be approximated from the Bode plot by using the following:

1. At low frequencies ($\omega/\omega_c \leq 0.1$) the voltage gain is approximately 0 dB with a phase shift of about 0°. This means that the output signal of the filter is very nearly equal to the input signal.

2. At the cutoff frequency, $\omega_c = 1/RC$ ($f_c = 1/2\pi RC$), the gain of the filter is −3 dB. This means that at the cutoff frequency, the circuit will deliver half the power that it would deliver at very low frequencies. Also, at the cutoff frequency, the output voltage will lag the input voltage by 45°.

3. As the frequency increases beyond the cutoff frequency, the amplitude of the output signal decreases approximately by a factor of ten for each ten fold increase in frequency; namely

Figure 22–15   *RL* low-pass filter.

the voltage gain is −20 dB per decade. For high frequencies ($\omega/\omega_c \geq 10$), the phase shift between the input and output voltage approaches −90°.

The *RL* Low-Pass Filter

A low-pass filter circuit may be made up of a resistor and an inductor as illustrated in Figure 22–15.

In a manner similar to that used for the *RC* low-pass filter, we may write the transfer function for the circuit of Figure 22–15 as follows:

$$\mathbf{TF} = \frac{\mathbf{V}_{out}}{\mathbf{V}_{in}}$$

$$= \frac{\mathbf{R}}{\mathbf{R} + \mathbf{X}_L} = \frac{R}{R + j\omega L}$$

Now, dividing the numerator and the denominator by *R*, we have the transfer function expressed as

$$\mathbf{TF} = \frac{1}{1 + j\omega\dfrac{L}{R}}$$

Since the cutoff frequency is found as $\omega_c = 1/\tau$, we have

$$\omega_c = \frac{1}{\tau} = \frac{1}{\dfrac{L}{R}} = \frac{R}{L}$$

and so

$$\mathbf{TF} = \frac{1}{1 + j\dfrac{\omega}{\omega_c}} \tag{22–16}$$

Notice that the transfer function for the *RL* low-pass circuit in Equation 22–16 is identical with the transfer function of an *RC* circuit in Equation 22–15. In each case, the cutoff frequency is determined as the reciprocal of the time constant.

■ **EXAMPLE 22–7**   Sketch the Bode plot and the actual response curve for the circuit of Figure 22–16. Show frequencies in hertz.

Figure 22–16

Solution   The cutoff frequency (in radians per second) for the circuit occurs at

$$\omega_c = \frac{1}{\tau} = \frac{1}{RC}$$

$$= \frac{1}{(10 \text{ k}\Omega)(2 \text{ nF})} = 50 \text{ krad/s}$$

which gives

$$f_c = \frac{\omega_c}{2\pi} = \frac{50 \text{ krad/s}}{2\pi} = 7.96 \text{ kHz}$$

In order to sketch the Bode plot, we begin with the asymptotes for the voltage gain response. The circuit will have a flat response until $f_c = 7.96$ kHz. Then the gain will drop at a rate of 20 dB for each decade increase in frequency. Therefore, the voltage gain at 79.6 kHz will be −20 dB, and at 796 kHz the voltage gain will be −40 dB. At the cutoff frequency for the filter, the actual voltage gain response will pass through a point which is 3 dB down from the intersection of the two asymptotes. The frequency response of the voltage gain is shown in Figure 22–17(a).

Figure 22–17

Next, we sketch the approximate phase shift response. The phase shift at 7.96 kHz will be −45°. At a frequency one decade below the cutoff frequency (at 796 Hz) the phase shift will be approximately equal to zero, while at a frequency one decade above the cutoff frequency (at 79.6 kHz) the phase shift will be near the maximum of −90°. The actual phase shift response will be a curve which varies slightly from the asymptotic response, as shown in Figure 22–17(b).

■ **EXAMPLE 22–8**   Consider the low-pass circuit of Figure 22–18:

Figure 22–18

a. Write the transfer function for the circuit.

b. Sketch the Bode plot.

Solution:

a. The transfer function of the circuit is found as

$$\mathbf{TF} = \frac{\mathbf{V}_{out}}{\mathbf{V}_{in}} = \frac{R_2}{R_2 + R_1 + j\omega L}$$

which becomes

$$\mathbf{TF} = \frac{R_2}{R_1 + R_2}\left(\frac{1}{1 + j\omega\dfrac{L}{R_1 + R_2}}\right)$$

b. From the transfer function of Part a), we see that the dc gain will no longer be 1 (0 dB) but rather is found as

$$A_v(dc) = 20\log\left(\frac{R_2}{R_1 + R_2}\right)$$

$$= 20\log\left(\frac{1}{10}\right)$$

$$= -20 \text{ dB}$$

The cutoff frequency occurs at

$$\omega_c = \frac{1}{\tau} = \frac{1}{\dfrac{L}{R_1 + R_2}}$$

$$\omega_c = \frac{R_1 + R_2}{L} = \frac{10 \text{ k}\Omega}{2 \text{ mH}}$$

$$= 5.0 \text{ Mrad/s}$$

The resulting Bode plot is shown in Figure 22–19. Notice that the frequency response of the phase shift is precisely the same as for other low-pass filters. However, the response of the voltage gain now starts at −20 dB and then drops at a rate of −20 dB/decade above the cutoff frequency, $\omega_c = 5$ Mrad/s.

Figure 22–19

Refer to the low-pass circuit of Figure 22–20:

Figure 22–20

a. Write the transfer function for the circuit.

b. Sketch the Bode plot.

a. Design a low-pass RC filter to have a cutoff frequency of 30 krad/s. Use a 0.01-μF capacitor.

b. Design a low-pass RL filter to have a cutoff frequency of 20 kHz and a dc gain of −6 dB. Use a 10-mH inductor. (Assume that the inductor has no internal resistance.)

22.5 The High-Pass Filter

The *RC* High-Pass Filter

As the name implies, the high-pass filter is a circuit which allows high-frequency signals to pass from the input to the output of the circuit while attenuating low-frequency signals. A simple RC high-pass filter circuit is illustrated in Figure 22–21.

Figure 22–21   *RC* high-pass filter.

At low frequencies, the reactance of the capacitor will be very large, effectively preventing any input signal from passing through to the output. At high frequencies, the capacitive reactance will approach a short-circuit condition, providing a very low impedance path for the signal from the input to the output.

The transfer function of the low-pass filter is determined as follows:

$$\mathbf{TF} = \frac{\mathbf{V}_{out}}{\mathbf{V}_{in}} = \frac{\mathbf{R}}{\mathbf{R} + \mathbf{X}_c}$$

$$= \frac{R}{R + \dfrac{1}{j\omega C}} = \frac{R}{\dfrac{j\omega RC + 1}{j\omega C}} = \frac{j\omega RC}{1 + j\omega RC}$$

Now, if we let $\omega_c = 1/\tau = 1/RC$, we have

$$\mathbf{TF} = \frac{j\dfrac{\omega}{\omega_c}}{1 + j\dfrac{\omega}{\omega_c}} \qquad (22\text{–}17)$$

Notice that the expression of Equation 22–17 is very similar to the expression for a low-pass filter, with the exception that there is an additional term in the numerator. Since the transfer function is a complex number which is dependent upon frequency, we may once again find the general expressions for voltage gain and phase shift as functions of frequency, ω.

The voltage gain may be found as

$$A_v = \frac{\dfrac{\omega}{\omega_c}}{\sqrt{1 + \left(\dfrac{\omega}{\omega_c}\right)^2}}$$

which, when expressed in decibels becomes

$$A_v(\text{dB}) = 20 \log \frac{\omega}{\omega_c} - 10 \log \left[1 + \left(\frac{\omega}{\omega_c}\right)^2\right] \qquad (22\text{--}18)$$

The phase shift of the numerator will be a constant 90°, since the term has only an imaginary component. The overall phase shift of the transfer function is then found as follows:

$$\theta = 90° - \arctan \frac{\omega}{\omega_c} \qquad (22\text{--}19)$$

In order to sketch the asymptotic response of the voltage gain we need to examine Equation 22–18 for frequencies around the cutoff frequency, ω_c.

For frequencies $\omega \leq 0.1\omega_c$, the second term of the expression will be essentially equal to zero, and so the voltage gain at low frequencies may be approximated as

$$A_v(\text{dB}) \cong 20 \log \frac{\omega}{\omega_c}$$

If we substitute some arbitrary values of ω into the above approximation, we arrive at a general statement. For example, by letting $\omega = 0.01\omega_c$, we have the voltage gain as

$$A_v = 20 \log(0.01) = -40 \text{ dB}$$

and by letting $\omega = 0.1\omega_c$, we have

$$A_v = 20 \log(0.1) = -20 \text{ dB}$$

In general, we see that the expression $A_v = 20 \log (\omega/\omega_c)$ may be represented as a straight line on a semilogarithmic graph. The straight line will intersect the 0-dB axis at the cutoff frequency, ω_c, and have a slope of +20 dB/decade.

For frequencies $\omega \gg \omega_c$, Equation 22–18 may be expressed as

$$A_v(\text{dB}) \cong 20 \log \frac{\omega}{\omega_c} - 10 \log \left[\left(\frac{\omega}{\omega_c}\right)^2\right]$$

$$= 20 \log \frac{\omega}{\omega_c} - 20 \log \frac{\omega}{\omega_c}$$

$$= 0 \text{ dB}$$

For the particular case when $\omega = \omega_c$, we have

$$A_v(\text{dB}) = 20 \log 1 - 10 \log 2 = -3.0 \text{ dB}$$

which is exactly the same result that we would expect, since the actual response will be 3 dB down from the asymptotic response.

Examining Equation 22–19 for frequencies $\omega \leq 0.1\omega_c$, we see that the phase shift for the transfer function will be essentially constant at 90°, while for frequencies $\omega \geq 10\omega_c$ the phase shift will be approximately constant at 0°. At $\omega = \omega_c$ we have $\theta = 90° - 45° = 45°$.

Figure 22–22 shows the normalized Bode plot and the actual response of the high-pass circuit of Figure 22–21.

(a)

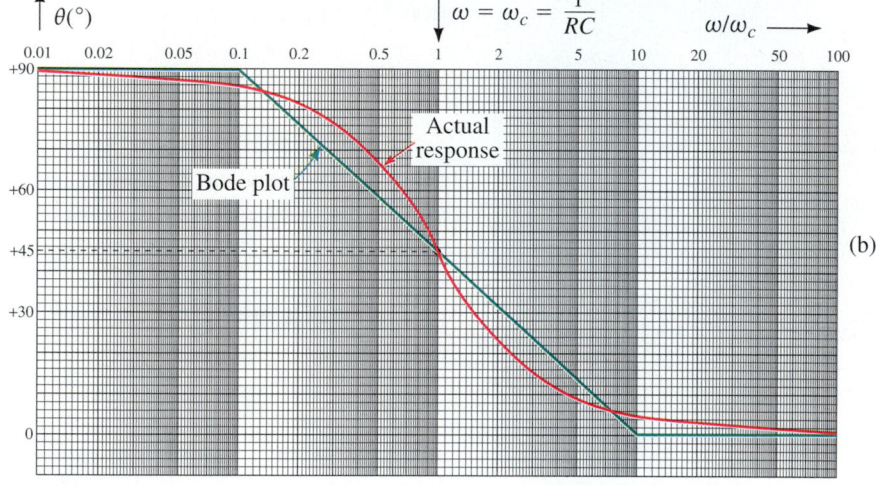

(b)

Figure 22–22   Normalized frequency response for an *RC* high-pass filter. (a) Voltage gain response. (b) Phase shift response.

The *RL* High-Pass Filter

A typical *RL* high-pass filter circuit is shown in Figure 22–23.

Figure 22–23   *RL* high-pass filter.

At low frequencies, the inductor is effectively a short circuit, which means that the output of the circuit is essentially zero at low frequencies. Inversely, at high frequencies, the reactance of the inductor approaches infinity and greatly exceeds the resistance, effectively preventing current. The voltage across the inductor is therefore very nearly equal to the applied input voltage signal. The transfer function for the high-pass RL circuit is derived as follows:

$$\mathbf{TF} = \frac{\mathbf{X}_L}{\mathbf{R} + \mathbf{X}_L}$$

$$= \frac{j\omega L}{R + j\omega L} = \frac{j\omega \dfrac{L}{R}}{L + j\omega \dfrac{L}{R}}$$

Now, letting $\omega_c = 1/\tau = R/L$, we simplify the expression as

$$\mathbf{TF} = \frac{j\omega\tau}{1 + j\omega\tau}$$

Notice that the above expression is identical with the transfer function for a high-pass RC filter, with the exception that in this case we have $\tau = L/R$.

■ **EXAMPLE 22–9**   Design the RL high-pass filter circuit of Figure 22–24 to have a cutoff frequency of 40 kHz. (Assume that the inductor has no internal resistance.) Sketch the Bode plot of the circuit expressing the frequency in kilohertz.

Figure 22-24

Solution   The cutoff frequency, ω_c, in radians per second is

$$\omega_c = 2\pi f_c = 2\pi(40 \text{ kHz}) = 251.33 \text{ krad/s}$$

Now, since $\omega_c = R/L$, we have

$$R = \omega_c L = (251.33 \text{ krad/s}(100 \text{ mH}) = 25.133 \text{ k}\Omega$$

The resulting Bode plot is shown in Figure 22–25.

Figure 22–25

practice
problem
VI

Consider the high-pass circuit of Figure 22–26:

Figure 22–26

a. Write the transfer function of the filter.

b. Sketch the Bode plot for the filter. Show the frequency in radians per second. (Hint: The frequency response of the filter is a step response.)

1. Use a 0.05-μF capacitor to design a high-pass filter having a cutoff frequency of 25 kHz. Sketch the Bode plot for the filter.

2. Use a 25-mH inductor to design a high-pass filter circuit having a cutoff frequency of 80 krad/s and a high-frequency gain of −12 dB. Sketch the Bode plot for the filter.

Figure 22–27   Block diagram of a band-pass filter.

Figure 22–28

22.6 The Band-Pass Filter

A band-pass filter will permit frequencies within a certain range to pass from the input of a circuit to the output. All frequencies which fall outside the desired range will be attenuated and so will not appear with appreciable power at the output. Such a filter circuit is easily constructed by using a low-pass filter cascaded with a high-pass circuit as illustrated in Figure 22–27.

Although the low-pass and high-pass blocks may consist of various combinations of elements, one possibility is to construct the entire filter network from resistors and capacitors as shown in Figure 22–28.

The bandwidth of the resulting band-pass filter will be approximately equal to the difference between the two cutoff frequencies, namely,

$$\text{BW} \cong \omega_2 - \omega_1 \quad (\text{rad/s}) \tag{22–20}$$

The above approximation will be most valid if the cutoff frequencies of the individual stages are separated by at least one decade.

■ **EXAMPLE 22–10**   Write the transfer function for the circuit of Figure 22–29. Sketch the resulting Bode plot and determine the expected bandwidth for the band-pass filter.

Figure 22–29

Solution   Although the transfer function for the circuit may be written by using circuit theory, it is easier to recognize that the circuit consists of two stages: one a low-pass stage and the other a high-pass stage. If the cutoff frequencies of each stage are separated by more than one decade, then we may assume that the impedance of one stage will not adversely affect the operation of the other stage. (If this is not the case, the analysis is complicated and is outside the scope of this textbook.) Based on the previous assumption, the transfer function of the first stage is determined as

$$\mathbf{TF}_1 = \frac{\mathbf{V}_1}{\mathbf{V}_{in}} = \frac{j\omega R_1 C_1}{1 + j\omega R_1 C_1}$$

and for the second stage as

$$\mathbf{TF}_2 = \frac{\mathbf{V}_{\text{out}}}{\mathbf{V}_1} = \frac{1}{1 + j\omega R_2 C_2}$$

Combining the above results, we have

$$\mathbf{TF} = \frac{\mathbf{V}_{\text{out}}}{\mathbf{V}_{\text{in}}} = \frac{(\mathbf{TF}_2)(\mathbf{V}_1)}{\dfrac{\mathbf{V}_1}{\mathbf{TF}_1}} = \mathbf{TF}_1 \mathbf{TF}_2$$

which, when simplified, becomes

$$\mathbf{TF} = \frac{\mathbf{V}_{\text{out}}}{\mathbf{V}_{\text{in}}} = \frac{j\omega\tau_1}{(1 + j\omega\tau_1)(1 + j\omega\tau_2)} \qquad (22\text{–}21)$$

where $\tau_1 = R_1 C_1 = 2.0$ ms and $\tau_2 = R_2 C_2 = 50$ μs. The corresponding cutoff frequencies are $\omega_1 = 500$ rad/s and $\omega_2 = 20$ krad/s. The transfer function of Equation 22–21 has three separate terms which, when taken separately, result in the approximate responses illustrated in Figure 22–30.

Figure 22–30

The resulting Bode plot is determined by the summation of the individual responses as shown in Figure 22–31.

Figure 22–31

From the Bode plot, we determine that the bandwidth of the resulting filter is

practice
problem
VII

Refer to the band-pass filter of Figure 22–32:

Figure 22–32

a. Calculate the cutoff frequencies and the approximate bandwidth.

b. Sketch the Bode plot for the filter.

Given a 0.1-μF capacitor and a 0.04-μF capacitor, design a band-pass filter having a bandwidth of 30 krad/s and a lower cutoff frequency of 5 krad/s. Sketch the Bode plot.

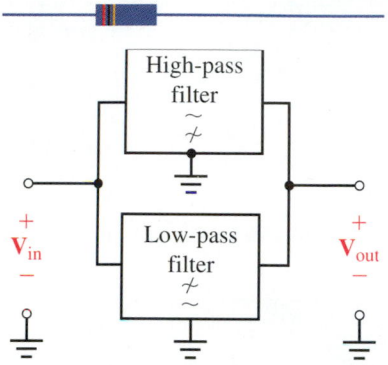

Figure 22–33 Block diagram of a band-reject filter.

22.7 The Band-Reject Filter

The band-reject filter has a response which is opposite to that of the band-pass filter. This filter passes all frequencies with the exception of a narrow band which is greatly attenuated. The band-reject filter effectively consists of a low-pass filter in shunt with a high-pass filter as illustrated in Figure 22–33.

A band-reject filter constructed of resistors, inductors, and capacitors is shown in Figure 22–34.

Figure 22–34 Notch filter.

Notice that the circuit uses a resonant tank circuit as part of the overall design. As we saw in the previous chapter, the combination of the inductor and the capacitor results in a very high tank impedance at the resonant frequency. Therefore, for any signals occurring at the resonant frequency, the output voltage is effectively zero. Because the filter circuit effectively removes any signal occurring at the resonant frequency, the circuit is often referred to as a *notch filter*. The voltage gain of the notch filter is shown in Figure 22–35.

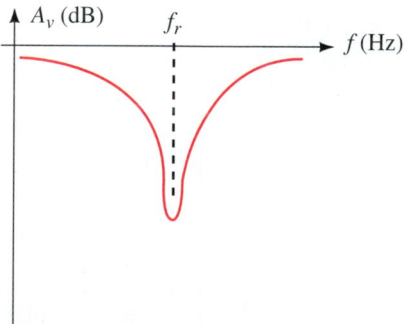

Figure 22–35 Frequency response of voltage gain for a notch filter.

For low-frequency signals, the inductor provides a low-impedance path from the input to the output, allowing these signals to pass from the input and appear across the resistor with minimal attenuation. Conversely, at high frequencies the capacitor provides a low-impedance path from the input to the output. Although the complete analysis of the notch filter is outside the scope of this textbook, the transfer function of the filter is determined by employing the same techniques as those previously developed.

$$\mathbf{TF} = \frac{\mathbf{R}}{\mathbf{R} + \mathbf{X}_L \| \mathbf{X}_C}$$

$$= \frac{R}{R + \dfrac{(j\omega L)\left(\dfrac{1}{j\omega C}\right)}{j\omega L + \dfrac{1}{j\omega C}}}$$

$$= \frac{R}{R + \dfrac{j\omega L}{1 - \omega^2 LC}}$$

$$= \frac{R(1 - \omega^2 LC)}{R - \omega^2 RLC + j\omega L}$$

$$\mathbf{TF} = \frac{1 - \omega^2 LC}{1 - \omega^2 LC + j\omega \dfrac{L}{R}} \tag{22--22}$$

Notice that the transfer function for the notch filter is significantly more complicated than for the previous filter circuits. Due to the presence of the complex quadratic in the denominator of the transfer function, this type of filter circuit is called a *second-order filter*. The design of such filters is a separate field in electronics engineering.

Actual filter design often involves using operational amplifiers to provide significant voltage gain in the passband of the filter. In addition, such active filters have the advantage of providing very high input impedance to prevent loading effects. Many excellent textbooks are available for assistance in filter design.

22.8 Circuit Analysis Using Computers

Despite the complexity of designing a filter for a particular application, the analysis of the filter is a relatively simple process when done by computer. We have already seen the ease with which PSpice may be used to examine the frequency response of resonant circuits. In this chapter we will again use the PROBE postprocessor of PSpice to plot the frequency characteristics of a particular circuit. We find that with some minor adjustments, the program is able to simultaneously plot both the voltage gain (in decibels) and the phase shift (in degrees) of any filter circuit.

■ **EXAMPLE 22–11**   Write the PSpice input file for the circuit of Figure 22–29. Use the PROBE postprocessor to observe the transfer functions over a frequency range from 1 Hz to 100 kHz.

Solution   The circuit has been redrawn in Figure 22–36 showing a voltage source and the node labels corresponding to the required PSpice format.

Figure 22–36

A supply voltage of 1 V is used as a reference input voltage since PSpice does not actually calculate the voltage gain in decibels, but rather determines the output voltage level in dBV (referenced to 1 V_{rms}).

The input file is given below:

```
Input File:
* Figure 22-36 Bandpass filter
Vs     1    0     AC     1V    0
C1     1    2     1uF
R1     2    0     2kohms
R2     2    3     500ohms
C2     3    0     .1uF
.AC    DEC  101   1Hz 100kHz
.PROBE
.OPTIONS NOPAGE
.END
```

In order to obtain the desired response curves, several adjustments are made once the program is in the PROBE postprocessor. An additional plot is obtained by using the *Plot_control* menu and selecting the *Add_plot* option. Then the output voltage is expressed in decibels by using the trace of VDB(3). The next plot is selected by once again using the *Plot_control* menu. The phase of the voltage is displayed in the second field by calling for the trace of VP(3). Because we know that the phase angle varies between 0° and 90°, the range of the Y-axis may be adjusted by using the *Y_axis* and the *Set_range* menu items. The range is set by entering "–90 90".

The resulting display is shown in Figure 22–37.

Figure 22–37

practice
problem
VIII

a. Write the PSpice input file for the circuit of Figure 22–32.

b. Use the PROBE postprocessor to observe the transfer functions over a frequency range from 1 Hz to 100 kHz.

c. From the display, determine the cutoff frequencies and use the cursors to determine the bandwidth.

d. Compare the results to those obtained in Practice Problem VII.

Problems

22.1   The Decibel

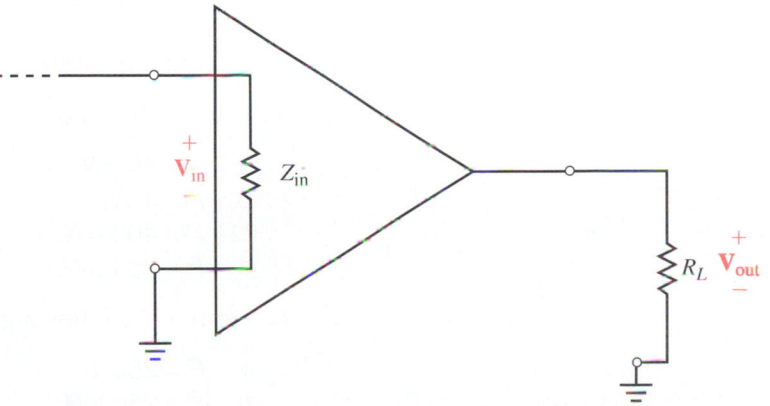

Figure 22–38

1. Refer to the amplifier shown in Figure 22–38. Determine the power gain both as a ratio and in decibels for the following power values.

   a. $P_{in} = 1.2$ mW, $P_{out} = 2.4$ W.
   b. $P_{in} = 3.5$ μW, $P_{out} = 700$ mW.
   c. $P_{in} = 6.0$ pW, $P_{out} = 12$ μW.
   d. $P_{in} = 2.5$ mW, $P_{out} = 1.0$ W.

2. If the amplifier of Figure 22–38 has $Z_{in} = 600$ Ω and $R_L = 2$ kΩ, find P_{in}, P_{out}, and A_p(dB) for the following voltage levels:

   a. $V_{in} = 20$ mV, $V_{out} = 100$ mV.
   b. $V_{in} = 100$ μV, $V_{out} = 400$ μV
   c. $V_{in} = 320$ mV, $V_{out} = 600$ mV.
   d. $V_{in} = 2$ μV, $V_{out} = 8$ V.

3. The amplifier of Figure 22–38 has $Z_{in} = 2$ kΩ and $R_L = 10$ Ω. Find the voltage gain and power gain both as a ratio and in dB for the following conditions:

   a. $V_{in} = 2$ mV, $P_{out} = 100$ mW.
   b. $P_{in} = 16$ μW, $V_{out} = 40$ mV.
   c. $V_{in} = 3$ mV, $P_{out} = 60$ mW.
   d. $P_{in} = 2$ pW, $V_{out} = 80$ mV.

4. The amplifier of Figure 22–38 has an input voltage of $V_{in} = 2$ mV and an output power of $P_{out} = 200$ mW. Find the voltage gain and power gain both as a ratio and in dB for the following conditions:

   a. $Z_{in} = 5$ kΩ, $R_L = 2$ kΩ.
   b. $Z_{in} = 2$ kΩ, $R_L = 10$ kΩ.
   c. $Z_{in} = 300$ kΩ, $R_L = 1$ kΩ
   d. $Z_{in} = 1$ kΩ $R_L = 1$ kΩ.

5. The amplifier of Figure 22–38 has an input impedance of 5 kΩ and a load resistance of 250 Ω. If the power gain of the amplifier is 35 dB, and the input voltage is 250 mV, find P_{in}, P_{out}, V_{out}, A_v, and A_v(dB).

6. Repeat Problem 5 if the input impedance is increased to 10 kΩ. (All other quantities remain unchanged.)

7. Express the following powers in dBm and in dBW:

   a. $P = 50$ mW.
   b. $P = 1$ W.
   c. $P = 400$ nW.
   d. $P = 250$ pW.

8. Express the following powers in dBm and in dBW.

   a. $P = 250$ W.
   b. $P = 250$ kW.
   c. $P = 540$ nW.
   d. $P = 27$ mW.

9. Convert the following power levels into watts:

    a. $P = 23.5$ dBm.
    b. $P = -45.2$ dBW.
    c. $P = -83$ dBm.
    d. $P = 33$ dBW.

10. Convert the following power levels into watts:

    a. $P = 16$ dBm.
    b. $P = -43$ dBW.
    c. $P = -47.3$ dBm.
    d. $P = 29$ dBW.

22.2   Multistage Systems

11. Calculate the power levels (in dBm) at the output of each of the stages of the system shown in Figure 22–39. Solve for the output power (in watts).

Figure 22–39

12. Calculate the power levels (in dBm) at the indicated locations of the system shown in Figure 22–40. Solve for the input and output powers (in watts).

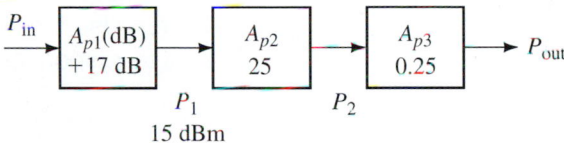

Figure 22–40

13. Given that power $P_2 = 140$ mW as shown in Figure 22–41, calculate the power levels (in dBm) at each of the indicated locations. Solve for the voltage across the load resistor, R_L.

Figure 22–41

14. Suppose that the system of Figure 22–42 has an output voltage of 2 V:

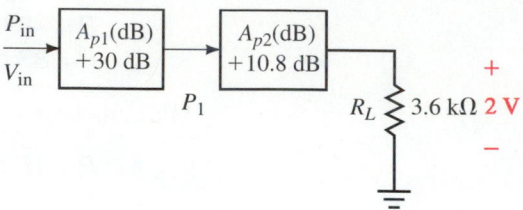

Figure 22–42

   a. Determine the power levels (in watts) at each of the indicated locations.

   b. Solve for the voltage, V_{in}, if the input impedance of the first stage is 1.5 kΩ.

   c. Convert V_{in} and V_L into voltage levels (in dBV).

   d. Solve for the voltage gain, A_v (in dB).

22.3   Simple *RC* and *RL* Transfer Functions

15. Consider the following transfer function:

$$\mathbf{TF} = \frac{200}{1 + j0.001\omega}$$

   a. Determine the cutoff frequency in radians per second and in hertz.

   b. Sketch the Bode plot showing the voltage gain and the phase shift responses on separate graphs. Label the abscissa in radians per second.

   c. Use the Bode plot to sketch the approximate actual responses.

16. Consider the following transfer function:

$$\mathbf{TF} = \frac{1 + j0.001\omega}{200}$$

   a. Determine the cutoff frequency in radians per second and in hertz.

   b. Sketch the Bode plot showing the voltage gain and the phase shift responses on separate graphs. Label the abscissa in radians per second.

   c. Use the Bode plot to sketch the approximate actual responses.

17. Consider the following transfer function:

$$\mathbf{TF} = \frac{1 + j0.02\omega}{1 + j0.001\omega}$$

a. Determine the cutoff frequencies in radians per second.
b. Sketch the Bode plot showing the voltage gain and the phase shift responses on separate graphs. Label the abscissa in radians per second.
c. Use the Bode plot to sketch the approximate actual responses.

18. Consider the following transfer function:

$$TF = \frac{1 + j0.04\omega}{(1 + j0.004\omega)(1 + j0.001\omega)}$$

a. Determine the approximate cutoff frequencies in radians per second.
b. Sketch the Bode plot showing the voltage gain and the phase shift responses on separate graphs. Label the abscissa in radians per second.
c. Use the Bode plot to sketch the approximate actual responses.

22.4 The Low-Pass Filter

19. Use a 4.0-μF capacitor to design a low-pass filter circuit having a cutoff frequency of 5 krad/s. Draw a schematic of your design and sketch the Bode plot showing the voltage gain and the phase shift responses.

20. Use a 100-mH inductor (assume $R_{coil} = 0\ \Omega$) to design a low-pass filter circuit having a cutoff frequency of 15 kHz. Draw a schematic of your design and sketch the Bode plot showing the voltage gain and the phase shift responses.

21. Refer to the low-pass circuit of Figure 22–43:
    a. Write the transfer function for the circuit.
    b. Sketch the Bode plot showing the voltage gain and phase shift responses.

22. Repeat Problem 21 for the circuit of Figure 22–44.

23. Use a 36-mH inductor to design a low-pass filter having a cutoff frequency of 36 kHz. Draw a schematic of your design and sketch the Bode plot showing the voltage gain and the phase shift responses.

24. Use a 5-μF capacitor to design a low-pass filter circuit having a cutoff frequency of 100 krad/s. Draw a schematic of your design and sketch the Bode plot showing the voltage gain and the phase shift responses.

Figure 22–43

Figure 22–44

22.5　The High-Pass Filter

25. Use a 0.05-μF capacitor to design a high-pass filter to have cutoff frequency of 100 krad/s. Draw a schematic of your design and sketch the Bode plot design showing the voltage gain and the phase shift responses.

26. Use a 16-mH inductor to design a high-pass filter circuit having a cutoff frequency of 250 kHz. Draw a schematic of your design and sketch the Bode plot showing the voltage gain and the phase shift responses.

22.6　The Band-Pass Filter

27. Refer to the filter of Figure 22–45.

Figure 22–45

a.　Determine the approximate cutoff frequencies and bandwidth of the filter. (Assume that the two stages of the filter operate independently.)

b.　Sketch the Bode plot showing the voltage gain and the phase shift responses.

28. Repeat Problem 27 for the circuit of Figure 22–46.

Figure 22–46

29. a.　Use two 0.01-μF capacitors to design a band-pass filter to have cutoff frequencies of 2 krad/s and 20 krad/s.

b.　Draw your schematic and sketch the Bode plot showing the voltage gain and the phase shift responses.

c.　Do you expect that the actual cutoff frequencies will occur at the designed cutoff frequencies? Explain.

30. a. Use two 10-mH inductors to design a band-pass filter to have cutoff frequencies of 25 krad/s and 40 krad/s.

   b. Draw your schematic and sketch the Bode plot showing the voltage gain and the phase shift responses.

   c. Do you expect that the actual cutoff frequencies will occur at the designed cutoff frequencies? Explain.

22.8 Circuit Analysis Using Computers

31. Write a PSpice input file to analyze the circuit of Figure 22–47 for frequencies of 100 Hz to 1 MHz. Use the PROBE postprocessor to obtain a display showing the voltage gain (in dBV) and the corresponding phase shift of the output.

L
250 mH
$V_{in} = 1\text{ V}\angle 0°$ $R \lessgtr 15\text{ k}\Omega \ V_{out}$

Figure 22–47

32. Write a PSpice input file to analyze your design of Problem 20, for frequencies of 100 Hz to 1 MHz. Use the PROBE postprocessor to obtain a display showing the voltage gain (in dBV) and the corresponding phase shift of the output.

33. Write a PSpice input file to analyze your design of Problem 29. Select an appropriate range of sweep frequencies. Use the PROBE postprocessor to obtain a display showing the voltage gain (in dBV) and the corresponding phase shift of the output. Determine the actual cutoff frequencies and compare these values to the design values.

34. Write a PSpice input file to analyze your design of Problem 30. Select an appropriate range of sweep frequencies. Use the PROBE postprocessor to obtain a display showing the voltage gain (in dBV) and the corresponding phase shift of the output. Determine the actual cutoff frequencies and compare these values to the design values.

Three-Phase Systems

CHAPTER PREVIEW

So far, we have looked only at single-phase systems. In this chapter, we consider three-phase systems. (Three-phase systems differ from single-phase systems in that they use a set of three voltages instead of one.) Three-phase systems are used for the generation and transmission of bulk electrical power. All commercial ac power systems, for example, are three-phase systems. Not all loads connected to a three-phase system need be three-phase, however—for example, the electric lights and appliances used in our homes require only single-phase ac. To get single-phase ac from a three-phase system, we simply tap off one of its phases.

Three-phase systems may be *balanced* or *unbalanced*. If a system is balanced, it can be analyzed by considering just one of its phases. (This is because, once you know the solution for one phase, you can write down the solutions to the other two phases without further computation.) This is significant because it makes the analysis of balanced systems only slightly more complex than the analysis of single-phase systems. Since most systems operate close to balance, many practical problems can be dealt with by assuming balance. This is the approach used in practice.

Three-phase systems possess economic and operating advantages over single-phase systems. For example, for the same power output, three-phase generators cost less than single-phase generators, produce uniform power rather than pulsating power, and operate with less vibration and noise.

We begin the chapter with a look at three-phase voltage generation.

Double circuit, three phase steel tower transmission line. *(Courtesy Bob Hibbert, Manitoba Hydro.)*

OBJECTIVES

After studying this chapter, you will be able to

- describe three-phase voltage generation,

- represent three-phase voltages and currents in phasor form,

- describe standard three-phase load connections,

- analyze balanced three-phase circuits,

- compute real power, reactive power, and apparent power in a three-phase system,

- measure power using the two-wattmeter method and the three-wattmeter method,

- analyze simple, unbalanced three-phase circuits,

- apply PSpice to three-phase problems.

23.1 Three-Phase Voltage Generation

Three-phase generators have three sets of windings and thus produce three ac voltages instead of one. To get at the idea, consider first the elementary single-phase generator of Figure 23–1. As coil AA' rotates, it produces a sinusoidal waveform $e_{AA'}$ as indicated in (b). This voltage can be represented by phasor $\mathbf{E}_{AA'}$ as shown in (c).

(a) Basic ac generator.   (b) Voltage waveform.   (c) Phasor.

Figure 23–1   A basic single-phase generator. (This is the same generator we looked at in Chapter 15.)

With two more windings added as in Figure 23–2, two additional voltages are generated. Since these windings are identical with AA' (except for their position on the rotor), they produce identical voltages. However, since coil BB' is placed 120° behind coil AA', voltage $e_{BB'}$ lags $e_{AA'}$ by 120°; similarly, coil CC', which is placed ahead of coil AA' by 120°, produces voltage $e_{CC'}$ that leads by 120°. Waveforms are shown in (b) and phasors in (c). As indicated, the generated voltages are equal in

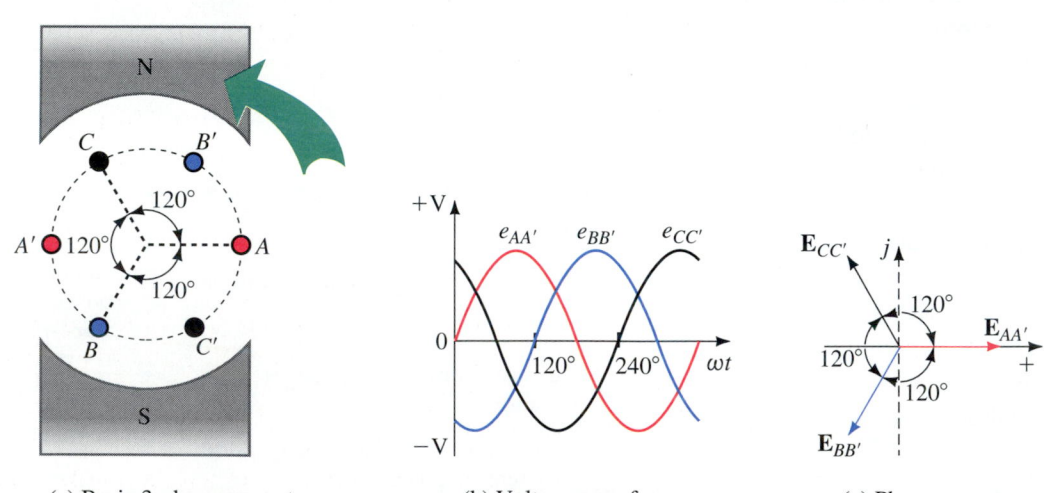

(a) Basic 3-phase generator.   (b) Voltage waveforms.   (c) Phasors.

Figure 23–2   A three-phase generator. Three sets of coils are used instead of one and thus, three sets of voltages are generated. Voltage $e_{BB'}$ lags $e_{AA'}$ by 120° while voltage $e_{CC'}$ leads $e_{AA'}$ by 120°. Three-phase generators always produce balanced voltages.

magnitude and phase displaced by 120°. Thus, if $\mathbf{E}_{AA'}$ is at 0°, then $\mathbf{E}_{BB'}$ will be at −120° and $E_{CC'}$ will be at +120°. Assuming an rms value of 120 V and a reference position of 0° for phasor $\mathbf{E}_{AA'}$ for example, yields $\mathbf{E}_{AA'}$ = 120 V∠0°, $\mathbf{E}_{BB'}$ = 120 V∠−120° and $\mathbf{E}_{CC'}$ = 120 V∠120°. Such a set of voltages is said to be *balanced*. Because of this fixed relationship between balanced voltages, you can, if you know one voltage, easily determine the other two.

practice problem 1

a. If $\mathbf{E}_{AA'}$ = 277 V∠0°, what are $\mathbf{E}_{BB'}$ and $\mathbf{E}_{CC'}$?
b. If $\mathbf{E}_{BB'}$ = 347 V∠−120°, what are $\mathbf{E}_{AA'}$ and $\mathbf{E}_{CC'}$?
c. If $\mathbf{E}_{CC'}$ = 120 V∠150°, what are $\mathbf{E}_{AA'}$ and $\mathbf{E}_{BB'}$?

Sketch the phasors for each set.

A Comment on Generator Construction

Except for small generators, most three-phase generators do not actually use the construction of Figure 23–2. Instead, they use a fixed set of windings and a rotating magnetic field. However, the voltages produced by both designs are the same. The design of Figure 23–2 was chosen to illustrate three-phase voltage generation because it is easier to visualize.

23.2　Basic Three-Phase Circuit Connections

The generator of Figure 23–2 has three independent windings: *AA'*, *BB'*, and *CC'*. As a first thought, you might try connecting loads using six wires as in Figure 23–3(a). This will work, although it is not a scheme that is used in practice. Nonetheless, some useful insights can be gained from it. To illustrate, assume a voltage of 120 V for each coil and a 12-ohm resistive load. With $\mathbf{E}_{AA'}$ as reference, Ohm's law applied to each circuit yields

$$\mathbf{I}_A = \mathbf{E}_{AA'}/R = 120 \text{ V}\angle0°/12\ \Omega = 10 \text{ A}\angle0°$$

$$\mathbf{I}_B = \mathbf{E}_{BB'}/R = 120 \text{ V}\angle-120°/12\ \Omega = 10 \text{ A}\angle-120°$$

$$\mathbf{I}_C = \mathbf{E}_{CC'}/R = 120 \text{ V}\angle120°/12\ \Omega = 10 \text{ A}\angle120°$$

These currents form a balanced set, as shown in Figure 23–3(b).

Four-Wire and Three-Wire Systems

Each load in Figure 23–3(a) has its own return wire. What if you replace them with a single wire as in (c)? By Kirchhoff's current law, the current in this wire (which we call the *neutral*) is the phasor sum of $\mathbf{I}_A$, $\mathbf{I}_B$, and $\mathbf{I}_C$. For the balanced 12-ohm load,

$$\mathbf{I}_N = \mathbf{I}_A + \mathbf{I}_B + \mathbf{I}_C = 10 \text{ A}\angle0° + 10 \text{ A}\angle-120° + 10 \text{ A}\angle120°$$

$$= (10 \text{ A} + j0) + (-5 \text{ A} - j8.66 \text{ A}) + (-5 \text{ A} + j8.66 \text{ A}) = 0 \text{ amps}$$

Thus, the return wire carries no current at all! (This result is always true regardless of load impedance, provided the load is balanced, i.e., all

(a) First thoughts.   (b) Currents form a balanced set.   (c) 4-Wire system. The return wire is called the neutral.

Figure 23–3   Evolution of three-phase connections.

phase impedances are the same.) In practice, power systems are normally operated close to balance. Thus, the return current, while not necessarily zero, will be quite small, and the neutral wire can be made smaller than the other three conductors. This configuration is called a *four-wire system* and is one of the systems used in practice.

The outgoing lines of Figure 23–3(c) are called *line* or *phase* conductors. They are the conductors that you see suspended by insulators on transmission lines as shown in the chapter opener photograph. (The smaller overhead wire is the neutral.)

Symbology

Having joined points A', B', and C' in Figure 23–3(c), we now drop the A', B', and C' notation and simply call the common point N. The voltages are then renamed $\mathbf{E}_{AN}$, $\mathbf{E}_{BN}$, and $\mathbf{E}_{CN}$. They are known as *line-to-neutral voltages*.

Standard Representation

Three-phase circuits are not usually drawn as in Figure 23–3. Rather, they are usually drawn as in Figure 23–4. (Figure 23–4(a), for example, shows Figure 23–3(c) redrawn in standard form.) Note that coil symbols are used to represent generator windings rather than the circle symbol that we use for single phase.

As Figure 23–4(a) shows, the circuit that we have been looking at is a *four-wire, wye-wye (Y-Y) circuit*. A variation, the *three-wire wye-wye circuit*, is shown in (b). Three-wire wye-wye circuits cay be used if the load can be guaranteed to remain balanced, since under balanced conditions the neutral conductor carries no current. However, for practical reasons (discussed in Section 23.7), most wye-wye systems use four wires.

(a) 4-wire Y-Y system. This is Figure 23-3(c) redrawn in standard form.

(b) 3-wire Y-Y system.

Figure 23–4   Conventional representation of three-phase circuits. These are both Y–Y systems.

Figure 23–5   A delta-connected generator. For practical reasons, delta generators are seldom used.

Delta-Connected Generators

Now consider Δ connection of the generator windings. Theoretically, this is possible; as indicated in Figure 23–5. However, there are practical difficulties. For example, when generators are loaded, distortions occur in the coil voltages due to magnetic fluxes produced by load currents. In Y-connected generators, these distortions cancel, but in Δ-connected generators, they do not. These distortions create a third harmonic current that circulates within the windings of the Δ-connected generator, lowering its efficiency. (You will learn about third harmonics in Chapter 25). For this and other reasons, Δ-connected generators are seldom used and will not be discussed in this book.

Neutral-Neutral Voltage in a Wye-Wye Circuit

In a balanced Y-Y system, neutral current is zero because line currents sum to zero. As a consequence, the voltage between neutral points is zero. To see why, consider again Figure 23–4(a). Assume that the wire joining points n and N has impedance $\mathbf{Z}_{nN}$. This yields voltage $\mathbf{V}_{nN} = \mathbf{I}_N \times \mathbf{Z}_{nN}$. But since $\mathbf{I}_N = 0$, $\mathbf{V}_{nN} = 0$, regardless of the value of $\mathbf{Z}_{nN}$. Even if the neutral conductor is absent as in (b), $\mathbf{V}_{nN}$ is still zero. Thus, *in a balanced Y-Y system, the voltage between neutral points is zero.*

■ **EXAMPLE 23–1**   Assume the circuits of Figure 23–4(a) and (b) are balanced. If $\mathbf{E}_{AN} = 247$ V∠0°, what are $\mathbf{V}_{an}$, $\mathbf{V}_{bn}$, and $\mathbf{V}_{cn}$?

Solution   In both cases, the voltage between neutral points is zero. Thus, by KVL, $\mathbf{V}_{an} = \mathbf{E}_{AN} = 247$ V∠0°. Since the system is balanced, $\mathbf{V}_{bn} = 247$ V∠−120 and $\mathbf{V}_{cn} = 247$ V∠120°.

Phase Sequence

Phase sequence refers to the order in which three-phase voltages are generated. Consider again Figure 23–2. As the rotor turns in the counterclockwise direction, voltages are generated in the sequence $e_{AA'}$, $e_{BB'}$, and $e_{CC'}$ and the system is said to have an *ABC phase sequence*. This is clearly evident when you think in terms of the phasor set (c). On the other hand, if the direction of rotation were reversed, the sequence would be *ACB*. Sequence *ABC* is called the *positive phase sequence* and is the only sequence considered in this book.

While voltages are generated in the sequence *ABC*, the order of voltages applied to a load depends on how you connect it to the source. For three-phase motors, the order is important, since if you reverse any pair of wires, the direction of the motor's rotation will reverse.

23.3 Basic Three-Phase Relationships

In three-phase systems, there are many voltages and currents to consider. To keep track of them, we use the symbols and notations of Figure 23–6. Capital letters are used at the source and lowercase letters at the load. As usual, E is used for source voltage and V for voltage drops.

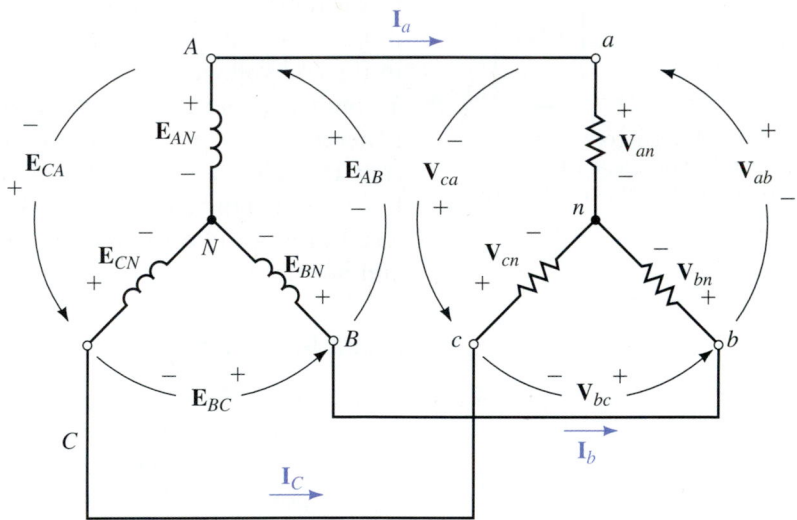

(a) For a Y, phases are defined from line to neutral.

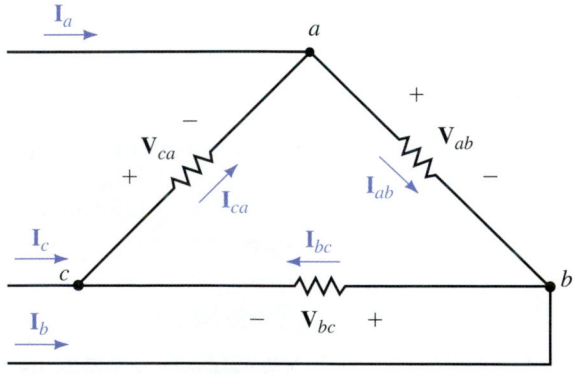

(b) For a Δ, phases are defined from line to line.

Figure 23–6   Symbols and notation. Capital letters are used at the generator and lowercase letters at the load.

Definitions

Line (also called *line-to-line*) voltages are voltages between lines. Thus, $\mathbf{E}_{AB}$, $\mathbf{E}_{BC}$, and $\mathbf{E}_{CA}$ are line-to-line voltages at the generator, while $\mathbf{V}_{ab}$, $\mathbf{V}_{bc}$, and $\mathbf{V}_{ca}$ are line-to-line voltages at the load.

Phase voltages are voltages across phases. For a Y load, phases are defined from line to neutral as indicated in (a); thus, $\mathbf{V}_{an}$, $\mathbf{V}_{bn}$, and $\mathbf{V}_{cn}$ are phase voltages for a Y load. For a Δ load, phases are defined from line to line as shown in (b); thus, $\mathbf{V}_{ab}$, $\mathbf{V}_{bc}$, and $\mathbf{V}_{ca}$ are phase voltages for a Δ. As you can see, for a Δ load, phase voltages and line voltages are the same thing. For the generator, $\mathbf{E}_{AN}$, $\mathbf{E}_{BN}$, and $\mathbf{E}_{CN}$ are phase voltages.

Line currents are the currents in the line conductors. Only a single subscript is needed. You can use either $\mathbf{I}_a$, $\mathbf{I}_b$, and $\mathbf{I}_c$ as in Figure 23–6 or $\mathbf{I}_A$, $\mathbf{I}_B$, and $\mathbf{I}_C$ as in Figure 23–4.

Phase currents are currents through phases. For the Y load Figure 23–6(a), $\mathbf{I}_a$, $\mathbf{I}_b$, and $\mathbf{I}_c$ pass through phase impedances and are therefore phase currents. For the Δ load (b), $\mathbf{I}_{ab}$, $\mathbf{I}_{bc}$, and $\mathbf{I}_{ca}$ are phase currents. As you can see, for a Y load, phase currrents and line currents are the same thing.

Phase impedances in a balanced system are equal. When we wish to refer to a phase impedance in general terms, we use $\mathbf{Z}_\phi$; when we wish to refer to a particular impedance, we use $\mathbf{Z}_{an}$, $\mathbf{Z}_{bn}$, or $\mathbf{Z}_{cn}$ as applicable for a Y load or $\mathbf{Z}_{ab}$, $\mathbf{Z}_{bc}$, or $\mathbf{Z}_{ca}$ as applicable for a Δ load.

Sometimes you need only the magnitude of a quantity. To represent the magnitude of a line voltage, use E_L at the generator and V_L at the load. For line currents, use I_L. For phase quantities, use E_ϕ, V_ϕ, and I_ϕ.

Line and Phase Voltages for a Wye Circuit

We now need the relationship between line and phase voltages for a Y circuit. Consider Figure 23–6(a). By KVL, $\mathbf{V}_{ab} - \mathbf{V}_{an} + \mathbf{V}_{bn} = 0$. Thus,

$$\mathbf{V}_{ab} = \mathbf{V}_{an} - \mathbf{V}_{bn} \qquad (23\text{–}1)$$

Now, assume a magnitude V for each phase voltage and take $\mathbf{V}_{an}$ as reference. Thus, $\mathbf{V}_{an} = V\angle 0°$ and $\mathbf{V}_{bn} = V\angle -120°$. Substitute these two into Equation 23–1:

$$\mathbf{V}_{ab} = V\angle 0° - V\angle -120° = V(1 + j0) - V(-0.5 - j0.866)$$

$$= V(1.5 + j0.866) = 1.732\ V\angle 30° = \sqrt{3}V\angle 30°$$

But $\mathbf{V}_{an} = V\angle 0°$. Thus,

$$\mathbf{V}_{ab} = \sqrt{3}\ \mathbf{V}_{an}\angle 30° \qquad (23\text{–}2)$$

Equation 23–2 shows that the magnitude of $\mathbf{V}_{ab}$ is $\sqrt{3}$ times the magnitude of $\mathbf{V}_{an}$ and that $\mathbf{V}_{ab}$ leads $\mathbf{V}_{an}$ by 30°. This is shown in Figure 23–7(a). Similar relationships hold for the other two phases. This is shown in (b). Thus, *for a balanced Y system, the magnitude of line-to-line voltage is $\sqrt{3}$ times the magnitude of the phase voltage and each line-to-line voltage leads its corresponding phase voltage by 30°.* From this you can see that *the line-to-line voltages also form a balanced set.* (Although we devel-

(a) $\mathbf{V}_{ab} = \mathbf{V}_{an} - \mathbf{V}_{bn}$.

(b) Line and phase voltages each form a balanced set.

Figure 23–7   Voltages for a balanced Y load. As soon as you know one, the other five can be determined by inspection.

oped these relationships with $\mathbf{V}_{an}$ in the 0° reference position, they are true regardless of the choice of reference.) They also hold at the source. Thus,

$$\mathbf{E}_{AB} = \sqrt{3}\ \mathbf{E}_{AN}\angle 30° \tag{23–3}$$

■ **EXAMPLE 23–2**

a. Given $\mathbf{V}_{an} = 120\ \text{V}\angle{-45°}$. Determine $\mathbf{V}_{ab}$ using Equation 23–2.
b. Verify $\mathbf{V}_{ab}$ by direct substitution of $\mathbf{V}_{an}$ and $\mathbf{V}_{bn}$ into equation 23–1.

Solution

a. $\mathbf{V}_{ab} = \sqrt{3}\mathbf{V}_{an}\angle 30° = \sqrt{3}\ (120\text{V}\angle{-45°})(1\angle 30°) = 207.8\angle{-15°}$.
b. $\mathbf{V}_{an} = 120\ \text{V}\angle{-45°}$. Thus, $\mathbf{V}_{bn} = 120\ \text{V}\angle{-165°}$.
$\mathbf{V}_{ab} = \mathbf{V}_{an} - \mathbf{V}_{bn} = (120\ \text{V}\angle{-45°}) - (120\ \text{V}\angle{-165°})$
$\quad = (84.853\ \text{V} - j84.853\ \text{V}) - (-115.911\ \text{V} - j31.058\ \text{V})$
$\quad = 200.764\ \text{V} - j53.795\ \text{V} = 207.8\ \text{V}\angle{-15°}$ as before

Nominal Voltages

While Example 23–2 yields 207.8 V for line-to-line voltage, we generally round this to 208 V and refer to the system as a 120/208-V system. These are nominal values. Other sets of nominal voltages used in practice are 277/480-V and 347/600-V.

■ **EXAMPLE 23–3**   For the circuits of Figure 23–4, suppose $\mathbf{E}_{AN} = 120\ \text{V}\angle 0°$.

a. Determine the phase voltages at the load.
b. Determine the line voltages at the load.
c. Show all voltages on a phasor diagram.

Solution

a. From Example 23–1, $\mathbf{V}_{an} = \mathbf{E}_{AN}$. Thus, $\mathbf{V}_{an} = 120\ \text{V}\angle0°$. Since the system is balanced, $\mathbf{V}_{bn} = 120\ \text{V}\angle{-120°}$ and $\mathbf{V}_{cn} = 120\ \text{V}\angle120°$.

b. $\mathbf{V}_{ab} = \sqrt{3}\ \mathbf{V}_{an}\angle30° = \sqrt{3} \times 120\ \text{V}\angle(0° + 30°) = 208\ \text{V}\angle30°$. Since line voltages form a balanced set, $\mathbf{V}_{bc} = 208\ \text{V}\angle{-90°}$ and $\mathbf{V}_{ca} = 208\ \text{V}\angle150°$.

c. The phasors are shown in Figure 23–8.

Figure 23–8

Equations 23–2 and 23–3 permit you to calculate line voltages from phase voltages. Rearranging them yields equation 23–4 which permits you to calculate phase voltage from line voltage.

$$\mathbf{V}_{an} = \frac{\mathbf{V}_{ab}}{\sqrt{3}\angle30°} \qquad \mathbf{E}_{AN} = \frac{\mathbf{E}_{AB}}{\sqrt{3}\angle30°} \qquad (23\text{–}4)$$

For example, if $\mathbf{E}_{AB} = 480\ \text{V}\angle45°$, then

$$\mathbf{E}_{AN} = \frac{\mathbf{E}_{AB}}{\sqrt{3}\angle30°} = \frac{480\ \text{V}\angle45°}{\sqrt{3}\angle30°} = 277\ \text{V}\angle15°$$

A Milestone

You have now reached an important milestone. *Given any voltage at a point in a balanced, three-phase Y system, you can, with the aid of Equation 23–2 or 23–4, determine the remaining five voltages by inspection*, i.e., by simply shifting their angles appropriately. This is the approach that we use in practice.

practice problem II

For a balanced Y generator, $\mathbf{E}_{AB} = 480\ \text{V}\angle20°$.

a. Determine the generator line voltages.
b. Determine the generator phase voltages.
c. Sketch the phasors.

Currents for a Wye Circuit

As you saw earlier, for a Y load, line currents are the same as phase currents. Consider Figure 23–9. As indicated in (b),

$$\mathbf{I}_a = \mathbf{V}_{an}/\mathbf{Z}_{an} \qquad (23\text{–}5)$$

Similarly for $\mathbf{I}_b$ and $\mathbf{I}_c$. Since $\mathbf{V}_{an}$, $\mathbf{V}_{bn}$, and $\mathbf{V}_{cn}$ form a balanced set, *line currents $\mathbf{I}_a$, $\mathbf{I}_b$, and $\mathbf{I}_c$ also form a balanced set.* Thus, if you know one, you can determine the remaining five by inspection.

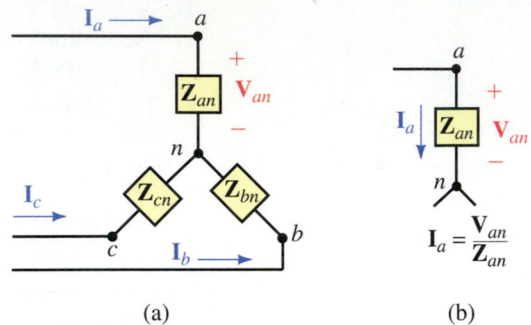

(a)   (b)

Figure 23–9   Determining currents for a Y load.

■ **EXAMPLE 23–4**   For Figure 23–10, suppose $\mathbf{V}_{an} = 120\ \text{V}\angle 0°$.

a. Compute $\mathbf{I}_a$, then determine $\mathbf{I}_b$ and $\mathbf{I}_c$ by inspection.

b. Verify by direct computation.

Figure 23–10

Solution

a. $\mathbf{I}_a = \dfrac{\mathbf{V}_{an}}{\mathbf{Z}_{an}} = \dfrac{120\angle 0°}{12 - j9} = \dfrac{120\angle 0°}{15\angle -36.87°} = 8.0\ \text{A}\angle 36.87°$

$\mathbf{I}_b$ lags $\mathbf{I}_a$ by 120°. Thus, $\mathbf{I}_b = 8\ \text{A}\angle -83.13°$.
$\mathbf{I}_c$ leads $\mathbf{I}_a$ by 120°. Thus, $\mathbf{I}_c = 8\ \text{A}\angle 156.87°$.

b. Since $\mathbf{V}_{an} = 120\ \text{V}\angle 0°$, $\mathbf{V}_{bn} = 120\ \text{V}\angle -120°$, and $\mathbf{V}_{cn} = 120\ \text{V}\angle 120°$. Thus,

$$\mathbf{I}_b = \frac{\mathbf{V}_{bn}}{\mathbf{Z}_{bn}} = \frac{120\angle-120°}{15\angle-36.87°} = 8.0\ \text{A}\angle-83.13°$$

$$\mathbf{I}_c = \frac{\mathbf{V}_{cn}}{\mathbf{Z}_{cn}} = \frac{120\angle120°}{15\angle-36.87°} = 8.0\ \text{A}\angle156.87°$$

These agree with the results obtained in (a).

1. If $\mathbf{V}_{ab} = 600\ \text{V}\angle0°$ for the circuit of Figure 23–10, what are $\mathbf{I}_a$, $\mathbf{I}_b$, and $\mathbf{I}_c$?

2. If $\mathbf{V}_{bc} = 600\ \text{V}\angle-90°$ for the circuit of Figure 23–10, what are $\mathbf{I}_a$, $\mathbf{I}_b$, and $\mathbf{I}_c$?

Line and Phase Currents for a Delta Load

Consider the delta load of Figure 23–11. Phase current $\mathbf{I}_{ab}$ can be found as in (b).

$$\mathbf{I}_{ab} = \mathbf{V}_{ab}/\mathbf{Z}_{ab} \qquad (23\text{–}6)$$

Similar relationships hold for $\mathbf{I}_{bc}$ and $\mathbf{I}_{ca}$. Since line voltages are balanced, phase currents are also balanced. Now consider again Figure 23–11(a). KCL at node a yields

$$\mathbf{I}_a = \mathbf{I}_{ab} - \mathbf{I}_{ca} \qquad (23\text{–}7)$$

After some manipulation, this reduces to

$$\mathbf{I}_a = \sqrt{3}\mathbf{I}_{ab}\angle-30° \qquad (23\text{–}8)$$

Thus, the magnitude of $\mathbf{I}_a$ is $\sqrt{3}$ times the magnitude of $\mathbf{I}_{ab}$, and $\mathbf{I}_a$ lags $\mathbf{I}_{ab}$ by 30°. Similarly, for the other two phases. Thus, *in a balanced Δ, the magnitude of the line current is $\sqrt{3}$ times the magnitude of the*

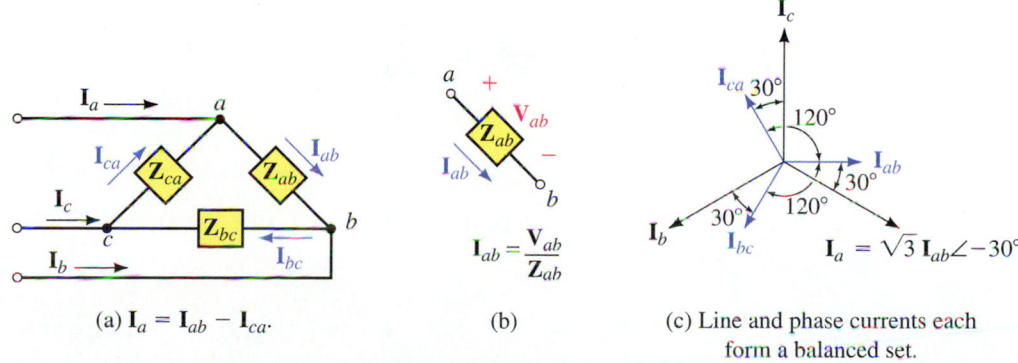

(a) $\mathbf{I}_a = \mathbf{I}_{ab} - \mathbf{I}_{ca}$.

(b)

(c) Line and phase currents each form a balanced set.

Figure 23–11   Currents for a balanced Δ load. As soon as you know one, the other five can be determined by inspection.

phase current, and each line current lags its corresponding phase current by 30°. Since phase currents are balanced, line currents are also balanced. This is shown in (c).

Equation 23–8 permits you to calculate line current from phase current. Rearranging, we get Equation 23–9, which permits you to calculate phase current from line current:

$$\mathbf{I}_{ab} = \frac{\mathbf{I}_a}{\sqrt{3}\angle -30°} \tag{23–9}$$

A Second Milestone

You have reached a second milestone. *Given any current in a balanced, three-phase Δ load, you can, with the aid of Equations 23–8 or 23–9, determine all remaining currents by inspection.*

■ **EXAMPLE 23–5**   Suppose $\mathbf{V}_{ab} = 240\ V\angle 15°$ for the circuit of Figure 23–12.

    a. Determine the phase currents.

    b. Determine the line currents.

    c. Sketch the phasor diagram.

Figure 23–12

Solution

a. $\mathbf{I}_{ab} = \dfrac{\mathbf{V}_{ab}}{\mathbf{Z}_{ab}} = \dfrac{240\angle 15°}{10 + j3} = 23.0\ A\angle -1.70°$

Thus,

$$\mathbf{I}_{bc} = 23.0\ A\angle -121.7° \quad \text{and} \quad \mathbf{I}_{ca} = 23.0\ A\angle 118.3°$$

b. $\mathbf{I}_a = \sqrt{3}\mathbf{I}_{ab}\angle -30° = 39.8\ A\angle -31.7°$

Thus,

$$\mathbf{I}_b = 39.8\ A\angle -151.7° \quad \text{and} \quad \mathbf{I}_c = 39.8\ A\angle 88.3°$$

c. Phasors are shown in Figure 23–13.

Figure 23–13

practice
problems
___IV___

1. For the circuit of Figure 23–12, if $\mathbf{I}_a = 17.32$ A$\angle 20°$, determine
   a. $\mathbf{I}_{ab}$
   b. $\mathbf{V}_{ab}$

2. For the circuit of Figure 23–12, if $\mathbf{I}_{bc} = 5$ A$\angle -140°$, what is $\mathbf{V}_{ab}$?

The Single-Phase Equivalent

By now it should be apparent that if you know the solution for one phase of a balanced system, you effectively know the solution for all three phases. We will now formalize this viewpoint by developing the *single-phase equivalent* approach to solving balanced systems. Consider a Y-Y system with line impedance. The system may be either a three-wire system or a four-wire system with neutral conductor impedance. In either case, since the voltage between neutral points is zero, you can join points n and N with a *zero-impedance conductor* without disturbing voltages or currents elsewhere in the circuit. This is illustrated in Figure 23–14(a). Phase a can now be isolated as in (b). Since $V_{nN} \equiv 0$ as before, the equa-

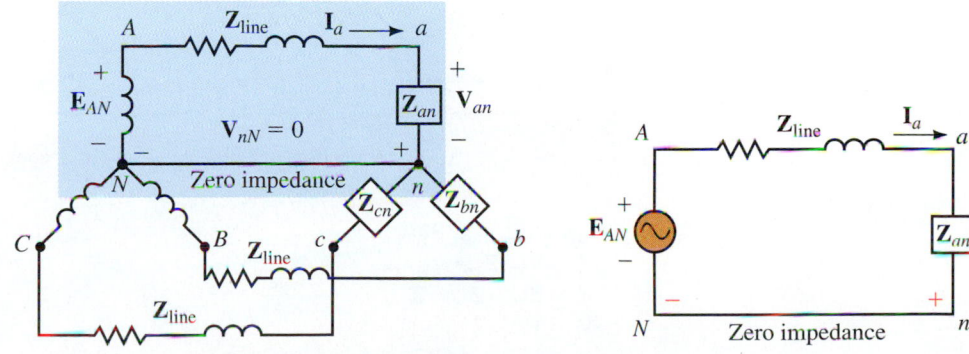

(a) Original circuit: $\mathbf{E}_{AN} = \mathbf{I}_a \mathbf{Z}_{\text{line}} + \mathbf{V}_{an}$ (b) Single phase equivalent: $\mathbf{E}_{AN} = \mathbf{I}_a \mathbf{Z}_{\text{line}} + \mathbf{V}_{an}$

Figure 23–14   Reduction to single-phase equivalent.

tion describing phase *a* in circuit (b) is the same as that describing phase *a* in the original circuit; thus, circuit (b) can be used to solve the original problem. If there are Δ loads present, convert them to Y loads using the Δ–Y conversion formula for balanced loads: $\mathbf{Z}_Y = \mathbf{Z}_\Delta/3$, from Chapter 8. This procedure is valid regardless of the configuration or complexity of the circuit. We will look at its use in Section 23.4.

Selecting a Reference

Before you solve a three-phase problem, you need to select a reference. For Y circuits, we normally choose $\mathbf{E}_{AN}$ or $\mathbf{V}_{an}$; for Δ circuits, we normally choose $\mathbf{E}_{AB}$ or $\mathbf{V}_{ab}$.

Summary of Basic Three-Phase Relationships

Table 23–1 summarizes the relationships developed so far. Note that in balanced systems (Y or Δ), <u>all</u> voltages and <u>all</u> currents are balanced.

in process
l e a r n i n g
check I

1. For Figure 23–4(a), if $\mathbf{E}_{AN} = 277$ V∠−20°, determine all line and phase voltages, source and load.

2. For Figure 23–4(a), if $\mathbf{V}_{bc} = 208$ V∠−40°, determine all line and phase voltages, source and load.

3. For Figure 23–10, if $\mathbf{I}_a = 8.25$ A∠35°, determine $\mathbf{V}_{an}$ and $\mathbf{V}_{ab}$.

4. For Figure 23–12, if $\mathbf{I}_b = 17.32$ A∠−85°, determine all voltages.

TABLE 23–1 Summary of Relationships (Balanced System). All Voltages and Currents Are Balanced

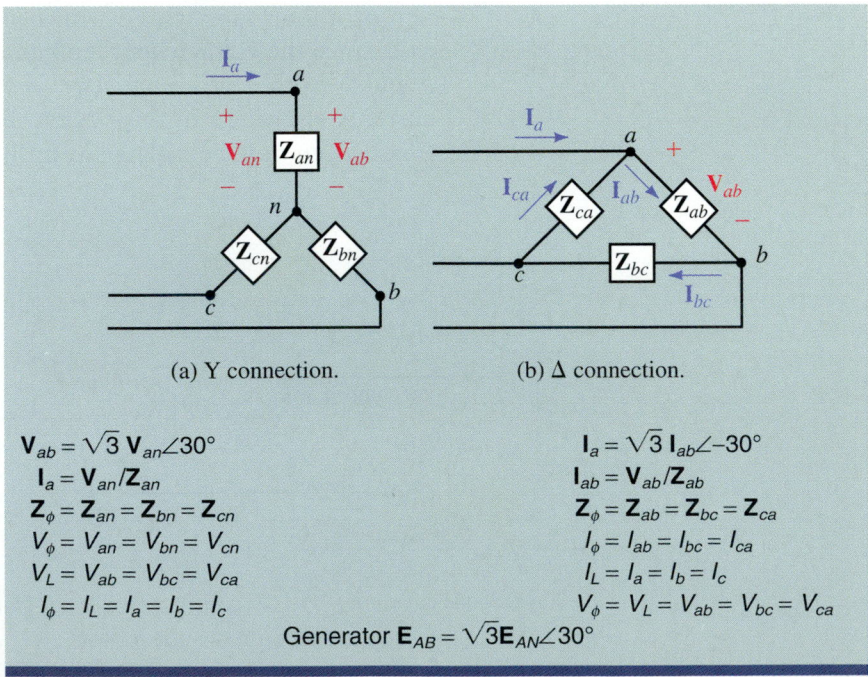

(a) Y connection. (b) Δ connection.

$\mathbf{V}_{ab} = \sqrt{3}\,\mathbf{V}_{an}\angle 30°$ $\qquad$ $\mathbf{I}_a = \sqrt{3}\,\mathbf{I}_{ab}\angle -30°$

$\mathbf{I}_a = \mathbf{V}_{an}/\mathbf{Z}_{an}$ $\qquad$ $\mathbf{I}_{ab} = \mathbf{V}_{ab}/\mathbf{Z}_{ab}$

$\mathbf{Z}_\phi = \mathbf{Z}_{an} = \mathbf{Z}_{bn} = \mathbf{Z}_{cn}$ $\qquad$ $\mathbf{Z}_\phi = \mathbf{Z}_{ab} = \mathbf{Z}_{bc} = \mathbf{Z}_{ca}$

$V_\phi = V_{an} = V_{bn} = V_{cn}$ $\qquad$ $I_\phi = I_{ab} = I_{bc} = I_{ca}$

$V_L = V_{ab} = V_{bc} = V_{ca}$ $\qquad$ $I_L = I_a = I_b = I_c$

$I_\phi = I_L = I_a = I_b = I_c$ $\qquad$ $V_\phi = V_L = V_{ab} = V_{bc} = V_{ca}$

Generator $\mathbf{E}_{AB} = \sqrt{3}\mathbf{E}_{AN}\angle 30°$

23.4 Examples

Generally, there are several ways to solve most problems. We usually try to use the simplest approach. Thus, sometimes we use the single-phase equivalent method and sometimes we solve the problem in its three-phase configuration. Generally, if a circuit has line impedance, we use the single-phase equivalent method; otherwise, we solve it directly.

■ **EXAMPLE 23–6**   For Figure 23–15, $E_{AN} = 120 \text{ V} \angle 0°$.

a. Solve for the line currents.
b. Solve for the phase voltages at the load.
c. Solve for the line voltages at the load.

(a) $Z_Y = 6 \, \Omega + j \, 8 \, \Omega$

(b) Single phase equivalent. The neutral conductor impedance has no effect on the solution since it carries no current.

Figure 23–15   A Y–Y problem.

Solution

a. Reduce the circuit to its single-phase equivalent as shown in (b).

$$I_a = \frac{E_{AN}}{Z_T} = \frac{120 \angle 0°}{(0.2 + j0.2) + (6 + j8)} = 11.7 \text{ A} \angle -52.9°$$

Therefore,

$$\mathbf{I}_b = 11.7 \text{ A}\angle{-172.9°} \quad \text{and} \quad \mathbf{I}_c = 11.7 \text{ A}\angle{67.1°}$$

b. $\mathbf{V}_{an} = \mathbf{I}_a \times \mathbf{Z}_{an} = (11.7\angle{-52.9°})(6 + j8) = 117 \text{ V}\angle{0.23°}$

Thus,

$$\mathbf{V}_{bn} = 117 \text{ V}\angle{-119.77°} \quad \text{and} \quad \mathbf{V}_{cn} = 117 \text{ V}\angle{120.23°}$$

c. $\mathbf{V}_{ab} = \sqrt{3}\mathbf{V}_{an}\angle{30°} = \sqrt{3} \times 117\angle{(0.23° + 30°)} = 202.6 \text{ V}\angle{30.23°}$

Thus,

$$\mathbf{V}_{bc} = 202.6 \text{ V}\angle{-89.77°} \quad \text{and} \quad \mathbf{V}_{ca} = 202.6 \text{ V}\angle{150.23°}$$

Note the phase shift and voltage drop across the line impedance. Note also that the impedance of the neutral conductor plays no part in the solution, since no current passes through it when the system is balanced.

practice problem V

For the circuit of Figure 23–16, $\mathbf{V}_{an} = 120 \text{ V}\angle{0°}$.

a. Find the line currents.
b. Verify that the neutral current is zero.
c. Determine generator voltages $\mathbf{E}_{AN}$ and $\mathbf{E}_{AB}$.

Figure 23–16

■ **EXAMPLE 23–7**   For the circuit of Figure 23–17, $\mathbf{E}_{AB} = 208 \text{ V}\angle{30°}$.

a. Determine the phase currents.
b. Determine the line currents.

Figure 23–17   A Y–Δ problem.

Solution

a. Since this circuit has no line impedance, the load connects directly to the source and $\mathbf{V}_{ab} = \mathbf{E}_{AB} = 208 \text{ V}\angle 30°$. Current $\mathbf{I}_{ab}$ can be found as

$$\mathbf{I}_{ab} = \frac{\mathbf{V}_{ab}}{\mathbf{Z}_{ab}} = \frac{208\angle 30°}{9 + j12} = \frac{208\angle 30°}{15\angle 53.1°} = 13.9 \text{ A}\angle -23.13°$$

Thus,

$$\mathbf{I}_{bc} = 13.9 \text{ A}\angle -143.13° \text{ and } \mathbf{I}_{ca} = 13.9 \text{ A}\angle 96.87°$$

b. $\mathbf{I}_a = \sqrt{3}\mathbf{I}_{ab}\angle -30° = \sqrt{3}(13.9)\angle(-30° -23.13°) = 24 \text{ A}\angle -53.13°$

Thus,

$$\mathbf{I}_b = 24 \text{ A}\angle -173.13° \quad \text{and} \quad \mathbf{I}_c = 24 \text{ A}\angle 66.87°$$

■ **EXAMPLE 23–8**   For the circuit of Figure 23–18(a), the magnitude of the line voltage at the generator is 208 volts. Solve for the line voltage $\mathbf{V}_{ab}$ at the load.

Solution   Since points A-a and B-b are not directly joined, $\mathbf{V}_{ab} \neq \mathbf{E}_{AB}$ and we cannot solve the circuit as in Example 23–7. Use the single-phase equivalent.

Phase voltage at the source is $208/\sqrt{3} = 120$ V. Choose $\mathbf{E}_{AN}$ as reference: $\mathbf{E}_{AN} = 120 \text{ V}\angle 0°$.

$$\mathbf{Z}_Y = \mathbf{Z}_\Delta/3 = (9 + j12)/3 = 3 \ \Omega + j4 \ \Omega.$$

The single-phase equivalent is shown in (b). Now use the voltage divider rule to find $\mathbf{V}_{an}$:

$$\mathbf{V}_{ab} = \left(\frac{3 + j4}{3.1 + j4.1}\right) \times 120\angle 0° = 117 \text{ V}\angle 0.22°$$

$$\mathbf{V}_{ab} = \sqrt{3}\mathbf{V}_{an}\angle 30° = \sqrt{3}(117 \text{ V})\angle 30.22° = 203 \text{ V}\angle 30.22°$$

(a) $\mathbf{Z}_Y = \dfrac{\mathbf{Z}_\Delta}{3} = 3\ \Omega + j\,4\ \Omega$.

(b) Single phase equivalent.

Figure 23–18   Y–Δ circuit with line impedance.

■ **EXAMPLE 23–9**   For the circuit of Figure 23–18(a), the generator phase voltage is 120 volts. Find the Δ currents.

Solution   From the single-phase equivalent (b),

$$\mathbf{I}_a = \frac{\mathbf{E}_{AN}}{\mathbf{Z}_T} = \frac{120\angle 0°}{(0.1 + j0.1) + (3 + j4)} = 23.3\ \text{A}\angle{-52.9°}$$

From Equation 23–9, Δ phase current $\mathbf{I}_{ab}$ is

$$\mathbf{I}_{ab} = \frac{\mathbf{I}_a}{\sqrt{3}\angle{-30°}} = \frac{23.3\angle{-52.9°}}{\sqrt{3}\angle{-30°}} = 13.5\ \text{A}\angle{-22.9°}$$

Thus,

$$\mathbf{I}_{bc} = 13.5\ \text{A}\angle{-142.9°} \quad \text{and} \quad \mathbf{I}_{ca} = 13.5\ \text{A}\angle 97.1°$$

■ **EXAMPLE 23–10**   A Y load and a Δ load are connected in parallel as in Figure 23–19(a). Line voltage magnitude at the generator is 208 V.

a. Find the phase voltages at the loads.
b. Find the line voltages at the loads.

(a)

(b) Δ load converted to equivalent Y.

Figure 23–19

Solution   Convert the Δ load to a Y load. Thus, $\mathbf{Z'}_Y = \frac{1}{3}\mathbf{Z}_\Delta = 30\ \Omega\angle 0°$ as in (b). Now join the neutral points N, n, and n' by a zero-impedance conductor to get the single-phase equivalent, which is shown in Figure 23–20(a). The parallel load resistors can be combined as in (b).

a. Phase voltage is 208 V/$\sqrt{3}$ = 120 V. Select $\mathbf{E}_{AN}$ as reference; $\mathbf{E}_{AN}$ = 120 V∠0°. Using the voltage divider rule yields

$$\mathbf{V}_{an} = \left(\frac{12}{12.5 + j0.5}\right) \times 120\angle 0° = 115.1\ V\angle{-2.29°}$$

Thus,

$$\mathbf{V}_{bn} = 115.1\ V\angle{-122.29°} \quad \text{and} \quad \mathbf{V}_{cn} = 115.1\ V\angle{117.71°}$$

b. $\mathbf{V}_{ab} = \sqrt{3}\mathbf{V}_{an}\angle 30° = \sqrt{3}(115.1\ V)\angle(-2.29° + 30°) = 199$ V∠27.71°

Thus,

$$\mathbf{V}_{bc} = 199\ V\angle{-92.29°} \quad \text{and} \quad \mathbf{V}_{ca} = 199\ V\angle{147.71°}$$

These are the line voltages for both the Y and Δ loads.

(a) Single phase equivalent of Figure 23-19   (b) Reduced circuit.

Figure 23–20

1. Repeat Example 23–7 using the single-phase equivalent. Compare the answers.

2. Determine Δ phase currents for the circuit of Figure 23–19. Hint: Find $\mathbf{I}_{a'}$ from Figure 23–20, then use Equation 23–9.

23.5   Power in a Balanced System

To find total power in a balanced system, determine power to one phase, then multiply by three. Per phase quantities can be found using the formulas of Chapter 17.

Real Power to a Balanced Wye Load

First, consider a Y load (Figure 23–21). The power to any phase as indicated in (b) is the product of the magnitude of the voltage V_{ϕ} across the phase impedance times the magnitude of the current I_{ϕ} through it times the cosine of the angle θ_{ϕ} between them. Since the angle between voltage and current is always the angle of the load impedance, the power per phase is

$$P_{\phi} = V_{\phi}I_{\phi}\cos\theta_{\phi} \quad \text{(W)} \tag{23–10}$$

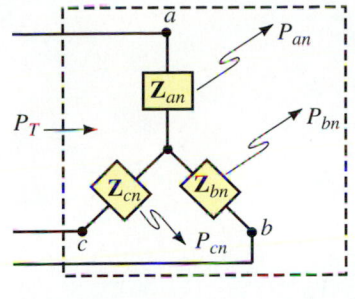

(a) $P_T = P_{an} + P_{an} + P_{cn} = 3\,P_\phi$

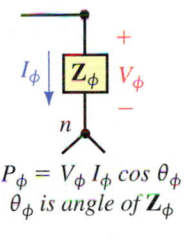

$P_\phi = V_\phi\,I_\phi\,\cos\,\theta_\phi$
θ_ϕ is angle of $\mathbf{Z}_\phi$

(b)

Figure 23-21   For a balanced system, $P_\phi = P_{an} = P_{bn} = P_{cn}$.

where θ_ϕ is the angle of $\mathbf{Z}_\phi$. Multiplying by three yields

$$P_T = 3P_\phi = 3V_\phi I_\phi \cos\,\theta_\phi \quad \text{(W)} \tag{23–11}$$

It is also handy to have a formula for power in terms of line quantities. For a Y load, $I_\phi = I_L$ and $V_\phi = V_L/\sqrt{3}$, where I_L is the magnitude of the line current and V_L is the magnitude of the line-to-line voltage. Substituting these relations into Equation 23–11 and noting that $3/\sqrt{3} = \sqrt{3}$ yields

$$P_T = \sqrt{3}\ V_L I_L \cos\,\theta_\phi \quad \text{(W)} \tag{23–12}$$

This is a very important formula and one that is widely used. Note carefully, however, that θ_ϕ is the angle of the load impedance and not the angle between V_L and I_L.

Power per phase can also be expressed as

$$P_\phi = I_\phi^2 R_\phi = V_R^2/R_\phi \tag{23–13}$$

where R_ϕ is the resistive component of the phase impedance and V_R is the voltage across it. Total power is thus

$$P_T = 3I_\phi^2 R_\phi = 3V_R^2/R_\phi \tag{23–14}$$

Reactive Power to a Balanced Wye Load

Equivalent expressions for reactive power are

$$Q_\phi = V_\phi I_\phi \sin\,\theta_\phi \quad \text{(VAR)} \tag{23–15}$$

$$= I_\phi^2 X_\phi = V_X^2/X_\phi \quad \text{(VAR)} \tag{23–16}$$

$$Q_T = \sqrt{3}V_L I_L \sin\,\theta_\phi \quad \text{(VAR)} \tag{23–17}$$

Apparent Power

$$S_\phi = V_\phi I_\phi \quad \text{(VA)} \tag{23–18}$$

$$S_T = \sqrt{3}V_L I_L \quad \text{(VA)} \tag{23–19}$$

Power Factor

$$F_p = \cos\,\theta_\phi = P_T/S_T = P_\phi/S_\phi \tag{23–20}$$

■ **EXAMPLE 23–11**   For Figure 23–22, the phase voltage is 120 V.

a. Compute real power to each phase and total power using each equation of this section.

b. Repeat (a) for reactive power.

c. Repeat (a) for apparent power.

Solution   Since we want to compare answers for the various methods, we will use 207.8 V for the line voltage rather than the nominal value of 208 V.

Figure 23–22

$$\mathbf{Z}_\phi = 9 - j12 = 15\Omega\angle{-53.13°}. \text{ Thus, } \theta_\phi = -53.13°.$$

$$V_\phi = 120 \text{ V and } I_\phi = V_\phi/Z_\phi = 120 \text{ V}/15 \ \Omega = 8.0 \text{ A}.$$

a. $P_\phi = V_\phi I_\phi \cos \theta_\phi = (120)(8)\cos(-53.13°) = 576 \text{ W}$

$P_\phi = I_\phi^2 R_\phi = (8^2)(9) = 576 \text{ W}$

$P_\phi = V_R^2/R_\phi = (72)^2/9 = 576 \text{ W}$

where $V_R = I_\phi R_\phi = (8)(9) = 72 \text{ V}.$

$P_T = 3P_\phi = 3(576) = 1728 \text{ W}$

$P_T = \sqrt{3} \ V_L I_L \cos \theta_\phi = \sqrt{3} \ (207.8)(8)\cos(-53.13°) = 1728 \text{ W}$

b. $Q_\phi = V_\phi I_\phi \sin \theta_\phi = (120)(8)\sin(-53.13°) = -768 \text{ VAR}$

$\qquad = 768 \text{ VAR (cap.)}$

$Q_\phi = I_\phi^2 X_\phi = (8)^2(12) = 768 \text{ VAR (cap.)}$

$Q_\phi = V_R^2/X_\phi = (96)^2/12 = 768 \text{ VAR (cap.)}$

$Q_T = 3Q_\phi = 3(768) = 2304 \text{ VAR (cap.)}$

$Q_T = \sqrt{3}V_L I_L \sin \theta_\phi = \sqrt{3} \ (207.8)(8)\sin(-53.13°) = -2304 \text{ VAR (cap.)}$

c. $S_\phi = V_\phi I_\phi = (120)(8) = 960 \text{ VA}$

$S_T = 3S_\phi = 3(960) = 2880 \text{ VA}$

$S_T = \sqrt{3} \ V_L I_L = \sqrt{3}(207.8)(8) = 2880 \text{ VA}$

Thus, all approaches yield the same answers.

Power to a Balanced Delta Load

For a Δ load [Figure 23–23(a)],

$$P_\phi = V_\phi I_\phi \cos \theta_\phi \quad \text{(W)} \tag{23–21}$$

where θ_ϕ is the angle of the Δ impedance. Note that this formula is identical with Equation 23–10 for the Y load. Similarly for reactive power, apparent power, and power factor. Thus, all power formulas are the same.

Results are tabulated in Table 23–2. *Note: In all of these formulas, θ_ϕ is the angle of the load impedance, i.e., the angle of* $\mathbf{Z}_{an}$ *for Y loads and* $\mathbf{Z}_{ab}$ *for* Δ *loads.*

 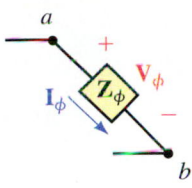

$$P_\phi = V_\phi I_\phi \cos \theta_\phi$$
θ_ϕ is angle of load impedance

(a) $P_T = P_{ab} + P_{bc} + P_{ca} = 3\,P_\phi.$

Figure 23–23   For a balanced system, $P_\phi = P_{ab} = P_{bc} = P_{ca}.$

TABLE 23–2   Power Formulas for Balanced Wye and Delta Circuits

Real power	$P_\phi = V_\phi I_\phi \cos \theta_\phi = I_\phi^2 R_\phi = \dfrac{V_R^2}{R_\phi}$
	$P_T = \sqrt{3}\, V_L I_L \cos \theta_\phi$
Reactive power	$Q_\phi = V_\phi I_\phi \sin \theta_\phi = I_\phi^2 X_\phi = \dfrac{V_X^2}{X_\phi}$
	$Q_T = \sqrt{3}\, V_L I_L \sin \theta_\phi$
Apparent power	$S_\phi = V_\phi I_\phi$
	$S_T = \sqrt{3}\, V_L I_L$
Power factor	$F_p = \cos \theta_\phi = \dfrac{P_T}{S_T} = \dfrac{P_\phi}{S_\phi}$

■ **EXAMPLE 23–12**   Determine per phase and total power (real, reactive, and apparent) for Figure 23–24. Use $V_\phi = 207.8$ V in order to compare results.

Figure 23–24

Solution

$$\mathbf{Z}_\phi = 27 - j36 = 45\ \Omega\angle{-53.13°}, \quad \text{so} \quad \theta_\phi = -53.13°$$

$$V_\phi = 207.8\ \text{V} \quad \text{and} \quad I_\phi = V_\phi/Z_\phi = 207.8\ \text{V}/45\ \Omega = 4.62\ \text{A}$$

$$P_\phi = V_\phi I_\phi \cos \theta_\phi = (207.8)(4.62)\cos(-53.13°) = 576\ \text{W}$$

$$Q_\phi = V_\phi I_\phi \sin \theta_\phi = (207.8)(4.62)\sin(-53.13°) = -768 \text{ VAR}$$

$$= 768 \text{ VAR (cap.)}$$

$$S_\phi = V_\phi I_\phi = (207.8)(4.62) = 960 \text{ VA}$$

$$P_T = 3P_\phi = 3(576) = 1728 \text{ W}$$

$$Q_T = 3Q_\phi = 3(768) = 2304 \text{ VAR (cap.)}$$

$$S_T = 3S_\phi = 3(960) = 2880 \text{ VA}$$

Note that the results here are the same as for Example 23–11. This is to be expected since the load of Figure 23–22 is the Y equivalent of the Δ load of Figure 23–24.

Power from the Single-Phase Equivalent

You can also use the single-phase equivalent to determine power. Here, all single-phase real, reactive, and apparent power formulas apply. The power that you get is of course just the power for one phase. To get the totals, multiply by three.

■ **EXAMPLE 23–13**   The total power to the balanced load of Figure 23–25 is 6912 W. The phase voltage at the load is 120 V. Compute the generator voltage $\mathbf{E}_{AB}$, magnitude and angle.

Figure 23–25

Solution   Consider the single-phase equivalent in Figure 23–26.

$$P_{an} = P_T/3 = \frac{1}{3}(6912) = 2304 \text{ W}$$

$$V_{an} = 120 \text{ V}$$

$$\theta_{an} = \cos^{-1}(0.8) = 36.87°$$

$$P_{an} = V_{an}I_a\cos\theta_{an}$$

Therefore,

$$I_a = \frac{P_{an}}{V_{an}\cos\theta_{an}} = \frac{2304}{(120)(0.8)} = 24.0 \text{ A}$$

Select as $\mathbf{V}_{an}$ reference. $\mathbf{V}_{an} = 120$ V$\angle 0°$. Thus, $\mathbf{I}_a = 24$ A$\angle -36.87°$ (since power factor was given as lagging).

$$\mathbf{E}_{AN} = \mathbf{I}_a \times \mathbf{Z}_{line} + \mathbf{V}_{an}$$

$$= (24\angle -36.87)(0.12 + j0.09) + 120\angle 0° = 123.6 \text{ V}\angle 0°$$

$$\mathbf{E}_{AB} = \sqrt{3}\ \mathbf{E}_{AN}\angle 30° = 214.1 \text{ V}\angle 30°$$

$\mathbf{Z}_{line} = 0.12\ \Omega + j\,0.09\ \Omega$

$\mathbf{I}_a = 24.0$ A$\angle -36.87°$

$\mathbf{E}_{AN}$

$\mathbf{V}_{an} = 120$ V$\angle 0°$

$P_\phi = 2304$ W

Figure 23–26

23.6   Measuring Power in Three-Phase Circuits

The Three-Wattmeter Method

Power to each phase of a three-phase circuit may be measured using the same technique used for single-phase circuits: You connect each wattmeter so that its current coil carries the current for the phase of interest while its voltage coil is connected across the phase. This is illustrated in Figure 23–27 for a four-wire Y circuit. For wattmeter W_1, voltage across its potential coil is V_{an} and current through its current coil is I_a. Thus,

$$P_1 = V_{an}I_a\cos\theta_{an}$$

which is power to phase an. Similarly, W_2 indicates power to phase bn and W_3 to phase cn. Loads may be balanced or unbalanced. The total power is

$$P_T = P_1 + P_2 + P_3 \tag{23–22}$$

CC = Current coil
PC = Potential coil

Figure 23–27   Three-wattmeter connection for a Y load.

The Two-Wattmeter Method

While three wattmeters are required for a four-wire system, for a three-wire system, only two are needed. The connection is shown in Figure 23–28. Loads may be Y or Δ, balanced or unbalanced. The meters may be connected in any pair of lines with the voltage coils connected to the third line. The total power is the algebraic sum of the meter readings.

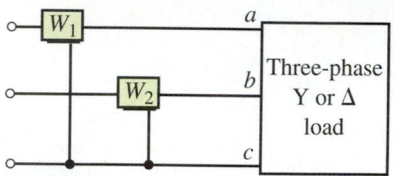

Figure 23–28   Two-wattmeter method. Load may be balanced or unbalanced.

Determining Wattmeter Readings

To determine a wattmeter reading, multiply the magnitude of the voltage across its voltage coil times the magnitude of the current through its current coil times the cosine of the angle between them. For each meter, you must carefully determine what this angle is. This is illustrated next.

■ **EXAMPLE 23–14**   For Figure 23–29, $\mathbf{V}_{an} = 120$ V∠0°. Compute the readings of each meter, then sum to determine total power. Compare P_T to the P_T found in Example 23–11.

Figure 23–29 θ_1 = angle between $\mathbf{V}_{ab}$ and $\mathbf{I}_a$ = 23.13°.
θ_2 = angle between $\mathbf{V}_{cb}$ and $\mathbf{I}_c$ = 83.13°.

Solution $\mathbf{V}_{an} = 120$ V∠0°. Thus, $\mathbf{V}_{ab} = 208$ V∠30° and $\mathbf{V}_{bc} = 208$ V∠−90°.

$\mathbf{I}_a = \mathbf{V}_{an}/\mathbf{Z}_{an} = 120$ V∠0°/(9 − j12) Ω = 8 A∠53.13°. Thus, $\mathbf{I}_c = 8$ A∠173.13°.

For W_1: $P_1 = V_{ab}I_a\cos\theta_1$, where θ_1 is the angle between $\mathbf{V}_{ab}$ and $\mathbf{I}_a$. $\mathbf{V}_{ab}$ has an angle of 30° and $\mathbf{I}_a$ has an angle of 53.13°. Thus, $\theta_1 = 23.13°$ and $P_1 = (208)(8)\cos 23.13° = 1530$ W.

For W_2: The voltage across meter 2 is $\mathbf{V}_{cb}$ and the current through it is $\mathbf{I}_c$. But $\mathbf{V}_{cb} = -\mathbf{V}_{bc} = 208\ V\angle 90°$ and $\mathbf{I}_c = 8\ A\angle 173.13°$. The angle between $\mathbf{V}_{cb}$ and $\mathbf{I}_c$ is thus $173.13° - 90° = 83.13°$. Therefore, $P_2 = V_{cb}I_c\cos\theta_2 = (208)(8)\cos 83.13° = 199\ W$.

$P_T = P_1 + P_2 = 1530 + 199 = 1729\ W$. This agrees well with the answer of 1728 W that we obtained in Example 23–11.

Change the load impedances of Figure 23–29 to $15\ \Omega\angle 70°$. Repeat Example 23–14. (As a check, total power to the load is 985 W. *Hint:* One of the meters reads backwards.)

As Practice Problem VII shows, the low-reading wattmeter may read backward. (If this happens, reverse either its CC or its PC connection to make it read upscale.) Its reading must then be subtracted. Thus, if P_h and P_ℓ are the high-reading and low-reading meters respectively,

$$P_T = P_h - P_\ell \qquad (23\text{–}23)$$

note

Watts Ratio Curve

***T**o understand why a wattmeter may read backward, recall that it indicates the product of the magnitude of the voltage across its potential coil times the magnitude of current through its current coil times the cosine of the angle between them. This angle is not the angle θ_ϕ of the load impedance; Rather, it can be shown that, for a balanced load, one meter will indicate $V_\phi I_\phi\cos(\theta_\phi - 30°)$ while the other will indicate $V_\phi I_\phi\cos(\theta_\phi + 30°)$. If the magnitude of either angle exceeds 90°, its cosine will be negative and the corresponding meter will read backward.*

The power factor for a balanced load can be obtained using Equation 23–20. It can also be obtained from the wattmeter readings using a simple curve called the *watts ratio curve*, shown in Figure 23–30. (To see where this curve comes from, see end-of-chapter Problem 46.)

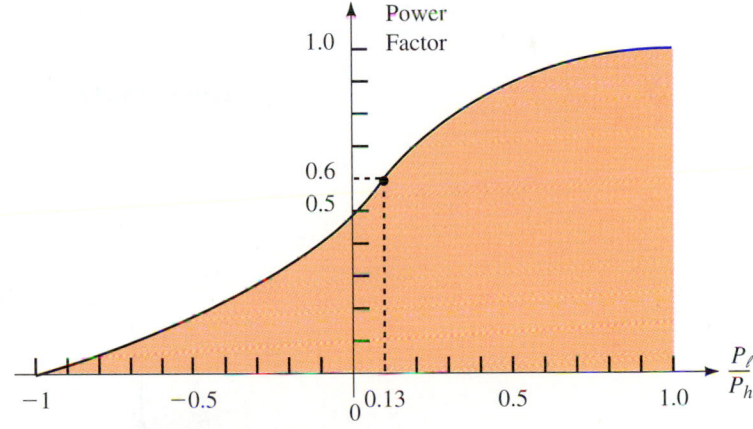

Figure 23–30   Watts ratio curve.

■ **EXAMPLE 23–15**   Consider again Figure 23–29.

a. Determine the power factor from the load impedance.

b. Using the meter readings of Example 23–14, determine the power factor from the watts ratio curve.

Solution

a. $F_p = \cos\theta_\phi = \cos 53.13° = 0.6$.

b. $P_\ell = 199\ W$ and $P_h = 1530\ W$. Therefore, $P_\ell/P_h = 0.13$. From Figure 23–30, $F_p = 0.6$.

The problem with the watts ratio curve is that values are difficult to determine accurately from the graph. However, it can be shown that

$$\tan \theta_\phi = \sqrt{3}\left(\frac{P_h - P_\ell}{P_h + P_\ell}\right) \qquad (23\text{–}24)$$

From this, you can determine θ_ϕ, then you can compute power factor from $F_p = \cos \theta_\phi$.

23.7 Unbalanced Systems

For unbalanced loads, none of the balanced circuit relationships apply. Each problem must therefore be treated as a three-phase problem. We will look at a few examples that can be handled by fundamental circuit techniques such as Kirchhoff's laws and mesh analysis. Assume balanced source voltages.

Unbalanced Wye Loads

Unbalanced four-wire Y systems without line impedance are easily handled using Ohm's law. However, for three-wire systems or four-wire systems with line and neutral impedance, you generally have to use mesh equations. One of the problems with unbalanced three-wire Y systems is that you get different voltages across each phase of the load and a voltage between neutral points. This is illustrated next.

■ **EXAMPLE 23–16**   For Figure 23–31(a), the line-to-line voltage at the generator is 208 V. Select $\mathbf{E}_{AB}$ as reference and determine line currents and load voltages.

(a)

Figure 23–31 (Continues)

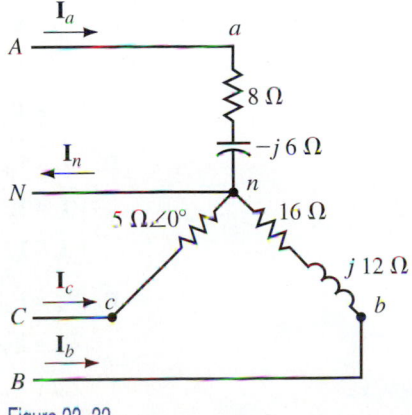

(b) $\mathbf{I}_a = \mathbf{I}_1$; $\mathbf{I}_b = \mathbf{I}_2 - \mathbf{I}_1$; $\mathbf{I}_c = -\mathbf{I}_3$

Figure 23–31   (Continued)

Solution   Use mesh analysis. The circuit has been redrawn in (b) to help set up the equations. $\mathbf{E}_{AB} = 208$ V$\angle 0°$ and $\mathbf{E}_{BC} = 208$ V$\angle -120°$.

$$\textit{Loop 1: } (8 + j4)\mathbf{I}_1 - (3 + j4)\mathbf{I}_2 = 208 \text{ V}\angle 0°$$

$$\textit{Loop 2: } -(3 + j4)\mathbf{I}_1 + (9 - j4)\mathbf{I}_2 = 208 \text{ V}\angle -120°$$

These equations can be solved using standard techniques such as determinants. The solutions are

$$\mathbf{I}_1 = 29.9 \text{ A}\angle -26.2° \quad \text{and} \quad \mathbf{I}_2 = 11.8 \text{ A}\angle -51.5°$$

$$\text{KCL: } \mathbf{I}_a = \mathbf{I}_1 = 29.9 \text{ A}\angle -26.2°$$

$$\mathbf{I}_b = \mathbf{I}_2 - \mathbf{I}_1 = 19.9 \text{ A}\angle 168.5°$$

$$\mathbf{I}_c = -\mathbf{I}_2 = 11.8 \text{ A}\angle 128.5°$$

$$\mathbf{V}_{an} = \mathbf{I}_a \mathbf{Z}_{an} = (29.9 \text{A}\angle -26.2°)\,(5) = 149.5 \text{ V}\angle -26.2°$$

Similarly,

$$\mathbf{V}_{bn} = 99 \text{ V}\angle -138.4° \quad \text{and} \quad \mathbf{V}_{cn} = 118 \text{ V}\angle 75.4°$$

practice problems VIII

1. Using KVL and the results from Example 23–16, compute the voltage between neutral points n and N of Figure 23–31(a).

2. For the circuit of Figure 23–32, $\mathbf{E}_{AN} = 120$ V$\angle 0°$. Compute the line currents, power to each phase, and total power. Do not use mesh analysis here.

Figure 23–32

Unbalanced Delta Loads

Systems without line impedance are easily handled since the source voltage is applied directly to the load. For systems with line impedance, use mesh equations.

■ **EXAMPLE 23–17** For the circuit of Figure 23–33, the line voltage is 240 V. Take $\mathbf{V}_{ab}$ as reference and do the following:

a. Determine the phase currents and sketch their phasor diagram.
b. Determine the line currents.
c. Determine the total power to the load.

(a)

$\mathbf{I}_{ca} = 15\ A\angle150°$

$\mathbf{I}_{bc} = 20\ A\angle-180°$ $\mathbf{I}_{ab} = 9.6\ A\angle0°$

(b) Phase currents

Figure 23–33   Unbalanced Δ load.

Solution

a. $\mathbf{I}_{ab} = \mathbf{V}_{ab}/\mathbf{Z}_{ab} = (240\ V\angle0°)/25\ \Omega = 9.6\ A\angle0°$

$\mathbf{I}_{bc} = \mathbf{V}_{bc}/\mathbf{Z}_{bc} = (240\ V\angle-120°)/(12\ \Omega\angle60°) = 20\ A\angle-180°$

$\mathbf{I}_{ca} = \mathbf{V}_{ca}/\mathbf{Z}_{ca} = (240\ V\angle120°)/(16\ \Omega\angle-30°) = 15\ A\angle150°$

b. $\mathbf{I}_a = \mathbf{I}_{ab} - \mathbf{I}_{ca} = 9.6\ A\angle0° - 15\ A\angle150° = 23.8\ A\angle-18.4°$

$\mathbf{I}_b = \mathbf{I}_{bc} - \mathbf{I}_{ab} = 20\ A\angle-180° - 9.6\ A\angle0° = 29.6\ A\angle180°$

$\mathbf{I}_c = \mathbf{I}_{ca} - \mathbf{I}_{bc} = 15\ A\angle150° - 20\ A\angle-180° = 10.3\ A\angle46.9°$

c. $P_{ab} = V_{ab}I_{ab}\cos\theta_{ab} = (240)(9.6)\cos 0° = 2304\ W$

$P_{bc} = V_{bc}I_{bc}\cos\theta_{bc} = (240)(20)\cos 60° = 2400\ W$

$P_{ca} = V_{ca}I_{ca}\cos\theta_{ca} = (240)(15)\cos 30° = 3118\ W$

$P_T = P_{ab} + P_{bc} + P_{ca} = 7822\ W$

■ **EXAMPLE 23–18**   A pair of wattmeters are added to the circuit of Figure 23–33 as illustrated in Figure 23–34. Determine wattmeter readings and compare them to the total power calculated in Example 23–17.

Figure 23–34

Solution $P_1 = V_{ac}I_a\cos\theta_1$, where θ_1 is the angle between V_{ac} and I_a. From Example 23–17, $I_a = 23.8$ A$\angle-18.4°$ and $V_{ac} = -V_{ca} = 240$ V$\angle-60°$. Thus, $\theta_1 = 60° - 18.4° = 41.6°$. Therefore,

$$P_1 = (240)(23.8)\cos 41.6° = 4271 \text{ W}$$

$P_2 = V_{bc}I_b\cos\theta_2$, where θ_2 is the angle between V_{bc} and I_b. $V_{bc} = 240$ V$\angle-120°$ and $I_b = 29.6$ A$\angle180°$. Thus, $\theta_2 = 60°$. So

$$P_2 = (240)(29.6)\cos 60° = 3552 \text{ W}$$

$P_T = P_1 + P_2 = 7823$ W, which compares well with our earlier solution of 7822 W.

23.8 Power System Loads

Before we leave this chapter, we look briefly at how both single-phase and three-phase loads may be connected to a three-phase system. (This is necessary because residential and business customers require only single-phase power, while industrial customers sometimes require both single-phase and three-phase power.) Figure 23–35 shows how this may be done. (Even this is simplified, as real systems contain transformers. However, the basic principles are correct.) Two points should be noted here.

1. In order to approximately balance the system, the utility tries to connect one third of its single phase loads to each phase. Three-phase loads generally are balanced.

2. Real loads are seldom expressed in terms of resistance, capacitance, and inductance. Instead, they are described in terms of power, power factor, and so on. This is because most loads consist of electric lights, motors, and the like which are never described in terms of impedance. (For example, you purchase light bulbs as 60-W bulbs, 100-W bulbs, etc., and electric motors as ½ horsepower, etc. You never ask for a 240-ohm light!) The circuit of Figure 23–35 can be analyzed using the principles of this chapter.

Figure 23–35   Combined three-phase and single-phase loads.

23.9 PSpice For Three-Phase Circuits

PSpice may be used to analyze three-phase circuits, balanced or unbalanced, wye or delta, and so on. Consider Figure 23–36. Nodes have been numbered in a straightforwad way, with the neutral of the generator selected as node 0. However, PSpice cannot solve this circuit unless we make a minor change. Because of its emulation technique, PSpice needs a dc path from every node to the reference node. Because of the capacitors, it sees node 7 as floating. A simple way around this problem is to place a high value resistance (called R7 in the input file) between nodes 7 and 0. A value of 100 kΩ has been arbitrarily chosen. Current $\mathbf{I}_a$ is the current in R_1 and so on. The input file is shown below.

```
Input File:
 * Figure 23-36: Balanced Y Load
VA   1   0   AC   120    0
VB   2   0   AC   120   -120
VC   3   0   AC   120    120
R1   1   4   3
R2   2   5   3
R3   3   6   3
C1   4   7   663.13uF
C2   5   7   663.13uF
C3   6   7   663.13uF
R7   7   0   100kohms
.AC LIN 1 60Hz    60Hz
.PRINT AC IM(R1) IP(R1) IM(R2) IP(R2) IM(R3) IP(R3)
.END
```

Figure 23–36   Circuit for PSpice analysis.

When you run this example, you get currents of $\mathbf{I}_a = 24 \text{ A}\angle 53.13°$, $\mathbf{I}_b = 24 \text{ A}\angle{-66.87°}$, and $\mathbf{I}_c = 24 \text{ A}\angle 173.13°$. These can be easily verified by manual solution.

Problems

23.2   Basic Three-Phase Circuit Connections

1. As long as the loads and voltages of Figure 23–3 are balanced (regardless of their actual value), currents $\mathbf{I}_A$, $\mathbf{I}_B$, and $\mathbf{I}_C$ will sum to zero. To illustrate, change the load impedance from 12 Ω to 15 $\Omega\angle30°$ and for $\mathbf{E}_{AA'} = 120$ V$\angle0°$, do the following:
   a. Compute currents $\mathbf{I}_A$, $\mathbf{I}_B$, and $\mathbf{I}_C$.
   b. Show that $\mathbf{I}_A + \mathbf{I}_B + \mathbf{I}_C = 0$.

2. For Figure 23–3(c), $\mathbf{E}_{AN} = 277$ V$\angle-15°$.
   a. What are $\mathbf{E}_{BN}$ and $\mathbf{E}_{CN}$?
   b. If the resistance is 5.54 Ω, compute currents $\mathbf{I}_A$, $\mathbf{I}_B$, and $\mathbf{I}_C$.
   c. Show that $\mathbf{I}_N = 0$.

3. If the generator of Figure 23–2 is rotated in the clockwise direction, what will be the phase sequence of the generated voltages? Sketch and label the waveforms.

23.3   Basic Three-Phase Relationships

4. For the generators of Figure 23–4, $\mathbf{E}_{AN} = 7620$ V$\angle-18°$.
   a. What are the phase voltages $\mathbf{E}_{BN}$ and $\mathbf{E}_{CN}$?
   b. Determine the line-to-line voltages.
   c. Sketch their phasor diagram.

5. For the loads of Figure 23–4, $\mathbf{V}_{bc} = 208$ V$\angle-75°$.
   a. Determine the line-to-line voltages $\mathbf{V}_{ab}$ and $\mathbf{V}_{ca}$.
   b. Determine the phase voltages.
   c. Sketch their phasor diagram.

6. Repeat Problem 5 if $\mathbf{V}_{ca} = 208$ V$\angle90°$.

7. For the load of Figure 23–37, $\mathbf{V}_{an} = 347$ V$\angle15°$. Determine all line currents.

Figure 23–37

8. For the load of Figure 23–37, if $\mathbf{I}_a = 7.8$ A$\angle-10°$, determine the phase voltages and line voltages.

9. For the circuit of Figure 23–38, $\mathbf{V}_{ab} = 480$ V$\angle0°$. Find the phase and line currents.

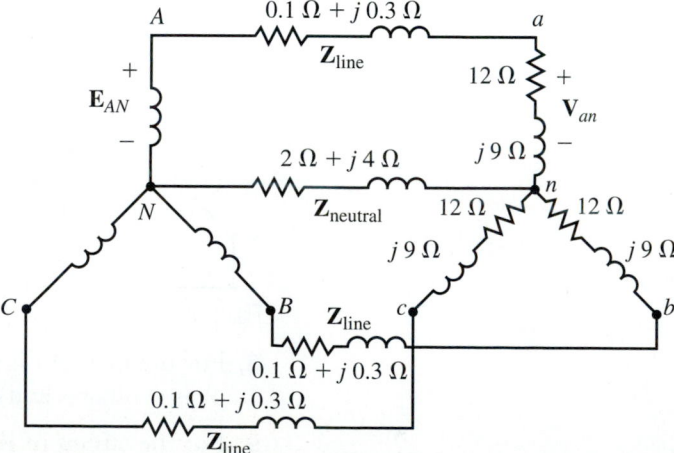

Figure 23–38

10. For the circuit of Figure 23–38, if $\mathbf{I}_a = 41.0$ A$\angle$–46.7°, find all the phase currents.

11. For the circuit of Figure 23–38, if $\mathbf{I}_{ab} = 10$ A$\angle$–21°, determine all the line voltages.

12. For the circuit of Figure 23–38, if $\mathbf{I}_a = 11.0$ A$\angle$30°, find all the line voltages.

13. a. For a certain balanced Y load, $\mathbf{V}_{ab} = 208$ V$\angle$30°, $\mathbf{I}_a = 24$ A$\angle$40° and $f = 60$ Hz. Determine the load (R and L or C).
    b. Repeat (a) if $\mathbf{V}_{bc} = 208$ V$\angle$–30° and $\mathbf{I}_c = 12$ A$\angle$140°.

14. Consider Figure 23–11(a). Show that $\mathbf{I}_a = \sqrt{3}\mathbf{I}_{ab}\angle$–30°.

15. At 60 Hz, a balanced Δ load has current $\mathbf{I}_{bc} = 4.5$ A$\angle$–85°. The line voltage is 240 volts and $\mathbf{V}_{ab}$ is taken as reference.
    a. Find the other phase currents.
    b. Find the line currents.
    c. Find the resistance R and capacitance C of the load.

16. A Y generator with $\mathbf{E}_{AN} = 120$ V$\angle$0° drives a balanced Δ load. If $\mathbf{I}_a = 43.6$ A$\angle$–37.5°, what are the load impedances?

23.4   Examples

17. For Figure 23–39, $\mathbf{V}_{an} = 120$ V$\angle$0°.
    a. Find phase voltage $\mathbf{E}_{AN}$, magnitude and angle.
    b. Find line voltage $\mathbf{E}_{AB}$, magnitude and angle.

Figure 23–39

18. For Figure 23–39, $\mathbf{E}_{AN} = 120\ \text{V}\angle 20°$.
    a.   Find phase voltage $\mathbf{V}_{an}$, magnitude and angle.
    b.   Find line voltage $\mathbf{V}_{ab}$, magnitude and angle.

19. For Figure 23–38, $\mathbf{E}_{AN} = 120\ \text{V}\angle -10°$. Find the line currents using the single-phase equivalent method.

20. Repeat Problem 19 if $\mathbf{E}_{BN} = 120\ \text{V}\angle -100°$.

21. For Figure 23–38, assume the lines have impedance $\mathbf{Z}_{\text{line}}$ of $0.15\Omega + j0.25\ \Omega$ and $\mathbf{E}_{AN} = 120\ \text{V}\angle 0°$. Convert the Δ load to a Y and use the single-phase equivalent to find the line currents.

22. For Problem 21, find the phase currents in the Δ network.

23. For the circuit of Figure 23–39, assume $\mathbf{Z}_{\text{line}} = 0.15\ \Omega + j0.25\ \Omega$ and $\mathbf{V}_{ab} = 600\ \text{V}\angle 30°$. Determine $\mathbf{E}_{AB}$.

24. For Figure 23–40, $\mathbf{Z}_Y = 12\ \Omega + j9\ \Omega$ and $\mathbf{Z}_\Delta = 27\ \Omega + j36\ \Omega$. At the Y load, $\mathbf{V}_{an} = 120\ \text{V}\angle 0°$.
    a.   Draw the single-phase equivalent.
    b.   Find generator voltage $\mathbf{E}_{AN}$.

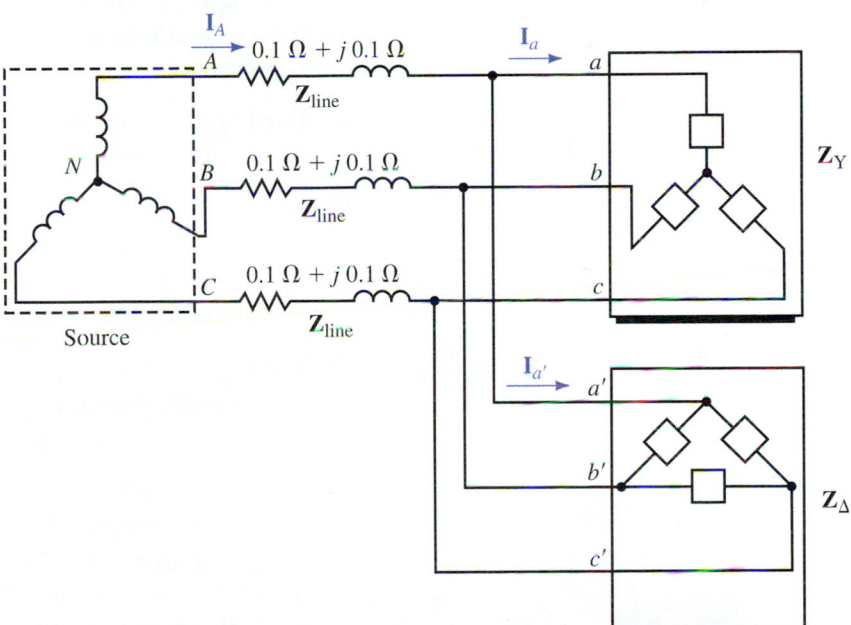

Figure 23–40

25. Same as Problem 24 except that the phase voltage at the Δ load is $\mathbf{V}_{a'b'} = 480\ \text{V}\angle 30°$. Find generator voltage $\mathbf{E}_{AB}$, magnitude and angle.

26. For Figure 23–40, $\mathbf{Z}_Y = 12\ \Omega + j9\ \Omega$ and $\mathbf{Z}_\Delta = 36\ \Omega + j27\ \Omega$. Line current $\mathbf{I}_A$ is $46.2\ \text{A}\angle -36.87°$. Find the phase currents for both loads.

27. For Figure 23–40, $Z_Y = 15\ \Omega + j20\ \Omega$, $Z_\Delta = 9\ \Omega - j12\ \Omega$ and $I_{a'b'} = 40\ A\angle73.13°$. Find Y phase voltage V_{an}, magnitude and angle.

23.5   Power in a Balanced System

28. For Figure 23–37, $E_{AN} = 120$ volts.
   a. Determine per phase real, reactive, and apparent power.
   b. Multiply per phase quantities to get total quantities.

29. For Figure 23–37, compute the total real, reactive, and apparent power using the P_T, Q_T, and S_T formulas of Table 23–2. (Use $V_L = 207.8$ V rather than the nominal value of 208 V here.) Compare these to the results of Problem 28.

30. For Figure 23–38, $E_{AB} = 208$ volts.
   a. Determine per phase real, reactive, and apparent power.
   b. Multiply per phase quantities to get total quantities.

31. For Figure 23–38, $E_{AB} = 208$ V. Compute the total real, reactive, and apparent power using the P_T, Q_T, and S_T formulas of Table 23–2. Compare to the results of Problem 30.

32. For Figure 23–10, if $V_{an} = 277$ V, determine the total active power, total reactive power, total apparent power, and power factor.

33. For Figure 23–12, if $V_{ab} = 600$ V, determine the total power, total reactive power, total apparent power, and power factor.

34. For Figure 23–16, if $V_{an} = 120$ V, determine the total power, total reactive power, and total apparent power,
   a. supplied to the load.
   b. output by the source.

35. For Figure 23–17, if $V_{ab} = 480$ V, determine the total power, total reactive power, total apparent power, and power factor.

36. For Figure 23–40, let $Z_{line} = 0\ \Omega$, $Z_Y = 20\ \Omega\angle0°$, $Z_\Delta = 30\ \Omega\angle10°$, and $E_{AN} = 120$ V.
   a. Find the total real power, reactive power, and apparent power to the Y load.
   b. Repeat (a) for the Δ load.
   c. Determine total watts, VARs, and VA using the results of (a) and (b).

37. $V_{ab} = 208$ V for a balanced Y load, $P_T = 1200$ W, and $Q_T = 750$ VAR (ind.). Choose V_{an} as reference and determine I_a.

38. A motor (delivering 100 hp to a load) and a bank of power factor capacitors are connected as in Figure 23–41. The capacitors are rated $Q_C = 45$ kVAR. Reduce the problem to its single-phase equivalent, then compute the resultant power factor of the system.

Figure 23–41

39. The capacitors of Figure 23–41 are connected in Y and each has the value $C = 120 \ \mu F$. Compute the resultant power factor. Frequency is 60 Hz.

23.6 Measuring Power in Three-Phase Circuits

40. Sketch a circuit showing how to measure power to a Δ load using three wattmeters.

41. For Figure 23–37, the generator phase voltage is 120 volts.
   a. Sketch three wattmeters correctly into the circuit.
   b. Compute the reading of each wattmeter.
   c. Sum the readings and compare this to the result of 2304 W obtained in Problem 28.

42. Figures 23–28 and 23–29 show two ways that two wattmeters may be connected to measure power in a three-phase, three-wire circuit. There is one more way. Sketch it.

43. For the circuit of Figure 23–42, $\mathbf{V}_{ab} = 208 \ V\angle 30°$.
   a. Determine the magnitude and angle of the currents.
   b. Determine power per phase and total power, P_T.
   c. Compute the reading of each wattmeter.
   d. Sum the meter readings and compare the result to P_T of (b).

Figure 23–42

44. Two wattmeters measure power to a load. Readings are $P_h = 1000$ W and $P_\ell = -400$ W. Determine the power factor of the load from Equation 23–24.

45. Consider the circuit of Figure 23–42.
   a. Compute the power factor from the angle of the phase impedances.
   b. In Problem 43, wattmeter readings $P_h = 1163$ W and $P_\ell = 870$ W were determined. Substitute these values into Equation 23–24 and compute the power factor of the load. Compare the results to (a).

46. For a balanced load, it can be shown that

$$\tan \theta_\phi = \sqrt{3}\left(\frac{P_h - P_\ell}{P_h + P_\ell}\right)$$

Dividing the top and bottom of the right hand side by P_h yields

$$\tan \theta_\phi = \sqrt{3}\left(\frac{1-x}{1+x}\right) \quad \text{where } x = \frac{P_\ell}{P_h}$$

   a. Vary x from 1 to -1 in increments of 0.2 to complete Table 23–3.
   b. From Table 23–3, plot a watts ratio curve.
   c. From this curve, determine F_p for the wattmeter readings of Problem 45.

TABLE 23–3

x	θ	F_p
1.0	0	1.0
0.8	10.9°	0.98
0.6		
0.4		
0.2		
0		
−0.2		
−0.4		
−0.6		
−0.8		
−1.0	90°	0

23.7 Unbalanced Systems

47. For Figure 23–43, $R_{ab} = 60\ \Omega$, $\mathbf{Z}_{bc} = 80\ \Omega + j60\ \Omega$. Compute the following:
   a. Phase currents, magnitudes and angles.
   b. Line currents, magnitudes and angles.
   c. Power to each phase.
   d. Total power to the load.

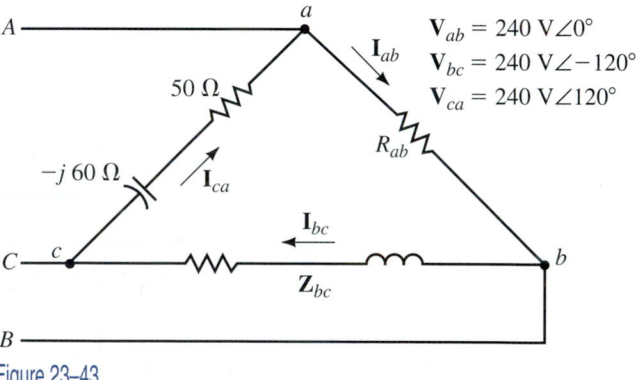

$\mathbf{V}_{ab} = 240\ \text{V}\angle 0°$
$\mathbf{V}_{bc} = 240\ \text{V}\angle -120°$
$\mathbf{V}_{ca} = 240\ \text{V}\angle 120°$

Figure 23–43

48. Repeat Problem 47 if $P_{ab} = 2400$ W and $\mathbf{Z}_{bc} = 50\ \Omega\angle 40°$.

49. For Figure 23–44, compute the following:
   a. Line currents, magnitudes and angles.
   b. The neutral current.
   c. Power to each phase.
   d. Total power to the load.

$\mathbf{E}_{AN} = 120 \text{ V}\angle 0°$

Figure 23–44

50. Remove the neutral conductor from the circuit of Figure 23–44 and compute the line currents. *Hint:* Use mesh equations.

51. From Problem 50, $\mathbf{I}_a = 1.94 \text{ A}\angle-0.737°$, $\mathbf{I}_b = 4.0 \text{ A}\angle-117.7°$ and $\mathbf{I}_c = 3.57 \text{ A}\angle 91.4°$. Compute the following:
   a. Voltages across each phase of the load.
   b. Voltage between the neutral of the load and the neutral of the generator.

23.9 PSpice For Three-Phase Circuits

52. The circuit of Figure 23–45 is balanced with $R_1 = R_2 = R_3 = 3 \text{ }\Omega$, $C_1 = C_2 = C_3 = 663.13 \text{ }\mu F$, and $f = 60$ Hz. Use PSpice to solve for currents $\mathbf{I}_{ab}$, $\mathbf{I}_{bc}$, and $\mathbf{I}_{ca}$.

Figure 23–45

53. The circuit of Figure 23–45 is unbalanced with $R_1 = 3 \text{ }\Omega$, $C_1 = 663.13 \text{ }\mu F$, $R_2 = 12 \text{ }\Omega$, $C_2 = 884.17 \text{ }\mu F$, $R_3 = 6 \text{ }\Omega$, $C_3 = 663.13 \text{ }\mu F$, and $f = 60$ Hz. Use PSpice to solve for currents $\mathbf{I}_{ab}$, $\mathbf{I}_{bc}$, and $\mathbf{I}_{ca}$. Sketch their phasors.

24

Transformers and Coupled Circuits

CHAPTER PREVIEW

In our study of induced voltage, in Chapter 13, we found that the changing magnetic field produced by current in one coil induced a voltage in a second coil wound on the same core. A device built to utilize this effect is the *transformer*.

Transformers have many applications. They are used in electrical power systems to step up voltage for long-distance transmission and then to step it down again to a safe level for use in our homes and offices. They are used in electronic equipment power supplies to raise or lower voltages, in audio systems to match speaker loads to amplifiers, in telephone, radio, and TV systems to couple signals, and so on.

In this chapter, we look at transformer fundamentals and the analysis of circuits containing transformers. We discuss transformer action, types of transformers, voltage and current ratios, applications, and so on. Both iron-core and air-core transformers are covered.

KEY TERMS

Air-Core Transformer
Autotransformer
Coefficient of Coupling
Copper Loss
Core Loss
Coupled Circuit
Current Ratio
Dot Convention
Ferrite Core
Ideal Transformer
Impedance Matching
Iron Core
Leakage Flux
Loosely Coupled
Magnetizing Current
Mutual Inductance
Open-Circuit Test
Primary
Reflected Impedance
Secondary
Short-Circuit Test
Step-Up/Step-Down
Tightly Coupled
Transformer
Turns Ratio
Voltage Ratio

Transformers are used in electronic equipment power supplies to step down the incoming line voltage. *(Courtesy Condor DC Power Supplies Inc.)*

OBJECTIVES

After studying this chapter, you will be able to

- describe how a transformer couples energy from its primary to its secondary via a changing magnetic field,

- describe basic transformer construction,

- use the dot convention to determine transformer phasing,

- determine voltage and current ratios from the turns ratio for iron-core transformers,

- compute voltage and currents in circuits containing iron-core and air-core transformers,

- use transformers to impedance match loads,

- describe some basic transformer applications,

- determine transformer equivalent circuits,

- compute iron-core transformer efficiency,

- use PSpice to solve circuits with transformers and coupled circuits.

24.1 Introduction

A transformer is a magnetically *coupled circuit*, i.e., a circuit in which the magnetic field produced by time-varying current in one circuit induces voltage in another. A basic iron-core transformer is shown in Figure 24–1. It consists of two coils wound on a common core. Alternating current in one winding establishes a flux which links the other winding and induces a voltage in it. Power thus flows from one circuit to the other via the medium of the magnetic field, with no electrical connection between the two sides. The winding to which we supply power is called the *primary*, while the winding from which we take power is called the *secondary*. Power can flow in either direction, as either winding can be used as the primary or the secondary.

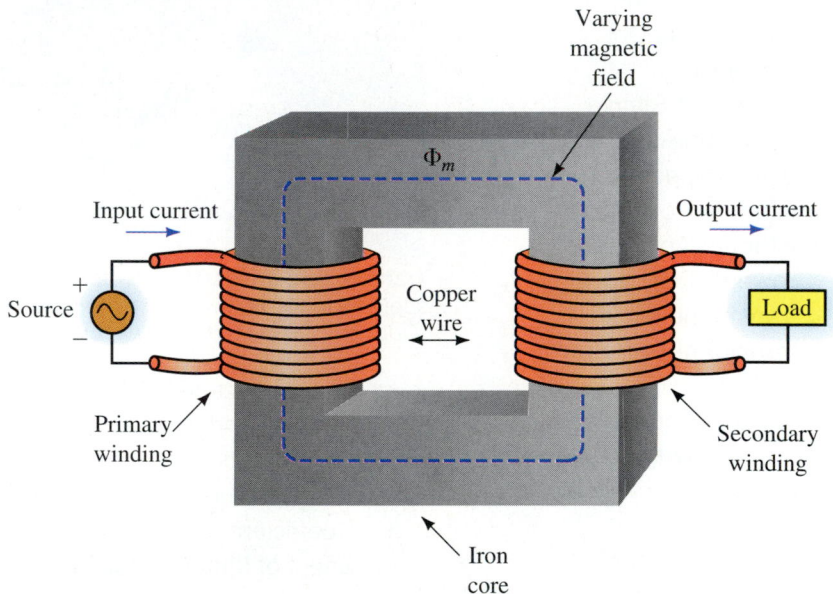

Figure 24–1   Basic iron-core transformer. Energy is transferred from the source to the load via the transformer's magnetic field with no direct electrical connection between the two sides.

Transformers fall into two broad categories: *iron-core* and *air-core*. Regardless of their physical construction, however, transformer action is described by Faraday's law. The resulting details, however, differ dramatically, as you will soon see.

Transformer Construction

Transformers used for audio- or power-frequency applications have iron cores. Figure 24–2 shows a transformer of the type used in electronic equipment, while Figure 24–3 shows a cutaway view of a transformer of the type used by electric utilities to distribute power to residential and commercial users.

Two principal types of construction are used for iron core transformers: the *core type* and the *shell type* (Figure 24–4). In both cases, cores

Figure 24–2   Iron-core transformer of the type used in electronic equipment. *(Courtesy Transformer Manufacturers Inc.)*

Figure 24–3   Distribution transformer, cutaway view. The tank is filled with oil to improve insulation and to remove heat from the core and windings. *(Courtesy Carte International Inc.)*

are made from laminated sheet steel rather than from blocks of solid iron. The sheets are insulated from each other by a thin coating of ceramic or other material to keep eddy current losses low (Section 24.6). The iron increases the coupling between coils by providing an easy path for magnetic flux.

Figure 24–4 Two basic iron-core configurations. For the core type, windings are placed on separate legs, while for the shell type, both windings are on the same leg. *(Adapted with permission from Perozzo, Practical Electronics Troubleshooting, © 1985 by Delmar Publishers Inc., Albany, New York)*

Because iron has considerable power loss due to hysteresis and eddy currents at high frequencies, iron cores are usable only at power and audio frequencies. For high-frequency applications (such as in radio circuits), *air-core* and *ferrite-core* types are used. Figure 24–5 shows a ferrite-core device. Ferrite greatly increases coupling between coils (compared with air) while maintaining low losses. Circuit symbols for transformers are shown in Figure 24–6.

Figure 24–5 A slug-tuned rf (radio frequency) ferrite-core transformer. Coupling between coils is varied by a ferrite slug that is positioned by means of an adjusting screw. (The slug has been removed and is shown in the foreground.)

(a) Iron-core

(b) Air-core

(c) Ferrite-core

Figure 24–6   Transformer schematic symbols.

Winding Directions

One of the advantages of a transformer is that it may be used to change the polarity of an ac voltage. This is illustrated in Figure 24–7. For the transformer of (a) the primary and secondary voltages are in phase (for reasons to be discussed later), while for (b) they are 180° out of phase.

(a) 0° phase shift

(b) 180° phase shift

Figure 24–7   Transformer phase shift depends on the relative direction of the windings.

Tightly Coupled and Loosely Coupled Circuits

If most of the flux produced by one coil links the other, the coils are said to be *tightly coupled*. Thus, iron-core transformers are tightly coupled (since close to 100% of the flux is confined to the core). For air- and ferrite-core transformers, however, much less than 100% of the flux links both windings. They are therefore *loosely coupled*. Since air-core and ferrite-core devices are loosely coupled, the same principles of analysis apply to both.

Faraday's Law

All transformer operation is described by Faraday's law. Faraday's law (in MKS units) states that the voltage induced in a circuit by a changing magnetic field is equal to the rate at which the flux linking the circuit is changing. When Faraday's law is applied to iron-core and air-core transformers, however, the results that emerge are quite different: Iron-core transformers are found to be characterized by their turns ratios, while air-core transformers are characterized by self- and mutual inductances. We begin with iron-core transformers.

24.2 Iron-Core Transformers: The Ideal Model

At first glance, iron-core transformers appear quite difficult to analyze because they have a number of characteristics such as winding resistance, core loss, and leakage flux that appear difficult to handle. Fortunately, these effects are small and often can be neglected. The result is the *ideal transformer.* Once you know how to analyze an ideal transformer, however, it is relatively easy to go back and add in the nonideal effects. This is the approach we use here.

To idealize a transformer, 1) neglect the resistance of its coils, 2) neglect its core loss, 3) assume all flux is confined to its core, and 4) assume that negligible current is required to establish its core flux. (Well-designed iron-core transformers are very close to this ideal.)

We now apply Faraday's law to the ideal transformer. Before we do this, however, we need to determine flux linkages. The flux linking a winding (as determined in Chapter 13) is the product of the flux that passes through the winding times the number of turns through which it passes. For flux Φ passing through N turns, flux linkage is $N\Phi$. Thus, for the ideal transformer (Figure 24–8), the primary flux linkage is $N_p\Phi_m$, while the secondary flux linkage is $N_s\Phi_m$, where the subscript "m" indicates mutual flux, i.e., flux that links both windings.

Figure 24–8   Ideal transformer. All flux is confined to the core and links both windings. This is a "tightly coupled" transformer.

Voltage Ratio

Now apply Faraday's law. Since the flux linkage equals $N\Phi$ and since N is constant, the induced voltage is equal to N times the rate of change of Φ, i.e., $e = Nd\Phi/dt$. Thus, for the primary,

$$e_p = N_p \frac{d\Phi_m}{dt} \tag{24–1}$$

while for the secondary

$$e_s = N_s \frac{d\Phi_m}{dt} \tag{24–2}$$

Dividing Equation 24–1 by Equation 24–2 and cancelling $d\Phi_m/dt$ yields

$$\frac{e_p}{e_p} = \frac{N_p}{N_s} \tag{24–3}$$

Equation 24–3 states that *the ratio of primary voltage to secondary voltage at any instant is equal to the ratio of primary turns to secondary turns.* This ratio is called the *transformation* ratio (or *turns ratio*) and is given the symbol *a*. Thus,

$$a = N_p/N_s \tag{24–4}$$

and

$$e_p/e_s = a \tag{24–5}$$

For example, a transformer with 1000 turns on its primary and 250 turns on its secondary therefore has a turns ratio of 1000/250 = 4. This is referred to as a 4:1 ratio.

Since the ratio of two instantaneous sinusoidal voltages is the same as the ratio of their effective values, Equation 24–5 can also be written as

$$E_p/E_s = a \qquad\qquad (24\text{–}6)$$

As noted earlier, e_p and e_s are either in phase or 180° out of phase, depending on the relative direction of coil windings. We can therefore also express the ratio of voltages in terms of phasors as

$$\mathbf{E}_p/\mathbf{E}_s = a \qquad\qquad (24\text{–}7)$$

where the relative polarity (in phase or 180° out of phase) is determined by the direction of the coil windings.

Step-Up and Step-Down Transformers

A *step-up* transformer is one in which the secondary voltage is higher than the primary voltage, while a *step-down* transformer is one in which the secondary voltage is lower. Since $a = E_p/E_s$, a step-up transformer has $a < 1$, while for a step-down transformer, $a > 1$. If $a = 1$, the transformer has a *unity turns ratio* and the secondary voltage is equal to the primary voltage.

■ **EXAMPLE 24–1**   Suppose the transformer of Figure 24–7(a) has 500 turns on its primary and 1000 turns on its secondary.

    a. Determine its turns ratio. Is it step-up or step-down?
    b. If its primary voltage is $e_p = 25 \sin \omega t$ V, what is its secondary voltage?
    c. Sketch the waveforms.

Solution

    a. The turns ratio is $a = N_p/N_s = 500/1000 = 0.5$. This is a step-up transformer.

    b. From Equation 24–5, $e_s = e_p/a = (25 \sin \omega t)/0.5 = 50 \sin \omega t$ V.

    c. Primary and secondary voltages are in phase as noted earlier. Figure 24–9 shows the waveforms.

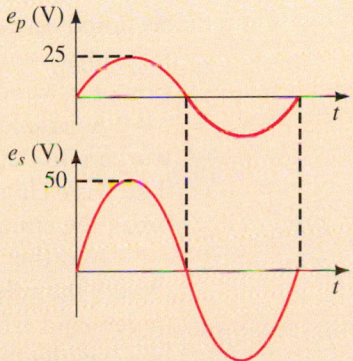

Figure 24–9

■ **EXAMPLE 24–2**   If the transformers of Figure 24–7 have 600 turns on their primaries and 120 turns on their secondaries, and $E_p = 120$ V$\angle 0°$, what is E_s for each case?

Solution   The turns ratio is $a = 600/120 = 5$. For transformer (a), E_s is in phase with E_p. Therefore, $E_s = E_p/5 = (120$ V$\angle 0°)/5 = 24$ V$\angle 0°$. For transformer (b), E_s is 180° out of phase with E_p. Therefore, $E_s = 24$ V$\angle 180°$.

practice problem 1

Repeat Example 24–1 for the circuit of Figure 24–7(b) if $N_p = 1200$ turns and $N_s = 200$ turns.

Current Ratio

Because we have neglected power loss in the ideal transformer, its efficiency is 100% and thus power in equals power out. Consider again Figure 24–8. At any instant, $p_{in} = e_p i_p$ and $p_{out} = e_s i_s$. Thus,

$$e_p i_p = e_s i_s \tag{24–8}$$

and

$$\frac{i_p}{i_s} = \frac{e_s}{e_p} = \frac{1}{a} \tag{24–9}$$

since $e_s/e_p = 1/a$. (This means that if voltage is stepped up, current is stepped down, and vice versa.) Since the ratio of instantaneous currents is the same as the ratio of current phasors and the ratio of current magnitudes, we can also write

$$\frac{I_p}{I_s} = \frac{I_p}{I_s} = \frac{1}{a} \tag{24–10}$$

For example, for a transformer with $a = 4$, and $I_p = 2$ A$\angle -20°$, $I_s = aI_p = 4(2$ A$\angle -20°) = 8$ A$\angle -20°$.

Polarity of Induced Voltage: The Dot Convention

As noted earlier, a transformer's secondary voltage is either in phase with its primary voltage or 180° out of phase, depending on the relative direction of its windings. We will now demonstrate why.

A simple test, called the *kick test* (sometimes used by electrical workers to determine transformer polarity), can help establish the idea. The basic circuit is shown in Figure 24–10. A switch is used to make and break the circuit (since voltage is induced only while flux is changing).

For the winding directions shown, at the instant the switch is closed, the voltmeter needle "kicks" upscale, then settles back to zero. To understand why, we need to consider magnetic fields. Before we start, however, let us place a dot on one of the primary terminals; in this case, we arbitrarily choose the top terminal. Let us also replace the voltmeter by its equivalent resistance (Figure 24–11).

Figure 24–10   The "kick test." For the winding directions shown, the meter kicks upscale at the instant the switch is closed. (This is the same transformer shown in Figure 24–7a.)

Figure 24–11   The dot convention for coupled circuits.

At the instant the switch is closed, the polarity of the dotted primary terminal is positive with respect to the undotted primary terminal (because the + end of the source is directly connected to it). As current in the primary builds, it creates a flux in the upward direction as indicated by the blue arrow (recall the right-hand rule). According to Lenz's law, the effect that results must oppose the cause that produced it. The effect is a voltage induced in the secondary winding. The resulting current in the secondary produces a flux which, according to Lenz's law, must *oppose the buildup of the original flux*, i.e., it must be in the direction of the purple arrow. Applying the right-hand rule, we see that secondary current must be in the direction indicated by i_s. Placing a plus sign at the tail of this arrow shows that the top end of the resistor is positive. This means that the top end of the secondary winding is also positive. Place a dot here. Dotted terminals are called *corresponding terminals*.

As you can see, corresponding terminals are positive (with respect to their companion undotted terminals) at the instant the switch is closed. If you perform a similar analysis at the instant the switch is opened, you will find that both dotted terminals are negative. Thus, *dotted terminals have the same polarity at all instants of time*. What we have developed here is known as the *dot convention for coupled circuits*.

practical

notes

1. *While we developed the dot convention using a switched dc source, it is valid for ac as well. In fact, we will use it mostly for ac.*

2. *The resistor of Figure 24–11 is required only to work through the physics to establish the secondary voltage polarity. You can now remove it without affecting the resulting dot position.*

3. *In practice, corresponding terminals may be marked with dots, by color coded wires, or by special letter designations such as H1 (for the high-voltage side) and X1 (for the low-voltage side).*

■ **EXAMPLE 24–3**   Determine the waveform for e_s in the circuit of Figure 24–12(a).

(a)   (b)

Figure 24–12

Solution   Dotted terminals have the same polarity (with respect to their undotted terminals) at all instants. During the first half-cycle, the dotted end of the primary coil is positive. Therefore, the dotted end of the secondary is also positive. During the second half-cycle, both are negative. The polarity marks on e_s mean that we are looking at the polarity of the top end of the secondary coil with respect to its bottom end. Thus, e_s is positive during the first half-cycle and negative during the second half cycle. It is therefore in phase with e_p as indicated in (b).

practice
problems
II

1. Determine the waveform for e_s in the circuit of Figure 24–13(b).

2. Repeat Problem 1 for Figure 24–13(c).

3. Using the "kick test" technique, determine dotted ends for the transformer of Figure 24–7(b).

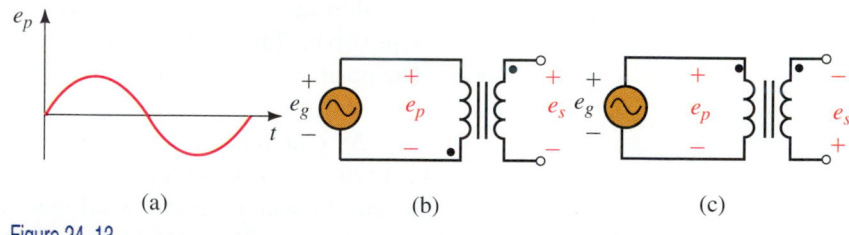

(a)   (b)   (c)

Figure 24–13

Analysis of Simple Transformer Circuits

Simple transformer circuits may be analyzed using the relationships described so far, namely $\mathbf{E}_p = a\mathbf{E}_s$, $\mathbf{I}_p = \mathbf{I}_s/a$, and $P_{\text{in}} = P_{\text{out}}$. This is illustrated in Example 24–4. More complex problems require some additional ideas.

■ **EXAMPLE 24–4** For Figure 24–14(a), $\mathbf{E}_g = 120 \text{ V} \angle 0°$, the turns ratio is 5:1 and $\mathbf{Z}_L = 4 \, \Omega \angle 30°$. Find

a. the load voltage,

b. the load current,

c. the generator current,

d. the power to the load,

e. the power output by the generator.

$\mathbf{Z}_L = 4 \, \Omega \angle 30°$

(a)

(b)

Figure 24–14

Solution

a. $\mathbf{E}_p = \mathbf{E}_g = 120 \text{ V} \angle 0°$

$\mathbf{V}_L = \mathbf{E}_s = \mathbf{E}_p/a = (120 \text{ V} \angle 0°)/5 = 24 \text{ V} \angle 0°$

b. $\mathbf{I}_L = \mathbf{V}_L/\mathbf{Z}_L = (24 \text{ V} \angle 0°)/(4 \, \Omega \angle 30°) = 6 \text{ A} \angle -30°$

c. $\mathbf{I}_g = \mathbf{I}_p$. But $\mathbf{I}_p = \mathbf{I}_s/a = \mathbf{I}_L/a$. Thus,

$\mathbf{I}_g = (6 \text{ A} \angle -30°)/5 = 1.2 \text{ A} \angle -30°$

Values are shown on Figure 24–14(b).

d. $P_L = V_L I_L \cos \theta_L = (24)(6)\cos 30° = 124.7 \text{ W}$.

e. $P_g = E_g I_g \cos \theta_g$, where θ_g is the angle between $\mathbf{E}_g$ and $\mathbf{I}_g$. $\theta_g = 30°$. Thus, $P_g = (120)(1.2)\cos 30° = 124.7 \text{ W}$, which agrees with (d) as it must since the transformer is lossless.

1. A transformer has a turns ratio of 1:8. Is it step-up or step-down? If $E_p = 25$ V, what is E_s?

2. For the transformers of Figure 24–7, if $a = 0.2$ and $\mathbf{E}_s = 600$ V∠−30°, what is $\mathbf{E}_p$ for each case?

3. For each of the transformers of Figure 24–15, sketch the secondary voltage, showing both phase and amplitude.

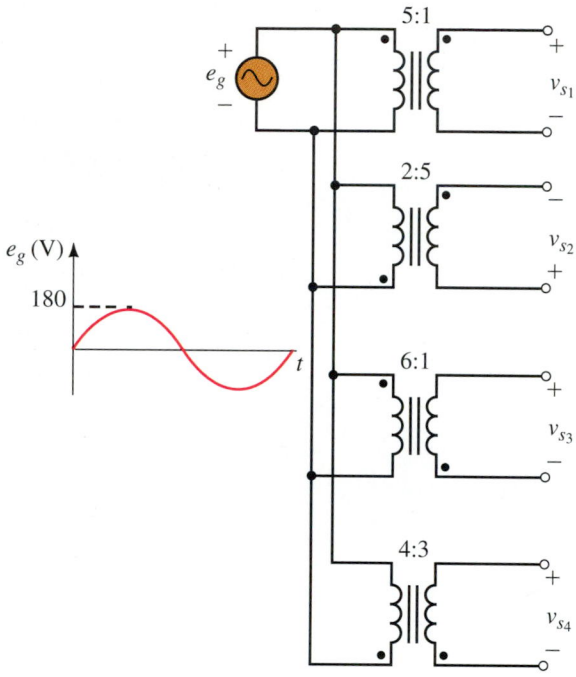

Figure 24–15

4. For Figure 24–16, determine the position of the missing dot.

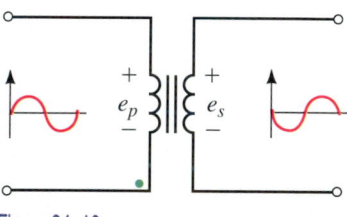

Figure 24–16

5. Figure 24–17 shows another way to determine dotted terminals. First, arbitrarily mark one of the primary terminals with a dot. Next, connect a jumper wire and voltmeter as indicated. From the readings, you can determine the secondary dotted terminal. For the two cases indicated, where should the secondary dot be? (*Hint:* Use KVL.)

Figure 24–17   Each transformer is rated 120/60 V.

24.3 Reflected Impedance

When an impedance $\mathbf{Z}_L$ is connected directly to a source, the impedance seen by the source is $\mathbf{Z}_L$. However, if a transformer is connected between them as in Figure 24–18(a), the impedance seen by the source is now quite different. Let this equivalent impedance be denoted as $\mathbf{Z}_p$. Then, $\mathbf{I}_p = \mathbf{E}_g/\mathbf{Z}_p$. Rearranging terms yields $\mathbf{Z}_p = \mathbf{E}_g/\mathbf{I}_p$. But $\mathbf{E}_g = \mathbf{E}_p$, $\mathbf{E}_p = a\mathbf{E}_s$, and $\mathbf{I}_p = \mathbf{I}_s/a$. Thus,

$$\mathbf{Z}_p = \frac{\mathbf{E}_p}{\mathbf{I}_p} = \frac{a\mathbf{E}_s}{\left(\dfrac{\mathbf{I}_s}{a}\right)} = a^2\frac{\mathbf{E}_s}{\mathbf{I}_s} = a^2\frac{\mathbf{V}_L}{\mathbf{I}_L}$$

However,

$$\frac{\mathbf{V}_L}{\mathbf{I}_L} = \mathbf{Z}_L$$

Thus,

$$\mathbf{Z}_p = a^2\mathbf{Z}_L \tag{24–11}$$

This means that $\mathbf{Z}_L$ now looks to the source like the transformer's turns ratio squared times the load impedance. The term $a^2\mathbf{Z}_L$ is referred to as the load's *reflected impedance*. Note that it retains the load's characteristics, that is, a capacitive load still looks capacitive, an inductive, load still looks inductive, and so on.

Figure 24–18   Concept of reflected impedance. Through the a:1 transformer, load $\mathbf{Z}_L$ looks to the source like an impedance $a^2\mathbf{Z}_L$ with voltage $a\mathbf{V}_L$ across it.

Equation 24–11 shows that a transformer can make a load look larger or smaller to the source, depending on its turns ratio. To illustrate, consider Figure 24–19. If the 1-Ω load resistor were connected directly to the source, it would look like a 1-Ω resistor and the generator current would be $100 A\angle 0°$. However, when connected to a 10:1 transformer, it looks like a $(10)^2(1\ \Omega) = 100$-Ω resistor and the generator current is only $1\ A\angle 0°$.

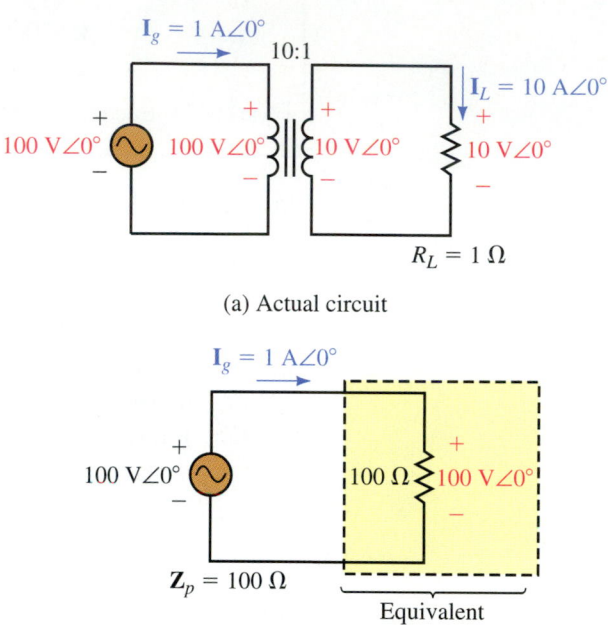

(a) Actual circuit

(b) Load as seen by the source

Figure 24–19   Illustrating reflected impedance. The 1-Ω resistor looks like 100 Ω through the 10:1 transformer.

The concept of reflected impedance is useful in a number of ways. It permits us to match loads to sources (such as amplifiers) as well as provides a better way to solve complex problems. Several of these ideas are considered in the next few sections.

■ EXAMPLE 24–5   Use the reflected impedance idea to solve for primary and secondary currents and the load voltage for the circuit of Figure 24–20(a). (Note: This is the same circuit we solved earlier in Example 24–4. While there is no advantage to the reflected impedance idea over our previous solution for a problem as simple as this, the advantages are considerable for complex problems.)

Solution

$$\mathbf{Z}_p = a^2\mathbf{Z}_L = (5)^2(4\ \Omega\angle 30°) = 100\ \Omega\angle 30°.$$

The equivalent circuit is shown in (b).

$$\mathbf{I}_g = \mathbf{E}_g / \mathbf{Z}_p = (120 \text{ V} \angle 0°) / (100 \ \Omega \angle 30°) = 1.2 \text{ A} \angle -30°$$

$$\mathbf{I}_L = a\mathbf{I}_p = a\mathbf{I}_g = 5(1.2 \text{ A} \angle -30°) = 6 \text{ A} \angle -30$$

$$\mathbf{V}_L = \mathbf{I}_L \mathbf{Z}_L = (6 \text{ A} \angle -30°)(4 \ \Omega \angle 30°) = 24 \text{V} \angle 0°$$

The answers are the same as before.

(a) Turns ratio = 5.

$$\mathbf{I}_g = 1.2 \text{ A} \angle -30°$$

Load as it looks
to the source.

(b)

Figure 24–20

24.4 Transformer Power Ratings

Transformers are rated in terms of voltage and apparent power (rather than in terms of real power as discussed in Chapter 17). Rated current can be determined from these ratings. Thus a transformer rated 2400/120 volt, 48 kVA, has a current rating of 48 000 VA/2400 V = 20 A on its 2400-V side and 48 000 VA/120 V = 400 A on its 120-V side (Figure 24–21). This transformer can handle a 48-kVA load, regardless of power factor.

20 A   400 A

2400 V   Load   120 V

$S_{in} = 48 \text{ kVA}$ $S_{out} = 48 \text{ kVA}$

Figure 24–21   Transformers are rated by how much apparent power they can handle.

Figure 24–22   A power supply transformer. The secondary is tapped at various voltages from 6.3 V to 30 V.

24.5 Transformer Applications

Power Supply Transformers

On electronic equipment, *power supply transformers* are used to convert the incoming 120 Vac to the voltage levels required for internal circuit operation. A variety of commercial transformers are made for this purpose. The transformer of Figure 24–22, for example, has a multi-tapped secondary winding, each tap providing a different output voltage. It is intended for laboratory supplies, test equipment, or experimental power supplies.

Figure 24–23 illustrates a typical use of a power supply transformer. First, the incoming line voltage is stepped down, then a rectifier circuit (a circuit that uses diodes to convert ac to dc using a process called rectification) converts the ac to pulsating dc, a filter smooths it, and finally, a voltage regulator (an electronic device used to maintain a constant output voltage) regulates it to the required dc value.

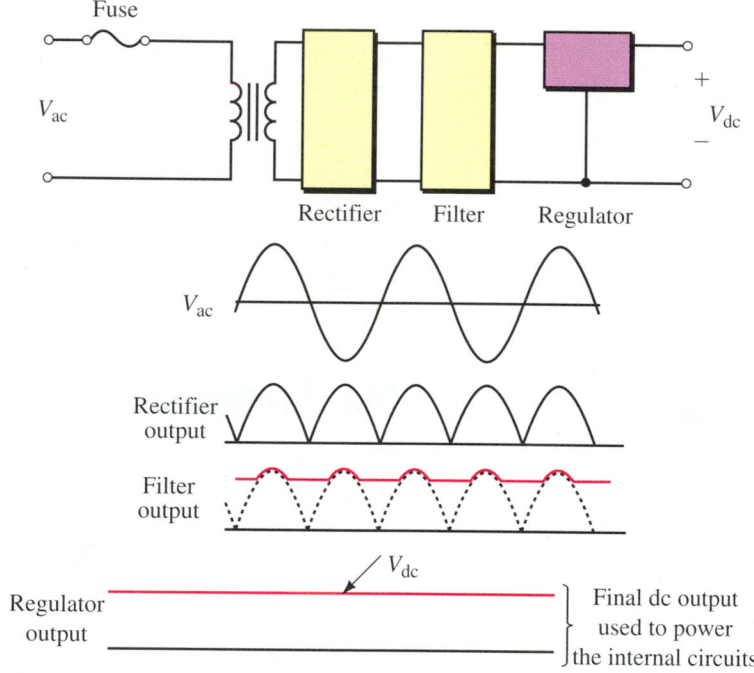

Figure 24–23   Transformer in a power supply application. Power supplies such as this are used in electronic equipment.

Transformers in Power Systems

Transformers are one of the key elements that have made commercial ac power systems possible. Transformers are used at generating stations to raise voltage for long-distance transmission. This lowers the transmitted current and hence the I^2R power losses in the transmission line. At the user end, transformers reduce the voltage to a safe level for everyday use. A typical residential connection is shown in Figure 24–24. The taps on the primary permit the electric utility company to compensate for line voltage drops. Transformers located far from substations, for example, have lower input voltages than those close to substations due to voltage drops on distribution lines. Taps permit the turns ratio to be changed to compensate. Note also the split secondary. It permits 120-V and 240-V loads to be supplied from the same transformer.

Figure 24–24   Typical distribution transformer supplying residential loads.

The transformer of Figure 24–24 is a single-phase unit (since residential customers require only single phase). By connecting its primary from line to neutral (or line to line), the required single-phase input is obtained from a three-phase line.

Isolation Applications

Transformers are sometimes used to isolate equipment for safety or other reasons as shown in Figure 24–26. If a piece of equipment has its frame or chassis connected to the grounded neutral of Figure 24–24, for example, the connection is perfectly safe as long as the connection is not changed. If, however, connections are inadvertently reversed as in Figure 24–25(b) (due to faulty installation for example), a dangerous situation results. A transformer used as in Figure 24–26 eliminates this danger by ensuring that the chassis is never directly connected to the "hot" wire. Isolation transformers are made for this purpose.

(a) Chassis is safe

(b) Chassis is at 120 V

Figure 24–25   If line connections are inadvertently reversed as in (b), you could get a shock if you touch the chassis.

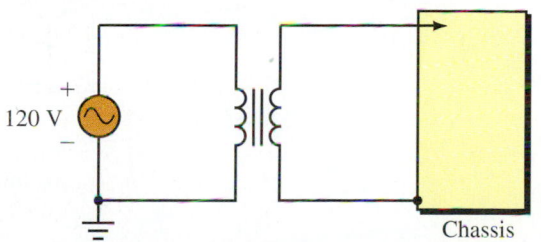

Figure 24–26   With a transformer, no shock hazard exists, even if the supply lines are reversed.

Impedance Matching

As you learned earlier, load impedance $\mathbf{Z}_L$, when viewed through a transformer, has a reflected value of $\mathbf{Z}_p = a^2 \mathbf{Z}_L$. This means that a transformer can be used to raise or lower the apparent impedance of a load by choice of turns ratio. This is referred to as *impedance matching*. Impedance matching is sometimes used to match loads to amplifiers to achieve maximum power transfer. If the load and source are not matched, a transformer can be inserted between them as illustrated next.

■ **EXAMPLE 24–6**   Figure 24–27(a) shows the schematic of a multi-tap sound distribution transformer with various taps that permit the matching of speakers to amplifiers. Over their design range, speakers are basically resistive. If the speaker of Figure 24–27(b) has a resistance of 4Ω, what transformer ratio should be chosen?

Solution   Make the reflected resistance of the speaker equal to the internal (Thévenin) resistance of the amplifier. Thus, $Z_p = 400\ \Omega = a^2 Z_L = a^2(4\ \Omega)$, and

$$a = \sqrt{\frac{400}{4}} = \sqrt{100} = 10$$

Thus, choose the tap that provides a 10:1 ratio.

(a) Schematic of a multi-tap sound distribution
transformer used to couple a speaker
to an amplifier.

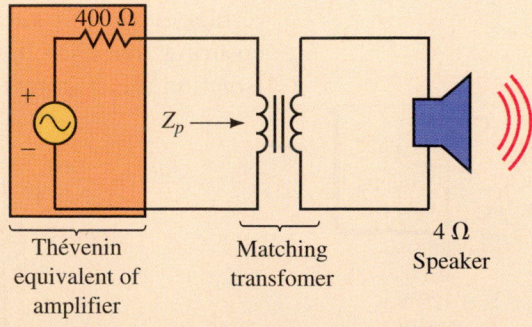

(b) An example

Figure 24–27   With a 10:1 matching transformer, the 4-Ω speaker looks like 400 Ω to the amplifier.

Transformers with Multiple Secondaries

For a transformer with multiple secondaries (Figure 24–28), each secondary voltage is governed by the appropriate turns ratio, that is, $\mathbf{E}_1/\mathbf{E}_2 = N_1/N_2$ and $\mathbf{E}_1/\mathbf{E}_3 = N_1/N_3$. Loads are reflected in parallel. That is, $\mathbf{Z}'_2 = a_2^2\mathbf{Z}_2$ and $\mathbf{Z}'_3 = a_2^2\mathbf{Z}_3$ appear in parallel in the equivalent circuit, (b).

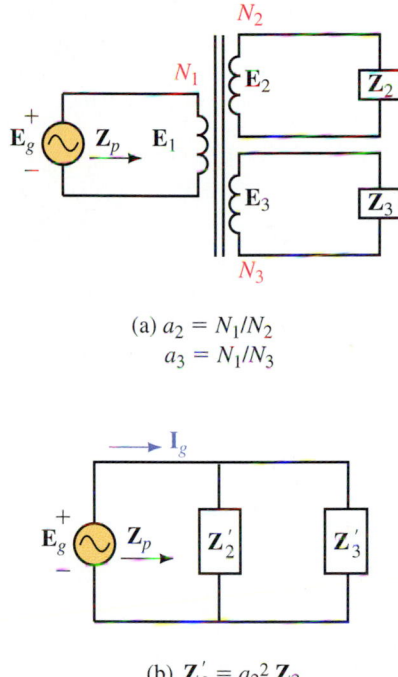

(a) $a_2 = N_1/N_2$
$a_3 = N_1/N_3$

(b) $\mathbf{Z}'_2 = a_2{}^2 \mathbf{Z}_2$
$\mathbf{Z}'_3 = a_3{}^2 \mathbf{Z}_3$

Figure 24–28   Multiple loads are reflected in parallel.

■ **EXAMPLE 24–7**   For the circuit of Figure 24–29(a),

a. determine the equivalent circuit,
b. determine the generator current,
c. show that apparent power in equals apparent power out.

Solution

a. See Figure 24–29(b).

b. $\mathbf{I}_g = \dfrac{\mathbf{E}_g}{\mathbf{Z}'_2} + \dfrac{\mathbf{E}_g}{\mathbf{Z}'_3} = \dfrac{100\angle 0°}{10} + \dfrac{100\angle 0°}{j10} = 10 + j10 = 14.14\ \text{A}\angle 45°$

c. $\mathbf{S}_{\text{in}} = \mathbf{E}_g\mathbf{I}_g{}^* = (100\angle 0°)(14.14\angle{-45°}) = 1414\ \text{VA}\angle{-45°}$ where $\mathbf{I}^*_g$ is the conjugate of $\mathbf{I}_g$.

Output apparent power can be determined by a power triangle. For the resistor load, $P = (100\ \text{V})^2/10\ \Omega = 1000\ \text{W}$. For the capacitor, $Q = (100\ \text{V})^2/10\ \Omega = 1000\ \text{VAR}$. The power triangle is shown in (c). Note that $\mathbf{S}_{\text{out}} = 1000 - j1000 = 1414\ \text{VA}\angle{-45°}$, the same as $\mathbf{S}_{\text{in}}$.

Figure 24-29

Autotransformers

An important variation of the transformer is the *autotransformer* (Figure 24–30). While autotransformers use two sets of windings, they are unusual in that their primary circuit is not electrically isolated from their secondary. However, they are smaller and cheaper than conventional transformers for the same rated load kVA since only part of the load power is transferred inductively. (The remainder is transferred by direct conduction.) Figure 24–30 shows several autotransformer connections. The transformer of (c) is variable by means of a slider, typically, from 0% to 110%.

For analysis, an autotransformer may be viewed as a standard two-winding transformer reconnected as in Figure 24–31(b). Voltage and current relationships between windings hold as they do for the standard

(a) Step-up

(b) Step-down

(c) Variable

Figure 24–30   Autotransformers.

connection. Thus, if you apply the rated voltage to the primary winding, you will obtain the rated voltage across the secondary winding. Finally, since we are assuming an ideal transformer, apparent power out equals apparent power in.

■ **EXAMPLE 24–8**   A 240/60-V, 3-kVA transformer is reconnected as an autotransformer to supply 300 volts to a load from a 240-V supply [Figure 24–31(b)].

a. Determine the rated primary and secondary currents.

b. Determine the maximum apparent power that can be delivered to the load.

c. Determine the supply current.

(a) 3 kVA transformer.

(b) Used as an autotransformer.

Figure 24–31

Solution

a. Rated current = rated kVA/rated voltage. Thus,

$$I_p = 3 \text{ kVA}/240 \text{ V} = 12.5 \text{ A} \quad \text{and} \quad I_s = 3 \text{ kVA}/60 \text{ V} = 50 \text{ A}$$

b. Load current = secondary current. At rated $I_s = 50$ A,

$$S_L = V_L I_L = (300 \text{ V})(50 \text{ A}) = 15 \text{ kVA}$$

This is five times the rated kVA of the transformer.

c. Apparent power in = apparent power out:

$$240 I_1 = 15 \text{ kVA}.$$

Thus, $I_1 = 62.5$ A. Current directions are as shown.

As a check, KCL at the junction of the two coils yields

$$I_1 = I_p + I_L = 12.5 + 50 = 62.5 \text{ A}$$

1. For each of the circuits of Figure 24–32, determine the required answer.

$\mathbf{Z}_p = $?

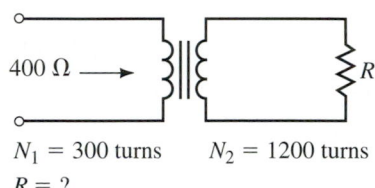

$N_1 = 300$ turns $N_2 = 1200$ turns

$R = $?

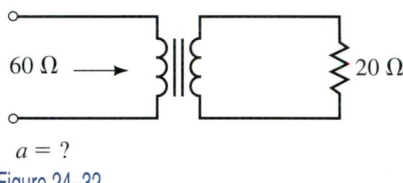

$a = $?

Figure 24–32

2. For Figure 24–33, if $a = 5$ and $\mathbf{I}_p = 5\,\text{A}\angle-60°$, what is $\mathbf{E}_g$?

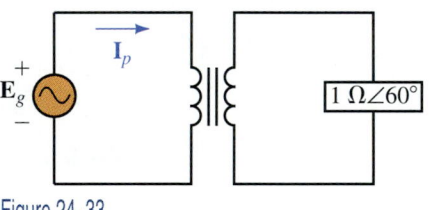

Figure 24–33

3. For Figure 24–33, if $\mathbf{I}_p = 30\,\text{mA}\angle-40°$ and $\mathbf{E}_g = 240\,\text{V}\angle20°$, what is a?

4. a. How many amps can a 24-kVA, 7200/120-V transformer supply to a 120-V unity power factor load? To a 0.75 power factor load?

b. How many watts can it supply to each load?

5. For the transformer of Figure 24–34, between position 2 and 0 there are 2000 turns. Between taps 1 and 2 there are 200 turns, and between taps 1 and 3 there are 300 turns. What will the output voltage be when the supply is connected to tap 1? To tap 2? To tap 3?

Figure 24–34

6. For the circuit of Figure 24–35, what is the power delivered to a 4-ohm speaker? What is the power delivered if an 8-ohm speaker is used instead? Why is the power to the 4-ohm speaker larger?

Figure 24–35

7. The autotransformer of Figure 24–36 has a 58% tap. The apparent power of the load is 7.2 kVA. Calculate the following:

  a. The load voltage and current.

  b. The source current.

  c. The current in each winding and its direction.

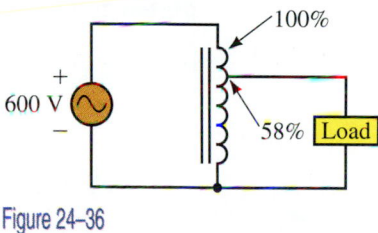

Figure 24–36

24.6 Practical Iron-Core Transformers

In Section 24.2, we idealized the transformer. We now add back the effects that we ignored.

Leakage Flux

While most flux is confined to the core, a small amount (called *leakage flux*) passes outside the core and through air at each winding as in Figure

24–37(a). The effect of this leakage can be modeled by inductances L_p and L_s as indicated in (b). The remaining flux, the *mutual flux* Φ_m, links both windings and is accounted for by the ideal transformer as previously.

(a) Flux leakage

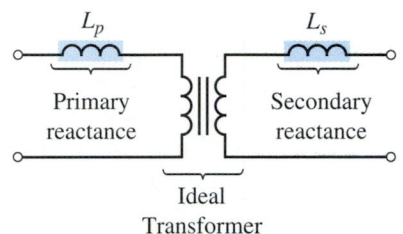

(b) Adding the effect of leakage
flux to the model.

Figure 24–37 Leakage flux has the same effect as L_p and L_s.

Winding Resistance

The effect of coil resistance can be approximated by resistances R_p and R_s as shown in Figure 24–38. The effect of these resistances is to cause a slight power loss and hence a reduction in efficiency as well as a small voltage drop. (The power loss associated with coil resistance is called *copper loss* and varies as the square of the load current.)

Figure 24–38 Adding winding resistance.

Core Loss

Losses occur in the core because of *eddy currents* and *hysteresis*. First, consider eddy currents. Since iron is a conductor, voltage is induced in it as core flux varies. This voltage creates currents that circulate as "eddies" within the core itself. One way to reduce these currents is to break their path of circulation by constructing the core from thin laminations of steel rather than using a solid block of iron. Laminations are insulated from each other by a coat of ceramic, varnish, or similar insulating material. (Although this does not eliminate eddy currents, it greatly reduces them.) Power and audio transformers are built this way (Figure 24–4). Another way to reduce eddy currents is to use powdered iron held together by an insulating binder. Ferrite cores are made like this.

Now consider hysteresis. Because flux constantly reverses, magnetic domains in the core steel constantly reverse as well. This takes energy. However, this energy is minimized by using special grain-oriented transformer steel.

The sum of hysteresis and eddy current loss is called *core loss* or *iron loss*. In a well-designed transformer, it is small, typically one to two percent of the transformer rating. The effect of core loss can be modeled as a resistor, R_c in Figure 24–39. Core losses vary approximately as the square of applied voltage. As long as voltage is constant (which it normally is), core losses remain constant.

Other Effects

We have also neglected *magnetizing current*. In a real transformer, however, some current is required to magnetize the core. To account for this, add path L_m as shown in Figure 24–39. Stray capacitances also exist between various parts of the transformer. They can be approximated by lumped capacitances as indicated.

The Full Equivalent

Figure 24–39 shows the final equivalent with all effects incorporated. How good is it? Calculations based on this model agree exceptionally well with measurements made on real transformers. However, the circuit is complex and awkward to use. In practice, therefore, only those elements of the model that affect a given application are actually retained. For example, at power frequencies, the effect of capacitance is negligible and the capacitors are omitted for power frequency analysis.

Figure 24–39   Final iron-core transformer equivalent circuit.

Voltage Regulation

Because of the internal impedance of a transformer, i.e., $R_p + jX_p$ and $R_s + jX_s$, voltage drops occur inside the transformer under load. Therefore, the output voltage of a transformer under load is different from its voltage under no load. This change in voltage (expressed as a percentage of full-load voltage) is termed *regulation*. For regulation analysis, parallel branches R_c and L_m and stray capacitance have negligible effect and can be neglected. This yields the simplified circuit of Figure 24–40(a). Even greater simplification is achieved by reflecting secondary impedances into the primary. This yields the circuit of (b). The reflected load voltage is $a\mathbf{V}_L$ and the reflected load current is $\mathbf{I}_L/a$. Regulation calculations are performed using this simplified circuit.

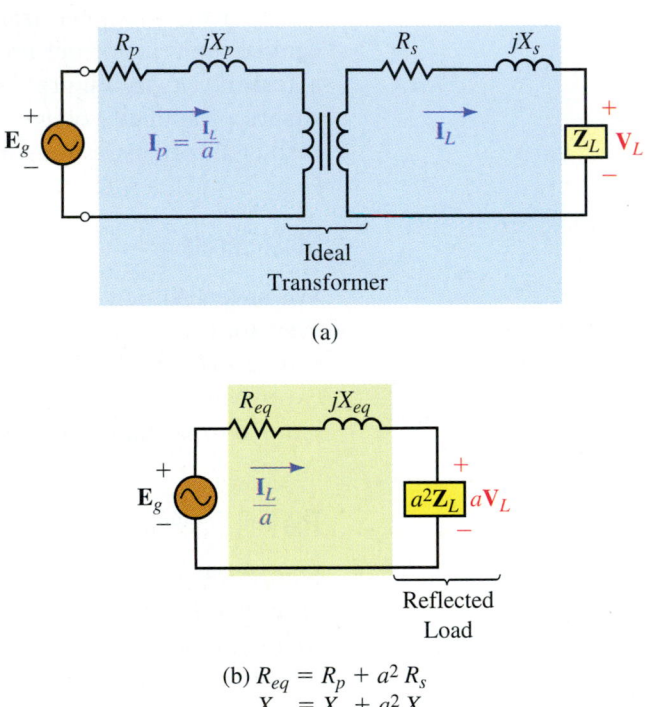

(a)

(b) $R_{eq} = R_p + a^2 R_s$
$X_{eq} = X_p + a^2 X_s$

Figure 24–40 Simplifying the equivalent circuit.

■ **EXAMPLE 24–9** A 10:1 transformer has primary and secondary resistance and reactance of $4\,\Omega + j4\,\Omega$ and $0.04\,\Omega + j0.04\,\Omega$ respectively as in Figure 24–41(a).

a. Determine its equivalent circuit.

b. If $\mathbf{V}_L = 120\ \text{V}\angle 0°$ and $\mathbf{I}_L = 20\ \text{A}\angle -30°$, what is the supply voltage, $\mathbf{E}_g$?

c. Determine the regulation.

Solution

a. $R_{eq} = R_p + a^2 R_s = 4\,\Omega + (10)^2(0.04\,\Omega) = 8\,\Omega$

$X_{eq} = X_p + a^2 X_s = 4\,\Omega + (10)^2(0.04\,\Omega) = 8\,\Omega$

(a)

(b) Equivalent circuit

Figure 24–41

Thus, $\mathbf{Z}_{eq} = 8\ \Omega + j8\ \Omega$ as shown in (b).

b. $a\mathbf{V}_L = (10)(120\ \text{V}\angle 0°) = 1200\ \text{V}\angle 0°$ and $\mathbf{I}_L/a = (20\ \text{A}\angle -30°)/10 = 2\ \text{A}\angle -30°$. From KVL, $\mathbf{E}_g = (2\ \text{A}\angle -30°)(8\ \Omega + j8) + 1200\ \text{V}\angle 0° = 1222\ \text{V}\angle 0.275°$.

Thus, there is a phase shift of 0.275° across the transformer's internal impedance and a drop of 22 V, requiring that the primary be operated slightly above its rated voltage. (This is normal.)

c. Now consider the no-load condition (Figure 12–42). Let V_{NL} be the no-load voltage. As indicated, $aV_{NL} = 1222$ V. Thus, $V_{NL} = 1222/a = 1222/10 = 122.2$ volts and

$$\text{regulation} = \frac{V_{NL} - V_{FL}}{V_{FL}} \times 100 = \frac{122.2 - 120}{120} \times 100 = 1.83\%$$

Note that only magnitudes are used in determining regulation.

Figure 24–42   No-load equivalent circuit.

practical

notes

1. *From Figure 24–42,* a = E_g/V_{NL}. *This means that the turns ratio is the ratio of input voltage to output voltage at no load.*

2. *The voltage rating of a transformer (such as 1200/120 V) is referred to as its nominal rating. The ratio of nominal voltages is the same as the turns ratio. Thus, for a nonloaded transformer, if nominal voltage is applied to the primary, nominal voltage will appear at the secondary.*

3. *Transformers are normally operated close to their nominal voltages. However, because of internal voltage drops under load, if the primary is at rated voltage, the secondary will not be, and vice versa. This is normal.*

practice
problem
III

A transformer used in an electronic power supply has a nominal rating of 120/12 volts and is connected to a 120-Vac source. Its equivalent impedance as seen from the primary is 10 Ω + j10 Ω. What is the magnitude of the load voltage if the load is 5 ohms resistive? Determine the regulation.

Transformer Efficiency

Efficiency is the ratio of output power to input power.

$$\eta = \frac{P_{out}}{P_{in}} \times 100\% \tag{24–12}$$

But $P_{in} = P_{out} + P_{loss}$. For a transformer, losses are due to I^2R losses in the windings (called copper losses) and losses in the core (called core losses). Thus,

$$\eta = \frac{P_{out}}{P_{out} + P_{loss}} \times 100\% = \frac{P_{out}}{P_{out} + P_{copper} + P_{core}} \times 100\% \tag{24–13}$$

Large power transformers are exceptionally efficient, of the order of 98 to 99 percent. The efficiencies of smaller transformers are around 95 percent or better. Losses may be measured experimentally. This is considered next.

24.7 Transformer Tests

Two important transformer tests, used mainly in the electrical power industry, are the *short-circuit test* and the *open-circuit test*. These tests provide the data needed to determine a transformer's equivalent circuit and to compute its efficiency.

The Short-Circuit Test

Figure 24–43 shows the test setup for the short-circuit test. Starting at 0 V, gradually increase E_g until the ammeter indicates the rated current. (This occurs at about 5 percent rated input voltage.) Since core losses are proportional to the square of voltage, at 5 percent rated voltage, core losses are negligible. The losses that you measure are therefore only copper losses.

Figure 24–43   Short-circuit test.

■ **EXAMPLE 24–10** Measurements on the high side of a 240/120-volt, 4.8-kVA transformer yield $V = 11.5$ V and $W = 172$ W at the rated current of $I = 4.8$ kVA/240 = 20 A. Determine $\mathbf{Z}_{eq}$.

Solution See Figure 24–44. Since $Z_L = 0$, the only impedance in the circuit is $\mathbf{Z}_{eq}$. Thus, $\mathbf{Z}_{eq} = V/I = 11.5$ V/20 A = 0.575 Ω. Further, $R_{eq} = W/I^2 = 172$ W/20^2 A^2 = 0.43 Ω. Therefore,

$$X_{eq} = \sqrt{\mathbf{Z}_{eq}^2 - R_{eq}^2} = \sqrt{(0.575)^2 - (0.43)^2} = 0.382 \text{ Ω}$$

Thus, $\mathbf{Z}_{eq} = 0.43$ Ω $+ j0.382$ Ω as shown in (b).

(a)

(b)

Figure 24–44   Determining the equivalent circuit.

The Open-Circuit Test

The setup for the open-circuit test is shown in Figure 24–45. Apply the full rated voltage. Since the load current is zero, only exciting current results. Since the exciting current is small, power loss in the coil resistance is negligible, and the power that you measure is just core loss.

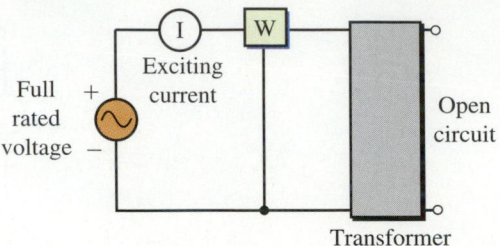

Figure 24–45   Open-circuit test.

■ **EXAMPLE 24–11**   An open-circuit test on the transformer of Example 24–10 yields a core loss of 106 W. Determine this transformer's efficiency when supplying the full, rated VA to a load at unity power factor.

Solution   Since the transformer is supplying the rated VA, its current is the full, rated current. From the short-circuit test, the copper loss at full rated current is 172 W. Thus,

$$\text{copper loss} = 172 \text{ W}$$

$$\text{core loss} = 106 \text{ W}$$

$$\text{output} = 4800 \text{ W}$$

Thus

$$\text{input} = 5078 \text{ W}$$

and

$$\eta = P_{\text{out}}/P_{\text{in}} = (4800 \text{ W}/5078 \text{ W}) \times 100 = 94.5\%$$

Copper loss varies as the square of load current. Thus, at half the rated current, the copper loss is $(\frac{1}{2})^2 = \frac{1}{4}$ of its value at full rated current, and so on. Core loss remains constant since applied voltage remains constant.

practice
problem
IV

For the transformer of Example 24–11, determine efficiency at half the rated VA output, unity power factor.

24.8 Voltage and Frequency Effects

Iron-core transformer characteristics vary with frequency and voltage. To determine why, we start with Faraday's law, $e = Nd\Phi/dt$. Specializing this to the sinusoidal ac case, it can be shown that

$$E_p = 4.44fN_p\Phi_m \tag{24–14}$$

where Φ_m is the mutual core flux.

Effect of Voltage

First, assume constant frequency. Since $\Phi_m = E_p/4.44fN_p$, the core flux is proportional to the applied voltage. Thus, if the applied voltage is increased, the core flux increases. Since magnetizing current is required to produce this flux, magnetizing current increases as well. An examination of the B-H curves of Chapter 12 shows that magnetizing current increases dramatically when flux density rises above the knee of the curve; in fact, the effect is so pronounced that the magnetizing current is no longer negligible, but may exceed the load current! For this reason, power transformers should be operated only at or near their rated voltage.

Effect of Frequency

Audio transformers must operate over a range of frequencies. Consider again $\Phi_m = E_p/4.44fN_p$. As this indicates, decreasing the frequency has the same effect as increasing the voltage—it increases the core flux and hence the magnetizing current. This large current causes internal voltage drops and hence a decrease in output voltage as indicated in Figure 24–46. Now consider increasing the frequency. As frequency increases, leakage inductance and shunt capacitance cause the voltage to fall off. To compensate for this, audio transformers are sometimes designed so that their internal capacitances resonate with their inductances to extend the operating range. This is what causes the peaking at the high-frequency end of the curve.

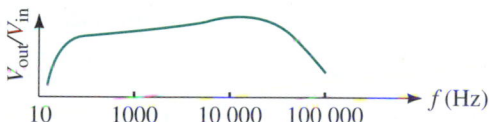

Figure 24–46   Frequency response curve, audio transformer.

in process
l e a r n i n g
check III

A transformer with a nominal rating of 240/120 V, 60 Hz, has its load on the 120-V side. Suppose $R_p = 0.4\ \Omega$, $L_p = 1.061$ mH, $R_s = 0.1\ \Omega$, and $L_s = 0.2653$ mH.

  a. Determine its equivalent circuit as per Figure 24–40(b).
  b. If $\mathbf{E}_g = 240\ \text{V}\angle0°$ and $\mathbf{Z}_L = 3 + j4\ \Omega$, what is $\mathbf{V}_L$?
  c. Compute the regulation.

24.9 Loosely Coupled Circuits

We now turn our attention to coupled circuits that do not have iron cores. For such circuits, only a portion of the flux produced by one coil links another and the coils are said to be *loosely coupled*. Loosely coupled circuits cannot be characterized by turns ratios; rather, as you will see, they are characterized by self- and mutual inductances. Air-core transformers, ferrite-core transformers, and general inductive circuit coupling fall into this category. In this section, we develop the main ideas.

Magnetic field

Figure 24–47   The plus sign goes at the tail of the current arrow.

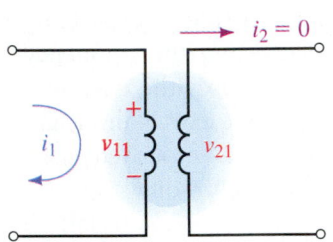

(a) v_{21} = voltage in coil 2 due to current i_1. v_{11} is voltage in coil 1 due to its own current.

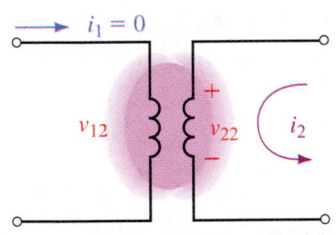

(b) v_{12} = voltage in coil 1 due to current i_2. v_{22} is voltage in coil 2 due to its own current.

Figure 24–48   To determine the effect of both coils energized simultaneously, use superposition.

Voltages in Air-Core Coils

To begin, consider the isolated (non-coupled) coil of Figure 24–47. As shown in Chapter 13, the voltage across this coil is given by $v_L = L di/dt$, where i is the current through the coil and L is its inductance. Note carefully the polarity of the voltage; the plus sign goes at the tail of the current arrow. Because the coil's voltage is created by its own current, it is called a *self-induced voltage*.

Now consider a pair of coupled coils (Figure 24–48). When coil 1 alone is energized as in (a), it looks just like the isolated coil of Figure 24–47; thus its voltage is

$$v_{11} = L_1 di_1/dt \qquad \text{(self-induced in coil 1)}$$

where L_1 is the inductance of coil 1 alone, and the subscripts indicate that v_{11} is the voltage across coil 1 due to its own current. Similarly, when coil 2 alone is energized as in (b), its self induced voltage is

$$v_{22} = L_2 di_2/dt \qquad \text{(self-induced in coil 2)}$$

For both of these self-voltages, note that the plus sign goes at the tail of their respective current arrows.

Mutual Voltages

Consider again Figure 24–48(a). When coil 1 is energized, some of the flux that it produces links coil 2, inducing voltage v_{21} in coil 2. Since the flux here is due to i_1 alone, v_{21} is proportional to the rate of change of i_1. Let the constant of proportionality be M. Then,

$$v_{21} = M di_1/dt \qquad \text{(mutually induced in coil 2)}$$

v_{21} is the *mutually induced voltage* in coil 2 and M is the *mutual inductance* between the coils. It has units of henries. Similarly, when coil 2 alone is energized as in (b), the voltage induced in coil 1 is

$$v_{12} = M di_2/dt \qquad \text{(mutually induced in coil 1)}$$

When both coils are energized, the voltage of each coil can be found by superposition; *in each coil, the induced voltage is the sum of its self-voltage plus the voltage mutually induced due to the current in the other coil.* The sign of the self term for each coil is straightforward: It is determined by placing a plus sign at the tail of the current arrow for the coil as shown in Figures 24–48(a) and (b). The polarity of the mutual term, however, depends on whether the mutual voltage is additive or subtractive.

Additive and Subtractive Voltages

Whether self- and mutual voltages add or subtract depends on the direction of currents through the coils relative to their winding directions. This is best described in terms of the dot convention. Consider Figure 24–49(a). Comparing the coils here to Figure 24–11, you can see that their top ends correspond and thus can be marked with dots. Now let

(a)   (b) Sign of mutual term is plus.

Figure 24–49   Fluxes are additive. Therefore, self and mutual terms add.

currents enter both coils at dotted ends. Using the right-hand rule, you can see that their fluxes add. The total flux linking coil 1 is therefore the *sum* of that produced by i_1 and i_2; therefore, the voltage across coil 1 is the sum of that produced by i_1 and i_2. That is,

$$v_1 = L_1 \frac{di_1}{dt} + M \frac{di_2}{dt} \qquad (24\text{--}15a)$$

Similarly, for coil 2,

$$v_2 = M \frac{di_2}{dt} + L_2 \frac{di_2}{dt} \qquad (24\text{--}15b)$$

Now consider Figure 24–50. Here, the fluxes oppose and the flux linking each coil is the *difference* between that produced by its own current and that produced by the current of the other coil. Thus, the sign in front of the mutual voltage terms will be negative.

(a)   (b) Sign of mutual term is negative.

Figure 24–50   Fluxes oppose. Therefore, self and mutual terms subtract.

The Dot Rule

As you can see from the above, the sign of the mutual voltage term is positive when both currents enter dotted terminals, but is negative when one current enters a dotted terminal and the other enters an undotted terminal. Stated another way, *the sign of the mutual voltage is the same as the sign of its self voltage when both currents enter dotted (or undotted) terminals, but is opposite when one current enters a dotted terminal and the other enters an undotted terminal.* This observation provides us with a procedure for determining voltage polarities in coupled circuits.

1. Assign a direction for currents i_1 and i_2.

2. Place a plus sign at the tail of the current arrow for each coil. This is the polarity of the self-induced component of voltage.

3. If both currents enter (or both leave) dotted terminals, make the sign of the mutually induced voltage the same as the sign of the self-induced voltage.

4. If one current enters a dotted terminal and the other leaves, make the sign of the mutually induced voltage opposite to the sign of the self-induced voltage.

■ **EXAMPLE 24–12**   Write equations for v_1 and v_2 of Figure 24–51(a).

(a)   (b)

Figure 24–51

Solution   One current enters an undotted terminal and the other enters a dotted terminal. Thus,

$$v_1 = L_1 \frac{di_1}{dt} - M \frac{di_2}{dt}$$

$$v_2 = -M \frac{di_1}{dt} + L_2 \frac{di_2}{dt}$$

practice
problem
V

Write equations for v_1 and v_2 of Figure 24–51(b).

Coefficient of Coupling

For loosely coupled coils (Figure 24–48), not all of the flux produced by one coil links the other. To describe the degree of coupling between coils, we introduce a *coefficient of coupling*, *k*. Mathematically, *k* is defined as the ratio of the flux that links the companion coil to the total flux produced by the energized coil. For iron-core transformers, almost all the flux is confined to the core and links both coils; thus, *k* is very close to 1. At the other extreme, where no flux linkage occurs, *k* = 0. Thus, $0 \leq k < 1$. Mutual inductance depends on *k*. It can be shown that mutual induc-

(a) $L'_1 = L_1 + M; L'_2 = L_2 + M$

(b) $L_T{}^+ = L_1 + L_2 + 2M$

Figure 24–52   Coils in series with additive mutual coupling.

tance, self-inductance, and the coefficient of coupling are related by the equation

$$M = k \sqrt{L_1 L_2} \qquad (24\text{–}16)$$

Thus, the larger the coefficient of coupling, the larger the mutual inductance.

Effect of Mutual Coupling on Inductors

Mutual coupling can alter a coil's apparent inductance. To illustrate, consider Figure 24–52, which shows a pair of inductors with inductances L_1 and L_2 respectively. If mutual coupling occurs, the coil inductances will no longer be L_1 and L_2. To see why, consider the voltage induced in each winding—it is the sum of the coil's own self-voltage plus the voltage mutually induced from the other coil. Since current is the same for both coils, $v_1 = L_1 di/dt + M di/dt = (L_1 + M)di/dt$, which means that coil 1 has an effective inductance of $L'_1 = L_1 + M$. Similarly, $v_2 = (L_2 + M)di/dt$, giving coil 2 an effective inductance of $L'_2 = L_2 + M$. The effective inductance of the series combination is then

$$L_T^+ = L_1 + L_2 + 2M \quad \text{(henries)} \qquad (24\text{–}17)$$

If coupling is subtractive, $L'_1 = L_1 - M$, $L'_2 = L_2 - M$, and

$$L_T^- = L_1 + L_2 - 2M \quad \text{(henries)} \qquad (24\text{–}18)$$

■ **EXAMPLE 24–13**   Three inductors are connected in series (Figure 24–53). Coils 1 and 2 interact, but coil 3 does not.

    a. Determine the effective inductance of each coil.
    b. Determine the total inductance of the series connection.

Figure 24–53

Solution

a. $L'_1 = L_1 - M = 2 \text{ mH} - 0.4 \text{ mH} = 1.6 \text{ mH}$

    $L'_2 = L_2 - M = 3 \text{ mH} - 0.4 \text{ mH} = 2.6 \text{ mH}$

    L'_1 and L'_2 are in series with L_3. Thus,

b. $L_T = 1.6 \text{ mH} + 2.6 \text{ mH} + 2.7 \text{ mH} = 6.9 \text{ mH}$

For the circuit of Figure 24–54, different symbols are used to represent coupling between sets of coils.

   a. Determine the effective inductance of each coil.
   b. Determine the total inductance of the series connection.

$$L_1 \qquad L_2 \qquad L_3$$

$L_1 = 10$ mH $M_{12} = 2$ mH (Mutual inductance between coils 1 and 2)
$L_2 = 40$ mH $M_{23} = 1$ mH (Mutual inductance between coils 2 and 3)
$L_3 = 20$ mH $M_{31} = 0.6$ mH (Mutual inductance between coils 3 and 1)

Figure 24–54

The effect of unwanted mutual inductance can be minimized by physically separating coils or by orienting their axes at right angles. The latter technique is used where space is limited and coils cannot be spaced widely. While it does not eliminate coupling, it can help minimize its effects.

24.10 Magnetic Coupling in Network Analysis

When coupling occurs between various parts of a circuit (whether wanted or not), the foregoing principles may be applied. Since it is difficult to continue the analysis in terms of general time-varying quantities, we will change over to steady state ac. This will permit us to look for the main ideas. In general, we will use the mesh approach. To use the mesh approach, 1) write mesh equations using KVL, 2) use the dot convention to determine the signs of the induced voltage components, and 3) solve the resulting equations in the usual manner using determinants (Appendix B) or a computer program.

Sinusoidal AC Excitation

To specialize to the sinusoidal ac case, convert voltages and currents to phasor form. Thus, voltage v_1 becomes $\mathbf{V}_1$ and voltage across an inductance $v_L = L di/dt$ becomes $\mathbf{V}_L = j\omega L \mathbf{I}$. The only new term is $M di/dt$ but it has the same form as $L di/dt$ and so becomes $j\omega M \mathbf{I}$. The sign associated with $j\omega M \mathbf{I}$ is determined by the dot convention. To illustrate, consider the air-core transformer of Figure 24–55. For this circuit,

$$\mathbf{V}_1 = j\omega L_1 \mathbf{I}_1 + j\omega M \mathbf{I}_2$$
$$\mathbf{V}_2 = j\omega M \mathbf{I}_1 + j\omega L_2 \mathbf{I}_2$$

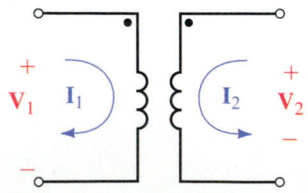

Figure 24–55   Coupled coils with sinusoidal ac excitation.

where we have used the plus sign in front of $j\omega M$ since both currents enter dotted terminals.

■ **EXAMPLE 24–14**   For Figure 24–56, write the mesh equations and solve for $\mathbf{I}_1$ and $\mathbf{I}_2$. Let $\omega = 100$ rad/s, $L_1 = 0.1$ H, $L_2 = 0.2$ H, and $M = 0.08$ H.

$R_1 = 15\ \Omega$
$R_2 = 20\ \Omega$

Figure 24–56   Air-core transformer.

Solution $\omega L_1 = (100)(0.1) = 10\ \Omega$, $\omega L_2 = (100)(0.2) = 20\ \Omega$, and $\omega M = (100)(0.08) = 8\ \Omega$. Since one current enters a dotted terminal and the other leaves, the sign of the mutual term is opposite to the sign of the self term. KVL yields

Loop 1: $\mathbf{E}_1 - R_1\mathbf{I}_1 - j\omega L_1\mathbf{I}_1 + j\omega M\mathbf{I}_2 = 0$

Loop 2: $\mathbf{E}_2 - j\omega L_2\mathbf{I}_2 + j\omega M\mathbf{I}_1 - R_2\mathbf{I}_2 = 0$

Thus,

$$(15 + j10)\mathbf{I}_1 - j8\mathbf{I}_2 = 100\angle 0°$$

$$-j8\mathbf{I}_1 + (20 + j20)\mathbf{I}_2 = 150\angle 30°$$

These can be solved by the usual methods such as determinants or by computer. The answers are $\mathbf{I}_1 = 6.36\angle{-6.57}°$ and $\mathbf{I}_2 = 6.54\angle{-2.24}°$.

■ **EXAMPLE 24–15**   For the circuit of Figure 24–57, determine $\mathbf{I}_1$ and $\mathbf{I}_2$.

$L_1 = 0.1$ H, $L_2 = 0.2$ H, $M = 80$ mH, $\omega = 100$ rad/s

Figure 24–57

Solution $\omega L_1 = 10\ \Omega$, $\omega L_2 = 20\ \Omega$, and $\omega M = 8\ \Omega$.

Loop 1: $100\angle 0° - 15\mathbf{I}_1 -j10\mathbf{I}_1 + j8\mathbf{I}_2 - 10\mathbf{I}_1 + 10\mathbf{I}_2 = 0$

Loop 2: $120\angle 30° + 20\mathbf{I}_2 + j20\mathbf{I}_2 - j8\mathbf{I}_1 + 10\mathbf{I}_2 - 10\mathbf{I}_1 = 0$

Thus:

$$(25 + j10)\mathbf{I}_1 - (10 + j8)\mathbf{I}_2 = 100\angle 0°$$

$$-(10 + j8)\mathbf{I}_1 + (30 + j20)\mathbf{I}_2 = -120\angle 30°$$

Solution yields $\mathbf{I}_1 = 3.05A\angle -38.1°$ and $\mathbf{I}_2 = 2.44A\angle -171°$.

practice problem VII

Refer to the circuit of Figure 24–58.

a. Determine the mesh equations.

b. Solve for currents $\mathbf{I}_1$ and $\mathbf{I}_2$.

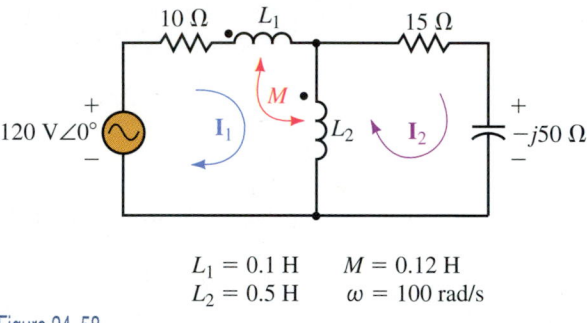

$L_1 = 0.1\ H$ $M = 0.12\ H$
$L_2 = 0.5\ H$ $\omega = 100\ rad/s$

Figure 24–58

24.11 Coupled Impedance

Earlier, we determined that an impedance $\mathbf{Z}_L$ in the secondary circuit of an iron-core transformer, when viewed from the primary, looks like $\mathbf{Z}_p = a^2\mathbf{Z}_L$. Note that $\mathbf{Z}_p$ is called the reflected load impedance and retains the characteristics of $\mathbf{Z}_L$. A somewhat similar situation occurs for the loosely coupled transformer. In this case, however, the reflected impedance does not retain the characteristics of the original impedance; instead, the sign of its reactive component is reversed, making an inductive load appear capacitive and vice versa. We now show this. Consider Figure 24–59.

KVL for the circuit of Figure 24–59 yields

Loop 1: $\mathbf{E}_g - R_1\mathbf{I}_1 - j\omega L_1\mathbf{I}_1 - j\omega M\mathbf{I}_2 = 0$

Loop 2: $-j\omega L_2\mathbf{I}_2 - j\omega M\mathbf{I}_1 - \mathbf{Z}_2\mathbf{I}_2 - \mathbf{Z}_L\mathbf{L}_2 = 0$

which reduces to

$$\mathbf{E}_g = \mathbf{Z}_p\mathbf{I}_1 + j\omega M\mathbf{I}_2 \tag{24–19a}$$

$$0 = j\omega M\mathbf{I}_1 + (\mathbf{Z}_s + \mathbf{Z}_L)\mathbf{I}_2 \tag{24–19b}$$

Figure 24–59

where $\mathbf{Z}_p = \mathbf{R}_1 + j\omega L_1$ and $\mathbf{Z}_s = \mathbf{Z}_2 + j\omega L_2$. Solving Equation 24–19b for $\mathbf{I}_2$ and substituting this into Equation 24–19a yields, after some manipulation,

$$\mathbf{E}_g = \mathbf{Z}_p \mathbf{I}_1 + \frac{(\omega M)^2}{\mathbf{Z}_s + \mathbf{Z}_L} \mathbf{I}_1$$

Now, divide both sides by $\mathbf{I}_1$, and define $\mathbf{Z}_{in} = \mathbf{E}_g/\mathbf{I}_1$. Thus,

$$\mathbf{Z}_{in} = \mathbf{Z}_p + \frac{(\omega M)^2}{\mathbf{Z}_s + \mathbf{Z}_L} \tag{24-20}$$

The term $(\omega M)^2/(\mathbf{Z}_s + \mathbf{Z}_L)$ which reflects the secondary impedances into the primary, is called the *coupled impedance* for the circuit. Note that since secondary impedances appear in the denominator, they reflect into the primary with reversed reactive parts. Thus, capacitance in the secondary circuit looks inductive to the source and vice versa.

■ **EXAMPLE 24–16**   For Figure 24–59, let $L_1 = L_2 = 10$ mH, $M = 9$ mH, $\omega = 1000$ rad/s, $R_1 = 5\ \Omega$, $\mathbf{Z}_2 = 1\ \Omega - j5\ \Omega$, $\mathbf{Z}_L = 1\ \Omega + j20\ \Omega$ and $\mathbf{E}_g = 100$ V∠0°. Determine $\mathbf{Z}_{in}$ and $\mathbf{I}_1$.

Solution

$\omega L_1 = 10\ \Omega$. Thus, $\mathbf{Z}_p = R_1 + j\omega L_1 = 5\ \Omega + j10\ \Omega$.
$\omega L_2 = 10\ \Omega$. Thus, $\mathbf{Z}_s = \mathbf{Z}_2 + j\omega L_2 = (1\ \Omega - j5\ \Omega) + j10\ \Omega = 1\ \Omega + j5\ \Omega$.
$\omega M = 9\ \Omega$ and $\mathbf{Z}_L = 1\ \Omega + j20\ \Omega$. Thus,

$$\mathbf{Z}_{in} = \mathbf{Z}_p + \frac{(\omega M)^2}{\mathbf{Z}_s + \mathbf{Z}_L} = (5 + j10) + \frac{(9)^2}{(1 + j5) + (1 + j20)}$$

$$= 8.58\ \Omega\ \angle 52.2°$$

$$\mathbf{I}_1 = \mathbf{E}_g/\mathbf{Z}_{in} = (100\angle 0°)/(8.58\angle 52.2°) = 11.7\ A\angle -52.2°$$

The equivalent circuit is shown in Figure 24–60.

Figure 24–60

practice
problem
VIII

Verify the current of Example 24–16 by solving for $\mathbf{I}_1$ using mesh analysis.

24.12   Circuit Analysis Using Computers

PSpice can be used to solve coupled circuits. It includes a coefficient of coupling function (called the k function) that permits us to specify coupling. To represent coupled coils, you specify their location in the circuit, their inductance values, the coupling between them, and the position of their dots. (The dot goes at the first-named node of each inductor.) For example, the statements

```
L1     2    0    2H
L2     0    3    1.5H
K12    L1   L2   0.5
```

specify that coil 1 is located between nodes 2 and 0, has a value of 2 H, and that the end connected to node 2 is dotted, while coil 2 is located between nodes 0 and 3, has a value of 1.5 H and that the end connected to node 0 is dotted. The coefficient of coupling between them is 0.5.

■ **EXAMPLE 24–17**   For the circuit of Figure 24–61, $L_1 = 0.1$ H, $L_2 = 0.2$ H, $M = 0.08$ H, and $f = 60$ Hz. Compute currents $\mathbf{I}_1$ and $\mathbf{I}_2$.

Solution   The coefficient of coupling is

$$k = \frac{M}{\sqrt{L_1 L_2}} = \frac{0.08}{\sqrt{(0.1)(0.2)}} = 0.5657$$

Thus,

```
Input File:
* Figure 24-61: Loosely Coupled Circuit
V1 1 0 AC 100 0
```

Figure 24–61   PSpice circuit. Note that PSpice requires that the bottom ends of the coils be joined here.

```
R1 1 2 15
R2 3 4 20
L1 2 0 0.1H
L2 3 0 0.2H
K12 L1 L2 0.5657
C1 4 0 10uF
.AC LIN 1 60 60
.PRINT AC IM(R1) IP(R1) IM(R2) IP(R2)
.OPTIONS NOPAGE
.END
```

When you run this simulation, you will obtain the answers

$$\mathbf{I}_1 = 2.21 \text{ A}\angle{-69.9°} \quad \text{and} \quad \mathbf{I}_2 = 0.35 \text{ A}\angle{104.1°}.$$

Problems

24.1   Introduction

1. For the transformers of Figure 24–62, sketch the missing waveforms.

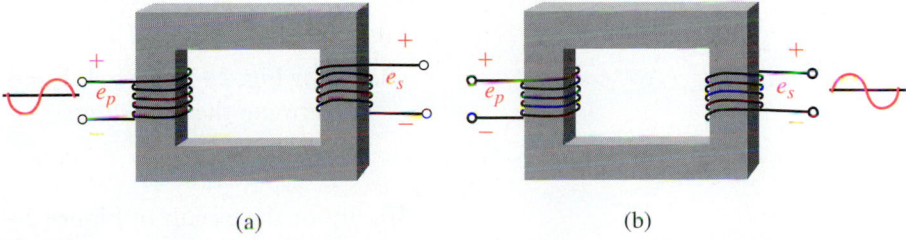

(a)   (b)

Figure 24–62

24.2   Iron-Core Transformers: The Ideal Model

2. List the four things that you neglect when you idealize an iron-core transformer.

3. An ideal transformer has $N_p = 1000$ turns and $N_s = 4000$ turns.
  a. Is it step-up or step-down?
  b. If $e_p = 100 \sin \omega t$, what is e_p?
  c. If $E_s = 24$ volts, what is E_p?
  d. If $\mathbf{E}_p = 24$ V$\angle 0°$, what is $\mathbf{E}_s$ when wound as in Figure 24–7(a)?
  e. If $\mathbf{E}_p = 800$ V$\angle 0°$, what is $\mathbf{E}_s$ when wound as in Figure 24–7(b)?

4. A 3:1 step-down transformer has a secondary current of 6 A. What is its primary current?

5. For Figure 24–63, sketch waveforms for v_1, v_2, and v_3.

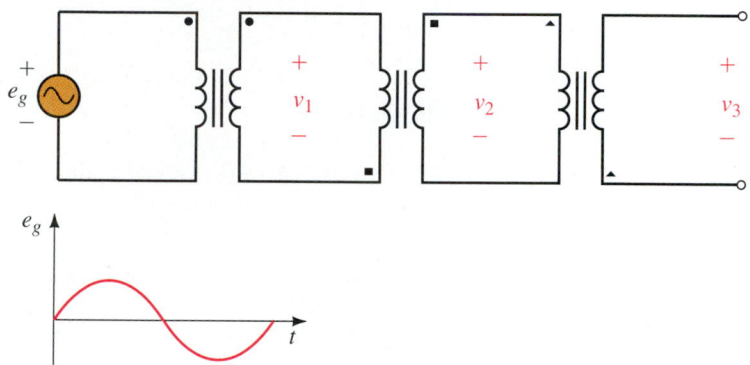

Figure 24–63

6. If, for Figure 24–64, $\mathbf{E}_g = 240$ V$\angle 0°$, $a = 2$, and $\mathbf{Z}_L = 8\,\Omega - j6\,\Omega$, determine the following:
  a. $\mathbf{V}_L$ b. $\mathbf{I}_L$ c. $\mathbf{I}_g$

7. If, for Figure 24–64, $\mathbf{E}_g = 240$ V$\angle 0°$, $a = 0.5$, and $\mathbf{I}_g = 2$ A$\angle 20°$, determine the following:
  a. $\mathbf{I}_L$ b. $\mathbf{V}_L$ c. $\mathbf{Z}_L$

8. If, for Fig. 24–64, $a = 2$, $\mathbf{V}_L = 40$ V$\angle 0°$, and $\mathbf{I}_g = 0.5$ A$\angle 10°$, determine $\mathbf{Z}_L$.

9. If, for Fig. 24–64, $a = 4$, $\mathbf{I}_g = 4$ A$\angle 30°$, and $\mathbf{Z}_L = 6\,\Omega - j8\,\Omega$, determine the following:
  a. $\mathbf{V}_L$.   b. $\mathbf{E}_g$.

10. If, for the circuit of Figure 24–64, $a = 3$, $\mathbf{I}_L = 4$ A $\angle 25°$, and $\mathbf{Z}_L = 10\,\Omega\angle -5°$, determine the following:
  a. Generator current and voltage.
  b. Power to the load.
  c. Power output by the generator.
  d. Does $P_{out} = P_{in}$?

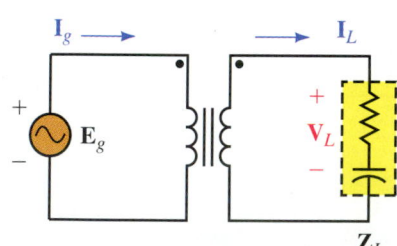

Figure 24–64

24.3 Reflected Impedance

11. For each circuit of Figure 24–65, determine Z_p.

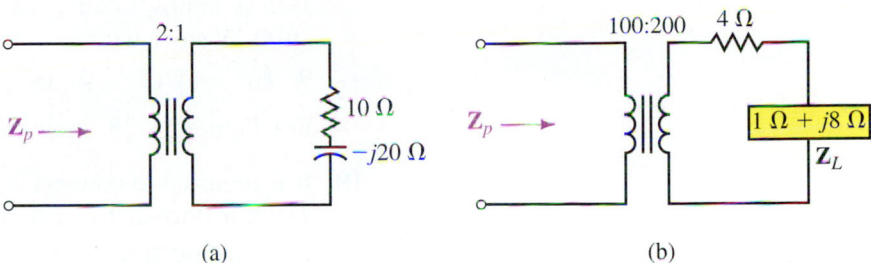

(a)   (b)

Figure 24–65

12. For each circuit of Figure 24–65, if $E_g = 120 \text{ V}\angle 40°$ is connected, determine the following, using the reflected impedance of Problem 11.

   a. I_g.   b. I_L.   c. V_L.

13. For each circuit of Figure 24–66, determine Z_T.

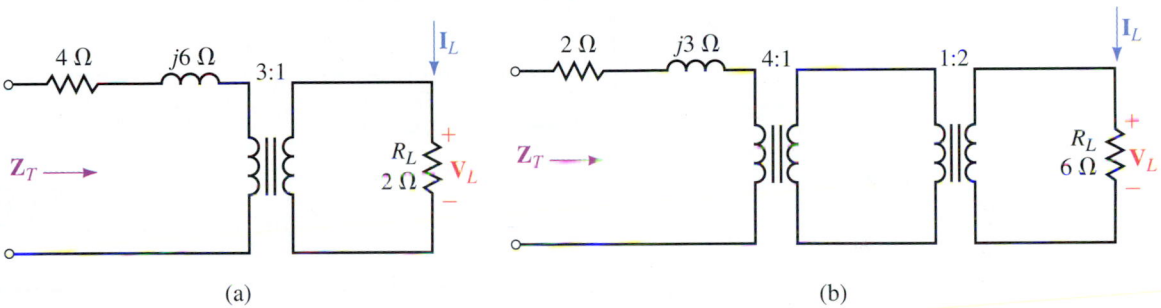

(a)   (b)

Figure 24–66

14. For each circuit of Figure 24–66, if $E_g = 120 \text{ V}\angle -40°$ is connected, determine the following:

   a. I_g.   b. I_L.   c. V_L.

24.4 Transformer Power Ratings

15. A transformer has a rated primary voltage of 7.2 kV, $a = 0.2$, and a secondary rated current of 3 A. What is its kVA rating?

16. Consider a 48 kVA, 1200/120-V transformer.
   a. What is the maximum kVA load that it can handle at $F_p = 0.8$?
   b. What is the maximum power that it can supply to a 0.75 power factor load?
   c. If the transformer supplies 45 kW to a load at 0.6 power factor, is it overloaded?

24.5   Transformer Applications

17. The transformer of Figure 24–24 has a 7200-V primary and center-tapped 240-V secondary. If Load 1 consists of twelve 100-W lamps, Load 2 is a 1500-W heater, and Load 3 is a 2400-W stove with $F_p = 1.0$, determine

a. I_1,   b. I_2,   c. I_N,   d. I_p.

18. For Figure 24–28, show that $\mathbf{Z}_2' = a_2^2\mathbf{Z}_2$ and $\mathbf{Z}_3' = a_3^2\mathbf{Z}_3$.

19. It is required to connect a 5-kVA, 120/240-V transformer as an autotransformer to a 120-V source to supply 360 V to a load.

a. Draw the circuit.
b. What is the maximum current that the load can draw?
c. What is the maximum load kVA that can be supplied?
d. How much current is drawn from the source?

24.6   Practical Iron-Core Transformers

20. For Figure 24–67, $\mathbf{E}_g = 1220\text{ V}\angle 0°$.

a. Draw the equivalent circuit,
b. Determine $\mathbf{I}_g$, $\mathbf{I}_L$, and $\mathbf{V}_L$.

Figure 24–67

21. For Figure 24–67, if $\mathbf{V}_L = 118\text{ V}\angle 0°$, draw the equivalent circuit and determine

a. $\mathbf{I}_L$,   b. $\mathbf{I}_g$,   c. $\mathbf{E}_g$,
d. no-load voltage,   e. regulation.

22. A transformer delivering $P_{\text{out}} = 48$ kW has a core loss of 280 W and a copper loss of 450 W. What is its efficiency at this load?

24.7   Transformer Tests

23. A short-circuit test (Figure 24–43) at rated current yields a wattmeter reading of 96 W, and an open-circuit test (Figure 24–45) yields a core loss of 24 W.

(a)

(b)

(c)

Figure 24–68

$L_1 = 1$ H $L_2 = 6$ H
$L_4 = 4$ H $L_3 = 1.5$ H
$M_{14} = 1$ H $M_{23} = 2$ H

Figure 24–70

a. What is the transformer's efficiency when delivering the full, rated output of 5 kVA at unity F_p?
b. What is its efficiency when delivering one quarter the rated kVA at 0.8 F_p?

24.9 Loosely Coupled Circuits

24. For Figure 24–68,

$$v_1 = L_1 \frac{di_1}{dt} \pm M \frac{di_2}{dt}, \qquad v_2 = \pm \frac{M di_1}{dt} + L_2 \frac{di_2}{dt}$$

For each circuit, indicate whether the sign to use with M is plus or minus.

25. For Figure 24–56, $R_1 = 10$ Ω, $R_2 = 30$ Ω, $L_1 = 100$ mH, $L_2 = 200$ mH, $M = 25$ mH, and $f = 31.83$ Hz. Write the mesh equations.

26. For the circuit of Question 25, what is the coefficient of coupling?

27. For a set of coils, $L_1 = 250$ mH, $L_2 = 0.4$ H, and $k = 0.85$, what is M?

24.10 Magnetic Coupling in Network Analysis

28. For each circuit of Figure 24–69, determine L_T.

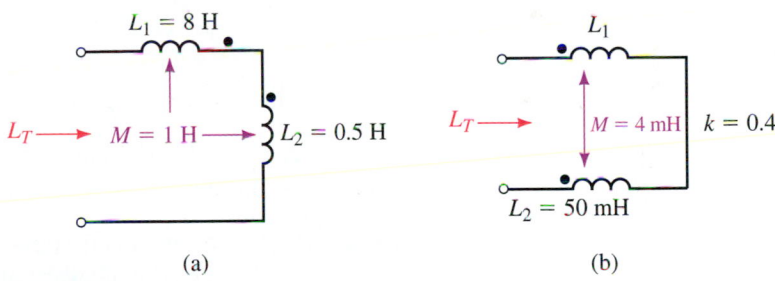

(a)   (b)

Figure 24–69

29. For Figure 24–70, determine L_T.

30. For the circuit of Figure 24–71, determine **I**.

$L_1 = 1.0$ H $k = 0.8$
$L_2 = 4.0$ H $f = 60$ Hz

Figure 24–71

31. For the circuit of Figure 24–72, determine I_1 and I_2.

$\omega L_1 = 40\ \Omega \quad \omega L_2 = 20\ \Omega \quad \omega M = 5\ \Omega$

Figure 24–72

24.11   Coupled Impedance

32. For the circuit of Figure 24–73,
    a. determine Z_{in}
    b. determine I_g. Compare to I_1 of Example 24–17.

$L_1 = 0.1\ \text{H};\ L_2 = 0.2\ \text{H};\ M = 0.08\ \text{H};\ f = 60\ \text{Hz}.$

Figure 24–73

24.12   Circuit Analysis Using Computers

33. Solve for the currents of Figure 24–57 using PSpice. Compare these to the answers of Example 24–15.

34. Solve for the currents of Figure 24–58 using PSpice. Compare these to the answers of Practice Problem VII.

35. Solve Example 24–16 for current I_1 using PSpice. Compare answers.

Nonsinusoidal
Waveforms

CHAPTER PREVIEW

In our analysis of ac circuits, we have dealt primarily with sinusoidal waveforms. Although the sinusoidal wave is the most common waveform in electronic circuits, it is by no means the only type of signal used in electronics. In previous chapters we observed how sinusoidal signals were affected by the characteristics of components within a circuit. For instance, if a 1-kHz sinusoid is applied to a low-pass filter circuit having a cutoff frequency of 3 kHz, we know that the signal appearing at the output of the filter will be essentially the same as the signal applied to the input. This effect is illustrated in Figure 25–1.

One would naturally expect that the 1-kHz low-pass filter would allow any other 1-kHz signal to pass from input to the output without being distorted. Unfortunately, this is not the case.

In this chapter we will find that any periodic waveform is composed of numerous sinusoidal waveforms, each of which has a unique amplitude and frequency. As we have already seen, circuits such as the low-pass filter and the resonant tank circuit do not allow all sinusoidal frequencies to pass from the input to the output in the same manner. As a result, the output signal may be dramatically different from the signal applied at the input. For example, if we were to apply a 1-kHz square wave to a low-pass filter having a 3-kHz cutoff, the output would appear as shown in Figure 25–2 (see page 960).

Although the frequency of the square wave is less than the 3-kHz cutoff frequency of the filter, we will find that the square wave has many high-frequency components which are well above the cutoff frequency. It is these components which are affected in the filter and this results in the distortion of the output waveform.

Once a periodic waveform is reduced to the summation of sinusoidal waveforms, it is a fairly simple matter to determine how various frequency components of the original signal will be affected by the circuit. The overall response of the circuit to a particular waveform can then be found.

KEY TERMS

Fourier Series
Frequency Spectrum
Fundamental Frequency
Distortion
Harmonic Frequency
Spectrum Analyzer

Typical recording studio control room. *(Photo courtesy JBL Professional)*

OBJECTIVES

After studying this chapter, you will be able to

- solve for the coefficients of the Fourier series of a simple periodic waveform, using integration,

- use tables to write the Fourier equivalent of any simple periodic waveform,

- sketch the frequency spectrum of a periodic waveform, giving the amplitudes of various harmonics in either volts, watts, or dBm,

- calculate the power dissipated when a complex waveform is applied to a resistive load,

- determine the output of a filter circuit given the frequency spectrum of the input signal and the frequency response of the filter,

- use PSpice to observe the actual response of a filter circuit to a nonsinusoidal input signal.

Figure 25–1

Figure 25–2

Jean Baptiste Joseph Fourier

ourier was born in Auxerre, Yonne, France on March 21, 1768. As a youth, Fourier reluctantly studied for the priesthood in the monastery of Saint-Benoit-sur Loire. However, his interest was in mathmatics.

In 1798, Fourier accompanied Napoleon to Egypt, where he was made a governor. After returning to France, Fourier was particularly interested in the study of heat transfer between two points of different temperature. He was appointed joint secretary of the Academy of Sciences in 1822.

In 1807, Fourier announced the discovery of a theorem which made him famous. Fourier's theorem states that any periodic waveform can be written as the summation of a series of simple sinusoidal functions.

Using his theorem, Fourier was able to develop important theories in heat transfer, which were published in 1822 in a book titled Analytic Theory of Heat.

Although still used to describe heat transfer, Fourier's theorem is used today to predict how filters and various other electronic circuits operate when subjected to a non-sinusoidal periodic function.

Fourier died in Paris on May 16, 1830 as the result of a fall down the stairs.

25.1 Fourier Series

In 1826, Baron Jean Baptiste Joseph Fourier developed a branch of mathematics which is used to express any periodic waveform as an infinite series of sinusoidal waveforms. Although it seems as though we are turning a simple waveform into a more complicated form, you will find that the resulting expression actually simplifies the analysis of many circuits which respond differently to signals of various frequencies. Using Fourier analysis, any periodic waveform can be written as a summation of sinusoidal waveforms as follows:

$$f(t) = a_0 + a_1 \cos \omega t + a_2 \cos 2\omega t + \cdots + a_n \cos n\omega t + \cdots$$
$$+ b_1 \sin \omega t + b_2 \sin 2\omega t + \cdots + b_n \sin n\omega t + \cdots \qquad (25\text{–}1)$$

The coefficients of the individual terms of the Fourier series are found by integrating the original function over one complete period. The coefficients are determined as

$$a_0 = \frac{1}{T} \int_{t_1}^{t_1 + T} f(t)\, dt \qquad (25\text{–}2)$$

$$a_n = \frac{2}{T} \int_{t_1}^{t_1 + T} f(t) \cos n\omega t \, dt \qquad (25\text{--}3)$$

$$b_n = \frac{2}{T} \int_{t_1}^{t_1 + T} f(t) \sin n\omega t \, dt \qquad (25\text{--}4)$$

Notice that Equation 25–1 indicates that the Fourier series of a periodic function may contain both a sine and a cosine component at each frequency. These individual components can be combined to give a single sinusoidal expression as follows:

$$a_n \cos nx + b_n \sin nx = a_n \sin(nx + 90°) + b_n \sin nx$$

$$= c_n \sin(nx + \theta)$$

where

$$c_n = \sqrt{a_n^2 + b_n^2} \qquad (25\text{--}5)$$

and

$$\theta = \tan^{-1}\left(\frac{a_n}{b_n}\right) \qquad (25\text{--}6)$$

Therefore, the Fourier equivalent of any periodic waveform may be simplified as follows:

$$f(t) = a_0 + c_1 \sin(\omega t + \theta_1) + c_2 \sin(2\omega t + \theta_2) + \cdots$$

The a_0 term is a constant which corresponds to the average value of the periodic waveform and the c_n coefficients give the amplitudes of the various sinusoidal terms. Notice that the first sinusoidal term ($n = 1$) has the same frequency as the original waveform. This component is referred to as the *fundamental frequency* of the given waveform. All other frequencies are integer multiples of the fundamental frequency and are called the *harmonic frequencies*. When $n = 2$, the resulting term is called the second harmonic; when $n = 3$, we have the third harmonic, etc. Using Equations 25–2 to 25–6, it is possible to derive the Fourier series for any periodic function.

■ **EXAMPLE 25–1**   Write the Fourier series for the pulse waveform shown in Figure 25–3.

$$V(t) = \begin{cases} 1, & 0 < t < \dfrac{T}{2} \\ 0, & \dfrac{T}{2} < t < T \end{cases}$$

Figure 25–3

Solution   The various coefficients are calculated by integrating as follows:

$$a_0 = \frac{1}{T}\int_0^{T/2} (1)\, dt + \frac{1}{T}\int_{T/2}^{T} (0)\, dt = \frac{1}{2}$$

$$a_n = \frac{2}{T}\int_0^{T/2} (1)\cos n\omega t\, dt + \frac{2}{T}\int_{T/2}^{T} (0)\, dt$$

$$= \frac{2}{T}\left[\left(\frac{1}{n\omega}\right)\sin n\omega t\right]_0^{T/2}$$

$$= \frac{1}{n\pi}\sin n\pi = 0$$

Notice that all $a_n = 0$ since $\sin n\pi = 0$ for all n.

$$b_1 = \frac{2}{T}\int_0^{T/2} (1)\sin \omega t\, dt + \frac{2}{T}\int_{T/2}^{T} (0)\, dt$$

$$= \frac{2}{T}\left[-\left(\frac{1}{\omega}\right)\cos \omega t\right]_0^{T/2}$$

$$= -\frac{1}{\pi}\left[\cos\left(\frac{2\pi t}{T}\right)\right]_0^{T/2}$$

$$= -\frac{1}{\pi}[(-1) - (1)] = \frac{2}{\pi}$$

$$b_2 = \frac{2}{T}\int_0^{T/2} (1)\sin 2\omega t\, dt + \frac{2}{T}\int_{T/2}^{T} (0)\, dt$$

$$= \frac{2}{T}\left[-\left(\frac{1}{2\omega}\right)\cos 2\omega t\right]_0^{T/2}$$

$$= -\frac{1}{2\pi}\left[\cos\left(\frac{4\pi t}{T}\right)\right]_0^{T/2}$$

$$= -\frac{1}{2\pi}[(1) - (1)] = 0$$

$$b_3 = \frac{2}{T}\int_0^{T/2} (1)\sin 3\omega t\, dt + \frac{2}{T}\int_{T/2}^{T} (0)\, dt$$

$$= \frac{2}{T}\left[-\left(\frac{1}{3\omega}\right)\cos 3\omega t\right]_0^{T/2}$$

$$= -\frac{1}{3\pi}\left[\cos\left(\frac{6\pi t}{T}\right)\right]_0^{T/2}$$

$$= -\frac{1}{3\pi}[(-1) - (1)] = \frac{2}{3\pi}$$

For all odd values of n, we have $b_n = 2/n\pi$ since $\cos n\pi = -1$. Even values of n give $b_n = 0$ since $\cos n\pi = 1$.

The general expression of the Fourier series for the given pulse wave is therefore written as

$$v(t) = \frac{1}{2} + \frac{2}{\pi}\sum_1^\infty \frac{\sin n\omega t}{n} \qquad n = 1, 3, 5, \ldots \qquad (25\text{–}7)$$

By examining the general expression for the pulse wave of Figure 25–3, several important characteristics are observed. The Fourier series confirms that the wave shown has an average value of $a_0 = 0.5$. In addition, the pulse wave has only odd harmonics. In other words, a pulse wave having a frequency of 1 kHz would have harmonic components occurring at 3 kHz, 5 kHz, etc. Although the given wave consists of an infinite number of sinusoidal components, the amplitudes of successive terms decrease as n increases.

If we were to consider only the first four nonzero frequency components of the pulse wave, we would have the following expression:

$$v(t) = 0.5 + \frac{2}{\pi}\sin \omega t + \frac{2}{3\pi}\sin 3\omega t + \frac{2}{5\pi}\sin 5\omega t + \frac{2}{7\pi}\sin 7\omega t \qquad (25\text{-}8)$$

The graphical representation of the above expression is shown in Figure 25–4.

Figure 25–4

Although this waveform is not identical with the given pulse wave, we see that the first four nonzero harmonics provide a reasonable approximation of the original waveform.

practice problem 1

Consider the ramp function shown in Figure 25–5.

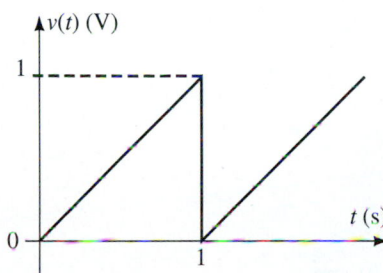

Figure 25–5

a. Use calculus to determine the Fourier expression for $v(t)$.
b. Verify that the a_0 term of the Fourier series is equal to the average value of the waveform.
c. From the Fourier expression, is the ramp function made up of odd harmonics, even harmonics, or all harmonic components? Briefly justify your answer.

in process
learning
check I

Without using calculus, determine a method of rewriting the expression of Equation 25–8 to represent a square wave having an amplitude of 1 V as illustrated in Figure 25–6. *Hint:* Notice that the square wave is similar to the pulse waveform with the exception that its average value is zero and that the peak-to-peak value is twice that of the pulse wave.

Figure 25–6

25.2 Fourier Series of Common Waveforms

All periodic waveforms can be converted into their Fourier equivalent series by using integration as shown in Section 25.1. Integration of common waveforms is time-consuming and prone to error. A more simple approach is to use tables such as Table 25-1, which gives the Fourier series of several common waveforms encountered in electrical circuits.

TABLE 25–1   Fourier Equivalents of Common Waveforms $(\omega = 2\pi/T)$

$$v(t) = \frac{4V}{\pi}\left(\sum_{1}^{\infty} \frac{\sin n\omega t}{n}\right) \qquad n = 1, 3, 5, \ldots$$

$$V(t) = \begin{cases} -V, -\dfrac{T}{2} < t < 0 \\ +V, 0 < t < \dfrac{T}{2} \end{cases}$$

Figure 25–7

$$v(t) = \frac{aV}{T} + \frac{2V}{\pi}\left(\sum_{1}^{\infty} (-1)^n \frac{\sin\left(\dfrac{\pi n a}{T}\right)\cos n\omega t}{n}\right) \qquad n = 1, 2, 3, \ldots$$

$$V(t) = \begin{cases} 0, -\dfrac{T}{2} + a/2 < t < \dfrac{T}{2} - a/2 \\ V, \dfrac{T}{2} - a/2 < t < \dfrac{T}{2} + a/2 \end{cases}$$

Figure 25–8

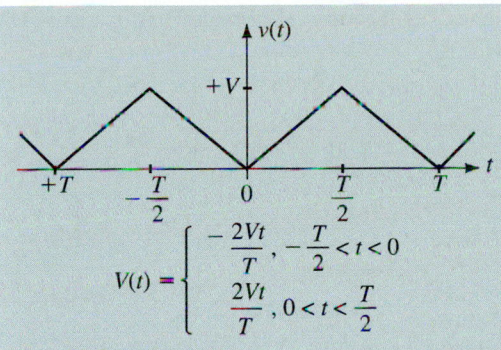

$$v(t) = \frac{V}{2} - \frac{4V}{\pi^2}\left(\sum_{1}^{\infty} \frac{\cos n\omega t}{n^2}\right) \qquad n = 1, 3, 5, \ldots$$

$$V(t) = \begin{cases} -\dfrac{2Vt}{T}, & -\dfrac{T}{2} < t < 0 \\[2mm] \dfrac{2Vt}{T}, & 0 < t < \dfrac{T}{2} \end{cases}$$

Figure 25–9

$$v(t) = -\frac{2V}{\pi}\left(\sum_{1}^{\infty} (-1)^n \frac{\sin n\omega t}{n}\right) \qquad n = 1, 2, 3, \ldots$$

$$V(t) = \frac{2Vt}{T}, \quad -\frac{T}{2} < t < \frac{T}{2}$$

Figure 25–10

$$v(t) = \frac{V}{2} - \frac{V}{\pi}\left(\sum_{1}^{\infty} \frac{\sin n\omega t}{n}\right) \qquad n = 1, 2, 3, \ldots$$

$$V(t) = \frac{Vt}{T}, \quad 0 < t < T$$

Figure 25–11

$$v(t) = \frac{2V}{\pi} - \frac{4V}{\pi}\left(\frac{\cos 2\omega t}{1 \cdot 3} + \frac{\cos 4\omega t}{3 \cdot 5} + \frac{\cos 6\omega t}{5 \cdot 7} + \cdots\right)$$

$$V(t) = |V \sin \omega t|, \quad -\frac{T}{2} < t < \frac{T}{2}$$

Figure 25–12

$$v(t) = \frac{V}{\pi} + \frac{V}{2}\sin \omega t - \frac{2V}{\pi}\left(\frac{\cos 2\omega t}{1 \cdot 3} + \frac{\cos 4\omega t}{3 \cdot 5} + \frac{\cos 6\omega t}{5 \cdot 7} + \cdots\right)$$

$$v(t) = \begin{cases} 0, & -\dfrac{T}{2} < t < 0 \\[2mm] V \sin \omega t, & 0 < t < \dfrac{T}{2} \end{cases}$$

Figure 25–13

The following example illustrates how a given waveform is converted into its Fourier series equivalent.

■ **EXAMPLE 25–2** Use Table 25–1 to determine the Fourier series for the ramp function of Figure 25–14.

Figure 25–14

Solution The amplitude of the waveform is 10 V and the angular frequency of the fundamental is $\omega = 2\pi/(2 \text{ ms}) = 1000\pi$ rad/s. The resulting series is determined from Table 25–1 as

$$v(t) = \frac{10}{2} - \frac{10}{\pi}\sin 1000\pi t - \frac{10}{2\pi}\sin 2000\pi t$$

$$- \frac{10}{3\pi}\sin 3000\pi t - \frac{10}{4\pi}\sin 4000\pi t - \cdots$$

$$= 5 - 3.18\sin 1000\pi t - 1.59\sin 2000\pi t$$

$$-1.06\sin 3000\pi t - 0.80\sin 4000\pi t - \cdots \quad \text{(V)}$$

If a given waveform is similar to one of the types shown in Table 25–1 but is shifted along the time axis, it is necessary to include a phase shift with each of the sinusoidal terms. The phase shift is determined as follows:

1. Determine the period of the given waveform.

2. Compare the given waveform with the figures appearing in Table 25–1 and select which of the waveforms in the table best describes the given wave.

3. Determine whether the given waveform leads or lags the selected figure of Table 25–1. Calculate the amount of the phase shift, as a fraction, t, of the total period. Since one complete cycle is equivalent to 360°, the phase shift is determined as

$$\theta = \frac{t}{T} \times 360°$$

4. Write the resulting Fourier expression for the given waveform. If the given waveform leads the selected figure of Table 25–1, then add the angle θ to each term. If the given waveform lags the selected figure, then subtract the angle θ from each term.

■ **EXAMPLE 25–3**   Write the Fourier expression for the first four nonzero sinusoidal terms of the waveform shown in Figure 25–15.

Figure 25–15

Solution   **Step 1:** The period of the given waveform is $T = 8.0$ ms, which gives a frequency of $f = 125$ Hz or an angular frequency of $\omega = 250\pi$ rad/s.

Step 2: From Table 25–1, we see that the given waveform is similar to the square wave of Figure 25–7.

Step 3: The waveform of Figure 25–15 leads the square wave of Figure 25–7 by an amount equivalent to $t = 2$ ms. This corresponds to a phase shift of

$$\theta = \frac{2 \text{ ms}}{8 \text{ ms}} \times 360° = 90°$$

Step 4: The Fourier expression for the first four terms of the waveform of Figure 25–15 is now written as follows:

$$v(t) = \frac{4(4)}{\pi} \sin(250\pi t + 90°) + \frac{4(4)}{3\pi} \sin(750\pi t + 90°) +$$

$$\frac{4(4)}{5\pi} \sin(1250\pi t + 90°) + \frac{4(4)}{7\pi} \sin(1750\pi t + 90°)$$

The above expression may be left as the summation of sine waves. However, since the cosine wave leads the sine wave by 90°, the expression may be simplified as a summation of cosine waves without any phase shift. This results in the following:

$$v(t) = 5.09 \cos 250\pi t + 1.70 \cos 750\pi t + 1.02 \cos 1250\pi t + 0.73 \cos 1750\ \pi t$$

practice problem *II*

Write the Fourier expression for the first four nonzero sinusoidal terms of the waveform shown in Figure 25–16. Express each term as a sine wave rather than as a cosine wave.

Figure 25–16

The waveforms of Table 25–1 provide most of the commonly observed waveforms. Occasionally, however, a particular waveform consists of a combination of a dc voltage together with a waveform or the combination of several simple waveforms. In such a case, it is generally easiest if we first redraw the original waveform as the summation of two or more recognizable waveforms. Then the Fourier series of each of the individual component waves is determined. Finally, the resultant is expressed as the summation of the two series.

■ **EXAMPLE 25–4**   Write the first four nonzero sinusoidal terms of the Fourier series for the waveform of Figure 25–17.

Figure 25–17

Solution   The waveform of Figure 25–17 is made up of a combination of waves as illustrated in Figure 25–18.

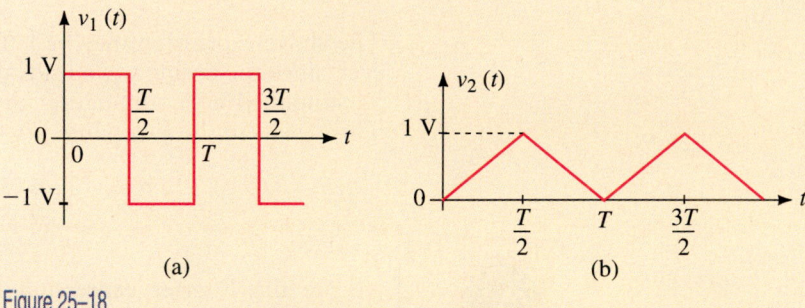

Figure 25–18

The Fourier series of each of the waveforms is determined from Table 25–1 as

$$v_1(t) = \frac{4}{\pi} \sin \omega t + \frac{4}{3\pi} \sin 3\omega t + \frac{4}{5\pi} \sin 5\omega t + \cdots$$

and

$$v_2(t) = \frac{1}{2} - \frac{4}{\pi^2} \cos \omega t - \frac{4}{3^2\pi^2} \cos 3\omega t - \frac{4}{5^2\pi^2} \cos 5\omega t - \cdots$$

When these series are added algebraically, we get

$$v(t) = v_1(t) + v_2(t)$$

$$= 0.5 + 1.27 \sin \omega t - 0.41 \cos \omega t$$

$$+ 0.42 \sin 3\omega t - 0.05 \cos 3\omega t$$

$$+ 0.25 \sin 5\omega t - 0.02 \cos 5\omega t$$

$$+ 0.18 \sin 7\omega t - 0.01 \cos 7\omega t$$

$$\vdots$$

The above series may be further simplified by using Equations 25–5 and 25–6 to provide a single coefficient and phase shift for each frequency. The resultant waveform is written as

$$v(t) = 0.5 + 1.34 \sin(\omega t - 17.7°) + 0.43 \sin(3\omega t - 6.1°)$$

$$+ 0.26 \sin(5\omega t - 3.6°) + 0.18 \sin(7\omega t - 2.6°)$$

practice
problem
III

A composite waveform is made up of the summation of the waveforms illustrated in Figure 25–19.

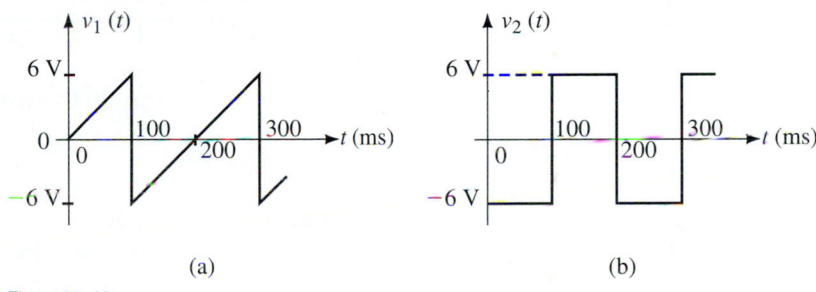

Figure 25–19

a. Sketch the composite waveform showing all voltage levels and time values.
b. Write the Fourier expression for the resultant waveform, $v(t)$.

in process
learning
check II

Consider the composite waveform of Figure 25–20:

Figure 25–20

a. Separate the given waveform into two recognizable waveforms appearing in Table 25–1.
b. Use Table 25–1 to write the Fourier series for each of component waveforms of (a).
c. Combine the results to determine the Fourier series for $v(t)$.

(a) Time-domain display
of a 1 kHz sine-wave.

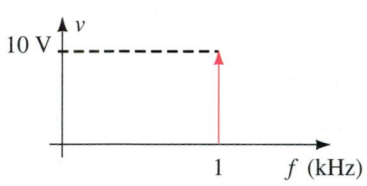

(b) Frequency-domain display
of a 1 kHz sine-wave.

Figure 25–21

25.3 Frequency Spectrum

Most waveforms that we have observed were generally shown as a function of time. However, they may also be shown as a function of frequency. In such cases, the amplitude of each harmonic is indicated at the appropriate frequency. Figure 25–21 shows the display of a 1-kHz sine wave in both the time domain and the frequency domain, while Figure 25–22 shows the corresponding displays for a pulse waveform.

(a) Time domain display
of a pulse-wave.

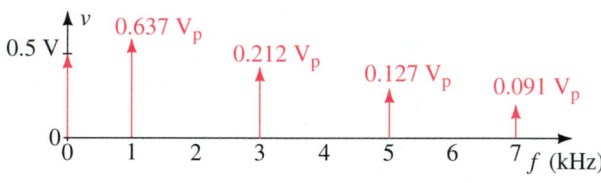

(b) Frequency domain display
of a pulse-wave.

Figure 25–22

The frequency spectrum of the pulse wave shows the average value (or dc value) of the wave at a frequency of 0 kHz and illustrates the absence of even harmonics. Notice that the amplitude of successive harmonic components decreases fairly quickly.

If the pulse wave were applied to a resistive element, power would be dissipated as if each frequency component had been applied independently. The total power is determined by the summation of the individual contributions of each frequency. In order to calculate the power dissipated at each sinusoidal frequency, we need to first convert the voltages into rms values. The frequency spectrum may then be represented in terms of power rather than as voltage.

■ **EXAMPLE 25–5**　Determine the total power dissipated by a 50-Ω resistor if the pulse waveform of Figure 25–22 is applied to the resistor. Consider the dc component and the first four nonzero harmonics. Indicate the power levels (in watts) on a frequency distribution curve.

Solution　The power dissipated by the dc component is determined as

$$P_0 = \frac{V_0^2}{R_L} = \frac{(0.5\text{ V})^2}{50\ \Omega} = 5.0\text{ mW}$$

The power dissipated by any resistor subjected to a sinusoidal frequency is determined as

$$P = \frac{V_{rms}^2}{R_L} = \frac{\left(\frac{V_P}{\sqrt{2}}\right)^2}{R_L} = \frac{V_P^2}{2R_L}$$

For the pulse wave of Figure 25–22, the power due to each of the first four nonzero sinusoidal components is found as follows:

$$P_1 = \frac{\left(\frac{2\,V}{\pi}\right)^2}{(2)(50\,\Omega)} = 4.05\text{ mW}$$

$$P_3 = \frac{\left(\frac{2\,V}{3\pi}\right)^2}{(2)(50\,\Omega)} = 0.45\text{ mW}$$

$$P_5 = \frac{\left(\frac{2\,V}{5\pi}\right)^2}{(2)(50\,\Omega)} = 0.16\text{ mW}$$

$$P_7 = \frac{\left(\frac{2\,V}{7\pi}\right)^2}{(2)(50\,\Omega)} = 0.08\text{ mW}$$

Figure 25–23 shows the power levels (in milliwatts) as a function of frequency.

Figure 25–23

Using only the dc component and the first four nonzero harmonics, the total power dissipated by the resistor is $P_T = 9.74$ mW. From Chapter 15, the actual rms voltage of the pulse waveform is found to be

$$V_{rms} = \sqrt{\frac{(1\text{ V})^2(0.5\text{ ms})}{1.0\text{ ms}}} = 0.707\text{ V}$$

Therefore, using the rms voltage, the power dissipated by the resistor is found as

$$P = \frac{(0.707\text{ V})^2}{50\,\Omega} = 10.0\text{ mW}$$

Although the pulse waveform has power contained in components with frequencies above the seventh harmonic, we see that more than 97% of the total power of a pulse waveform is contained in only the first seven harmonics.

Power levels and frequencies of the various harmonics of a periodic waveform may be measured with an instrument called a *spectrum analyzer*, shown in Figure 25–24.

Figure 25–24   Spectrum analyzer. *(Courtesy of Tektronix Inc.)*

Some spectrum analyzers are able to display either voltage levels or power levels in the frequency domain, while most others display only power levels (in dBm). When displaying power levels, the spectrum analyzer usually uses a reference 50-Ω load. Figure 25–25 shows the display of a 1.0-V pulse waveform as it would appear on a typical spectrum analyzer.

Vertical : 5 dB/division
Horizontal : 1 kHz/division
Reference : +20 dBm

Figure 25–25   Spectrum analyzer display of a 1-kHz pulse wave having a peak of 1 V.

Notice that the spectrum analyzer has a reference of +20 dBm and that this reference is shown at the top of the display rather than at the bottom. The vertical scale of the instrument is measured in decibels, where each vertical division corresponds to a difference of 5 dB. The horizontal axis for the spectrum analyzer is scaled in hertz, where each division in Figure 25–25 corresponds to 1 kHz.

practical

notes

Spectrum analyzers are very sensitive instuments. As a result, great care must be taken to ensure that the input power never exceeds the rated maximum. When in doubt, it is best to insert an extra attenuator to lower the amount of power entering the spectrum analyzer.

■ **EXAMPLE 25–6**   A spectrum analyzer with a 50-Ω input is used to display the power levels in dBm of the Fourier series components of the ramp waveform shown in Figure 25–26.

Figure 25–26

Determine the voltage and power levels of the various components and sketch the resultant display as it would appear on a spectrum analyzer. Assume that the spectrum analyzer has the same vertical and horizontal settings as those shown in Figure 25–25.

Solution   The Fourier series of the given waveform is determined from Table 25–1 as

$$v(t) = \frac{2}{2} - \frac{2}{\pi}\sin\omega t - \frac{2}{2\pi}\sin 2\omega t - \frac{2}{3\pi}\sin 3\omega t - \cdots$$

Because the fundamental frequency occurs at $f = 2$ kHz, we see that harmonic frequencies will occur at 4 kHz, 6 kHz, etc. However, because the spectrum analyzer is able to display only up to 10 kHz, we need go no further.

The dc component will have an average value of $v_0 = 1.0$ V, as expected. The rms values of the harmonic sinusoidal waveforms are determined as

$$V_{\text{rms}} = \frac{V_p}{\sqrt{2}}$$

which gives the following:

$$V_{1(\text{rms})} = \frac{2}{\pi\sqrt{2}} = 0.450 \text{ V}$$

$$V_{2(\text{rms})} = \frac{2}{2\pi\sqrt{2}} = 0.225 \text{ V}$$

$$V_{3(rms)} = \frac{2}{3\pi\sqrt{2}} = 0.150 \text{ V}$$

$$V_{4(rms)} = \frac{2}{4\pi\sqrt{2}} = 0.113 \text{ V}$$

$$V_{5(rms)} = \frac{2}{5\pi\sqrt{2}} = 0.090 \text{ V}$$

The above rms voltages are used to calculate the powers (and power levels in dBm) of the various harmonic components.

$$P_0 = \frac{(1.0 \text{ V})^2}{50 \text{ }\Omega} = 20.0 \text{ mW} \equiv 10 \log \frac{20 \text{ mW}}{1 \text{ mW}} = 13.0 \text{ dBm}$$

$$P_1 = \frac{(0.450 \text{ V})^2}{50 \text{ }\Omega} = 4.05 \text{ mW} \equiv 10 \log \frac{4.05 \text{ mW}}{1 \text{ mW}} = 6.08 \text{ dBm}$$

$$P_2 = \frac{(0.225 \text{ V})^2}{50 \text{ }\Omega} = 1.01 \text{ mW} \equiv 10 \log \frac{1.01 \text{ mW}}{1 \text{ mW}} = 0.04 \text{ dBm}$$

$$P_3 = \frac{(0.150 \text{ V})^2}{50 \text{ }\Omega} = 0.45 \text{ mW} \equiv 10 \log \frac{0.45 \text{ mW}}{1 \text{ mW}} = -3.5 \text{ dBm}$$

$$P_4 = \frac{(0.113 \text{ V})^2}{50 \text{ }\Omega} = 0.25 \text{ mW} \equiv 10 \log \frac{0.25 \text{ mW}}{1 \text{ mW}} = -6.0 \text{ dBm}$$

$$P_5 = \frac{(0.090 \text{ V})^2}{50 \text{ }\Omega} = 0.16 \text{ mW} \equiv 10 \log \frac{0.16 \text{ mW}}{1 \text{ mW}} = -7.9 \text{ dBm}$$

A spectrum analyzer would indicate a display similar to that shown in Figure 25–27.

Vertical : 5 dB/division
Horizontal : 1 kHz/division
Reference : +20 dBm

Figure 25–27

practice problem IV

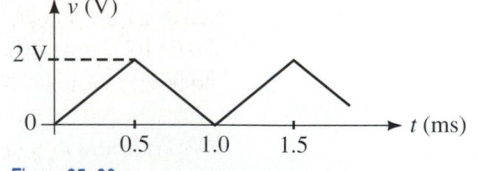

Figure 25–28

The sawtooth waveform of Figure 25–28 is applied to a 50-Ω spectrum analyzer. Sketch the display that would be observed, assuming that the spectrum analyzer has the same vertical and horizontal settings as those shown in Figure 25–25.

25.4 Circuit Response to a Nonsinusoidal Waveform

We have determined that all periodic, nonsinusoidal waveforms are comprised of numerous sinusoidal components together with a dc component. In Chapter 22 we observed how various frequencies were affected when they were applied to a given filter. We will now examine how the frequency components of a waveform will be modified when applied to the input of a given filter.

Consider what happens when a pulse waveform is applied to a bandpass filter tuned to the third harmonic, as shown in Figure 25–29.

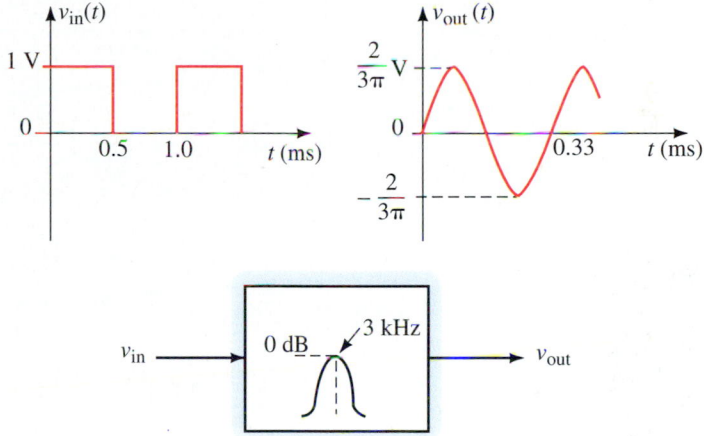

Figure 25–29

Since the band-pass filter is tuned to the third harmonic, only the frequency component corresponding to this harmonic will be passed from the input of the filter to the output. Further, because the filter has a voltage gain of 0 dB at the center frequency, the amplitude of the resulting sinusoidal output will be the same voltage level as the amplitude of the original third harmonic. All other frequencies, including the dc component, will be attenuated by the filter, and so are effectively eliminated from the output.

This method is used extensively in electronic circuits to provide frequency multiplication, since any distorted waveform will be rich in harmonics. The desired frequency component is easily extracted by using a tuned filter circuit. Although any integer multiplication is theoretically possible, most frequency multiplier circuits are either frequency doublers or frequency triplers since higher-order harmonics have much lower amplitudes.

In order to determine the resulting waveform after it passes through any other filter, it is necessary to determine the amplitude and phase shift of numerous harmonic components.

■ EXAMPLE 25–7   The circuit of Figure 25–30 has the frequency response shown in Figure 25–31.

Figure 25–30

Figure 25–31

a. Determine the dc component at the output of the low-pass filter.

b. Calculate the amplitude and corresponding phase shift of the first four nonzero sinusoidal output components.

Solution

a. From previous examples, we have determined that the given waveform is expressed by the following Fourier series:

$$v(t) = 0.5 + \frac{2}{\pi} \sin \omega t + \frac{2}{3\pi} \sin 3\omega t + \frac{2}{5\pi} \sin 5\omega t + \frac{2}{7\pi} \sin 7\omega t$$

Since the circuit is a low-pass filter, we know that the dc component will pass from the input to the output without being attenuated. Therefore

$$V_{0(out)} = V_{0(in)} = 0.5 \text{ Vdc}$$

b. By examining the frequency response of Figure 25–29, we see that all sinusoidal components will be attenuated and phase shifted. From the graphs we have the following:

$$
\begin{aligned}
&1 \text{ kHz:} \quad A_{V1} = -1.0 \text{ dB,} \quad \Delta\theta_1 = -26.6° \\
&3 \text{ kHz:} \quad A_{V3} = -5.1 \text{ dB,} \quad \Delta\theta_3 = -56.3° \\
&5 \text{ kHz:} \quad A_{V5} = -8.6 \text{ dB,} \quad \Delta\theta_5 = -68.2° \\
&7 \text{ kHz:} \quad A_{V7} = -11.2 \text{ dB,} \quad \Delta\theta_7 = -74.1°
\end{aligned}
$$

The amplitudes of the various harmonics at the output of the filter are determined as follows:

$$V_{1(out)} = \left(\frac{2}{\pi}\right) 10^{-1.0/20} = 0.567 \text{ V}_p$$

$$V_{3(out)} = \left(\frac{2}{3\pi}\right) 10^{-5.1/20} = 0.118 \text{ V}_p$$

$$V_{5(out)} = \left(\frac{2}{5\pi}\right) 10^{-8.6/20} = 0.047 \text{ V}_p$$

$$V_{7(out)} = \left(\frac{2}{7\pi}\right) 10^{-11.2/20} = 0.025 \text{ V}_p$$

The Fourier series of the output waveform is now approximated as

$$v(t) = 0.5 + 0.567 \sin (\omega t - 26.6°) + 0.118 \sin (3\omega t - 56.3°)$$

$$+ 0.047 \sin (5\omega t - 68.2°) + 0.025 \sin (7\omega t - 74.1°)$$

Various computer-aided design (CAD) and mathematical applications programs are able to generate a time-domain display of a waveform from a mathematical expression. When the above waveform is plotted in the time domain, it appears as illustrated in Figure 25–32.

Figure 25–32

The voltage waveform of Figure 25–33 is applied to a high-pass filter circuit having the frequency response shown in Figure 25–34.

Figure 25–33

(a) Frequency response of voltage gain for a high-pass filter

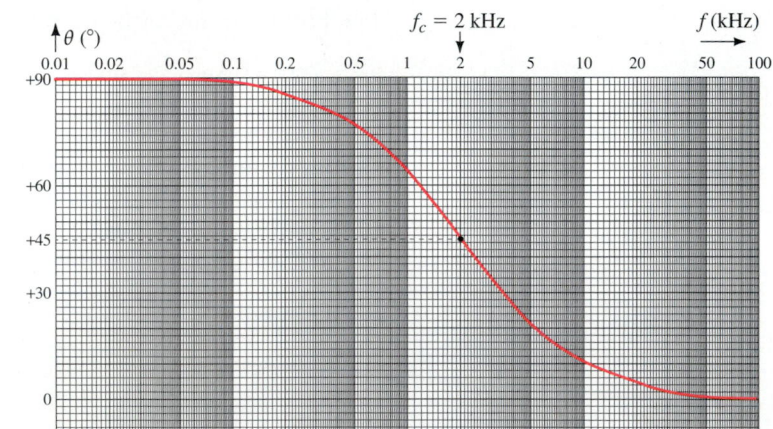

(b) Frequency response of phase shift for a high-pass filter

Figure 25–34

a. Determine the dc component at the output of the high-pass filter.

b. Calculate the amplitude and corresponding phase shift of the first four nonzero sinusoidal output components.

25.5 Circuit Analysis Using Computers

PSpice and PROBE can be used to help visualize the frequency spectrum at the input and the output of a given circuit. By comparing the input and the output, we are able to observe how a given circuit distorts the waveform due to attenuation and phase shift of the various frequency components which make up the input signal. In this example we will use a pulse waveform. The PSpice statement for a square wave is given below:

Vxxx   n+   n−   PULSE(min max [delay] [t_r] [t_f] [width] [period])

where

> **Vxxx** is the unique name of the voltage source.
> **n+** is the node to which the positive terminal of the voltage source is connected.
> **n−** is the node to which the negative terminal of the voltage source is connected.
> **min** is the minimum voltage level of the pulse.
> **max** is the maximum voltage level of the pulse.
> **delay** is the length of time the start of the pulse is delayed [optional].
> **t_r** is the rise time of the pulse [optional].
> **t_f** is the fall time of the pulse [optional].
> **width** is the width of the pulse.
> **period** is the period of the waveform.

In addition to performing an ac analysis, we will also perform a Fourier series analysis of the given input and output waveforms. The PSpice statements required for performing the Fourier analysis are given as

.TRAN   tstep   tstop

where

> **.TRAN** performs a transient analysis.
> **.tstep** is the time interval between plotted points.
> **.tstop** is the time over which the transient is analyzed. If this time is several periods in length, the resulting Fourier display has a better resolution than if the time is only one or two periods long. However, the time required by the computer to perform the analysis may be excessively long.

.FOUR   freq   vname

where

> **freq** is the fundamental frequency of the input waveform.
> **vname** if the list of the variables for which the Fourier series coefficients are to be found.

■ EXAMPLE 25–8 Write the PSpice input file to find the Fourier series for both the input and output waveforms for the circuit of Figure 25–35. Use the PROBE postprocessor to obtain both a time-domain and frequency-domain display of the output waveform.

Figure 25–35

Solution The input file and the pertinent data from the output file are given below:

```
Input File:
Response to a nonsinusoidal input
Vs    1    0    PULSE(0    1    0    .1us    .1us    .5ms    1ms)
R1    1    2    53.05ohms
C1    2    0    1uF
.TRAN    0.5us    2ms
.FOUR    1kH    V(1)    V(2)
.PROBE
.OPTIONS NOPAGE
.END
Output File:
FOURIER COMPONENTS OF TRANSIENT RESPONSE V(1)
DC COMPONENT =    5.000000E-01
```

HARMONIC NO	FREQUENCY (HZ)	FOURIER COMPONENT	NORMALIZED (DEG)	PHASE PHASE (DEG)	NORMALIZED
1	1.000E+03	6.366E-01	1.000E+00	-9.000E-02	0.000E+00
2	2.000E+03	3.910E-09	6.142E-09	9.364E+01	9.373E+01
3	3.000E+03	2.122E-01	3.333E-01	-2.700E-01	-1.800E-01
4	4.000E+03	3.905E-09	6.135E-09	9.728E+01	9.737E+01
5	5.000E+03	1.273E-01	2.000E-01	-4.500E-01	-3.600E-01
6	6.000E+03	3.898E-09	6.123E-09	1.009E+02	1.010E+02
7	7.000E+03	9.095E-02	1.429E-01	-6.300E-01	-5.400E-01
8	8.000E+03	3.887E-09	6.106E-09	1.046E+02	1.047E+02
9	9.000E+03	7.074E-02	1.111E-01	-8.100E-01	-7.200E-01

```
TOTAL HARMONIC DISTORTION =    4.287984E+01 PERCENT
FOURIER COMPONENTS OF TRANSIENT RESPONSE V(2)
DC COMPONENT =    5.000977E-01
```

HARMONIC NO	FREQUENCY (HZ)	FOURIER COMPONENT	NORMALIZED COMPONENT	PHASE (DEG)	NORMALIZED PHASE (DEG)
1	1.000E+03	6.052E-01	1.000E+00	-1.854E+01	0.000E+00
2	2.000E+03	1.562E-04	2.581E-04	6.343E+01	8.197E+01
3	3.000E+03	1.504E-01	2.484E-01	-4.602E+01	-2.748E+01
4	4.000E+03	1.348E-04	2.227E-04	4.768E+01	6.622E+01
5	5.000E+03	6.443E-02	1.064E-01	-6.064E+01	-4.210E+01
6	6.000E+03	1.249E-04	2.063E-04	4.353E+01	6.207E+01
7	7.000E+03	3.456E-02	5.710E-02	-6.788E+01	-4.933E+01
8	8.000E+03	1.586E-04	2.620E-04	2.426E+01	4.280E+01
9	9.000E+03	2.147E-02	3.548E-02	-7.152E+01	-5.298E+01

```
TOTAL HARMONIC DISTORTION =    2.785088E+01 PERCENT
```

The time-domain display of the input and output voltage waveforms are shown in Figure 25–36.

Figure 25–36

The corresponding frequency-domain displays of the input and output waveforms are shown in Figure 25–37.

Figure 25–37

Problems

25.1   Fourier Series

1. Use calculus to derive the Fourier series for the waveform shown in Figure 25–38.

Figure 25–38

2. Repeat Problem 1 for the waveform shown in Figure 25–39.

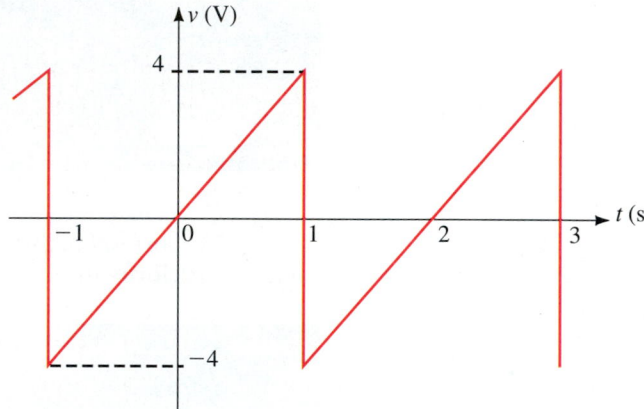

Figure 25–39

25.2   Fourier Series of Common Waveforms

3. Use Table 25–1 to determine the Fourier series for the waveform of Figure 25–40.

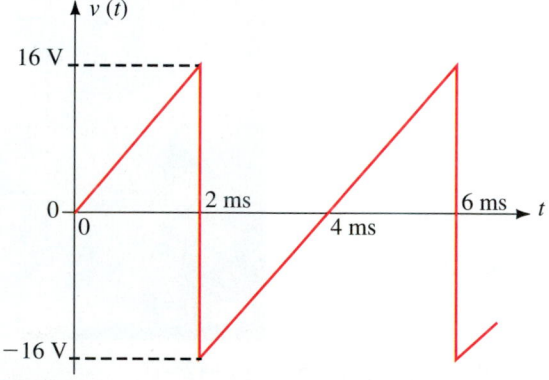

Figure 25–40

4. Repeat Problem 3 for the waveform of Figure 25–41.

5. Repeat Problem 3 for the waveform of Figure 25–42.

Figure 25–41

$$v(t) = \begin{cases} 10 \sin \omega t, \ 0 < t < 10 \ ms \\ 0, \ 10 \ ms < t < 20 \ ms \end{cases}$$

Figure 25–42

6. Repeat Problem 3 for the waveform of Figure 25–43.

7. Write the expression including the first four sinusoidal terms of the the Fourier series for the waveform of Figure 25–44.

$$v(t) = \left| 10 \sin \omega t \right|$$

Figure 25–43

Figure 25–44

8. Repeat Problem 7 for the waveform of Figure 25–45.

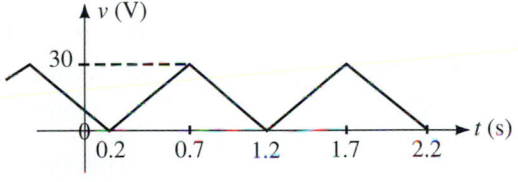

Figure 25–45

9. A composite waveform is made up of the two periodic waves shown in Figure 25–46.

(a)

(b)

Figure 25–46

    a. Sketch the resulting waveform.

    b. Write the Fourier series of the given waveforms.

    c. Determine the Fourier series of the resultant.

10. Repeat Problem 9 for the periodic waveforms shown in Figure 25–47.

Figure 25–47

11. A composite waveform is made up of the two periodic waves shown in Figure 25–48.

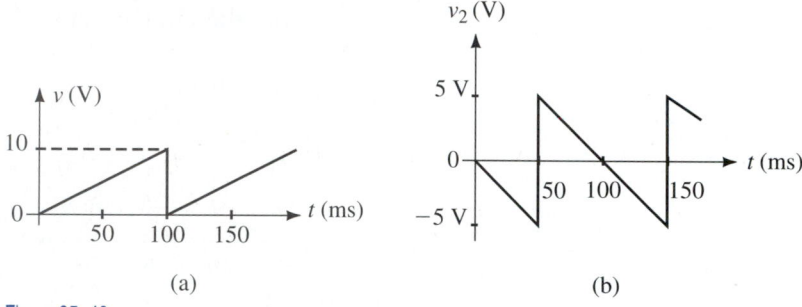

Figure 25–48

a. Sketch the resulting waveform.
b. Solve for the dc value of the resultant.
c. Write the Fourier series of the given waveforms.
d. Determine the Fourier series of the resultant.

12. Repeat Problem 11 for the waveforms shown in Figure 25–49.

Figure 25–49

Figure 25–50

13. The waveform of Figure 25–50 is made up of two fundamental waveforms from Table 25–1. Sketch the two waveforms and determine the Fourier series of the composite wave.

14. The waveform of Figure 25–51 is made up of a dc voltage combined with two fundamental waveforms from Table 25–1. Determine the dc voltage and sketch the two waveforms. Determine the Fourier series of the composite wave.

Figure 25–51

25.3   Frequency Spectrum

15. Determine the total power dissipated by a 50-Ω resistor, if the voltage waveform of Figure 25–40 is applied to the resistor. Consider the dc component and the first four nonzero harmonics. Indicate the power levels (in watts) on a frequency distribution curve.

16. Repeat Problem 15 for the waveform of Figure 25–41.

17. A spectrum analyzer with a 50-Ω input is used to measure the power levels in dBm of the Fourier series components of the waveform shown in Figure 25–42. Determine the power levels (in dBm) of the dc component and the first four nonzero harmonic components. Sketch the resultant display as it would appear on a spectrum analyzer.

18. Repeat Problem 17 if the waveform of Figure 25–43 is applied to the input of the spectrum analyzer.

25.4   Circuit Response to a Nonsinusoidal Waveform

19. The circuit of Figure 25–52 has the frequency response shown in Figure 25–53.

Figure 25–52

Figure 25–53   (*Continues*)

Figure 25–53   (*Continued*)

    a. Determine the dc component at the output of the filter.

    b. Calculate the amplitude and corresponding phase shift of the first four nonzero sinusoidal output components.

20. Repeat Problem 19 for Figures 25–54 and 25–55.

Figure 25–54

Figure 25–55   (*Continues*)

Figure 25–55   (*Continued*)

25.5   Circuit Analysis Using Computers

21. Write the PSpice input file to find the Fourier series for both the input and output waveforms for the circuit of Figure 25–52. Use the PROBE postprocessor to obtain both a time-domain and frequency-domain display of the output waveform. Compare your results to those obtained in Problem 19.

22. Repeat Problem 21 for the circuit of Figure 25–54. Compare your results to those obtained in Problem 20.

CIRCUIT ANALYSIS USING PSpice

PSpice is circuit simulation software designed to solve electrical and electronic problems from physical descriptions of circuits, rather than from a set of mathematical equations. The version of PSpice used in this book is the Design Center evaluation version from MicroSim Corporation. (It is available free to educational institutions for use on IBM PCs and compatibles or the Macintosh as noted on the software request form included with this text.) Although the evaluation version does not provide the full capabilities of the commercial version, it is able to solve all example problems in this book.

A.1 Using PSpice

To solve a problem using the evaluation version: 1) Create a circuit description file; 2) Call up an editor and key in the file; 3) Run the simulation; 4) View the results. For DOS machines, PSpice includes a menu-driven development environment called *Control Shell* that integrates these four steps. To use Control Shell, type *ps* at the DOS prompt. The general procedure to follow thereafter is outlined by the following examples. For non-Control Shell users, create your input file using a standard text editor. (A short overview for non-Control Shell users follows the Control Shell section.)

A Circuit Analysis Example

Consider Figure A–1. (This is the first PSpice example of the book, Section 4.8, Chapter 4.) Although it is simple, it will get you started.

Figure A–1

First, number all nodes, with the reference node designated node 0. Next, prepare an input file. The first line must be a heading (which PSpice uses as the title for its output). Since we are going to save the file on disk as *PROBA1.CIR*, we have noted its name on the title line. Here is the input file.

```
* Example A-1: Ohm's Law, (File Name: PROBA1.CIR)
Vs 1 0 12V
R1 1 2 10ohms
R2 2 0 15ohms
.END
```

Running the Example Under Control Shell

1. At the DOS prompt, type *ps* then press the Enter key.

2. The startup screen appears with *Files* highlighted. Press the Enter key. A drop-down menu appears with *Current File* highlighted. Press Enter. Type in the name of the file that you wish to create, in this case, *PROBA1.CIR*, then press Enter.

3. *Files* is again highlighted. Press Enter. The drop-down menu appears with *Edit* highlighted. Press Enter. PSpice's Editor appears. Key in the input file. Press Esc when done. You are asked if you want to save changes. Press Enter to Save (S). The drop-down menu goes away. (If it does not, you have errors. Go to Step 3A. Otherwise, go to Step 4.)

   a. Press key F6 to view error messages. Press Esc, then Enter to open the editor. Correct errors. Press Esc to save changes as in Step 3.

4. Use the arrow keys to move the highlight to *Analysis*, then press Enter. Select *Run PSpice* and press Enter. If there are no errors, the analysis will run to completion. If there are errors, return to the editor and correct them.) To view results, select *Files*, then *Browse Output*. Use PgDn or the arrow keys to scroll to the answers. Press Esc when done. (Note: PSpice computes current entering nodes. Because of the way we set up Example A–1, PSpice determined the current into the source (which is −0.48 A), rather than the current out (which is +0.48 A). A better solution is shown in Example A–2.)

5. If you want a hard copy, select *Quit* from the menu, press Enter, select *DOS Command* and press Enter. Key in *PRINT PROBA1.OUT* (or the appropriate file name), press Enter and follow screen instructions. Press any key to return to Control Shell.

6. To exit from PSpice, select *Quit*, press Enter, select *Exit to DOS* and press Enter. If asked to save files, press Enter to answer each question. This finally takes you back to DOS.

Some Points about Input Files

- The first line must be a title.

- The last line must be .END (with the period).

- You can use upper- or lowercase letters.

- You can insert comment lines to make the program more readable by using an * in the first column.

Improving the Example

Example A–1 performs what is called a *bias point analysis* where only node voltages and source currents are calculated. To determine currents and voltages in other parts of the circuit, we must include a source specification statement such as .DC as shown below, in Example A–2. As noted in Section A.2, the .DC statement permits us to specify a fixed dc source or to sweep the source voltage over a range of values. In this example, we use a single 12 V value. We can then use a .PRINT statement to print the voltages and currents that we want to see. The .PRINT DC I(R1) statement used here, for example, prints current I(R1). Enter this example as file name PROBA2.CIR and run it as before. The answer is found in the Output File. As expected, it is + 0.48 A.

```
Example A-2: Ohm's Law again (File Name: PROBA2.CIR)
Vs 1 0
R1 1 2 10ohms
R2 2 0 15ohms
.DC Vs 12 12 1
.PRINT DC I(R1)
.END
```

Graphic Output

For problems requiring graphic output PSpice includes a graphic post-processor. To use it, place a .PROBE statement in your program as illustrated in Example A–3. (Here, PSpice computes the current for the circuit of Figure A–2 over a range of voltages from 0 to 25 V in 5-V steps.) You can get both a screen display and a hardcopy if desired.

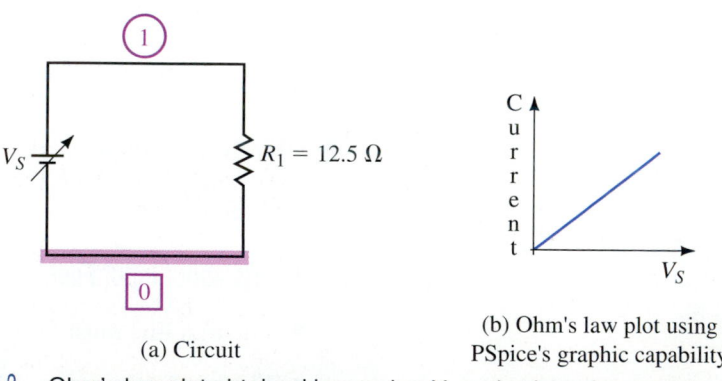

(a) Circuit

(b) Ohm's law plot using PSpice's graphic capability

Figure A–2   Ohm's law plot obtained by varying V_S and solving for current.

```
* Example A-3: Graph of Ohm's Law (File Name: PROBA3.CIR)
* Circuit Description
Vs 1 0
R1 1 0 12.5ohms
* DC Source Specification
.DC Vs 0 25V 5V
* Plot Results
.PROBE
.END
```

Upon completion of analysis, a set of empty axes appears on the screen. To get a graphic output, proceed as follows:

1. Move the cursor to *Add_trace*, then press Enter.

2. Type I(R1) and press Enter. A graph of current versus voltage appears as illustrated in Figure 4–31, Chapter 4.

3. Press Esc twice to return to Control Shell.

Displaying Multiple Traces

Multiple traces may be graphed with independent *y*-axis scales. For example, to display load voltage V(1) and current I(R1) for Example A–3, proceed as follows. When the empty axis appears,

Create the Voltage Trace and Label its Axis

1. Select *Add_trace* and press the Enter key.

2. Type *V(1)* and press the Enter key. The voltage trace appears on the screen.

3. Select *Y_axis* and press the Enter key.

4. Select *Change_title* and press the Enter key.

5. Type *Load Voltage* and press the Enter key. This labels the voltage axis.

Create the Current Trace and Label its Axis

1. Select *add_aXis* and press the Enter key.

2. Select *Change_title* and press the Enter key.

3. Type *Load Current* and press the Enter key. This labels the current axis.

4. Select *Exit* and press the Enter key.

5. Select *Add_trace* and press the Enter key.

6. Type *I(R1)* and press the Enter key. This adds the current trace.

Non-Control Shell Users

Use the following to adapt Control Shell operation to your system. First, prepare your circuit description file using any standard text editor (i.e., one that does not insert control characters into the file), then save with a .CIR extension. Next, call up PSpice by typing *pspice* at the system prompt. Key in the name of the input file. PSpice then computes and writes results to an output file using the same name as the input file, but with an .OUT extension. (For example, if your input file is *PROBA1.CIR*, the output file will be *PROBA1.OUT*.) Use your editor to view this file. If your program includes a .PROBE statement, PROBE runs automatically. In this case, to view graphic results, proceed as described earlier.

A.2 Syntax and Other Things

You can use standard engineering prefixes, Table A–1. For example, you may specify a 12 000 ohm resistor as 12000ohms or as 12kohms.

TABLE A–1 Standard Prefixes used by PSpice

SYMBOL	SCALE	NAME
T or t	10^{12}	tera
G or g	10^{9}	giga
MEG or meg	10^{6}	mega
K or k	10^{3}	kilo
M or m	10^{-3}	milli
U or u	10^{-6}	micro
N or n	10^{-9}	nano
P or p	10^{-12}	pico

Reference Conventions

Current   PSpice computes current into the node named first in a statement. Thus, R1 1 0 10ohms causes PSpice to compute current into node 1 of R1.

Voltage   Voltage is referenced at the first node with respect to the second. Thus, $V(1,2) = 20V$ means that node 1 is 20 V positive with respect to node 2.

Devices

All devices need a type designation such as R for resistors, C for capacitors, etc., Table A–2. If a capacitor or inductor has an initial value, use IC = value at the end of its statement. Initial values on capacitors use the reference convention for voltage while initial currents for inductors use the reference convention for current. Thus, C1 1 2 5uF IC = 10V specifies a 5-μF capacitor between nodes 1 and 2 with an initial voltage of 10 V, node 1 positive with respect to node 2.

TABLE A–2 Devices Encountered in This Book. PSpice Includes Many Others.

CLASS	TYPE	DESCRIPTION
Passive	R	Resistor
	C	Capacitor IC = value (optional)
	L	Inductor IC = value (optional)
	K	Inductor coupling
Sources	V	Voltage source
	I	Current source
Ideal Switch	S	Voltage controlled switch

DC Voltages

Voltage sources use the reference convention noted above. Thus, Vs 1 0 45V specifies a 45 V dc source between nodes 1 and 0 with its positive end connected to node 1.

AC Voltages

Vs 1 0 AC 70.7V 60deg means an ac voltage source connected between nodes 1 and 0 with 70.7 V amplitude and 60° phase angle.

.DC Statement

The .DC statement permits you to specify a single voltage or a range of voltages. For example, .DC Vs 0 25V 5V steps Vs from 0 to 25 V in increments of 5 V. For a fixed source, make the initial and final voltages the same. Thus, .DC Vs 12 12 1 specifies a fixed 12 V source as we saw in Example A–2. (Note: the V is optional. Thus, use 12 or 12V.)

.AC Statement

The .AC statement permits you to specify a single frequency or a range of frequencies. Frequency may be swept linearly, by octave or by decade. Thus, .AC LIN 101 100Hz 200Hz computes 101 evenly spaced points (i.e., a linear spacing) from 100 Hz to 200 Hz. For analysis at a single frequency, set the number of points to 1 and make the initial and final frequencies the same. Thus, .AC LIN 1 2kHz 2kHz specifies a single frequency of 2 kHz.

.TRAN Statement

The .TRAN TSTEP TSTOP statement causes PSpice to perform a transient analysis with time t varied from 0 to TSTOP. TSTEP is the print increment. Thus, the statement .TRAN 200ms 2s displays values every 20 ms from 0 to 2 s. Keyword UIC (Use Initial Conditions) placed at the end of the .TRAN statement causes sources to be applied as step sources.

Pulse Sources

Pulse sources are created by a PULSE declaration. For example, PULSE (V1 V2 td tr tf pw per) defines a pulse with an initial value of V1, an amplitude of V2, a delay to start of td, rise and fall times of tr and tf respectively, a pulse width of pw and a period of per. (Not all values need be specified.) Thus, the statement Vs 1 0 PULSE(0 20V 0 1ns) defines Vs as a 20 V pulse source connected between nodes 1 and 0 with a delay of 0 to start and a rise time of 1 ns.

.PRINT Statement

The .PRINT statement writes the results of a PSpice analysis to an output file for viewing. It includes three basic parts, the command .PRINT, the type of analysis being performed, and a list of variables. Thus, for a dc analysis, use .PRINT DC (e.g., .PRINT DC I(R1) as in Example A–2), for an ac analysis, use .PRINT AC while for a transient analysis, use .PRINT TRAN. If a table of data is generated by the analysis, .PRINT creates a table of results. For instance, if you add the statement .PRINT DC I(R1) to Example A–3, a table of current versus voltage will be created (as V_S is varied from 0 to 25 V) that can be used to plot the Ohm's law graph manually. Similarly, the .PRINT TRAN statement used in transient analysis creates a table of values versus time. When used with .PROBE, both an on-screen plot and a table of values are created.

Other Comments

1) PSpice does not permit voltage sources or inductors to be connected in loops. To get around this, insert a small resistor in the loop. 2) PSpice requires a dc path from all nodes to the reference node. If no path exists (because of capacitors for example), add a high value resistor from the node to the reference.

PSpice Documentation

We have included only a brief overview here. MicroSim sells a set of documentation manuals that provides full details.

SOLUTION OF SIMULTANEOUS LINEAR EQUATIONS

Simultaneous linear equations appear often in the solution of problems in electrical/electronics technology. The solutions of these equations has been simplified by a branch of mathematics called linear algebra. Although the actual theorems and proofs are well outside the scope of this textbook, we will use some of the principles of linear algebra to solve simple linear equations. The following is a set of N simultaneous linear equations in N unknowns:

$$a_{11}x_1 + a_{12}x_2 + \ldots + a_{1N}x_N = b_1$$
$$a_{21}x_1 + a_{22}x_2 + \ldots + a_{2N}x_N = b_2$$
$$\cdot$$
$$\cdot$$
$$\cdot$$
$$a_{N1}x_1 + a_{N2}x_2 + \ldots + a_{NN}x_N = b_N$$

The above equations may also be expressed in matrix form as

$$\mathbf{AX} = \mathbf{B}$$

where

$$\mathbf{A} = \begin{bmatrix} a_{11} a_{12} \ldots a_{1N} \\ a_{21} a_{22} \ldots a_{2N} \\ \cdot \quad \quad \cdot \\ \cdot \quad \cdot \quad \cdot \\ \cdot \quad \cdot \quad \ldots \\ a_{N1} a_{N2} \ldots a_{NN} \end{bmatrix}, \quad \mathbf{X} = \begin{bmatrix} x_1 \\ x_2 \\ \cdot \\ \cdot \\ \cdot \\ x_N \end{bmatrix}, \quad \mathbf{B} = \begin{bmatrix} b_1 \\ b_2 \\ \cdot \\ \cdot \\ \cdot \\ b_N \end{bmatrix}$$

Substitution

Although simultaneous linear equations may be expressed in several unknowns, we begin with the most simple, namely two simultaneous linear equations in two unknowns. Consider the equations below:

$$a_{11}x_1 + a_{12}x_2 = b_1 \qquad \text{(B–1)}$$

$$a_{21}x_1 + a_{22}x_2 = b_2 \qquad \text{(B–2)}$$

If we multiply Equation B–1 by a_{22} and Equation B–2 by a_{12}, we have

$$a_{11}a_{22}x_1 + a_{12}a_{22}x_2 = a_{22}b_1$$

$$a_{12}a_{21}x_1 + a_{12}a_{22}x_2 = a_{12}b_2$$

Subtracting, we obtain

$$a_{11}a_{22}x_1 - a_{12}a_{21}x_1 = a_{22}b_1 - a_{12}b_2$$

which gives

$$x_1 = \frac{a_{22}b_1 - a_{12}b_2}{a_{11}a_{22} - a_{12}a_{21}}$$

Similarly, we solve for the unknown x_2 as

$$x_2 = \frac{a_{11}b_2 - a_{21}b_1}{a_{11}a_{22} - a_{12}a_{21}}$$

■ **EXAMPLE B–1**   Use substitution to find the solutions for the following following linear equations:

$$2x_1 + 8x_2 = -2$$

$$x_1 + 2x_2 = 5$$

Solution   Rewriting the first equation, we have

$$2x_1 = -2 - 8x_2$$

$$x_1 = -1 - 4x_2$$

Now, substituting the above expression into the second equation, we have

$$(-1 - 4x_2) + 2x_2 = 5$$

$$-2x_2 = 6$$

$$x_2 = -3$$

Finally, we have

$$x_1 = -1 - 4(-3) = 11$$

Determinants

While substitution may be used for solving simultaneous linear equations in two variables, it is lengthy and particularly complicated when solving for more than two unknowns. An easier method used for solving simultaneous linear equations involves using *determinants*. We begin by expressing the simultaneous linear equations (B–1) and (B–2) as a product of matrices:

$$\begin{bmatrix} a_{11} & a_{12} \\ a_{21} & a_{22} \end{bmatrix} \begin{bmatrix} x_1 \\ x_2 \end{bmatrix} = \begin{bmatrix} b_1 \\ b_2 \end{bmatrix} \tag{B–3}$$

Column 1   Column 2   Column 3

A determinant is a set of coefficients which has the same number of rows and columns and which may be expressed as a single value. The number of rows (or columns) defines the *order* of a determinant. The second-order determinant corresponding to the coefficients of the matrix equation (B–3) consists of the elements in columns 1 and 2 and is expressed as

$$D = \begin{vmatrix} a_{11} & a_{12} \\ a_{21} & a_{22} \end{vmatrix}$$

The value of the second-order determinant is found by taking the product of the upper left term and the lower right term (elements of the principal diagonal) and then subtracting the product of the lower left term and the upper right term (elements of the secondary diagonal). The result is given as

$$D = a_{11}a_{22} - a_{12}a_{21}$$

The unknowns of the simultaneous linear equations are found by using a technique called *Cramer's rule*. In applying this rule, we need to solve the following determinants:

$$x_1 = \frac{\begin{vmatrix} b_1 & a_{12} \\ b_2 & a_{22} \end{vmatrix}}{\begin{vmatrix} a_{11} & a_{12} \\ a_{21} & a_{22} \end{vmatrix}} \quad \frac{a_{22}b_1 - a_{12}b_2}{a_{11}a_{22} - a_{21}a_{12}}$$

and

$$x_2 = \frac{\begin{vmatrix} a_{11} & b_1 \\ a_{21} & b_2 \end{vmatrix}}{\begin{vmatrix} a_{11} & a_{12} \\ a_{21} & a_{22} \end{vmatrix}} \quad \frac{a_{11}b_2 - a_{21}b_1}{a_{11}a_{22} - a_{21}a_{12}}$$

The application of Cramer's rule gives the solution for each unknown by first placing the determinant of the coefficient matrix in the denominator. The numerator is then developed by using the same determinant with the exception that the coefficients of the variable to be found are replaced by the coefficients of the solution matrix. The resulting solutions are precisely those found when we used substitution.

■ **EXAMPLE B–2** Use determinants to find solutions for the following linear equations:

$$2x_1 + 8x_2 = -2$$

$$x_1 + 2x_2 = 5$$

Solution The determinant of the denominator is found as

$$D = \begin{vmatrix} 2 & 8 \\ 1 & 2 \end{vmatrix} = (2)(2) - (1)(8) = -4$$

The variables are now calculated as

$$x_1 = \frac{\begin{vmatrix} -2 & 8 \\ 5 & 2 \end{vmatrix}}{-4} = \frac{(-2)(2) - (5)(8)}{-4} = \frac{-44}{-4} = 11$$

and

$$x_2 = \frac{\begin{vmatrix} 2 & -2 \\ 1 & 5 \end{vmatrix}}{-4} = \frac{(2)(5) - (1)(-2)}{-4} = \frac{12}{-4} = -3$$

The solution of third-order simultaneous linear equations is similar to the method used for solving second-order equations. Consider the following third-order simultaneous linear equation:

$$a_{11}x_1 + a_{12}x_2 + a_{13}x_3 = b_1$$

$$a_{21}x_1 + a_{22}x_2 + a_{23}x_3 = b_2$$

$$a_{31}x_1 + a_{32}x_2 + a_{33}x_3 = b_3$$

The corresponding matrix equation is shown as follows:

$$\begin{bmatrix} a_{11} & a_{12} & a_{13} \\ a_{21} & a_{22} & a_{23} \\ a_{31} & a_{32} & a_{33} \end{bmatrix} \begin{bmatrix} x_1 \\ x_2 \\ x_3 \end{bmatrix} = \begin{bmatrix} b_1 \\ b_2 \\ b_3 \end{bmatrix}$$

The value of the third-order determinant may be found in one of several ways. The first method works for only third-order determinants, while the second method is a more general approach which evaluates any order of determinant.

Method I. This method works only for third-order determinants:

1. Begin by writing the original columns of the third-order determinant.

2. Copy the first two columns, placing them to the right of the original determinant.

3. Add the product of the elements of the principal diagonal to the products of the adjacent two parallel diagonals to the right of the principal diagonal.

4. Subtract the product of the elements of the secondary diagonal and also subtract the products of the elements along the two other parallel diagonals.

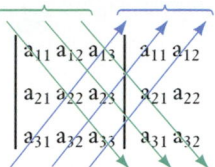

The resultant determinant is written as

$$D = a_{11}a_{22}a_{33} + a_{12}a_{23}a_{31} + a_{13}a_{21}a_{32} - a_{31}a_{22}a_{13} - a_{32}a_{23}a_{11} - a_{33}a_{21}a_{12}$$

■ **EXAMPLE B–3**   Evaluate the following determinant:

$$D = \begin{vmatrix} 3 & 1 & -2 \\ 1 & -2 & 3 \\ 2 & 3 & 2 \end{vmatrix}$$

Solution   We begin by rewriting the first two columns as follows:

$$D = \begin{vmatrix} 3 & 1 & -2 \\ 1 & -2 & 3 \\ 2 & 3 & 2 \end{vmatrix} \begin{matrix} 3 & 1 \\ 1 & -2 \\ 2 & 3 \end{matrix}$$

Now, adding the products of the principal diagonal and adjacent diagonals and subtracting the products of the secondary diagonal and adjacent diagonals, we have

$$D = (3)(-2)(2) + (1)(3)(2) + (-2)(1)(3)$$
$$- (2)(-2)(-2) - (3)(3)(3) - (2)(1)(1)$$
$$= -49$$

METHOD II.   This evaluation of determinants is achieved by expansion by minors. The *minor* of an element is the determinant which remains after deleting the row and the column in which the element lies. The value of any *n*th-order determinant is found as follows:

1. For any row or column, find the product of each element and the determinant of its minor.

2. A product is given a positive sign if the sum of the row and the column of the element is even. The product is given a negative sign if the sum is odd.

3. The value of the determinant is the sum of the resulting terms.

As before, Cramer's rule is used to solve for the unknowns, x_1, x_2, and x_3, by using determinants and replacing the appropriate terms of the numerator with the terms of the solution matrix. The resulting determinants and solutions are given as follows:

$$x_1 = \frac{\begin{vmatrix} b_1 & a_{12} & a_{13} \\ b_2 & a_{22} & a_{23} \\ b_3 & a_{32} & a_{33} \end{vmatrix}}{\begin{vmatrix} a_{11} & a_{12} & a_{13} \\ a_{21} & a_{22} & a_{23} \\ a_{31} & a_{32} & a_{33} \end{vmatrix}}$$

By expansion of minors, the determinant of the denominator is found as

$$D = + a_{11} \begin{vmatrix} a_{22} & a_{23} \\ a_{32} & a_{33} \end{vmatrix} - a_{21} \begin{vmatrix} a_{12} & a_{13} \\ a_{32} & a_{33} \end{vmatrix} + a_{31} \begin{vmatrix} a_{12} & a_{13} \\ a_{22} & a_{23} \end{vmatrix}$$

$$= a_{11}(a_{22}a_{33} - a_{23}a_{32}) - a_{21}(a_{12}a_{33} - a_{13}a_{32}) + a_{31}(a_{12}a_{23} - a_{13}a_{22})$$

The solution for x_1 is now found to be

$$x_1 = \frac{+ b_1 \begin{vmatrix} a_{22} & a_{23} \\ a_{32} & a_{33} \end{vmatrix} - b_2 \begin{vmatrix} a_{12} & a_{13} \\ a_{32} & a_{33} \end{vmatrix} + b_3 \begin{vmatrix} a_{12} & a_{13} \\ a_{22} & a_{23} \end{vmatrix}}{D}$$

$$= \frac{b_1(a_{22}a_{33} - a_{23}a_{32}) - b_2(a_{12}a_{33} - a_{13}a_{32}) + b_3(a_{12}a_{23} - a_{13}a_{22})}{D}$$

Similarly, for x_2, we get

$$x_2 = \frac{\begin{vmatrix} a_{11} & b_1 & a_{13} \\ a_{21} & b_2 & a_{23} \\ a_{31} & b_3 & a_{33} \end{vmatrix}}{\begin{vmatrix} a_{11} & a_{12} & a_{13} \\ a_{21} & a_{22} & a_{23} \\ a_{31} & a_{32} & a_{33} \end{vmatrix}}$$

$$= \frac{-b_1(a_{21}a_{33} - a_{23}a_{31}) + b_2(a_{11}a_{33} - a_{13}a_{31}) + b_3(a_{11}a_{23} - a_{13}a_{21})}{D}$$

and for x_3 we have

$$x_3 = \frac{\begin{vmatrix} a_{11} & a_{12} & b_1 \\ a_{21} & a_{22} & b_2 \\ a_{31} & a_{32} & b_3 \end{vmatrix}}{\begin{vmatrix} a_{11} & a_{12} & a_{13} \\ a_{21} & a_{22} & a_{23} \\ a_{31} & a_{32} & a_{33} \end{vmatrix}}$$

$$= \frac{b_1(a_{21}a_{32} - a_{22}a_{31}) - b_2(a_{11}a_{32} - a_{12}a_{31}) + b_3\,(a_{11}a_{22} - a_{12}a_{21})}{D}$$

■ **EXAMPLE B–4**   Solve for x_1 in the following system of linear equations using minors.

$$3x_1 + x_2 - 2x_3 = 1$$

$$x_1 - 2x_2 + 3x_3 = 11$$

$$2x_1 + 3x_2 + 2x_3 = -3$$

Solution   The determinant of the denominator is evaluated as follows:

$$D = \begin{vmatrix} 3 & 1 & -2 \\ 1 & -2 & 3 \\ 2 & 3 & 2 \end{vmatrix}$$

$$= +(3)\begin{vmatrix} -2 & 3 \\ 3 & 2 \end{vmatrix} - (1)\begin{vmatrix} 1 & -2 \\ 3 & 2 \end{vmatrix} + (2)\begin{vmatrix} 1 & -2 \\ -2 & 3 \end{vmatrix}$$

$$= (3)(-4 - 9) - (2 + 6) + (2)(3 - 4)$$

$$= -49$$

and so the unknown x_1 is calculated to be

$$x_1 = \frac{\begin{vmatrix} 1 & 1 & -2 \\ 11 & -2 & 3 \\ -3 & 3 & 2 \end{vmatrix}}{-49}$$

$$= \frac{+(1)\begin{vmatrix} -2 & 3 \\ 3 & 2 \end{vmatrix} - (11)\begin{vmatrix} 1 & -2 \\ 3 & 2 \end{vmatrix} + (-3)\begin{vmatrix} 1 & -2 \\ -2 & 3 \end{vmatrix}}{-49}$$

$$= \frac{(-4-9) - (11)(2 + 6) - (3)(3 - 4)}{-49}$$

$$= 2$$

practice problem

Use expansion by minors to solve for x_2 and x_3 in the Example B–4.

ANSWERS TO IN-PROCESS LEARNING CHECKS AND PRACTICE PROBLEMS

Chapter 1

In-Process Learning Check I

1. One

2. 1 280 cm

3. m/s

4. a. 4.27 s   b. 1.53 mV   c. 1.23 ms

5. a. $1.8 \times 10^4 = 18 \times 10^3$
   b. $36 \times 10^4 = 3.6 \times 10^5 = 0.36 \times 10^6$
   c. $4.45 \times 10^2 = 44.5 \times 10^1$
   d. $27 \times 10^{-6} = 2.7 \times 10^{-5}$

6. a. 0.752 mA   b. 980 μV   c. 400 μs = 0.4 ms

Practice Problems I

1. 0.1297 m^2
2. a. 9.144 m/s   b. 32.92 km/h

Practice Problems II

   a. 2.71×10^3 d. 25.9×10^3
   b. 2.93×10^6 e. 6.75
   c. $15.7 \times 10^2 = 157 \times 10^1$

Practice Problems III

1. 1.8 MV

Practice Problems IV

1. 196

2. 2.34

3. 195.83287;   2.3390710

4. 14.3 A

Chapter 2

In-Process Learning Check I

1. An atom consists of a nucleus of protons and neutrons orbited by electrons. The nucleus is positive because protons are positive, but the atom is neutral because it contains the same number of electrons as protons, and their charges cancel.

2. The valence shell is the outermost shell, and the electrons in this shell are called valence electrons.

3. The force between charged particles is proportional to the product of their charges and inversely proportional to the square of their spacing. Since force decreases as the square of the spacing, electrons far from the nucleus experience little force of attraction.

4. If a loosely bound electron gains sufficient energy, it may break free from its parent atom and wander throughout the material. Such an electron is called a free electron. For materials like copper, heat (thermal energy) can give an electron enough energy to dislodge it from its parent atom.

5. A normal atom is neutral because it has the same number of electrons as protons and their charges cancel. An atom that has lost an electron is called a positive ion, while an atom that has gained an electron is called a negative ion.

6. The electrons remain in the material.

7. Good conductors have many free electrons while good insulators have few.

In-Process Learning Check II

1. $Q_A = 13.74 \ \mu C$ (pos.); $Q_B = 13.57 \ \mu C$ (neg.)

2. Chemical action creates an excess of electrons at the zinc and a deficiency of electrons at the carbon electrode. Because one pole is positive and the other negative, a voltage exists between them.

3. Motion is random. Since the net movement in all directions is zero, current is zero.

4. 80 A

5. 3195 s

Practice Problems I

1. 2160 N (repulsive)

2. 32.8 nC, both (+) or both (−)

3. 14 μC (+)

Practice Problems II

1. 204 J

2. 2 mC

Practice Problems III

1. 0.615 mA

2. 720 C

Practice Problems IV

1. 3 mA

2. 0.15 mA

Chapter 3

In-Process Learning Check I

1. Iron wire will have approximately seven times more resistance than copper.

2. The longer wire will have twice the resistance.

3. The wire having the greater diameter will have one quarter of the resistance.

In-Process Learning Check II

1. 200 A

2. The actual value is less than the theoretical value. Since only the surface of the cable is able to dissipate heat, the current must be decreased to prevent heat build-up.

In-Process Learning Check III

A = 63.7 CM

A = 3.23×10^{-8} m^2 = 0.0323 mm^2

In-Process Learning Check IV

Positive temperature coefficient means that resistance of a material increases as temperature increases. Negative temperature coefficient means that resistance of a material decreases as temperature increases. Aluminum has a positive temperature coefficient.

Practice Problem I

$R = 698 \ \Omega$

Practice Problems II

1. AWG19: $d = 0.912$ mm $A = 0.653$ mm^2
   AWG30: $d = 0.254$ mm $A = 0.0507$ mm^2

2. AWG16: $A = 1.306$ mm^2
   AWG40: $A = 0.0051$ mm^2

3. AWG16: $A = 1.31$ mm^2
   AWG40: $A = 0.00487$ mm^2

Practice Problems III

1. $R = 42.4 \ \Omega$
2. $R = 42.5 \ \Omega$
3. $R = 27.2 \ \Omega$

Practice Problem IV

$R_{-40°C} = 20.8 \ \Omega$
$R_{+40°C} = 29.3 \ \Omega$

Practice Problem V

Brown   Black   Violet   Gold

Practice Problem VI

a. $R = 0 \ \Omega$ (short circuit)
b. $R = \infty \ \Omega$ (open circuit)

Practice Problem VII

a. $590 \ \Omega$
b. $250 \ \Omega$
c. $120 \ \Omega$

Practice Problems VIII

1. $R = 200 \ \Omega$
2. Resistance is decreased to half. (R = 100 Ω)

Practice Problem IX

    a. $G = 5$ pS

    b. $R = 200$ GΩ

Chapter 4

In-Process Learning Check I

1.

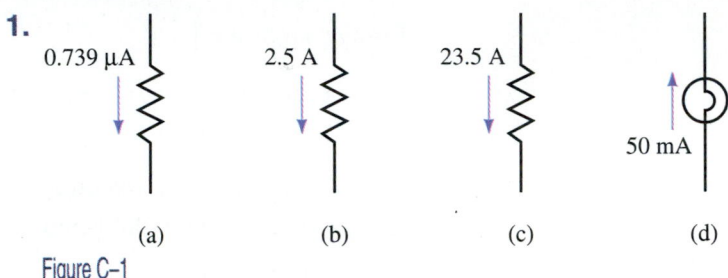

     (a)   (b)   (c)   (d)

Figure C–1

2. 25 V

3. 0 A; No

4. a. 1.25 A   **b.** 0.625 A   **c.** 12.5 A   **d.** 5 A

5. a. 30 V   **b.** −72 V   **c.** −90 V   **d.** 160 V

Practice Problems I

1. a. When the voltage is halved to 5 V, $I = 5V/10\ \Omega = 0.5$ A

    b. When the resistance is halved to 6 Ω, current doubles to 36 V/6 Ω = 6 A

2. a. 2.4 A   **b.** 2.0 V   **c.** 25 kΩ

Practice Problem II

    + 1000 V −   − 450 V +   + 120 V −   − 1.0 V +

     (a)   (b)   (c)   (d)

Figure C–2

Practice Problem III

    a. $P = I^2 R$. Thus, $I^2 = P/R$ and $I = \sqrt{P/R}$

    b. $P = V^2/R$. Thus, $V^2 = PR$ and $V = \sqrt{P/R}$

    c. 1.3 A   d. 27 V   e. $P_1 = 5$ W; $P_2 = 25$ W

Chapter 5

In-Process Learning Check I

1. Two elements are connected at only one node.

2. No current-carrying element is connected to the common node.

In-Process Learning Check II

The summation of voltage drops and rises around any closed loop is equal to zero.

or

The summation of voltage rises is equal to the summation of voltage drops around a closed loop.

In-Process Learning Check III

$R_1 = 4.2 \text{ k}\Omega$
$R_2 = 12.6 \text{ k}\Omega$
$R_3 = 25.2 \text{ k}\Omega$

In-Process Learning Check IV

$E_{\text{CELL}} = 2.18 \text{ V}$

In-Process Learning Check V

$R_1 = 18.8 \text{ k}\Omega$
$R_2 = 28.2 \text{ k}\Omega$

In-Process Learning Check VI

The chassis of the oven is grounded when it is connected to the electrical outlet.

In-Process Learning Check VII

$V_{ab} = 7.5 \text{ V}$
$V_{ca} = -10.0 \text{ V}$
$V_{bc} = 2.5 \text{ V}$

In-Process Learning Check VIII

$R_T = 2 \text{ M}\Omega$

Practice Problem I

$2 \text{ V} - 4 \text{ V} + 4 \text{ V} - 3.5 \text{ V} - 1.5 \text{ V} + 3 \text{ V} = 0$

Practice Problem II

a. $R_T = 90 \ \Omega$

b. $I = 1.33$ A   counterclockwise

c.

$$V_1 = 26.7 \text{ V} \qquad V_2 = 53.3 \text{ V} \qquad V_3 = 40 \text{ V}$$

Figure C-3

d. $P_1 = 35.6$ W $P_2 = 71.1$ W $P_3 = 53.3$ W

e. $P_T = 160$ W

f. $P_T = P_1 + P_2 + P_3 = 160$ W

Practice Problem III

i. $V_{10\Omega} \cong 0$ $V_{10\Omega} = 10 \ \mu\text{V}$

$V_{5\text{M}\Omega} \cong 5$ V $V_{5\text{M}\Omega} = 5$ V

$V_{15\text{M}\Omega} \cong 15$ V $V_{15\text{M}\Omega} = 15$ V

ii. $V_{5\Omega} \cong 0$ $V_{10\Omega} = 0.30$ mV

$V_{10\Omega} \cong 0$ $V_{5\text{M}\Omega} = 0.60$ mV

$V_{1\text{M}\Omega} \cong 60$ V $V_{1\text{M}\Omega} = 60$ V

Practice Problem IV

$V_{ab} = -8.00$ V

Practice Problem V

$R_{S_1} = 1.245 \ \text{M}\Omega$

$R_{S_2} = 4.995 \ \text{M}\Omega$

Chapter 6

In-Process Learning Check I

a. $I = 2.0$ A

b. $I_1 = 0.267$ A $= I_2 = I_3$, $I_4 = 1.200$ A

c. $(0.267 \text{ A})(3) + 1.20 \text{ A} = 2.0$ A (as required)

In-Process Learning Check II

a. $I_{1\Omega} = 30$ A $I_{3\Omega} = 10$ A $I_{4\Omega} = 7.5$ A

b. $I_T = 53.5$ A

c. $R_T = 0.561 \ \Omega$ (Same result as in part a.)

In-Process Learning Check III

The voltmeter with the smaller internal resistance would load the circuit more, since more of the circuit current would enter the instrument.

Practice Problem I

$\Sigma I_{in} = 3$ mA $+ 6$ mA $+ 1$ mA $= 10$ mA
$\Sigma I_{out} = 4$ mA $+ 4$ mA $+ 2$ mA $= 10$ mA
$\Sigma I_{in} = \Sigma I_{out}$

Practice Problem II

$I_2 = I_1 = 500$ μA
$I_4 = -700$ μA

Practice Problem III

$G_T = 0.125$ S
$R_T = 8$ Ω

Practice Problem IV

i. $R_T = 12$ Ω
ii. $R_T = 240$ Ω

Practice Problem V

a. $I_1 = 100$ A $I_2 = 150$ A
b. $I_1 = 4.5$ A $I_2 = 13.5$ A $I_3 = 9.0$ A
c. $I_1 = 3.0$ mA $I = 5.0$ mA

Practice Problem VI

Refer to Figure 6–33. $R_{shunt} = 25.13$ Ω. $R_m = 5$ kΩ.

Chapter 7

In-Process Learning Check I

$P_T = 115.2$ mW $P_1 = 69.1$ mW $P_2 = 36.9$ mW $P_3 = 9.2$ mW
$P_1 + P_2 + P_3 = 115.2$ mW as required.

In-Process Learning Check II

$P_T = 30.0$ mW $P_{R1} = 20$ mW $P_{R2} = 2.5$ mW $P_Z = 7.5$ mW
$P_{R1} + P_{R2} + P_Z = 30$ mW as required.

In-Process Learning Check III

$R_L = 2\ \text{k}\Omega$:

    a. N = 1/4: $V_L = 8.70\%$ of V_{in}.

    b. N = 1/2: $V_L = 14.29\%$ of V_{in}.
       N = 3/4: $V_L = 26.09\%$ of V_{in}.

$R_L = 200\ \text{k}\Omega$:

    c. N = 1/4: $V_L = 24.5\%$ of V_{in}.
       N = 1/2: $V_L = 48.8\%$ of V_{in}.
       N = 3/4: $V_L = 73.6\%$ of V_{in}.

    d. If $R_L \gg R_1$, then the loading effect is minimal.

Practice Problem I

$$R_T = R_1 + R_2\|[(R_3\|R_5) + (R_4\|R_6)]$$

Practice Problem II

    a. $R_T = 7.2\ \text{k}\Omega$

    b. $I_T = 1.11\ \text{mA}$

    c. $I_2 = 0.444\ \text{mA}$
      $I_1 = 0.133\ \text{mA}$

    d. $V_{ab} = -0.800\ \text{V}$

Practice Problem III

$$V_G = 0\ \text{V} \qquad I_D = 3.0\ \text{mA} \qquad V_{DS} = +6.0\ \text{V}$$

Practice Problem IV

    a. $R_1 = 10\ \text{k}\Omega$ $I_1 = 1.0\ \text{mA}$ $I_2 = 0.5\ \text{mA}$
      $I_Z = 0.5\ \text{mA}$

    b. $R_1 = 30\ \text{k}\Omega$ $I_Z = 0$ $I_1 = I_2 = 0.375\ \text{mA}$

Practice Problem V

    a. $V_{bc} = 0 \rightarrow 10\ \text{V}$

    b. $V_{bc} = 0 \rightarrow 57.1\ \text{V}$

    c. When R_L is large in comparison to the potentiometer resistance, the output voltage will better approximate the unloaded voltage. In this example, the unloaded voltage is $V_{bc} = 0 \rightarrow 60\ \text{V}$.

Practice Problem VI

V_1:   Loading effect = 94.3%
V_2:   Loading effect = 94.3%

Practice Problem VII

$I_{T(RDG)} = 19.35$ mA   Loading effect $= 19.4\%$
$I_{1(RDG)} = 3.846$ mA   Loading effect $= 3.85\%$
$I_{2(RDG)} = 16.67$ mA   Loading effect $= 16.7\%$

Chapter 8

In-Process Learning Check I

1. A voltage source consisting of E in series with R is equivalent to a current source having an ideal current source, $I = E/R$ in parallel with the same resistance, R.

2. Current sources are never connected in series.

In-Process Learning Check II

1. Voltage is zero.

2. Current is zero.

3. R_5 can be replaced with either a short circuit or an open circuit.

Practice Problems I

1. a. 3-A current source in parallel with a 12-Ω resistor.
   b. 5-μA current source in parallel with a 50-kΩ a resistor.
2. a. 750-V voltage source in series with a 30-Ω resistor.
   b. 6.25-V voltage source in series with a 50-kΩ resistor.

Practice Problem II

$I_1 = 4.40$ A $I_2 = 3.54$ A $I_3 = -0.860$ A

Practice Problem III

$I_1 = 3.25$ A $I_2 = -1.95$ A $I_3 = 1.72$ A

Practice Problem IV

$V_1 = 23.5$ V $V_2 = 25.2$ V $V_3 = 0.133$ V

Practice Problem V

$R_1 = 4.8$ kΩ $R_2 = 3.6$ kΩ $R_3 = 2.4$ kΩ

Practice Problems VI

1. $R_4 = 20$ Ω
2. a. $R_4 = 0$ Ω: $I = 0.286$ A
   b. $R_4 = 50$ Ω: $I = -0.0526$ A

Chapter 9

In-Process Learning Check I

$P_{R_1} = 304$ mW $\qquad P_{R_2} = 76$ mW $\qquad P_{R_3} = 276$ mW

Assuming that superposition applies for power:

$P_{R_1}(1) = 60$ mW $\qquad P_{R_1}(2) = 3.75$ mW $\qquad P_{R_1}(3) = 60$ mW

$P_{R_1} = 123.75$ mW $\neq 304$ mW

In-Process Learning Check II

$R_1 = 133$ Ω

In-Process Learning Check III

1. Refer to Figure 9–27.
2. $I_N = 200$ μA $\qquad R_N = 500$ Ω
3. $E_{Th} = 0.2$ V $\qquad R_{Th} = 20$ kΩ

In-Process Learning Check IV

a. $\eta = 50\%$
b. $\eta = 33.3\%$
c. $\eta = 66.7\%$

Practice Problem I

$V_{R_1} = 27$ V $\qquad V_{R_3} = 21$ V

Practice Problem II

$R_{Th} = 16$ Ω $\qquad E_{Th} = 52$ V

Practice Problem III

$R_{Th} = 120$ Ω $\qquad E_{Th} = 2$ V $\qquad I_L = 10$ mA (upward)

Practice Problem IV

$R_N = 17.6$ Ω $\qquad I_N = 50$ mA $\qquad E_{Th} = 0.88$ V

Practice Problem V

$R_N = 42$ kΩ:	$I_N = 1.00$ mA
$R_L = 0$:	$I_L = 1.00$ mA
$R_L = 10$ kΩ:	$I_L = 0.808$ mA
$R_L = 50$ kΩ:	$I_L = 0.457$ mA
$R_L = 100$ kΩ:	$I_L = 0.296$ mA

Practice Problem VI

$R_L = R_N = 42$ kΩ:	$P_L = 10.5$ mW
when $R_L = 25$ kΩ:	$P_L = 9.82$ mW
when $R_L = 50$ kΩ:	$P_L = 10.42$ mW

Chapter 10

In-Process Learning Check I

1. It becomes six times larger.
2. 1.1 nF
3. Mica

Practice Problems I

1. a. Doubles;   b. Increases 1.5 times
2. 36.8 pF

Practice Problem II

5000 V

Practice Problem III

Maximum = 4.5 μF;   Minimum = 2 μF

Practice Problems IV

1. 1000 μF
2. $C_1 = 20\ \mu$F $C_2 = 40\mu$F $C_3 = 10\ \mu$F

Practice Problems V

1. $V_1 = 16.7$ V $V_2 = 8.3$ V $V_3 = 25$ V
2. $V_1 = 10$ V $V_2 = V_3 = 10$ V $V_4 = V_5 = 30$ V

Chapter 11

In-Process Learning Check I

1. 0.2 A

2. a. 50 ms

   b.

t (m$_c$)	i_c (mA)
0	50
25	30.3
50	18.4
75	11.2
100	6.8
500	0.0

3. a. 20 ms

b.

t (ms)	v_C (V)
0	0
25	71.3
50	91.8
75	97.7
100	99.3
500	100

4. $40(1 - e^{-5t})$ V; $200e^{-5t}$ mA

5. a. $v_C(0^+) = v_C(0^-) = 0$ **b.** $i_C(0^-) = 0$; $i_C(0^+) = 10$ mA

6. 80 V   40 kΩ   0.2 μF

In-Process Learning Check II

1. a. $40 + 40e^{-5t}$ V. v_C starts at 80 V and decays exponentially to 40 V.

b. There is no transient since initial value = final value.

c. $40 - 100e^{-5t}$ V. v_C starts at –60 V and climbs exponentially to 40 V.

2. 0.1833 s

3. a. $100e^{-10t}$ V; $-10e^{-10t}$ mA   **b.** $-100e^{-10t}$ V; $10e^{-10t}$ mA

4. a., b., and c. Same as Example 11–11
d. $-30 + 37.14e^{-250t}$ V   **e.** $-37.14e^{-250t}$ mA

f.

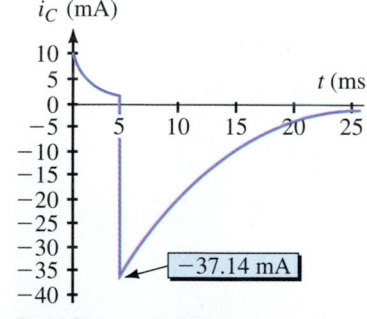

Figure C–4

5. $6(1 - e^{-25t})$ V; $150e^{-25t}$ mA

Practice Problems II

1.

t (ms)	v_C (V)	i_C (mA)
0	0	10
100	63.2	3.68
200	86.5	1.35
300	95.0	0.498
400	98.2	0.183
500	99.3	0.067

2. $80(1 - e^{-50t})$ V $20e^{-50t}$ mA

t (ms)	v_C (V)	i_C (mA)
0	0	20
20	50.6	7.36
40	69.2	2.70
60	76.0	0.996
80	78.6	0.366
100	79.4	0.135

Practice Problem III

a. $40 - 190e^{-5t}$ V   b. $0.95e^{-5t}$ A   c. -75.2 V;   0.576 A

Practice Problems IV

1. $75(1 - e^{-250t})$ V; $0.375e^{-250t}$ A
2. $40 + 20e^{-100t}$ V; $-0.1e^{-100t}$ A
3. $75 - 125e^{-250t}$ V; $0.625e^{-250t}$ A

Practice Problems V

1. $20e^{-20t}$ V; $-0.2e^{-20t}$ A
2. $12.6e^{-25t}$ V; $-6.3e^{-25t}$ mA
3. 0.208 A;   1.67 A

Practice Problems VI

1. 544 kΩ
2. 55.5 s

Chapter 12

In-Process Learning Check I

1. Force
2. Strength, direction
3. Iron, nickel, cobalt
4. North, south

5. Same except direction of flux reversed
6. Same as Figure 12–6
7. Perpendicular
8. 10 cm^2

In-Process Learning Check II

1. While flux is the same throughout, the effective area of each section differs.

2. $B_{iron} = 1.1$ T; $B_{steel} = 1.38$ T; $B_g = 1.04$ T

3. $\Phi_2 = \Phi_3 = 0.5$ mWb

In-Process Learning Check III

1. 3.5 A
2. a. 2.1 A   b. 0.525 A
3. 1650 At/m
4. Same as Figure 12–21(b) except add $H_{g_2}\ell_{g_2}$.
5. a. $NI = H_1\ell_1 + H_2\ell_2$ b. $0 = H_2\ell_2 - H_3\ell_3$
6. 0.32 m

Practice Problems I

1. a. 2×10^{-4} Wb   b. 90 μWb
2. 0.5 T; 1.0 T

Practice Problem II

a. 1.35×10^{-3} m^2
b. 1.04 T

Practice Problem III

1.81 mWb

Practice Problem IV

13.5 mWb

Practice Problem V

6.4 A

Chapter 13

In-Process Learning Check I

1. Both a and b. Current cannot change instantaneously.
2. 600 Wb–turns

3. 30 V

4. −300 V

In-Process Learning Check II

1. 28 mH

2. 9 times

In-Process Learning Check III

1. 87.5 V

2. Rate of change of current is the same;   20 V

Practice Problems I

1. 25 mH

2. 25×10^3 A/s

Practice Problems II

1. v_L is a square wave, oscillating between 15 V and −15 V.

2. 7 mH

Practice Problem III

3 H

Practice Problem IV

$I = 10.7$ A $V_{C_1} = 63$ V $V_{C_2} = 31.5$ V

Chapter 14

In-Process Learning Check I

1. a. $20e^{-100\,000t}$ V; $2(1 - e^{-100\,000t})$ mA

b.

$t\,(\mu s)$	$v_L\,(V)$	$i_L\,(mA)$
0	20	0
10	7.36	1.26
20	2.71	1.73
30	0.996	1.90
40	0.366	1.96
50	0.135	1.99

2. 4 V

3. 3.88 A

Practice Problem I

Figure C–5

Practice Problem II

a. $i = 16(1 - e^{-2 \times 10^6 t})$ mA; $v_L = 80e^{-2 \times 10^6 t}$ V;
$v_R = 80(1 - e^{-2 \times 10^6 t})$

b.

t (µs)	v_L (V)	i_L (mA)	v_R (V)
0	80	0	0
0.5	29.4	10.1	50.6
1.0	10.8	13.8	69.2
1.5	3.98	15.2	76.0
2.0	1.47	15.7	78.5
2.5	0.539	15.9	79.5

c. Yes

Practice Problem III

The switch voltage is identical to Figure 14–16. Voltage access R_2 is 120 V (constant) when the switch is closed. At the instant the switch is opened, v_{R_2} drops to –2400 V, then climbs exponentially to zero.

Practice Problems IV

1. a. $160(1 - e^{-2000t})$ mA   b. $80e^{-2000t}$ V
2. 7.21 H

Chapter 15

In-Process Learning Check I

1. 16.7 ms

2. Frequency doubles, period halves

3. 50 Hz;   20 ms

4. 0.25 ms;   4 ms

5. (c) and (d); Since current is directly proportional to the voltage, it will have the same waveshape.

6. 250 Hz

7. $f_1 = 100$ Hz; $f_2 = 33.3$ Hz

8. 50 kHz

9. 22.5 Hz

10. At 12 ms, direction →; at 37 ms, direction ←; at 60 ms, →

11. At 75 ms, $i = -5$ A

In-Process Learning Check II

1.

α (deg)	0	45	90	135	180	225	270	315	360
i (mA)	0	10.6	15	10.6	0	−10.6	−15	−10.6	0

2. a. 0.349   c. 2.09
  b. 0.873   d. 4.36

3. 30°;   120°;   450°

4. Same as Figure 15–17 with $T = 10$ s and amplitude = 50 mA.

5. 1.508×10^6 rad/s

6. $i = 8 \sin 157t$ A

7. $i = 6 \sin 69.81t$ A

8. a. $i = 250 \sin(251t - 30°)$ A
  b. $i = 20 \sin(62.8t + 45°)$ A
  c. $v = 40 \sin(628t - 30°)$ V
  d. $v = 80 \sin(314 \times 10^3 t + 36°)$ V

9. a. 2.95 ms; 7.05 ms; 22.95 ms; 27.05 ms
  b. 11.67 ms; 18.33 ms; 31.67 ms; 38.33 ms

10.

(a)

(b)

Figure C–6   (*Continues*)

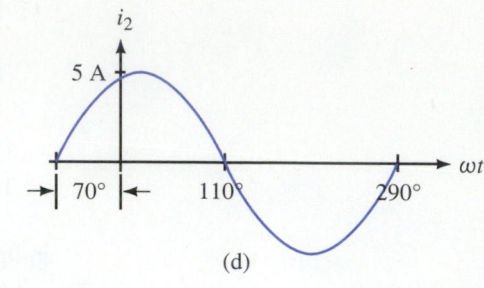

(c)

(d)

Figure C–6   (*Continued*)

11. a. 2.25 s   **b.** 1.33 s

12. b. 125°   **c.** v_1 leads

13. b. 20°   **c.** i_1 leads

Practice Problems I

1.

t (ms)	i (mA)
0	0
0.5	8
1.0	14
1.5	16
3.0	0
4.5	−16
5.25	−11.2

2.

t (μs)	i (mA)
1.5	0.25
3.0	0.5
7.5	−0.25
9.0	−0.5

Practice Problem III

21 600°/s

Practice Problems IV

1. 1031.4 cycles
2. 100.5×10^3 rad/s

Practice Problem V

a. $i = 10 \sin 52.36t$ A

b.

t (ms)	i (mA)
0	0
10	5
20	8.66
30	10
40	8.66
50	5
60	0
70	−5
80	−8.66
90	−10
100	−8.66
110	−5
120	0

c. • Mark the end of the cycle as $T = 120$ ms, the half-cycle point as 60 ms, the quarter-cycle point as 30 ms, and the 3/4-cycle point as 90 ms.

• Mark +10 A and −10 A on the vertical grid.

• Sketch a sinewave which reaches 10 A at $T/4$ (30 ms), crosses zero at $T/2$ (60 ms), reaches −10 A at $3T/4$ (90 ms), and ends at T (120 ms).

Practice Problem VI

80 ms;   100 ms

Practice Problems VII

1. 1.64 A

Practice Problem IX

a. 1.43 A   b. 6.67 V

Practice Problems X

1. Area = 1.855;   Exact = 1.866;   Error = 0.6%
2. Area = 67.9;   Exact = 68.3;   Error = 0.6%

Practice Problem XI

$V_{avg} = -5$ V;   Positive Peak = 10 V;   Negative Peak = −20 V

Practice Problems XII

1. 1.73 A

2. 20 V

Chapter 16

In-Process Learning Check I

1. a. $6\angle90°$ d. $7.21\angle-56.3°$ g. $3.61\angle-123.7°$
   b. $4\angle-90°$ e. $9.43\angle122.0°$
   c. $4.24\angle45°$ f. $2.24\angle-63.4°$

2. a. $j4$ d. $3.83 + j3.21$ g. $-1.64 - j0.599$
   b. $3 + j0$ e. $-3 + j5.20$
   c. $-j2$ f. $2.35 - j0.855$

3. $12\angle40°$

4. $88 - j7$; $-16 + j15$

5. $288\angle-100°$; $2\angle150°$

6. a. $14.70 + j2.17$ b. $-8.94 + j7.28$

7. $18.0 \sin(\omega t - 56.3°)$

In-Process Learning Check II

1. $50 \sin(\omega t + 30°)$ A

2. $50 \sin(\omega t - 60°)$ A

3. $50 \sin(\omega t + 120°)$ A

4. 0.637 H

5. 3.98 μF

6. a. Voltage and current are in phase. Therefore, R;
   b. Current leads. Therefore, C;
   c. Current leads. Therefore, C;
   d. Current lags. Therefore, L.

Practice Problems I

1. Since the numbers have the same angle, their sum also has the same angle and thus, their magnitudes simply add (or subtract).

2. a. $12 - j6$;   b. $-13 + j13$

Practice Problem II

ωt	e_1	e_2	$e_1 + e_2$	v
0	0	13	13	13
30°	5	15	20	20
60°	8.66	13	21.7	21.7
90°	10	7.5	17.5	17.5
120°	8.66	0	8.66	8.66
150°	5	−7.5	−2.5	−2.5
180°	0	−13	−13	−13
210°	−5	−15	−20	−20
240°	−8.66	−13	−21.7	−21.7
270°	−10	−7.5	−17.5	−17.5
300°	−8.66	0	−8.66	−8.66
330°	−5	7.5	2.5	2.5
360°	0	13	13	13

Practice Problems III

1. a. $e = 707\sin(\omega t - 20°)$ mV   b. $i = 113\sin(\omega t + 40°)$ A

2. $e_1 = 221 \sin(\omega t - 99.8°)$ V

Practice Problems IV

1. $i_R = 6 \cos \omega t$ mA

2. $i_R = 0.5 \sin(\omega t + 30°)$ mA

Practice Problems V

1. $i_L = 1.99 \sin(\omega t - 90°)$ mA

2. $i_L = 83.3 \sin(2400t - 30°)$ mA

Practice Problem VI

200 Ω

Practice Problems VII

1. v_C is a negative cosine wave.

Practice Problem VIII

$i = 0.942 \sin(62.8t + 90°) = 0.942 \cos 62.8t$ A
See Figure 16–27 for waveforms.

Practice Problems IX

1. 628 mV∠30°

2. 0.170 $\sin(800t + 90°)$ A

Chapter 17

In-Process Learning Check I

1. 12.37 V

Figure C–7

2. 9.96 A

3. Fig. 17–7: 0.6 (lag);   Fig. 17–8: 0.97 (lead);
Fig. 17–9: 0.97 (lead)

4. Yes. ($S = 128$ kVA)

Practice Problems I

1. Outward from the load.

2. a. 50 W;   25 W
b. $V_{\text{rms}} = 7.07$ V, $I_{\text{rms}} = 3.54$ A, $P = 25$ W

3. a. yes   b. yes

Practice Problem II

300 W;   400 VAR

Practice Problem III

$F_p = 0.883$;   No correction needed

Practice Problem IV

750 W

Chapter 18

In-Process Learning Check I

1. Current and voltage are in phase.
2. Current leads voltage by 90°.
3. Voltage leads current by 90°.

In-Process Learning Check II

1. The phasor sum of voltage drops and rises around a closed loop is equal to zero.

2. All voltages must be expressed as phasors.
ie. $\mathbf{V} = V\angle\theta = V\cos\theta + jV\sin\theta$.

In-Process Learning Check III

1. The phasor sum of currents entering a node is equal to the phasor sum of currents leaving the same node.

2. All currents must be expressed as phasors.
   ie. $\mathbf{I} = I\angle\theta = I\cos\theta + jI\sin\theta$.

In-Process Learning Check IV

1. At $f = 0$ Hz, $Z = \infty$(open)
   As $f \to \infty$, $Z = R$

2. At $f = 0$ Hz, $Z = 0$(short)
   As $f \to \infty$, $Z = R$

In-Process Learning Check V

a. $R_P = 795\ \Omega$ $L_P = 10.1$ mH

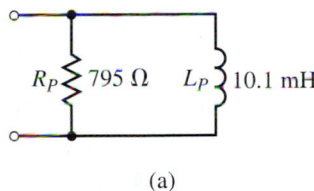

(a)

b. $R_P = 790$ MΩ $L_P = 10.0$ mH

(b)

Figure C–8

c. R_P will increase. L_P will not change. i.e., $L_P = L_S$

Practice Problem I

a. $\mathbf{I} = 0.2$ A$\angle{-60°}$

c. $e = 14.14 \sin(\omega t + 30°)$ (V)
   $i = 0.2828 \sin(\omega t - 60°)$ (A)

Practice Problem II

a. $\mathbf{I} = 0.5$ A$\angle{120°}$

c. $e = 14.14 \sin(\omega t + 30°)$ (V)
   $i = 0.7071 \sin(\omega t + 120°)$ (A)

Practice Problem III

a. $\mathbf{I} = 1.28$ A$\angle{-25.19°}$

b. $F_P = 0.6402$

c. $P_T = 41.0$ W

d. $\mathbf{V}_R = 32$ V$\angle{-25.19°}$ $\mathbf{V}_L = 51.2$ V$\angle{64.81°}$
   $\mathbf{V}_C = 12.8$ V$\angle{-115.19°}$

Practice Problem IV

a. $\mathbf{V}_R = 32$ V$\angle-25.19°$ $\mathbf{V}_L = 51.2$ V$\angle64.81°$
   $\mathbf{V}_C = 12.8$ V$\angle-115.19°$

b. $\Sigma\mathbf{V} = 50.0$ V$\angle25° = \mathbf{E}$

Practice Problem V

a. $\mathbf{V} = 0.2495$ V$\angle61.93°$

b. $F_P = 0.4705$

c. $P_R = 2.49$ mW $= P_T$

d. $\mathbf{I}_R = 9.98$ mA$\angle61.93°$ $\mathbf{I}_C = 6.24$ mA$\angle151.93°$
   $\mathbf{I}_L = 24.95$ mA$\angle-28.07°$

Practice Problem VI

a. $X_L = 8.89$ Ω $X_C = 26.7$ Ω
b. $I_L = 2.25$ A $I_C = 0.75$ A

Practice Problem VII

a. $\mathbf{I}_L = 175.6$ μA$\angle-69.44°$ $\mathbf{I}_R = 234.1$ μA$\angle20.56°$
   $\mathbf{I}_C = 87.8$ μA$\angle110.57°$

b. $\Sigma\mathbf{I}_{out} = 250$ μA$\angle0° = \Sigma\mathbf{I}_{in}$

Practice Problem VIII

a. $\mathbf{Z}_T = 18.9$ Ω$\angle-45°$

b. $\mathbf{I} = 1.06$ A$\angle45°$

c. $\mathbf{I}_1 = 0.3536$ A$\angle135°$
   $\mathbf{I}_2 = 1.12$ A$\angle26.57°$

d. $F_P(1) = 0.4472$ leading $F_P(2) = 0.7071$ lagging

e. $F_P = 0.7071$ leading

f. $P_T = 15.0$ W $P_1 = 2.50$ W $P_2 = 12.5$ W
   $P_1 + P_2 = 15.0$ W $= P_T$

Practice Problem IX

$\omega_c = 96.71$ rad/s $f_c = 15.39$ Hz
At $\omega = \omega_c/10$, $Z = 472.3$ kΩ
At $\omega = \omega_c$, $Z = 66.5$ kΩ
At $\omega = 10\omega_c$, $Z = 47.2$ kΩ

Practice Problem X

a. $R_S = 12.9$ Ω $L_S = 1.54$ H
b. $X_{LP} = 51.76$ kΩ $L_P = 1.65$ H

Chapter 19

In-Process Learning Check I

a. $E = 1000\ V\angle30°$
b. $E = 4\ V\angle-90°$
c. $E = 1024\ V\angle-38.66°$

In-Process Learning Check II

1. Convert current sources to voltage sources.
2. Redraw circuit.
3. Assign clockwise currents to each loop.
4. Write loop equations using KVL.
5. Solve the resulting simultaneous linear equations.

In-Process Learning Check III

1. Convert voltage sources to current sources.
2. Redraw the circuit.
3. Label all nodes, including the reference.
4. Write nodal equations using KVL.
5. Solve the resulting simultaneous linear equations.

In-Process Learning Check IV

$$Z_1 = 150\ \Omega\angle90° \qquad Z_2 = 150\ \Omega\angle-90° \qquad Z_3 = 150\ \Omega\angle0°$$

In-Process Learning Check V

$$R_x = \frac{R_2}{\omega^2 R_1^2 C_1^2 C_3^2} + \frac{R_2 C_1}{C_3} \qquad C_x = \frac{R_1 C_3}{R_2(1 + \omega^2 R_1^2 C_1^2)}$$

Practice Problem I

a. $V_R = 2.4\ mV\angle180°$
b. $V_R = 480\ \mu V\angle0°$
c. $V_R = 7.2\ mV\angle-90°$

Practice Problem II

a. $I = 312.5\ \mu A\angle60° \| Z = 16\ k\Omega\angle-30°$
b. $I = 1.61\ mA\angle93.43° \| Z = 4.472\ k\Omega\angle-63.43°$

Practice Problem III

$$I_1 = 1.19\ A\angle1.59° \qquad I_2 = 1.28\ A\angle-46.49° \qquad I = 1.01\ A\angle72.16°$$

Practice Problem IV

$$I = 1.18\ A\angle1.25°$$

Practice Problem V

$\mathbf{Z}_1 = 2.4\ \Omega + j1.2\ \Omega$ $\mathbf{Z}_2 = -1.2\ \Omega + j2.4\ \Omega$
$\mathbf{Z}_3 = 2.4\ \Omega + j4.8\ \Omega$ $\mathbf{Z}_T = 2\ \Omega + j6\ \Omega$

Practice Problem VI

$R_1 = 7916\ \Omega$ $R_3 = 199.6\ \Omega$

Chapter 20

In-Process Learning Check I

   a. $\mathbf{I} = 3.39\ \text{A}\angle 25.34°$
   b. $P_T = 230.4\ \text{W}$
   c. $P_1 + P_2 + P_3 = 145\ \text{W} \neq P_T = 230.4\ \text{W}$
     Superposition does not apply for power.

In-Process Learning Check II

1. Remove the inductor from the circuit. Label the remaining terminals as "a" and "b."
2. Set the voltage source to zero by removing it from the circuit and replacing it with a short circuit.
3. Determine the values of the impedance using the given frequency. Calculate the Thévenin impedance between terminals "a" and "b."
4. Convert the voltage source into its equivalent phasor form. Solve for the open-circuit voltage between terminals "a" and "b."
5. Sketch the resulting Thévenin equivalent circuit.

In-Process Learning Check III

1. Remove the load resistor from the circuit. Label the remaining terminals as "a" and "b."
2. Set the voltage source to zero by removing it from the circuit and replacing it with a short circuit.
3. Calculate the Thévenin impedance between terminals "a" and "b."
4. Solve for the short-circuit current between terminals "a" and "b."
5. Sketch the resulting Norton equivalent circuit.

In-Process Learning Check IV

   a. $R_L = 40\ \Omega$
   b. $P_L = 2.78\ \text{W}$
   c. $\omega = 3333\ \text{rad/s}$
   d. $P_L = 4.9\ \text{W}$

Practice Problem I

$I = 2.52$ A$\angle$$-25.41°$ $V = 4.45$ V$\angle 104.18°$

Practice Problem II

$V = 2.73$ V$\angle 180°$

Practice Problem III

    a. $Z_{Tb} = 1.5$ k$\Omega - j2.0$ kΩ $E_{Tb} = 3.16$ V$\angle -63.43°$
    b. $V_L = 1.75$ V$\angle 60.26°$
    c. $v_L = 2.48 \sin(2000t + 60.26°)$ (V)

Practice Problem IV

$Z_N = 13.5\ \Omega + j4.5\ \Omega$ $I_N = 0.333$ A$\angle 0°$ $I_L = 0.091$ A$\angle 14.04°$

Practice Problem V

    a. $E_{Tb} = 3.33$ V$\angle 0°$ $Z_{Tb} = 0.667\ \Omega$
    b. $R_L = 0$: $I_L = 5.0$ A   (upward)
        $R_L = 20\ \Omega$: $I_L = 0.161$ A   (upward)

Practice Problem VI

$V(1) = 12$ V $V(2) = -12$ V $V_{ab} = +20$ V
$I_{ab}(1) = -3.08$ mA $I_{ab}(2) = +4.62$ mA $I_{ab} = +1.54$ mA
$Z_{Tb} = 13$ kΩ

Practice Problem VII

$P_L = 100$ mW

Practice Problem VIII

    a. $R_L = 390\ \Omega$
    b. $P_L = 11.1$ W

Chapter 21

In-Process Learning Check I

    a. $L = 320\ \mu$H

    b. $R = 20.1\ \Omega$

    c. $I = 0.995$ mA$\angle 0°$ $V_L = 1.20$ V$\angle 90°$ $P = 20.0\ \mu$W

    d. $f_1 = 595$ kHz $f_2 = 605$ kHz

    e. $I = 0.700$ mA$\angle 45.28°$ $P_1 = 9.85\ \mu$W
        $P_1/P = 0.492 \cong 0.5$

In-Process Learning Check II

a. $\omega_s = 10$ krad/s $P = 500\ \mu W$

b. $Q = 10$ BW $= 1$ krad/s $\omega_1 = 9.5$ krad/s
$\omega_2 = 10.5$ krad/s

d. $\omega_s = 10$ krad/s $P = 1000\ \mu W$ $Q = 20$
BW $= 0.5$ krad/s $\omega_1 = 9.75$ krad/s
$\omega_2 = 10.25$ krad/s

e. As resistance decreases, selectivity increases.

In-Process Learning Check III

a. $X_L = 200\ \Omega$ $R_P = 1600\ \Omega$
$X_C = 80\ \Omega$ $R_P = 40\ \Omega$

b. $X_{LS} = 197\ \Omega$ $R_S = 24.6\ \Omega$
$X_{CS} = 16\ \Omega$ $R_S = 32\ \Omega$

In-Process Learning Check IV

a. $L = 180\ \mu H$
b. $R_P = 28.9$ kΩ
c. $\mathbf{V} = 57.9$ mV$\angle 0°$ $\mathbf{I}_L = 64.0\ \mu A\angle -90°$ $P = 115$ nW
d. $f_1 = 788$ kHz $f_2 = 813$ kHz
e. $\mathbf{V} = 41.1$ mV$\angle 44.72°$ $P = 58$ nW

Practice Problem I

a. $\omega = 102$ krad/s $f = 16.2$ kHz
b. $\mathbf{Z} = 55.0\ \Omega\angle 0°$
c. $\mathbf{I} = 205.7$ mA$\angle 0°$ $\mathbf{V}_L = 46$ V$\angle 90°$ $\mathbf{V}_C = 46$ V$\angle -90°$
d. $Q_C = 9.46$ vars cap. $Q_L = 9.46$ vars ind.
e. $Q = 4.065$

Practice Problem II

a. $P_{max} = 2.33$ W
b. BW $= 25.0$ krad/s   or   BW $= 3.98$ kHz
$\omega_1 = 89.14$ krad/s $\omega_2 = 114.14$ krad/s
c. $\omega_1 = 89.91$ krad/s $\omega_2 = 114.91$ krad/s
The approximation assumes that the power-frequency curve is symmetrical around ω_R. This is in fact, not exactly true.
d. $P = 1.16$ W (half-power)

Practice Problem III

a. $Q_s = 2.5$ W $Q_s = 1.0$

b. $R_P = 725\ \Omega$ $X_{LP} = 290\ \Omega$

    $R_P = 4\ k\Omega$ $X_{LP} = 4\ k\Omega$

c. $Q_s = 12.5$ $R_P = 15.725\ k\Omega,$ $X_{LP} = 1258\ \Omega$

    $Q_s = 5.0$ $R_P = 52\ k\Omega,$ $X_{LP} = 10.4\ \Omega$

Practice Problem IV

a. $Q_P = 1.257$ $Q_s = 12.732$

b. $R_s = 388\ \Omega,$ $X_{CS} = 487\ \Omega$

    $R_P = 816\ \Omega,$ $X_{CP} = 64.1\ \Omega$

c. $Q_P = 251$ $R_S = 0.0158\ \Omega,$ $X_{CS} = 3.97\ \Omega$

    $Q_S = 0.0637$ $R_P = 5.02\ \Omega,$ $X_{CP} = 78.9\ \Omega$

Practice Problem V

a. $\omega_P = 2.5$ Mrad/s $f_P = 398$ kHz

b. $Q_{cct} = 75$

c. BW $= 33.3$ krad/s

d. $\mathbf{V} = 7.5\ V\angle 180°$

Chapter 22

In-Process Learning Check I

1. a. -1.99 dBm   b. -15 dBm   c. -66.0 dBm

2. a. $0.561\ V_{rms}$ b. $0.224\ V_{rms}$ c. $0.354\ V_{rms}$

In-Process Learning Check II

a. $R = 3333\ \Omega$ in series with $C = 0.01\ \mu F$ (Output across C).

b. $R_1 = 630\ \Omega$ in series with $L = 10$ mH and $R_2 = 630\ \Omega$ (Output across R_2).

In-Process Learning Check III

1. TF $= \dfrac{j\dfrac{\omega}{157.1\ \text{krad/s}}}{1 + j\dfrac{\omega}{157.1\ \text{krad/s}}}$

$C = 0.05\ \mu F$ in series with $R = 127.3\ \Omega$ (Output across R).

2. TF $= \dfrac{j\dfrac{\omega}{320\ \text{krad/s}}}{1 + j\dfrac{\omega}{80\ \text{krad/s}}}$

$R_1 = 8\ k\Omega$ in series with $L = 25$ mH$\|R_2 = 2.67\ k\Omega$ (Output across the parallel combination).

In-Process Learning Check IV

Figure C–9

Practice Problem I

$A_v = 66.02$ dB $A_P = 92.04$ dB

Practice Problem II

a. 33.01 dBm;   3.01 dBW
   −37.0 dBm;   −67.0 dBm

b. 20 dBV;   −34.0 dBV

Practice Problem III

$P_1 = -62.0$ dBm $P_2 = -72.0$ dBm $P_3 = -45.0$ dBm

Practice Problem IV

(a)

(b)

Figure C–10   (*Continues*)

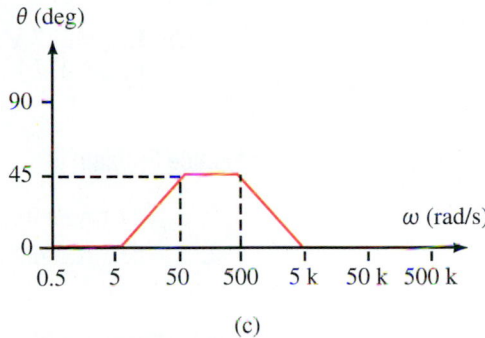

(c)

Figure C–10 (*Continued*)

Practice Problem V

a. $TF = \dfrac{0.25}{1 + j0.0015\omega}$

Practice Problem VI

a. $TF = \dfrac{0.25\left(1 + j\dfrac{\omega}{200\ \text{krad/s}}\right)}{1 + j\dfrac{\omega}{800\ \text{krad/s}}}$

Practice Problem VII

a. $\omega_1 = 1\ \text{krad/s}$ $\omega_2 = 10\ \text{krad/s}$ $BW = 9\ \text{krad/s}$

Chapter 23

In-Process Learning Check I

1. $\mathbf{V}_{an} = \mathbf{E}_{AN} = 277\ \text{V}\angle{-20°};$ $\mathbf{V}_{bn} = \mathbf{E}_{BN} = 277\ \text{V}\angle{-140°};$
   $\mathbf{V}_{cn} = \mathbf{E}_{CN} = 277\ \text{V}\angle{100°};$ $\mathbf{V}_{ab} = \mathbf{E}_{AB} = 480\ \text{V}\angle{10°};$
   $\mathbf{V}_{bc} = \mathbf{E}_{BC} = 480\ \text{V}\angle{-110°};$ $\mathbf{V}_{ca} = \mathbf{E}_{CA} = 480\ \text{V}\angle{130°}$

2. $\mathbf{V}_{an} = \mathbf{E}_{AN} = 120\ \text{V}\angle{50°};$ $\mathbf{V}_{bn} = \mathbf{E}_{BN} = 120\ \text{V}\angle{-70°};$
   $\mathbf{V}_{cn} = \mathbf{E}_{CN} = 120\ \text{V}\angle{170°};$ $\mathbf{V}_{ab} = \mathbf{E}_{AB} = 208\ \text{V}\angle{80°};$
   $\mathbf{V}_{bc} = \mathbf{E}_{BC} = 208\ \text{V}\angle{-40°};$ $\mathbf{V}_{ca} = \mathbf{E}_{CA} = 208\ \text{V}\angle{-160°}$

3. $\mathbf{V}_{an} = 124\ \text{V}\angle{-1.87°};$ $\mathbf{V}_{ab} = 214\ \text{V}\angle{28.13°}$

4. $\mathbf{V}_{ab} = 104\ \text{V}\angle{81.7°};$ $\mathbf{V}_{bc} = 104\ \text{V}\angle{-38.3°};$ $\mathbf{V}_{ca} = 104\ \text{V}\angle{-158.3°}$

Practice Problem I

a. $\mathbf{E}_{BB'} = 277 \text{ V}\angle-120°$; $\mathbf{E}_{CC'} = 277 \text{ V}\angle120°$

b. $\mathbf{E}_{AA'} = 347 \text{ V}\angle0°$; $\mathbf{E}_{CC'} = 347 \text{ V}\angle120°$

c. $\mathbf{E}_{AA'} = 120 \text{ V}\angle30°$; $\mathbf{E}_{BB'} = 120 \text{ V}\angle-90°$

Practice Problem II

a. $\mathbf{E}_{BC} = 480 \text{ V}\angle-100°$; $\mathbf{E}_{CA} = 480 \text{ V}\angle140°$

b. $\mathbf{E}_{AN} = 277 \text{ V}\angle-10°$; $\mathbf{E}_{BN} = 277 \text{ V}\angle-130°$;
   $\mathbf{E}_{CN} = 277 \text{ V}\angle110°$

Practice Problem III

1. $\mathbf{I}_a = 23.1 \text{ A}\angle6.9°$; $\mathbf{I}_b = 23.1 \text{ A}\angle-113.1°$; $\mathbf{I}_c = 23.1 \text{ A}\angle126.9°$
2. $\mathbf{I}_a = 23.1 \text{ A}\angle36.9°$; $\mathbf{I}_b = 23.1 \text{ A}\angle-83.1°$; $\mathbf{I}_c = 23.1 \text{ A}\angle156.9°$

Practice Problems IV

1. a. $10 \text{ A}\angle50°$

   b. $104 \text{ V}\angle66.7°$

2. $52.2 \text{ V}\angle-3.30°$

Practice Problem V

a. $\mathbf{I}_a = 24 \text{ A}\angle36.9°$; $\mathbf{I}_b = 24 \text{ A}\angle-83.1°$; $\mathbf{I}_c = 24 \text{ A}\angle156.9°$

b. $24\angle36.9° + 24\angle-83.1° + 24\angle156.9° = 0$

c. $\mathbf{E}_{AN} = 121 \text{ V}\angle3.18°$; $\mathbf{E}_{AB} = 210 \text{ V}\angle33.18°$

Practice Problem VI

1. Same

2. $\mathbf{I}_{a'b'} = 2.22 \text{ A}\angle27.7°$; $\mathbf{I}_{b'c'} = 2.22 \text{ A}\angle-92.3°$;
   $\mathbf{I}_{c'a'} = 2.22 \text{ A}\angle147.7°$

Practice Problem VII

$W_1 = -289 \text{ W}$; $W_2 = 1275 \text{ W}$; $W_T = 986 \text{ W}$

Practice Problem VIII

1. $30.8 \text{ V}\angle168.8°$

2. $\mathbf{I}_a = 12 \text{ A}\angle36.9°$; $\mathbf{I}_b = 6 \text{ A}\angle-156.9°$; $\mathbf{I}_c = 24 \text{ A}\angle120°$;
   $P_a = 1152 \text{ W}$; $P_b = 576 \text{ W}$; $P_c = 2880 \text{ W}$; $P_T = 4608 \text{ W}$

Chapter 24

In-Process Learning Check I

1. Step-up; 200 V

2. a. $120\ V\angle-30°$ **b.** $120\ V\angle150°$

3.

Figure C–11

4. Secondary, upper terminal.

5. a. Terminal 4   **b.** Terminal 4

In-Process Learning Check II

1. $\mathbf{Z}_p = 18.75\ \Omega - j25\ \Omega$; $R = 6400\ \Omega$; $a = 1.73$

2. $125\ V\angle0°$

3. 89.4

4. a. 200 A; 200 A   **b.** 24 kW; 18 kW

5. Tap 1: 109.1 V;   Tap 2: 120 V;   Tap 3: 126.3 V

6. 0.81 W; 0.72 W;   Maximum power is delivered when $R_s = a^2R_L$.

7. a. 348 V;   20.7 A   **b.** 12 A   **c.** 12 A;   8.69 A

In-Process Learning Check III

1. a. $\mathbf{Z}_{eq} = 0.8\ \Omega + j0.8\ \Omega$

   b. $113.6\ V\angle0.434°$

   c. 5.63%

Practice Problem I

   a. $a = 6$;   Step-down

   b. $e_s = 4.17\sin(\omega t + 180°)\ V$

Practice Problems II

1. e_s is 180° out of phase with e_p.

2. e_s is 180° out of phase with e_p.

3. Primary: Top end is dotted.
Secondary: Bottom end is dotted.

Practice Problem III

11.8 V;   2.04%

Practice Problem IV

94.2%

Practice Problem V

$$v_1 = L_1 \frac{di_1}{dt} + M \frac{di_2}{dt}$$

$$v_2 = M \frac{di_1}{dt} + L_2 \frac{di_2}{dt}$$

Practice Problem VI

a. $L_1' = 7.4$ mH; $L_2' = 39$ mH; $L_3' = 20.4$ mH

b. 66.8 mH

Practice Problem VII

a. $(10 + j84) \mathbf{I}_1 - j62 \mathbf{I}_2 = 120\angle0°$
   $-j62 \mathbf{I}_1 + 15 \mathbf{I}_2 = 0$

b. $\mathbf{I}_1 = 0.43$ A$\angle-17.5°$; $\mathbf{I}_2 = 1.78$ A$\angle72.5°$

Chapter 25

In-Process Learning Check I

The waveform of Figure 25–6 has an average value of zero. Therefore, $a_0 = 0$. The waveform also has a peak-to-peak value which is double that of Figure 25–3, which means that the amplitude of each harmonic must be doubled.

$$v(t) = \frac{4}{\pi} \sin(\omega t) + \frac{4}{3\pi} \sin(3\omega t) + \frac{4}{5\pi} \sin(5\omega t) + \ldots$$

In-Process Learning Check II

a. Figure 25–9 with $V_m = 25$ V and $T = 20$ ms

b. Figure 25–10 with $V_m = 25$ V and $T = 20$ ms.

c. $v(t) = 12.5 + 18.9 \sin(100\pi t - 32.48°)$
   $-7.96 \sin(200\pi t) + 5.42 \sin(300\pi t - 11.98°)$
   $-3.98 \sin(400\pi t)$

Practice Problem I

a. $v(t) = 0.5 - \dfrac{1}{\pi}\sin(2\pi t) - \dfrac{1}{2\pi}\sin(4\pi t) - \dfrac{1}{3\pi}\sin(6\pi t) \; \cdots$

b. $V_{av} = 0.5$ V

c. All harmonics since $b_n = \dfrac{-1}{(n\pi)}$ for all n.

Practice Problem II

$v(t) = 4.863 \sin(250\pi t - 135°) + 0.540 \sin(750\pi t - 135°)$
$+ 0.195 \sin(1250\pi t - 135°) + 0.099 \sin(1750\pi t - 135°)$.

Practice Problem III

b. $v(t) = -\dfrac{12}{\pi}\sin(10\pi t) - \dfrac{12}{2\pi}\sin(20\pi t) - \dfrac{12}{3\pi}\sin(30\pi t) - \; \cdots$

Practice Problem IV

$V_{dc} = 1.0$ V	$P_{dc} = +13.0$ dBm
$V_{1kHz} = 0.811$ V	$P_{1kHz} = +8.2$ dBm
$V_{3kHz} = 0.901$ V	$P_{3kHz} = -10.9$ dBm
$V_{5kHz} = 0.032$ V	$P_{5kHz} = -19.8$ dBm
$V_{7kHz} = 0.017$ V	(All other components are less than
$V_{9kHz} = 0.010$ V	-20 dBm and so will not appear).

Practice Problem V

a. dc:

b. 0 V

1 kHz:	0.142 V_P	$\theta = -116°$
2 kHz:	0.113 V_P	$\theta = -135°$
3 kHz:	0.088 V_P	$\theta = -147°$
4 kHz:	0.071 V_P	$\theta = -154°$

Appendix B

Practice Problem I

$x_2 = -3.0 \qquad x_3 = +1.0$

Answers to Selected Odd-Numbered Questions

Chapter 1

1. a. 9.29 m^2
    b. $0.000\ 645 \text{ m}^2$
    c. 0.84 m^2
    d. 209.1 N
    e. $26\ 110 \text{ W}$
    f. $138°/\text{s}$

3. $94\ 608\ 000 \text{ s}$

5. 1.63m; 163cm

7. 3.175 s

9. 1.28 s

11. a. 8.675×10^3
    b. 8.72×10^{-3}
    c. 1.24×10^3
    d. 3.72×10^{-1}
    e. 3.48×10^2
    f. 2.15×10^{-7}
    g. 1.47×10^1

13. 1.179; 4.450; Direct computation is less work for these examples.

15. 6.24×10^{18}

17. 1.16 s

19. a. mA
    b. kV
    c. MW
    d. μs
    e. μm
    f. ms
    g. nA

21. a. 150 μV
    b. 0.156 V
    c. $47\ 000 \text{ W}$
    d. 57 kW
    e. 35 kV
    f. 35.7 μA

23. Radio signal, 16.68 ms; Telephone signal, 33.33 ms; The radio signal by 16.66 ms.

25. 134 V (3 digits)

27. a. 0.25 mm b. 0.65 mm

Chapter 2

1. K holds 2; L holds 8; M holds 18

3. It decreases by a factor of nine.

5. The material with one valence electron.

7. It has an excess or deficiency of electrons.

9. 2 μC; (Attraction)

11. 0.333 μC, 1.67 μC; both (+) or both (−)

13. 30.4 μC

15. 1.294 μC

17. When you slide off the chair, you strip electrons from their parent atoms. This leaves you with a net charge. Since the other person is uncharged, a potential difference exists and a shock results.

19. 4.25 mJ

21. 3 V

23. 4.75 C

25. 28 mC

27. 3 mA

29. *DC current* is current that always flows in one direction, while *dc voltage* is voltage whose polarity does not change. *AC current* is current that periodically alters the direction of its flow, while *ac voltage* is voltage whose polarity cycles alternately between positive and negative.

31. 50 h

33. 24 V

35. (c) Both

37. 0.045 mA

39. The voltmeter and ammeter are interchanged.

41.

SW1	SW2	Lamp
open	open	off
open	closed	off
closed	open	off
closed	closed	on

43. If you exceed a fuse's voltage rating, it may arc over when it "blows."

Chapter 3

1. a. 3.6 Ω   b. 0.90 Ω   c. 36.0 kΩ   d. 36.0 mΩ

3. 0.407 inch

5. 300 m = 986 feet

7. 982×10^{-8} Ω · m (Resistivity is less than for carbon.)

9. 2.26×10^{-8} Ω · m (This alloy is not as good a conductor as copper.)

11. AWG 22: 4.86 Ω

AWG 19: 2.42 Ω

Diameter of AWG 19 is 1.42 times the diameter of AWG 22. The resistance of AWG 19 is half the resistance of an equal length of AWG 22.

13. AWG 19 should be able to handle 4 A.
AWG 30 can handle about 0.30 A.

15. 405 meters

17. a. 256 CM   **b.** 6200 CM   **c.** 1910 MCM

19. a. 16.2 Ω **b.** 0.668 Ω **c.** 2.17×10^{-3} Ω

21. a. 4148 CM = 3260 sq mil   **b.** 0.0644 inch

23. a. 1600 CM = 1260 sq mil   **b.** 1930 feet

25. $R_{-30°C} = 40.2$ Ω $R_{0°C} = 46.1$ Ω $R_{200°C} = 85.2$ Ω

27. a. Positive temperature coefficient
b. 0.00385 $(°C)^{-1}$
c. $R_{0°C} = 18.5$ Ω $R_{100°C} = 26.2$ Ω

29. 16.8 Ω

31. a. $R_{ab} = 10$ kΩ $R_{bc} = 0$ Ω
b. $R_{ab} = 8$ kΩ $R_{bc} = 2$ kΩ
c. $R_{ab} = 2$ kΩ $R_{bc} = 8$ kΩ
d. $R_{ab} = 0$ Ω $R_{bc} = 10$ kΩ

33. a. 150 kΩ $\pm$ 10%
b. 2.8 Ω $\pm$ 5% with a reliability of 0.001%
c. 47 MΩ $\pm$ 5%
d. 39 Ω $\pm$ 5% with a reliability of 0.1%

35. Connect the ohmmeter between the two terminals of the light bulb. If the ohmmeter indicates an open circuit, the light bulb is burned out.

37. AWG 24 has a resistance of 25.7 Ω/1000 ft. Measure the resistance between the two ends and calculate the length as

$$\ell = \frac{R}{0.0257 \text{ } \Omega/\text{ft}}$$

39. a. 380 Ω

b. 180 Ω

c. Negative temperature coefficient. Resistance decreases as temperature increases.

41. a. 4.0 S   **b.** 2.0 mS   **c.** 4.0 μS   **d.** 0.08 μS

43. 2.93 mS

Chapter 4

1. a. 2 A   **c.** 5 mA   **e.** 3 mA
b. 7.0 A   **d.** 4 μA   **f.** 6 mA

3. a. 40 V   **b.** 0.3 V   **c.** 400 V   **d.** 0.36 V

5. 9.6 Ω

7. a. 20 mA   b. 18.2 mA; 22.2 mA
9. a. 4 Ω   b. 25 A
11. 400 V
13. 3.78 mA
15. a. 45 V   b. −60 V   c. 90 V   d. −105 V
17. 3.19 J/s; 3.19 W
19. 36 W
21. 14.1 A
23. 47.5 V
25. 37.9 A
27. 2656 W
29. 23.2 V; 86.1 mA

31. a. 48 W ($\rightarrow$)   c. 128 W ($\leftarrow$)
    b. 30 W ($\leftarrow$)   d. 240 W ($\rightarrow$)

33. a. 1.296×10^6 J   b. 360 Wh   c. 2.88 cents
35. $175.20
37. 26 cents
39. 5 cents
41. 82.7%
43. 2.15 hp

45. Consider two devices as in Figure 4–26(b). For device
1, $\eta_1 = \dfrac{P_{\text{out}_1}}{P_{\text{in}}}$. For device 2, $\eta_2 = \dfrac{P_{\text{out}}}{P_{\text{in}_2}}$. But $P_{\text{out}_1} = P_{\text{in}_2}$.

Thus, $\eta_1 \times \eta_2 = \dfrac{P_{\text{out}_1}}{P_{\text{in}}} \times \dfrac{P_{\text{out}}}{P_{\text{in}_2}} = \dfrac{P_{\text{out}}}{P_{\text{in}}} = \eta_T$.

Continuing in this manner, $\eta_T = \eta_1 \times \eta_2 \times \ldots \eta_n$.

47. 1.97 hp
49. $137.45
51. a. 10 Ω   b. 13.3 Ω

Chapter 5

1. a. 7 V b. $V_2 = 4$ V $V_1 = 4$ V
3. $V_3 = 12$ V $V_4 = 2$ V
5. a. 10 kΩ   b. 2.94 MΩ   c. 23.4 kΩ
7. a. 1650 Ω, 6.06 mA   b. 18.15 kΩ, 16.5 mA

9. a. 10 mA

    b. 13 kΩ

    c. Current is clockwise.

    d. 5 kΩ

    e. $V_{3\text{-}k\Omega} = 30$ V $V_{4\text{-}k\Omega} = 40$ V $V_{1\text{-}k\Omega} = 10$ V
      $V_R = 50$ V

f. $P_T = 1300$ mW $P_{3\text{-}k\Omega} = 300$ mW $P_{4\text{-}k\Omega} = 400$ mW
   $P_R = 500$ mW

11. a. 40 mA   b. $V_{R_1} = 12$ V $V_2 = V_{R_3} = 10$ V   c. 26 V

13. a. 8.5 V   b. 1.02 mA   c. 7.32 kΩ

15. a. 457 Ω

b. 78.8 mA

c. $V_1 = 9.45$ V $V_2 = 3.07$ V $V_3 = 6.14$ V
   $V_4 = 17.33$ V

d. $V_T = 36$ V

e. $P_1 = 0.745$ W $P_2 = 0.242$ W $P_3 = 0.484$ W
   $P_4 = 1.365$ W

f. R_1: 1 W R_2: 1/4 W R_3: 1/2 W R_4: 2 W

g. 2.836 W

17. a. 0.15 A   b. 0.115 mA

19. a. $R_T = 24$ Ω $V_{6\text{-}\Omega} = 6$ V $V_{3\text{-}\Omega} = 3$ V
   $V_{5\text{-}\Omega} = 5$ V $V_{8\text{-}\Omega} = 8$ V $V_{2\text{-}\Omega} = 2$ V
   $V_T = 24$ V

b. $V_{4.3\text{-}k\Omega} = 21.6$ V $V_{2.7\text{-}k\Omega} = 13.6$ V
   $V_{7.8\text{-}k\Omega} = 39.2$ kΩ $V_{9.1\text{-}k\Omega} = 45.7$ V $V_T = 120$ V

21. Circuit (a):
   a. $R_1 = 0.104$ kΩ $R_2 = 0.365$ kΩ $R_3 = 0.730$ kΩ
   b. $V_1 = 2.09$ V $V_2 = 7.30$ V $V_3 = 14.61$ V
   c. $P_1 = 41.7$ mW $P_2 = 146.1$ mW $P_3 = 292.2$ mW

Circuit (b):
   a. $R_1 = 977$ Ω $R_2 = 244$ Ω $R_3 = 732$ Ω
   b. $V_1 = 25.0$ V $V_2 = 6.25$ V $V_3 = 18.75$ V
   c. $P_1 = 640$ mW $P_2 = 160$ mW $P_3 = 480$ mW

23. a. 0.2 A
   b. 5.0 V
   c. 1 W
   d. $R_T = 550\Omega$ $I = 0.218$A $V = 5.45$V $P = 1.19$W
   e. Life expectancy decreases.

25. Circuit (a): $V_{ab} = 9.39$ V $V_{bc} = 14.61$ V
   Circuit (b): $V_{ab} = -25.0$ V $V_{bc} = -6.25$ V

27. Circuit (a):

$V_{3k\text{-}\Omega} = 9$ V $V_{9k\text{-}\Omega} = 27$ V $V_{6k\text{-}\Omega} = 18$ V $V_a = 45$ V

Circuit (b):

$V_{330\text{-}\Omega} = 2.97$ V $V_{670\text{-}\Omega} = 6.03$ V $V_a = 3.03$ V

29. a. $109 \, \Omega$ b. $9.20 \, V$

31. a. $9 \, k\Omega$

     b. $49 \, k\Omega$

     c. $249 \, \Omega$

     d. A 1.0-V range is not possible since the meter movement requires a minimum of 2 V to give full scale deflection.

33. $R_S = 3.5 \, k\Omega$

     25% deflection: $R_x = 13.5 \, k\Omega$
     50% deflection: $R_x = 4.5 \, k\Omega$
     75% deflection: $R_x = 1.5 \, k\Omega$
     100% deflection: $R_x = 0$

35. Circuit (a):
     $I_{actual} = 0.375 \, mA$ $I_{measured} = 0.3745 \, mA$
     loading error $= 0.125\%$

     Circuit (b):
     $I_{actual} = 0.375 \, mA$ $I_{measured} = 0.3333 \, mA$
     loading error $= 11.1\%$

Chapter 6

1. a. A and B are in series; D and E are in series; C and F are parallel

     b. B, C, and D are parallel

     c. A and B are parallel; D and F are parallel; C and E are in series

     d. A, B, C, and D are parallel

5. a. $I_1 = 3 \, A$ $I_2 = -1 \, A$
     b. $I_1 = 7 \, A$ $I_2 = 2 \, A$ $I_3 = -7 \, A$
     c. $I_1 = 4 \, mA$ $I_2 = 20 \, mA$

7. a. $I_1 = 1.25 \, A$ $I_2 = 0.0833 \, A$ $I_3 = 1.167 \, A$ $I_4 = 1.25 \, A$
     b. $R_3 = 4.29 \, \Omega$

9. a. $I_1 = 200 \, mA$ $I_2 = 500 \, mA$ $I_3 = 150 \, mA$
     $I_4 = 200 \, mA$

     b. $2.5 \, V$

     c. $R_1 = 12.5 \, \Omega$ $R_3 = 16.7 \, \Omega$ $R_4 = 50 \, \Omega$

11. a. $R_T = 2.4 \, \Omega$ $G_T = 0.417 \, s$
     b. $R_T = 32 \, k\Omega$ $G_T = 31.25 \, \mu s$
     c. $R_T = 4.04 \, k\Omega$ $G_T = 247.6 \, \mu s$

13. a. $2.0 \, M\Omega$ b. $450 \, \Omega$

15. a. $R_1 = 1250\ \Omega$ $R_2 = 5\ k\Omega$ $R_3 = 250\ \Omega$
   b. $I_{R_1} = 0.40\ A$ $I_{R_2} = 0.10\ A$
   c. 2.5 A

17. a. 900 mV   b. 4.5 mA

19. a. $240\ \Omega$ b. $9.392\ k\Omega$ c. $1.2\ k\Omega$

21. $I_1 = 0.235\ mA$ $I_2 = 0.706\ mA$ $I_3 = 1.059\ mA$
   $R_1 = 136\ k\Omega$ $R_2 = 45.3\ k\Omega$ $R_3 = 30.2\ k\Omega$

23. a. $12.5\ k\Omega$ b. 0   c. $75\ \Omega$

25. $I = 0.2\ A$ $I_1 = 0.1\ A = I_2$

27. a. $I_1 = 2\ A$ $I_2 = 8\ A$ b. $I_1 = 4\ mA$ $I_2 = 12\ mA$

29. a. $I_1 = 6.48\ mA$ $I_2 = 9.23\ mA$ $I_3 = 30.45\ mA$
   $I_4 = 13.84\ mA$

   b. $I_1 = 60\ mA$ $I_2 = 30\ mA$ $I_3 = 20\ mA$
   $I_4 = 40\ mA$ $I_5 = 110\ mA$

31. $12\ \Omega$

33. a. $8\ \Omega$
   b. 1.50 A
   c. $I_1 = 0.50\ A$ $I_2 = 0.25\ A$ $I_3 = 0.75\ A$
   d. $\Sigma I_{in} = \Sigma I_{out} = 1.50\ A$

35. a. $25\ \Omega$

   b. $I = 9.60\ A$ $I_1 = 4.0\ A$ $I_2 = 2.40\ A$
   $I_3 = 3.20\ A$ $I_4 = 5.60\ A$

   c. $\Sigma I_{in} = \Sigma I_{out} = 9.60\ A$

   d. $P_1 = 960\ W$ $P_2 = 576\ W$ $P_3 = 768\ W$
   $P_T = 2304\ W = P_1 + P_2 + P_3$

37. a. $I_1 = 1.00\ A$ $I_2 = 2.00\ A$ $I_3 = 5.00\ A$
   $I_4 = 4.00\ A$

   b. 12.00 A

   c. $P_1 = 20\ W$ $P_2 = 40\ W$ $P_3 = 100\ W$
   $P_4 = 80\ W$

39. a. $R_3 = 4\ k\Omega$ $R_4 = 6\ k\Omega$ $R_2 = 8\ k\Omega$ $R_1 = 2\ k\Omega$
   b. $I_{R_1} = 24\ mA$ $I_{R_2} = 6\ mA$ $I_{R_4} = 8\ mA$
   c. $I_1 = 20\ mA$ $I_2 = 50\ mA$
   d. $P_2 = 288\ mW$ $P_3 = 576\ mW$ $P_4 = 384\ mW$

41. $I_1 = 8.33\ A$ $I_2 = 5.00\ A$ $I_3 = 2.50\ A$ $I_4 = 7.50\ A$
   $I_T = 15.83\ A$

   The rated current of the fuse will be exceeded; the fuse
   will "blow."

43. $25.1 \ \Omega$

45. $I_{range} = 10 \ A \qquad I = 6.2 \ A$

47. a. $V_{measured} = 20 \ V$ \qquad b. loading effect $= 33.3\%$

49. $25.2 \ V$

Chapter 7

1. a. $R_T = R_1 + R_5 + [(R_2 + R_3)\|R_4]$
   b. $R_T = (R_1\|R_2) + (R_3\|R_4)$

3. a. $R_{T_1} = R_1 + [(R_3 + R_4)\|R_2] + R_5$
   $R_{T_2} = R_5$

   b. $R_{T_1} = R_1 + (R_2\|R_3\|R_5)$
   $R_{T_2} = R_5\|R_3\|R_2$

7. a. $1500 \ \Omega$ \qquad b. $2.33 \ k\Omega$

9. $R_{ab} = 140 \ \Omega \qquad R_{cd} = 8.89 \ \Omega$

11. a. $R_T = 314 \ \Omega$

   b. $I_T = 63.7 \ mA \qquad I_1 = 19.2 \ mA \qquad I_2 = 44.5 \ mA$
   $I_3 = 34.1 \ mA \qquad I_4 = 10.4 \ mA$

   c. $V_{ab} = 13.6 \ V \qquad V_{bc} = -2.9 \ V$

13. a. $I_1 = 5.19 \ mA \qquad I_2 = 2.70 \ mA \qquad I_3 = 1.081 \ mA$
   $I_4 = 2.49 \ mA \qquad I_5 = 1.621 \ mA \qquad I_6 = 2.70 \ mA$

   b. $V_{ab} = 12.43 \ V \qquad V_{cd} = 9.73 \ V$

   c. $P_T = 145.3 \ mW \qquad P_1 = 26.9 \ mW \qquad P_2 = 7.3 \ mW$
   $P_3 = 3.5 \ mW \qquad P_4 = 30.9 \ mW \qquad P_5 = 15.8 \ mW$
   $P_6 = 7.0 \ mW \qquad P_7 = 53.9 \ mW$

15. Circuit (a):

   a. $I_1 = 4.5 \ mA \qquad I_2 = 4.5 \ mA \qquad I_3 = 1.5 \ mA$

   b. $V_{ab} = -9.0 \ V$

   c. $P_T = 162 \ mW \qquad P_{6\text{-}k\Omega} = 13.5 \ mW$
   $P_{3\text{-}k\Omega} = 27.0 \ mW \qquad P_{2\text{-}k\Omega} = 40.5 \ mW$
   $P_{4\text{-}k\Omega} = 81.0 \ mW$

   Circuit (b):

   a. $I_1 = 0.571 \ A \qquad I_2 = 0.365 \ A \qquad I_3 = 0.122 \ A$
   $I_4 = 0.449 \ A$

   b. $V_{ab} = -1.827 \ V$

   c. $P_T = 5.14 \ W \qquad P_{10\text{-}\Omega} = 3.26 \ W \qquad P_{16\text{-}\Omega} = 0.68 W$
   $P_{5\text{-}\Omega} = 0.67 \ W \qquad P_{6\text{-}\Omega} = 0.36 \ W \qquad P_{8\text{-}\Omega} = 0.12 \ W$
   $P_{4\text{-}\Omega} = 0.06 \ W$

17. $I_1 = 93.3 \ mA \qquad I_2 = 52.9 \ mA \qquad I_Z = 40.4 \ mA$
   $P_T = 2240 \ mW \qquad P_1 = 1306.7 \ mW \qquad P_2 = 109.2 \ mW$
   $P_3 = 419.9 \ mW \qquad P_Z = 404.2 \ mW$

19. $R = 31.1\ \Omega \rightarrow 3900\ \Omega$

21. $I_C = 1.70$ mA $V_B = -1.87$ V $V_{CE} = -8.10$ V

23. a. $I_D = 3.6$ mA   b. $R_S = 556\ \Omega$ c. $V_{DS} = 7.6$ V

25. a. $V_L = 0 \rightarrow 7.2$ V   b. $V_L = 2.44$ V   c. $V_{ab} = 3.0$ V

27. $V_{bc} = 7.45$ V $V_{ab} = 16.55$ V

29. a. $V_{\text{out(min)}} = 0$ V $V_{\text{out(max)}} = 40$ V

   b. $R_2 = 3.82$ kΩ

31. a. 11.33 V   b. 8.95 V   c. 44.1%   d. 1.333 V

33. a. Break the circuit between the 5.6-Ω resistor and the voltage source. Insert the ammeter at the break, connecting the red (+) lead of the ammeter to the positive terminal of the voltage source and the black (−) lead to the 5.6-Ω resistor.

   b. $I_{1(\text{loaded})} = 19.84$ mA $I_{2(\text{loaded})} = 7.40$ mA
      $I_{3(\text{loaded})} = 12.22$ mA

   c. loading effect (I_1) = 19.9%
      loading effect (I_2) = 18.0%
      loading effect (I_3) = 22.3%

Chapter 8

1. 38 V

3. a. 12 mA   b. $V_S = 4.4$ V $V_1 = 2.0$ V

5. $I_1 = 400\ \mu$A $I_2 = 500\ \mu$A

7. $P_T = 7.5$ mW $P_{50\text{-k}\Omega} = 4.5$ mW $P_{150\text{-k}\Omega} = 1.5$ mW
   $P_{\text{current source}} = 1.5$mW

   Note: The current source is absorbing energy from the circuit rather than providing energy.

11. a. 7.2 A   b. $E = 3600$ V $I_L = 7.2$ A

13. b. 6.06 mA   c. 0.606 V

15. $V_2 = -80$ V $I_1 = -26.7$ mA

19. Circuit (a):

   $(30\ \Omega)I_1 + (6\ \Omega)I_2 + 0\ I_3 = 15$ V
   $0\ I_1 - (6\ \Omega)I_2 + (20\ \Omega)I_3 = 5$ V
   $I_1 - I_2 - I_3 = 0$
   $I_1 = 0.467$ A $I_2 = 0.167$ A $I_3 = 0.300$ A

   Circuit (b):

   $(8\ \Omega)I_1 + 0\ I_2 + (4\ \Omega)I_3 = 3$ V
   $0\ I_1 + (4\ \Omega)I_2 + (4\ \Omega)I_3 = 5$ V

$I_1 + I_2 - I_3 = 0$

$I_1 = 0.050$ A $I_2 = 0.600$ A $I_3 = 0.650$ A

21. $-(5\ \Omega)I_1 - (14.4\ \Omega)I_2 + 0\ I_3 = 10.5$ V

$(5\ \Omega)I_1 + 0\ I_2 + (4.7\ \Omega)I_3 = 10$ V

$I_1 - I_2 - I_3 = 0$

$I_2 = -0.931$ A

23. a. $(8\ \Omega)I + 0\ I_2 - (10\ \Omega)I_3 = 24$ V

$0\ I + (4\ \Omega)I_2 + (10\ \Omega)I_3 = 16$ V

$I - I_2 + I_3 = 0$

b. $I = 3.26$ A

c. $V_{ab} = -13.89$ V

25. Circuit (a):

$(36\ \Omega)I_1 - (6\ \Omega)I_2 = 15$ V

$-(6\ \Omega)I_1 + (26\ \Omega)I_2 = 5$ V

$I_1 = 0.467$ A $I_2 = 0.300$ A

Circuit (b):

$(12\ \Omega)I_1 - (4\ \Omega)I_2 = 3$ V

$-(4\ \Omega)I_1 + (8\ \Omega)I_2 = -5$ V

$I_1 = 0.050$ A $I_2 = -0.600$ A

27. $(19.4\ \Omega)I_1 - (5\ \Omega)I_2 = 10.5$ V

$-(5\ \Omega)I_1 + (9.7\ \Omega)I_2 = 10$ V

$I = -0.931$ A

29. $(100\ \Omega)I_1 - (60\ \Omega)I_2 = -10$ V

$-(60\ \Omega)I_1 + (140\ \Omega)I_2 = 20$ V

$I_1 = -19.23$ mA $V_{ab} = 2.77$ V

31. $(10\ \Omega)I_1 - (1\ \Omega)I_2 - (6\ \Omega)I_3 = -6$ V

$-(1\ \Omega)I_1 + (8\ \Omega)I_2 - (3\ \Omega)I_3 = 10$ V

$-(6\ \Omega)\ I_1 - (3\ \Omega)I_2 + (9\ \Omega)I_3 = 5$ V

$I_1 = 0.495$ A $I_2 = 1.879$ A $I_3 = 1.512$ A

33. Circuit (a):

$(0.700\ \text{S})V_1 - (0.200\ \text{S})V_2 = -5$ A

$-(0.200\ \text{S})V_1 + (0.450\ \text{S})V_2 = 2$ A

$V_1 = -6.73$ V $V_2 = 1.455$ V

Circuit (b):

$(0.5\ \text{mS})V_1 - 0\ V_2 = -3$ mA

$-0\ V_1 + (0.300\ \text{mS})V_2 = 6$ mA

$V_1 = -6.0$ V $V_2 = 20$ V

35. Circuit (a):

$(0.417 \text{ S})V_1 - (0.250 \text{ S})V_2 = 5 \text{ A}$
$-(0.250 \text{ S})V_1 + (0.750 \text{ S})V_2 = 2 \text{ A}$
$V_1 = 17.0 \text{ V} = V_{6\,\Omega}$

Circuit (b):

$(0.422 \text{ S})V_1 - (0.111 \text{ S})V_2 - (0.111 \text{ S})V_3 = 4 \text{ A}$
$-(0.111 \text{ S})V_1 + (0.389 \text{ S})V_2 - (0.111 \text{ S})V_3 = 0$
$-(0.111 \text{ S})V_1 - (0.111 \text{ S})V_2 + (0.556 \text{ S})V_3 = 0$
$V_2 = 4.14 \text{ V} = V_{6\,\Omega}$

41. Circuit (a): $I = 6.67 \text{ mA}$
Circuit (b): $I = 0.15 \text{ A}$

43. a. The bridge is not balanced.

b. $(18 \,\Omega)I_1 - (12 \,\Omega)I_2 - (6 \,\Omega)I_3 = 15 \text{ V}$

$-(12 \,\Omega)I_1 + (54 \,\Omega)I_2 - (24 \,\Omega)I_3 = 0$

$-(6 \,\Omega)I_1 - (24 \,\Omega)I_2 + (36 \,\Omega)I_3 = 0$

c. $I = 38.5 \text{ mA}$

d. $V_{R_5} = 0.923 \text{ V}$

45. $I_{R_5} = 0 \qquad I_{RS} = 60 \text{ mA} \qquad I_{R_1} = I_{R_3} = 45 \text{ mA}$
$I_{R_2} = I_{R_4} = 15 \text{ mA}$

Chapter 9

1. $I_{R_1} = 75 \text{ mA (up)} \qquad I_{R_2} = 75 \text{ mA (to the right)}$
$I_{R_3} = 87.5 \text{ mA (down)} \qquad I_{R_4} = 12.5 \text{ mA (to the right)}$

3. $V_a = -3.11 \text{ V} \qquad I_1 = 0.1889 \text{ A}$

5. $E = 30 \text{ V} \qquad I_L(1) = 2.18 \text{ mA} \qquad I_L(2) = 2.82 \text{ mA}$

7. a. $R_{Tb} = 20 \,\Omega \qquad E_{Tb} = 10 \text{ V} \qquad V_{ab} = 6.0 \text{ V}$
b. $R_{Tb} = 2.02 \text{ k}\Omega \qquad E_{Tb} = 1.20 \text{ V} \qquad V_{ab} = -0.511 \text{ V}$

9. a. $R_{Tb} = 16 \,\Omega \qquad E_{Tb} = 5.6 \text{ V}$

b. When $R_L = 20 \,\Omega$: $V_{ab} = 3.11 \text{ V}$
When $R_L = 50 \,\Omega$: $V_{ab} = 4.24 \text{ V}$

11. Circuit (a):

b. $I = 0.75 \text{ A}$

Circuit (b):

b. $I = 13.21 \text{ mA}$

13. Circuit (a):

a. $R_{Tb} = 60 \text{ k}\Omega$

b. $E_{Tb} = 25$ V
$R_L = 0$: $I = -0.417$ mA
$R_L = 10$ kΩ: $I = -0.357$ mA
$R_L = 50$ kΩ: $I = -0.227$ mA

Circuit (b):

a. $R_{Tb} = 14.40$ kΩ

b. $E_{Tb} = 1.948$ V
$R_L = 0$: $I = 0.1353$ mA
$R_L = 10$ kΩ: $I = 0.0798$ mA
$R_L = 50$ kΩ: $I = 0.0302$ mA

15. $R_{Tb} = 7.2$ Ω $E_{Tb} = 4.56$ V

17. Circuit (a):

$I_L = 0.200$ A

Circuit (b):

$I_L = 0.341$ mA

19. b. $R_L = 20$ Ω: $I_L = 0.156$ A
$R_L = 50$ Ω: $I_L = 0.085$ A

21. Circuit (a):

a. $R_N = 50$ Ω b. $I_N = 1.5$ A

Circuit (b):

a. $R_N = 3.8$ kΩ b. $I_N = 13.16$ mA

23. Circuit (a):

$R_N = 60$ kΩ $I_N = 0.417$ mA

Circuit (b):

$R_N = 14.4$ kΩ $I_N = 0.1353$ mA

25. $R_N = 7.2$ Ω $I_N = 633$ mA

27. a. 60 kΩ b. $P_{max} = 2.60$ mW

29. Circuit (a):

a. $R = 31.58$ Ω b. $P_{max} = 7.81$ mW

Circuit (b):

a. $R = 3$ kΩ b. $P_{max} = 313$ nW

33. Circuit (a):

$I = 0.054$ A $P_L = 0.073$ W

Circuit (b):

$I = 0.284$ mA $P_L = 0.807$ mW

35. a. $I = 0.24$ A
b. $I = 0.24$ A
c. Reciprocity does apply.

Chapter 10

1. 1.6 mC

3. 0.4 μF

5. 73 pF

7. 5.65×10^{-4} m^2

9. 16.2 nF

11. 150 kV/m

13. 2.03 kV

15. 1328 V

17. No. When the gap with the lowest rating breaks down, it arcs over. The voltage across the arc will always be lower than the voltage needed to break down the second gap.

19. 532 ppm/°C, positive.

21. 12 μF

23. 40 μF

25. The 10–μF capacitor is in parallel with the series combination of the 1–μF and 1.5–μF capacitors.

27. a. $V_1 = 60$ V; $V_2 = V_3 = 40$ V

b. $V_1 = 50$ V; $V_2 = V_3 = 25$ V; $V_4 = 25$ V;
$V_5 = 8.3$ V; $V_6 = 16.7$ V

29. 30 μF

31. −50 mA from 0 to 1 ms; 50 mA from 1 ms to 4 ms; 0 mA from 4 ms to 6 ms; 50 mA from 6 ms to 7 ms; −75 mA from 7 ms to 9 ms.

33. $-23.5 \, e^{-0.05t} \, \mu$A

35. 0 mJ, 0.25 mJ, 1.0 mJ, 1.0 mJ, 2.25 mJ, 0 mJ

Chapter 11

1. a. 0 V;   5 A   b. 20 V;   0 A

3. a. Short circuit

b. Voltage source

c. Open circuit

d. $i(0^-)$ = current just before t = 0 s; $i(o^+)$ = current just after t = 0 s

5. a. $45(1-e^{-80t})$ V

b. $90e^{-80t}$ mA

c.

t (ms)	v_C (V)	i_C (mA)
0	0	90
20	35.9	18.2

40	43.2	3.67
60	44.6	0.741
80	44.93	0.150
100	44.98	0.030

7. $\tau = RC = \left(\dfrac{V}{I}\right)\left(\dfrac{Q}{V}\right) = \dfrac{Q}{I} = \dfrac{It}{I} = t$ (i.e., seconds)

9. $8\ \mu F$

11. $25\ k\Omega;\qquad 8\mu F$

13. $45\ V;\qquad 4.5\ k\Omega;\qquad 0.222\ \mu F$

15. a. $20 + 10e^{-25\,000t}\ V$

  b. $-2.5e^{-25\,000t}\ A$

  c. v_C starts at 30 V and decays exponentially to 20 V in 200 μs. i_C is 0 A at $t = 0^-$, -2.5 A at $t = 0^+$, and decays exponentially to zero in 200 μs.

17. a. $50e^{-2t}\ V$

  b. $-2e^{-2t}\ mA$

  c. $0.5\ s$

  d. v_C: 50 V, 18.4 V, 6.77 V, 2.49 V, 0.916 V, 0.337 V

  i_C: -2 mA, -0.736 mA, -0.271 mA, -0.0996 mA, -0.0366 mA, -0.0135 mA

19. $5\ k\Omega;\qquad 0.2\ \mu F$

21. a. $60(1-e^{-500t})\ V$ b. $1.5e^{-500t}\ A$

23. $90\ V;\qquad 15\ k\Omega;\qquad 100\ \mu F$

25. $V_{C_1} = 65\ V;\qquad V_{C_2} = 10\ V;\qquad V_3 = 55\ V;\qquad I_T = 0.5\ A$

27. $14.0\ \mu F$

29. a. 16.1 s   b. 11.8 s;   24.0 s

31. a. 4 ms   b. 75%   c. 250 pulses per second

33. a. Similar to v_R of Figure 11–38.
  b. Similar to v_R of Figure 11–37.

35. 6.6 ns

37. Sample points: $t = 0$, $v_C = 0$ V, $i_C = 200$ mA
  $t = 10$ ms, $v_C = 25.3$ V, $i_C = 73.6$ mA

39. Same

Chapter 12

1. a. A_1 b. 1.4 T

3. 0.50 T

5. $1.23 \times 10^{-3}\ Wb$

7. 0.45 T;   0.9 T

9. 0.625 T

11. $B = \dfrac{\Phi}{A} = \dfrac{\mathscr{F}/\mathscr{R}}{A} = \dfrac{\mathscr{F}}{\mathscr{R}A} = \dfrac{NI}{\left(\dfrac{\ell}{\mu A}\right)A} = \mu\left(\dfrac{NI}{\ell}\right) = \mu H$

13. $N_1 I_1 = H_1 \ell_1 + H_2 \ell_2; \qquad N_2 I_2 = H_2 \ell_2 - H_3 \ell_3$

15. 0.47 A

17. 0.89 A

19. 0.53 A

21. 0.83 A

23. 4.49 A

25. 4.43×10^{-4} Wb

27. 1.06×10^{-4} Wb

Chapter 13

1. 225 V

3. 6.0 V

5. 150 mV

7. 0.111 s

9. 79.0 μH

11. $L = \dfrac{N\Phi}{I} = \dfrac{N(B_g A_g)}{I} = \dfrac{N(\mu_0 H_g)A_g}{I}$

$\qquad = \dfrac{N\mu_0\left(\dfrac{NI}{\ell_g}\right)A_g}{I} = \dfrac{\mu_0 N^2 A_g}{\ell_g}$

13. 4 H

15. a. 4 H   c. 5 A

17. a. 20 H   b. 4 H   c. 4 mH

19. 3 H

21. 1.6 H, in series with 6 H∥4 H

23. 1.2 H in series with 8 H∥12 H

25. $\dfrac{1}{L_T} = \dfrac{1}{L_1} + \dfrac{1}{L_2} = \dfrac{L_2 + L_1}{L_1 L_2}; \qquad$ Inverting $L_T = \dfrac{L_1 L_2}{L_1 + L_2}$

27. 0.32 J

29. The path containing L_1 and L_2 is open.

Chapter 14

1. 0.75 A

3. 2.75 A

5. a. 250 ms

   b. $3(1 - e^{-20t})$ A; $180 e^{-20t}$ V

c.

t	i_L(A)	v_L(V)
0	0	180
τ	1.90	66.2
2τ	2.59	24.4
3τ	2.85	8.96
4τ	2.95	3.30
5τ	2.98	1.21

7. 80 V;   20 Ω;   2 H

9. 40 V;   4 kΩ;   2 H

11. a. 2 ms

b. $5e^{-500t}$ A; $-1250e^{-500t}$ V

c.

t	i_L(A)	v_L(V)
0	5	−1250
τ	1.84	−460
2τ	0.677	−169
3τ	0.249	−62.2
4τ	0.092	−22.9
5τ	0.034	−8.42

13. $R_1 = 20\ \Omega$; $R_2 = 30\ \Omega$

15. a. 20 μs; $5(1 - e^{-t/20\mu s})$ mA; $90\ e^{-t/20\mu s}$ V
     b. 33.1 V;   3.16 mA

17. a. 10 ms
     b. $90\ (1 - e^{-t/10ms})$ mA; $36\ e^{-t/10ms}$ V
     c. 2.96 V;   82.6 mA

19. 103.3 mA;   3.69 V

21. At 0.2 s, $v_C = -7.36$ V and $i_C = -0.632$ A
     At 0.4 s, $v_C = -2.71$ V and $i_C = -0.865$ A

Chapter 15

1. It does not change polarity.
3. 1200 rpm
5. 0 V, 10 V, 20 V, 20 V, 20 V, 0 V, − 20 V, −20 V, 0 V

7. 0 mA, 2.5 mA, 5 mA, 5 mA, 5 mA, 0 mA, −5 mA, −5 mA, 0 mA. When current is positive, it is in the direction of the arrow; when negative, it is opposite to the arrow.

9. a. 2 Hz   b. 10 Hz   c. 62.5 kHz
11. 149 919 cycles
13. 100 μs;   10 kHz;   150 V
15. 100 V

17. a. 0.1745   **c.** 1.3963   **e.** 6.1087
    b. 0.4363   **d.** 2.618   **f.** 10.821

19. 43.3 A;   25 A;   −49.2 A;   −50 A

21. $V_m = 85.1$ V. Waveform is like Figure 15–25 except $T = 50\ \mu s$.

23. a. 62.83×10^6 rad/s
    b. 188.5 rad/s
    c. 157.1 rad/s
    d. 314.2 rad/s
    e. 1571 rad/s

25. a. $v = 170 \sin 377t$ V
    b. $i = 40 \sin 628t\ \mu A$
    c. $v = 17 \sin 52.4 \times 10^3 t$ V

27. $v = 20 \sin 125.7t$ V

29. 35 ms; 55 ms

31. a. $T = 27$ ms. Same as Figure 15–31(a) with $\theta = 40°$ (i.e., 3 ms on the time scale) and $V_m = 100$ V.

    b. $T = 50$ ms. Same as Figure 15–31(b) with $\theta = 60°$ (i.e., 8.33 ms on the time scale) and V_m replaced by $I_m = 20$ mA.

33. −16.9 A

35. a. $100 \sin(\omega t + 70°)$ V
    b. $20 \sin(\omega t + 140°)$ A
    c. $10 \sin(\omega t − 35°)$ mA

37. a. A leads by 90°   **b.** A leads by 150°

39. a. 1.1 A   **b.** −5 V   **c.** 1.36 A

41. $y = \sin x$. Compute y at $\pi/12$ (i.e., 15°) intervals and substitute. Thus, $y_0 = \sin 0° = 0$, $y_1 = \sin 15° = 0.2588$, $y_2 = \sin 30° = 0.5$, etc.
Area $= (y_0/2 + y_1 + y_2 + \ldots + y_{12}/2) = (7.5958)\,(\pi/12) \approx 2$

43. a. Similar to Figure 15–59(b), except positive peak is 40 V, negative peak is −10 V, and $V_{avg} = 15$ V. $T = 120$ ms.

    b. 40 V;   −10 V

    c. 27.5 V;   36.7 V;   2.5 V;   −6.65 V

    d. 15 V

45. a. 70.7 V
    b. 5.66 A
    c. 28.3 V
    d. 84.8 mA

47. 2.074 A;   30 A

49. 26.7 A

Chapter 16

1. a. $13\angle 67.4°$

  b. $10.8\angle -33.7°$

  c. $17\angle 118.1°$

  d. $10.8\angle -158.2°$

5. a. $7 + j6$

  b. $1 + j10$

  c. $15.4 - j3.50$

  d. $-4.64 + j1.86$

  e. $-7.2 - j12.1$

7. a. $14.2\angle -23.8°$

  b. $1.35\angle -69.5°$

  c. $5.31\angle 167.7°$

9. a. $10 \sin(\omega t + 30°)$ V

  b. $15 \sin(\omega t - 10°)$ V

11. a. 10 V$\angle 30°$; 15 V$\angle -20°$

  b. 11.5 V$\angle 118.2°$

  c. $11.5 \sin(\omega t + 118.2°)$ V

13. a. 35.4 A$\angle 60°$; 14.1 A$\angle -30°$

  b. 38.1 A$\angle 81.8°$

  c. $53.8 \sin(\omega t + 81.8°)$ A

15. a. $5 \sin(\omega t - 30°)$ A

  b. $5 \sin(\omega t - 105°)$ A

  c. $10 \sin(\omega t + 120°)$ V

  d. $10 \sin(\omega t + 100°)$ V

17. a. $0.5231 \sin(377t - 90°)$ A

  b. $31.8 \sin(6283t - 90°)$ mA

  c. $0.4 \sin(500t - 90°)$ A

19. a. $530.5\ \Omega$ b. $31.83\ \Omega$ c. $400\ \Omega$

21. a. $48\ \Omega\angle 0°$ b. $j37.7\ \Omega$ c. $-j50\ \Omega$

23. a. $\mathbf{V}_R = 25$ V$\angle 0°$; $\mathbf{V}_L = 12.5$ V$\angle 90°$; $\mathbf{V}_C = 5$ V$\angle -90°$

  b. $v_R = 35.4 \sin\omega t$ V; $v_L = 17.7 \sin(\omega t + 90°)$ V;
    $v_C = 7.07 \sin(\omega t - 90°)$ V

25. a. 39.8 Hz   b. $6.37\ \mu$F

27. See Figure 16–26. Check point: at $f = 200$ Hz, $X_L = 251.3\ \Omega$.

Chapter 17

1. Real or active power is power that does work. Reactive power, on the other hand, does no work — it represents energy that shuttles back and forth between the source and load. Reactive power has an average value of zero.

3. a. $P_1 = 1000$ W; $P_2 = 500$ W   b. 1500 W

5. a. 10 Ω   b. 6 Ω   c. 265 μF

7. a. 20 Ω   b. 6 Ω   c. 8 Ω   d. 1.2 H

9. a. 125 VA   b. 82.5 VA   c. 1237 VA

11. 2.36 Ω

13. 120 Ω

15. a. 721 W;   83 VAR (cap.);   726 VA

b. $I = 6.05$ A;   No

17. a. Across the load   b. 73.9 μF

19. 57.3 kW

Chapter 18

1. Network (a): 0.125 sin ωt
Network (b): 1.866×10^{-3} sin($\omega t + 30°$)

3. Network (a): 1.36 sin ($\omega t - 90°$)
Network (b): 1333 sin($2000\pi t + 30°$)

5. Network (a): 62.5 sin (10 000t – 90°)
Network (b): 67.5 sin(20 000t – 160°)

7. a. 31.6 Ω∠18.43°   b. 8.29 kΩ∠−29.66°

9. Network (a): 42.0 Ω∠19.47°
Network (b): 62.8 Ω∠47.15°

11. $R = 15$ Ω $C = 1.932$ μF

13. a. $\mathbf{Z}_T = 50$ Ω∠−36.87° $\mathbf{I} = 2.4$ A∠36.87°
$\mathbf{V}_R = 96$ V∠36.87° $\mathbf{V}_L = 48$ V∠126.87°
$\mathbf{V}_C = 120$ V∠−53.13°

c. $P_R = 230.4$ W

d. $P_T = 230.4$ W

15. a. $\mathbf{Z}_T = 45$ Ω∠−36.87°

b. $i = 0.533$ sin($\omega t + 36.87°$)
$v_R = 19.20$ sin($\omega t + 36.87°$)

$$v_C = 25.1 \sin(\omega t - 53.13°)$$
$$v_L = 10.67 \sin(\omega t + 126.87°)$$

e. $P_R = 5.12$ W

f. $P_T = 5.12$ W

17. Network (a):

　　a. $\mathbf{V}_R = 9.49$ V$\angle$–18.43°　　　$\mathbf{V}_L = 11.07$ V$\angle$71.57°
　　　$\mathbf{V}_C = 7.91$ V$\angle$–108.43°

　　b. $\Sigma\,\mathbf{V} = 10.00$ V$\angle$0°

　　Network (b):

　　a. $\mathbf{V}_{R_1} = 3.62$ V$\angle$29.66°　　　$\mathbf{V}_{R_2} = 5.07$ V$\angle$29.66°
　　　$\mathbf{V}_{XL} = 3.98$ V$\angle$119.66°　　　$\mathbf{V}_{XC_1} = 1.810$ V$\angle$–60.34°
　　　$\mathbf{V}_{XC_2} = 7.12$ V$\angle$–60. 34°

　　b. $\Sigma\,\mathbf{V} = 10.00$ V$\angle$0°

19. a. $\mathbf{V}_C = 317$ V$\angle$–30°　　　$\mathbf{V}_L = 99.8$ V$\angle$150°　　　b. 25 Ω

21. a. $\mathbf{V}_C = 6.0$ V$\angle$–110°　　　　c. 69.4 $\Omega\angle$79.92°
　　b. $\mathbf{V}_Z = 13.87$ V$\angle$59.92°　　　d. 1.286 W

23. Network (a): $\mathbf{Z}_T = 199.9$ $\Omega\angle$–1.99°
　　Network (b): $\mathbf{Z}_T = 485$ $\Omega\angle$–14.04°

25. a. $\mathbf{Z}_T = 3.92$ k$\Omega\angle$–78.69°　　　$\mathbf{I}_T = 2.55$ mA $\angle$78.69°
　　　$\mathbf{I}_1 = 0.5$ mA$\angle$0°　　　$\mathbf{I}_2 = 10.0$ mA$\angle$–90°
　　　$\mathbf{I}_3 = 12.5$ mA$\angle$90°

　　d. $P_R = 5.00$ mW

　　e. $F_p = 0.1961$ leading $P_T = 5.00$ mW

27. a. 5.08 k$\Omega\angle$23.96°

　　b. 152.3 V$\angle$23.96°

29. 2.55 $\Omega\angle$81.80°

31. Network (a):

　　$\mathbf{I}_R = 10.00$ mA$\angle$–31.99°　　　$\mathbf{I}_L = 4.00$ mA$\angle$–121.99°
　　$\mathbf{I}_C = 4.35$ mA$\angle$58.01°

　　Network (b):

　　$\mathbf{I}_R = 9.70$ mA$\angle$–44.04°　　　$\mathbf{I}_{C_1} = 1.617$ mA$\angle$45.96°
　　$\mathbf{I}_{C_2} = 0.808$ mA$\angle$45.96°

33. $\mathbf{I}_L = 2.83$ mA$\angle$–135°　　　$\mathbf{I}_C = 3.54$ mA$\angle$45°
　　$\mathbf{I}_R = 0.707$ mA$\angle$–45°　　　$\Sigma\mathbf{I}_{out} = \Sigma\mathbf{I}_{in} = 1.00$ mA$\angle$0°

35. a. $\mathbf{I}_R = 6.25$ A$\angle$90°
　　b. $R = 72.06$ Ω
　　c. $\mathbf{I} = 8.00$ A$\angle$51.32°

37. a. $\mathbf{Z}_T = 31.38\ \Omega\angle67.52°$ $\mathbf{I}_L = 3.82\ A\angle-67.52°$
   $\mathbf{I}_C = 3.42\ A\angle-40.96°$ $\mathbf{I}_R = 1.71\ A\angle-130.96°$

c. $P_R = 175.4\ W$

d. $P_T = 175.4\ W$

39. a. $\mathbf{Z}_T = 10.53\ \Omega\angle10.95°$ $\mathbf{I}_T = 1.900\ A\angle-10.95°$
   $\mathbf{I}_1 = 2.28\ A\angle-67.26°$ $\mathbf{I}_2 = 2.00\ A\angle60.61°$

b. $\mathbf{V}_{ab} = 8.87\ V\angle169.06°$

41. a. $\mathbf{Z}_T = 7.5\ k\Omega$ $\mathbf{I}_1 = 0.75\ mA\angle0°$ $\mathbf{I}_2 = 0.75\ mA\angle90°$
   $\mathbf{I}_3 = 0.791\ mA\angle60.61°$

b. $\mathbf{V} = 7.12\ V\angle-161.57°$

43. $\omega_C = 2000\ rad/s$

45. $f_C = 3.39\ Hz$

49. $\omega = 1\ krad/s$: $\mathbf{Y}_T = 0.01\ S + j0$ $\mathbf{Z}_T = 100\ \Omega$
   $\omega = 10\ krad/s$: $\mathbf{Y}_T = 0.01\ S + j0$ $\mathbf{Z}_T = 100\ \Omega$

Chapter 19

1. Circuit (a):

a. $5.00\ V\angle180°$

b. $12.50\ V\angle0°$

c. $15.00\ V\angle-120°$

Circuit (b):

a. $3.20\ mV\angle180°$

b. $8.0\ mV\angle0°$

c. $9.60\ mV\angle-120°$

3. $7.80\ V\angle-120°$

5. Circuit (a):

$\mathbf{E} = 54\ V\angle0°$ $\mathbf{V}_L = 13.5\ V\angle0°$

Circuit (b):

$\mathbf{E} = 450\ mV\angle-60°$ $\mathbf{V}_L = 439\ mV\angle-47.32°$

7. a. $4.69\ V\angle180°$

b. $\mathbf{E} = (7.5\ M\Omega)\mathbf{I}$ $\mathbf{V} = 4.69\ V\angle180°$

9. a. $(4\ \Omega + j2\ \Omega)\mathbf{I}_1 - (4\ \Omega)\mathbf{I}_2 = 20\ V\angle0°$
   $-(4\ \Omega)\ \mathbf{I}_1 + (6\ \Omega + j4\ \Omega)\mathbf{I}_2 = 48.4\ V\angle-161.93°$

b. $\mathbf{I}_1 = 2.39\ A\angle72.63°$ $\mathbf{I}_2 = 6.04\ A\angle154.06°$

c. $\mathbf{I} = 6.15\ A\angle-3.33°$

11. a. $(12\ \Omega - j16\ \Omega)\mathbf{I}_1 + (j15\ \Omega)\mathbf{I}_2 = 13.23\ \text{V}\angle-79.11°$
$(j15\ \Omega)\mathbf{I}_1 + 0\ \mathbf{I}_2 = 10.27\ \text{V}\angle-43.06°$

   b. $\mathbf{I}_1 = 0.684\ \text{A}\angle-133.06°$ $\mathbf{I}_2 = 1.443\ \text{A}\angle-131.93°$

   c. $\mathbf{V} = 11.39\ \text{V}\angle-40.91°$

13. $\mathbf{V} = 27.8\ \text{V}\angle6.79°$ $\mathbf{I} = 6.95\ \text{mA}\angle6.79°$

15. a. $(0.417\ \text{S}\angle36.87°)\ \mathbf{V}_1 - (0.25\ \text{S}\angle90°)\ \mathbf{V}_2 = 3.61\ \text{A}\angle56.31°$
$-(0.25\ \text{S}\angle90°)\ \mathbf{V}_1 + (0.083\ \text{S}\angle90°)\ \mathbf{V}_2 = 7.00\ \text{A}\angle90°$

   b. $\mathbf{V}_1 = 30.1\ \text{V}\angle139.97°$ $\mathbf{V}_2 = 60.0\ \text{V}\angle75.75°$

   c. $\mathbf{I} = 13.54\ \text{V}\angle-44.31°$

17. a. $(0.0893\ \text{S}\angle22.08°)\mathbf{V}_1 - (0.04\ \text{S}\angle90°)\mathbf{V}_2 = 0.570\ \text{A}\angle93.86°$
$-(0.04\ \text{S}\angle90°)\mathbf{V}_1 + (0.06\ \text{S}\angle90°)\mathbf{V}_2 = 2.00\ \text{A}\angle180°$

   b. $\mathbf{V}_1 = 17.03\ \text{V}\angle18.95°$ $\mathbf{V}_2 = 31.5\ \text{V}\angle109.91°$

   c. $11.39\ \text{V}\angle-40.91°$

19. $(0.372\ \mu\text{S}\angle-5.40°)\mathbf{V} = 10.33\ \text{mA}\angle1.39°$
$\mathbf{V} = 27.8\ \text{V}\angle6.79°$ $\mathbf{I} = 6.95\ \text{mA}\angle6.79°$

21. Network (a):

   $\mathbf{Z}_1 = 284.4\ \Omega\angle-20.56°$ $\mathbf{Z}_2 = 94.8\ \Omega\angle69.44°$
   $\mathbf{Z}_3 = 31.6\ \Omega\angle159.44°$

   Network (b):

   $\mathbf{Z}_1 = 11.84\ \text{k}\Omega\angle9.46°$ $\mathbf{Z}_2 = 5.92\ \text{k}\Omega\angle-80.54°$
   $\mathbf{Z}_3 = 2.96\ \text{k}\Omega\angle-80.54°$

23. $\mathbf{I}_T = 0.337\ \text{A}\angle-2.82°$

25. a. $\mathbf{Z}_T = 3.03\ \Omega\angle-76.02°$
   b. $\mathbf{I} = 5.28\ \text{A}\angle76.02°$ $\mathbf{I}_1 = 0.887\ \text{A}\angle-15.42°$

27. a. $\mathbf{Z}_2 = 1\ \Omega - j7\ \Omega = 7.07\ \Omega\angle-81.87°$
   b. $\mathbf{I} = 142.5\ \text{mA}\angle52.13°$

29. $\mathbf{Z}_1\mathbf{Z}_4 = \mathbf{Z}_2\mathbf{Z}_3$ as required
33. $R_3 = 50.01\ \Omega$ $R_1 = 253.3\Omega$

Chapter 20

1. $\mathbf{I} = 4.12\ \text{A}\angle50.88°$

3. a. $\mathbf{V}(1) = 2.75\ \text{V}\angle15.95°$ $\mathbf{V}(2) = 13.23\ \text{V}\angle40.71°$
   $\mathbf{V} = 15.77\ \text{V}\angle36.52°$

   b. $P(1) + P(2) = 1.826\ \text{W} \neq P_{100-\Omega} = 2.49\ \text{W}$

5. a. $\mathbf{V}_L = 1.265 \text{ V}\angle 161.57°$ **b.** $\mathbf{V}_L = 6.32 \text{ V}\angle 161.57°$
7. $\mathbf{V}_L = 9.88 \text{ V}\angle 0°$
11. a. $\mathbf{Z}_{Th} = 37.2 \text{ }\Omega\angle 57.99°$ **b.** $\mathbf{E}_{Th} = 9.63 \text{ V}\angle 78.49°$

13. a. $\mathbf{Z}_{Th} = 20.6 \text{ }\Omega\angle 34.94°$ $\mathbf{E}_{Th} = 10.99 \text{ V}\angle 13.36°$
b. $P_L = 1.607 \text{ W}$

15. $\mathbf{Z}_N = 1.875 \text{ k}\Omega\angle -53.13°$ $\mathbf{I}_N = 17.09 \text{ mA}\angle 163.69°$

17. a. $\mathbf{Z}_N = 109.9 \text{ }\Omega\angle -28.44°$ $\mathbf{I}_N = 0.132 \text{ A}\angle -63.16°$
b. $\mathbf{I}_L = 36.2 \text{ mA}\angle -84.09°$
c. $P_L = 0.394 \text{ W}$

19. a. $\mathbf{Z}_N = 14.13 \text{ }\Omega\angle 85.41°$ $\mathbf{I}_N = 0.1807 \text{ A}\angle 29.91°$
b. $\mathbf{I}_L = 0.0746 \text{ A}\angle 90.99°$

21. a. $\mathbf{Z}_{Th} = 16 \text{ k}\Omega - j8 \text{ k}\Omega$ $\mathbf{E}_{Th} = 1.789 \text{ V}\angle 153.43°$
b. $\mathbf{I}_L = 31.6 \text{ }\mu\text{A}\angle 161.57°$
c. $P_L = 40 \text{ }\mu\text{W}$

25. a. $\mathbf{Z}_L = 8 \text{ }\Omega\angle 22.62°$ **b.** $P_{max} = 40.17 \text{ W}$
27. a. $\mathbf{Z}_L = 2.47 \text{ }\Omega\angle 21.98°$ **b.** $P_{max} = 1.039 \text{ W}$
29. $\mathbf{Z}_L = 4.15 \text{ }\Omega\angle 85.24°$

31. a. $\mathbf{Z}_L = 37.2 \text{ }\Omega\angle -57.99°$ **c.** $P_L = 1.176 \text{ W}$
b. $R_L = 19.74 \text{ }\Omega$

Chapter 21

1. a. $\omega_S = 3835 \text{ rad/s}$ $f_S = 610.3 \text{ Hz}$

b. $\mathbf{I} = 153.8 \text{ mA}\angle 0°$

c. $\mathbf{V}_C = 59.0 \text{ V}\angle -90°$ $\mathbf{V}_L = 59.03 \text{ V}\angle 87.76°$
$\mathbf{V}_R = 7.69 \text{ V}\angle 0°$

d. $P_L = 0.355 \text{ W}$

3. a. $R = 25.0 \text{ }\Omega$ $C = 4.05 \text{ nF}$

b. $P = 15.6 \text{ mW}$

c. $X_C = 1.57 \text{ k}\Omega$ $\mathbf{V}_C = 39.3 \text{ V}\angle -90°$
$\mathbf{V}_L = 39.3 \text{ V}\angle +90°$ $\mathbf{V}_R = \mathbf{E} = 0.625 \text{ V}\angle 0°$

d. $v_C = 55.5 \sin(50{,}000\pi t - 90°)$
$v_L = 55.5 \sin(50{,}000\pi t + 90°)$
$v_R = 0.884 \sin(50{,}000\pi t)$

5. a. $\omega_S = 500 \text{ krad/s}$ $f_S = 79.6 \text{ kHz}$

b. $\mathbf{Z}_T = 200 \text{ }\Omega$

c. $\mathbf{I} = 10 \text{ mA}\angle 0°$

d. $\mathbf{V}_R = 1 \text{ V}\angle 0°$ $\mathbf{V}_L = 50.01 \text{ V}\angle 88.85°$
$\mathbf{V}_C = 50 \text{ V}\angle -90°$

e. $P_T = 20$ mW $Q_C = 0.5$ VAR (cap)
$Q_L = 0.5$ VAR (ind)

f. $Q_S = 25$

7. a. $C = 0.08$ μF $R = 6.4$ Ω
b. $P_T = 0.625$ W
c. $\mathbf{V}_L = 62.5$ V∠90°

9. a. $\omega_S = 8000$ rad/s

11. a. $\omega_S = 3727$ rad/s $Q = 7.45$ BW = 500 rad/s

b. $P_{max} = 144$ W

c. $\omega_1 \approx 3477$ rad/s $\omega_2 \approx 3977$ rad/s

d. $\omega_1 = 3485.16$ rad/s $\omega_2 = 3985.16$ rad/s

e. The results are close, although the approximation will yield some error if used in further calculations. The error would be less if the Q were larger.

f. $\mathbf{I}_1 = 0.849$ A∠45° $P = 72$ W

13. a. $R = 1005$ Ω $C = 63.325$ pF
b. $P = 0.625$ W $\mathbf{V}_L = 312.5$ V∠90°
c. $v_{out} = 442 \sin (400 \pi \times 10^3 t + 90°)$

15. Circuit (a):

a. $Q = 24$
b. $R_P = 5770$ Ω $X_{LP} = 240$ Ω
c. $Q = 240$ $R_P = 576$ kΩ $X_{LP} = 2400$ Ω

Circuit (b):

a. $Q = 1$
b. $R_P = 200$ Ω $X_{LP} = 200$ Ω
c. $Q = 10$ $R_P = 10.1$ kΩ $X_{LP} = 1.01$ kΩ

Circuit (c):

a. $Q = 12.5$
b. $R_P = 314.5$ Ω $X_{LP} = 25.16$ Ω
c. $Q = 125$ $R_P = 31.25$ kΩ $X_{LP} = 250$ Ω

17. Circuit (a):

$Q = 15$ $R_P = 4500$ Ω $X_{CP} = 300$ Ω

Circuit (b):

$Q = 2$ $R_P = 225$ Ω X_{CP} 112.5 Ω

Circuit (c):

$Q = 5$ $R_P = 13$ kΩ $X_{CP} = 2600$ Ω

19. Circuit (a):

$Q = 4 \qquad R_S = 4 \text{ k}\Omega \qquad X_{LS} = 16 \text{ k}\Omega$

Circuit (b):

$Q = 0.333 \qquad R_S = 3.6 \text{ k}\Omega \qquad X_{CS} = 1.2 \text{ k}\Omega$

Circuit (c):

$Q = 100 \qquad R_S = 10 \text{ }\Omega \qquad X_{LS} = 1 \text{ k}\Omega$

21. $L_S = 1.2 \text{ mH} \qquad L_P = 2.4 \text{ mH}$

23. a. $\omega_P = 20 \text{ krad/s}$

25. a. $\omega_P = 39.6 \text{ krad/s} \qquad f_P = 6310 \text{ Hz}$

b. $Q = 6.622$

c. $\mathbf{V} = 668.2 \text{ V}\angle 0° \qquad \mathbf{I}_R = 11.14 \text{ mA}\angle 0°$
$\mathbf{I}_L = 668.2 \text{ m}\angle -82.36° \qquad \mathbf{I}_C = 662.2 \text{ mA}\angle 90°$

d. $P_T = 66.82 \text{ W}$

e. $\text{BW} = 5.98 \text{ krad/s} \qquad \text{BW} = 952 \text{ Hz}$

27. $R_1 = 41.94 \text{ k}\Omega \qquad C = 405 \text{ pF} \qquad \mathbf{I}_L = 5.0 \text{ mA}\angle -89.27°$

Chapter 22

1. a. 2000 (33.0 dB)
b. 200,000 (53.0 dB)
c. 2×10^6 (63.0 dB)
e. 400 (26.0 dB)

3. a. $A_p = 50 \times 10^6$ (77.0 dB) $\qquad A_v = 500$ (54.0 dB)
b. $A_p = 10$ (10.0 dB) $\qquad A_v = 0.224$ (−13.0 dB)
c. $A_p = 13.3 \times 10^6$ (71.3 dB) $\qquad A_v = 258$ (48.2 dB)
d. $A_p = 320 \times 10^6$ (85.1 dB) $\qquad A_v = 1265$ (62.0 dB)

5. $P_{\text{in}} = 12.5 \text{ μW} \qquad P_{\text{out}} = 39.5 \text{ mW}$
$V_{\text{out}} = 3.14 \text{ V} \qquad A_v = 12.6$

$A_v = 22.0 \text{ dB}$

7. a. 17.0 dBm (−13.0 dBW)
b. 30.0 dBm (0.0 dBW)
c. −34.0 dBm (−64.0 dBW)
d. −66.0 dBm (−96.0 dBW)

9. a. 0.224 W $\qquad$ c. 5.01 pW
b. 30.2 μW $\qquad$ d. 1995 W

11. $P_1 = 5.05 \text{ dBm} \qquad P_2 = 2.05 \text{ dBm}$
$P_3 = 14.09 \text{ dBm} \qquad P_0 = 25.6 \text{ mW}$

13. $P_1 = 25.5 \text{ dBm} \qquad P_{\text{in}} = -14.5 \text{ dBm} \qquad V_{\text{out}} = 52.8 \text{ V}$

15. a. $\omega_C = 1000$ rad/s $f_C = 159.2$ Hz

17. a. $\omega_1 = 50$ rad/s $\omega_2 = 1000$ rad/s

21. a. T.F. $= \dfrac{1 + j0.00003\,\omega}{1 + j0.00006\omega}$

27. a. Low-pass filter: $\omega_C = 500$ krad/s
High-pass filter: $\omega_C = 25$ krad/s

BW $= 475$ krads/s

c. The actual cutoff frequencies will be close to the designed values since the break frequencies are separated by more than one decade.

Chapter 23

1. a. 8 A$\angle$–30°;   8 A$\angle$–150°;   8 A$\angle$90°

3. ACB. The waveforms are identical to those of Figure 23-2(b) if you interchange labels e_{BB}' and e_{CC}'.

5. a. $\mathbf{V}_{ab} = 208$ V$\angle$45°; $\mathbf{V}_{ca} = 208$ V$\angle$165°

b. $\mathbf{V}_{an} = 120$ V$\angle$15°; $\mathbf{V}_{bn} = 120$ V$\angle$–105°;
$\mathbf{V}_{cn} = 120$ V$\angle$135°

7. $\mathbf{I}_a = 23.1$ A$\angle$–21.9°; $\mathbf{I}_b = 23.1$ A$\angle$–141.9°;
$\mathbf{I}_c = 23.1$ A$\angle$98.1°

9. $\mathbf{I}_{ab} = 19.2$ A$\angle$36.9°; $\mathbf{I}_{bc} = 19.2$ A$\angle$–83.1°;
$\mathbf{I}_{ca} = 19.2$ A$\angle$156.9°; $\mathbf{I}_a = 33.3$ A$\angle$6.9°;
$\mathbf{I}_b = 33.3$ A$\angle$–113.1°; $\mathbf{I}_c = 33.3$ A$\angle$126.9°

11. $\mathbf{V}_{ab} = 250$ V$\angle$–57.9°; $\mathbf{V}_{bc} = 250$ V$\angle$–177.9°;
$\mathbf{V}_{ca} = 250$ V$\angle$62.1°.

13. a. $R = 3.83\ \Omega$; $C = 826\ \mu$F
b. $R = 7.66\ \Omega$; $L = 17.1$ mH

15. a. $\mathbf{I}_{ab} = 4.5$ A$\angle$35°; $\mathbf{I}_{ca} = 4.5$ A$\angle$155°
b. $\mathbf{I}_a = 7.79$ A$\angle$5°; $\mathbf{I}_b = 7.79$ A$\angle$–115°; $\mathbf{I}_c = 7.79$ A$\angle$125°
c. $R = 43.7\ \Omega$; $C = 86.7\ \mu$F

17. a. 122.1 V$\angle$0.676°
b. 212 V$\angle$30.676°

19. $\mathbf{I}_a = 14.4$ A$\angle$26.9°; $\mathbf{I}_b = 14.4$ A$\angle$–93.1°; $\mathbf{I}_c = 14.4$ A$\angle$146.9°
21. $\mathbf{I}_a = 14.4$ A$\angle$34.9°
23. 598 V$\angle$32.0°
25. b. 489 V$\angle$30°
27. 346 V$\angle$–10°
29. 2303 W;   1728 VAR (ind.);   2879 VA
31. 4153 W;   3115 VAR;   5191 VA
33. 99.1 kW;   29.7 kVAR (ind.);   103 kVA;   0.958

35. 27.7 kW;   36.9 kVAR (ind.);   46.1 kVA; 0.60

37. 3.93 A∠−32°

39. 0.909

41. a. Same as Figure 23–27   b. 768 W   c. 2304 W

43. a. $\mathbf{I}_a = 5.82$ A∠−14.0°; $\mathbf{I}_b = 5.82$ A∠−134.0°;
$\mathbf{I}_c = 5.82$ A∠106°

b. $P_\phi = 678$ W; $P_T = 2034$ W

c. $W_1 = 1163$ W; $W_2 = 870$ W

d. 2033 W

45. a. 0.970   b. 0.970

47. a. $\mathbf{I}_{ab} = 4$ A∠0°; $\mathbf{I}_{bc} = 2.4$ A∠−156.9°; $\mathbf{I}_{ca} = 3.07$ A∠170.2°

b. $\mathbf{I}_a = 7.04$ A∠−4.25°; $\mathbf{I}_b = 6.28$ A∠−171.4°;
$\mathbf{I}_c = 1.68$ A∠119.2°

c. $P_{ab} = 960$ W; $P_{bc} = 461$ W; $P_{ca} = 471$ W

d. 1892 W

49. a. $\mathbf{I}_a = 6.67$ A∠0°; $\mathbf{I}_b = 2.68$ A∠−93.4°; $\mathbf{I}_c = 2.4$ A∠66.9°
b. 7.47 A∠−3.59°
c. $P_{an} = 800$ W; $P_{bn} = 287$ W; $P_{cn} = 173$ W
d. 1260 W

51. a. $\mathbf{V}_{an} = 34.9$ V∠−0.737°; $\mathbf{V}_{bn} = 179$ V∠−144°;
$\mathbf{V}_{cn} = 178$ V∠145°

b. $\mathbf{V}_{nN} = 85.0$ V∠0.302°

53. $\mathbf{I}_{ab} = 41.57$ A∠83.13°; $\mathbf{I}_{bc} = 16.8$ A∠−75.96°;
$\mathbf{I}_{ca} = 28.82$ A∠−176.3°

Chapter 24

1. a. e_s is in phase with e_p.
b. e_p is 180° out of phase with e_s.

3. a. Step-up   d. 96 V∠0°
b. 25 sinωt V   e. 3200 V∠180°
c. 6 V

5. v_1 is in phase with e_g. v_2 is 180° out of phase with v_1. v_3 is 180° out of phase with v_2.

7. a. 1 A∠20°   b. 480 V∠0°   c. 480 Ω∠−20°

9. a. 160 V∠−23.1°   b. 640 V∠−23.1°

11. a. 40 Ω − j80 Ω   b. 1.25 Ω + j2 Ω

13. a. 22 Ω + j6 Ω   b. 26 Ω + j3 Ω

15. 108 kVA

17. a. 20 A   **b.** 22.5 A   **c.** −2.5 A   **d.** 0.708 A
19. b. 20.83 A   **c.** 7500 VA   **d.** 62.5 A

21. a. 20.86 A∠−45°   **d.** 119.8 V
 b. 2.086 A∠−45°   **e.** 1.53%
 c. 1198 V∠0°

23. a. 97.7%   **b.** 97.1%

25. $(10 + j20)\,\mathbf{I}_1 - j5\mathbf{I}_2 = 100\angle 0°$
 $-j5\mathbf{I}_1 + (30 + j40)\mathbf{I}_2 = 150\angle 30°$

27. 0.269 H
29. 10.5 H
31. $\mathbf{I}_1 = 11.3$ A∠−63.2°; $\mathbf{I}_2 = 12.2$ A∠−116.8°
33. Answers are the same as Example 24–15.
35. $\mathbf{I}_1 = 11.66$ A∠−52.2° (same).

Chapter 25

1. $v(t) = 1 + \dfrac{2}{\pi}\sin \omega t + \dfrac{2}{2\pi}\sin(2\omega t) + \dfrac{2}{3\pi}\sin(3\omega t) + \ldots\ldots$

3. $v(t) = \dfrac{32}{\pi}\sin(500\pi t) - \dfrac{32}{2\pi}\sin(1000\pi t) + \dfrac{32}{3\pi}\sin(1500\pi t) -$

$\dfrac{32}{4\pi}\sin(2000\pi t) + \ldots\ldots$

5. $v(t) = \dfrac{10}{\pi} + 5\sin(\omega t) - \dfrac{20}{\pi}\left[\dfrac{\cos(2\omega t)}{3} + \right.$

$\left. \dfrac{\cos(4\omega t)}{15}\ \ldots\ldots\right]$

7. $v(t) = \dfrac{32}{\pi}\sin(\omega t + 30°) + \dfrac{32}{3\pi}\sin(3\omega t + 30°) +$

$\dfrac{32}{5\pi}\sin(5\omega t + 30°) + \dfrac{32}{7\pi}\sin(7\omega t + 30°) + \ldots\ldots$

9. b. $v_1 = -\dfrac{16}{\pi}\sin\omega t - \dfrac{16}{3\pi}\sin(3\omega t) - \dfrac{16}{5\pi}\sin(5\omega t) - \ldots$

$v_2 = \dfrac{8}{\pi}\sin\omega t - \dfrac{8}{2\pi}\sin(2\omega t) + \dfrac{8}{3\pi}\sin(3\omega t) - \ldots$

 c. $v = -\dfrac{8}{\pi}\sin\omega t - \dfrac{8}{2\pi}\sin(2\omega t) - \dfrac{8}{3\pi}\sin(3\omega t) - \ldots$

11. b. $V_{avg} = 5$ V

 c. $v_1 = 5 - \dfrac{10}{\pi}\sin\omega t - \dfrac{10}{2\pi}\sin(2\omega t) - \dfrac{10}{3\pi}\sin(2\omega t) - \ldots$

$v_2 = -\dfrac{10}{\pi}\sin\omega t + \dfrac{10}{2\pi}\sin(2\omega t) - \dfrac{10}{3\pi}\sin(3\omega t) - \ldots$

d. $v_1 + v_2 = 5 - \dfrac{20}{\pi}\sin\omega t - \dfrac{20}{3\pi}\sin(3\omega t) - \dfrac{20}{5\pi}\sin(5\omega t) - \dots$

13. $v_1 + v_2 = \dfrac{16}{\pi}\sin\omega t - \dfrac{4}{2\pi}\sin(2\omega t) + \dfrac{16}{3\pi}\sin(3\omega t) - \dots$

15. $P = 1.477$ W

17. $P_0 = 23.1$ dBm $P_1 = 24.0$ dBm $P_2 = 26.1$ dBm
$P_3 = 2.56$ dBm $P_4 = -4.80$ dBm

19. a. $V_0 = 0.5$ V

b. $V_1 = 0.90\ V_p$ $\theta_1 = -45°$
$V_3 = 0.14\ V_p$ $\theta_3 = -63°$
$V_5 = 0.05\ V_p$ $\theta_5 = -79°$
$V_7 = 0.03\ V_p$ $\theta_7 = -82°$

Glossary

ac Abbreviation for alternating current; used to denote periodically varying quantities such as ac current, ac voltage, and so on.

admittance (Y) A vector quantity (measured in siemens, s) which is the reciprocal of impedance. $\mathbf{Y} = 1/\mathbf{Z}$.

alternating current Current that periodically reverses in direction, commonly called an ac current.

alternating voltage Voltage that periodically changes in polarity, commonly called an ac voltage. The most common ac voltage is the sine wave.

American Wire Gage (AWG) An American standard for classifying wire and cable.

ammeter An instrument that measures current.

ampere (A or amp) The SI unit of electrical current, equal to a rate of flow of one coulomb of charge per second.

ampere-hour (Ah) A measure of the storage capacity of a battery.

angular frequency (ω) Frequency of an ac waveform in radians/s. $\omega = 2\pi f$ where f is frequency in Hz.

apparent power (S) The power that apparently flows in an ac circuit. It has components of real power and reactive power, related by the power triangle. The magnitude of apparent power is equal to the product of effective voltage times effective current. Its unit is the VA (volt-amp).

atom The basic building block of matter. In the Bohr model, an atom consists of a nucleus of positively charge protons and uncharged neutrons, surrounded by negatively charged orbiting electrons. An atom normally consists of equal numbers of electrons and protons and is thus uncharged.

attenuation The amount that a signal decreases as it passes through a system. The attenuation is usually measured in decibels, dB.

audio frequency A frequency in the range of human hearing, which is typically from about 15 Hz. to 20 kHz.

autotransformer A type of transformer with a partially common primary and secondary winding. Part of its energy is transferred magnetically and part conductively.

average of a waveform The mean value of a waveform, obtained by algebraically summing the areas above and below the zero axis of the waveform, divided by the cycle length of the waveform. It is equal to the dc value of the waveform as measured by an ammeter or a voltmeter.

balanced (1) For a bridge circuit, the voltage between midpoints on its arms is zero. (2) In three-phase systems, a load that is identical for all three phases.

band-pass filter A circuit that permits signals within a range of frequencies to pass through a circuit. Signals of all other frequencies are prevented from passing through the circuit.

band-stop filter (or notch filter) A circuit designed to prevent signals within a range of frequencies from passing through a circuit. Signals of all other frequencies freely pass through the circuit.

bandwidth (BW) The difference between the half-power frequencies for any resonant, band-pass, or band-stop filter. The bandwidth may be expressed in either hertz or radians per second.

Bode plot A straight line approximation that shows how the voltage gain of a circuit changes with frequency.

branch A portion of a circuit that occurs between two nodes (or terminals).

branch current The current through a branch of a circuit.

capacitance A measure of charge storage capacity, for example, c a capacitor. A circuit with capacitance opposes a change in voltag Unit is the farad (F).

capacitor A device that stores electrical charges on conductive "plates" separated by an insulating material called a dielectric.

cascade Two stages of a circuit are said to be in a cascade conr tion when the output of one stage is connected to the input of next stage.

CGS system A system of units based on centimeters, grams, seconds.

charge (1) The electrical property of electrons and protons causes a force to exist between them. Electrons are negative charged while protons are positively charged. Charge is der by Q and is defined by Coulomb's law. (2) An excess or defi of electrons on a body. (3) To store electricity as in to charg pacitor or charge a battery.

choke Another name for an inductor.

circuit A system of interconnected components such as resistors, capacitors, inductors, voltages sources, and so on.

circuit breaker A resettable circuit protection device that trips a set of contacts to open the circuit when current reaches a preset value.

circuit common The reference point in a electrical circuit from which voltages are measured.

circular mil (CM) A unit used to specify the cross-section area of a cable or wire. The circular mil is defined as the area contained in a circle having a diameter of 1 mil (0.001 inch).

coefficient of coupling (k) A measure of the flux linkage between circuits such as coils. If $k = 0$, there is no linkage; if $k = 1$, all of the flux produced by one coil links another. The mutual inductance M between coils is related to k by the relationship $M = k\sqrt{L_1 L_2}$, where L_1 and L_2 are the self-inductances of the coils.

coil A term commonly used to denote inductors or windings on transformers.

conductance (G) The reciprocal of resistance. Unit is the siemens (S).

conductor A material through which charges move easily. Copper is the most common metallic conductor.

copper loss The I^2R power loss in a conductor due to its resistance, for example the power loss in the windings of a transformer.

core The form or structure around which an inductor or the coils of a transformer are wound. The core material affects the magnetic properties of the device.

core loss Power loss in the core of a transformer or inductor due to hysteresis and eddy currents.

coulomb (C) The SI unit of electrical charge, equal to the charge carried by 6.24×10^{18} electrons.

Coulomb's law An experimental law which states that the force (in Newtons) between charged particles is $F = Q_1 Q_2 / 4\pi \epsilon r^2$, where Q_1 and Q_2 are the charges (in coulombs), r is the distance between their centers in meters, and ϵ is the permittivity of the medium. For air, $\epsilon = 8.85 \times 10^{-12} F/m$.

critical temperature The temperature below which a material becomes a superconductor.

current (I or i) The rate of flow of electrical charges in a circuit, measured in amperes.

current source A practical current source can be modeled as an ideal current source in parallel with an internal impedance.

cutoff frequency, f_c or ω_c The frequency at which the output power of a circuit is reduced to half of the maximum output power. The cutoff frequency may be measured in either hertz, (Hz) or radians per second, (rad/s).

cycle One complete variation of an ac waveform.

decade A tenfold change in frequency.

decibel (dB) A logarithmic unit used to represent an increase (or decrease) in power levels or sound intensity.

delta (Δ) A small change (increment or decrement) in a variable. For example, if current changes a small amount from i_1 to i_2, its increment is $\Delta i = i_2 - i_1$, while if time changes a small amount from t_1 to t_2, its increment is $\Delta t = t_2 - t_1$.

delta load A configuration of circuit components connected in the shape of a Δ (Greek letter delta). Sometimes called a pi (π) load.

derivative The instantaneous rate of change of a function. It is the slope of the tangent to the curve at the point of interest.

dielectric An insulating material. The term is commonly used with reference to the insulating material between the plates of a capacitor.

dielectric constant ($\in$) A common name for permittivity.

differentiator A circuit whose output is proportional to the derivative of its input.

diode A two-terminal component made of semiconductor material, which permits current in one direction while preventing current in the opposite direction.

direct current (dc) Unidirectional current such as that from a battery.

DMM A digital multimeter that displays results on a numeric readout. In addition to voltage, current, and resistance, some dmms measure other quantities such as frequency and capacitance.

duty cycle The ratio of on time to the duration of a pulse waveform, expressed in percent.

eddy current A small circulating current. Usually refers to the unwanted current that is induced in the core of an inductor or transformer by changing core flux.

effective resistance Resistance defined by $R = P/I^2$. Effective resistance is larger than dc resistance due to skin effect and other effects such as power losses.

effective value An equivalent dc value of a time varying waveform, hence, that value of dc that has the same heating effect as the given waveform. Also called *rms* (root mean square) value. For sinusoidal current, $I_{\text{eff}} = 0.707 I_m$, where I_m is the amplitude of the ac waveform.

efficiency (η) The ratio of output power to input power, expressed as a percentage. $\eta = P_{out}/P_{in} \times 100\%$.

electron A negatively charged atomic particle. *See* atom.

energy (W) The ability to do work. Its SI unit is the joule; electrical energy is also measured in kilowatt-hours (kWh).

fall time (t_f) The time it takes for a pulse or step to change from its 90% value to its 10% value.

farad (F) The SI unit of capacitance, named in honor of Michael Faraday.

ferrite A magnetic material made from powdered iron oxide. Provides a good path for magnetic flux and has low enough eddy current losses that it is used as a core material for high frequency inductors and transformers.

field A region in space where a force is felt, hence a force field. For example, magnetic fields exist around magnets and electric fields exist around electric charges.

field intensity The strength of a field.

filter A circuit that passes certain frequencies while rejecting all other frequencies.

flux A way of representing and visualizing force fields by drawing lines that show the strength and direction of a field at all points in space. Commonly used to depict electric or magnetic fields.

free electron An electron that is weakly bound to its parent atom and is thus easily broken free. For materials like copper, there are billions of free electrons per cubic centimeter at room temperature. Since these electrons can break free and wander from atom to atom, they form the basis of an electric current.

frequency (f) The number of times that a cycle repeats itself each second. Its SI unit is the hertz (Hz).

gain The ratio of output voltage, current, or power to the input. Power gain for an amplifier is defined as the ratio of ac output power to ac input power, $A_p = P_{out}/P_{in}$. Gain may also be expressed in decibels. In the case of power gain, $A_p(dB) = 10 \log P_{out}/P_{in}$.

gauss The unit of magnetic flux density in the CGS system of units.

giga (G) A prefix with a value of 10^9.

ground (1) An electrical connection to earth. (2) A circuit common. (*See* circuit common.) (3) A short to ground, such as a ground fault.

harmonics Integer multiples of a frequency.

henry (H) The SI unit of inductance, named in honor of Joseph Henry.

hertz (Hz) The SI unit of frequency, named in honor of Heinrich Hertz. One Hz equals one cycle per second.

high-pass filter A circuit which readily permits frequencies above the cutoff frequency to pass from the input to the output of the circuit, while attenuating frequencies below the cutoff frequency. (*See* cutoff frequency).

hysteresis loss Power loss in a ferromagnetic material caused by the reversal of magnetic domains in a time varying magnetic field.

ideal current source A current source having an infinite shunt (parallel) impedance. An ideal current source is able to provide the same current to all loads (except an open circuit). The voltage across the current source is determined by the value of the load impedance.

ideal transformer A transformer having no losses and characterized by its turns ratio $a = N_p/N_s$. For voltage, $\mathbf{E}_p/\mathbf{E}_s = a$, while for current $\mathbf{I}_p/\mathbf{I}_s = 1/a$.

ideal voltage source A voltage source having zero series impedance. An ideal voltage source is able to provide the same voltage across all loads (except a short circuit). The current through the voltage source is determined by the value of the load impedance.

impedance (Z) Total opposition that a circuit element presents to sinusoidal ac in the phasor domain. $\mathbf{Z} = \mathbf{V}/\mathbf{I}$ ohms, where $\mathbf{V}$ and $\mathbf{I}$ are voltage and current phasors respectively. Impedance is a complex quantity with magnitude and angle.

induced voltage Voltage produced by changing magnetic flux linkages.

inductance (L) That property of a coil (or other current-carrying conductor) that opposes a change in current. The SI unit of inductance is the henry.

inductor A circuit element designed to posses inductance, e.g., a coil of wire wound to increase its inductance.

instantaneous value The value of a quantity (such as voltage or current) at some instant of time.

insulator A material such as glass, rubber, bakelite, and so on, that does not conduct electricity.

integrator A circuit whose output is proportional to the integral of its input.

internal impedance The impedance that exists internally in a device such as a voltage source.

ion An atom that has become charged. If it has an excess of electrons, it is a negative ion, while if it has a deficiency, it is a positive ion.

joule (J) The SI unit of energy, equal to one newton-meter.

kilo A prefix with the value of 10^3.

kilowatt-hour (kWh) A unit of energy equal to 1000 W times one hour and commonly used by electrical utilities.

Kirchhoff's current law An experimental law which states that the sum of the currents entering a junction is equal to the sum leaving.

Kirchhoff's voltage law An experimental law that states that the algebraic sum of voltages around a closed path in a circuit is zero.

lagging load A load in which current lags voltage (e.g., an inductive load).

leading load A load in which current leads voltage (e.g., a capacitive load).

linear circuit A circuit in which relationships are proportional. In a linear circuit, current is proportional to voltage.

load (1) The device that is being driven by a circuit. Thus, the lamp in a flashlight is the load. (2) The current drawn by a load.

low-pass filter A circuit that permits frequencies below the cutoff frequency to pass through from the input to the output of the circuit, while attenuating frequencies above the cutoff frequency. (*See* cutoff frequency.)

magnetic flux density *(B)* The number of magnetic flux lines per unit area, measured in the SI system in tesla (T), where one T = one Wb/m^2.

magnetomotive force (mmf) The flux producing ability of a coil. In the SI system, the mmf of a coil of N turns with current I is NI ampere-turns.

maxwell (Mx) The CGS unit of magnetic flux Φ.

mega (M) A prefix with the value of 10^6.

micro (μ) A prefix with the value of 10^{-6}.

milli (m) A prefix with the value of 10^{-3}.

multimeter A multifunction meter used to measure a variety of electrical quantities such as voltage, current, and resistance. Its function and range is selected by a switch. (*See also* DMM.)

mutual inductance *(M)* The inductance between circuits (such as coils) measured in henries. The voltage induced in one circuit by changing current in another circuit is equal to M times the rate of change of current in the first circuit.

nano (n) A prefix with the value of 10^{-9}.

neutron An atomic particle with no charge. (*See* atom.)

node A junction where two or more components connect in an electric circuit.

ohm (Ω) The SI unit of resistance. Also used as the unit for reactance and impedance.

ohmmeter An instrument for measuring resistance.

open circuit A discontinuous circuit, hence one that does not provide a complete path for current.

oscilloscope An instrument that electronically displays voltage waveforms on a screen. The screen is ruled with a scaled grid to permit measurement of the waveform's characteristics.

parallel Elements or branches are said to be in a parallel connection when they have exactly two nodes in common. The voltage across all parallel elements or branches is exactly the same.

peak The maximum instantaneous value (positive or negative) of a waveform.

peak-to-peak The difference between a waveform's maximum and minimum values.

period (T) The time for a waveform to go through one cycle. $T = 1/f$ where f is frequency in Hz.

periodic Repeating at regular intervals.

permeability (μ) A measure of how easy it is to magnetize a material. $B = \mu H$, where B is the resulting flux density and H is the magnetizing force that creates the flux.

permittivity (ϵ) A measure of how easy it is to establish electric flux in a material. (*See also* relative dielectric constant and Coulomb's law.)

phase shift The angular difference by which one waveform leads or lags another, hence the relative displacement between time varying waveforms.

phasor A way of representing the magnitude and angle of a sine wave graphically or by a complex number. The magnitude of the phasor represents the rms value of the ac quantity and its angle represents the waveform's phase.

pico (p) A prefix with the value of 10^{-12}.

potentiometer A three-terminal resistor consisting of a fixed resistance between two end terminals and a third terminal that is connected to a movable wiper arm. When the end terminals are connected to a voltage source, the voltage between the wiper and either of the other terminals is adjustable.

power (P, p) The rate of doing work, with units of watts, where one watt equals one joule per second. Also called real or active power.

power factor The ratio of active power to apparent power, equal to cos θ, where θ is the angle between the voltage and the current.

power triangle A way to represent the relationship between real power, reactive power, and apparent power using a triangle.

primary The winding of a transformer to which we connect the source.

proton A positively charged atomic particle. (*See* atom).

pulse A short duration voltage or current that abruptly changes from one value to another, then back again.

pulse width The duration of a pulse. For non-ideal pulses, it is measured at the 50% amplitude point.

quality factor (Q) (1) A figure of merit. Q for a coil is the ratio of its reactive power to its real power. The higher the Q, the more closely the coil approaches the ideal.(2) A measure of the selectivity of a resonant circuit. The higher the Q, the narrower the bandwidth.

reactance (X) The opposition that a reactive element (capacitance or inductance) presents to sinusoidal ac, measured in ohms.

reactive power A component of power that alternately flows into then out of a reactive element, measured in VARs (volt-amps reactive). Reactive power has an average value of zero and is sometimes called "wattless" power.

rectifier A circuit, generally consisting of a least one diode, which permits current in only one direction.

regulation The change in voltage from no-load to full-load expressed as a percentage of full load voltage.

relative dielectric constant (ϵ_r) The ratio of the dielectric constant of a material to that of a vacuum.

relay A switching device that is opened or closed by an electrical signal. May be electromechanical or electronic.

reluctance The opposition of a magnetic circuit to the establishment of flux.

resistance (R) The opposition to current that results in power dissipation. Thus, $R = P/I^2$ ohms. For a dc circuit, $R = V/I$, while for an ac circuit containing reactive elements, $R = V_R/I$, where V_R is the component of voltage across the resistive part of the circuit.

resistor A circuit component designed to posses resistance.

resonance, resonant frequency The frequency at which the output power of an L-R-C circuit is at a maximum. $f = 1/(2\pi LC)$

rheostat A variable resistor connected so that current through the circuit is controlled by the position of the wiper.

rise time (t_r) The time that it takes for a pulse or step to change from its 10% value to its 90% value.

rms value The root-mean-square value of a time varying waveform. (*See* effective value.)

saturation The condition of a ferromagnetic material where it is fully magnetized. Thus, if the magnetizing force (current in a coil for example) is increased, no significant increase in flux results.

schematic diagram A circuit diagram that uses symbols to represent physical components.

secondary winding The output winding of a transformer.

selectivity A measure of the ability of a resonant circuit to select a very narrow band of frequencies and reject all others. The higher the Q, the narrower the bandwidth and hence, the greater the selectivity.

semiconductor A material such as silicon from which transistors, diodes, and the like are made.

series circuit A closed loop of elements where two elements have no more than one common terminal. In a series circuit, there is only one current path and all series elements have the same current.

short circuit A short circuit occurs when two terminals of an element or branch are connected together by a low-resistance conductor. When a short circuit occurs, very large currents may result in sparks or a fire, particularly when the circuit is not protected by a fuse or circuit breaker.

SI System The international system of units used in science and engineering. It is a metric system and includes the standard units for length, mass, and time (e.g., meters, kilograms, and seconds), as well as the electrical units (e.g., volts, amperes, ohms, and so on).

siemens (S) A unit of measure for conductance, admittance, and susceptance. The siemens is the reciprocal of ohm.

sine wave A periodic waveform that is described by the trigonometric sine function. It is the principle waveform used in ac systems.

skin effect At high frequencies, the tendency of current to travel in a thin layer near the surface of a conductor.

steady state The condition of operation of a circuit after transients have subsided.

step An abrupt change in voltage or current, as for example when a switch is closed to connect a battery to a resistor.

superconductor A conductor that has no internal resistance. Current will continue unimpeded through a superconductor even though there is no externally applied voltage or current source.

susceptance The reciprocal of reactance. Unit is the siemens.

tank circuit A circuit consisting of an inductor and capacitor connected in parallel. Such an *L-C* circuit is used in oscillators and receivers to provide maximum signal at the resonant frequency. (*See* selectivity.)

temperature coefficient (1) The rate at which resistance changes as the temperature changes. A material has a positive temperature coefficient if the resistance increases with an increase in temperature. Conversely, a negative temperature coefficient means that resistance decreases as temperature is increased. (2) Similarly for capacitance. The change in capacitance is due to changes in the characteristics of its dielectric with temperature.

tesla (T) The SI unit of magnetic flux density. One T = one Wb/m^2.

time constant (τ) A measure of how long a transient lasts. For example, during charging, capacitor voltage changes by 63.2% in one time constant, and for all practical purposes, charges fully in five time constants, For an *RC* circuit, $\tau = RC$ seconds and for an *RL* circuit, $\tau = L/R$ seconds.

transformer A device with two or more coils in which energy is transferred from one winding to the other by electromagnetic action.

transient A temporary or transitional voltage or current.

turns ratio (*a*) The ratio of primary turns to secondary turns; $a = N_p/N_s$.

valence shell The outermost shell of an atom.

volt The unit of voltage in the SI system.

voltage (*V, v, E, e*) Potential difference created when charges are separated, as for example by chemical means in a battery. If one joule of work is required to move a charge of one coulomb from one point to another, the potential difference between the points is one volt.

voltage source A practical voltage source can be modeled as an ideal voltage source in a series with an internal impedance.

watt (W) The SI unit of active power. Power is the rate at which work is done; one watt equals one joule/s.

watthour (Wh) A unit of energy, equal to one watt times one hour. One Wh = 3600 joules.

waveform The variation versus time of a time varying signal, hence, the shape of a signal.

weber (Wb) The SI unit of magnetic flux.

work (W) The product of force times distance, measured in joules in the SI system, where one joule equals one newton-meter.

wye load A configuration of circuit components connected in the shape of a Y. Sometimes called a star or T load.

INDEX